CHEMICAL PROPERTIES OF MATERIAL SURFACES

CHEMICAL PROPERTIES OF MATERIAL SURFACES

Marek Kosmulski

Technical University of Lublin
Lublin, Poland

CRC Press
Taylor & Francis Group
Boca Raton London New York

CRC Press is an imprint of the
Taylor & Francis Group, an **informa** business

CRC Press
Taylor & Francis Group
6000 Broken Sound Parkway NW, Suite 300
Boca Raton, FL 33487-2742

First issued in paperback 2019

© 2001 by Taylor & Francis Group, LLC
CRC Press is an imprint of Taylor & Francis Group, an Informa business

No claim to original U.S. Government works

ISBN-13: 978-0-8247-0560-2 (hbk)
ISBN-13: 978-0-367-39711-1 (pbk)

Visit the Taylor & Francis Web site at
http://www.taylorandfrancis.com

and the CRC Press Web site at
http://www.crcpress.com

To the memory of my father
Zdzislaw Kosmulski
1922–1998

Preface

Adsorption phenomena at solid–electrolyte solution interfaces at room temperature and at atmospheric pressure are reviewed in this book with a special emphasis on the mutual relationship between adsorption and surface charging. This relationship is particularly significant for adsorption of inorganic ions on silica, metal oxides and hydroxides, certain salts, e.g. silicates, and clay minerals. The models of surface ionization and complexation originally developed in colloid chemistry are widely used in different fields, including catalysis, ceramics, corrosion science, environmental sciences, geology, mineral processing, nuclear waste management, and soil science. Association with one of the traditional branches of science (and thus preference for particular journals or conferences) is only one of many factors that have split the scientists interested in adsorption at solid–solution interfaces into many insulated groups. For example, Western papers have rarely been cited in former Soviet Union papers and vice versa even though English translations of the leading Russian scientific journals are readily available. Each group has its own goals and methods, and specific systems of interest, and often ignore other systems and methods. The generalizations formulated by different groups are frequently based on a selective approach. They are not necessarily applicable in other systems and sometimes contradict each other. The aim of this book is to systematize the existing knowledge and to facilitate the exchange of ideas between different parts of the scientific community.

The point of zero charge (PZC) is a central concept in adsorption of charged species. This well-known term has been given very different meanings. The relationship between the zero points obtained by different methods and at different conditions is discussed in this book. An up-to-date compilation of values of the points of zero charge of various materials obtained by different methods is presented. These materials range from simple to very complex and from well- to ill-defined. Collections of zero points compiled by different authors are compared, and the correlation between these zero points and other physical quantities is analyzed.

Methods used in studies of adsorption of ions, their advantages and limitations, the meaning of results, and possibilities to combine results obtained by different methods are discussed. A large compilation of adsorption data is presented. The results obtained in simple adsorption systems (with one specifically adsorbed species) are sorted by the adsorbent and then by the adsorbate. Then, more complex systems are discussed with many specifically adsorbing species.

Many materials show a certain degree of chemical dissolution. The present survey is limited to materials whose solubility is low. This does not imply that the solubility is always negligible in the systems of interest.

Kinetics of adsorption and adsorption of surfactants and macromolecular species are broad fields, with their own methodologies, theories, and literature. Only selected topics directly related to the main subject of the present book are briefly treated.

Many recently published review articles, book chapters, and even entire books are devoted to adsorption of ionic species. Usually they cover one adsorbent (or a group of related adsorbents) or specific method(s). Some of these publications were very helpful during the preparation of this book, but current original papers were the main source of information.

ACKNOWLEDGMENTS

Substantial parts of this book were prepared at North Carolina State University, Raleigh, NC, and at Forschungszentrum Rossendorf, Dresden, Germany. Professors Robert A. Osteryoung and Thomas Fanghänel are acknowledged for their hospitality. A grant from the Alexander von Humboldt Foundation, for Chapter 5, is gratefully acknowledged.

Marek Kosmulski

Contents

1

Introduction

Properties of numerous adsorbents have been reported in the literature. This presentation is confined to materials

- Having the same bulk and surface chemical composition.
- Sparingly soluble in water.
- Showing pH dependent surface charging.

The above terms are relative, for example, some changes in surface composition due to solvation or selective leaching are unavoidable, but this is a part of the adsorption process. Adsorbents prepared by grafting or by adsorption of thin films (one or a few molecular layers) of substances whose properties are completely different from those of the support constitute an example of essential difference between the surface and bulk properties. Such materials combine high mechanical resistance, high surface area, and low cost of the support and desired sorption properties of the film and they are widely applied in different fields, but they are out of scope of this book. On the other hand, the surface properties of composite materials with external layer at least 10 nm thick are closely related to bulk properties of the coating, and a few examples of such materials will be discussed.

Solubility is another issue that requires some explanation. Materials more soluble than silica, i.e. about 10^{-3} mol dm^{-3} (this is an arbitrary choice) will not be discussed here. There is no sharp border between "soluble" and "insoluble", e.g. dissolution of relatively soluble materials is often sufficiently slow to allow

1

completion of sorption process before significant amount of the adsorbent is dissolved, moreover, the presence of certain adsorbates can substantially depress the dissolution rate and/or the equilibrium solubility. On the other hand, the solubility of materials reputed insoluble, can be significantly enhanced in the presence of certain complexing and/or redox agents. Kinetics of dissolution of silica is related to surface charging and depends on pH and ionic strength [1], and so is the solubility of many other materials. Chemical dissolution of oxides has been reviewed by Blesa et al. [2, 3] It should be emphasized that points of zero charge of relatively soluble materials, e.g. BaO and SrO have been reported in the literature.

Gel like materials containing sufficient amount of water are much more penetrable to ions than crystalline materials. Materials penetrable to ions are often characterized using the methods and terminology of ion-exchange, i.e. by their exchange capacities and diffusion coefficients of particular ions, and the affinity of particular ions to such materials is expressed in terms of selectivity coefficients. Misak [4] reviewed sorption properties of hydrous oxides from such a perspective, but in some other publications the uptake of ions by gel like materials is treated as adsorption. It is rather difficult to compare results interpreted in terms of the "ion exchange" approach on the one hand and adsorption approach on the other. Finally, in some publications dealing with materials not penetrable to ions a terminology borrowed from ion exchange is used: the adsorbents are "in hydrogen form", "in sodium form", etc. This can be translated into the language used in adsorption, e.g. "adsorbent in calcium form" is considered as adsorbent with pre-adsorbed calcium. Specific approach is required to describe surface charging of and adsorption on clay minerals on the one hand and zeolites on the other.

Most materials discussed in this book are electrical insulators, and their surface charge is regulated by sorption processes. However, a few oxides show sufficient degree of electronic conductivity that makes it possible to polarize the surface using an external battery. For example, the charging curves of IrO_2 can be plotted as a function of pH (when the oxide is polarized to constant potential) or as a function of polarizing potential at constant pH [5]. Properties and preparation of oxide electrodes (often termed DSA, dimensionally stable anodes) were reviewed by Trasatti [6]. Also adsorption properties of some sulfides, e.g. natural chalcocite [7] can be modified by polarization by external electric potentials.

The "dry" surface chemistry, i.e. chemistry of solid-gas interfaces has its own methodology and language. A substantial difference between wet and dry surface chemistry is that adsorption from solution is always an exchange, the "empty" surface is in fact occupied by solvent. In spite of an obvious relationship between dry and wet surfaces, only wet surface chemistry will be discussed here, although some quantities (e.g. the BET surface area) and relationships involve results obtained for dry surfaces. In particular, certain adsorbates considered show substantial vapor pressure at room temperature, and sorption of their vapors has been studied. Such results, albeit related to sorption of the same species from aqueous solution are beyond the scope of the book.

With a huge amount of information, that has been published on sorption properties of materials of interest, and in view of broad spectrum of goals and viewpoints of the authors of the cited publications and of potential readers of this book, it is not easy to organize the entire material. The grouping of data is based on concepts of colloid chemistry. Many results are compiled in tabular form. They are sorted by the formal chemical formula of the adsorbent and then by the adsorbate.

Chapter 2, which is not directly related to surface properties presents physical and chemical properties of the materials of interest. Not all materials described in Chapter 2 are then directly referred to in subsequent chapters. For example mixed oxides are not only adsorbents, but also potential products of surface reactions involving simple oxides: crystallographic data are helpful in identifying such products, and thermochemical data in predicting the direction of the reaction. Chapter 2 also shows how many well-defined and potentially interesting materials have not been studied as adsorbents and may stimulate further research. The present author has once submitted a manuscript "Let us measure points of zero charge of exotic oxides", but a reviewer was against such an intriguing title and finally the title was changed it into a more usual one. Now there is a chance to broadcast and extend this idea: there are so many important and well-defined materials whose points of zero charge are unknown. On the other hand for some materials that have been extensively used as adsorbents crystallographic or thermochemical data are incomplete. Availability of physical data for oxides, aluminates and silicates is illustrated in Figs. 1.1–1.3, respectively. Symbols of elements whose simple or mixed oxides are considered as adsorbents in this book are printed in boldface; black background: crystallographic and thermochemical data available, gray background: only crystallographic data available.

The organization scheme of the crystallographic and thermochemical tables presented below is also used in the next chapters. Simple oxides, hydroxides and oxohydroxides are listed first, in alphabetical order of the symbol of the electropositive element and then from low to high oxidation state, and then from low to high degree of hydration. Then results for mixed oxides (all component oxides are insoluble with one exception of CO_2) are listed according to the symbol of the most acidic element, and then according to the symbols of less acidic elements. Aluminosilicates are listed after aluminates as a separate group, followed by clay minerals (listed alphabetically by name). A few basic carbonates are also included in spite of solubility of CO_2 in water. Namely, basic carbonates are potential products of reaction of certain (hydr)oxides with atmospheric CO_2. Also other sparingly soluble basic salts of water soluble acids can be formed from (hydr)oxides at sufficiently high concentrations of certain anions, so the example of carbonates is not unique, but physical properties of the other basic salts are not reported in Chapter 2. For derivatives of less common oxides the crystallographic data are not given explicitly, only formulae of salts for which such data exist are reported.

The presentation of adsorption data follows the rule "from the simplest to the most complicated". This was achieved by organizing the adsorbates into the following categories (listed from the simplest to the most complicated)

- H^+/OH^-
- Inert electrolytes
- Small ions and neutral organic molecules that tend to be specifically adsorbed
- Surfactants
- Polymers

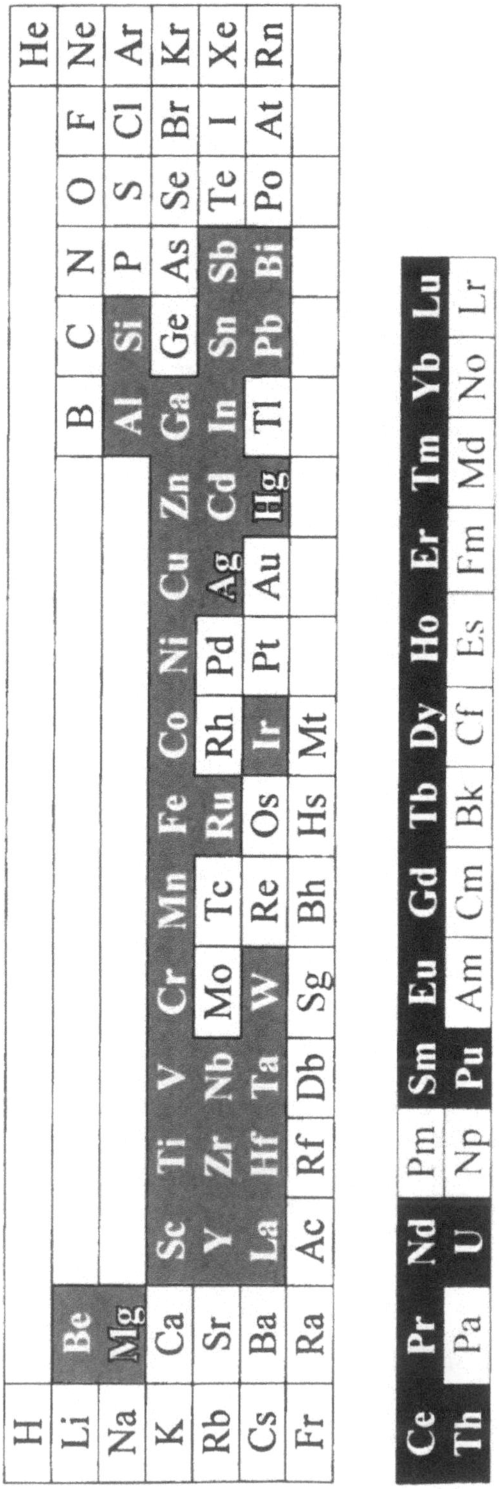

H																	He
Li	Be											B	C	N	O	F	Ne
Na	Mg											Al	Si	P	S	Cl	Ar
K	Ca	Sc	Ti	V	Cr	Mn	Fe	Co	Ni	Cu	Zn	Ga	Ge	As	Se	Br	Kr
Rb	Sr	Y	Zr	Nb	Mo	Tc	Ru	Rh	Pd	Ag	Cd	In	Sn	Sb	Te	I	Xe
Cs	Ba	La	Hf	Ta	W	Re	Os	Ir	Pt	Au	Hg	Tl	Pb	Bi	Po	At	Rn
Fr	Ra	Ac	Rf	Db	Sg	Bh	Hs	Mt									

Ce	Pr	Nd	Pm	Sm	Eu	Gd	Tb	Dy	Ho	Er	Tm	Yb	Lu
Th	Pa	U	Np	Pu	Am	Cm	Bk	Cf	Es	Fm	Md	No	Lr

FIG. 1.1 Availability of physical data for oxides; black background: crystallographic and thermochemical data available.

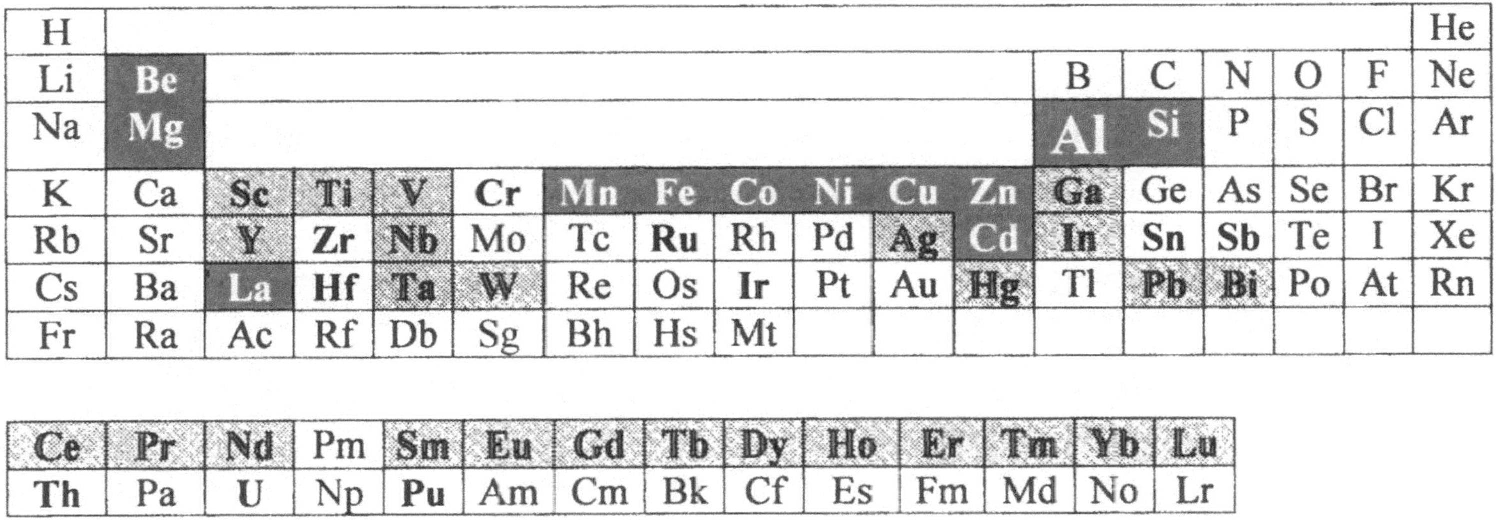

H																	He
Li	Be											B	C	N	O	F	Ne
Na	Mg											Al	Si	P	S	Cl	Ar
K	Ca	Sc	Ti	V	Cr	Mn	Fe	Co	Ni	Cu	Zn	Ga	Ge	As	Se	Br	Kr
Rb	Sr	Y	Zr	Nb	Mo	Tc	Ru	Rh	Pd	Ag	Cd	In	Sn	Sb	Te	I	Xe
Cs	Ba	La	Hf	Ta	W	Re	Os	Ir	Pt	Au	Hg	Tl	Pb	Bi	Po	At	Rn
Fr	Ra	Ac	Rf	Db	Sg	Bh	Hs	Mt									

Ce	Pr	Nd	Pm	Sm	Eu	Gd	Tb	Dy	Ho	Er	Tm	Yb	Lu
Th	Pa	U	Np	Pu	Am	Cm	Bk	Cf	Es	Fm	Md	No	Lr

FIG. 1.2 Availability of physical data for aluminates; black background: crystallographic and thermochemical data available, gray background: only crystallographic data available.

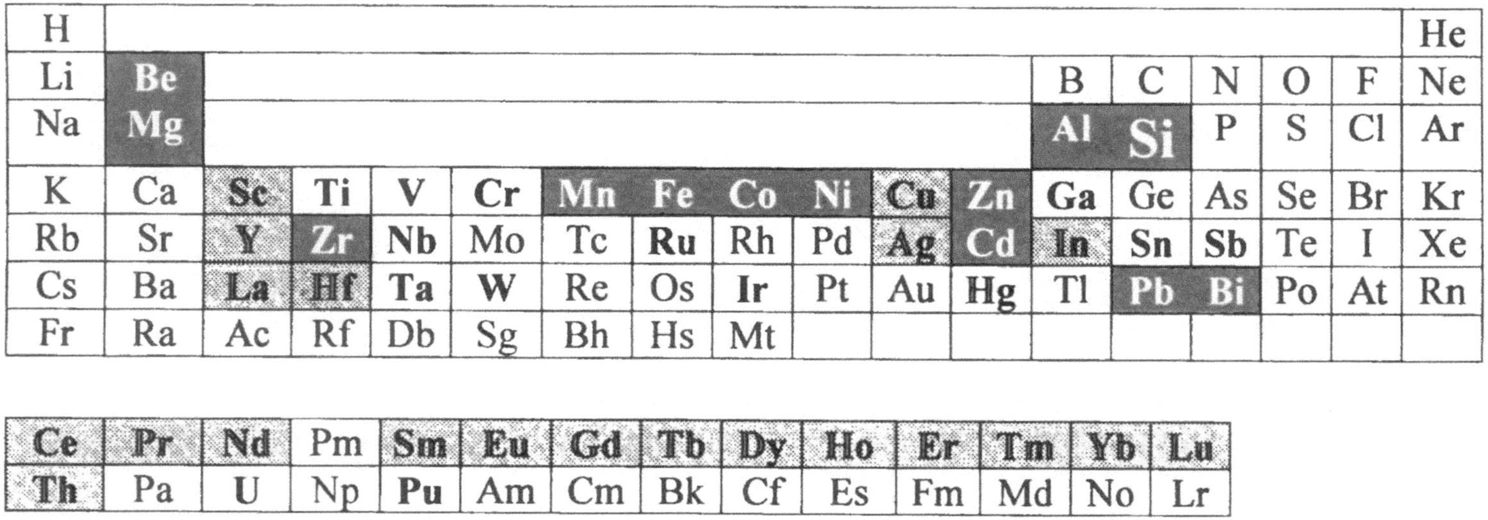

H																	He
Li	Be											B	C	N	O	F	Ne
Na	Mg											Al	Si	P	S	Cl	Ar
K	Ca	Sc	Ti	V	Cr	Mn	Fe	Co	Ni	Cu	Zn	Ga	Ge	As	Se	Br	Kr
Rb	Sr	Y	Zr	Nb	Mo	Tc	Ru	Rh	Pd	Ag	Cd	In	Sn	Sb	Te	I	Xe
Cs	Ba	La	Hf	Ta	W	Re	Os	Ir	Pt	Au	Hg	Tl	Pb	Bi	Po	At	Rn
Fr	Ra	Ac	Rf	Db	Sg	Bh	Hs	Mt									

Ce	Pr	Nd	Pm	Sm	Eu	Gd	Tb	Dy	Ho	Er	Tm	Yb	Lu
Th	Pa	U	Np	Pu	Am	Cm	Bk	Cf	Es	Fm	Md	No	Lr

FIG. 1.3 Availability of physical data for silicates; black background: crystallographic and thermochemical data available, gray background: only crystallographic data available.

gulation involving small and large particles is often considered as adsorption of particles on particles. This phenomenon plays an important role in transport of toxic waste in nature but it will not be discussed here. Adsorption of particles on particles was recently reviewed by Ryan and Elimelech [8].

In most publications a clear distinction is made between sorption (interaction between sorbate and pre-formed particles of the adsorbent) and coprecipitation (particles are formed in the presence of foreign species). Sometimes, however, these two phenomena are confused, e.g. in [9, 10] the terms "adsorption" and "sorption", respectively, are used in the title of the paper, abstract, text and figure captions, although the description of the experimental procedure clearly indicates that in fact coprecipitation was studied. Studies reporting only coprecipitation were not taken into account in the literature survey.

The quantities, which have been directly measured are referred to rather than those which were derived from the measured data. For example when the adsorption of counterions was assumed to entirely balance the surface charge (measured directly), the results are listed under "charge" rather than under "adsorption of counterions", even if the later was used in the title of the paper, abstract and figure captions.

In the studies conducted in aqueous solutions, the presence and thus sorption of H^+/OH^- ions cannot be avoided, also some concentration of inert electrolyte is usually present, even if such electrolyte was not added on purpose. To avoid repetitions, the studies involving multiple categories of adsorbates are listed only under the most complex one, e.g. coadsorption of surfactants and small anions is only listed in the section on surfactants but not in the section on sorption of anions.

The above classification of adsorbates is relative and subjective, it is practical, but not generally accepted. As a matter of fact, the same adsorbate can belong to different categories. This depends on specific system (adsorbent, solvent), concentration, solid to liquid ratio and other experimental conditions. Moreover, some results (e.g. Hoffmeister series) suggest that there is rather a continuum than a sharp border between strongly and weakly adsorbing ions. In the ion-exchange approach (*vide ultra*) protons are treated as any other cation. In spite of some shortcomings of the proposed classification, it is expedient to associate an adsorbate with certain category typical for it and then to consider a few exceptions. By no means is this classification original. The terms "potential determining ions", "specifically adsorbing ions", and "inert electrolyte" are widely used in the literature. Lyklema [11] introduced a term "generic adsorption" as opposite to "specific adsorption".

Chapter 3 presents data on points of zero charge obtained by different methods in the absence of strongly adsorbing species, usually at low concentrations (on the order of 10^{-2} mol dm^{-3}) of inert electrolytes, i.e. alkali nitrates V, chlorates VII and halides. Also corresponding salts of ammonium and its short-chain tetraalkyl derivatives are considered as inert electrolytes. The definition of the zero point is anything but trivial. For materials that are not penetrable to ions, the interfacial region as a whole is electrically neutral, but usually there is some excess of positive or negative charge at the surface due to adsorption/dissociation of protons, which is balanced by counterions (chiefly ions of the inert electrolyte whose sign is opposite to the sign of the surface charge) in the layer of solution next to the surface. The distribution of the counterions is governed primarily by the electrostatics. With purely electrostatic interaction the adsorption of coions (ions of the inert electrolyte whose sign is like the sign of the surface charge) should be negative. The countercharge is distributed over a thin layer of solution next to the surface, and a sufficiently thin layer of solution carries net positive or negative charge, despite the entire system is neutral. One way of defining the zero points is as the pH, at which certain layer of solution next to the surface has a zero charge. The charge of such thin layers cannot be measured directly, so this kind of definition must be based on some assumptions; specific examples are discussed in Chapter 3. To avoid assumptions an empirical definition may be used, linking the zero point with the results obtained by means of specified experimental procedure. In many cited publications more than one method was used to determine the zero point of the same material.

The presence of equal amounts of positively and negatively charged sites more or less evenly distributed over the surface results in zero net charge. It is often expedient to assume that the surface charge behaves as smeared out homogeneous charge. However in some phenomena discreteness of charge may play a significant role, especially if the surface shows a patchwise heterogeneity.

The materials in the compilation of zero point values are sorted by their formal chemical formulae and then the entries are sorted by the year of publication. This approach makes it somewhat difficult to compare results corresponding to given crystalline modification. On the other hand in many publications mixtures of different phases were used or the crystallographic data were not reported. Finally, different compositions have been alleged for apparently the same commercial product. Thus, it would be rather difficult to sort the data (within the same formal chemical formula) by the structure. Commercial materials are often characterized by trade names rather than by their crystalline structure.

Reference herein to any specific commercial product by trade name, trademark, manufacturer, or otherwise, does not constitute or imply its recommendation, or favoring. The difference in the point of zero charge PZC between different crystallographic forms of the same compound has been widely discussed, e.g. according to Parfitt [12] the zero point of rutile is about 1 pH unit below that of anatase. Many collections of selected PZC have been published. These collections are summarized and briefly discussed after presentation of original data.

In addition to sparingly soluble metal (hydr)oxides, salt type materials involving two such oxides or more, and clay minerals, whose crystallographic and thermochemical data are presented in Chapter 2, the zero points of zeolites, clays, and glasses are listed (in this order) after mixed oxides. Soils and other complex and ill-defined materials are on the end of the list. It should be emphasized that the terms "soil", "sediment", etc. have somewhat different meanings in different scientific and technical disciplines. This may lead to confusion, e.g. terms "kaolin" (clay) and "kaolinite" (clay mineral) are treated as synonyms in some publications. The zero points obtained for composite materials with a layer structure (core covered by coating) are listed separately from those in which the distribution of components is more uniform.

The salts involving relatively soluble acid or base may show very low solubility. A few examples of pH dependent surface charging of such materials are discussed in a separate section. These salts behave differently from the salts produced by sparingly soluble acid and base, that are listed as mixed oxides. In the table of zero points the salts are sorted according to the chemical symbol of the more acidic element, and then according to the symbol of the basic element. Sulfides are discussed separately, and their zero points are not listed in the table.

Correlations between the PZC and other measured quantities are discussed and illustrated by experimental results. In certain studies of surface charging or electrokinetic behavior no zero point has been found, often because the zero point was beyond the experimental range. Such results are collected in two separate tables (surface charging and ζ potentials). Selected values of ζ potentials and surface charge density σ_0 are presented in graphical form. They show the range of values reported in different sources, and are not supposed to be the "most reliable" ones. In many experimental studies the inert electrolyte concentrations are integer powers of ten, and only such data were used in the graphs. With other concentrations it was difficult to find another set of data obtained at the same ionic strength for comparison.

The "inert" electrolytes are, indeed indifferent at low concentrations and near the PZC. For example the values of ζ potentials and σ_0 are often symmetrical with respect to the PZC and insensitive to the nature of these salts. However certain results, e.g. Hoffmeister's series, studies of uptake of ions from solution, and even shifts in the isoelectric point IEP and PZC suggest non-electrostatic interactions of these ions and surfaces of materials, especially at high concentrations and far from the PZC. These results are shown and briefly discussed next.

The compilation of PZC in this book involves results obtained in temperature range 15–40°C. In some studies the temperature was not controlled or measured (room temperature) or at least the temperature is not reported. Detailed discussion of temperature effects on the PZC is not intended but a few examples of studies reporting the temperature dependence of the PZC are presented to show the general trends. Surface charging in mixed solvents and nonaqueous media, especially in polar solvents and water-organic mixtures is not much different from that in aqueous solution. A few results of such studies are presented as the last section of Chapter 3.

Strongly adsorbing species are discussed in Chapter 4. Specific adsorption of each class of compounds (small ions, surfactants, polymers) has specific terminology and methodology. First, methods used to study specific adsorption of small ions are discussed and the results are presented for cations and anions separately. The results are organized according to the adsorbent in the same order as the values of PZC. There are many publications in which more than one adsorbate was studied at basically the same conditions. There is only one table entry for each set of such data. The papers reporting data for multiple adsorbates are listed after the publications dealing with single adsorbate for the same type of adsorbent (in terms of formal chemical formula). Some readers may be interested in data for specific adsorbate rather than specific adsorbent, e.g. when looking for the best scavenger for certain element or species. For their convenience the Tables have indices of entries corresponding to specific adsorbates (sorted by adsorbate). The term "neutral organic molecules" used as a title of the next section is again conventional, namely, weak carboxylic acids are discussed in the section on specific adsorption of anions, although at experimental conditions their degree of dissociation is often low. In the section on neutral organic molecules the adsorption data for amines, phenols, amino acids, dyes, and humic substances are reported. These substances show certain degree of acidic or basic dissociation at the experimental conditions and their sorption behavior is similar to that of weak acids and bases. On the other hand, surface charging in the presence of lower alcohols, ethers, ketones, and amides is discussed in the section on mixed solvents. Separate sections are devoted to adsorption of surfactants and polymers, in view of their complex solution chemistry. The section on adsorption competition presents some results obtained in the systems with more than one adsorbate representing the same class, e.g. adsorption from solution containing two or more specifically adsorbing anions. Not necessarily does a real competition occur in such systems; sometimes the presence of other solutes even leads to a synergistic effect. On the other hand, simultaneous adsorption of adsorbates representing different classes, e.g. specific sorption of a cation in the presence of specifically adsorbing anion(s) or at different concentrations of inert electrolyte is not considered in the section on adsorption competition. The experimental results reported in Chapter 4 were obtained for different equilibration times ranging from a few minutes to over one year. The choice of equilibration time was often based on preliminary studies of the kinetics. Some results are presented as a conclusion of Chapter 4. Studies of adsorption kinetics can be also a source of information about the sorption mechanism.

Models of adsorption are discussed in detail in Chapter 5, but some terms used in adsorption modeling appear already in the chapters presenting the experimental data. The concepts and models originally developed for crystalline inorganic materials were also used to describe sorption properties of organic materials. A few examples are presented in Chapter 6.

The phenomena presented in this book were discussed in many reviews. For example, Schwarz [13] discussed methods used to characterize the acid base properties of catalysts. The review on sorption on solid – aqueous solution interface by Parks [14] includes also principles of surface science. The book *Environmental Chemistry of Aluminum* edited by Sposito reviews the solution and surface chemistry of aluminum compounds. Chapter 3 [15] provides thermochemical data for aluminum compounds. Chapter 5 [16] lists the points of zero charge of aluminum oxides, oxohydroxides and hydroxides with many references on adsorption of metal cations and various anions on these materials. Unlike the present book, which is confined to sorption from solution at room temperature, publications on coprecipitation and adsorption from gas phase or at elevated temperatures are also cited there. Brown et al. [17] reviewed on dry and wet surface chemistry of metal oxides. Stumm [18] reviewed sorption of ions on iron and aluminum oxides. The review by Schindler and Stumm [19] is devoted to surface charging and specific adsorption on oxides. Schindler [19] published a review on similar topic in German. Many other reviews related to specific topics are cited in respective chapters.

REFERENCES

1. M. Löbbus, W. Vogelsberger, J. Sonnefeld, and A. Seidel. Langmuir 14: 4386–4396 (1998).
2. M. A. Blesa, P. J. Morando and A. E. Regazzoni, Chemical Dissolution of Metal Oxides, CRC Press, Boca Raton 1994.
3. J. A. Salfity, A. E. Regazzoni and M. A. Blesa. In Interfacial Dynamics (N. Kallay, ed.) Marcel Dekker New York 1999, pp. 513–540.
4. N. Z. Misak. Adv. Colloid Interf. Sci. 51: 29–135 (1994).
5. O. A. Petrii, Electrochem. Acta 41: 2307–2312 (1996).

6.   S. Trasatti, Croat. Chem. Acta 63: 313–329 (1990).
7.   J. S. Hanson, and D. W. Fuerstenau, Colloids Surf. 26: 133–140 (1987).
8.   J. N. Ryan, and M. Elimelech, Colloids Surf. A. 107: 1–56 (1996).
9.   T. Aoki, and M. Munemori. Water Res. 16: 793–796 (1982).
10. S. Music. J. Radioanal. Nucl. Chem. 99: 161–170 (1986).
11. J. Lyklema. In Adsorption from Solution at Solid Liquid Interface (G.D. Parfitt and C. H. Rochester, eds.) Academic Press, New York 1983, pp. 223–246.
12. G. D. Parfitt, Pure Appl. Chem. 48: 415–418 (1976).
13. J. A. Schwarz, J. Colloid Interf. Sci. 218: 1–12 (1999).
14. G. A. Parks, Rev. Mineral. 23: 133–175 (1990).
15. B. S. Hemingway, and G. Sposito. In The Environmental Chemistry of Aluminum (G. Sposito, ed.), CRC Press 1996, pp. 81–116.
16. S. Goldberg, J. A. Davis, and J. D. Hem. In The Environmental Chemistry of Aluminum (G. Sposito, ed.), CRC Press 1996, pp. 271–331.
17. G. E. Brown, V. E. Henrich, W. H. Casey, D. L. Clark, C. Eggleston, A. Felmy, D. W. Goodman, M. Gratzel, G. Maciel, M. I. McCarthy, K. H. Nealson, D. A. Sverjensky, M. F. Toney and J. M. Zachara, Chem. Rev. 99: 77–174 (1999).
18. W. Stumm. Colloids Surf. A 73: 1–18 (1993).
19. P. W. Schindler and W. Stumm. In Aquatic Surface Chemistry: Chemical Processes at the Particle–Water Interface (W. Stumm, ed.) John Wiley 1987, pp. 83–110.
20. W. Schindler, Oster. Chem. Z. 86: 141 (1985).

2

Physical Properties of Adsorbents

I. CRYSTALLOGRAPHIC DATA

The surface properties of various phases corresponding to the same chemical formula can substantially differ from one crystallographic form to another. This is well known, and the information about the crystallographic form of the adsorbents is specified in many publications on adsorption, especially when the compound of interest forms more than one phase showing a sufficient stability at the experimental conditions. Since the present survey is devoted to adsorption studies performed at room temperature and atmospheric pressure the phases existing only at very high pressures or at very high (or very low) temperatures are not considered.

Usually there is no doubt which phase is meant in a publication, but also no normalized or generally accepted system has been established to name different phases having the same chemical formula. For example, in many publications γ and δ Al_2O_3 are considered two different phases, bayerite is called $\alpha\text{-}Al(OH)_3$, and goethite is called α-FeOOH. However, the following excerpt: "... $\delta\text{-}Al_2O_3$ (sometimes called γ -Al_2O_3) ..." was found in a recent publication in a highly ranked scientific journal and this was supported by three literature references. In another recent publication "$\gamma\text{-}Al(OH)_3$ (alumina gel, bayerite)" was studied. Apparently these three descriptions of the adsorbent are treated as synonyms. In another recent paper goethite is called "β-FeOOH". These three examples of inconsistent nomenclature are only a tip of the iceberg.

12

There are a few examples of two or more different names for the same crystallographic form or the same name shared by different crystallographic forms or even different chemical compounds. Specific examples of such inconsistencies can be found in Table 2.1. Probably the most accurate way to distinguish one phase from another would be to specify the space group, the lengths of axes of the elementary cell and the angles between them. This, however, is not practiced; sometimes the crystal system is specified.

Crystal, system	axes lengths	axial angles
triclinic	$a \neq b \neq c$	$\alpha \neq \beta \neq \gamma$
monoclinic	$a \neq b \neq c$	$\alpha = \gamma = 90° \neq \beta$
orthorhombic	$a \neq b \neq c$	$\alpha = \gamma = \beta = 90°$
tetragonal	$a = b \neq c$	$\alpha = \gamma = \beta = 90°$
trigonal	$a = b = c$	$120° > \alpha = \gamma = \beta \neq 90°$
hexagonal	$a = b \neq c$	$\alpha = \beta = 90°; \gamma = 120°$
cubic	$a = b = c$	$\alpha = \gamma = \beta = 90°$

The above choice of axes for particular systems is the most common one but it is not unique, e.g. for the trigonal and hexagonal systems it is convenient to use four axes: three of them are in one plane symmetrically spread to 120 and the fourth axis is perpendicular to this plane.

Table 2.1 lists adsorbents of interest, i.e. their chemical formulas and own names (and/or Greek letters used to distinguish given phase from the other polymorphs) and their crystallographic data: space group number (1–230; groups 1 and 2 belong to triclinic system, 3–15 to monoclinic, 16–74 to orthorhombic, 75–142 to tetragonal, 143–167 to trigonal, 168–194 to hexagonal and 195–230 to cubic system), space group according to Schoenflies, space group according to Hermann–Mauguin (sometimes the group number is unknown, in such instance the crystallographic system is given in Table 2.1, P = primitive; F = face centered; R = rhombohedral), the number of formula units in one cell Z, experimentally determined specific density d (in kg m^{-3} at 25°C or at the temperature given in brackets in °C), cell axes lengths a, b, c (in nm), and axial angles α, β, γ (decimal fractions are used rather than angular minutes). The empty cells in Table 2.1 have the following meanings: for axial angles empty cell = 90^c; for cell axes empty b or c means that $b=a$ or $c=a$, respectively; empty Z or d cell: data not available; there are also many phases which to the best knowledge of the present author do not have any names, sometimes the "name" cell contains other informations (e.g. about thermodynamic stability).

Table 2.1 reports crystallographic data of simple oxides, hydroxides and oxohydroxides, and then for mixed oxides (as discussed in Chapter 1). The principle of organization of Table 2.1 has been described in the Introduction. For derivatives of less common oxides, the compounds having known crystallographic structure are only listed, without specific information on this structure.

Reference [1] was the sole source of this data. Attempts to find a more up to date and equally comprehensive compilation failed. A relatively recent compilation [2] covers only rock-forming materials and most of the cited literature dates back to the sixties and early seventies. Reference [3] presents a collection of crystallographic data based on older compilations. The data from [1–3] match for a vast majority of

(*Text continues on pg. 49*)

TABLE 2.1 Crystallographic Data

Simple Oxides, Oxohydroxides and Hydroxides

Formula	Name	Number	Schoenflies	Hermann–Mauguin	Z	d	a	b	c	α	β	γ
Ag₂O		224	Oh⁴	Pn3m	2	7520	0.4736					
Al₂O₃	corundum, α	167	D3d⁶	R(-3)c	6	3990	0.47589		1.2991			
Al₂O₃	γ	227	Oh⁷	Fd3m	32/3	3619	0.7911					
Al₂O₃	δ	tetragonal			16	3420	0.796		1.170			
Al₂O₃	κ	hexagonal			28	3250	0.971		1.786			
Al₂O₃	κ'	186	C6v⁴	P6₃mc	16/3	3680	0.5544		0.9024			
5Al₂O₃·H₂O	tohdite	186	C6v⁴	P6₃mc	1	3720	0.5575		0.8761			
4Al₂O₃·H₂O	akdalaite	178	D6²	P6,22	18	3680	1.287		1.497			
AlOOH	diaspore, α	62	D2h¹⁶	Pbnm	4	3400	0.4401	0.9421	0.2845			
AlOOH	boehmite, γ	63	D2h¹⁷	Cmcm	4	2977; 3020	0.2868	1.2227	0.3700			
Al(OH)₃	gibbsite, hydrargillite, γ	14	C2h⁵	P2₁/n	8	2420 (20)	0.8676	0.5070	0.9721		94.57	
Al(OH)₃	bayerite, α	14	C2h⁵	P2₁/a	4	2529 (20)	0.5062	0.8671	0.4713		90.27	
Al(OH)₃	nordstrandite	2	Ci¹	I(-1)	8	2430	0.8752	0.5069	1.0244	109.326	97.662	88.340
BeO	bromellite	186	C6v⁴	P6₃mc	2	3015 (0)	0.26979		0.43772			
Be(OH)₂	behoite, β	19	D2⁴	P2₁2₁2₁	4	1924	0.4620	0.7039	0.4535			
Be(OH)₂	α (metastable)	tetragonal			16		1.088		0.783			
Bi₂O₃	bismite	14	C2h⁵	P2₁/c	4	9200	0.58486	0.81661	0.75097		113.00	
Bi₂O₃	sillenite, γ	197	T³	I23	12	8400..9200	1.0268					
Bi₂O₃	β	117	D2d⁷	P(-4)b2	8	9195	1.095		0.563			
CdO	monteponite	225	Oh⁵	Fm3m	4	8150	0.46951					
Cd(OH)₂	β	164	D3d³	P(-3)ml	1	4790(15)	0.3500		0.4710			
Cd(OH)₂	γ	8	Cs³	Im	4	4814 (20)	0.567	1.025	0.341		91.4	
CeO₂(-x)		225	Oh⁵	Fm3m	4	7132 (23)	0.5409					
CoO		225	Oh⁵	Fm3m	4	6470	0.42581					
CoO		12	C2h³	C2/m	2		0.5183	0.3015	0.3017		125.55	
Co(OH)₂	β	164	D3d³	P(-3)ml	1	3597 (15)	0.3173		0.4640			
3Co(OH)₂·2H₂O	α	162	D3d¹	P(-3)1m	1	2890	0.540		0.807			

Formula	Mineral	No.	Schoenflies	Space group	Z	d (kg/m³)	a (nm)	b (nm)	c (nm)	β (°)
Co_3O_4		227	O_h^7	Fd3m	8	6070	0.80835			
CoOOH	heterogenite	166	D_{3d}^5	$R\bar{3}m$	3	4720	0.2855		1.3157	
Cr_2O_3	eskolaite	167	D_{3d}^6	$R\bar{3}c$	6	5215	0.495762	0.27407	1.35874	
CrOOH	bracewellite	62	D_{2h}^{16}	Pbnm	4		0.449	0.986	0.297	
CrOOH	guyanaite	58	D_{2h}^{12}	Pnnm	2	4110	0.4861	0.4292	0.2960	
CrOOH	grimaldiite	166	D_{3d}^5	$R\bar{3}m$	3		0.2984		1.340	
Cu_2O	cuprite	224	O_h^4	$Pn\bar{3}m$	2	6000	0.4268			
CuO	tenorite	15	C_{2h}^6	C2/c	4	6450	0.46837	0.54226	0.51288	95.54
CuO	Paratenorit (Ger)	141	D_{4h}^{19}	$I4_1/amd$	16	6040	0.584		0.990	
$Cu(OH)_2$		63	D_{2h}^{17}	Cmcm	4	3850	1.059	0.2949	0.5256	
Dy_2O_3		206	T_h^7	Ia3	16	8120	1.0667			
Dy_2O_3		12	C_{2h}^3	C2/m	6		1.397	0.3519	0.8661	100.00
DyOOH		11	C_{2h}^2	$P2_1/m$	2	7170	0.598	0.364	0.429	109
DyOOH		113	D_{2d}^3	$P\bar{4}2_1/m$	4		0.5573		0.5446	
$Dy(OH)_3$		176	C_{6h}^2	$P6_3/m$	2	5780	0.6280		0.3569	
Er_2O_3		206	T_h^7	Ia3	16	8640	1.0550			
ErOOH		11	C_{2h}^2	$P2_1/m$	2	7590	0.594	0.352	0.430	109.3
ErOOH		113	D_{2d}^3	$P\bar{4}2_1/m$	4		0.5517		0.5383	
$Er(OH)_3$		176	C_{6h}^2	$P6_3/m$	2	5800	0.6232		0.3518	
Eu_2O_3		206	T_h^7	Ia3	16	7060	1.08670			
EuOOH		11	C_{2h}^2	$P2_1/m$	2	5850	0.6109	0.3748	0.4347	108.6
$Eu(OH)_3$		176	C_{6h}^2	$P6_3/m$	2	5100	0.6365		0.3645	
$Fe_{1-x}O$	wustite	225	O_h^5	Fm3m	4		0.43125			
$Fe(OH)_2$		164	D_{3d}^3	$P\bar{3}m1$	1		0.3262		0.4596	
$Fe_5O(OH)_9$	green rust	hexagonal R			1		0.3198		2.42	
$Fe_5O(OH)_9$	green rust	hexagonal P			1/2		0.3174		1.094	
$Fe_5O(OH)_9$		orthorhombic					0.5305	0.3160	0.7834	
Fe_3O_4	magnetite	227	O_h^7	Fd3m	8	5128	0.83940			
Fe_2O_3	hematite, α	167	D_{3d}^6	$R\bar{3}c$	6	5200..5300	0.50340		1.3752	
Fe_2O_3	β	206	T_h^7	Ia3	16		0.940			
Fe_2O_3	maghemite, γ	198	T^4	$P2_13$	32/3	4590	0.8339			
Fe_2O_3	δ	hexagonal			2		0.509		0.441	
Fe_2O_3	ε	monoclinic			20	4780	1.297	1.021	0.844	95.33
FeOOH	goethite, α	62	D_{2h}^{16}	Pbnm	4	3800..4400	0.462	0.995	0.301	
FeOOH	akaganeite, β	87	C_{4h}^5	I4/m	8	3100	1.048		0.3023	
FeOOH	lepidocrocite, γ	63	D_{2h}^{17}	Amam	4	3950	0.387	1.252	0.307	

TABLE 2.1 (Continued)

Formula	Name	Number	Schoenflies	Hermann Mauguin	Z	d	a	b	c	α	β	γ
FeOOH	δ	149	D_3^1	$P312$	1	3950	0.295		0.453			
Fe(OH)$_3$		14	C_{2h}^5	$P2_1/n$	8	3120	0.899	0.512	0.992		93.43	
Ga$_2$O$_3$	β	12	C_{2h}^3	$C2/m$	4	5950	1.223	0.304	0.580		103.7	
Ga$_2$O$_3$	α	167	D_{3d}^6	$R\bar{3}c$	6	6480	0.49825		1.3433			
GaOOH	α	62	D_{2h}^{16}	$Pbnm$	4		0.451	0.975	0.2965			
Ga(OH)$_3$	sohngeite	62	C_{2v}^7	$Pmn2_1$	8	3840	0.74865	0.74379	0.74963			
Gd$_2$O$_3$		206	T_h^7	$Ia3$	16	7600	1.0818					
Gd$_2$O$_3$		12	C_{2h}^3	$C2/m$	6	8220	1.4061	0.3566	0.8760		100.10	
GdOOH		11	C_{2h}^2	$P2_1/m$	2		0.606	0.371	0.434		108.9	
GdOOH		113	D_{2d}^3	$P\bar{4}2_1/m$	4	5540	0.5656		0.5552			
Gd(OH)$_3$		176	C_{6h}^2	$P6_3/m$	2		0.630		0.361			
HfO$_2$		14	C_{2h}^5	$P2_1/c$	4	10050	0.51156	0.51722	0.52948		99.18	
HgO	montroydite	62	D_{2h}^{16}	$Pnma$	4	11080(28)	0.66129	0.55208	0.35219			
HgO		152,154	$D_3^{4,6}$	$P3_{1,2}21$	3	11000	0.3577		0.8681			
Ho$_2$O$_3$		206	T_h^7	$Ia3$	16	8310	1.0606					
HoOOH		11	C_{2h}^2	$P2_1/m$	2	7150	0.596	0.364	0.431		109.1	
HoOOH		113	D_{2d}^3	$P\bar{4}2_1/m$	4		0.5541		0.5410			
Ho(OH)$_3$		176	C_{6h}^2	$P6_3/m$	2	5700	0.6255		0.3545			
In$_2$O$_3$		206	T_h^7	$Ia3$	16	7040	1.0117					
InOOH		31	C_{2v}^7	$P2_1nm$	2	7000	0.526	0.456	0.327			
In(OH)$_3$	djalindite (dzhalindite)	204	T_h^5	$Im3$	8	4410	0.7939					
IrO$_2$		136	D_{4h}^{14}	$P4_2/mnm$	2		0.44983		0.31544			
La$_2$O$_3$		206	T_h^7	$Ia3$	16		1.136					
La$_2$O$_3$		194	D_{6h}^4	$P6_3/mmc$	1	6510	0.39373		0.61299			
LaOOH		11	C_{2h}^2	$P2_1/m$	2	5360	0.6572	0.3929	0.4417		112.5	
La(OH)$_3$		176	C_{6h}^2	$P6_3/m$	2	4453	0.6531		0.3853			
Lu$_2$O$_3$		206	T_h^7	$Ia3$	16	9170	1.0390					
LuOOH		11	C_{2h}^2	$P2_1/m$	2		0.5955	0.355	0.423		112.5	
LuOOH		113	D_{2d}^3	$P\bar{4}2_1/m$	4		0.5464		0.5332			
MgO	periclase	225	O_h^5	$Fm3m$	4	3576	0.42117					

Formula	Mineral	System / No.	Schoenflies	Space group	Z	Density	a	b	c	β
$Mg(OH)_2$	brucite	164	D_{3d}^{3}	$P\bar{3}m1$	1	2360 (15)	0.3142		0.4766	
$Mg(OH)_2$		orthorhombic					0.3792	1.016	0.2842	
Mn_3O_4	hausmannite	141	D_{4h}^{19}	$I4_1/amd$	4	4720;4800	0.5763		0.9456	
$MnOOH$	groutite, α	62	D_{2h}^{16}	Pbnm	4	4140	0.4560	1.0700	0.2870	
$MnOOH$	feitknechtite, β	tetragonal (?)					0.86		0.93	
$MnOOH$	manganite, γ	14	C_{2h}^{5}	$B2_1/d$	8	4300	0.886	0.524	0.570	90
Mn_2O_3OH	Groutellit (Ger)	orthorhombic			2		0.461	0.954	0.288	
$Mn_7O_{13}\cdot5H_2O$	"δ-MnO₂"	hexagonal P			3	3660 (20)	0.284		0.727	
MnO_2	cryptomelane, α	87	C_{4h}^{5}	I4/m	8	4230	0.9815		0.2847	
MnO_2	pyrolusite, β	136	D_{4h}^{14}	$P4_2/mnm$	2	5120	0.4388		0.2865	
MnO_2	nsutite, γ	orthorhombic			4		0.443	0.936	0.285	
MnO_2	ε	194	D_{6h}^{4}	$P6_3/mmc$	1		0.2786		0.4412	
MnO_2	ramsdellite, ρ	62	D_{2h}^{16}	Pbnm	4		0.4533	0.927	0.2866	
Nb_2O_5		33	C_{2v}^{9}	$Pna2_1$	48	4870	1.470	0.9376	3.136	
Nb_2O_5		hexagonal			0.5	4860	0.3607		0.3925	
Nd_2O_3		206	T_h^{7}	Ia3	16	6490	1.1078			
$Nd_2O_3\cdot1\,3\,H_2O$		206	T_h^{7}	Ia3	16		1.1080			
$Nd_2O_3\cdot2\,3\,H_2O$		206	T_h^{7}	Ia3	16		1.1518			
$NdOOH$		11	C_{2h}^{2}	$P2_1\,m$	2	5840	0.6242	0.3809	0.4391	108.18
$Nd(OH)_3$		176	C_{6h}^{2}	$P6_3\,m$	2	4920	0.6425		0.3739	
NiO	bunsenite	hexagonal-R			3	6700..6900	0.29549		0.7227	
$4\,Ni(OH)_2\cdot NiOOH$		hexagonal			3/4	2950	0.3071		2.32	
$Ni(OH)_2$	β	164	D_{3d}^{3}	$P\bar{3}m1$	1	3850;3650	0.3126		0.4605	
$3Ni(OH)_2\cdot2H_2O$	α	162	D_{3d}^{1}	$P\bar{3}1m$	1	2520	0.534		0.750	
$Ni_3O_2(OH)_4$		hexagonal			1	3330	0.304		1.460	
Ni_2O_3		cubic-F			1		0.4186			
$NiOOH$		164	D_{3d}^{3}	$P\bar{3}m1$		4150 (20)	0.281		0.484	
PbO	massicot, β	57	D_{2h}^{11}	Pbma	4	9520	0.5489	0.4755	0.5891	
PbO	litharge, α	129	D_{4h}^{7}	P4/nmm	2	9280	0.39759		0.5023	
$Pb(OH)_2$		hexagonal			4		0.526		1.47	
Pb_3O_4	minium	135	D_{4h}^{13}	$P4_2/mbc$	4	9100	0.8815		0.6565	
PbO_2	plattnerite, β	136	D_{4h}^{14}	$P4_2/mnm$	2	9390..9490	0.4955		0.3383	
PbO_2	α (metastable)	60	D_{2h}^{14}	Pbcn	4	9530	0.4988	0.5958	0.5465	
Pr_2O_3		164	D_{3d}^{3}	$P\bar{3}m1$	1	6870..7070	0.3859		0.6008	
$Pr_2O_{3(x)}$ x)		206	T_h^{7}	Ia3	16		1.1152			

TABLE 2.1 (Continued)

Formula	Name	Number	Schoenflies	Hermann–Mauguin	Z	d	a	b	c	α	β	γ
PrOOH		11	$C2h^2$	$P2_1/m$	2		0.6287	0.3861	0.4397		107.81	
Pr(OH)$_3$		176	$C6h^2$	$P6_3/m$	2	4730	0.6454		0.3772			
PuO$_2$		225	Oh^5	Fm3m	4		0.53960					
RuO$_2$		136	$D4h^{14}$	$P4_2/mnm$	2	7200	0.44910		0.31064			
Sb$_2$O$_3$	valentinite	56	$D2h^{10}$	Pccn	4	5780 (20)	0.4914	1.2468	0.5421			
Sb$_2$O$_3$	senarmontite	227	Oh^7	Fd3m	16	5190	1.1152					
Sc$_2$O$_3$		206	Th^7	Ia3	16	3860	0.9849					
ScOOH		63	$D2h^{17}$	Cmcm	4	3020	0.324	1.301	0.401			
Sc(OH)$_3$	γ	204	Th^5	Im3	8	2650	0.7898					
SiO$_2$	quartz α	152,154	$D3^{4,6}$	$P3_{1,3}21$	3	2650	0.49138		0.54052			
SiO$_2$	tridymite α	15	$C2h^6$	C2/c	48	2270	1.854	0.501	2.579		117.67	
SiO$_2$	cristobalite α	92,96	$D4^{4,8}$	$P4_{1,3}2_11$	4	2331	0.4971		0.6918			
H$_2$Si$_2$O$_5$		46	$C2v^{22}$	Ibm2	4		0.612	1.540	0.496			
H$_2$Si$_2$O$_5$		37	$C2v^{13}$	Ccc2	24		0.57	1.46	2.976			
H$_2$Si$_2$O$_5$		62	$D2h^{16}$	Pmnb	4		0.747	1.194	0.491			
H$_2$Si$_2$O$_5$		13	$C2h^4$	P2/c	8	2060	1.1286	0.9905	0.8377		103.78	
H$_4$Si$_4$O$_{10}$·0.5H$_2$O		130	$D4h^8$	$P4/ncc$	4	2050	0.764		1.510			
Sm$_2$O$_3$		206	Th^7	Ia3	16	7310	1.0927					
Sm$_2$O$_3$		12	$C2h^3$	C2/m	6	7680	1.4177	0.3633	0.8847		99.96	
SmOOH		11	$C2h^2$	$P2_1/m$	2		0.613	0.377	0.438		108.6	
SmOOH		113	$D2d^3$	$P(-4)2_1/m$	4		0.5710		0.5599			
Sm(OH)$_3$		176	$C6h^2$	$P6_3/m$	2	5210	0.636		0.366			
Sm(OH)$_3$·H$_2$O		hexagonal			2		0.636		0.366			
SnO$_2$	cassiterite	136	$D4h^{14}$	$P4_2/mnm$	2	6720	0.47374		0.318638			
Ta$_2$O$_5$		orthorhombic			11		0.6198	4.0290	0.3888			
H$_2$Ta$_2$O$_6$·H$_2$O		227	Oh^7	Fd3m	8	5380	1.0580					
Tb$_2$O$_3$		206	Th^7	Ia3	16		1.07281					
Tb$_2$O$_3$		12	$C2h^3$	C2/m	6		1.403	0.3536	0.8717		100.10	
TbOOH		11	$C2h^2$	$P2_1/m$	2	6980	0.604	0.369	0.433		109.0	
TbOOH		113	$D2d^3$	$P(-4)2_1/m$	4		0.5623		0.5498			
Tb(OH)$_3$		176	$C6h^2$	$P6_3/m$	2	5660	0.6312		0.3601			

Formula	Mineral name	No.	Schoenflies	System	Space group	Z	ρ	a	b	c	β
ThO_2	thorianite	225	O_h^5		$Fm3m$	4	9870	0.55972			
TiO_2	brookite	61	D_{2h}^{15}		$Pbca$	8	4130	0.91819	0.54558	0.51429	
TiO_2	anatase	141	D_{4h}^{19}		$I4_1/amd$	4	4060	0.37852		0.95139	
TiO_2	rutile	136	D_{4h}^{14}		$P4_2/mnm$	2	4249.3	0.4593659		0.2958682	
Tm_2O_3		206	T_h^7		$Ia3$	16	8470	1.0488			
$TmOOH$		11	C_{2h}^2		$P2_1/m$	2		0.6015	0.359	0.425	112.5
$TmOOH$		113	D_{2d}^3		$P\bar{4}2_1/m$	4		0.5507		0.5364	
$Tm(OH)_3$		176	C_{6h}^2		$P6_3/m$	2		0.6233		0.3501	
UO_2	uraninite	225	O_h^5		$Fm3m$	4		0.54690			
U_4O_9		220	T_d^6		$I\bar{4}3d$	64	10730..10820	2.17632			
V_2O_3	karelianite	167	D_{3d}^6		$R\bar{3}c$	6	11159 (20)	0.49515		1.4003	
$VOOH$	montroseite	62	D_{2h}^{16}		$Pbnm$	4	4720..4780	0.454	0.997	0.303	
$V_3O_4(OH)_4$	doloresite	12	C_{2h}^3		$C2/m$	2	4090	1.964	0.299	0.483	103.92
V_2O_5	Vanadinocker (Ger)	59	D_{2h}^{13}		$Pmmm$	2	3270..3330	1.1510	0.43763	0.35677	
$V_2O_5 \cdot 3H_2O$	navajoite			monoclinic		6	3357	1.743	0.365	1.225	97
$W_{18}O_{49}$	γ WO_3	10	C_{2h}^1		$P2/m$	1	2560	1.832	0.379	1.404	115.03
$W_{20}O_{58}$	β WO_3	10	C_{2h}^1		$P2/m$	1	7720	1.205	0.3767	2.359	95.72
$W_{20}O_{59}$		3	C_2^1		$P2$	2	7150	1.19	0.385	4.94	106.07
$W_{25}O_{74}$	α WO_3	13	C_{2h}^4		$P2/c$	2	7140	1.190	0.3826	5.964	98.4
WO_3		14	C_{2h}^5		$P2_1/n$	8	7126	0.7306	0.7540	0.7692	90.881
$WO_2(OH)_2 \cdot H_2O$	hydrotungstite	10	C_{2h}^1		$P2_1/m$	2	7286	0.745	0.692	0.372	90
$H_8W_{12}O_{40} \cdot 5H_2O$		224	O_h^4		$Pn3m$	2		1.215			
Y_2O_3		206	T_h^7		$Ia3$	16	5010	1.0604			
$YOOH$		11	C_{2h}^2		$P2_1/m$	2	4620	0.595	0.365	0.430	109.1
$Y(OH)_3$		176	C_{6h}^2		$P6_3/m$	2	3810	0.6241		0.3539	
$Y(OH)_3$		14	C_{2h}		$P2_1/c$			0.625	0.601	1.540	97.5
Yb_2O_3		206	T_h^7		$Ia3$	16	9170	1.04342			
$YbOOH$		11	C_{2h}^2		$P2_1/m$	2	7270	0.587	0.358	0.427	109.3
$YbOOH$		113	D_{2d}^3		$P\bar{4}2_1/m$	4		0.5482		0.5326	
$Yb(OH)_3$		176	C_{6h}^2		$P6_3/m$	2		0.620		0.346	
ZnO	zincite	186	C_{6v}^4		$P6_3mc$	2	5642	0.3249858		0.5206619	
$Zn(OH)_2$	β	44	C_{2v}^{20}	orthorhombic	$Imm2$	42	3354 (20)	1.317	0.642	2.41	
$Zn(OH)_2$	γ	19	D_2^4		$P2_12_12_1$	12	3234 (20)	2.307	0.804	0.330	
$Zn(OH)_2$	ε					4	3030	0.5170	0.8547	0.4930	
ZrO_2	baddeleyite	14	C_{2h}^5		$P2_1/c$	4	5820:5730	0.5138	0.5204	0.5313	99.2

Aluminates

Formula	Name	Number	Schoenflies	Hermann–Mauguin	Z	d	a	b	c	β
$Ag_8Al_{22}O_{34}$		194	$D6h^4$	$P6_3/mmc$	1	3813	0.5595		2.2488	
$AgAlO_2$		166	$D3d^5$	$R(-3)m$	3		0.2890		1.827	
$BeAl_2O_4$	chrysoberyl	62	$D2h^{16}$	Pnma	4	3600..3860	0.9404	0.5476	0.4427	
$BeAl_6O_{10}$		57	$D2h^{11}$	Pcam	8	3740	0.955	1.3757	0.8909	
$Bi_2Al_4O_9$		55	$D2h^9$	Pbam	2	6220	0.7721	0.8099	0.5689	
$Cd_8Al_2O_{11}\cdot 11H_2O$	hexagonal				3/10		0.331		2.285	
$CdAl_2O_4$		148	$C3i^2$	$R(-3)$	18		1.4212		0.9581	
$CdAl_4O_7$		15	$C2h^6$	$C2/c$	4		1.268	0.886	0.540	105.98
$CeAlO_3$		167	$D3d^6$	$R(-3)c$	6	6170	0.5348		1.3021	
$CoAl_2O_4$		227	Oh^7	$Fd3m$	8	4370 (18)	0.8160			
$Co_4Al(OH)_{11}$	hexagonal				1		0.313		2.37	
$CuAl_2O_4$		227	Oh^7	$Fd3m$	8	4532 (20)	0.8078			
$Dy_3Al_5O_{12}$		230	Oh^{10}	$Ia3d$	8		1.20381			
$Dy_3Sc_2Al_3O_{12}$		230	Oh^{10}	$Ia3d$	8		1.2360			
$Dy_4Al_2O_9$		14	$C2h^5$	$P2_1/c$	4		0.7432	1.0533	1.1089	108.16
$DyAlO_3$		194	$D6h^4$	$P6_3/mmc$	2		0.3700		1.050	
$Er_3Al_5O_{12}$		230	Oh^{10}	$Ia3d$	8		1.1981			
$Er_3Sc_2Al_3O_{12}$		230	Oh^{10}	$Ia3d$	8		1.2300			
$ErAlO_3$		194	$D6h^4$	$P6_3/mmc$	2		0.3660		1.050	
$Eu_3Sc_2Al_3O_{12}$		230	Oh^{10}	$Ia3d$	8		1.2460			
$Eu_4Al_2O_9$		14	$C2h^5$	$P2_1/c$	4		0.7608	1.0616	1.1101	108.50
$EuAlO_3$		62	$D2h^{16}$	Pbnm	4		0.5267	0.5294	.7459	
$EuAlO_3$		194	$D6h^4$	$P6_3/mmc$	2		0.3760		1.052	
$FeAl_2O_4$	hercynite	227	Oh^7	$Fd3m$	8		0.8150			
$Gd_3Al_5O_{12}$		230	Oh^{10}	$Ia3d$	8		1.2113			
$Gd_4Al_2O_9$		14	$C2h^5$	$P2_1/c$	4		0.759	1.061	1.112	108.5
$GdAlO_3$		194	$D6h^4$	$P6_3/mmc$	2		0.3730		1.051	
$HgAl_{12}O_{19}$		194	$D6h^4$	$P6_3/mmc$	2		0.558		2.268	
$Ho_3Al_5O_{12}$		230	Oh^{10}	$Ia3d$	8		1.2011			
$Ho_3Sc_2Al_3O_{12}$		230	Oh^{10}	$Ia3d$	8		1.2324			
$Ho_4Al_2O_9$		14	$C2h^5$	$P2_1/c$	4		0.737	1.050	1.099	108.1

$HoAlO_3$		194	D_{6h}^4	$P6_3/mmc$	2		0.3670		1.051	
$LaAlO_3$		167	D_{3d}^6	$R\bar{3}c$	6	5840	0.5365		1.311	
$Lu_3Al_5O_{12}$		230	O_h^{10}	$Ia3d$	8		1.1906			
$LuAlO_3$		194	D_{6h}^4	$P6_3/mmc$	2		0.3614		1.0500	
$Mg_4Al_2(OH)_{12}CO_3 \cdot 3H_2O$	hydrotalcite	166	D_{3d}^5	$R\bar{3}m$	1/2	2090	0.3054		2.281	
$Mg_6Al_2O_9 \cdot 13H_2O$	meixnerite	166	D_{3d}^5	$R\bar{3}m$	3/8	1900	0.30463		2.293	
$MgAl_2O_4$	spinel	227	O_h^7	$Fd3m$	8	3570	0.80831			
$AlMg_{64}(OH)_{11}$		hexagonal			1		0.310		2.37	
$MnAl_2O_4$	galaxite	227	O_h^7	$Fd3m$	8	4120	0.8258			
$Nd_4Al_2O_9$		14	C_{2h}^5	$P2_1/c$	4		0.7725	1.0846	1.1306	109.5
$NdAlO_3$		167	D_{3d}^6	$R\bar{3}c$	6	7030	0.5322		1.2916	
$NiAl_2O_4$		227	O_h^7	$Fd3m$	8		0.8061			
$Ni_4Al(OH)_{11}$		hexagonal			1		0.306		2.34	
$Pb_2Al_2O_5$		Cubic P (?)			17	6650	1.324			
$PbAl_{12}O_{19}$		194	D_{6h}^4	$P6_3/mmc$	2	4590	0.5563		2.2033	
$PbAl_2O_4$		4	C_2^2	$P2_1$	2	5640	0.527	0.846	0.507	118.8
$PrAlO_3$		167	D_{3d}^6	$R\bar{3}c$	6		0.5334		1.297	
$ScAlO_3$		hexagonal R			8		1.307		1.705	
$Sm_3Sc_2Al_3O_{12}$		230	O_h^{10}	$Ia3d$	4		1.2495			
$Sm_4Al_2O_9$		14	C_{2h}^5	$P2_1/c$	8		0.762	1.068	1.115	108.5
$SmAlO_3$		62	D_{2h}^{16}	$Pbnm$	8	6120	0.52912	0.52904	0.74740	
$Tb_3Al_5O_{12}$		230	O_h^{10}	$Ia3d$	2		1.2074			
$TbAlO_3$		194	D_{6h}^4	$P6_3/mmc$	8		0.3710		1.051	
$Tm_3Al_5O_{12}$		230	O_h^{10}	$Ia3d$	8		1.1957			
$Tm_3Sc_2Al_3O_{12}$		230	O_h^{10}	$Ia3d$	8		1.2220			
$TmAlO_3$		194	D_{6h}^4	$P6_3/mmc$	2		0.364		1.0510	
$Y_3Al_5O_{12}$		230	O_h^{10}	$Ia3d$	8	4650	1.2008			
$Y_3Sc_2Al_3O_{12}$		230	O_h^{10}	$Ia3d$	8		1.2324			
$Y_4Al_2O_9$		14	C_{2h}^5	$P2_1/c$	4		0.7373	1.0467	1.1121	108.53
$Y_4Al_4O_{12}$		230	O_h^{10}	$Ia3d$	8	4530	1.1989			
$YAlO_3$		194	D_{6h}^4	$P6_3/mmc$	2		0.3678		1.052	
$Yb_3Al_5O_{12}$		230	O_h^{10}	$Ia3d$	8		1.19295			
$Yb_3Sc_2Al_3O_{12}$		230	O_h^{10}	$Ia3d$	8		1.2162			
$YbAlO_3$		194	D_{6h}^4	$P6_3/mmc$	2		0.3625		1.0500	
$Zn_8Al_2O_{11} \cdot 11H_2O$		166	D_{3d}^5	$R\bar{3}m$	3,10		0.311		2.34	
$ZnAl_2O_4$	gahnite	227	O_h^7	$Fd3m$	8	4602 (20)	0.80848			

Aluminosilicates

Formula	Name	Number	Schoenflies	Hermann–Mauguin	Z	d	a	b	c	α	β	γ
$Al_4Ti_2SiO_{12}$	Pseudoanatas (Ger.)	141	D_{4h}^{19}	$I4_1/amd$	2/3		0.3777		0.9460			
$Be_3Al_2Si_6O_{18}$	beryl	192	D_{6h}^2	P6/mcc	2	2781	0.9212		0.9236			
$BeAlSiO_4(OH)$	euclase	14	C_{2h}^5	$P2_1/a$	4	3095	0.4763	1.429	0.4618		100.2	
$Cu_2AlSi_3O_9(OH)\cdot 2H_2O$	ajoite	11	C_{2h}^2	$P2_1/m$			1.5218	2.4712	1.3632		92.91	
$Fe_2Al_4(SiO_4)_2O_2(OH)_4$	chloritoid	2	C_i^1	P(-1)	2	3580	0.950	0.548	0.916	96.88	101.82	90.03
$Fe_2Al_3Si_5O_{18}$	Fe-indialite	192	D_{6h}^2	P6/mcc	2		0.9860		0.9285			
$Fe_2Al_5Si_5O_{18}$	Fe-cordierite	66	D_{2h}^{20}	Cccm	4		1.7065	0.9726	0.9287			
$Fe_2Al_9O_7(SiO_4)_4(OH)$	staurolite	12	C_{2h}^3	C2/m	2	3770	0.7879	1.6635	0.5664		90	
$Fe_3Al_2(SiO_4)_3$	almandine	230	O_h^{10}	$Ia3d$	8	4325 (20)	1.1526					
$Fe_5Al_3Si_3O_{11}(OH)_2$	ferrogedrite	62	D_{2h}^{16}	Pnma	4	3566	1.8514	1.7945	0.53158			
$Mg_2Al(Al_8O_7)(SiO_4)_4(OH)$		12	C_{2h}^3	C2/m	2		0.7887	1.6552	0.5635		90	
$Mg_2Al_4Si_5O_{18}$	indialite β	192	D_{6h}^2	P6/mcc	2		0.9770		0.9352			
$Mg_2Al_4Si_5O_{18}$	cordierite	66	D_{2h}^{20}	Cccm	4	2632	1.7083	0.9738	0.9335			
$Mg_2Al_4Si_5O_{18}\cdot 0.5H_2O$	cordierite	66	D_{2h}^{20}	Cccm	4		1.7079	0.9730	0.9356			
$Mg_{3.5}Al_9Si_{1.5}O_{20}$	sapphirine	14	C_{2h}^5	$P2_1/a$	4	3490	1.1266	1.4401	0.9929		125.46	
$Mg_3Al_2(SiO_4)_3$	pyrope	230	O_h^{10}	$Ia3d$	8	3708	1.1456					
$Mg_6Al_2Si_7O_{22}(OH)_2$		62	D_{2h}^{16}	Pnma	4		1.8433	1.7807	0.5270			
$MgAl_2O_4\cdot 3SiO_2$		180,181	$D_6^{4,5}$	$P6_{2,4}22$	1		0.5182		0.5360			
$MgAl_2Si_4O_{12}$		192	D_{6h}^2	P6/mcc	6		1.012		1.436			
$Mn_2Al_2(SiO_4)_3$	spessartine	230	O_h^{10}	$Ia3d$	8	4152	1.16207					
$Mn_2Al_4Si_5O_{18}$	Mn-indialite	192	D_{6h}^2	P6/mcc	2		0.9925		0.9297			
$Mn_2Al_4Si_5O_{18}$	Mn-cordierite	66	D_{2h}^{20}	Cccm	4		1.728	0.995	0.928			
$MnAl_2(Si_2O_6)(OH)_4$		68	D_{2h}^{22}	Ccca	8	3040	1.3831	2.0296	0.5121			
$Pb_3Al_{10}SiO_{20}$		12	C_{2h}^3	$I2/m$	2		1.434	1.139	0.496		90.0	
$Pb_4Al_2Si_2O_{11}$		hexagonal			2		0.880		0.950			
$Pb_6Al_6Si_6O_{21}$		139	D_{4h}^{17}	I4/mmm	2	5360	1.1723		0.8044			
$PbAl_2Si_2O_8$		15	C_{2h}^6	I2/c	8		0.8408	1.3038	1.4358		115.11	

Clay Minerals

Formula	Name	Number	Schoenflies	Hermann Mauguin	Z	d	a	b	c	α	β	γ
$Cu_4H_4(Si_4O_{10})(OH)_8$	chrysocola	orthorhombic					0.57	0.885	0.67			
$Al_2Si_2O_5(OH)_4$	dickite	9	Cs^4	Cc	4		0.5150	0.8940	1.4424		96.73	
$Al_4Si_4O_{10}(OH)_8 \cdot 4H_2O$	halloysite	8	Cs^3	Cm	1	2100..2600	0.520	0.892	1.025		100.2	
$Al_2Si_2O_5(OH)_4$	kaolinite	2	Ci^1	P(-1)	2		0.514	0.893	0.737	91.8	104.5..105	90
$Al_2Si_2O_5(OH)_4$	kaolinite	9	Cs^4	Cc	6	2100..2600	0.513	0.890	2.15		90	
$Al_4Si_4O_{10}(OH)_8$	metahalloysite	8	Cs^3	Cm	1		0.515	0.89	0.757		100	
$Na_{0.33}Mg_{0.33}Al_{1.67}$ $Si_4O_{10}(OH)_2 \cdot 4H_2O$	montmorillonite	12	$C2h^3$	C2/m	2		0.517	0.894	1.52		90	
$Al_2Si_2O_5(OH)_4$	nacrite	9	Cs^4	Cc	4		0.8909	0.5146	1.5697		113.7	

Basic Carbonates

Formula	Name	Number	Schoenflies	Hermann–Mauguin	Z	d	a	b	c	α	β	γ
$Al_{14}(OH)_{36}(CO_3)_3$	scarbroite	triclinic			4	2170	0.994	1.488	2.647	98.7	96.5	89
$Bi_2O_2CO_3$		139	$D4h^{17}$	I4/mmm	2	7300..8200	0.3867		1.3686			
$Cu_2(OH)_2(CO_3)$	malachite	14	$C2h^5$	P2₁/a	4	4050	0.9502	1.1974	0.3240		98.75	
$Cu_2Mg_2(OH)_6(CO_3) \cdot 2H_2O$	callaghanite	15	$C2h^6$	C2/c	2	2710	1.006	1.180	0.824		107.3	
$Cu_3(OH)_2(CO_3)_2$	azurite	14	$C2h^5$	P2₁/c	2	3773	0.500	0.585	1.035		92.33	
$Dy_2O_2CO_3$		139	$D4h^{17}$	I4/mmm	2		0.393		1.290			
$Er_2O_2CO_3$		139	$D4h^{17}$	I4/mmm	2		0.392		1.275			
$Fe_6(OH)_{12}(CO_3) \cdot 3H_2O$		hexagonal			3/8		0.317		2.28			
$La_2O_2CO_3$		194	$D6h^4$	P6₃/mmc	4		0.776		0.947			
$LaOHCO_3$		190	$D3h^4$	P(-6)2c	1		0.4214		0.5041			
$Mg_{10}Fe_2(OH)_{24}(CO_3)_2 \cdot 2H_2O$	coalingite	166	$D3d^5$	R(-3)m	1/2	2320..2330	0.312		3.74			
$Mg_2(OH)_2CO_3 \cdot 3H_2O$	artinite	12	$C2h^3$	C2/m	4	2040 (16)	1.6561	0.5298	0.6220		99.15	

Basic Carbonates (Continued)

Formula	Name	Number	Schoenflies	Hermann–Mauguin	Z	d	a	b	c	α	β	γ
$Mg_4Al_2(OH)_{12}CO_3 \cdot 3H_2O$	hydrotalcite	166	D_{3d}^5	R(-3)m	1/2	2090	0.3054		2.281			
$Mg_5(OH)_2(CO_3)_4 \cdot 4H_2O$	hydromagnesite	monoclinic			4	2236	1.858	0.906	0.842		90	
$Mg_8Al_2(OH)_{16}(CO_3) \cdot 4H_2O$	manasseite	194	D_{6h}^4	$P6_3/mmc$	1	2050	0.613		1.537			
$Mg_6Cr_2(OH)_{16}(CO_3) \cdot 4H_2O$	barbertonite	hexagonal			1	2100	0.618		1.555			
$Mg_6Cr_2(OH)_{16}(CO_3) \cdot 4H_2O$	stichtite	166	D_{3d}^5	R(-3)m	3		0.619		4.647			
$Mg_6Fe(OH)_{13}(CO_3) \cdot 4H_2O$	brugnatellite	hexagonal P			1	2140	0.547		1.597			
$Nd_2O(CO_3)_2 \cdot 2H_2O$		orthorhombic					0.817	0.850	0.745			
$Nd_2O_2CO_3$		194	D_{6h}^4	$P6_3/mmc$	2	5890	0.3974		1.5703			
$Ni_3(OH)_4(CO_3) \cdot 4H_2O$	zaratite	cubic P			1	2570..2650	0.616					
$Ni_6Fe_2(OH)_{16}(CO_3) \cdot 4H_2O$	reevesite	166	D_{3d}^5	R(-3)m	3	2870..2900	0.6164		4.554			
$Pb_{10}(OH)_6O(CO_3)_6$	plumbonacrite	hexagonal			3	7070	0.9076		2.496			
$Pb_2Al_4(OH)_8(CO_3)_4 \cdot 3H_2O$	dundasite	51	D_{2h}^5	Pbmm	3	3550	0.905	1.635	0.561			
$Pb_3(OH)_2(CO_3)_2$	hydrocerussite	hexagonal R			3	6820	0.5239		2.365			
$Pb_3O(CO_3)_2$		1	C_1^1	P1	1	5910 (20)	0.866	0.515	0.789	61.08	57.3	116.35
$Pr_2O(CO_3)_2 \cdot 2H_2O$		orthorhombic					0.816	0.860	0.743			
$Pr_2O_2CO_3$		194	D_{6h}^4	$P6_3/mmc$	2		0.4012		1.5693			
$PrOHCO_3$		190	D_{3h}^4	P(-6)2c	1		0.4146		0.4986			
$Tb_4O(CO_3)_5 \cdot 7H_2O$		orthorhombic					0.928	1.142	0.769			
$YOHCO_3$		19	D_2^4	$P2_12_12_1$	4		0.4809	0.6957	0.8466			
$Zn_4(OH)_2(CO_3)_3 \cdot 4H_2O$		hexagonal					1.332		0.7537			
$Zn_5(OH)_6(CO_3)_2$	hydrozincite	12	C_{2h}^3	C2/m	2	4000	1.3479	0.6320	0.5368		95.5	

Cobaltates

$AgCoO_2$, $BiCoO_3$, Cu_2CoO_3, $CuCo_2O_4$, $CuCoO_2$, $DyCoO_3$, $ErCoO_3$, $EuCoO_3$, $FeCo_2O_4$, $GdCoO_3$, $HoCoO_3$, $LaCoO_3$, $MgCo_2O_4$, $MnCo_2O_4$, $MnCoO_3$, $NdCoO_3$, $NiCo_2O_4$, $PrCoO_3$, $SmCoO_3$, $TbCoO_3$, $YCoO_3$, $ZnCo_2O_4$, $Co_3Zn_2(OH)_{10} \cdot 2H_2O$.

Chromates (III)

Formula	Name	Number	Schoenflies	Hermann–Mauguin	Z	d	a	b	c	α	β	γ
$AgCrO_2$		166	$D3d^5$	R(-3)m	3	6850	0.29843		1.8511			
$BeCr_2O_4$		62	$D2h^{16}$	Pnma	4	4420	0.9792	0.5663	0.4555			
$BiCrO_3$		pseudotriclinic			1		0.3906	0.3870	0.3906	90.55	89.15	90.55
$CdCr_2O_4$		227	Oh^7	Fd3m	8	5790 (17)	0.8597					
$CrCd_4(OH)_{11}$		hexagonal			1		0.333		2.37			
$CeCrO_3$		62	$D2h^{16}$	Pbnm	4	5143	0.5473	0.5473	0.7742			
$CoCr_2O_4$		227	Oh^7	Fd3m	8	5580	0.83299					
$CuCr_2O_4$		122	$D2d^{12}$	I(-4)2d	8	5490	$a\sqrt{2} = .8532$		0.7788			
$CuCrO_2$	mcconnellite	166	$D3d^5$	R(-3)m	3		0.29747		1.71015			
$CuGaCr_2O_4$		227	Oh^7	Fd3m	8		0.8305					
$DyCrO_3$		62	$D2h^{16}$	Pbnm	4	7850	0.5265	0.5520	0.7559			
$ErCrO_3$		62	$D2h^{16}$	Pbnm	4		0.5223	0.5516	0.7519			
$EuCrO_3$		62	$D2h^{16}$	Pbnm	4		0.5340	0.5515	0.7622			
$FeCr_2O_4$	chromite	227	Oh^7	Fd3m	8	5100	0.8377					
$GdCrO_3$		62	$D2h^{16}$	Pbnm	4		0.5312	0.5514	0.7611			
$HgCr_2O_4$		cubic-F			8	7540 (20)	0.865					
$HoCrO_3$		62	$D2h^{16}$	Pbnm	4		0.5243	0.5519	0.7538			
$InCrO_3$		62	$D2h^{16}$	Pbnm	4		0.5170	0.5355	0.7543			
$LaCrO_3$		62	$D2h^{16}$	Pbnm	4		0.5477	0.5514	0.7755			
$LuCrO_3$		62	$D2h^{16}$	Pbnm	4		0.5176	0.5497	0.7475			
$MgCr_2O_4$	magnesiochromite	227	Oh^7	Fd3m	8	4415	0.8333					
$MgInCrO_4$		227	Oh^7	Fd3m	8		0.8588					
$MnCr_2O_4$		227	Oh^7	Fd3m	8	4870	0.8437					
$NdCrO_3$		62	$D2h^{16}$	Pbnm	4		0.5412	0.5494	0.7695			
$NiCr_2O_4$		227	Oh^7	Fd3m	8		0.8328					
$NiCr_2O_4$		141	$D4h^{19}$	$I4_1/amd$	8		$a\sqrt{2} = .8253$		0.8441			
$PrCrO_3$		62	$D2h^{16}$	Pbnm	4	4000	0.5444	0.5484	0.7710			
Sc_3CrO_6		148	$C3i^2$	R(-3)	6		0.882		1.022			
$SmCrO_3$		62	$D2h^{16}$	Pbnm	4		0.5372	0.5502	0.7650			
$TbCrO_3$		62	$D2h^{16}$	Pbnm	4		0.5291	0.5513	0.7557			

Chromates (III) (Continued)

Formula	Name	Number	Schoenflies	Hermann–Mauguin	Z	d	a	b	c	α	β	γ
$TmCrO_3$		62	$D2h^{16}$	Pbnm	4		0.5209	0.5508	0.7500			
$YbCrO_3$		62	$D2h^{16}$	Pbnm	4		0.5195	0.5510	0.7490			
$YCrO_3$		62	$D2h^{16}$	Pbnm	4	5780	0.5247	0.5518	0.7540			
$ZnCr_2O_4$		227	Oh^7	Fd3m	8	5290 (13)	0.83275					

Ferrates

Formula	Name	Number	Schoenflies	Hermann–Mauguin	Z	d	a	b	c
$AgFeO_2$		166	$D3d^5$	R(-3)m	3	6390	0.30391		1.8590
$AlFeO_3$		33	$C2v^9$	$Pc2_1n$	8	4410..4450 (27)	0.860	0.925	0.497
$Bi_{24}Fe_2O_{39}$		tetragonal					0.8633		0.9605
$Bi_{24}Fe_2O_{39}$		197	T^3	I23	1	9080	1.018		
$Bi_2Fe_4O_9$		55	$D2h^9$	Pbam	2	6530	0.7950	0.8428	0.6005
$Bi_{40}Fe_2O_{48}$		197	T^3	I23			1.01		
$Bi_{40}Fe_2O_{63}$		tetragonal					1.044		1.354
$Bi_{46}Fe_2O_{72}$		197	T^3	I23	1/2	9290	1.019		
$Bi_4Fe_2O_9$		tetragonal					1.2102		1.7865
$BiFeO_3$		161	$C3v^6$	R3c	6	8310	0.55876		1.3867
$CeFeO_3$		62	$D2h^{16}$	Pbnm	4		0.5519	0.5536	0.7819
$CoFe_2O_4$		227	Oh^7	Fd3m	8	5190	0.83919		
$Co_4Fe(OH)_{11}$		hexagonal			1		0.311		2.31
$CuFe_2O_4$		141	$D4h^{19}$	$I4_1/amd$	8		$a\sqrt{2}=.8222$		0.872
$Dy_3Fe_5O_{12}$		230	Oh^{10}	Ia3d	8	6080	1.2405		
$DyFeO_3$		62	$D2h^{16}$	Pbnm	4		0.5302	0.5598	0.7623
$Er_3Fe_5O_{12}$		230	Oh^{10}	Ia3d	8	6330	1.2347		

Formula	Mineral	No./System	Schoenflies	Space group	Z	Density	a	b	c
ErFeO$_3$		62	D_{2h}^{16}	Pbnm	4		0.5267	0.5581	0.7593
Eu$_3$Fe$_5$O$_{12}$		230	O_h^{10}	Ia3d	8		1.2498		
EuFeO$_3$		62	D_{2h}^{16}	Pbnm	4		0.5372	0.5606	0.7685
GaFeO$_3$		33	C_{2v}^9	Pc2$_1$n	8	5330	0.87512	0.93993	0.50806
Gd$_3$Fe$_5$O$_{12}$		230	O_h^{10}	Ia3d	8		1.2478937		
GdFeO$_3$		62	D_{2h}^{16}	Pbnm	4		0.5349	0.5611	0.7669
Ho$_3$Fe$_5$O$_{12}$		230	O_h^{10}	Ia3d	8		1.2380		
HoFeO$_3$		62	D_{2h}^{16}	Pbnm	4	6050	0.5278	0.5591	0.7602
LaFeO$_3$		62	D_{2h}^{16}	Pbnm	4		0.5553	0.5563	0.7867
Lu$_3$Fe$_5$O$_{12}$		230	O_h^{10}	Ia3d	8		1.2277		
LuFeO$_3$		62	D_{2h}^{16}	Pbnm	4	4436	0.5213	0.5547	0.7565
MgFe$_2$O$_4$	magnesioferrite	227	O_h^7	Fd3m	8		0.8380	1.003	0.991
MgLaFeO$_4$		36	C_{2v}^{12}	Bm2$_1$b	4		0.388		
FeMg$_4$(OH)$_{11}$		hexagonal			1		0.314		2.34
FeMn$_4$(OH)$_{11}$		hexagonal			1		0.321		2.37
MnFe$_2$O$_4$	jacobsite	227	O_h^7	Fd3m	8	4760	0.85050		
Nd$_3$Fe$_5$O$_{12}$		230	O_h^{10}	Ia3d	8		1.2596		
NdFeO$_3$		62	D_{2h}^{16}	Pbnm	4		0.5453	0.5584	0.7768
NiFe$_2$O$_4$	trevorite	227	O_h^7	Fd3m	8	5268	0.8339		
Pb$_2$Fe$_2$O$_5$		tetragonal	D_{2d}^6	$P\bar{4}c2$	8	8180	0.779		1.585
Pb$_3$Fe$_2$O$_6$		116			1		0.391		1.528
Pb$_4$Fe$_2$O$_{11}$		tetragonal			4		1.083		1.586
PbFe$_{12}$O$_{19}$	magnetoplumbite	194	D_{6h}^4	P6$_3$/mmc	2	5517 (20)	0.5889		2.311
PbFe$_2$O$_4$		198	T^4	P2$_1$3	4		0.7830		
PbFe$_4$O$_7$	plumboferrite	149	D_3^1	P312	42	6070	1.186		4.714
PbFe$_6$O$_{10}$		hexagonal			42		1.180		2.315
PbFe$_8$O$_{13}$		hexagonal			2	6050	0.662		1.019
PbZn$_2$Fe$_{16}$O$_{27}$		194	D_{6h}^4	P6$_3$/mmc	2		0.588		3.282
PrFeO$_3$		62	D_{2h}^{16}	Pbnm	4		0.5482	0.5578	0.7786
Sm$_3$Fe$_5$O$_{12}$		230	O_h^{10}	Ia3d	8		1.2540		
SmFeO$_3$		62	D_{2h}^{16}	Pbnm	4		0.5400	0.5597	0.7711
Tb$_3$Fe$_5$O$_{12}$		230	O_h^{10}	Ia3d	8		1.2437471		
TbFeO$_3$		62	D_{2h}^{16}	Pbnm	4		0.5326	0.5602	0.7635
Tm$_3$Fe$_5$O$_{12}$		230	O_h^{10}	Ia3d	8		1.2325		
TmFeO$_3$		62	D_{2h}^{16}	Pbnm	4		0.52491	0.55716	0.75824
Y$_3$Fe$_5$O$_{12}$		230	O_h^{10}	Ia3d	8		1.2376		

Ferrates (Continued)

Formula	Name	Number	Schoenflies	Hermann–Mauguin	Z	d	a	b	c
YFeO$_3$		62	D2h^{16}	Pbnm	4		0.5283	0.5592	0.7603
Yb$_3$Fe$_5$O$_{12}$		230	Oh^{10}	Ia3d	8	6080	1.2291		
YbFeO$_3$		62	D2h^{16}	Pbnm	4		0.5233	0.5557	0.7570
ZnFe$_2$O$_4$	franklinite	227	Oh^7	Fd3m	8		0.84411		

Gallates (attention, the same name is used for salts of 3, 4, 5-trihydroxybenzoic acid)

AgGaO$_2$, AlGaO$_3$, BeGa$_4$O$_7$, BeNd$_2$Ga$_2$O$_7$, BeSm$_2$Ga$_2$O$_7$, Bi$_2$Ga$_4$O$_9$, Bi$_3$Sb$_2$GaO$_{11}$, CdGa$_2$O$_4$, CeGaO$_3$, CoGa$_2$O$_4$, CrMnGaO$_4$, CuGa$_2$O$_4$, CuGa$_5$O$_8$, CuGaO$_2$, CuMnGaO$_4$, Dy$_3$Ga$_5$O$_{12}$, DyGaO$_3$, Er$_3$Ga$_5$O$_{12}$, ErGaO$_3$, Eu$_3$Ga$_5$O$_{12}$, EuGaO$_3$, FeGa$_2$O$_4$, Gd$_3$Ga$_5$O$_{12}$, GdGaO$_3$, Ho$_3$Ga$_5$O$_{12}$, HoGaO$_3$, La$_3$Lu$_2$Ga$_3$O$_{12}$, La$_3$Sc$_2$Ga$_3$O$_{12}$, La$_3$Tm$_2$Ga$_3$O$_{12}$, La$_3$Yb$_2$Ga$_3$O$_{12}$, LaGaO$_3$, Lu$_3$Ga$_5$O$_{12}$, LuGaO$_3$, MgGa$_2$O$_4$, MnFeGaO$_4$, MnGa$_2$O$_4$, Nd$_3$Er$_2$Ga$_3$O$_{12}$, Nd$_3$Ga$_5$O$_{12}$, Nd$_3$Lu$_2$Ga$_3$O$_{12}$, Nd$_3$Tm$_2$Ga$_3$O$_{12}$, Nd$_3$Yb$_2$Ga$_3$O$_{12}$, Nd$_4$Ga$_2$O$_9$, Nd$_4$Ga$_4$O$_{12}$, NdGaO$_3$, NiGa$_2$O$_4$, Pb$_2$Ga$_2$O$_5$, PbGa$_{12}$O$_{19}$, PbGa$_2$O$_4$, Pr$_3$Er$_2$Ga$_3$O$_{12}$, Pr$_3$Ga$_5$O$_{12}$, Pr$_3$Lu$_2$Ga$_3$O$_{12}$, Pr$_3$Tm$_2$Ga$_3$O$_{12}$, Pr$_3$Yb$_2$Ga$_3$O$_{12}$, PrGaO$_3$, ScGaO$_3$, Sm$_3$Ga$_5$O$_{12}$, Sm$_4$Ga$_4$O$_{12}$, SmGaO$_3$, Tb$_3$Ga$_5$O$_{12}$, TbGaO$_3$, Tm$_3$Ga$_5$O$_{12}$, TmGaO$_3$, Y$_3$Ga$_5$O$_{12}$, Y$_3$Sc$_2$Ga$_3$O$_{12}$, Y$_4$Ga$_4$O$_{12}$, Yb$_3$Ga$_5$O$_{12}$, YbGaO$_3$, YGaO$_3$, ZnGa$_2$O$_4$, ZnMnGaO$_4$.

Hafnates

CdHfO$_3$, Ce$_2$Hf$_2$O$_7$, Er$_6$HfO$_{11}$, Eu$_2$Hf$_2$O$_7$, Gd$_2$Hf$_2$O$_7$, Gd$_6$HfO$_{11}$, Ho$_2$Hf$_2$O$_7$, La$_2$Hf$_2$O$_7$, La$_2$ZnHfO$_6$, Mg$_2$Hf$_5$O$_{12}$, Nd$_2$Hf$_2$O$_7$, PbHfO$_3$, Pr$_2$Hf$_2$O$_7$, Sc$_2$Hf$_2$O$_7$, Sc$_2$Hf$_5$O$_{13}$, Sc$_4$Hf$_3$O$_{12}$, Sc$_2$Hf$_3$O$_{12}$, Sm$_2$Hf$_2$O$_7$, TiHfO$_4$.

Indates

AgInO$_2$, AlInO$_3$, BeIn$_2$O$_4$, CdAlInO$_4$, CdGaInO$_4$, CdIn$_2$O$_4$, CoAlInO$_4$, CoGaInO$_4$, Cu$_2$In$_2$O$_5$, CuAlInO$_4$, CuGaInO$_4$, DyInO$_3$, ErInO$_3$, EuInO$_3$, GaInO$_3$, GdInO$_3$, HoInO$_3$, LaInO$_3$, LuInO$_3$, MgAlInO$_4$, MgGaInO$_4$, MgIn$_2$O$_4$, MnAlInO$_4$, MnIn$_2$O$_4$, NdInO$_3$, NiAlInO$_4$, NiGaInO$_4$, SmInO$_3$, Tb$_3$InO$_6$, TbInO$_3$, Y$_3$Ga$_3$In$_2$O$_{12}$, YInO$_3$, Zn$_2$In$_2$O$_6$, Zn$_2$In$_2$O$_5$, Zn$_3$In$_2$O$_6$, Zn$_4$In$_2$O$_7$, Zn$_5$In$_2$O$_8$, Zn$_7$In$_2$O$_{10}$, ZnAlInO$_4$, ZnGaInO$_4$, ZnIn$_2$O$_4$.

Manganates

Formula	Name	Number	Schoenflies	Hermann–Mauguin	Z	d	a	b	c	α	β	γ
AgMn$_2$O$_4$		227	Oh^7	Fd3m	8		0.8650	0.8534	0.5766			
Bi$_2$Mn$_4$O$_{10}$		55	D2h^9	Pbam	2	7133	0.7540	0.3989	0.3935			
BiMnO$_3$		pseudotriclinic			1		0.3935			91.47	90.93	91.47
Cd$_2$Mn$_3$O$_8$		12	C2h^3	C2/m	2		1.0806	0.5808	0.4932		109.53	
CdMn$_2$O$_4$		141	D4h^{19}	I4$_1$/amd	4	5520	0.5832		0.9754			
CeMnO$_3$		62	D2h^{16}	Pbnm	4		0.5537	0.5557	0.7812			
CoMn$_2$O$_4$		141	D4h^{19}	I4$_1$/amd	8		a√2 = .804		0.904			
CrMn$_2$O$_4$		141	D4h^{19}	I4$_1$/amd	8		a√2 = .8355		0.8752			
CuCrMnO$_4$		227	Oh^7	Fd3m	8		0.831					
CuGaMnO$_4$		227	Oh^7	Fd3m	8		0.8365					
CuMn$_2$O$_4$		227	Oh^7	Fd3m	8	5270	0.8362					
CuMnO$_2$	crednerite	12	C2h^3	C2/m	2	5380	0.5530	0.2884	0.5898		104.6	
DyMn$_2$O$_5$		55	D2h^9	Pbam	4	6500	0.72940	0.85551	0.56875			
DyMnO$_3$		62	D2h^{16}	Pbnm	4		0.5279	0.5843	0.7378			
ErMn$_2$O$_5$		55	D2h^9	Pbam	4		0.724	0.840	0.566			
EuMn$_2$O$_5$		55	D2h^9	Pbam	4		0.737	0.856	0.570			
EuMnO$_3$		62	D2h^{16}	Pbnm	4		0.5535	0.5853	0.7448			
FeMn$_2$O$_4$		141	D4h^{19}	I4$_1$/amd	8		a√2 = .8402		0.8799			
GdMn$_2$O$_5$		55	D2h^9	Pbam	4		0.736	0.852	0.569			
GdMnO$_3$		62	D2h^{16}	Pbnm	4		0.5317	0.5863	0.7433			
HoMn$_2$O$_5$		55	D2h^9	Pbam	4		0.733	0.846	0.568			
HoMnO$_3$		62	D2h^{16}	Pbnm	4		0.5255	0.5831	0.7354			
La$_2$MgMnO$_6$		221	Oh^1	Pm3m	1/2		0.3882					
LaMn$_2$O$_5$		55	D2h^9	Pbam	4		0.767	0.866	0.572			
LaMnO$_3$		62	D2h^{16}	Pbnm	4		0.5537	0.5743	0.7695			
LuMn$_2$O$_5$		55	D2h^9	Pbam	4		0.723	0.842	0.566			
LuMnO$_3$		185	C6v^3	P6$_3$cm	6		0.60455		1.1394			
Mg$_2$MnO$_4$		227	Oh^7	Fd3m	8	3630	0.838					
Mg$_6$MnO$_8$		225	Oh^5	Fm3m	4		0.8381					
MgMn$_2$O$_4$		141	D4h^{19}	I4$_1$/amd	4		0.5806		0.8915			

Manganates (Continued)

Formula	Name	Number	Schoenflies	Hermann–Mauguin	Z	d	a	b	c	α	β	γ
MgMnO$_3$		148	C3i^2	R(-3)	6		0.4945		1.373			
NdMn$_2$O$_5$		55	D2h^9	Pbam	4		0.754	0.863	0.570			
NdMnO$_3$		62	D2h^{16}	Pbnm	4		0.5414	0.5829	0.7551			
Ni$_6$MnO$_8$		225	Oh^5	Fm3m	4		0.8312					
NiCrMnO$_4$		227	Oh^7	Fd3m	8		0.827					
NiMn$_2$O$_4$		227	Oh^7	Fd3m	8		0.8399					
PbMn$_2$O$_2$OH	quenselite	13	C2h^4	P2/a	4	6842	0.561	0.570	0.915		93.0	
PrMn$_2$O$_5$		55	D2h^9	Pbam	4		0.754	0.866	0.570			
PrMnO$_3$		62	D2h^{16}	Pbnm	4		0.5545	0.5787	0.7575			
ScMnO$_3$		185	C6v^3	P6$_3$cm	6	4350	0.5830		1.1179			
SmMn$_2$O$_5$		55	D2h^9	Pbam	4		0.741	0.856	0.570			
SmMnO$_3$		62	D2h^{16}	Pbnm	4		0.5359	0.5843	0.7482			
TbMn$_2$O$_5$		55	D2h^9	Pbam	4		0.734	0.851	0.569			
TbMnO$_3$		62	D2h^{16}	Pbnm	4		0.5297	0.5831	0.7403			
TmMn$_2$O$_5$		55	D2h^9	Pbam	4		0.724	0.840	0.567			
TmMnO$_3$		185	C6v^3	P6$_3$cm	6		0.6062		1.140			
YbMn$_2$O$_5$		55	D2h^9	Pbam	4		0.723	0.840	0.567			
YbMnO$_3$		185	C6v^3	P6$_3$cm	6		0.6062		1.140			
YMn$_2$O$_5$		55	D2h^9	Pbam	4		0.727	0.845	0.566			
YMnO$_3$		185	C6v^3	P6$_3$cm	6		0.6125		1.141			
Zn$_2$Mn$_3$O$_8$		212,213	O^{6,7}	P4$_{3,1}$32	4		0.8220					
ZnCrMnO$_4$		141	D4h^{19}	I4$_1$/amd	8	5180	a$\sqrt{2}$ = .825					
ZnMn$_2$O$_4$	hetaerolite	141	D4h^{19}	I4$_1$/amd	4		0.5722		0.863			
ZnMn$_3$O$_7$·3H$_2$O	chalkophanite	2	Ci^1	P(-1)	2	3980	0.754	0.754	0.9236	90	117.2	120
ZnMnO$_3$		148	C3i^2	R(-3)	6		0.4965		1.380			

Niobates

$AgNbO_3$, $AlNb_{11}O_{29}$, $AlNbO_4$, $Bi_2Nb_{10}O_{28}$, Bi_3TiNbO_9, $Bi_5Nb_3O_{15}$, $Bi_6Nb_{34}O_{94}$, $Bi_6Ti_8Nb_2O_{30}$, $Bi_8Nb_{18}O_{57}$, $BiNbO_4$, Cd_2CrNbO_6, $Cd_2Nb_2O_7$, Cd_2SbNbO_6, $CdNb_2O_6$, Ce_3NbO_7, $CeNbO_4$, $CeNb_5O_{14}$, $CeTiNbO_6$, $Co_3Nb_4O_{14}$, $Co_4Nb_2O_9$, $CoNb_2O_6$, $CrNb_{49}O_{124}$, $CuNb_2O_6$, Dy_3NbO_7, $DyNbO_4$, $DyTiNbO_6$, Er_3NbO_7, $ErNbO_4$, $EuNbO_4$, Eu_3NbO_7, $EuTiNbO_6$, $Fe_4Nb_2O_9$, $Fe_8Nb_{10}O_{37}$, $FeNb_{11}O_{29}$, $FeNb_2O_6$, $FeNb_{49}O_{124}$, $FeNbO_4$, $GaNb_{11}O_{29}$, $GaNbO_4$, Gd_3NbO_7, $GdNbO_4$, $GdTiNbO_6$, $Hg_2Nb_2O_7$, Ho_3NbO_7, $HoNbO_4$, $InNbO_4$, $InTiNbO_6$, $La_3Co_2NbO_9$, La_3NbO_7, $LaNb_3O_9$, $LaNb_5O_{14}$, $LaTiNbO_6$, $LuNbO_4$, $LuTiNbO_6$, $Mg_4Nb_2O_9$, $Mg_5Nb_4O_{15}$, $MgFeNbO_5$, $MgNb_2O_6$, $Mn_4Nb_2O_9$, $MnNb_2O_6$, $MnNbO_4$, $Nd_2BiFe_2Nb_3O_{15}$, $Nd_3Fe_2Nb_3O_{15}$, Nd_3NbO_7, $NdNb_3O_9$, $NdNb_5O_{14}$, $NdNbO_4$, $NdTiNbO_6$, $Ni_4Nb_2O_9$, $NiNb_2O_6$, $NiSbNbO_6$, Pb_2AlNbO_6, Pb_2CrNbO_6, Pb_2GaNbO_6, Pb_2HoNbO_6, Pb_2InNbO_6, Pb_2LuNbO_6, $Pb_2Nb_2O_7$, $Pb_2Nd_4Fe_3Nb_7 O_{30}$, Pb_2NiNbO_6, Pb_2SbNbO_6, Pb_2VNbO_6, Pb_2YbNbO_6, Pb_2YNbO_6, $Pb_3Nb_2O_8$, $Pb_3Nb_4O_{13}$, $Pb_3NiNb_2O_9$, $Pb_5Nb_4O_{15}$, $PbBi_2Nb_2O_9$, $PbBi_3Ti_2NbO_{12}$, $PbLaTiNbO_6$, $PbNb_2O_6$, $P_5Ti_3Nb_4O_{17}$, $PbTiNb_2O_8$, $PrNb_3O_9$, $PrNb_5O_{14}$, $PrNbO_4$, $PrTiNbO_6$, $SbNbO_4$, Sc_3NbO_7, $ScNbO_4$, $ScTiNbO_6$, Sm_2DyNbO_7, Sm_2ErNbO_7, Sm_2EuNbO_7, Sm_2GdNbO_7, Sm_2HoNbO_7, Sm_2InNbO_7, Sm_2LuNbO_7, Sm_2TmNbO_7, Sm_2YbNbO_7, Sm_2YNbO_7, $Sm_3 NbO_7$, $SmNbO_4$, $SmTiNbO_6$, Tb_3NbO_7, $TbTiNbO_6$, $Ti_2Nb_{10}O_{29}$, $Ti_3Nb_{34}O_{91}$, $Ti_5Nb_{44}O_{120}$, $TiNb_{24}O_{62}$, $TiNb_2O_7$, $TiNb_{38}O_{97}$, $TiNb_{52}O_{132}$, $TmNbO_4$, $TmTiNbO_6$, $V_2Nb_{23}O_{62}$, $V_3Nb_{17}O_{50}$, V_3NbO_{10}, VNb_9O_{25}, Y_3NbO_7, Yb_3NbO_7, $YbTiNbO_6$, $YNbO_4$, $YTiNbO_6$, $ZnNb_2O_6$, $ZrNb_{10}O_{27}$, $ZrNb_{14}O_{37}$, $ZrNb_{24}O_{62}$.

Niclates

$AgNiO_2$, $BiNiO_3$, $CuNi_2O_4$, $DyNiO_3$, $ErNiO_3$, $EuNiO_3$, $GdNiO_3$, $HoNiO_3$, La_2MnNiO_6, $LaNiO_3$, $LuNiO_3$, $MnNiO_3$, $NdNiO_3$, $SmNiO_3$, $TmNiO_3$, $YbNiO_3$, $YNiO_3$.

Plumbates

Ag_2PbO_2, $Ag_5Pb_2O_6$, $Bi_{12}PbO_{20}$, $Cd_2Pb_2O_3(OH)_6$, Cd_2PbO_4, $CdPb(OH)_6$, $CdPbO_2(OH)_2$, $CdPbO_3$, Cu_6PbO_8, $Dy_2Pb_2O_7$, $Er_2Pb_2O_7$, $Eu_2Pb_2O_7$, $Gd_2Pb_2O_7$, $Ho_2Pb_2O_7$, $La_2Pb_2O_7$, $Nd_2Pb_2O_7$, $Pr_2Pb_2O_7$, $Sm_2Pb_2O_7$, $Tb_2Pb_2O_7$, $Y_2Pb_2O_7$, Zn_2PbO_4.

Ruthenates

$Bi_2Ru_2O_7$, Co_2RuO_2, $Dy_2Ru_2O_7$, $Er_2Ru_2O_7$, $Eu_2Ru_2O_7$, $Gd_2Ru_2O_7$, $GdBiRu_2O_7$, $Ho_2Ru_2O_7$, $Lu_2Ru_2O_7$, $Nd_2Ru_2O_7$, $NdBiRu_2O_7$, $PbRuO_3$, $Pr_2Ru_2O_7$, $Sm_2Ru_2O_7$, $Tb_2Ru_2O_7$, $Tm_2Ru_2O_7$, $Y_2Ru_2O_7$, $Yb_2Ru_2O_7$.

Silicates

Formula	Name	Number	Schoenflies	Hermann Mauguin	Z	d	a	b	c	α	β	γ
$Ag_2Si_2O_5$		60	D_{2h}^{14}	$Pcnb$	4	5620	0.653	1.568	0.497			
Ag_2SiO_3		orthorhombic			4	6070	1.00	0.713	0.455			
Ag_4SiO_4		2	C_i^1	$P\bar{1}$	2	5950	0.674	0.924	0.582	94.0	95.1	115.5
$Ag_6Si_2O_7$		13	C_{2h}^4	$P2/n$	2	6580	1.0264	0.5259	0.8052		110.5	
$Al_2Si_4O_{10}(OH)_2$	pyrophyllite	15	C_{2h}^6	$C2/c$	4	2650..2900	0.5172	0.8958	1.8676		100.0	
$Al_2Si_4O_{10}(OH)_2$	pyrophyllite	2	C_i^1	$P\bar{1}$	2		0.51614	0.89576	0.93511	91.03	100.37	89.75
$Al_3Si_4O_{11}$		monoclinic			4		0.5173	0.9114	1.8995		100.2	
$Al_2Si_4O_{11}$		triclinic			2		0.5140	0.9116	0.9504	91.2	100.0	90.0
Al_2SiO_5	sillimanite	62	D_{2h}^{16}	$Pbnm$	4	3230..3240	0.74856	0.76738	0.57698			
Al_2SiO_5	andalusite	58	D_{2h}^{12}	$Pnnm$	4	3140..3150	0.77942	0.78985	0.5559			
Al_2SiO_5	cyanite, disthene	2	C_i^1	$P\bar{1}$	4	3560..3670	0.71192	0.78473	0.55724	89.977	101.121	106.006
$Al_3Si_2O_7(OH)_3$	piezotite	2	C_i^1	$P\bar{1}$	2		0.7287	0.7720	0.9729	70.32	128.7	115.72
$Al_6Si_2O_{13}$	mullite	55	D_{2h}^9	$Pbam$	3/4	3160	0.7583	0.7681	0.28854			
$H_{50}Al_{50}Si_{133}O_{384}$	H-faujasite	227	O_h^7	$Fd3m$	1		2.4756					
$Be_2FeY_2(SiO_4)_2O_2$	gadolinite	14	C_{2h}^5	$P2_1/c$	2	4000..4500	0.471	0.752	0.989		90.55	
Be_2SiO_4	phenakite	148	C_{3i}^2	$R\bar{3}$	18	2980	1.2474		0.8251			
$Be_2Y_2SiO_7$		113	D_{2d}^3	$P\bar{4}2_1m$	2	4280	0.7283		0.4755			
$Be_2Yb_2NiSi_2O_{10}$		14	C_{2h}^5	$P2_1/c$	2		0.4664	0.7385	0.9866		90.02	
$Be_4Si_2O_7(OH)_2$	bertrandite	36	C_{2v}^{12}	$Ccm2_1$	4	2599 (20)	1.531	0.873	0.458			
$Bi_{12}SiO_{20}$		197	T^3	$I23$	2		1.0098					
Bi_2SiO_5		36	C_{2v}^{12}	$Cmc2_1$	4		1.5195	0.54680	0.53148			
$Bi_4(SiO_4)_3$	eulytite	220	T_d^6	$I\bar{4}3d$	4	6750	1.0294					
$Cd_2La_8(SiO_4)_6O_2$		176	C_{6h}^2	$P6_3/m$	1	5860	0.964		0.709			
$Cd_2Nd_8(SiO_4)_6O_2$		176	C_{6h}^2	$P6_3/m$	1	5800	0.9562		0.7075			
Cd_2SiO_4		70	D_{2h}^{24}	$Fddd$	8		0.9796	1.1797	0.6007			
$Cd_3Y_8(SiO_4)_6O_2$		176	C_{6h}^2	$P6_3/m$	1		0.939		0.677			
Cd_3SiO_5		129	D_{4h}^7	$P4/nmm$	2		0.6835		0.4954			
$Cd_3V_2(SiO_4)_3$		230	O_h^{10}	$Ia3d$	8		1.203					
$Cd_3(SiO_4)_2(OH)_2$	Cd-chondrodite	14	C_{2h}^5	$P2_1/c$	4	5400	1.795	0.528	1.150		109	
$CdSiO_3$		14	C_{2h}^5	$P2_1/a$	12	5100	1.504	0.71	0.696		94	

Formula	Mineral	System	No.	Point group	Space group	Z		a	b	c	α	β	γ
$Ce_3(SiO_4)_3$		hexagonal	176		$P6_3/m$	1		1.136		0.471			
$Ce_{9.333}(SiO_4)_6O_2$			161	C_{6h}^{2}	$R3c$	6		0.96586		0.71207			
$Co_2La_7Si_6O_{23}(OH)_3$			62	C_{3v}^{6}	$Pbnam$	4		1.078	1.0302	3.840			
Co_2SiO_4			161	D_{2h}^{16}	$R3c$	6		0.4782	0.922	0.6003			
$Co_2Sm_7Si_6O_{23}(OH)_3$		orthorhombic / hexagonal		C_{3v}^{6}		1		1.06		3.68			
$Co_6Si_4O_{10}(OH)_8$	Co-antigorite		20	D_2^{5}	$C222_1$	16		0.536	1.309	0.720			
$Co_7Si_4O_{15} \cdot 8H_2O$			61	D_{2h}^{15}	$Pbca$	8		0.538	0.8927	0.738			
$CoSiO_3$			14	C_{2h}^{5}	$P2_1/c$	4		1.228	0.8933	0.259		108.53	
$CoSiO_3$			61	D_{2h}^{15}	$Pcab$	3	4110	1.8302	1.9832	0.5205			
$CoSiO_3$			148	C_{3i}^{2}	$R\bar{3}$	4	3280	0.9655		0.5199			
$Cu_5(SiO_3)_4(OH)_2$	shattuckite		2	C_1^{1}	$P\bar{1}$	8		0.9885	0.6691	0.53825	94.03	91.69	122.2
$Cu_6Si_6O_{18} \cdot 6H_2O$	dioptase		15	C_{2h}^{6}	$B2/b$	1	6060 (20)	1.4566	1.037	0.7778		89.81	
$Dy_2Si_2O_7$			176	C_{6h}^{2}	$P6_3/m$	1		0.6639		1.2152			
Dy_2SiO_5			176	C_{6h}^{2}	$P6_3/m$	1		1.4350		0.671			
$Dy_8Mn_2(SiO_4)_6O_2$			176	C_{6h}^{2}	$P6_3/m$	4		0.933		0.671			
$Dy_8Pb_2(SiO_4)_6O_2$			2	C_1^{1}	$P\bar{1}$	8		0.947	0.671	0.687	94.50	91.97	122.3
$Dy_{9.333}(SiO_4)_6O_2$			15	C_{2h}^{6}	$B2/b$	2		0.9373	0.687	0.6784		90.57	
$Er_2Si_2O_7$			11	C_{2h}^{2}	$P2_1/m$	1	6280 (20)	0.6583	0.6784	1.2000		107.2	
Er_2SiO_5			176	C_{6h}^{2}	$P6_3/m$	1	6870	1.432		0.669			
$Er_4PbSi_5O_{17}$			176	C_{6h}^{2}	$P6_3/m$	1		0.5534		0.6960			
$Er_6Pb_3(SiO_4)_6$			176	C_{6h}^{2}	$P6_3/m$	4		0.964		0.6780			
$Er_8Mn_2(SiO_4)_6O_2$			2	C_1^{1}	$P\bar{1}$	4		0.930	0.6609	0.665	94.36	91.75	
$Er_{9.333}(SiO_4)_6O_2$			14	C_{2h}^{5}	$P2_1/c$	1		0.9324	1.035	0.6686		90.02	
$Eu_2Si_2O_7$			176	C_{6h}^{2}	$P6_3/m$	2	5540 (20)	0.6716		1.2321			
Eu_2SiO_5			8	C_s^{3}	Cm	6		0.9142	1.058	0.6790		107.53	
$Eu_{9.333}(SiO_4)_6O_2$			161	C_{3v}^{6}	$R3c$	2		0.9472		0.6905			
$Fe_2Bi(SiO_4)_2(OH)$	bismutoferrite		8	C_s^{3}	Cm	8		0.521	0.6762	0.774		100.67	
$Fe_2La_7Si_6O_{23}(OH)_3$			161	C_{3v}^{6}	$R3c$	4	3750	1.082	0.7054	3.845			
$Fe_2Sb(SiO_4)_2(OH)$	chapmanite		8	C_s^{3}	Cm	4		0.519	0.902	0.770		100.67	
$Fe_2Si_2O_6$	orthoferrosilite		61	D_{2h}^{15}	$Pbca$	2		1.8418	0.899	0.52366			
$Fe_2Si_3O_6$	clinoferrosilite		14	C_{2h}^{5}	$P2_1/c$	1		0.97085	0.9078	0.52284		108.432	
Fe_2SiO_4	fayalite		62	D_{2h}^{16}	$Pbnm$	1/2	4296 (18)	0.4818	0.90872	0.6086			
Fe_7SiO_{10}			11	C_{2h}^{2}	$P2_1/m$	2	5020	2.14	1.0471	0.588		98	
$Fe_8Si_2O_{10}(OH)_8$	cronstedtite		8	C_s^{3}	Cm			0.549	0.306	0.732		104.52	
$Fe_8Si_2O_{10}(OH)_8$	cronstedtite		157	C_{3v}^{2}	$P31m$	1/2		0.549	0.951	0.7085			
$Fe_8Si_2O_{10}(OH)_8$	cronstedtite		9	C_s^{4}	Cc	2		0.549	0.951	1.429		97.37	

Silicates (Continued)

Formula	Name	Number	Schoenflies	Hermann–Mauguin	Z	d	a	b	c	α	β	γ
$Fe_9Si_6O_{20}(OH)_5$	deerite	14	$C2h^5$	$P2_1/a$	4	3837	1.0786	1.888	0.9564		107.45	
$Ga_2Pb_2Si_2O_9$		20	$D2^5$	$C222_1$	4		0.695	1.090	0.991			
$Gd_2Si_2O_7$		2	Ci^1	$P(-1)$	4	5820 (20)	0.6624	0.6679	1.2132	94.10	89.79	91.60
Gd_2SiO_5		14	$C2h^5$	$P2_1/c$	4	6550	0.9131	0.7045	0.6749		107.52	
$Gd_8Mn_2(SiO_4)_6O_2$		176	$C6h^2$	$P6_3/m$	1		0.934		0.679			
$Gd_8Pb_2(SiO_4)_6O_2$		176	$C6h^2$	$P6_3/m$	1		0.954		0.695			
$Gd_{9.333}(SiO_4)_6O_2$		176	$C6h^2$	$P6_3/m$	1		0.9431		0.6873			
$HfSiO_4$		141	$D4h^{19}$	$I4_1/amd$	4	6340	0.6581		0.5967			
$Ho_2Si_2O_7$		2	Ci^1	$P(-1)$	4	6110 (20)	0.6612	0.6669	1.2085	85.81	89.38	88.57
Ho_2SiO_5		15	$C2h^6$	$B2/b$	8		1.435	1.037	0.671			122.2
$Ho_{9.333}(SiO_4)_6O_2$		176	$C6h^2$	$P6_3/m$	1		0.9346		0.6744			
$In_2Pb_2Si_2O_9$		20	$D2^5$	$C222_1$	4		0.702	1.139	1.054			
$In_2Si_2O_7$		12	$C2h^3$	$C2/m$	2	5112	0.66238	0.85958	0.47023		102.94	
$La_2Si_2O_7$		76	$C4^2$	$P4_1$	8		0.67945		2.4871			
La_2SiO_5		14	$C2h^5$	$P2_1/c$	4	5720	0.9420	0.7398	0.7028		108.21	
$La_4(SiO_4)_3$		hexagonal			2	5310	1.125		0.4683			
$La_4(SiO_4)_3$		176	$C6h^2$	$P6_3/m$	2		0.969		0.717			
$La_8Mn_2(SiO_4)_6O_2$		176	$C6h^2$	$P6_3/m$	1		0.963		0.708			
$La_8Pb_2(SiO_4)_6O_2$		176	$C6h^2$	$P6_3/m$	1		0.971		0.720			
$La_{9.333}(SiO_4)_6O_2$		176	$C6h^2$	$P6_3/m$	1		0.9713		0.7194			
$Lu_2Si_2O_7$		12	$C2h^3$	$C2/m$	2	5340 (20)	0.67655	0.88369	0.47121		102.00	
Lu_2SiO_5		15	$C2h^6$	$B2/b$	8	6020 (20)	1.4254	1.0241	0.6641			122.20
$Lu_{9.333}(SiO_4)_6O_2$		176	$C6h^2$	$P6_3/m$	1		0.9260		0.6621			
$Mg_2Dy_7Si_6O_{23}(OH)_3$		161	$C3v^6$	$R3c$			1.047		3.632			
$Mg_2Dy_8(SiO_4)_6O_2$		176	$C6h^2$	$P6_3/m$	1		0.931		0.669			
$Mg_2Er_8(SiO_4)_6O_2$		176	$C6h^2$	$P6_3/m$	1		0.928		0.658			
$Mg_2Gd-Si_6O_{23}(OH)_3$		161	$C3v^6$	$R3c$			1.049		3.664			
$Mg_2Gd_8(SiO_4)_6O_2$		176	$C6h^2$	$P6_3/m$	1		0.933		0.675			
$Mg_2Ho_7Si_6O_{23}(OH)_3$		161	$C3v^6$	$R3c$			1.044		3.592			
$Mg_2La_4Ti_3Si_4O_{22}$	perrierite	14	$C2h^5$	$P2_1/a$	2		1.3818	0.5677	1.1787		113.85	

Formula	Mineral	No.	Schoenflies	Space group	Z		a	b	c	angle
Mg$_2$La$_7$Si$_6$O$_{23}$(OH)$_3$		161	C3v^6	R3c	1		1.077		3.840	
Mg$_2$La$_8$(SiO$_4$)$_6$O$_2$		176	C6h^2	P6$_3$/m	4		0.959		0.705	
Mg$_2$Mn$_5$(SiO$_4$)$_3$(OH)$_2$	manganesehumite	62	D2h^{16}	Pbnm	2		0.4815	1.0580	2.1448	
Mg$_2$Nd$_4$Ti$_3$Si$_4$O$_{22}$		14	C2h^5	P2$_1$/a			1.3328	0.5727	1.0971	100.91
Mg$_2$Nd$_7$Si$_6$O$_{23}$(OH)$_3$		161	C3v^6	R3c	1		1.064		3.763	
Mg$_2$Nd$_8$(SiO$_4$)$_6$O$_2$		176	C6h^2	P6$_3$/m	2		0.945		0.686	
Mg$_2$Pr$_4$Ti$_3$Si$_4$O$_{22}$		14	C2h^5	P2$_1$/a	2		1.357	0.5643	1.166	113.1
Mg$_2$SiO$_4$	forsterite	62	D2h^{16}	Pbnm	4	3220..3260	0.47535	1.01943	0.59807	
Mg$_2$Sm$_7$Si$_6$O$_{23}$(OH)$_3$		161	C3v^6	R3c	1		1.062		3.68	
Mg$_2$Sm$_8$(SiO$_4$)$_6$O$_2$		176	C6h^2	P6$_3$/m	1		0.938		0.680	
Mg$_2$Y$_3$Mg$_{0.5}$Si$_{2.5}$O$_{12}$		230	Oh^{10}	Ia3d	8		1.2125			
Mg$_2$Y$_8$(SiO$_4$)$_6$O$_2$		176	C6h^2	P6$_3$/m	1		0.9298		0.6635	
Mg$_3$Cr$_2$(SiO$_4$)$_3$	knorringite	230	Oh^{10}	Ia3d	8	3756	1.1659			
Mg$_3$Si$_4$O$_{10}$(OH)$_2$	talc	15	C2h^6	C2/c	4	2600..2800	0.5287	0.9158	1.895	99.5
Mg$_3$Si$_4$O$_{10}$(OH)$_2$	talc	2	Ci^1	P(-1)	2		0.5293	0.5299	0.9469	98.91 86.06 119.99
Mg$_4$Si$_6$O$_{15}$(OH)$_2$·2H$_2$O	sepiolite anhydride	14	C2h^5	P2$_1$/n	4		1.09	2.33	0.528	90
Mg$_4$Si$_6$O$_{15}$(OH)$_2$·6H$_2$O	sepiolite	52	D2h^6	Pncn	4	2080	0.5255	2.697	1.350	
Mg$_4$Y$_6$(SiO$_4$)$_6$O		176	C6h^2	P6$_3$/m	1		0.930		0.663	
Mg$_5$Er$_6$Si$_5$O$_{24}$		230	Oh^{10}	Ia3d	4		1.208			
Mg$_5$Ho$_6$Si$_5$O$_{24}$		230	Oh^{10}	Ia3d	4		1.212			
Mg$_5$Lu$_6$Si$_5$O$_{24}$		230	Oh^{10}	Ia3d	4		1.199			
Mg$_5$Yb$_6$Si$_5$O$_{24}$		230	Oh^{10}	Ia3d	4		1.202			
Mg$_6$Si$_2$O$_6$(OH)$_{12}$·11H$_2$O		hexagonal					0.895		0.810	
Mg$_6$Si$_4$O$_{10}$(OH)$_8$	antigorite	8	Cs^3	Cm	1		0.530	0.920	0.746	91.40
Mg$_6$Si$_4$O$_{10}$(OH)$_8$	antigorite	185	C6v^3	P6$_3$cm	2	2560	0.5322	0.9219	1.453	
Mg$_6$Si$_4$O$_{10}$(OH)$_8$	lizardite	monoclinic pseudohexagonal			1		0.5301	0.9186	0.7281	93.27
Mg$_6$Si$_4$O$_{10}$(OH)$_8$	chrysotile	monoclinic			2		0.534	0.92	1.465	
MgMnSi$_2$O$_6$	kanoite	14	C2h^5	P2$_1$/c	4	3660	0.9739	0.8939	0.5260	108.56
MgSiO$_3$	clinoenstatite	14	C2h^5	P2$_1$/c	8		0.96065	0.88146	0.51688	108.335
Mn$_2$La$_7$Si$_6$O$_{23}$(OH)$_3$		161	C3v^6	R3c	1		1.088		3.85	
Mn$_2$SiO$_4$	tephroite	62	D2h^{16}	Pbnm	4	4113	0.48968	1.0590	0.6250	
Mn$_3$V$_2$(SiO$_4$)$_3$	yamatoite	230	Oh^{10}	Ia3d	8		1.1953			
Mn$_4$Dy$_6$(SiO$_4$)$_6$(OH)$_2$		176	C6h^2	P6$_3$/m			0.933		0.668	
Mn$_4$Er$_6$(SiO$_4$)$_6$(OH)$_2$		176	C6h^2	P6$_3$/m			0.928		0.663	
Mn$_4$Gd$_6$(SiO$_4$)$_6$(OH)$_2$		176	C6h^2	P6$_3$/m			0.938		0.678	
Mn$_4$La$_6$(SiO$_4$)$_6$(OH)$_2$		176	C6h^2	P6$_3$/m			0.966		0.705	

Silicates (Continued)

Formula	Name	Number	Schoenflies	Hermann–Mauguin	Z	d	a	b	c	α	β	γ
$Mn_4Nd_6(SiO_4)_6(OH)_2$		176	$C6h^2$	$P6_3/m$			0.950		0.690			
$Mn_4Sm_6(SiO_4)_6(OH)_2$		176	$C6h^2$	$P6_3/m$			0.943		0.682			
$Mn_4Y_6(SiO_4)_6(OH)_2$		176	$C6h^2$	$P6_3/m$			0.931		0.665			
$Mn_5Si_4O_{10}(OH)_6$	bementite	19	$D2^4$	$P2_1 2_1 2_1$	18	2980	1.450	1.750	2.912			
$Mn_7(SiO_4)_3(OH)_2$	leukophoenicite	14	$C2h^5$	$P2_1/a$	2	3620..3930	1.0842	0.4826	1.1324		103.93	
Mn_7SiO_{11}	langbanite	164	$D3d^3$	$P(-3)m1$	2	4600..4920	0.677		1.112			
Mn_7SiO_{12}	braunite	142	$D4h^{20}$	$I4_1/acd$	8	4750..4820	0.9408		1.984			
$MnSiO_3$	α	2	Ci^1	$P(-1)$	14		0.6717	0.7603	1.7448	113.83	82.35	94.72
$MnSiO_3$	β	2	Ci^1	$A(-1)$	24		1.606	0.711	1.368	90	90	90
$MnSiO_3$		14	$C2h^5$	$P2_1/c$	4		0.9864	0.9179	0.5298		108.22	
$Nd_2Si_2O_7$		76	$C4^2$	$P4_1$	8	5382 (20)	0.67405		2.4524			
Nd_2SiO_5		14	$C2h^5$	$P2_1/c$	4		0.9250	0.7268	0.6886		108.30	
$Nd_8Mn_2(SiO_4)_6O_2$		176	$C6h^2$	$P6_3/m$	1		0.947		0.691			
$Nd_8Pb_2(SiO_4)_6O_2$		176	$C6h^2$	$P6_3/m$	1		0.965		0.712			
$Nd_{9.33}(SiO_4)_6O_2$		176	$C6h^2$	$P6_3/m$	1	5700 (20)	0.9563		0.7029			
$Ni_2Ho_7Si_6O_{23}(OH)_3$		161	$C3v^6$	$R3c$	6		1.047		3.634			
Ni_2SiO_4		62	$D2h^{16}$	$Pbnm$	4		0.4726	1.0118	0.5913			
$Ni_2Sm_7Si_6O_{23}(OH)_3$		161	$C3v^6$	$R3c$	6		1.058		3.674			
$Ni_5Si_8O_{21}\cdot8H_2O$		hexagonal					0.534		0.726			
$Ni_6Si_4O_{10}(OH)_8$		185	$C6v^3$	$P6_3cm$	1		0.529	0.914	0.723			
$Pb_2Cr_2Si_2O_9$		20	$D2^5$	$C222_1$	4		0.688	1.084	1.001			
$Pb_2Mn_2(Si_2O_7)O_2$	kentrolite	20	$D2^5$	$C222_1$	4	6190	0.7006	1.1059	0.9997			
Pb_2SiO_4		10	$C2h^1$	$P2/m$	20	7600	2.220	1.536	0.642		93	
$Pb_3Si_2O_7$		167	$D3d^6$	$R(-3)c$	18	6840	1.01264		3.8678			
$Pb_4Ce_6(SiO_4)_6(OH)_2$		176	$C6h^2$	$P6_3/m$	1		0.977		0.719			
$Pb_4Dy_6(SiO_4)_6(OH)_2$		176	$C6h^2$	$P6_3/m$	1		0.970		0.690			
$Pb_4Er_6(SiO_4)_6(OH)_2$		176	$C6h^2$	$P6_3/m$	1		0.968		0.684			
$Pb_4Gd_6(SiO_4)_6(OH)_2$		176	$C6h^2$	$P6_3/m$	1		0.972		0.699			
$Pb_4La_6(SiO_4)_6(OH)_2$		176	$C6h^2$	$P6_3/m$	1		0.980		0.726			
$Pb_4Lu_6(SiO_4)_6(OH)_2$		176	$C6h^2$	$P6_3/m$	1		0.964		0.675			

Formula	Mineral	Ref	Class	Space group	Z	ρ	a	b	c	α	β	γ
$Pb_4Nd_6(SiO_4)_6(OH)_2$		176	C_{6h}^2	$P6_3/m$	1		0.976		0.713			
$Pb_4Sm_6(SiO_4)_6(OH)_2$		176	C_{6h}^2	$P6_3/m$	1		0.974		0.705			
$Pb_4Y_6(SiO_4)_6(OH)_2$		176	C_{6h}^2	$P6_3/m$	6		0.968		0.686			
$Pb_8Mn(Si_2O_7)_3$	barysilite	167	D_{3d}^6	$R\bar{3}c$	4/51	6720	0.9821		3.838			
$PbSi_{50}O_{101}$		92,96	$D_4^{4,8}$	$P4_13_22$	12		0.4980		0.6943			
$PbSiO_3$	alamosite	13	C_{2h}^4	$P2/n$	8	6488	1.123	0.708	1.226		113.25	
$Pr_2Si_2O_7$		76	C_4^2	$P4_1$	4	5266	0.67657		2.4608			
Pr_2SiO_5		14	C_{2h}^5	$P2_1/c$	2		0.9253	0.7301	0.6938		108.15	
$Pr_4Ni_2Ti_3Si_4O_{22}$		14	C_{2h}^5	$P2_1/a$	1		1.357	0.5655	1.170		113.34	
$Pr_6Pb_3(SiO_4)_6$		176	C_{6h}^2	$P6_3/m$	1		0.9662		0.7162			
$Pr_{9.333}(SiO_4)_6O_2$		176	C_{6h}^2	$P6_3/m$	4		0.9607		0.7073			
$Sc_2Pb_2Si_2O_9$		20	D_2^5	$C222_1$	2		0.700	1.130	1.042			
$Sc_2Si_2O_7$	thortveitite	12	C_{2h}^3	$C2/m$	2	3390	0.6508	0.8506	0.4677		102.72	
$ScFeSi_2O$		12	C_{2h}^3	$C2/m$	2		0.64414	0.83800	0.46697		103.06	
$ScGaSi_2O_7$		12	C_{2h}^3	$C2/m$	2		0.6469	0.8268	0.46591		104.30	
$ScInSi_2O_7$		12	C_{2h}^3	$C2/m$	8		0.65627	0.85499	0.46916		102.804	
$Sm_2Si_2O_7$		76	C_4^2	$P4_1$	4	5672 (20)	0.66933		2.4384			
Sm_2SiO_5		14	C_{2h}^5	$P2_1/c$	1	6550	0.9161	0.7112	0.6821		107.51	
$Sm_8Mn_2(SiO_4)_6O_2$		176	C_{6h}^2	$P6_3/m$	1		0.942		0.685			
$Sm_8Pb_2(SiO_4)_6O_2$		176	C_{6h}^2	$P6_3/m$	4		0.958		0.706			
$Tb_2Si_2O_7$		2	C_i^1	$P\bar{1}$	4	5930 (20)	0.6623	0.6684	1.2101	93.97	89.85	91.55
Tb_2SiO_5		14	C_{2h}^5	$P2_1/c$	1		0.9083	0.6990	0.6714		107.31	
$Tb_{9.333}(SiO_4)_6O_2$		176	C_{6h}^2	$P6_3 m$	4		0.9401		0.6825			
$ThSiO_4$	thorite	141	D_{4h}^{19}	$I4_1/amd$	4	6630	0.71328		0.63188			
$ThSiO_4$	huttonite	14	C_{2h}^5	$P2_1\,n$	8	7200	0.6784	0.6974	0.6500		104.92	
Tm_2SiO_5		15	C_{2h}^6	$B2\,b$	1		1.4302	1.0313	0.6662		122.21	
$Tm_{9.333}(SiO_4)_6O_2$		176	C_{6h}^2	$P6_3/m$	2		0.9300		0.6666			
$Y_2Si_2O_7$	yttrialite β	12	C_{2h}^3	$C2/m$	4	3690	0.6845	0.9139	0.4687		100.57	
$Y_2Si_2O_7$	yttrialite α	2	C_i^1	$P\bar{1}$	6	3980	0.6584	0.6643	1.2390	93.62	89.81	91.17
$Y_2Si_2O_7$	thalenite	14	C_{2h}^5	$P2_1/n$	8	4340	1.0343	1.1093	0.7294		96.92	
Y_2SiO_5		15	C_{2h}^6	$I2/a$	6	4490	1.0410	0.6721	1.2490		102.65	
$Y_3Si_3O_{10}OH$		14	C_{2h}^5	$P2_1/n$	8	4340	1.0343	0.7294	1.1093		96.92	
$Y_4(SiO_4)_3$		230	O_h^{10}	$Ia3d$			1.192					
$Y_8Mn_2(SiO_4)_6O_2$		176	C_{6h}^2	$P6_3/m$	1		0.932		0.669			
$Y_8Pb_2(SiO_4)_6O_2$		176	C_{6h}^2	$P6_3/m$	1		0.942		0.680			
$Y_{9.333}(SiO_4)_6O_2$		176	C_{6h}^2	$P6_3/m$	1	3500	0.940		0.670			

Silicates (Continued)

Formula	Name	Number	Schoenflies	Hermann–Mauguin	Z	d	a	b	c	α	β	γ
$HYSiO_4$	tombarthite	14	$C2h^5$	$P2_1/n$	4	3510	0.712	0.729	0.671		102.68	
$Yb_2Si_2O_7$		12	$C2h^3$	$C2/m$	2	6010 (20)	0.6789	0.9067	0.4681		101.84	
Yb_2SiO_5		15	$C2h^6$	$B2/b$	8	7150	1.428	1.028	0.6653			122.2
$Yb_{9.333}(SiO_4)_6O_2$		176	$C6h^2$	$P6_3/m$	1		0.9275		0.6636			
$Zn_2Dy_8(SiO_4)_6O_2$		176	$C6h^2$	$P6_3/m$	1	6340	0.930		0.671			
$Zn_2Gd_8(SiO_4)_6O_2$		176	$C6h^2$	$P6_3/m$	1	6150	0.937		0.680			
$Zn_2Ho_8(SiO_4)_6O_2$		176	$C6h^2$	$P6_3/m$	1	6560	0.928		0.667			
$Zn_2La_8(SiO_4)_6O_2$		176	$C6h^2$	$P6_3/m$	1	5190	0.964		0.709			
$Zn_3Mn(SiO_4)(OH)_2$	hodgkinsonite	14	$C2h^5$	$P2_1/a$	4	4010	0.8170	0.5316	1.1761		95.25	
$Zn_2Nd_8(SiO_4)_6O_2$		176	$C6h^2$	$P6_3/m$	1	5680	0.950		0.695			
Zn_2SiO_4	willemite	148	$C3i^2$	$R(-3)$	18	4250 (20)	1.3948		0.9315			
$Zn_2Sm_8(SiO_4)_6O_2$		176	$C6h^2$	$P6_3/m$	1	5960	0.942		0.685			
$Zn_2Y_8(SiO_4)_6O_2$		176	$C6h^2$	$P6_3/m$	1		0.943		0.679			
$Zn_4Si_2O_7(OH)_2$		44	$C2v^{20}$	$Imm2$	2		0.817	1.075	0.508			
$Zn_4Si_2O_7(OH)_2 \cdot H_2O$	hemimorphite	44	$C2v^{20}$	$Imm2$	2	3450	0.8367	1.0730	0.5115			
$ZnMgSi_2O_6$		61	$D2h^{15}$	$Pbca$	18		1.8201	0.8916	0.5209			
$ZnPb_2Si_2O_7$		113	$D2d^3$	$P(-4)2_1m$	2		0.7980		0.5252			
$ZnSiO_3$		61	$D2h^{15}$	$Pbca$	16		1.8204	0.9087	0.5278			
$ZrSiO_4$	zircon	141	$D4h^{19}$	$I4_1/amd$	4	4706 (22)	0.66164		0.60150			

Stannates

$Bi_{12}SnO_{20}$, Bi_2MgSnO_6, Bi_2ZnSnO_6, Cd_2SnO_4, $CdSn(OH)_6$, $CdSnO_3$, Co_2SnO_4, $CoSn(OH)_6$, $CuSn(OH)_6$, $Dy_2Sn_2O_7$, $Er_2Sn_2O_7$, $Eu_2Sn_2O_7$, Fe_2SnO_4, $FeSnO(OH)_6$, $Gd_2Sn_2O_7$, $Ho_2Sn_2O_7$, La_2MgSnO_6, $La_2Sn_2O_7$, $Lu_2Sn_2O_7$, Mg_2SnO_4, $MgCoSnO_4$, $MgMnSnO_4$, $MgSn(OH)_6$, $MgSnO_3$, Mn_2SnO_4, $MnCoSnO_4$, $MnSn(OH)_6$, $Nd_2Sn_2O_7$, $NiSn(OH)_6$, Pb_2SnO_4, $PbSnO_3$, $Pr_2Sn_2O_7$, $Sc_2Sn_2O_7$, $Sm_2Sn_2O_7$, $Tb_2Sn_2O_7$, $Tm_2Sn_2O_7$, $Y_2Sn_2O_7$, $Yb_2Sn_2O_7$, $Zn_2Sn_2O_7$, $ZnCoSnO_4$, $ZnNiSnO_4$, $ZnSn(OH)_6$.

Tantalates

$AgTaO_3$, $AlTaO_4$, $AlTiTaO_6$, Bi_3TiTaO_9, $BiTaO_4$, Cd_2CrTaO_6, Cd_2FeTaO_6, Cd_2MnTaO_6, Cd_2ScTaO_6, $Cd_2Ta_2O_7$, $CdTaO_3$, $CdTa_2O_6$, $CeTa_3O_9$, $CeTa_5O_{14}$, $CeTa_7O_{19}$, $CeTiTaO_6$, $Co_4Ta_2O_9$, $CoTa_2O_6$, $CuTaO_3$, Dy_3TaO_7, $DyTa_3O_9$, $DyTaO_4$, $DyTiTaO_6$, Er_3TaO_7, $ErTa_3O_9$, $ErTaO_4$, $ErTiTaO_6$, Eu_3TaO_7, $EuTa_7O_{19}$, $EuTaO_4$, $Fe_4Ta_2O_9$, $FeTaO_4$, $FeTi_5Ta_2O_{16}$, $GaTaO_4$, $GaTiTaO_6$, Gd_3TaO_7, $GdTa_3O_9$, $GdTiTaO_6$, $Hg_2Ta_2O_7$, Ho_3TaO_7, $HoTa_3O_9$, $HoTaO_4$, $InTaO_4$, La_3TaO_7, $LaTa_3O_9$, $LaTa_5O_{14}$, $LaTa_7O_{19}$, $LaTaO_4$, $LaTiTaO_6$, Lu_3TaO_7, $LuTaO_4$, $LuTiTaO_6$, $Mg_4Ta_2O_9$, $Mg_5Ta_4O_{15}$, $MgFeTaO_5$, $MgNbTaO_6$, $MgTa_2O_6$, $Mn_4Ta_2O_9$, $MnTa_2O_6$, Nd_3TaO_7, $NdTa_3O_9$, $NdTa_7O_{19}$, $NdTaO_4$, $NdTiTaO_6$, $NiSbTaO_6$, $NiTa_2O_6$, Pb_2AlTaO_6, Pb_2CoTaO_6, Pb_2CrTaO_6, Pb_2FeTaO_6, Pb_2GaTaO_6, Pb_2LuTaO_6, Pb_2MnTaO_6, Pb_2NiTaO_6, Pb_2PrTaO_6, Pb_2ScTaO_6, Pb_2SmTaO_6, $Pb_2Ta_2O_7$, Pb_2VTaO_6, Pb_2YbTaO_6, $Pb_3Ta_4O_{14}$, $PbTa_2O_6$, $PrTa_3O_9$, $PrTa_7O_{19}$, $PrTaO_4$, $SbTaO_4$, Sc_3TaO_7, $ScTaO_4$, $ScTiTaO_6$, Sm_3TaO_7, $SmTa_7O_{19}$, $SmTiTaO_6$, Tb_3TaO_7, $TbTa_3O_9$, $TbTaO_4$, $Th(TaO_3)_4$, $TiTa_2O_7$, Tm_3TaO_7, $TmTaO_4$, VTa_9O_{25}, $VTaO_4$, Y_3TaO_7, $YbTa_3O_9$, $YbTaO_4$, $YbTiTaO_6$, YTa_3O_9, $YTaO_4$, $YTiTaO_6$, $Zn_5Co_2Ta_2O_{12}$, $ZnTa_2O_6$.

Titanates

Formula	Name	Number	Schoenflies	Hermann–Mauguin	Z	d	a	b	c	α	β	γ
Al_2TiO_5	tialite	63	$D2h^{17}$	Cmcm	4	3770	0.35875	0.94237	0.96291			
$Bi_{12}TiO_{20}$		197	T^3	I23	2	9074	1.01760					
$Bi_4Ti_3O_{12}$		42	$C2v^{18}$	Fm2m	4	7850	0.5408	0.5444	3.2840			
$Bi_5FeTi_3O_{15}$		orthorhombic					0.5445	0.5455	4.131			
$Bi_6Fe_2Ti_3O_{18}$		orthorhombic					0.5490	0.5500	5.0185			
Bi_8TiO_{14}		197	T^3	I23	3	8930	1.019					
$Bi_9Fe_5Ti_3O_{27}$		orthorhombic			4		0.5491	0.5502	7.620			
$CdTiO_3$		33	$C2v^9$	$Pc2_1n$	4		0.5348	0.7615	0.5417			
$CeTi_2O_6$		12	$C2h^3$	C2/m	2		0.984	0.375	0.691		119.25	
Co_2TiO_4		227	Oh^7	Fd3m	8		0.8448					
$CoTiO_3$		148	$C3i^2$	R(-3)	6		0.50683		1.39225			
$CuCoTiO_4$		227	Oh^7	Fd3m	8		0.845					
$CuFeTiO_4$		tetragonal			8		0.837		0.878			

Titanates (Continued)

Formula	Name	Number	Schoenflies	Hermann–Mauguin	Z	d	a	b	c	α	β	γ
$CuZnTiO_4$		227	Oh^7	Fd3m	8		0.846					
$Dy_2Ti_2O_7$		227	Oh^7	Fd3m	8	6870	1.0119					
Dy_2TiO_5		62	$D2h^{16}$	Pnma	4		1.049	0.370	1.126			
$Er_2Ti_2O_7$		227	Oh^7	Fd3m	8	7000	1.00869					
$Eu_2Ti_2O_7$		227	Oh^7	Fd3m	8	6510	1.0195					
Eu_2TiO_5		62	$D2h^{16}$	Pnma	4		1.582	0.534	1.576		92.48	
$EuCrTiO_5$		55	$D2h^9$	Pbam	4		0.7475	0.860	0.5785			
$EuFeTiO_5$		55	$D2h^9$	Pbam	4		0.747	0.867	0.5825			
$Fe_2Ti_3O_9$	pseudorutile	hexagonal					0.2872		0.4594			
Fe_2TiO_4	ulvaspinel	227	Oh^7	Fd3m	8		0.8529					
Fe_2TiO_5	pseudobrookite	63	$D2h^{17}$	Cmcm	4	4390	0.37385	0.97954	0.99853			
$FeTi_2O_5$		63	$D2h^{17}$	Cmcm	4		0.3741	0.9798	1.0041			
$FeTiO_3$	ilmenite	148	$C3i^2$	R(-3)	6	4500..5000	0.5082		1.4027			
$Gd_2Ti_2O_7$		227	Oh^7	Fd3m	8	6660	1.0171					
Gd_2TiO_5		62	$D2h^{16}$	Pnma	4		1.04788	0.37547	1.1328			
$GdCrTiO_5$		55	$D2h^9$	Pbam	4		0.745	0.859	0.5785			
$GdFeTiO_5$		55	$D2h^9$	Pbam	4		0.743	0.865	0.582			
$Ho_2Ti_2O_7$		227	Oh^7	Fd3m	8	6940	1.00998					
$HoYTi_2O_7$		227	Oh^7	Fd3m	8		1.00961					
In_2TiO_5		62	$D2h^{16}$	Pnma	4	6280	0.7237	0.3429	1.486		92.48	
$La_2Bi_2Ti_3O_{12}$		42	$C2v^{18}$	Fm2m	4		0.5390	0.5410	3.282			
La_2CoTiO_6		221	Oh^1	Pm3m	1/2		0.3922					
La_2NiTiO_6		221	Oh^1	Pm3m	1/2		0.3918					
$La_2Ti_2O_7$		11	$C2h^2$	P2₁/m	4	5680	0.780	0.554	1.301		98.62	
La_2TiO_5		62	$D2h^{16}$	Pnam	4	5500	1.097	1.137	0.393			
$La_3BiTi_3O_{12}$		42	$C2v^{18}$	Fm2m	4		0.5390	0.5390	3.280			
$LaBi_4FeTi_3O_{15}$		orthorhombic					0.5442	0.5452	4.129			
$Lu_2Ti_2O_7$	metastable	227	Oh^7	Fd3m	8	7290	1.0011					
Mg_2TiO_4		227	Oh^7	Fd3m	8	3520	0.8441					
Mg_2TiO_4	stable	91,95	$D4^{3,7}$	P4₁,₃22	8		$a\sqrt{2} = .8455$		0.8412			

$MgFe_2Ti_3O_{10}$	kennedyite	63		D_{2h}^{17}	Cmcm	2	4070	0.373	0.977	0.995	
$MgMnTiO_4$		227		O_h^7	Fd3m	8		0.848			
$MgNd_2TiO_6$		221		O_h^1	Pm3m	1/2		0.390			
$MgTi_2O_5$	karrooite	63		D_{2h}^{17}	Cmcm	4	3910	0.37442	0.97363	0.99870	
$MgTiO_3$	geikielite	148		C_{3i}^2	R(-3)	6		0.5054		1.3898	
Mn_2TiO_4		91,95		$D_4^{3,7}$	$P4_{1,3}22$	4		0.6170		0.8564	
$MnCoTiO_4$		227		O_h^7	Fd3m	8		0.849			
$MnTiO_3$	pyrophanite	148		C_{3i}^2	R(-3)	6	4550	0.51374		1.4284	
$Nd_2Ti_2O_7$		11		C_{2h}^2	$P2_1/m$	4	5850	0.768	0.546	1.299	98.5
$NdCrTiO_5$		55		D_{2h}^9	Pbam	4		0.756	0.867	0.580	
$NdFeTiO_5$		55		D_{2h}^9	Pbam	4		0.754	0.871	0.585	
$NiTiO_3$		148		C_{3i}^2	R(-3)	6		0.5031		1.3785	
$Pb_2Bi_4Ti_5O_{18}$		42		C_{2v}^{18}	Fm2m	4		0.5461	0.5461	4.970	
$PbBi_4Ti_4O_{15}$		42		C_{2v}^{18}	Fm2m	4		0.5437	0.5437	4.131	
$PbBi_5FeTi_4O_{18}$			orthorhombic			4		0.5450	0.5460	5.0110	
$PbTi_3O_7$			monoclinic			2	5790	1.0732	0.3812	0.6578	98.08
$PbTiO_3$	macedonite	99		C_{4v}^1	P4mm	1	7520	0.38985		0.41511	
$Pr_2Bi_2Ti_3O_{12}$		42		C_{2v}^{18}	Fm2m	4		0.5375	0.5375	3.280	
$Pr_2Bi_4Fe_2Ti_3O_{18}$			orthorhombic			4		0.5424	0.5435	4.990	
$Pr_3BiTi_3O_{12}$		42		C_{2v}^{18}	Fm2m	4		0.5375	0.5375	3.280	
$PrBi_3Ti_3O_{12}$		42		C_{2v}^{18}	Fm2m	4		0.5365	0.5365	3.284	
$PrBi_4FeTi_3O_{15}$			orthorhombic			4		0.5443	0.5450	4.130	
$PrCrTiO_5$		55		D_{2h}^9	Pbam	4	5010	0.758	0.868	0.581	
$PrFeTiO_5$		55		D_{2h}^9	Pbam	4		0.757	0.872	0.5855	
$Sc_2Ti_2O_7$		225		O_h^5	Fm3m	1	3620	0.49042		1.0272	
Sc_2TiO_5		63		D_{2h}^{17}	Cmcm	4	4120	0.3843	1.0105	1.707	
$Sc_4Ti_3O_{12}$		148		C_{3i}^2	R(-3)	6		0.9094		3.280	
$Sm_2Bi_2Ti_3O_{12}$		42		C_{2v}^{18}	Fm2m	4	6060	0.5378	0.5378		
$Sm_2Ti_2O_7$		227		O_h^7	Fd3m	8		1.0211			
Sm_2TiO_5		62		D_{2h}^{16}	Pnma	4		1.059	0.3792	1.135	
$SmBi_3Ti_3O_{12}$		42		C_{2v}^{18}	Fm2m	4		0.5380	0.5380	3.280	
$SmCrTiO_5$		55		D_{2h}^9	Pbam	4		0.750	0.862	0.579	
$SmFeTiO_5$		55		D_{2h}^9	Pbam	4		0.750	0.869	0.583	
$Tb_2Ti_2O_7$		227		O_h^7	Fd3m	8	6700	1.0148			
$ThTi_2O_6$		15		C_{2h}^6	C2/c	4		1.0808	0.8580	0.5196	115.25
$Tm_2Ti_2O_7$		227		O_h^7	Fd3m	8	7010	1.0050			

Titanates (Continued)

Formula	Name	Number	Schoenflies	Hermann–Mauguin	Z	d	a	b	c	α	β	γ
$Y_2Ti_2O_7$		227	Oh^7	Fd3m	8	5030	1.00896					
Y_2TiO_5		62	$D2h^{16}$	Pnma	4		1.035	0.370	1.125			
$Yb_2Ti_2O_7$		227	Oh^7	Fd3m	8	7310	1.00298					
Yb_6TiO_{11}		triclinic			4		3.600	0.5131	0.5123	87.06	85.18	84.28
$Zn_2Ti_3O_8$		212,213	$O^{6,7}$	$P4_3,32$	4		0.8395					
Zn_2TiO_4	α	91,95	$D4^{3,7}$	$P4_1,22$	4		0.6005		0.8415			
$ZnTi_2O_5$		63	$D2h^{17}$	Cmcm	4		0.3741	0.980	0.992			
$ZnTiO_3$		227	Oh^7	Fd3m	4		0.8450					

Vanadates

Formula	Name	Number	Schoenflies	Hermann–Mauguin	Z	d	a	b	c	α	β	γ
$AgVO_3$		orthorhombic			4	5200	1.430	0.541	0.347			
$AlVO_3$		227	Oh^7	Fd3m	32/3		0.8422					
$BiVO_4$		15	$C2h^6$	I2/a	4		0.5195	1.1701	0.5098		90.38	
$BiVO_4$	pucherite	60	$D2h^{14}$	Pnca	4	6250	0.5332	0.5060	1.202			
$Cd_2V_2O_7$		12	$C2h^3$	C2/m	2	4850	0.7088	0.9091	0.4963		103.35	
CdV_2O_4		227	Oh^7	Fd3m	8	5500	0.8696					
CdV_2O_6		12	$C2h^3$	C2/m	2	4280	0.9794	0.3616	0.7018		103.77	
$CdVO_3$		orthorhombic			4		0.5345	0.5623	0.6638			
$CeVO_3$		62	$D2h^{16}$	Pbnm	4		0.5541	0.5541	0.7807			
$CeVO_4$		14	$D4h^{19}$	$I4_1/amd$	4	4730(22)	0.7398		0.6498			
Co_2VO_4		227	Oh^7	Fd3m	8		0.8379					
$Co_3(VO_4)_2$		64	$D2h^{18}$	Cmca	4	4500	0.6030	1.150	0.830			
CoV_2O_4		227	Oh^7	Fd3m	8	4950	0.8407					

Formula	Phase / mineral	No.	Schoenflies	H–M	Z	ρ	a	b	c	α	β	γ
$CrVO_3$		167	D_{3d}^{6}	R(-3)c	6	3890	0.4982		1.3752			
$CrVO_4$		63	D_{2h}^{17}	Cmcm	4	3530	0.5579	0.8224	0.5989			
$Cu_3V_2O_7(OH)_2 \cdot 2H_2O$	volborthite	12	C_{2h}^{3}	C2/m	2		1.060	0.586	0.721		95.08	
$Cu_5V_3O_{14}$	monoclinic						1.520	0.361	0.735		101.5	
$CuVO_3$		148	C_{3i}^{2}	R(-3)	6		1.2859		0.7186			
$DyVO_3$		62	D_{2h}^{16}	Pbnm	4		0.5302	0.5602	0.7601			
$DyVO_4$		141	D_{4h}^{19}	$I4_1/amd$	4		0.71434		0.6313			
$ErVO_3$		62	D_{2h}^{16}	Pbnm	4		0.5262	0.5604	0.7578			
$ErVO_4$		141	D_{4h}^{19}	$I4_1/amd$	4		0.7100		0.6279			
$EuVO_4$		141	D_{4h}^{19}	$I4_1/amd$	4		0.72365		0.63675			
Fe_2VO_4		227	O_h^{7}	Fd3m	8		0.8417					
FeV_2O_4	coulsonite	227	O_h^{7}	Fd3m	8	5170..5200	0.8453					
$FeVO_4$		2	C_i^{1}	P(-1)	6	3650	0.6719	0.8060	0.9254	96.65	106.57	101.60
$FeVO_4 \cdot H_2O$	schubnelite	2	C_i^{1}	P(-1)	2	3280	0.659	0.543	0.662	125	104	84.72
$GdVO_3$		62	D_{2h}^{16}	Pbnm	4		0.5345	0.5623	0.7638			
$GdVO_4$		141	D_{4h}^{19}	$I4_1/amd$	4		0.72126		0.63483			
$Hg_2V_2O_7$	α	62	D_{2h}^{16}	Pnma			0.7165	0.3636	2.152			
$Hg_2V_2O_7$	β, triclinic						0.8570	0.3580	0.591	126.43	85.22	100.38
HgV_2O_6	α	12	C_{2h}^{3}	C2/m	2		0.9580	0.3644	0.6655		107.23	
$HoVO_4$		141	D_{4h}^{19}	$I4_1/amd$	4		0.71214		0.62926			
$InVO_4$		63	D_{2h}^{17}	Cmcm	4		0.575	0.855	0.660			
$LaVO_3$		62	D_{2h}^{16}	Pbnm	4	5030	0.5546	0.5546	0.7827			
$LaVO_4$		14	C_{2h}^{5}	$P2_1/n$	4		0.7038	0.7269	0.6719		104.91	
$LuVO_4$		141	D_{4h}^{19}	$I4_1/amd$	4	3550	0.70243		0.62316			
Mg_2VO_4		227	O_h^{7}	Fd3m	8	4230	0.8403					
$Mg_3(VO_4)_2$		64	D_{2h}^{18}	Cmca	4		0.6053	1.1442	0.8330			
MgV_2O_4		227	O_h^{7}	Fd3m	8	3790	0.8416					
$Mn_2V_2O_7$		12	C_{2h}^{3}	C2/m	2		0.6710	0.8726	0.4970		103.57	
Mn_2VO_4		227	O_h^{7}	Fd3m	8		0.8575					
$Mn_2V_2O_4$		227	O_h^{7}	Fd3m	8	4700	0.8522					
$NdVO_3$		62	D_{2h}^{16}	Pbnm	4		0.5440	0.5589	0.7733			
$NdVO_4$		141	D_{4h}^{19}	$I4_1/amd$	4		0.73290		0.64356			
$Ni_3(VO_4)_2$		64	D_{2h}^{18}	Cmca	4		0.5906	1.138	0.824			
$Pb_{10}(VO_4)_6(OH)_2$		176	C_{6h}^{2}	$P6_3/m$	1	6300..6320	1.0165		0.7463			
$Pb_2V_2O_7$	chervetite	14	C_{2h}^{5}	$P2_1/a$	4	7100	1.3470	0.7326	0.6956		107.42	
$Pb_3(VO_4)_2$		14	C_{2h}^{5}	$P2_1/c$	2		0.7525	0.6109	0.9325		111.81	

Vanadates (Continued)

Formula	Name	Number	Schoenflies	Hermann–Mauguin	Z	d	a	b	c	α	β	γ
$Pb_3Bi(VO_4)_3$		220	Td[6]	I($\bar{4}$)3d	4		1.0733					
$PbCuVO_4OH$	α-mottramite	62	D2h[16]	Pnam	4		0.748	0.924	0.598			
$PbCuVO_4OH$	β-mottramite	19	D2[4]	P2₁2₁2₁	4		0.756	0.956	0.587			
$PbFe_2(VO_4OH)_2$	mounanaite	2	Ci[1]	P($\bar{1}$)	1	4850	0.555	0.766	0.556	111.02	112.12	94.15
$PbMnVO_4OH$	pyrobelonite	62	D2h[16]	Pnma	4	5820	0.7644	0.6182	0.9508			
$PrVO_3$		62	D2h[16]	Pbnm	4		0.5477	0.5545	0.7759			
$PrVO_4$		141	D4h[19]	I4₁/amd	4	4820 (22)	0.7367		0.6468			
$SbVO_4$		136	D4h[14]	P4₂/mnm	1	5980 (23)	0.4598		0.3078			
Sc_2VO_5		tetragonal			8		0.6954		1.4572			
$ScVO_3$		206	Th[7]	Ia3	16		0.9602					
$ScVO_4$		141	D4h[19]	I4₁/amd	4	3600	0.6779		0.6136			
$SmVO_3$		62	D2h[16]	Pbnm	4		0.5371	0.5625	0.7693			
$SmVO_4$		141	D4h[19]	I4₁/amd	4		0.72652		0.63894			
$TbVO_4$		141	D4h[19]	I4₁/amd	4		0.71772		0.63289			
$Th_3(VO_4)_4$		141	D4h[19]	I4₁/amd	1	6240	0.726		0.6474			
ThV_2O_7		58	D2h[12]	Pnnm	8	5150	0.7216	0.6964	2.280			
$TmVO_4$		141	D4h[19]	I4₁/amd	4		0.70712		0.62606			
$YbVO_4$		141	D4h[19]	I4₁/amd	4		0.70435		0.62470			
YVO_3		62	D2h[16]	Pbnm	4		0.5284	0.5605	0.7587			
YVO_4	wakefieldite	141	D4h[19]	I4₁/amd	4	4490	0.71192		0.62898			
Zn_2VO_4		227	Oh[7]	Fd3m	8	5200	0.8395					
$Zn_3(VO_4)_2$		64	D2h[18]	Cmca	4	5000	0.6088	1.1489	0.8280			
ZnV_2O_4		227	Oh[7]	Fd3m	8	5100	0.8409					
ZnV_2O_6		5	C2[3]	C2	2	4350 (20)	0.9242	0.3526	0.6574		111.55	
ZrV_2O_7		205	Th[6]	Pa3	4		0.876					

Tungstates

Formula	Name	Number	Schoenflies	Herman-Mauguin	Z	d	a	b	c	α	β	γ
$Al_2(WO_4)_3$		60	$D2h^{14}$	Pnca	4	5080	0.9139	1.2596	0.9060			
Bi_2WO_6	russellite	41	$C2v^{17}$	B2cb	4		0.5457	0.5436	1.6427			
$BiTaW_2O_{10}$		189	$D3h^3$	P(-6)2m	1	8300	0.7420		0.3881			
Cd_2TiWO_7		227	Oh^7	Fd3m	8		1.0191					
$CdWO_4$		13	$C2h^4$	P2/c	2	7770	0.5029	0.5859	0.5074		91.47	
$Ce_2(WO_4)_3$		15	$C2h^6$	C2/c	4	6770	0.7817	1.1724	1.1629		110.03	
$CoWO_4$		13	$C2h^4$	P2/c	2	8420 (20)	0.49478	0.56827	0.46694		90.0	
Cr_2WO_6		136	$D4h^{14}$	P4₂/mnm	2	6680	0.4582		0.8870			
Cu_3WO_6		205	Th^6	Pa3	8	6620	0.9797					
$CuWO_4$		2	Ci^1	P(-1)	2	7610..7650	0.47026	0.58389	0.48784	91.677	92.469	82.805
$Dy_2(WO_4)_3$		15	$C2h^6$	C2/c	4	7690	0.7601	1.1333	1.1293		109.47	
Dy_2WO_6		15	$C2h^6$	C2/c	8		1.656	1.106	0.5366		110.47	
Dy_6WO_{12}		cubic					1.064					
$Er_{22}W_6O_{51}$		hexagonal					1.947		0.9382			
Er_2WO_6		11	$C2h^2$	P2₁/m	4		1.131	0.5317	0.7551		104.37	
Er_6WO_{12}		148	$C3i^2$	R(-3)	3		0.9712		0.9264			
$Eu_2(WO_4)_3$		15	$C2h^6$	C2/c	4	7440	0.7676	1.1463	1.1396		109.63	
Eu_2WO_6		15	$C2h^6$	C2/c	8		1.673	1.1225	0.5447		110.65	
$Fe_2(WO_4)_3$		tetragonal I			4	7150	1.066		0.711			
Fe_2WO_6		62	$D2h^{16}$	Pnma	4	6430	0.4566	1.672	0.4954			
$FeWO_4$	ferberite	13	$C2h^4$	P2/c	2		0.4730	0.5703	0.4952		90	
$Gd_2(WO_4)_3$		15	$C2h^6$	C2/c	4	7470	0.7656	1.1413	1.1388		109.62	
Gd_2SiWO_8		5	$C2^3$	I2	2		0.492	1.128	0.512		92.5	
Gd_2WO_6		15	$C2h^6$	C2/c	8		1.669	1.119	0.5427		110.65	
Gd_6WO_{12}		cubic					1.074					
HfW_2O_8		208	O^2	P4₃32	4		0.913					
$HgWO_4$		15	$C2h^6$	C2/c	4	9 82	1.1375	0.6007	0.5145		113.20	
$Ho_{10}W_2O_{21}$		pseudotetragonal					0.5287		0.5268			
$Ho_{14}W_4O_{33}$		hexagonal R					0.9752		1.8796			
$Ho_2(WO_4)_3$		15	$C2h^6$	C2/c	4	7800	0.7576	1.1283	1.1256		109.42	
Ho_2WO_6		11	$C2h^2$	P2₁/m	4		1.135	0.5330	0.7568		104.45	

Tungstates (Continued)

Formula	Name	Number	Schoenflies	Herman–Mauguin	Z	d	a	b	c	α	β	γ
Ho_6WO_{12}		cubic					1.058					
$In_2(WO_4)_3$		60	$D2h^{14}$	Pnca	4		1.072	1.304	1.035			
In_6WO_{12}		148	$C3i^2$	R(-3)	3		0.9494		0.8952			
$La_2(WO_4)_3$		15	$C2h^6$	C2/c	4	6570	0.7876	1.1836	1.1717		110.10	
La_2SiWO_8		5	$C2^3$	I2	2		0.515	1.190	0.515		90.5	
$La_4W_3O_{15}$		137	$D4h^{15}$	P4$_2$/nmc	4	7160	1.006		1.263			
La_6WO_{12}		cubic					1.118					
Lu_2WO_6		11	$C2h^2$	P2$_1$/m	4		1.121	0.5264	0.7496		104.52	
Lu_6WO_{12}		148	$C3i^2$	R(-3)	3		0.9618		0.9145			
$MgWO_4$		13	$C2h^4$	P2/c	2	6841	0.46864	0.56755	0.49284		89.32	
$MgWO_4$		214	O^8	I4$_1$32	8		1.286					
$MnWO_4$	hubnerite	13	$C2h^4$	P2/c	2		0.4829	0.5759	0.4998		91.16	
$Nb_{12}W_{11}O_{63}$		18	$D2^3$	P2$_1$2$_1$2			1.2195	3.6740	0.3951			
$Nb_{18}W_{17}O_{95}$		4	$C2^2$	P2$_1$	1		1.2195	3.6740	0.3951		90.5	
$Nb_{26}W_4O_{77}$		5	$C2^3$	C2	2	4900	2.974	0.3824	2.597		92.3	
$Nb_2W_3O_8$		57	$D2h^{11}$	Pmab	4		1.6615	1.7616	0.3955			
$Nb_4W_7O_{31}$		tetragonal					2.4264		0.3924			
$Nb_6W_8O_{39}$		115	$D2d^5$	P(-4)m2	2	5300	1.2130		0.3936			
$Nb_8W_9O_{47}$		18	$D2^3$	P2$_1$2$_1$2	2	5950	1.2251		0.3943			
$Nd_2(WO_4)_3$		15	$C2h^6$	C2/c	4	7030	0.7751	1.1633	1.1512		109.65	
Nd_2WO_6		15	$C2h^6$	I2/c	8		1.592	1.139	0.5508		92	
$Nd_4W_3O_{15}$		137	$D4h^{15}$	P4$_2$/nmc	4	7240	0.992		1.250			
Nd_6WO_{12}		tetragonal			1/2		0.5470		0.5442			
$NdWO_4OH$		14	$C2h^5$	P2$_1$/c	4		0.538	1.277	0.697		114.5	
$NiWO_4$		13	$C2h^4$	P2/c	2	6885 (20)	0.4600	0.5665	0.4912		90	
Pb_2CdWO_6		monoclinic			1/2		0.416	0.407	0.416		91	
Pb_2CoWO_6		cubic			4		0.8017					
Pb_2CoWO_6		62	$D2h^{16}$	Pbnm	2		0.5669	0.5689	0.7956			
Pb_2FeWO_6		225	Oh^5	Fm3m	4		0.805					
Pb_2MgWO_6		20	$D2^5$	C222$_1$	64	9190	2.269	2.274	1.587			
Pb_2MnWO_6		monoclinic			1/2		0.4063	0.4033	0.4063		90.2	

Formula	Name	No.	Sch.	Space group	Z	ρ	a	b	c	β
Pb₂NiWO₆		225	Oh⁵	Fm3m	4		0.7977			
Pb₂NiWO₆		tetragonal			4		0.8006		0.7920	
Pb₂TiWO₇		227	Oh⁷	Fd3m	8		1.0381			
Pb₂ZnWO₆		tetragonal			4		0.8123		0.7910	
PbMg₂WO₆		cubic			4		0.8001			
PbWO₄	raspite	14	C2h⁵	P2₁/c	4	8460	0.558	0.500	1.364	107.55
PbWO₄	stolzite	88	C4h⁶	I4₁/a	4	8240	0.54619		1.2049	
Pr₂(WO₄)₃		15	C2h⁶	C2/c	4	6870	0.7781	1.1653	1.1573	109.88
Pr₂W₂O₉		14	C2h⁵	P2₁/c	4	7690	0.770	0.984	0.927	106.5
Pr₂WO₆		15	C2h⁶	C2/c	8		1.703	1.144	0.5544	110.82
Pr₄W₃O₁₅		137	D4h¹⁵	P4₂/nmc	4	7370	0.9962		1.2584	
Pr₆WO₁₂		cubic					1.100			
PrGd(WO₄)₃		15	C2h⁶	C2/c	8		0.776	1.58	2.156	90
PrSc(WO₄)₃		60	D2h¹⁴	Pnca	4		0.94	1.33	0.94	
Sc₂(WO₄)₃		60	D2h¹⁴	Pnca	4	4440	0.9596	1.3330	0.9512	
Sc₆WO₁₂		148	C3i²	R(-3)	3		0.921		0.873	
ScYb(WO₄)₃		60	D2h¹⁴	Pnca	4		0.94	1.31	0.94	
Sm₂(WO₄)₃		15	C2h⁶	C2/c	4	7230	0.7699	1.1492	1.1451	109.53
Sm₂WO₆		15	C2h⁶	C2/c	8		1.681	1.127	0.5464	110.65
Sm₆WO₁₂		cubic					1.080			
Ta₂₂W₄O₆₇		38	C2v¹⁴	C2mm	1		0.6136	4.740	0.384	
Ta₂WO₈		51	D2h⁵	Pmma	2		1.670	0.3877	0.8864	
Ta₃₀W₂O₈₁		orthorhombic					0.6172	5.8452	0.3850	
Ta₃₈WO₉₈		orthorhombic			1		0.6188	6.9570	0.388	
Tb₂(WO₄)₃		15	C2h⁶	C2/c	4	7590	0.7646	1.373	1.1356	109.70
Tb₂WO₆		15	C2h⁶	C2/c	8		1.660	1.10	0.5386	110.53
Th(WO₄)₂		187	C3h¹	P(-6)	9		1.761		0.6259	
Th(WO₄)₂		61	D2h¹⁵	Pcab	8		0.968	1.040	1.461	
Tm₂WO₆		11	C2h²	P2₁/m	4		1.125	0.5292	0.7537	104.45
Tm₆WO₁₂		148	C3i²	R(-3)	3		0.9679		0.9230	
V₂WO₆		136	D4h¹⁴	P4₂/mmm	2	6560	0.4629		0.8912	
VCrWO₆		136	D4h¹⁴	P4₂/mmm	2		0.4595		0.8884	
Y₂SiWO₈		88	C4h⁶	I4₁/a	2		0.500		1.109	
Y₂W₄O₁₄(OH)₂·2H₂O	yttrotungstite	11	C2h²	P2₁/m	1	5960	0.6954	0.8637	0.5771	104.93
Y₂WO₆		11	C2h²	P2₁/m	4	6630	1.135	0.5333	0.7589	104.43
Y₆WO₁₂		225	Oh⁵	Fm3m	4	6000	0.5276			

Tungstates (Continued)

Formula	Name	Number	Schoenflies	Hermann-Mauguin	Z	d	a	b	c	α	β	γ
Yb_2WO_6		11	$C2h^2$	$P2_1/m$	4		1.123	0.5280	0.7508		104.42	
Yb_6WO_{12}		148	$C3i^2$	R(-3)	3		0.9640		0.9188			
$ZnWO_4$		13	$C2h^4$	P2/c	2	7800	0.4691	0.5720	0.4925		89.36	
ZrW_2O_8		208	O^2	$P4_332$	4	5210	0.9159					

Zirconates

Formula	Number	Schoenflies	Hermann-Mauguin	Z	d	a	b	c
$Ce_2Zr_2O_7$	227	Oh^7	Fd3m	8	6180	1.0750		
Dy_2ZrO_{11}	hexagonal R			6		0.9902		1.8573
$Er_4Zr_3O_{12}$	148	$C3i^2$	R(-3)	3		0.970		0.908
$Eu_2Zr_2O_7$	227	Oh^7	Fd3m	8		1.0545		
$Gd_2Zr_2O_7$	227	Oh^7	Fd3m	8		1.0520		
Gd_6ZrO_{11}	hexagonal R			6		1.0032		1.8814
$Ho_2Zr_2O_7$	227	Oh^7	Fd3m	8		1.038		
$La_2Zr_2O_7$	227	Oh^7	Fd3m	8	5940	1.0786		
$Lu_4Zr_3O_{12}$	148	$C3i^2$	R(-3)	6		0.9631		1.789
$Nd_2Zr_2O_7$	227	Oh^7	Fd3m	8	6270	1.0678		
$PbZrO_3$	32	$C2v^8$	Pba2	8		0.5884	1.1768	0.8220
$Pr_2Zr_2O_7$	227	Oh^7	Fd3m	8	6270	1.0699		
$Sc_2Zr_5O_{13}$	148	$C3i^2$	R(-3)	6	5420	0.9532		1.7442
$Sm_2Zr_2O_7$	227	Oh^7	Fd3m	8	6490	1.0548		
$TiZrO_4$	60	$D2h^{14}$	Pbcn	2		0.4806	0.5447	0.5032
$Y_2Zr_2O_7$	227	Oh^7	Fd3m	8	5530	1.0426		
$Yb_2Zr_2O_7$	227	Oh^7	Fd3m	8	7410	1.0312		
$Yb_4Zr_3O_{12}$	148	$C3i^2$	R(-3)	3		0.965		0.902

chemical compounds, there are only a few exceptions. Updates on structures of inorganic compounds can be found in Structure Reports A. The last volume of this series was published in 1993 [4].

Different phases having the same formula can be distinguished in different ways. Some phases, especially those found as minerals have their names, e.g. rutile, anatase and brookite are different forms of TiO_2. Many names are of foreign origin and different English spellings appear in different publications. The alternative spelling is given in brackets. A few names of minerals are given in German (Ger), namely, for minerals which do not have well established English names.

Letters of the Greek alphabet can have two different functions. Gibbsite is also called γ-$Al(OH)_3$, i.e. these two names are synonyms and they are separated by comma in Table 2.1. In contrast "α quartz" (without comma) is one name (to be distinguished from β quartz, a high temperature modification).

The nomenclature can be sometimes misleading, e.g. so called α, β and γ-WO_3 have in fact a substantial deficit of oxygen, β-alumina is $NaAl_{11}O_{17}$ rather than Al_2O_3, birnessite (nominally $Na_4Mn_{14}O_{27} \cdot 9H_2O$) has been considered a form of MnO_2 in some publications, etc. Finally different phases corresponding to the same or to different chemical formulas can share the same name, e.g. three structures were reported for cronstedtite, for more examples, cf. Table 2.1.

The chemical formulae given in Table 2.1 are nominal or "idealized" ones and the real composition of native minerals and synthetic materials is often substantially different.

II. THERMOCHEMICAL DATA

The reader is assumed to know the principles of chemical thermodynamics, and how to use thermodynamic tables. The present nomenclature in thermochemistry based on the recent IUPAC recommendations [5] is different from that used in older publications, but the symbols used in mathematical equations remain unchanged.

A chemical reaction

$$a\,A + b\,B \rightarrow c\,C + d\,D \tag{2.1}$$

is spontaneous at isothermal and isobaric conditions (constant temperature and pressure) when the change in Gibbs energy of the system caused by this reaction is negative, i.e. the energy flows from the system to the surroundings (this is a generally accepted sign convention in thermochemistry). It should be emphasized that the spontaneity of chemical reactions is not directly related to the change in Gibbs energy in many systems of practical importance, namely, when the conditions are not isothermal and isobaric. Let us define an intensive quantity (independent of the amounts of reactants), namely, the standard Gibbs energy of reaction (1) ΔG° as the change in Gibbs energy when a moles of A react with b moles of B to form c moles of C and d moles of D, and A, B, C, and D are in their standard states. The standard state is defined as specific pressure (10^5 Pa), temperature (298.15 K), and composition of the system. ΔG° of the reaction depends on the way the equation of reaction is written, e.g.

$$H_2 + 1/2\,O_2 \rightarrow H_2O \tag{2.2}$$

$$2\,H_2 + O_2 \rightarrow 2\,H_2O \tag{2.3}$$

$$\Delta G^o \text{ (reaction 2.3)} = 2\,\Delta G^o \text{ (reaction 2.2)} \tag{2.4}$$

thus, talking about $\Delta G°$ of formation of water from elements we must specify whether reaction (2.2) or (2.3) is meant.

The Gibbs energy G is a thermodynamic function, i.e. it is unequivocally defined by the state of the system: p, T, and the composition. In other words for a cycle of processes which ends at the initial state $\Delta G=0$. The Gibbs energy of the entire system (an extensive quantity) can be split into a sum of contributions of particular species i.

$$G = \Sigma\, n_i\, G_i \tag{2.5}$$

where n_i is the number of moles of the species i, and

$$G_i = G_i° + RT \ln a_i, \tag{2.6}$$

where a is the activity. Activity is a dimensionless number expressing the ratio of activity at given state and in the standard state. The choice of the standard state, in which $a = 1$ by definition is a question of convention. For solids a pure compound is a convenient standard state. There is no generally accepted method or theory to determine the activities of components in a solid mixture.

Let us consider the following cyclic process under isothermal and isobaric conditions with all the reactants in their standard states:

$$a\,A + b\,B \rightarrow c\,C + d\,D \tag{2.1}$$
$$\text{elements } A \rightarrow a\,A \tag{2.7}$$
$$\text{elements } B \rightarrow b\,B \tag{2.8}$$
$$c\,C \rightarrow \text{elements } C \tag{2.9}$$
$$d\,D \rightarrow \text{elements } D \tag{2.10}$$

$$\text{elements } A + \text{elements } B \rightarrow \text{elements } C + \text{elements } D \tag{2.11}$$

Since we have the same number of atoms of particular elements on the both sides of reaction (2.1), the sum of reactions (2.1) and (2.7)–(2.10) formally written as reaction (2.11) has the same reagents and products, so the change in Gibbs energy equals zero. Let us define the Gibbs energy of formation of a compound A ΔG_f as the change in Gibbs energy when one mole of A is formed from elements (in their stable forms at given conditions). For a cyclic process (sum of reactions (2.1) and (2.7)–(2.10))

$$\Delta G° \text{ (reaction 2.1)} + a\,\Delta G_f(A) + b\,\Delta G_f(B) - c\,\Delta G_f(C) - d\,\Delta G_f(D) = 0 \tag{2.12}$$

or

$$\Delta G° \text{ (reaction)} = \Sigma\,\Delta G_f(\text{products}) - \Sigma\,\Delta G_f(\text{reactants}) \tag{2.13}$$

where the sums in eq. (2.13) take into account the stoichiometric coefficients. Equation (2.13) allows to calculate $\Delta G°$ of any reaction when ΔG_f of products and reactants are known. With gaseous and liquid compounds the ΔG_f is unequivocally defined by their formulas. In contrast for solids the ΔG_f can be considerably different for particular crystallographic forms. Only one of these forms is stable at given T but other metastable forms can also exist. Thus, difference in ΔG_f for a solid compound reported in different publications, not necessarily is a discrepancy.

It is not possible to calculate the absolute value of Gibbs energy (Eqs (2.5) and (2.6) without certain convention. Let us introduce another thermodynamic function, namely, enthalpy H (the heat produced or consumed, that can be measured calorimetrically). For elements in their most stable form

$$H° (298.15) = 0 \tag{2.14}$$

(by definition). Eq. (2.13) is valid not only for Gibbs energy but also for other thermodynamic functions. Thus,

$$\Delta H_f(298.15) = H° (298.15) \tag{2.15}$$

for any compound. The Gibbs energy is related to the enthalpy by the following equation

$$G = H - TS \tag{2.16}$$

The absolute entropy of a solid at 298.15 K (assuming that there is no phase transition between 0 and 298.15 K) can be calculated from the following equation

$$S°(298.15) = \int_0^{298.15} (C_p/T)\,dT \tag{2.17}$$

namely, according to the Nernst theorem the absolute entropy of a perfect crystal at 0 K equals zero. Since $S°$ (298.15) is positive (Eq. 2.17) this convention leads to

$$\Delta G_f(298.15) > G° (298.15) \tag{2.18}$$

It can be easily shown that

$$\Delta G° \text{ (reaction)} = \Sigma\, G° \text{ (products)} - \Sigma\, G° \text{ (reactants)} \tag{2.19}$$

so the $G°$ of the reactants can be used instead of their ΔG_f to predict the spontaneity of any reaction.

When ΔH is known the effect of temperature on ΔG can be estimated, namely

$$\delta(\Delta G/T)/\delta T = -\Delta H/T^2 \tag{2.20}$$

However, ΔH is also a function of T

$$\Delta H(T) = \Delta H°(298.15) + \int_{298.15}^{T} \Delta C_p\,dT \tag{2.21}$$

where

$$\Delta C_p = \Sigma\, C_p \text{ (products)} - \Sigma\, C_p \text{ (reactants)} \tag{2.22}$$

ΔC_p in turn also varies with T, and many empirical equations for the dependence C_p (T) have been proposed. For the systems discussed in the present book (room or nearly room temperature), Eq. (2.20) with the values of G^c and $H°$ (at 298.15 K) gives a sufficiently good estimate.

With solid reactants a spontaneous reaction will eventually lead to complete exhaustion of the reactant(s). In contrast reactions in gaseous phase and in solution always leave some amount of reactants in equilibrium with the products, although, when the $\Delta G°$ is sufficiently negative the equilibrium concentration of reactants can be beyond the range of common analytical methods. It is expedient to define the equilibrium constant of reaction (2.1)

$$K = a_C{}^c\, a_D{}^d\, a_A{}^{-a}\, A_B^{-b} \qquad\qquad (2.23)$$

K is a function of pressure and temperature but it is independent of the activities of reactants and it is related to the Gibbs energy of the reaction

$$\Delta G = -RT\ln K. \qquad\qquad (2.24)$$

The effect of pressure on ΔG is governed by the following equation

$$(\delta\Delta G/\delta p)_T = \Delta V \qquad\qquad (2.25)$$

Volume changes (ΔV) caused by reactions in condensed phases are often overlooked or deemed insignificant. This may be surprising for many chemists that the field of reaction volumes in solution has been extensively studied and the recent review [6] contains as many as 780 references. More work was summarized in two older reviews [7]. Volume change due to reaction in solution can be measured using a specially designed dilatometer. The principle of the measurement is very straightforward: two compartments contain two different solutions and the one on the top is connected with a capillary tube. Once the diaphragm between the two compartments is removed and the reaction is complete, the volume changes can be calculated from the height of the liquid in the capillary. The same apparatus can be used to determine the volume change due to adsorption. Certainly in that case one of the solutions is replaced by dispersion. After the measurement of the volume changes the dispersion can be analyzed to determine the uptake of the solutes. To the best knowledge of the present author no systematic studies of volume changes due to adsorption have been conducted until very recently [8]. Equation (2.25) will be discussed in more detail in the chapter devoted to sorption kinetics.

Many compilations of thermodynamic data partially covering the materials of interest have been published in the nineties. Table 2.2 is based on data from these compilations. Most compilations report also thermodynamic data at elevated temperatures (usually every 100°C). No attempt to update the thermodynamic data using recent publications or to verify the data in the compilations using original publications has been made. The thermodynamic functions are interrelated. Each compilation has its unique style of presentation, namely some functions are given *explicite*, and the other ones must be calculated. Some values reported in the Table 2.2 below were calculated by means of procedures explained below.

Reference [9] abbreviated as **K** was used as the primary source of thermodynamic data. This is because **K** contains more relevant information than any other up to date compilation. For most compounds thermodynamic data from different compilations are identical or almost identical, but there are a few clear discrepancies. Assessment of credibility of data from different compilations is beyond the scope of the present book.

The values of ΔG_f are not given directly in **K**, but they have been calculated using the following equation

$$\Delta G_f(\text{compound}) = G°\,(\text{compound}) - \Sigma\, G°\,(\text{elements}) \qquad\qquad (2.26)$$

TABLE 2.2 Thermochemical Data

Simple Oxides, Oxohydroxides and Hydroxides

		C_p J mol^{-1} K^{-1}	$H° = \Delta H°_f$ J mol^{-1}	$\Delta G°_f$ J mol^{-1}	$G°$ J mol^{-1}	Ref.
water		75.79	-285829	-237142	-306685	K
Ag$_2$O		66.31	-30898	-10781	-66811	K
Al$_2$O$_3$	corundum	78.83	-1675692	-1582276	-1690882	K
Al$_2$O$_3$	γ	82.706	-1656864	-1563850	-1672457	B
Al$_2$O$_3$	δ	81.362	-1666487	-1572947	-1681581	B
Al$_2$O$_3$	κ	80.73	-1666487	-1573847	-1682454	B
AlOOH	boehmite	60.42	-985265	-910162	-999236	K
AlOOH	diaspore	52.76	-1002068	-923491	-1012565	K
Al(OH)$_3$	hydrargillite	93.08	-1293274	-1155552	-1314169	K
Al(OH)$_3$	bayerite		-1288250	-1149000	-1307645	S
Al(OH)$_3$	amorphous	93.149	-1276120	-1138706	-1297327	B
BeO	α	24.98	-608399	-579092	-612505	K
Be(OH)$_2$	β	65.64	-905798	-817811	-920767	K
Be(OH)$_2$	α	65.674	-902907	-815933	-918874	B
Bi$_2$O$_3$		113.52	-573208	-492790	-618366	K
CdO		44.16	-258362	-228678	-274703	K
Cd(OH)$_2$		-118.8	-560702	-473753	-589324	B
CeO$_2$		61.52	-1090350	-1027054	-1108925	K
CoO		55.22	-237944	-214190	-253736	K
Co$_3$O$_4$		123.42	-918680	-802056	-951276	K
Co(OH)$_2$		97.06	-541338	-460067	-569156	K
Co(OH)$_2$	precipitated	97.064	-539698	-454168	-563251	B
Cr$_2$O$_3$		114.26	-1140558	-1058917	-1164759	K
Cr(OH)$_3$	cryst.			-848056		R
Cu$_2$O		62.543	-170707	-147886	-198245	C
CuO		42.25	-156059	-128287	-168756	K
Cu(OH)$_2$		87.85	-443086	-359018	-469033	B
Dy$_2$O$_3$		116.26	-1862716	-1770971	-1907375	K
Er$_2$O$_3$		108.5	-1897900	-1808700	-1944292	H
Eu$_2$O$_3$	cubic	124.67	-1662721	-1566361	-1704511	K
Eu$_2$O$_3$	monoclinic	121.74	-1657123	-1562634	-1700784	K
Fe$_{0.947}$O	wustite	48.116	-266270	-245143	-283440	C
FeO		49.915	-272044	-251429	-290158	C
Fe$_3$O$_4$	magnetite	151.78	-1115479	-1012351	-1159078	K
Fe$_2$O$_3$	hematite	104.77	-823411	-741471	-849483	K
FeOOH	goethite	74.32	-558145	-487369	-576146	K
Fe(OH)$_2$		86.31	-574044	-491983	-600241	K
Fe(OH)$_3$		92.58	-832616	-705482	-863802	K
Ga$_2$O$_3$		93.86	-1089095	-998328	-1114418	K
Ga(OH)$_3$			-964400	-831300	-994215	H
Gd$_2$O$_3$	cubic	105.51	-1826901	-1739548	-1871810	K
Gd$_2$O$_3$	monoclinic	106.62	-1819404	-1733921	-1866183	K
HfO$_2$		60.27	-1117546	-999946	-1135260	K
HgO	red	43.89	-90830	-58565	-111775	K
Ho$_2$O$_3$		114.96	-1880917	-1791593	-1928071	K

TABLE 2.2 (Continued)

		C_p J mol^{-1} K^{-1}	$H° = \Delta H°_f$ J mol^{-1}	$\Delta G°_f$ J mol^{-1}	G J mol^{-1}	Ref.
In$_2$O$_3$		100.39	-925919	-830756	-956980	K
IrO$_2$		55.64	-249366	-192836	-264585	K
La$_2$O$_3$		108.77	-1794936	-1707220	-1832896	K
Lu$_2$O$_3$		101.76	-1878197	-1788848	-1910982	K
MgO	periclase	37.26	-601701	-569409	-609733	K
MgO	microcrystalline	38.48	-597981	-565979	-606303	K
Mg(OH)$_2$		77.51	-924998	-833987	-943854	K
Mn$_3$O$_4$	hausmanite			-1282978		R
Mn$_2$O$_3$	bixbyite			-879271		R
MnOOH	manganite			-558527		R
MnO$_2$	birnessite			-453777		R
MnO$_2$	nsutite			-457129		R
MnO$_2$	pyrolusite			-466347		R
MnO$_2$		54.41	-522054	-467190	-537897	K
MoO$_3$		75.14	-744982	-667892	-768166	K
NbO$_2$		57.452	-794960	-739175	-811211	C
Nb$_2$O$_5$		132.13	-1899536	-1765821	-1940471	K
Nd$_2$O$_3$		111.33	-1807906	-1721051	-1855185	K
NiO		44.29	-239743	-211582	-251070	K
Ni(OH)$_2$			-529700	-447200	-555938	H
PbO	red	45.74	-220007	-189633	-239530	K
PbO	massicot	45.8	-217300	-187900	-237783	H
Pb$_3$O$_4$		154.94	-730672	-615467	-795740	K
PbO$_2$		60.99	-282784	-224509	-304988	K
Pr$_2$O$_3$		117.98	-1809664	-1720237	-1856069	B
PuO$_2$		66.25	-1055832	-999041	-1075548	K
RuO$_2$		52.69	-305013	-252807	-322478	K
Sb$_4$O$_6$	valentinite	202.77	-1417539	-1253109	-1490889	K
Sc$_2$O$_3$		93.94	-1908322	-1818873	-1931275	K
SiO$_2$	quartz α	44.43	-910856	-856442	-923219	K
SiO$_2$	tridimite α	44.71	-910053	-856200	-922977	K
SiO$_2$	cristobalite α	44.95	-908346	-854508	-921285	K
H$_2$SiO$_3$			-1188700	-1092400	-1228653	H
H$_4$SiO$_4$			-1481100	-1332900	-1538345	H
Sm$_2$O$_3$		115.82	-1827403	-1737380	-1870566	K
SnO$_2$		52.6	-580822	-520001	-596428	K
Ta$_2$O$_5$		131.48	-2045976	-1911035	-2088693	K
Tb$_2$O$_3$		115.05	-1865227	-1776551	-1912007	K
ThO$_2$		61.81	-1226414	-1168781	-1245862	K
TiO$_2$	rutile	55.1	-944747	-889506	-959841	K
TiO$_2$	anatase	54.02	-941400	-885947	-956282	K
Tm$_2$O$_3$		116.73	-1888657	-1794442	-1930322	K
UO$_2$		63.58	-1084994	-1031802	-1107960	K
U$_4$O$_9$		292.14	-4510398	-4274804	-4610018	K
V$_2$O$_3$		101.87	-1218799	-1137492	-1246492	K
V$_2$O$_5$		127.34	-1550172	-1418928	-1589092	K
WO$_{2.72}$		68.283	-781153	-708632	-801554	C

$WO_{2.9}$		71.379	-820064	-743513	-841940	C
$WO_{2.96}$		72.383	-834959	-757027	-857289	C
WO_3	α	72.8	-842908	-764054	-865537	K
H_2WO_4		104.09	-1131893	-1004029	-1175055	K
Y_2O_3		102.51	-1905309	-1816620	-1934862	K
Yb_2O_3		115.3	-1814517	-1726764	-1854186	K
ZnO		41.06	-350619	-320636	-363630	K
$Zn(OH)_2$			-641900	-553500	-666110	H
ZrO_2		56.21	-1100559	-1042788	-1115578	K

Mixed Oxides (Salts)

		C_p J mol^{-1} K^{-1}	$H = \Delta H°_f$ J mol^{-1}	ΔG_f J mol^{-1}	G J mol^{-1}	Ref.
Aluminates						
$BeO·Al_2O_3$		104.94	-2300781	-2178522	-2320541	K
$BeO·3Al_2O_3$		261.92	-5624133	-5317245	-5676476	K
$CdO·Al_2O_3$		129.97	-1918991	-1801659	-1956290	K
$CoO·Al_2O_3$		132.23	-1947108	-1828645	-1976797	K
$CuO·Al_2O_3$		127.09	-1822500	-1703988	-1853063	K
$FeO·Al_2O_3$		123.48	-1969471	-1853836	-2001157	K
$La_2O_3·Al_2O_3$		185.29	-3587780	-3404394	-3638676	K
$MgO·Al_2O_3$		115.96	-2299108	-2176624	-2325554	K
$MnO·Al_2O_3$		124.63	-2100368	-1982574	-2131305	K
$NiO·Al_2O_3$		131.57	-1921497	-1802719	-1950813	K
$ZnO·Al_2O_3$		119.36	-2071289	-1945636	-2097236	K
Aluminosilicates						
$BeAlSiO_4OH$	euclase		-2532910	-2370170	-2559473	S
$Be_3Al_2Si_6O_{18}$	beryl		-9006520	-8500360	-9109889	S
$FeAl_2Si_2O_6(OH)_4$	Fe-carpholite	284.1	-4433850	-4080260	-4500338	P
$FeAl_2SiO_5(OH)_2$	Fe-chloritoid	202.5	-3209170	-2973740	-3257471	P
$Fe_2Al_4Si_5O_{18}$	Fe-cordierite	464.4	-8436510	-7949440	-8578132	P
$Fe_{3.5}Al_9Si_{1.5}O_{20}$	Fe-sapphirine	575.2	-9834000	-9273620	-9998281	P
$Fe_4Al_{18}Si_{7.5}O_{48}H_4$	Fe-staurolite	1274	-23753880	-22282230	-24055012	P
$Fe_2Al_4Si_3O_{10}(OH)_8$	Fe-sudoite	605.6	-7912770	-7275660	-8049025	P
$Fe_5Al_2Si_3O_{10}(OH)_8$	daphnite	547.8	-7153990	-6535560	-7316482	P
$Fe_3Al_2Si_3O_{12}$	almandine	350.3	-5276340	-4951526	-5376608	C
$Fe_2Al_4Si_5O_{18}$	ferrocordierite	400.5	-8470600	-7964583	-8593110	C
$Al_4Mg_2Si_5O_{18}$	cordierite	452.31	-9161701	-8651336	-9283079	B
$Mg_4Al_4Si_2O_{10}(OH)_8$	amesite	513.6	-9052530	-8378380	-9168809	P
$MgAlSi_4O_{10}(OH)_2$	celadonite	325.8	-5844480	-5464110	-5930050	P
$Mg_5Al_4Si_6O_{22}(OH)_2$	gedrite	642.6	-12319880	-11584190	-12473428	P
$MgAl_2Si_2O_6(OH)_4$	Mg-carpholite	278	-4794700	-4430870	-4852542	P
$MgAl_2SiO_5(OH)_2$	Mg-chloritoid	196.3	-3559530	-3313560	-3598886	P
$Mg_4Al_{18}Si_{7.5}O_{48}H_4$	Mg-staurolite	1251	-25103090	-23595240	-25374407	P
$MgAl_2SiO_6$	Mg-Tschermak pyroxene	160.3	-3188830	-3012120	-3227888	P
$Mg_2Al_2Si_3O_{10}(OH)_2$	Tschermak-talc	317.8	-5987610	-5605600	-6064831	P
$Mg_4Al_8Si_2O_{20}$	sapphirine (442)	546.6	-11014080	-10415770	-11145266	P

Mixed Oxides (Salts) (Continued)

		C_p J mol^{-1} K^{-1}	$H° = \Delta H°_f$ J mol^{-1}	$\Delta G°_f$ J mol^{-1}	$G°$ J mol^{-1}	Ref.
$Mg_{3.5}Al_9Si_{1.5}O_{20}$	sapphirine (793)	544.7	-11067020	-10469450	-11199697	P
$Mg_3Al_2Si_3O_{12}$	pyrope	325.6	-6291540	-5941028	-6370937	C
$Mg_2Al_4Si_5O_{18} \cdot H_2O$	hydrous cordierite	506.9	-9430320	-8811292	-9512580	C
$Mg_5Al_2Si_3O_{10}(OH)_8$	clinochlorite	517.4	-8928250	-8265042	-9053771	C
$Mn_2Al_4Si_3O_{10}(OH)_8$	sudoite	593.5	-8634360	-7976770	-8753322	P
$Mn_5Al_2Si_3O_{10}(OH)_8$	Mn-chlorite	542.6	-7680490	-7065500	-7853417	P
$MnAl_2SiO_5(OH)_2$	Mn-chloritoid	196.3	-3330930	-3095300	-3380423	P
$Mn_2Al_4Si_5O_{18}$	Mn-cordierite	462.3	-8681710	-8191830	-8823332	P
$Mn_4Al_{18}Si_{7.5}O_{48}H_4$	Mn-staurolite	1271	-24206000	-22732930	-24511306	P
Clay Minerals						
$Al_2Si_2O_7 \cdot 2H_2O$	dickite	239.48	-4118299	-3795819	-4177067	B
$Al_2Si_2O_7 \cdot 2H_2O$	halloysite	246.26	-4101199	-3780565	-4161813	B
$Al_2O_3 \cdot 2SiO_2 \cdot 2H_2O$	kaolinite	245.26	-4095843	-3775099	-4156345	K
$KAl_3Si_3O_{10}(OH)_2$	muscovite	326	-5976300	-5600298	-6067653	C
Chromates (III)						
$CoO \cdot Cr_2O_3$		157.22	-1438333	-1330743	-1476131	K
$CuO \cdot Cr_2O_3$		148.08	-1293479	-1186089	-1332400	K
$FeO \cdot Cr_2O_3$		133.8	-1458605	-1357833	-1502390	K
$MgO \cdot Cr_2O_3$		126.79	-1777781	-1663176	-1809342	K
$NiO \cdot Cr_2O_3$		148.8	-1392435	-1285776	-1431106	K
$ZnO \cdot Cr_2O_3$		143.36	-1553937	-1439781	-1588617	K
Ferrates						
$CoO \cdot Fe_2O_3$		152.58	-1088676	-983657	-1131215	K
$CuFeO_2$		79.996	-512958	-460268	-539454	B
$CuO \cdot Fe_2O_3$	chalcopyrite	148.8	-966504	-861778	-1010259	K
$MgO \cdot Fe_2O_3$	magnesioferrite	137.78	-1440132	-1328721	-1477057	K
$MnO \cdot Fe_2O_3$		149.22	-1228840	-1126610	-1274747	K
$NiO \cdot Fe_2O_3$		159.17	-1084492	-974541	-1122041	K
$ZnO \cdot Fe_2O_3$		137.33	-1179051	-1073752	-1224758	K
Gallates						
$CdO \cdot Ga_2O_3$		138.5	-1355105	-1234655	-1396770	K
$CuO \cdot Ga_2O_3$		135.03	-1228004	-1115106	-1271665	K
Silicates						
$3Al_2O_3 \cdot 2SiO_2$	mullite	322.2	-6819208	-6441794	-6901166	K
$Al_2O_3 \cdot SiO_2$	andalusite	122.44	-2590314	-2442890	-2618273	K
$Al_2O_3 \cdot SiO_2$	sillimanite	121.93	-2587804	-2441070	-2616453	K
$Al_2O_3 \cdot SiO_2$	kyanite	120.01	-2594080	-2443880	-2619263	K
$Al_2O_3 \cdot 2SiO_2$	metakaolinite	224.09	-3341154	-3139676	-3381836	K
$Al_2SiO_4(OH)_2$	hydroxy-topaz	159.1	-2905000	-2689970	-2934965	P
$Al_2Si_4O_{10}(OH)_2$	pyrophyllite	292	-5640960	-5267080	-5712337	C
$HOSiO_3Al_2(OH)_3$	imogolite		-3192100	-2929200	-3243710	S
$2BeO \cdot SiO_2$	phenacite	92.47	-2142563	-2028134	-2161737	K
$CdO \cdot SiO_2$		88.82	-1189302	-1105565	-1218367	K
$2CoO \cdot SiO_2$		133.95	-1398711	-1300120	-1445989	K

$2FeO \cdot SiO_2$	fayalite	132.9	-1477073	-1376153	-1520360	K
Fe_2SiO_4	γ-fayalite	130.2	-1471500	-1369332	-1513539	C
Fe_2SiO_4	β-fayalite	136.1	-1468000	-1366071	-1510278	C
$FeSiO_3$		89.454	-1194950	-1117463	-1222954	B
$Fe_{18}Si_{12}O_{40}(OH)_{10}$	deerite	1552	-18348400	-16902080	-18840348	P
$Fe_7Si_8O_{22}(OH)_2$	Fe-anthophyllite	692.8	-9627760	-8968900	-9843919	P
$Fe_3Si_4O_{10}(OH)_2$	Fe-talc	340	-4799030	-4451060	-4903979	P
$Fe_7Si_8O_{22}(OH)_2$	grunerite	694.3	-9631500	-8969794	-9844558	C*
$2MgO \cdot SiO_2$		118.42	-2176935	-2057890	-2205315	K
Mg_2SiO_4	β-forsterite	110.7	-2142000	-2022303	-2169728	C
Mg_2SiO_4	γ-forsterite	106.6	-2133200	-2012310	-2159735	C
$MgSiO_3$	low clinoenstatite	81.8	-1548715	-1460457	-1567558	C
$MgO \cdot SiO_2$		81.9	-1548498	-1461631	-1568732	K
$MgSiO_3$	orthoenstatite	82.2	-1546290	-1458956	-1566057	C
$MgSiO_3$	garnet – Mg	79.9	-1513000	-1423490	-1530591	C
$MgSiO_3$	ilmenite – Mg	75.1	-1489500	-1400199	-1507300	C
$MgSiO_3$	perovskite – Mg	79.8	-1450000	-1360639	-1467740	C
$Mg_7Si_8O_{22}(OH)_2$	anthophyllite	628.99	-12086405	-11367041	-12253066	B
$Mg_3Si_4O_{10}(OH)_2$	talc	321.7	-5922498	-5542603	-6000225	B
$Mg_3Si_2O_5(OH)_2$	chrisotile	273.68	-4365598	-4037962	-4431577	B
$Mg_7Si_2O_8(OH)_6$	phase A	462.4	-7129870	-6609650	-7234223	P
$Mg_6Si_4O_{10}(OH)_8$	Al-free chlorite	521.3	-8744200	-8078470	-8865846	P
$Mg_9Si_4O_{16}(OH)_2$	clinohumite	549.2	-9637840	-9062780	-9762467	P
$Mg_{48}Si_{34}O_{85}(OH)_{62}$	antigorite	4435	-71377000	-66110242	-72472045	C
$Mg_7Si_8O_{22}(OH)_2$	cummingtonite	428.9	-12217100	-11475109	-12361136	C
$MnSiO_3$	pyroxmangite	86.63	-1322500	-1245180	-1352107	P
$MnO \cdot SiO_2$	rhodonite	86.39	-1318085	-1241746	-1348648	K
$2MnO \cdot SiO_2$	tephroite	129.87	-1725314	-1620700	-1767727	K
$2NiO \cdot SiO_2$		127.01	-1397234	-1284289	-1430042	K
$4PbO \cdot SiO_2$		229.73	-1801086	-1633432	-1899797	K
$2PbO \cdot SiO_2$		136.92	-1369004	-1258070	-1424641	K
$PbO \cdot SiO_2$		89.13	-1138089	-1054190	-1170864	K
$2ZnO \cdot SiO_2$	willemite	121.83	-1644730	-1531135	-1683900	K
$ZnSiO_3$		84.762	-1262572	-1179479	-1289268	B
$ZrO \cdot 2SiO_2$		98.57	-2035700	-1921186	-2060753	K
Titanates						
$Al_2O_3 \cdot TiO_2$		136.4	-2607046	-2460788	-2639729	K
$CdO \cdot TiO_2$		98.5	-1230807	-1145758	-1262118	K
$2CoO \cdot TiO_2$		160.48	-1447203	-1345175	-1494602	K
$CoO \cdot TiO_2$		107.77	-1207381	-1126378	-1236259	K
$FeO \cdot TiO_2$	ilmenite	99.5	-1239229	-1161740	-1270790	K
Fe_2TiO_4		142.3	-1515236	-1417863	-1565633	B
$2MgO \cdot TiO_2$		128.16	-2164383	-2047705	-2198688	K
$MgO \cdot TiO_2$		91.19	-1572556	-1484127	-1594786	K
$MgO \cdot 2TiO_2$		146.65	-2509354	-2368777	-2549771	K
$2MnO \cdot TiO_2$		144.59	-1749543	-1649480	-1800065	K
$MnO \cdot TiO_2$		99.82	-1354486	-1275587	-1386047	K
$NiO \cdot TiO_2$		99.25	-1201435	-1116225	-1226048	K
$PbO \cdot TiO_2$		104.4	-1194741	-1107878	-1228110	K
$2ZnO \cdot TiO_2$		137.32	-1649751	-1536590	-1692913	K
Vanadates						
$2MgO \cdot V_2O_5$		201.79	-2834660	-2643476	-2894288	K

Mixed Oxides (Salts) (Continued)

	C_p J mol^{-1} K^{-1}	$H° = \Delta H°_f$ J mol^{-1}	$\Delta G°_f$ J mol^{-1}	$G°$ J mol^{-1}	Ref.
MgO·V$_2$O$_5$	156.35	-2200784	-2038198	-2248686	K
Tungstates					
FeO·WO$_3$	114.37	-1184176	-1083211	-1223409	K
MgO·WO$_3$	109.89	-1517118	-1405437	-1547244	K
MnO·WO$_3$	124.08	-1305826	-1206133	-1347741	K
NiO·WO$_3$	121.6	-1127839	-1022046	-1163017	K
PbO·WO$_3$	119.64	-1121312	-1019955	-1171335	K
ZnO·WO$_3$	125.5	-1235367	-1133928	-1278405	K
Zirconates					
Nd$_2$O$_3$·2ZrO$_2$	223.39	-4121658	-3919287	-4199001	K
Sm$_2$O$_3$·2ZrO$_2$	224.33	-4130444	-3926651	-4205417	K
Y$_2$Zr$_2$O$_7$	215.03	-4121993	-3917872	-4181622	B

The values of $G°$ (elements) were taken from **K** and the sum in Eq. (2.26) takes into account numbers of atoms or molecules of particular elements in the compound, e.g.

$$\Delta G_f(Ag_2O) = G°\,(Ag_2O) - 2\,G°\,(Ag) - 1/2\,G°\,(O_2) \qquad (2.27)$$

If thermodynamic data for given compound were not available in **K**, [10] (**B**), [3] (**H**) and then [11] (**C**) were used. The values in Table 2.2 were taken directly from **B**, and no calculation was needed. The values of $G°$ are not given directly in **H** and **C**, but they have been calculated using the following equation

$$G° = H° - TS° \qquad (2.28)$$

The values of $S°$ were taken from **H** and **C**, and $T = 298.15$ K in Eq. (2.28). The C_p data for a few compounds are missing in **H**. A few entries are based on Ref. [12] abbreviated as **S**. The $G°$ values were calculated from Eq. (2.28) using the $S°$ taken from **S**.

A few ΔG_f values (chiefly Mn oxides) not available from **K**, **B**, **H**, or **S** were taken from Ref. [13] abbreviated as **R**. Only ΔG_f are listed there (no H or S data).

Only **K**, **B**, **H**, **S** and **R** were used as sources of data for simple oxides, oxohydroxides and hydroxides. Specialized databases created primarily for mineralogists and geologists report on thermodynamic data for materials of their special interest, e.g. natural silicates and aluminosilicates. These databases were used to complete data which were not available in **K**, **B**, **H**, **S** or **R**. Ref. [14] abbreviated as **C** does not directly list $G°$ or ΔG_f values. The $G°$ values were calculated from Eq. (2.28) using the $S°$ taken from **C**. Then, Eq. (2.26) was applied to calculate ΔG_f (compound) and the values of $G°$ (elements) were taken from **K**. Then, data from Ref. [15] abbreviated as **P** were used. A few chemical formulas in **P** have typographical errors, which have been corrected in Table 2.2. The $G°$ values were calculated from Eq. (2.28) using the S° taken from **P**. **P** does not directly report C_p values. They were calculated using the following equation

$$C_p = a + bT + c\,T^2 + d\,T^{1/2} \tag{2.29}$$

The empirical parameters a, b, c and d in Eq. (2.29) are given in **P** and $T = 298.15$ K. Table 2.2 shows as many decimal digits as the source compilations. Calculated values were rounded to 1 J mol^{-1} for the enthalpy and Gibbs energy and to 0.01 J mol^{-1} K^{-1} for C_p Considering the difference in data for the same compound reported in different compilations no more than three digits can be considered as significant for most compounds.

Some data in Table 2.2 refer to amorphous materials. The crystalline phases are well defined, but amorphous materials are not, thus, the thermochemical data can differ from one amorphous sample to another. The data in Table 2.2 illustrate a typical difference in Gibbs energy between the most stable crystalline phase and amorphous materials. There are other compilations of thermodynamic data, which were not used in Table 2.2. Some additional data for oxides and silicates is available from Ref. [16]. The old units (kcal/mol) were used in that book. A collection of thermodynamic data, chiefly for silicates can be found in Ref. [17].

A few examples of application of thermochemical tables are discussed below.

Example 1. The Most Stable Oxide

Corundum has a more negative G° (and ΔG_f) than any other form of aluminum (III) oxide (Table 2.2). Therefore, Gibbs energies of reactions

$$\gamma\text{-Al}_2\text{O}_3 \rightarrow \text{corundum} \tag{2.30}$$

$$\delta\text{-Al}_2\text{O}_3 \rightarrow \text{corundum} \tag{2.31}$$

etc. calculated from Eq. (2.13) or Eq. (2.19) are negative. This means that the reactions (2.30) and (2.31) are spontaneous at room temperature and atmospheric pressure.

Thus, corundum is the thermodynamically stable form of aluminum (III) oxide at room temperature. However, this result does not imply that γ-Al$_2$O$_3$ would actually convert into corundum in given experiment, namely, the reaction (2.30) can be very slow. This allows existence of many different forms of aluminum (III) oxide. Corundum is stable and the other forms are metastable. On the other hand, a spontaneous conversion of corundum into another form at room temperature is impossible. Gibbs energies of reactions (2.30) and (2.31) are obtained as a difference of two large and almost equal numbers. The value and even the sign of such a difference is very uncertain, especially when data for two crystalline forms of the same oxide are taken from two different sources.

Example 2. Molecular Formula Multiplied by 2

In Table 2.2 thermodynamic data for Al(OH)$_3$ (hydrargillite) from **K** are used. Thermodynamic data for the same compound are listed in **B** as Al$_2$O$_3$·3H$_2$O (gibbsite). Since the molecular formula in **B** is multiplied by two so are the values of thermodynamic functions, i.e.

$$C_p(Al_2O_3 \cdot 3H_2O, \text{ gibbsite}) = 2\,C_p(Al(OH)_3, \text{hydrargillite}) \tag{2.32}$$

$$H^o(Al_2O_3 \cdot 3H_2O, \text{ gibbsite}) = 2\,H^o(Al(OH)_3, \text{hydragillite}) \tag{2.33}$$

etc. As a matter of fact there is almost no difference between thermodynamic data for $Al_2O_3 \cdot 3H_2O$ (gibbsite) taken from **B** and doubled values for $Al(OH)_3$ (hydrargillite) taken from **K**.

Example 3.   The Most Stable Oxide/Hydroxide/Oxohydroxide

In aqueous media oxides can spontaneously turn into hydroxides or oxohydroxides and vice versa. Let us explore the hydration-dehydration reactions for Al(III). Example 1 shows that corundum is the most stable oxide. Likewise hydrargillite has a more negative negative G^o (and ΔG_f) than the amorphous hydroxide and thus the former from is stable (not a big surprise), and diaspore is more stable than boehmite. In order to find which of the three forms (corundum, hydrargillite and diaspore) is thermodynamically stable, let us calculate ΔG^o for the following reactions.

$$Al_2O_3 \text{ (corundum)} + 3\,H_2O \text{ (liquid)} \rightarrow 2\,Al(OH)_3 \text{ (hydrargillite)} \tag{2.34}$$

$$\Delta G^o(\text{reaction } 2.34) = 2G^o\,(Al(OH)_3(\text{hydragillite})) - G^o(Al_2O_3(\text{corundum}))$$

$$- 3\,G^o(H_2O) = 2 \times (-1314169) + 1690882 + 3 \times 306685 = -17401 \text{ J} \tag{2.35}$$

Since ΔG^o (reaction 2.34) is negative this reaction is spontaneous when the reactants are in their standard states. Likewise, for the reaction

$$AlO(OH) \text{ (diaspore)} + H_2O(\text{liquid}) \rightarrow Al(OH)_3 \text{ (hydrargillite)} \tag{2.36}$$

$$\Delta G^o(\text{reaction } 2.36) = G^o\,(Al(OH)_3(\text{hydrargillite})) - G^o\,(AlO(OH)\,(\text{diaspore}))$$

$$- G^o\,(H_2O) = 1314169 + 1012565 + 306685 = 5081 \text{ J} \tag{2.37}$$

Since ΔG^o (reaction 2.36) is positive this reaction is not spontaneous when the reactants are in their standard states. Moreover, hydrargillite is expected to dehydrate spontaneously to form diaspore, which is the most stable among aluminum (III) oxides, hydroxides and oxohydroxides. However the practice shows that processes of hydration and dehydration are rather slow at ambient conditions and the unstable forms show sufficient degree of metastability to remain unchanged over the time of typical laboratory experiments. The result obtained for aluminum (III) does not imply that other oxohydroxides are also more stable than oxides or hydroxides. Many oxides do not undergo spontaneous hydration, e.g.

$$Fe_2O_3 \text{ (hemattie)} + H_2O(\text{liquid}) \rightarrow 2\,FeO(OH)\,(\text{goethite}) \tag{2.38}$$

$$\Delta G^o \text{ (reaction } 2.38) = 2G^o\,(FeO(OH)\,(\text{goethite})) - G^o(Fe_2O_3\,(\text{hematite}))$$

$$- G^o\,(H_2O) = 2 \times (-576146) + 849483 + 306685 = 3876 \text{ J} \tag{2.39}$$

$$Fe_2O_3 \text{ (hematite)} + 3\,H_2O \text{ (liquid)} \rightarrow 2\,Fe(OH)_3 \tag{2.40}$$

$$\Delta G^o(\text{reaction } 2.40) = 2G^o\,(Fe(OH)_3) - G^o\,(Fe_2O_3\,(\text{hematite})) - 3\,G^o\,(H_2O)$$

$$= 2 \times (-863802) + 849483 + 3 \times 306685 = 41934 \text{ J} \tag{2.41}$$

Positive Gibbs energies indicate that reactions (2.38) and (2.40) are not spontaneous and hematite is more stable than goethite or $Fe(OH)_3$. In some handbooks of corrosion goethite has been considered as the most stable form of iron (III). This discrepancy illustrates the rule that the sign of results obtained as a difference of two large and almost equal numbers is not necessarily reliable. The value of Gibbs energy of reaction (2.38) is small compared with the standard Gibbs energies of reactants and products. The Gibbs energy of reaction (2.38) is a fraction of 1% of the standard Gibbs energies of hematite and goethite. In other words, an error in the value of standard Gibbs energy of hematite and goethite on the order of 0.5% could result in a negative Gibbs energy of reaction (2.38). Also a small change in experimental conditions (e.g. temperature) can cause a reversal of sign of Gibbs energy of reaction (2.38).

The result obtained with Fe(III) is not the only example of discrepancy between the results obtained using data from Table 2.2 and literature data. According to Ref. [18] $Ni(OH)_2$ is more stable than NiO, while the present data suggest an opposite, but the positive Gibbs energy of hydration calculated using the data from Table 2.2 is very small.

For some metals hydroxides are the most stable forms, e.g.

$$MgO(periclase) + H_2O \rightarrow Mg(OH)_2 \tag{2.42}$$

$$\Delta G^o(\text{reaction } 2.42) = G^o(Mg(OH)_2) - G^o(MgO(periclase)) - G^o(H_2O) =$$
$$- 943854 + 609733 + 306685 = -27434 \text{ J} \tag{2.43}$$

Reaction (2.42) is spontaneous. Moreover, the negative Gibbs energy of reaction (2.42) is large enough compared with the standard Gibbs energies of the reactants and products and there is no doubt about the significance of the result.

Example 4. Formation of a Mixed Oxide (Salt)

The data from Table 2.2 can be used to assess whether a mixed oxide (salt) is formed spontaneously. It has been already shown that $Mg(OH)_2$ is more stable than MgO. Let us explore the possibility of formation of magnesium titanate (Mg(II)–Ti(IV) mixed oxide) as a result of reaction of $Mg(OH)_2$ (e.g. formed by surface precipitation) and rutile.

$$Mg(OH)_2 + TiO_2 \text{ (rutile)} \rightarrow MgO \cdot TiO_2 + H_2O \tag{2.44}$$

$$\Delta G^o(\text{reaction } 2.44) = G^o(MgO \cdot TiO_2) + G^o(H_2O) - G^o(Mg(OH)_2)$$
$$- G^o(TiO_2(\text{rutile})) = -1594786 - 306685 + 943854 + 959841 = 2224 \text{ J} \tag{2.45}$$

$$2\,Mg(OH)_2 + TiO_2 \text{ (rutile)} \rightarrow 2MgO \cdot TiO_2 + 2H_2O \tag{2.46}$$

$$\Delta G^o(\text{reaction } 2.46) = G^o(2MgO \cdot TiO_2) + 2G^o(H_2O) - 2G^o(Mg(OH)_2)$$
$$- G^o(TiO_2(\text{rutile})) = -2198688 + 2 \times (- 306685) + 2 \times 943854$$
$$+ 959841 = 35491 \text{ J} \tag{2.47}$$

$$Mg(OH)_2 + 2TiO_2 \text{ (rutile)} \rightarrow MgO \cdot 2TiO_2 + H_2O \tag{2.48}$$

$$\Delta G^o(\text{reaction } 2.48) = G^o(MgO \cdot 2TiO_2) + G^o(H_2O) - G^o(Mg(OH)_2)$$
$$- 2G^o(TiO_2(\text{rutile})) = -2549771 - 306685 + 943854 + 2 \times 959841$$
$$= 7080 \text{ J} \tag{2.49}$$

The positive Gibbs energies of reactions (2.44, 2.46, and 2.48) show that these reactions are not spontaneous. This means that magnesium hydroxide precipitated on rutile will not form magnesium titanate.

There are many examples that double oxides are more stable than single oxides or hydroxides. Rutile is more stable than anatase and $Co(OH)_2$ is more stable than CoO (this can be easily shown using data from Table 2.2).

$$Co(OH)_2 + TiO_2 \text{ (rutile)} \rightarrow CoO \cdot TiO_2 + H_2O \tag{2.50}$$

$$\Delta G^o(\text{reaction 2.50}) = G^o(CoO \cdot TiO_2) + G^o(H_2O) - G^o(Co(OH)_2)$$

$$- G^o(TiO_2(\text{rutile})) = -1236259 - 306685 + 569156 + 959841 =$$

$$- 13947 \text{ J} \tag{2.51}$$

The Gibbs energy of synthesis of $CoO \cdot TiO_2$ from oxides would be even more negative for less stable reactants, e.g. anatase instead of rutile, or CoO instead of $Co(OH)_2$, thus, reaction (2.50) and analogous reactions involving less stable reactants to synthesize $CoO \cdot TiO_2$ are spontaneous. This does not mean however that surface precipitation of Co on TiO_2 will actually result in formation of $CoO \cdot TiO_2$. First, for kinetic reasons discussed above. Then, $CoO \cdot TiO_2$ is one of many products that can be potentially formed. Of those, thermochemical data is available only for 2 $CoO \cdot TiO_2$, but this does not exclude a possibility of formation of other products.

Example 5.   T≠295.15 K

The above calculations (examples 2–4) were performed for $T = 298.15$ K. Let us check if reaction (2.50) is also spontaneous at $T = 293.15$ K. Data from Table 2.2 can be used to calculate the $G(293.15$ K) for each reactant but it is easier to perform the calculations for the entire reaction.

$$\Delta H^o(\text{reaction 2.50}) = H^o(CoO \cdot TiO_2) + H^o(H_2O) - H^o(Co(OH)_2)$$

$$- H^o(TiO_2(\text{rutile})) = -1207381 - 285829 + 541338 + 944747 =$$

$$- 7125 \text{ J} \tag{2.52}$$

When the difference in $\Delta H/T^2$ (reaction 2.50) between 293.15 and 298.15 K is neglected, Eq. (2.20) for this reaction can be written as

$$\Delta G(293.15) = (293.15/298.15)G^o(298.15) - 293.15(293.15 - 298.15) \Delta H^o(298.15)$$

$$/298.15^2 = -13831 \text{ J} \tag{2.53}$$

This means that reaction (2.50) is also spontaneous at 293.15 K and the difference in ΔG^o (reaction 2.50) between 293.15 and 298.15 is small enough to justify the disregard of variability of $\Delta H/T^2$ with T (the error caused by this simplification is within the limit of the data accuracy). Generally with ΔH^o on the order of a few kJ and ΔT of a few K dramatic changes in ΔG are not expected.

Example 6.   Experimental Study of Actual Products of Aging of Fe(III) Precipitates

The fate of Fe(III) precipitates was studied as a function of pH, type of anion and concentration of Mg and Ca by X-ray diffraction and TEM [19]. In the presence of

Ca, ferrihydrite ($5Fe_2O_3 \cdot 9H_2O$) was initially formed and then gradually transformed into goethite and/or hematite. In presence of Mg at pH > 9 a multicomponent phase similar to the mineral pyroaurite $Mg_6Fe_2(OH)_{16}CO_3 \cdot 4H_2O$ was formed, which did not undergo transformation into goethite or hematite.

Ferrihydrite obtained by hydrolysis of $Fe_2(SO_4)_3$ is very slowly transformed into goethite at pH 7 [20]. After one year only 7% of the initially formed ferrihydrite was transformed, while for ferrihydrite obtained by hydrolysis of $FeCl_3$ or $Fe(NO_3)_3$ the degree of transformation into mixture of goethite and hematite was 30% at the same pH. On the other hand at pH 11 the transformation into goethite was almost complete after one year and the difference in the transformation kinetics between ferrihydrite obtained by hydrolysis of $Fe_2(SO_4)_3$, $FeCl_3$ and $Fe(NO_3)_3$ was less significant. At pH 8–10 substantial amount of hematite is present after a one year aging. These results show that in adsorption experiments with fresh precipitates we can deal with two different adsorbents in the beginning and in the end the experiment.

In Example 6 formation of new phase involving a dissolved species (in this instance Mg) was discussed without an example of calculations. Formation of new phases involving dissolved species is commonplace, however, thermochemical calculations involving such species are much more complicated than those involving only water and pure crystalline phases whose activities equal one by definition. The activities of solution species are interrelated by a set of equations analogous to (2.23), one for each species, and mass balance equations, one for each component, and the relationship between activity and concentration is anything by trivial. Many computer programs were developed to facilitate such calculations. A few examples are discussed in the section of sorption modeling.

REFERENCES

1. H. Landolt, and R. Börnstein. Zahlenwerte und Funktionen aus Physik, Chemie, Astronomie, Geophysik und Technik; in Gemeinschaft mit J. Bartels (et al.) und unter Vorbereitender Mitwirkung von D'Ans (et al.) hrsg. von Arnold Eucken, 6th Edn. Springer, Berlin 1977–1979. Vol. 3, Part 7b (oxides and hydroxides), 7c (basic carbonates), 7d (silicates, plumbates, stannates, aluminates, gallates, indates), 7e (titanates, zirconates, vanadates) 7f (chromates, tungstates, manganates, ferrates, cobaltates, niclates).
2. W. A. Deer, R. A. Howie, and J. Zusman. An introduction to the rock-forming minerals, Longman, Wiley, New York, 1992.
3. CRC Handbook of Chemistry and Physics. CRC Boca Raton, FL.
4. G. Ferguson and J. Trotter. Structure Reports A. Vol. 58 Kluwer, Dordrecht, 1993.
5. I. Mills, T. Cvitas, K. Homann, N. Kallay, and K. Kuchitsu. Quantities, Units, and Symbols in Physical Chemistry, Blackwell Scientific, Oxford 1993.
6. A. Drljaca, C. D. Hubbard, R. van Eldik, T. Asano, M. V. Basilevsky, and W. J. le Noble. Chem Rev. 98: 2167–2289 (1998).
7. T. Asano, and J. le Noble. Chem Rev. 78: 407–489 (1978); R. Van Eldik, T. Asano, and J. le Noble, Chem Rev, 89: 549–699 (1989).
8. N. U. Yamaguchi, M. Okazaki, and T. Hashitani. J. Colloid Interf. Sci. 209: 386–391 (1999).
9. O. Knacke, O. Kubaschewski, and K. Hesselmann, (eds). Thermochemical Properties of Inorganic Substances, Springer-Verlag Berlin, 1991.
10. I. Barin. Thermochemical Data of Pure Substances, Vols 1 and 2, VCH Weinheim, 1993.

11. M. W. Chase. NIST-JANAF Thermochemical Tables, 4th Edn, Part II, Cr-Zr. Monograph No. 9, Journal of Physical and Chemical Reference Data, ACS Woodbury NY, 1998.
12. B. S. Hemingway, and G. Sposito. In The Environmental Chemistry of Aluminum (G. Sposito, ed.), CRC Press 1996, pp. 81–116.
13. H. Ruppert, Chem. Erde. 39: 97–132 (1980).
14. S. K. Saxena, N. Chatterjee, Y. Feui, and G. Shen. Thermodynamic Data on Oxides and Silicates, Springer-Verlag, Berlin, 1993.
15. T. J. B. Holland, and R. Powell. J. Metamorph. Geol 16: 309–343 (1998).
16. V. I. Babushkin, G. M. Matveyev, and O. P. Mchedlov-Petrossyan. Thermodynamics of Silicates, Springer-Verlag, Berlin, 1985.
17. S. Saxena. In Modelling in Aquatic Chemistry. (I. Grenthe and I. Puigdomenech, eds.), Nuclear Energy Agency. OECD Paris, 1997, pp. 289–323.
18. M. A. Blesa, P. J. Morando, and A. E. Regazzoni. Chemical Dissolution of Metal Oxides, CRC Press, Boca Raton, 1994.
19. K. A. Baltpurvins, R. C. Burns, and G. A. Lawrance. Environ. Sci. Techn. 31: 1024–1032 (1997).
20. K. A. Baltpurvins, R. C. Burns, and G. A. Lawrance. Environ. Sci. Techn. 30: 939–944 (1996).

3

Surface Charging in Absence of Strongly Adsorbing Species

I. ZERO CHARGE CONDITIONS

A. Definitions

The point of zero charge PZC is defined as the point at which the surface charge equals zero. The isoelectric point IEP is defined as the point at which the electrokinetic potential equals zero. Simple, isn't it? Why then the literature is full of expressions like zero point of charge ZPC, point of zero salt effect PZSE, point of zero net charge PZNC, point of zero net proton charge PZNPC, isoelectric point of the solid IEPS, pristine point of zero charge PPZC, zero point of titration ZPT, etc. and do we really need them? The answer is "no", provided that we deal with a crystalline, insoluble oxide in absence of strongly absorbing species (or—more precisely—sparingly soluble oxide at sufficiently low concentrations of strongly adsorbing species). Unfortunately, most systems of interest do not belong to this category. However, it is expedient to start the discussion on zero points from the simplest case of the system crystalline oxide—dilute inert electrolyte solution. Typical experimental results, for a hypothetical oxide whose PZC falls at pH 7 are illustrated in Figs 3.1–3.3. Figure 3.1 presents a course of potentiometric titration of dispersion of this oxide (details on the experimental procedure are discussed in Section I.B) at constant solid to liquid ratio and at three different concentrations of $NaClO_4$. Raw data is shown, i.e. the pH of dispersion is plotted as a function of volume of acid or base added (base is plotted as "minus" acid). The curves have a

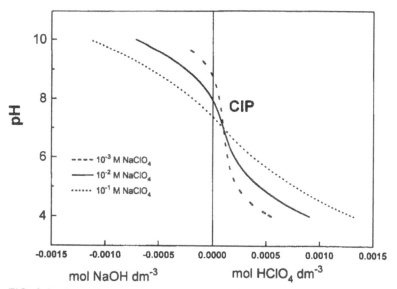

FIG. 3.1 Potentiometric titration of oxide dispersion at three different ionic strengths.

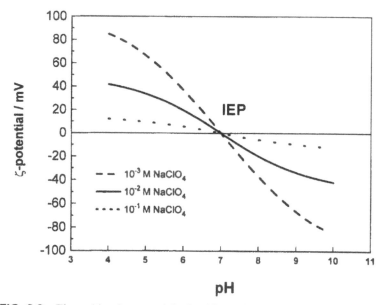

FIG. 3.2 Electrokinetic potential of oxide at three different ionic strengths.

common intersection point CIP, and not necessarily does this point correspond to a zero volume of acid or base added (vertical line). Only three titration curves are shown, but additional titrations at different ionic strengths would also produce curves going through this point. Figure 3.2 shows the electrokinetic potential of the same oxide as the function of pH at three different ionic strengths. Calculation of the ζ potential from experimental data can be difficult, cf. Section III, B, but a zero value of directly measured quantities, e.g. electrophoretic mobility corresponds to $\zeta = 0$, so

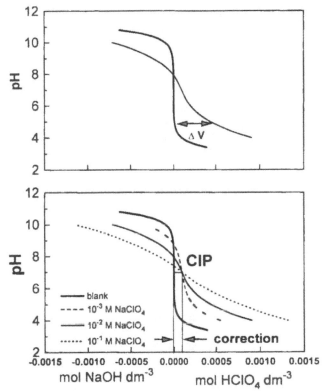

FIG. 3.3 Calculation of σ_0 from potentiometric titration.

the isoelectric point IEP can be determined experimentally without any model assumptions. For pure oxides CIP = IEP; moreover, this pH value corresponds to the minimum stability (maximum coagulation rate) and to maximum viscosity of dispersions, to onset of uptake of counterions (anions of inert electrolyte are adsorbed at pH below and cations at pH above this point), and to many other peculiarities in physical and chemical behavior. This characteristic pH value is also independent of the nature of inert electrolyte, and the literature is full of examples of CIP and IEP values determined for the same sample in the presence of different salts. For PbO_2 and RuO_2 [1] the CIP obtained with $NaClO_4$ and $LiNO_3$, $NaNO_3$ and $CsNO_3$ match, but $NaCl$ and $NaBr$ give a different value. In contrast for Co_3O_4 the CIP obtained with $LiNO_3$, $NaNO_3$, $CsNO_3$, $NaCl$ and $NaBr$ match, but $NaClO_4$ gives a different value.

A discrepancy between CIP of reagent grade hematite in $NaNO_3$ and $NaClO_4$ was reported, but a synthetic material gave practically identical CIP in these two electrolytes [2]. The CIP and IEP of rutile in KCl, $LiCl$, $CsCl$, KNO_3, $KClO_4$ match, although the positive branch of charging curves is affected by the nature of the anion and the negative branch of charging and electrokinetic curves depends on the nature of the cation [3]. These and many other examples suggest that the coincidence between IEP and CIP for many different salts is something more than fortuitous incident. Why should we give more credit to CIP as the characteristic pH value of the oxide than, e.g. to intersections of charging curves shown in Fig. 3.1 with the vertical line, i.e. the natural pH of the dispersion? First, this natural pH depends on the ionic strength as it can be seen in Fig. 3.1. It is also a function of the solid to liquid ratio while the CIP is independent of ionic strength (by definition) and it is rather insensitive to the solid to liquid ratio. The offset of the CIP from the vertical line is due to the base (in the example presented in Fig. 3.1) or acid (in such case the offset would be to the left) which was associated with the surface of dry solid. Presence of substantial amount of acid or base in solid oxides can be hardly avoided unless the PZC is known *a priori*. For example, an oxide separated from solution always has a film of solution on it, which cannot be removed by filtration or centrifugation. This film contains an excess of acid or base depending whether the pH is below or above the PZC. At typical drying conditions most water is evaporated, but the acid or base persists. When the dry oxide is brought into contact with solution, this acid or base is still there and it must be accounted for in the bookkeeping of the entire amount of acid or base added in potentiometric titrations. This is illustrated in Fig. 3.3. The blank curve represents titration of the solution without solid particles. The horizontal segment ΔV represents the difference in the volume of acid (base is counted as "minus" acid) which is necessary to bring the dispersion on the one hand and the solution on the other to the same pH. This volume is proportional to the amount of adsorbed acid. Assuming the protons of adsorbed acid to be associated with the surface and the anions to remain at some distance from the surface we can write:

$$\sigma_{0,\text{apparent}} = \Delta V c F / A \tag{3.1}$$

where c is the concentration of acid, and A is the specific surface area of the solid. Eq. (3.1) gives the apparent surface charge density σ_0, not corrected for the acid or base associated with the dry oxide. The principle of correcting the ΔV for the acid or base associated with the dry oxide is shown in Fig. 3.3. This correction leads to $\sigma_0 = 0$ at CIP, thus CIP = IEP = PZC. Typical charging curves (after correction) are shown in Fig. 3.4.

Most published results (cf. Section III) suggest that $|\sigma_0|$ at given pH always increases and $|\zeta|$ always decreases when the ionic strength increases. The ionic strength effects shown in Figs. 3.1–3.4 are purposely overemphasized, usually they are somewhat less pronounced. There are rather few publications, which did not confirm these trends. Almost ionic strength independent (10^{-3}–10^{-1}) charging curves of silica in the presence of $(C_2H_5)_4NCl$ were reported [4] while with alkali chlorides the absolute value of negative charge clearly increased with the ionic strength. The ζ potentials presented in Ref. [5] are ionic strength independent (10^{-4}–10^{-2} mol dm^{-3}). In Ref. [6] the absolute value of ζ potential at given pH even increases when ionic strength increases. Also sets of charging curves with $|\sigma_0|$ decreasing with ionic strength can be found in literature [7]. It is difficult to assess if such data represent typographical errors or real effects.

Figure 3.4 shows also typical ranges of PZC reported in the literature for four real materials. Examples of zero points for specific samples of materials are presented and discussed in Section I.C.

The above discussion refers to a somewhat idealized situation. For instance, what should we do when CIP≠IEP? First, the level of the experimental error should be considered as an important factor in the assessment whether or not CIP = IEP. The number of significant digits reported in the literature often exceeds the actual

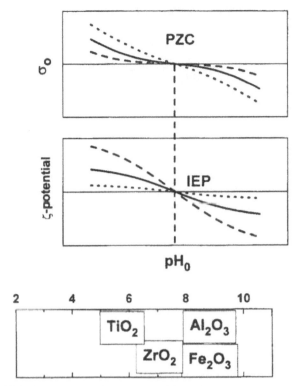

FIG. 3.4 Electrokinetic potential and σ_0 of oxide at three different ionic strengths.

accuracy of the measurements. Practical possibility to determine the CIP or IEP with an accuracy below 0.1 pH unit is questionable, thus a difference between these quantities below 0.2 pH units can be considered as insignificant. Very likely a substantial difference between CIP and IEP indicates the presence of strongly adsorbing impurities. The high purity of some commercial reagents can be misleading, because it reflects the bulk composition, while the surface can be relatively dirty. With low specific surface area the contribution of the surface impurities to the overall impurity level can be negligible. Presence of phosphates and sulfates is probably the most common cause of a low IEP, that has been reported for commercial and reagent grade titania samples. Properly designed washing procedure can reduce the amount of specifically adsorbed anions and shift the IEP to high pH values, but complete removal of preadsorbed phosphates is difficult [8]. In most publications reporting low IEP values for titania no attempt was made to remove phosphates or to determine their concentration. Washing with water alone is insufficient to remove strongly adsorbing anions but they can be partially desorbed by alkali treatment. Presence of specifically adsorbing ions can also result in hysteresis loops, i.e. acid and base titrations produce somewhat different results. Problems with reversibility and hysteresis in acid-base titration curves was addressed in several publications, e.g. [9]. The presumption that IEP and CIP obtained at the same temperature are compared seems too obvious to be pronounced. Surprisingly, an example can be found, that CIP was determined at 22 C and IEP at 25 C in the same paper, and despite a small difference between IEP and CIP they were considered "the same, within experimental error". Perhaps also other small discrepancies between CIP and IEP reported in literature were caused by difference in temperature or lack in temperature control (cf. Section IV). It seems also obvious that titrations at different ionic strengths should be performed at the same solid to liquid ratio, but, in a recent publication acidity constants were determined using data from titrations performed at two different ionic strengths, each data set at different solid to liquid ratio.

Potential pitfalls of specific experimental methods are discussed in more detail in Section I.B. A proper experimental procedure with sufficiently pure materials leads to CIP = IEP = PZC (±0.1 pH unit) for crystalline oxides in dilute solutions of some electrolytes. This equality has been challenged on grounds of different theoretical models (for detailed discussion of the models and their parameters cf. Chapter 5). TLM with unsymmetrical counterion binding gives PZC$\neq$IEP [10]. The following expressions for PZC and IEP in the TLM framework were proposed by Zhukov [11].

$$PZC = 0.5(pK_{a1} + pK_{a2}) - 0.5\log[(1 + K_X a_X)/(1 + K_Y a_Y)] \qquad (3.2)$$

$$IEP = 0.5(pK_{a1} + pK_{a2}) - (0.43F^2 N_s/C_1 RT)(K_Y a_Y - K_X a_X)/$$
$$[2 + (K_{a1}/K_{a2})^{1/2} + K_Y a_Y + K_X a_X] \qquad (3.3)$$

where K_{a1} and K_{a2} are the acidity constants of the surface equilibrium constants K_x and K_Y refer to reactions

$$\equiv SOH^- + X_s^+ = \equiv SO^- \cdot \cdot X_s^+ \qquad (3.4)$$

$$\equiv SOH_2^+ + Y_s^- = \equiv SOH_2^+ \cdot \cdot Y_s^- \qquad (3.5)$$

and a are activities of supporting electrolyte ions. The origin and meaning of these parameters will be discussed in more detail in Chapter 5. Another zero point, namely point of zero surface potential was introduced by Lützenkirchen [12]. Considerations based on TLM lead to conclusion that in this model both IEP and PZC can vary as a function of the ionic strength (although for typical values of model parameters and typical experimental conditions the differences do not exceed a few tenths of one pH unit).

Equations for mutual relationship between different zero points have been derived. The equality of PZC and IEP of pure oxides was also challenged by Charmas et al. [13] who wrote:

"...difference between pzc and iep values [is] observed more or less clearly in many adsorption systems. [...] Nowadays, the general feeling is growing that the inequality of pzc and iep is a fundamental feature..."

To avoid too many detailed descriptions of experimental methods it is expedient to arrange the zero points reported in the literature and presented in Table 3.1 into possibly few categories. These categories are named by lowercase abbreviations in quotation marks to be distinguished from similar abbreviations in their usual meaning (uppercase abbreviations, no quotation marks). The above discussion makes it possible to define the following types of zero points:

"cip"—CIP of at least three charging curves obtained at different ionic strengths.

"iep"—results obtained by electrophoresis, electroosmosis, and streaming potential (no results obtained by electroacoustic method are listed under this category).

"pH"—intersection of a titration curve of a suspension for one ionic strength with the blank titration curve (no correction for acid/base associated with the solid, thus the significance of such results can be challenged, but it will be discussed later that the assumption CIP = PZC is not always valid).

This classification seems obvious at the first glance, but analysis of specific descriptions of experimental procedures and the results reveals a few important issues to be addressed.

cip

In order to claim common intersection point charging data for at least three ionic strengths are necessary. With only two curves intersecting at some point, no necessarily will the charging curves for other ionic strength to through this point. For example the "STPT-ZPC" method which has been used in studies on a number of soils and clay minerals published in soil science journals in late eighties relies upon intersection point of the charging curves in water and one arbitrarily selected ionic strength. These results are shown in Table 3.3, but they are not listed as "cip" but as "intersection". A few examples of sets of charging curves without a sharp CIP can be found in literature, e.g. three intersection points with three ionic strengths, etc. Some of these results are still listed as "cip", namely when all intersection points are within a range below 0.2 pH unit, or when three or more charging curves obtained at low ionic strengths have CIP but charging curves at high ionic strengths do not go through this point. The latter situation can be due to specific interaction between the surface and monovalent ions. For instance, the charging curves of ZnO in four different 1–1 electrolytes at ionic strength below 0.02 mol dm^{-3} had a CIP, but the curves obtained at ionic strengths about 0.1 mol dm^{-3} did not go through this point [14]. The intersection of two charging curves can be also considered as a reliable indication of zero point when this result is supported by zero point determined by some independent method, e.g. electrokinetic data.

Some published results suggest that for silica instead of one sharp CIP there is a pH range (a few tenths of one pH unit) over which the charge density is zero and it is ionic strength independent while beyond this range an usual ionic strength effect is observed (Fig. 3.4). Since the apparent charge densities at pH < 3 are obtained as a difference between two large and almost equal numbers this hypothesis is difficult to verify but these results (pH ranges) are reported in Table 3.1 as "cip".

iep

Electrokinetic data, which did not lead to direct determination of the IEP are presented and discussed separately in Section III. For instance, the ζ potentials reported for silica and glasses are often negative over the entire covered pH range.

Many apparent IEP values obtained by extrapolation from such data sets were published, but in principle they are not shown in Table 3.1, namely at least one data point representing positive and negative electrokinetic potential is required to claim an "iep". This does not imply that these extrapolated IEP values are challenged. Extrapolation is very likely to produce the real IEP when the ζ potentials are linearly increasing (become less negative) when the pH decreases near the low pH edge of data. On the other hand, extrapolation of the high pH data (when all measured ζ potentials are negative) to produce an IEP above the high pH data edge is not recommended. The senselessness of such extrapolation is rather obvious, but examples of such "IEP" can be found in the literature. Also the IEP values derived without any explanation from a random mixture of positive and negative ζ potentials over the entire pH range can be found in literature but such data are not presented in Table 3.1. There are many reasons why the procedure described above for an idealized system does not work for most real systems. First, strongly adsorbing ions (added on purpose or present as impurities) affect the surface charge and electrokinetic potential even at low concentrations. Sorption of strongly adsorbed species is discussed in detail in Chapter 4. In contrast with the ions of inert electrolytes, these strongly adsorbed species can contribute to the surface charge. Thus, the assumption (which was the rationale for the discussion leading to IEP = CIP = PZC) that only adsorption and dissociation of protons contribute to the surface charge is false. The term pristine point of zero charge PPZC is often used to clearly distinguish between the particular case of inert electrolyte and the general case, where "pristine" means "in the presence of inert electrolyte" (no strongly adsorbing species). In such parlance we have IEP = CIP = PPZC (in absence of strongly adsorbing species), but generally IEP and PZC can be different and there is no simple relationship between them. IEP is measured experimentally and there is no need for a new definition when strongly adsorbing ions are present. However, in general case Eq. (3.1) with or without correction (*vide ultra*) is not valid any more, namely there are ions other than protons which can contribute to the surface charge. Contribution of protons can be still determined using a construction similar to this shown in Fig. 3.3, but it would be incorrect to identify this contribution with the entire surface charge. In particular the proton contribution can be equal zero when the entire charge is not and *vice versa*. Also in systems without strongly adsorbing ions mechanisms of surface charging different from proton adsorption/dissociation can be important and even prevailing. Clay minerals consist of aluminosilicate layers carrying negative charge, which is balanced by alkali metal cations in the interlayer spaces. In contact with solution these cations are mobile to some degree, and this results in negative charge of the clay mineral particles. This behavior is often referred to as permanent charge (as opposite to variable, i.e. pH dependent charge). Clay minerals show also some degree of pH dependent charging, but the contribution of the latter is not too significant. Moreover, the permanent negative charge is not necessarily located on the external surface of the particles, so this is not surface charge. Similar concerns refer to the charging of more complex materials containing clay minerals as constituents, and to zeolites except the cations in zeolites are located in a network of channels rather than in the interlayer spaces. The methods discussed above for oxides can be applied to study such materials (and many researchers do so), but the meaning of the results is different.

Several efforts to systematize the nomenclature referring to different zero points for the general case of materials carrying some degree of permanent charge, and in presence of strongly adsorbing species have been published. Five different zero points, namely point of zero charge PZC, point of zero proton charge PZNPC, point of zero net charge PZNC, isoelectric point IEP and point of zero salt effect PZSE are defined in Ref. [15] and listings of PZNPC, PZNC, IEP, and PZSE for different materials are presented. Procedures to obtain values for these four zero points are described in detail, but the discrepancies between PZNPC, PZNC, IEP, and PZSE reported for the same material are not explained. The definition of IEP is the same as used in the present work and PZSE (which is described as *"easy to measure but theoretically ambiguous"*) corresponds to CIP. The description of the procedure to obtain PZNPC begins with the following statement:

> "The most common procedure to *estimate* (underlined by MK) the PZNPC is to equate it with the pH value at the crossover point of potentiometric titration curves obtained at different ionic strengths..."

This seems to mean that PZNPC = PZSE = CIP. However, the further text suggests that apparent proton surface charge density obtained by assuming zero proton charge at CIP should be corrected. Unfortunately no clear description how to determine the proton surface charge density at CIP has been offered. Finally the PZNC is defined as the pH at which cation and anion exchange capacities determined for 1 mol dm^{-3} KCl using certain procedure are equal. The zero points based on uptake of ions different from protons are not reported in Table 3.1 for reasons discussed later in this section.

Definitions of zero points as the pH of zero electrophoretic mobility and CIP of acid-base titration curves were criticized by Sposito [16], who wrote about "... *illusory PZSE* (point of zero salt effect) *values inferred from the crossover point of titration ....*" Moreover, *"Other master variables than pH (e.g. the negative logarithm of the aqueous solution activity of any ion that adsorbs primarily through inner-sphere surface complexation) can be utilized to define a zero point of charge ..."*; and *"electrically neutral particle (ellipsoid with zero net charge because of mutually canceling patches of positive and negative surface charge) will have a nonzero electrophoretic mobility."* Importance of these concerns depends very much on the nature of systems under consideration.

Certainly PZC of surfaces carrying chiefly permanent charge, as clay minerals, clays and some soils is not equal to CIP, but with for many other materials the contribution of permanent charge to the total surface charge is negligible. The unique role of protons (as opposed to other strongly adsorbing species) in surface charging of oxides and many other materials is well documented, but activity of Ag^+ cations is better suited than pH to describe surface charging of AgI. Finally patchwise surface heterogeneity is possible for certain materials, but one can hardly find evidence of major shift in iep caused by such effect.

Many names for zero points that can be found in literature are synonyms of the terms discussed above, namely, some authors use their own nomenclature ignoring the well established terminology. In a relatively recent paper the isoelectric point was abbreviated as pI. What seems to be surface charge density was termed "capacity" in a relatively recent paper published one of the leading colloid journals. In some papers, e.g. Ref [17] a term equiadsorption point EAP is used. This terminology may suggest something different from point of zero charge, but considering how these quantities are calculated, the "apparent capacity" or "Cl$^-$ and Na$^+$ uptake" (terminology used in "EAP" publications) is nothing else but apparent surface charge density (without correction to obtain CIP = PZC) except for the units (meq/g can be easily converted into C/m^2 when the surface area of the adsorbent is known) and sign (apparent capacity and cation/anion uptake is always positive). Consequently the "EAP" is equal to the uncorrected PZC (*vide ultra*) and it is listed as "pH" in Table 3.1.

The term IEP has been used even quite recently for a zero point determined by "drift method" [18]. The principle of the method is as follows. A series of buffer solutions of equal volume and different pH (in this instance chloroacetic acid-sodium chloroacetate for acidic range and NH_3-NH_4NO_3 for basic range) is prepared. The same amount of powder is added to each solution and the pH of the slurry is measured. The instant change in pH (negative or positive) induced by addition of powder is plotted as the function of initial pH. The pH at which this change equals zero is taken as the zero point. This method is in fact a modified potentiometric titration without correction. Consequently such results are referred to as "pH" in Table 3.1. Moreover, weak acids often adsorb specifically and this affects the obtained zero point, thus pristine value can be only obtained in case of fortuitous coincidence using this method.

In Ref. [19] the IEP is termed "piep". In addition "siep" (solution IEP) is used but not defined; only the following explanation can be found "*the pH$_{siep}$ value for the mixed oxohydroxide was calculated for a solution of the same composition as the solid; because of low degree of substitution, the pH$_{siep}$ value was identical with that of pure ferric solution*". Only a few examples of peculiar terminology referring to the quantities discussed in this chapter were mentioned. For example interchange of the meanings of the terms PZC and IEP is commonplace.

The uptake of ions of inert electrolyte is discussed in detail in Section III, and the zero points derived solely from uptake of these ions are not reported in Table 3.1. At certain conditions the pH of equal uptake of anions and cations of inert electrolyte can match the point of zero charge, but generally the equiadsorption points depend on the nature and concentration of the electrolyte. Namely, at high concentrations of 1–1 electrolytes, at which such experiments are conducted, the ion binding becomes asymmetrical even for "inert" electrolytes. The cations tend to be preferentially adsorbed, and this is reflected in a substantial shift of the IEP (cf. Section III). Even less acceptable are results, (e.g. [20]) based on uptake of multivalent ions which tend to be specifically adsorbed at low concentrations and match between onset of their sorption and PZC can only happen by fortuitous coincidence. The literature is full of "surface charging" studies based solely on uptake of 1–1 electrolyte ions. Isoelectric point of titania was claimed on basis of surface excess of chlorates (VII) (positive branch) and tetramethyl ammonium (negative branch) measured by single internal reflection spectroscopy [21]. All results were plotted as positive numbers versus pH and the minimum of the best fit fourth degree polynomial was assumed to be the zero point. This method belongs to the same category as that discussed above PZNC based on anion and cation exchange capacities, and it is not suitable to be presented in Table 3.1.

A complicated method is claimed in Ref. [22] to give positive and negative components of surface charge ("PZC" = the positive and negative component are equal) but this method apparently gives rather the uptake of Na^+ and Cl^- ions. Uptake of sodium cations and nitrate anions [23] was used to characterize acid-base properties of different samples of hydrous ceria. Concentrations and pK of "more acidic sites" and "less acidic sites" and basic sites were reported without detailed description how were they obtained. Application of (seemingly) similar procedure led to only one type of acidic sites for hydrous SnO_2 [24].

Rather few studies of surface charging of metal oxides in the presence of controlled concentrations of ions of the same metal in solution (Al(III) for Al_2O_3, etc.) have been reported. Interestingly, while the IEP in 0.001 mol dm^{-3} KCl reported in Ref. [25] for some allegedly pure metal oxides are far from values reported by other authors, presence of 0.001 mol dm^{-3} of ions of the same metal shifts the IEP to values close to the IEP reported in other studies. On the other hand, for those metal oxides whose IEP in 0.001 mol dm^{-3} KCl reported in Ref. [25] are close to expected, addition of salts of the same metal did not cause any shift in the IEP.

B. Experimental Methods

1. Characterization and Pretreatment of Sample

Possible effects of strongly adsorbing impurities on the values of zero points have been briefly discussed in Section A. These impurities can be originally present in the material of interest or introduced during the experiment. Among the latter silica and CO_2 are the main concerns. An instructive example of possible effect of storage of suspensions in glassware on their electrokinetic properties is given in Ref. [26]. Alumina stored in glass for 10 days at pH 4.5 has the same IEP as that stored in polyethylene. The same alumina stored in glass at pH > 6.7 for one day or more has a much lower IEP due to adsorption of silicate. IEP of alumina at pH 4.25 reported in Ref. [27] is very likely another, although not intended example of the effect of silicate sorption. Such results are not reported in Table 3.1. The solid to liquid ratio is an important factor defining the effect of silicates in addition to the discussed above effects of pH and time of storage. An extensive study involving titania, alumina and latex shows that the effect of silicate is negligible when the solid to liquid ratio is sufficiently high [28]. The critical solid to liquid ratio (expressed as surface area per unit volume), below which storage in glassware affects the IEP strongly depends on the nature of the surface. Latex is relatively insensitive to sorption of silicates as compared to alumina. This is not surprising, namely, when the surface area is high enough, the solid surface is only partially covered by silicate and the overall surface properties are defined by the uncovered surface. The silicate adsorption affects the IEP obtained by microelectrophoresis when the solid to liquid ratio is low, but the silicate effect in potentiometric titration and related methods is negligible. On the other hand, electrokinetic measurements involving macroscopic specimens, e.g. single crystals are potentially even more affected by traces of silicates and other impurities than microelectrophoresis of powders. For example, in a recent paper an IEP of pulverized mica at pH 3 was reported, while a flat specimen had a high negative electrokinetic potential even at pH 2. This result can reflect a difference in electrokinetic properties between the basal and edge planes, but the flat specimen could be also influenced by trace amounts of impurities.

Protection of the sample from the atmospheric CO_2 by inert gas atmosphere (nitrogen, argon) during a potentiometric titration is a standard procedure, and most CIP values reported in Table 3.1 were obtained under such conditions. Carbon dioxide seems to have no particular effect on the IEP of goethite, but the CIP shifts to low pH values when CO_2 is not properly removed [29]. In an older publication [30] a substantial difference (1 pH unit) is reported between CIP in a CO_2 free system and crossover point of two charging curves of goethite suspensions equilibrated with atmospheric CO_2 in 0.01 and 0.1 mol dm^{-3} NaI. The later titrations were conducted under nitrogen and the suspensions were pre-equilibrated under nitrogen for > 4 h at pH 9, so the origin of the difference in charging behavior is not quite clear.

Many synthetic materials, especially monodispersed particles (for details cf. Section III.2) are prepared in the presence of strongly adsorbing ions. It is often impossible to entirely remove these ions from the final material, even after multiple washing cycles, dialysis, etc. For example, a substantial difference in surface charging behavior between indium hydroxide and oxide prepared in the presence of sulfate on the one hand and nitrate on the other has been reported [31].

Many publications report PZC of commercial and reagent grade materials, which underwent more or less sophisticated washing procedure to remove the impurities. An example how washing of a commercial material affects its IEP is presented in Ref. [32]. An extensive study of the effect of washing procedure on the zero point of commercial rutile was published by Cornell et al. [33]. In addition to electrokinetic and titration data the concentration and location (bulk, surface) of impurities is reported. Dietrich et al. [34] measured concentrations of Fe, Ni, Mg, and Al in chromatographic silicas and determined their IEP. The purest samples (below 2 ppm of Fe, Ni, Mg, and Al) had IEP at pH 2.44 and 2.9 and the samples whose IEP was about pH 2 showed substantial concentrations of Al (up to 540 ppm) and Fe (up to 154 ppm). As received Houdry 415 alumina gives CIP at pH 5, probably due to some impurities. However, the same alumina after impregnation with $NaNO_3$ solution and calcination gives cip at pH > 9. Most likely this effect is due to oxidation of surface active organic impurity, but is was interpreted as modification of alumina by sodium [35]. It would be interesting to check if such effect is also observed when another (not oxidizing) sodium salt is used instead of $NaNO_3$.

Storage of dispersion before measurements can lead to various effects. Colloidal magnetite (0.1 μm particles) was found to undergo transformation into maghemite in acidic medium [36] using x-ray diffraction and Mössbauer spectroscopy. Biscan et al. [37] observed a strong effect of ammonium acetate on electrokinetic behavior of PbO, namely, the ζ potentials were positive over a limited pH range in the center of the pH scale and the sign apparently reversed to negative at very low and very high pH. This effect was time dependent, and it was caused by dissolution of the oxide. The same electrolyte did not induce substantial changes in the electrokinetic behavior of TiO_2 or ZrO_2. Dissolution of different types of alumina at pH 2–6 (1 and 24 h equilibration times) was studied by Regalbuto et al. [38]. Effect of hydration time on electrokinetics of oxides has been studied by Watanabe and Seto [39].

The effect of heat treatment on the electrokinetic behavior of powders has been emphasized by Kittaka [40,41]. Janssen and Stein [42] compared electrokinetic and charging curves of two samples of titania which underwent H_2 and O_2 treatment at temperatures about 500°C and Soxhlet extraction (in different order) but no significant effects were observed.

The course of titration curves can be changed by exposure of dispersion to radiofrequency electric field [43]. The charging curves of treated suspensions often do not have a CIP.

2. Titration

The values of PZC referred to as "cip" and "pH" in Table 3.1 are obtained by acid-base titration. The experimental procedure is basically the same (although many variants have been described), and the substantial difference is in the method of selection of the zero point. Titration gives very accurate changes in σ_0 from one pH value to another (or at least the part of σ_0 due to proton adsorption/dissociation) without assumptions, but some assumptions are necessary to obtain the zero point. Existence of CIP does not prove the absence of specifically adsorbed ions. Lyklema [44] showed that in case of specific adsorption of metal cations the shift in the CIP to pH below the pristine PZC is more pronounced for metals having stronger affinity toward the surface (e.g. Pb > Ca) and the σ_0 at CIP is also more positive for more strongly bound cations.

The so-called batch equilibration technique is classified as "pH". In this method the final (after long shaking) pH of dispersion is plotted as a function of the initial pH at constant pH and solid to liquid ratio and the plateau of the curve indicates the PZC provided that the oxide is free of soluble acids and bases.

In case of materials showing some degree of solubility the problem how to subtract the "blank" is not trivial. Schulthess and Sparks proposed a back titration method to determine the surface charge density of such materials [45]. In this method a series of solutions of different pH is prepared and the solid is added to each solution. After certain time necessary to reach equilibrium a sample of supernatant is taken from each suspension and it is titrated back to the original pH value (before addition of solid). This method was designed to distinguish between surface charging and uptake/release of protons by soluble species. The volume of titrant is then substituted to Eq. (3.1) to obtain σ_0 as discussed above for the "pH" and "cip" results, and the results obtained by back titration method are listed as "pH" and "cip" in Table 3.1.

Pechenyuk [46] noticed that the apparent PZC (referred to as "pH") of hydrous oxides obtained by hydrolysis of metal salts at certain pH increased as the pH of precipitation increased. The charging curves at three ionic strengths did not intersect, but the plots of apparent PZC as a function of pH of precipitation at three ionic strengths usually showed a common intersection point at pH of precipitation≈apparent PZC. This pH value was termed "true PZC". Unfortunately, with data points every 1 pH unit or more the results depend on rather arbitrarily drawn curves connecting the points representing one ionic strength.

Certain materials give no CIP but the charging curves obtained at various ionic strengths merge at sufficiently low (or sufficiently high) pH. Ahmed and Maksimov report merge of positive branches of charging curves obtained in $NaClO_4$ and KNO_3 but with KCl positive charge increases with the ionic strength [47]. This type of behavior is marked as "merge" in the column "method" in Table 3.1.

Unsuccessful attempts to determine the PZC by means of potentiometric titration have been also reported [48].

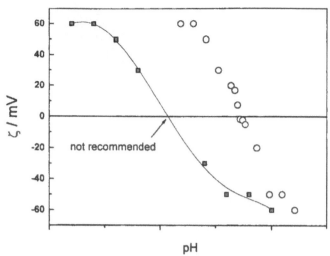

FIG. 3.5 Estimation of IEP from electrokinetic data. White symbols: many data points near the IEP (recommended). Black symbols: no data points near the IEP, the line is drawn arbitrarily (not recommended). One tick on pH axis corresponds to 1 pH unit.

3. Classical Electrokinetic Methods

Recent years witnessed an enormous progress in the experimental methods to measure the electrokinetic potential, especially by electrophoresis. In the sixties movement of single particles was observed in the eyepiece of microscope and the time was recorded manually using a stopwatch. Fast coagulation and sedimentation of particles often caused problems (too few particles). Nowadays fully automatic instruments are able to record velocities of hundreds of particles within a few seconds. These user-friendly designs do not, however, eliminate all problems and errors. Some concerns regarding sample preparation have been already discussed. In order to receive a reliable IEP it is recommended to have as many data points as possible for pH values slightly above and slightly below the IEP. Unfortunately, some published IEP values are based on lines arbitrarily drawn through data points, which are far from IEP. This is illustrated in Fig. 3.5. For example in Ref. [49] IEP for ZrO_2 at pH 7.4 is claimed without having any data point at 5.6 < pH < 8, and the curve on which this value is based has two minimums and two maximums.

Some types of electrophoretic cells are "stationary layer problem free", but in the other cells the electroosmotic flow can lead to erroneous results. The observed velocity of particles is a sum of the electroosmotic flow of the fluid and the velocity of particles with respect to the fluid. The latter is a function of the ζ potential of the particles and the former is a function of the position in the cell cross section. Hydrodynamic calculations make it possible to find the stationary levels, i.e. the positions in the cell cross section where the electroosmotic flow equals zero. Certainly the position of stationary levels in commercial electrophoretic cells can be found in the user's manual, and there is no need to perform any calculations. The fastest method to determine the electrophoretic mobility is from the velocity at one stationary level, but such a procedure can lead to substantial errors. For example, when the cell position is adjusted at room temperature and then measurements taken at elevated temperature, the results can be erroneous due to the shift in cell position on heating [50].

The hydrodynamic calculations upon which the theoretical stationary level is based assume that the electroosmotic velocity at cell walls is constant, and this is true when the walls are made of the same material. However, sedimentation of fine particles can change the electroosmotic velocity at the bottom of the cell, and thus the entire distribution of electroosmotic velocity in the cell, and this usually results in nonzero electroosmotic velocity at the calculated stationary level. This can be easily checked by comparison of results obtained at two stationary levels (e.g. in a rectangular cell they are located at the same distance from the upper and lower cell wall, respectively). When these two mobilities are not equal the distribution of electroosmotic velocity is different from theoretical, and results obtained from measurements at one stationary layer are erroneous. Unfortunately, it is usually impossible to check whether or not such precautions were exercised in a publication of interest. With an asymmetrical profile of electroosmotic velocity the actual mobility of particles can be still calculated for a flat cell (when the effect of the sidewalls is negligible, the actual mobility is an average of apparent mobilities at the two stationary levels) and estimated for a rectangular cell. The calculation of the profile from known ζ potentials of the walls is rather straightforward, but the calculation in opposite direction is difficult. The experiment is also tedious, because multiple measurements are required to receive one mobility value. The same algorithm, which is used to estimate the actual mobility of particles for an asymmetrical profile of electroosmotic flow can be utilized to determine the ζ potential of macroscopic samples [51]. One wall of a rectangular cell is replaced by the material of interest, and the profile of apparent velocity across the cell is determined. The ζ potentials of cell walls including the one made of material of interest and of particles can be calculated from such a profile.

The problem with the stationary layer can be also avoided by running a measurement with a standard colloid whose electrophoretic mobility is known prior to the measurement with a sample of interest. Such standards are often delivered with the instrument, some suggestions have also been published in the literature. Standard colloids have limited shelf time (usually about 1 year) due to slow processes of diffusion in solid and growth of microorganisms. Some standards have a pH close to the IEP of glass (usual cell material) and this is rather unfortunate from the point of view of confirmation of the stationary layer, namely, they give "reasonable" mobilities even when particles off the stationary layer are observed. In particular at IEP of the cell wall there is no electroosmotic flow at all and the observed velocity of the particles is not affected by the position in the cell. On the other hand when the cell walls carry a high positive or negative charge even small deviation from the stationary level causes a substantial error in the electrokinetic mobility.

Production of heat and electrolysis during the electrophoretic experiment can cause errors for long measurement times in certain types of electrophoretic cells. The temperature affects rather the absolute value of the mobility, but the presence of electrolysis products can result in an erroneous IEP. For example with KCl electrolyte and silver electrodes the overall reaction can be written as:

$$Ag + KCl + H_2O = AgCl + KOH + 1/2 H_2 \tag{3.6}$$

This results in more basic pH and thus more negative mobility. Usually the pH is measured before but not during or after the electrokinetic measurements. The electrolysis products can be in situ neutralized using buffer solutions rather than strong 1–1 electrolytes. However, such solutions often contain strongly adsorbing ions that shift the IEP.

The ζ potential of macroscopic grains can be conveniently measured by electroosmosis. A few commercial instruments are available, and many home made designs have been described in literature [52]. The apparatus described in Ref. [53] requires samples in form of two flat disks about 50 mm in diameter one of which has a central hole 6 mm in diameter to measure radial flow streaming potential.

Also results obtained by means of mass transport method are listed as "iep". In this method the mass of particles migrated through a orifice is directly measured, and this mass is related to the electrophoretic mobility by the following equation [54]:

$$m = c_{solid} E A t \, \mu_e (1-f)^{-1} \rho_s (\rho_s - \rho_d)^{-1} \tag{3.7}$$

where c_{solid} is the concentration of solid particles, E is the field strength, A is the cross section of the orifice, t is time, μ_e is the electrophoretic mobility, f is the volume fraction of the solid particles, ρ_s is the specific density and the subscripts s and d relate to the solid and dispersion, respectively. Commercial equipment using this concept is not available, but experiments with a home made instrument confirmed the proportionality of the mass transported to the amount of charge passed, and proportionality of the mass transport rate to the field strength.

Szymczyk et al. [55] published an equation for the streaming potential of the n-th layer of multilayer membrane when the streaming potential of the entire membrane and of a membrane composed of the remaining n-1 layers is known. A number of IEP values was obtained by means of this equation, but the results are not used in Tables 3.1 and 3.3.

4. Electroacoustic Method

In principle the results obtained using the electroacoustic method give the IEP. The recent developments in theory allow interpretation of electroacoustic signal without restrictions of dilute case ($<2\%$ v/v) [56]. However, many published values of ζ potentials were obtained at the times when the theory was not fully developed, using prototype versions of instruments, etc. This often resulted in different values of ζ potential and even different IEP for different solid to liquid ratios, and concerns about credibility of such data are fully understandable. Therefore they are marked as "acousto" in Table 3.1 to distinguish them from the "iep", i.e. results obtained using classical electrokinetic methods. On the other hand the agreement between electroacoustic and electrophoretic measurements for many systems is fully satisfactory [57]. In contrast with classical electrokinetic methods, electroacoustic measurements at relatively high ionic strengths are possible.

5. Coagulation

Coagulation is not considered as standalone method to determine the zero point, but as a valuable confirmation of results obtained by other methods. Coagulation is usually fast even far from IEP when the ionic strength is high. Moreover, the IEP can

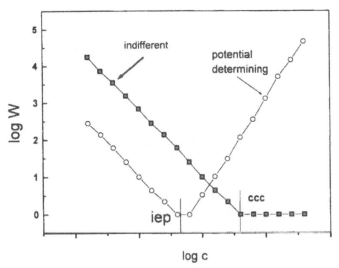

FIG. 3.6 Stability ratio for an oxide dispersion as the function of concentration of H ions (white symbols) and of inert electrolyte (black symbols).

be shifted at high ionic strengths in electrolytes, which are inert at low concentrations. Thus, zero points obtained from coagulation data can be taken as reliable only for ionic strengths 0.001 or below. The principle of this method is illustrated in Fig. 3.6. White symbols represent a typical pH dependence of the stability ratio W (rate of fast coagulation: rate of coagulation at given conditions) at low ionic strength. At higher ionic strengths the horizontal segment of the coagulation curve near the IEP is longer. Figure 3.6 shows also the effect of concentration of inert electrolyte on W at constant activity of potential determining ions (black symbols), and CCC is the critical coagulation concentration, which is pH sensitive for materials whose surface charging is pH dependent.

Dumont and Watillon [58] plotted coagulation concentrations of different anions as a function of pH and extrapolated the linear segments obtained for the anions having the lowest coagulation effect (high CCC) to intersect the pH axis. The pH value at intersection was claimed to be the PZC. Similar plot for cations did not confirm this result.

6. Inert Electrolyte Titration

Figure 3.4 shows that the absolute value of σ_0 increases with the ionic strength, except at the CIP. Thus addition of inert electrolyte (crystalline or as concentrated solution) to suspension at CIP does not result in any pH change. At any other initial pH the addition of inert electrolyte induces a shift of pH toward the CIP. This means that the pH increases on addition of inert electrolyte at pH < CIP and decreases at pH > CIP. A series of suspensions is prepared using the same amounts of solid and of dilute solution of inert electrolyte, the suspensions are adjusted to different pH, and allowed to equilibrate. The normal precautions as by classical acid-base titration (thermostating, protection from CO_2) have to be respected. Then constant amount of inert electrolyte is added to all samples, and the change in pH is recorded, and the results (ΔpH) are plotted as a function of pH. The intersection of the obtained curve with the pH axis (ΔpH = 0) represents the zero point. In this method the problem of slow pH drifts after addition of a portion of titrant, affecting classical titration is avoided, namely the results of classical titrations often depend on an arbitrarily chosen rate of titration. Inert electrolyte titration gives rather an intersection of two charging curves than CIP, but two or more series of experiments with different initial ionic strengths can be conducted.

In a recent modification of this method only one sample of solid is used to determine the CIP. The dispersion containing about 10^{-2} mol dm^{-3} of inert electrolyte is brought to pH$\approx$CIP (determined in a preliminary experiments or estimated from literature data) and a portion of saturated solution of inert electrolyte is added. At CIP there is no change in pH, otherwise addition of inert electrolyte induces a pH shift toward CIP. Then a small amount of acid or base is added (having common anion or cation with the inert electrolyte) to shift the pH further toward CIP and another portion of inert electrolyte solution is added after equilibration. This procedure can be repeated many times and a pH range in which the pH does not change on addition of inert electrolyte is determined. The width of this pH range depends on the experimental conditions (e.g. solid to liquid ratio) and type of pH meter and pH electrode.

Only the zero point is obtained by inert electrolyte titration (no σ_0 data).

7. Mass Titration

The method termed mass titration was originally described by Noh and Schwarz [59]. Dispersions containing 0.01, 0.1, 1, 5, 10 and 20% of solid in deionized and outgased water are equilibrated for one day under nitrogen atmosphere. The pH is measured and plotted versus log (mass% of solid) and the plateau in the curve indicates the PZC. Addition of inert electrolyte was recommended (but not applied) in the original paper and the problem of optimum concentration of inert electrolyte was not addressed. The agreement between results obtained by mass titration and the CIP was good for alumina but problematic for titania.

Although mass titration was originally designed as a method of determination of PZC, surface charging curves can also be calculated from mass titration data [60].

Mass titrations performed at different ionic strengths [61] with alumina give somewhat different results. While those performed at [NaCl] > 10^{-2} mol dm^{-3} were quite consistent and led to PZC similar to literature values, the titration in pure water gave pH 4.3 with 40% solid by weight.

Park and Regalbuto [62] argue that classical mass titration gives only good results for materials having their PZC in the pH range 3–11. Namely, the plateau at high solid loadings for such materials corresponds to the PZC, so this value can be directly read from experimental curves. For materials having their PZC at extreme pH values the plateau does not correspond to the PZC. The PZC can be calculated but this requires some model assumptions. To avoid a necessity of making such assumptions, a method termed EpHL (the equilibrium pH at high oxide loading) is recommended by these authors, which in fact is a modification of mass titration and all limitations discussed for classical mass titration apply. The principle of the method is as follows: a paste is made from oxide and electrolyte and its pH is measured using a pH electrode specially designed for paste like materials. This pH value is reported as the PZC of oxide. No special notation was used to distinguish results obtained by means this method from classical mass titration in Table 3.1. The mass titration method was modified by Wernet and Feke [63], who recorded the pH as a function of mass of powders added to solution and fitted the surface acidity constants (cf. Chapter 5) to get a match between experimental results and model curves. This procedure makes it possible to estimate the PZC even if a plateau at high solid loading is not reached.

A completely different method is referred to as "mass titration" in Ref. [64] Namely, a certain amount of powder is added to the solution of given pH and the dispersion is titrated with acid or base until the original pH value is reached. This procedure is very different from the mass titration discussed above, and it seems to be variation of potentiometric titration without correction for PZC = CIP, thus the results are reported as "pH" in Table 3.1.

Mass titration relies on the assumption that there is no acid or base associated with the solid surface, and this assumption is problematic for many specific materials (cf. the discussion concerning Figs. 3.1 and 3.3). Simulations of mass titration taking into account such acidic or basic impurities were carried out by Zalac and Kallay [65].

8. Inflection

The same authors used the inflection point of titration curves to find the PZC from a charging curve for one ionic strength. The $d\sigma_0/dpH$ is plotted versus pH and the maximum indicates the PZC. The point of zero charge corresponds to the inflection point of titration curves (second derivative of σ_0 versus pH = 0). [66]. Sometimes the σ_0-pH dependence is linear [67], and in such case the "infection point" method to find the PZC cannot be applied. An example of application of the inflection point method to authentic experimental data (In_2O_3 and $In(OH)_3$), Hamada et al., cf. Table 3.1) is given in Figs. 3.7 and 3.8 (first and second derivative). The match between originally claimed PZC and that obtained from the first and second derivatives of σ_0 versus pH is satisfactory.

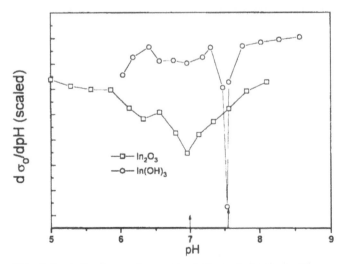

FIG. 3.7 Inflection point method (first derivative). The arrows indicate the PZC from titration. Calculated from uncorrected titration curves published by Hamada et al. (1990).

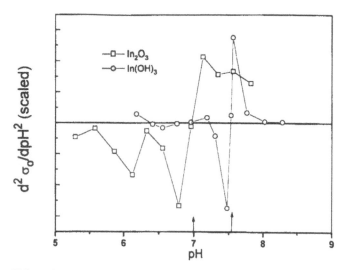

FIG. 3.8 Inflection point method (second derivative). The arrows indicate the PZC from titration. Calculated from uncorrected titration curves published by Hamada et al. (1990).

9. EMF

Tschapek et al. [68, 69] proposed a method to determine the position of the PZC of oxides and related materials in the presence of KCl based on the pH dependence of the transference number of cation t_{K+} in a plug or paste. KCl was chosen because the K^+ and Cl^- ions are nearly equitrasferent, i.e. another salt or salt mixtue with $t_{cation} \approx 0.5$ could be used as well. The transference number is determined from the potential of the following cell:

calomel electrode $|c_1,$ e.g. 2×10^{-3} mol dm^{-3} KCl|paste|$c_2,$ e.g. 10^{-3}

mol dm^{-3} KCl| calomel electrode   (3.8)

The principle of the method is that $t_{K+} = 0.5$ at PZC, $t_{K+} < 0.5$ for positively charged surfaces, and $t_{K+} > 0.5$ for negatively charged surfaces when [KCl] ≫[HCl] and [KCl]≫[KOH]. The results for common oxides agree with the CIP of potentiometric curves determined experimentally for the same samples.

The method was extended for the materials whose PZC falls at very low pH values, where classical electrokinetic methods fail, and the CIP cannot be observed. For such material a linear relationship between t_{K+} and pH over the pH range ≈4–7 was observed. Extrapolation of this straight line to $t_{K+} = 0.5$ gives the PZC. The data points obtained for low pH values show substantial deviations from linearity and they are not used in this extrapolation.

10. Adhesion

Adhesion method [70, 71] can be applied to macroscopic, conductive specimens, for which other methods are not efficient, but also to other materials. Many latexes have positive or negative ζ potentials which are rather insensitive to pH (cf. Chapter 6). Retention of such a positively charged latex on the one hand and negatively charged latex on the other, in a column containing particles of interest in studied as a

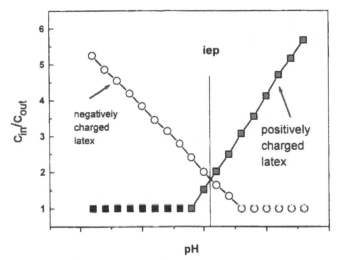

FIG. 3.9 Adhesion method. The crossover point corresponds to the IEP (schematically).

function of pH. When the charge of the studied surface and the latex have like signs there is no adhesion and the concentration of the latex in the effluent (c_{out}) is the same as in the influent (c_{in}). Once the signs become opposite the concentration of the latex in the effluent stepwise decreases.

The intersection of the c_{in}/c_{out}(pH) curves obtained for a positively and negatively charged latex corresponds to the IEP of the surface of interest (Fig. 3.9). Experimental results obtained by means of this method for 12 different metals (some of them underwent different surface treatment before the measurement) have been presented [72]. For metals having their IEP at pH < 4 only results with negatively charged latex particles were used to estimate the IEP. Figure 4 in this paper demonstrates a linear relationship between IEP of metals and corresponding oxides or hydroxides. This result indicates formation of thin layer of corrosion products whose IEP is actually determined. Selected IEP values for oxides from literature were used, which are sometimes (as in case of TiO_2) far from "recommended" values. Charging curves of fine metal particles were obtained by potentiometric titration, and the CIP is close to that of corresponding oxides or hydroxides [73]. Also IEP values for metal particles obtained by electrophoresis have been reported [74]. The metal surfaces unless covered by a thick layer of oxidation products are beyond the scope of this book, and they will not be discussed in further sections.

11. AFM

Tip-sample force measured by atomic force microscopy AFM at a distance beyond the range of van der Waals forces can be used to estimate the IEP of one material when the IEP of another is known [75]. The principle of this method is illustrated by the force-distance curves in Fig. 3. 10. For example the force between silica sample and Si_3N_4 tip in 10^{-3} mol dm^{-3} KNO_3 for a tip-sample distance of 17 nm (arbitrarily chosen) is attractive at pH > 6.2 and repulsive at pH < 6.2. Since the silica surface is known to be negatively charged in this pH range, this result suggests IEP of Si_3N_4 at pH 6.2. This is a rough estimate neglecting the attractive double layer interaction. A somewhat lower IEP value is obtained for Si_3N_4 when this interaction is considered.

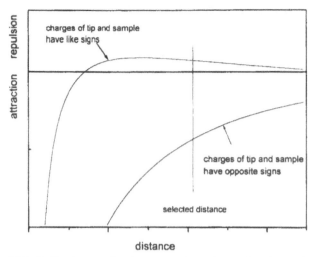

FIG. 3.10 Interaction between tip and sample whose charges have like and opposite signs (schematically).

An advantage of the AFM approach over macroscopic methods is that distribution of local IEP can be determined for a heterogeneous surface. This can be done when the difference in IEP between different parts of the surface is sufficiently large, and specific example of such results have not been reported. Arai et al. [76] observed two zero points in plots of force between two different surfaces measured at separation of half of Debye length as a function of pH and they argue that these points correspond to IEP of the two materials of interest. Indeed, their estimated values are close to IEP reported in literature. Phosphate buffer was used to establish the pH, so the results are probably affected by strong specific adsorption of phosphates.

In many studies good agreement between experimental AFM curves and DLVO theory was reported. On the other hand, Ducker et al. [77] suggest existence of additional "non DLVO" forces.

AFM was also used to study the interaction between silica tip and a single crystal semiconductor, which can be polarized to different potentials. For example, n-type TiO_2 at pH close to IEP of titania showed a repulsive interaction with silica when held at negative potentials (versus SCE) but at positive potentials the interaction became attractive [78]. With two identical surfaces the absolute value of ζ potential can be estimated by fitting the calculated force-distance curves to experimental ones [97]. Although the sign of ζ cannot be directly determined from such measurements the IEP can be found when sufficient number of data points at different pH is available, namely, the absolute value of the calculated ζ shows a minimum at the IEP. The surface charge density of different crystal faces can be quite different from the average, and this affects the AFM results. Hiemstra and van Riemsdijk [80] showed that the experimental-force-at-contact versus pH curve obtained for a hematite tip and the 001 face of cleaved hematite is very different from that expected from overall ζ potential of hematite, but it is in reasonable agreement with the prediction of surface charging of the 001 face of hematite in terms of their MUSIC model (Chapter 5).

Crystal face specific determination of IEP by means of SFM scanning force microscope was discussed by Eggleston and Jordan [87]. When tip and sample are made of the same material the force at contract shows a minimum at IEP. Preliminary results for silica and hematite showed that this minimum was rather broad, but the method is promising. One shortcoming in this method is a rather subjective definition where is the tip-sample contact on the force-distance curves.

12.   Contact Angle

Contact angles measured for a plate of boehmitized aluminum vertically immersed in KCl solutions have a maximum at the IEP [82]. In solutions containing sulfates the IEP shifts to low pH values and the maximum in the contact angle shifts accordingly. Contact angles at solid–hexadecane–aqueous solution interface as a function of pH go through a maximum, which is supposed to indicate the IEP of the solid [83]. The results reported for oxide covered aluminum are rather scattered and the maximum can be at any pH from 8 to 11. Indeed, IEP for alumina in this pH range has been reported. With oxide layer formed on chromium plated steel the maximum in contact angle falls at pH 5 which is clearly below IEP of chromium oxides and hydroxides reported in other sources. Finally, for oxidation product of tantalum the contact angle continuously increases as pH decreases, even below pH 0, which is again in clear contradiction with results of electrokinetic studies. Although the results reported for Ta_2O_5 in different sources are less consistent than those for chromium(III) oxides and hydroxides, the IEP for this material below pH 0 seems very unlikely. Thus, only one of three tested materials gave a result consistent with IEP measured by classical method, but it was rather a range 3 pH units broad than a single point on the pH scale, and the above method does not seem to be an attractive alternative to other methods.

13.   Other Methods

The logarithm of pseudo first-order rate constant for reduction of methyl viologen in the presence of TiO_2 is a linear function of pH and the slope is severely affected by the ionic strength [84]. The linear segments obtained at various ionic strength ($0.001-0.4$ mol dm^{-3}) have a common intersection point at pH 4.5 and pH 4.3 for two different samples of TiO_2 obtained by hydrolysis of $TiCl_4$ in aqueous ammonia. Brown and Darwent argue that this intersection point corresponds to the PZC of TiO_2.

An idea to use maximum of rate of filtration (or rather minimum of time necessary to filter certain volume of liquid) as an indication of the IEP has been suggested [85]. The result obtained by this method matches the intersection point of two charging curves at different ionic strengths.

PZC of V_2O_5 at pH 3 was derived [86] from pH dependence of rate of dissolution in chloric VII acid.

Bürgisser et al. proposed a chromatographic method at determine σ_0 of materials having low surface area [87]. The surface charge density is calculated from breakthrough curves. Agreement with results obtained by titration has been demonstrated for one ionic strength. Several problems, e.g. dissolution of the solid have been addressed. The chromatographic method is not suitable to obtain charging data at pH about 7.

Design and fabrication of ISFET was described in Ref. [88] The interest in ISFET arises chiefly from their application as pH and ion sensors. A graphical procedure to find PZC from capacitance–voltage characteristics of electrolyte–insulator–semiconductor and metal–insulator–semiconductor structures was discussed [89]. Due to the choice of electrolyte (2 mol dm^{-3} Na_2SO_4) the PZC values reported in this study (2.5 for SiO_2, 2.8 for Ta_2O_5 and 3–3.4 for Si_3N_4) are not likely to be the pristine values due to specific adsorption of anions.

Some results in Table 3.1 are marked as "authors' method", namely when

- The principle of the method have not been published in easily accessible literature.
- The present author had difficulties in understanding how does the method work (but there is no clear indication that the method is incorrect).

C. Simple Hydr(oxides)

In most studies reported in Table 3.1 determination of the zero point was not the main goal, but there is no reason to challenge the reliability of results. Table 3.1 contains the following information:

1st column: chemical formula.
2nd column: sample characterization. This may include: trade name (for commercial samples), brief description of preparation (for synthetic materials), crystallographic form, degree of hydration, major impurities.
3rd column: supporting electrolyte (formula and concentration or concentration range).
4th column: temperature in °C.
5th column: method, cf. section B.
6th column: pH_0.
7th column: source.

Some entries have empty columns 2, 3 or 4 because relevant information was not given explicitly. Some studies report only negative (or less often only positive) surface charge densities (or electrokinetic potentials) over the studied pH range. This suggests that the pH_0 is beyond the studied pH range or does not exist at all. Such results are summarized in separate tables in Section III.

With materials for which the PZC information is abundant, the data presentation in Table 3.1 is discriminating, i.e. many results of problematic accuracy or reliability were rejected (cf. Section B). In contrast, when data for given material are scarce, each piece of information is presented.

Very likely there is some duplication of information in Table 3.1, namely, when consecutive publications of the same author(s) present data for the same material, it is often difficult to assess if the material was re-synthesized (or at least the measurements were repeated) or the same results were republished. When the same results were clearly republished (the assessment was made intuitively) they are reported only once in Table 3.1, but unless appropriate referencing was provided no effort was made to find out which publication was the "original" one, i.e. the publication cited as the source is not necessarily the oldest one.

Some publications report experimental data and zero points calculated from this data using different models. The modeling is discussed in Chapter 5. Usually theoretical and experimental zero points match, but in a few instances the models did not properly reflect the experimental trend and the values given in Table 3.1 based on graphical interpolation of data are different from those published in original papers. For example, Ref. [90] reports only negative ζ potentials for Fe_2O_3 at $pH \geq 4$ but the "theoretical" IEP at pH 5 is reported.

The electrophoretic mobility was calculated as a linear combination of two terms corresponding to mobilities (sic!) for positively and negatively charged surface, and this model is erroneous. Apparently the idea to calculate the electrophotetic mobility of a particle as a linear combination of electrophoretic mobilities of different surface species has many subscribers. For example the electrophoretic mobility of alumina at three different ionic strengths was calculated as linear combination of mobilities of 11 surface species whose individual mobilities were supposed to range from -2.3 to $+ 3 \times 10^{-8} m^2V^{-1}s^{-1}$ [91].

If there is a serious doubt about the significance of the published zero point but the raw data shown in the publication could not be reinterpreted, the values are marked by a question mark (?) in Table 3.1.

This collection of PZC values is not supposed to show only "recommended" values. Some results presented in Table 3.1 are rather unusual and they are most likely due to impurities or some experimental errors. Such results can be a warning that certain routes of preparation lead to products with surface impurities, or that certain reagent grade chemicals contain substantial amounts of surface-active species.

Most authors who have recently reported unusual PZC values probably realized that their data was in conflict with results of other authors, very often this was explicitly stated in publications. Probably they also realized what potential pitfalls are encountered by attempts of PZC determination, thus, there is no reason to completely disregard such publications.

A few PZC values calculated from acidity constants (cf. Chapter 5) taken from original papers are reported in Table 3.1, namely, when PZC was not given explicitly.

The range of PZC reported in literature and the degree of their scatter can be easily comprehended from Table 3.1 when there are only a few entries referring to material of interest, but not with dozens of publications reporting PZC for the same material. The results for the most common materials, namely, Al_2O_3, AlOOH, $Al(OH)_3$, Cr oxides and hydroxides, Fe_3O_4, FeOOH, $Fe(OH)_3$, SiO_2, SnO_2, TiO_2, and ZrO_2 are summarized in Figs. 3.11–3.37. The possible effect of the crystalline form (α versus γ Al_2O_3, α versus γ Fe_2O_3, quartz versus amorphous silica, rutile versus anatase), origin of the sample (synthetic versus natural hematite) and experimental method ("iep", "cip", and "acousto" on the one hand and "pH" and mass titration on the other) on the PZC is also discussed. By no means is the discussion below intended as statistical analysis. The results corresponding to the same formal stoichiometry represent various samples whose PZC can be different because of their actual stoichiometry, hydration, level of impurities, etc. The significance of particular entries in Table 3.1 is very different. Some PZC values represent many hundreds of actual measurements, e.g. when the sample was studied using different methods, different types and concentrations of electrolytes, and by changing other experimental conditions, while some other entries represent only a few data points. The measurements were also conducted at different temperatures, and the temperature affects the PZC (cf. Section IV). Finally, some publications

(*Text continues on pg. 157*)

TABLE 3.1 Zero Points of Simple Hydr(Oxides) (See Appendix for the most recent results)

Oxide	Description	Salt	T	Method	pH_0	Source
Al_2O_3	cf. Table 3.9					
Al_2O_3	α, Aluminum Co. of Canada	$0.001-1$ mol dm^{-3} KNO$_3$, KCl, NaClO$_4$	25	pH, merge	$\sigma_0 = 0$ at pH 5-7	S.M. Ahmed. J.Phys.Chem. 73: 3546-3555 (1969).
Al_2O_3	chromatographic, Veb Lab.	10^{-4}-1 mol dm^{-3} KCl	25± 2	cip	8.9	H. Sadek, A.K. Helmy, V.M. Sabet, and Th.F. Tadros. J.Electroanal.Chem. 27: 257-266 (1970).
Al_2O_3	from nitrate, different T treatment	$0-0.005$ mol dm^{-3} KNO$_3$, KCl, KClO$_4$	30	iep	8.2-8.5 600 C 6.8-7.7 1200 C	H. Fukuda, and M. Miura. J.Sci.Hiroshima Univ.Ser.A. 36: 77-86 (1972).
Al_2O_3	chromatographic, Fisher	$0.005-0.1$ mol dm^{-3} KNO$_3$,	30	cip	9.06	P.H. Tewari, and A.W. McLean. J.Colloid Int.Sci.40: 267-272 (1972).
Al_2O_3				iep coagulation	8.9	T.W. Healy, G.R. Wiese, D.E. Yates, and B.E. Kavanagh. J.Colloid Int.Sci. 42: 647-649 (1973).
Al_2O_3	γ, Cabot	$10^{-3}-10^{-1}$ mol dm^{-3} NaCl	25	cip iep	8.5	C.P. Huang, and W. Stumm. J.Colloid Int.Sci. 43: 409-420 (1973).
Al_2O_3	α, Linde A	0.002 mol dm^{-3} NaCl	23	iep	9.1	K.N. Han, T.W. Healy, and D.W. Fuerstenau. J.Colloid Int.Sci. 44: 407-414 (1973).
Al_2O_3	T-60 Alcoa	KNO$_3$	25	iep	≈ 8	P. Somasundaran, and R.D. Kulkarni. J.Colloid Interf.Sci. 45: 591-600 (1973).
Al_2O_3	Fluka	$0.001-1$ mol dm^{-3} KCl		cip	7.8	M. Tschapek, L. Tcheichvili, and C. Wasowski. Clay Miner. 10: 219-229 (1974).
Al_2O_3	chromatography grade	KNO$_3$, KCl	22± 2	pH, iep	8.3	E.A. Nechaev, and T.B. Golovanova. Kolloid.Zh. 36: 889-894 (1974).
Al_2O_3	from nitrate, treated at different T	10^{-3} mol dm^{-3} NaCl		iep	9.4 (>1000 C) 11.4 (<800 C)	S. Kittaka. J.Colloid Int.Sci. 48: 327-333 (1974).
Al_2O_3	γ, high purity, Linde B	$0.0001-0.01$ mol dm^{-3} KNO$_3$	25	iep coagulation	8.9	G.R. Wiese, and T.W. Healy. J.Colloid Int.Sci. 51: 427-433 (1975).
Al_2O_3	99.5%, reagent	0.01 mol dm^{-3} KNO$_3$	30	iep	9.1	P.H. Tewari and W. Lee. J.Colloid Int.Sci. 52: 77-88 (1975).
Al_2O_3	reagent grade, "C", Degussa	$0.01-1$ mol dm^{-3} KCl		cip EMF	8.8 8.8	M. Tschapek, R.M. Torres Sanchez, and C. Wasowski. Colloid Polym.Sci. 254: 516-521 (1976).

Material	Sample	Electrolyte	t/°C	method	pH	Reference
Al_2O_3	reagent grade, Fluka	0.001 1 $mol\,dm^{-3}$ KCl		cip	7.8	M. Tschapek, R.M. Torres Sanchez, and C. Wasowski. Colloid Polym.Sci. 254: 516-521 (1976).
Al_2O_3	γ, Cabot	0.1 $mol\,dm^{-3}$ $NaClO_4$		EMF	7.8	H. Hohl, and W. Stumm. J.Colloid Int. Sci. 55: 281 288 (1976).
Al_2O_3	Degussa, γ	10^{-2}-1 $mol\,dm^{-3}$ LiCl; 10^{-2}-1 $mol\,dm^{-3}$ KCl		pH; cip	8.3; no cip	M. Tschapek, C. Wasowski, and R. M. Torres Sanchez. J.Electroanal.Chem. 74: 167-176 (1976).
Al_2O_3	synthetic sapphire, single crystal	10^{-3} $mol\,dm^{-3}$ KCl, LiCl; 10^{-3} $mol\,dm^{-3}$ NaCl	20–25	iep; iep	9.1; 9.1; 3.3	W. Smit, and H.S. Stein. J.Colloid Interf. Sci. 60: 299-307 (1977).
Al_2O_3	γ,δ,θ,α, also bayerite			iep	9.2	G.C. Bye, and G.T. Simpkin. J.Appl.Chem. Biotechnol 28: 116-118 (1978).
Al_2O_3	γ, Linde B			iep	8.5	H.A. Elliott, and C.P. Huang. J.Colloid Int. Sci. 70: 29-45 (1979).
Al_2O_3	activated, different washing procedures			iep	6.2; 7.3; 8.9	W.W. Choi, and K.Y. Chen. J.Am.Water Works Assoc. 71: 562 569 (1979)
Al_2O_3	δ or γ, Cabot	0.01 $mol\,dm^{-3}$ $NaClO_4$	25	iep	9.1	M.A. Anderson, and D.T. Malotky. J.Colloid Int Sci. 72: 413-427 (1979)
Al_2O_3	Carlo Erba, hydrous	0.001, 0.1 $mol\,dm^{-3}$ KNO_3		intersection	7.5	A.K. Helmy, E.A. Ferreiro, and S.G. de Bussetti. Z. Phys.Chem. (Leipzig) 261: 1065 1073 (1980).
Al_2O_3		0.01 $mol\,dm^{-3}$ KOH (?)		iep	8.3	E.A. Nechaev. Kolloidn. Zh. 42: 371-373 (1980).
Al_2O_3		0.01 $mol\,dm^{-3}$ KCl		iep	8.3	E.A. Nechaev, and G.V. Zvonareva. Kolloidn. Zh 42: 511-516 (1980).
Al_2O_3	α, Alcoa A-14	0.001 $mol\,dm^{-3}$ $NaClO_4$	20	iep	9	J.S Moya, J. Rubio, and J.A. Pask. Ceramic Bull. 59: 1198 1200 (1980).
Al_2O_3	γ, Degussa C	0.1 $mol\,dm^{-3}$ $NaClO_4$	22	pH	8.5	R. Kummert, and W. Stumm. J.Colloid Int. Sci. 75: 373-385 (1980).
Al_2O_3	γ, AE-11 Nishio	0.01 $mol\,dm^{-3}$ KCl	40	pH	7	H. Kita, N. Henmi, K. Shimazu, H. Hattori and K. Tanabe. J.Chem.Soc. Faraday Trans. 1 77: 2451-2463 (1981).
Al_2O_3	α, Linde B, Union Carbide	10^{-3} $mol\,dm^{-3}$ KNO_3		iep	8.9	D.N. Furlong, P.A. Freeman, and A.C.M. Lau. J.Colloid Int.Sci. 80: 20-31 (1981).
Al_2O_3	α, Linde A		23	pH	8.8	D.A. Griffiths, and D.W. Fuerstenau. J.Colloid Interf.Sci. 80: 271-283 (1981).
Al_2O_3	γ (fresh dehydrated sample)	0.001 $mol\,dm^{-3}$ KCl	22.5	iep	8.8; (7.3)	F.J. Gil-Llambias, and A.M. Escudey-Castro. J.Chem.Soc.Chem.Commun. 478-479 (1982).

TABLE 3.1 (Continued)

Oxide	Description	Salt	T	Method	pH₀	Source
Al_2O_3	γ, Cabot	10^{-2} mol dm^{-3} NaCl + 10^{-3} mol dm^{-3} NaHCO$_3$	25	iep	9.2	J.A. Davis. Geochim. Cosmochim. Acta 46: 2381 2393 (1982).
Al_2O_3	γ, POCh	NaBr		cip	8.75	R. Sprycha. J.Colloid Int.Sci. 96: 551–554 (1983).
Al_2O_3	Linde sintered at 1000 C, α			iep pH	8.8	R.T. Lowson, and J.V. Evans. Aust.J.Chem. 37: 2165- 2178 (1984).
Al_2O_3	from AlCl$_3$	10^{-3} 10^{-1} mol dm^{-3} NaNO$_3$		cip	7.2	J.A. Schwarz, C.T. Driscoll, and A.K. Bhanot. J.Colloid Int.Sci. 97: 55–61 (1984).
Al_2O_3	γ, Degussa C	0.1 mol dm^{-3} NaClO$_4$ 0.001 mol dm^{-3} NaClO$_4$		iep	9.7 9	A.R. Bowers, and C.P. Huang. J.Colloid Int. Sci. 105: 197–214 (1985).
Al_2O_3	γ	0.001 mol dm^{-3} KCl, KClO$_4$, NaCl, KNO$_3$, NH$_4$Cl	22.5	iep	8.55 8.8	M. Escudey, and F. Gil-Llambias. J.Colloid Int.Sci. 107: 272 275 (1985).
Al_2O_3	α, A 16 SG Alcoa	0.01–0.1 mol dm^{-3} NaCl	25	iep	8.7	D. Ballion, and N. Jaffrezic-Renault.J.Radioanal. Nucl.Chem. 92: 133 150 (1985).
Al_2O_3	γ, Degussa C, washed; different methods of subtracting the "blank"	10^{-3} 10^{-1} mol dm^{-3} NaClO$_4$		cip	8.6 7.5	C.P. Schulthess, and D.L. Sparks. Soil Sci.Soc.Am.J. 50: 1406 1411 (1986)
Al_2O_3	γ, Degussa C	0.1 mol dm^{-3} NaClO$_4$ 0.001 mol dm^{-3} NaClO$_4$	25± 2	iep	9.7 9.0	A.R. Bowers, and C.P. Huang. J.Colloid Interf.Sci. 110: 575–590 (1986).
Al_2O_3	γ, F-1 Alcoa	0.05 and 0.5 mol dm^{-3} NaClO$_4$		intersection	7	O.J. Hao, and C.P. Huang. J.Environ Eng. 112: 1054- 1069 (1986).
Al_2O_3	γ, F-1 Alcoa	0.1 mol dm^{-3} NaCl			7.7	M.M. Ghosh, and Y.P. Yang, US Environmental Protection Agency Report R809425-01-0 (1984), quoted after O.J. Hao, and C.P. Huang. J.Environ.Eng. 112: 1054- 1069 (1986).
Al_2O_3	γ, Houdry 415, 417, 425, dehydration of bauxite	0.001–0.1 mol dm^{-3} KNO$_3$		cip	3.55 3.6 3.61 5.3 5.5 9.9	L. Vordonis, P.G. Koutsoukos, and A. Lycourghiotis. Langmuir 2: 281–283 (1986).
Al_2O_3	A 16 SG Alcoa	0.001 mol dm^3 KCl	25	iep	9.2	E.M. de Liso, W.R. Cannon, and A.S. Rao. Mat.Res.Soc.Symp.Proc. 60: 43 50 (1986).

		electrolyte	T	method	pH	reference
Al_2O_3	γ, Houdry 415	0.001–0.1 mol dm^{-3} KNO$_3$	25	cip	5.4	K.Ch. Akratopulu, L. Vordonis, and A. Lycourghiotis. J.Chem.Soc. Faraday Trans. I. 82: 3697–3708 (1986).
Al_2O_3	γ, Houdry 415, calcined without and with NaNO$_3$	0.001–0.1 mol dm^{-3} KNO$_3$	25	cip	w o 5.3 with 9.4–10.1	L. Vordonis, P.G. Koutsoukos, and A. Lycourghiotis. J. Catalysis 98: 296–307 (1986).
Al_2O_3	$\gamma(+\delta)$, from butoxide, calcined at 300–800 C, different aging times	10^{-3} mol dm^{-3} NaCl	25	iep	7.8 10.6	B.I. Lee, and L.L. Hench. Colloids Surf. 23: 211–229 (1987).
Al_2O_3	$\alpha(+\theta)$ from butoxide, calcined at 1100–1400 C, different aging times, and one commercial sample CR6 from Baikalox	10^{-3} mol dm^{-3} NaCl	25	iep	7.2–9.2	B.I. Lee, and L.L. Hench. Colloids Surf. 23: 211 229 (1987).
Al_2O_3	Degussa C	0.15 mol dm^{-3} NaCl	30	iep	9.5	D. Bitting, and J.H. Harwell. Langmuir 3: 500–511 (1987).
Al_2O_3	γ, Degussa C, washed	0–0.05 mol dm^{-3} NaCl		cip	7.5–7.76	C.P. Schulthess, and D.L. Sparks. Soil.Sci. Soc.Am.J. 51: 1136–1144 (1987).
Al_2O_3	commercial, Alcoa	0.001 mol dm^{-3} KNO$_3$		iep	8.7	A.S. Rao. J.Disp.Sci.Technol. 8: 457–476 (1987).
Al_2O_3	γ, Cyanamid			mass titration	8.3	S. Subramanian, J.S. Noh, and J.A. Schwarz. J.Catalysis 114: 433 439 (1988).
Al_2O_3	Thiokol, washed	0.1 mol dm^{-3} NaCl		pH	8.2	J.F. Kuo, and T.F. Yen. J.Colloid Int.Sci. 121: 220–225 (1988).
Al_2O_3	Thiokol, washed	0.001 mol dm^{-3} NaCl		iep	8.1	J.F. Kuo, M.M. Sharma, and T.F. Yen. J.Colloid Int.Sci. 126: 537 546 (1988).
Al_2O_3	Houdry 415	10^{-3} and 10^{-1} mol dm^{-3} KCl		intersection	5.9	R.M. Torres Sanchez, E.F. Aglietti, and J.M. Porto-Lopez. Mater.Chem.Phys. 20: 27 38 (1938).
Al_2O_3	γ one sample α two samples			titration	8.8	W.C. Hasz. Thesis. quoted after W.M. Mullins, anc B.L. Averbach. Surface Sci. 206: 41–51 (1938).
Al_2O_3	α, A16 SG, Alcoa	NH$_4$Cl		iep	8.0	M. Hashiba, H. Okamoto, Y. Nurishi, and K. Hiramatsu. J.Mater.Sci. 23: 2893–2896 (1938).
Al_2O_3	POCh, washed	0.001–0.1 mol dm^{-3} NaCl		cip	8.5	W. Janusz, and J. Szczypa. Mater.Sci. Forum 25–26: 427–430 (1988).
Al_2O_3	γ, Degussa C, washed	0–0.007 mol dm^{-3} NaCl		iep	9.5–9.8	C.P. Schulthess, and D.L. Sparks. Soil.Sci. Soc.Am.J. 52: 92–97 (1988).

TABLE 3.1 (Continued)

Oxide	Description	Salt	T	Method	pH_0	Source
Al_2O_3		0.01 mol dm⁻³ KNO₃	25	iep	8.7	L. Vordonis, C. Kordulis, and A. Lycourghiotis. J.Chem.Soc. Faraday Trans. I 84: 1593–1601 (1988).
Al_2O_3	γ, Cyanamid		23	mass titration	7.2	S. Subramanian, J.A. Schwarz, and Z. Hejase. J. Catalysis 117: 512–518 (1989).
Al_2O_3	γ, POCh	0.0001 1 mol dm⁻³ NaCl		pH / iep / cip	7.8 / 8.1	R. Sprycha. J.Colloid Int.Sci. 127: 1 11 (1989).
Al_2O_3	γ, Cyanamid	10⁻³–10⁻¹ mol dm⁻³ NaCl		cip / mass titration	7.5 / 7.4	J.S. Noh, and J.A. Schwarz. J.Colloid Interf. Sci. 130: 157–164 (1989).
Al_2O_3	γ, Degussa C	0.01 mol dm⁻³ NaClO₄		iep	9.4	C.P. Huang, and E.A. Rhoads. J.Colloid Int. Sci. 131: 289–306 (1989).
Al_2O_3	α	0.1 mol dm⁻³ KNO₃		from acidity constants	7.35	B.I. Lobov, L.A. Rubina, I.G. Vinogradova, and I. F. Mavrin. Zh. Neorg. Khim. 34: 2495–2498 (1989).
Al_2O_3	Degussa C	0.001–0.1 mol dm⁻³ KCl	25	iep / cip	8.6 / 8.5	F. Thomas, J.Y. Bottero, and J.M. Cases. Colloids Surf. 37: 281–294 (1989).
Al_2O_3	Merck 90	0.001–0.1 mol dm⁻³ KCl	25	iep / cip	8.4 / 8.5	F. Thomas, J.Y. Bottero, and J.M. Cases. Colloids Surf. 37: 281 294 (1989).
Al_2O_3	γ(+α) reagent grade, washed	0.001–0.1 mol dm⁻³ NaCl	20	cip / pH	8.1 / 7 7.6	M. Kosmulski. J.Colloid Int.Sci. 135: 590 593 (1990).
Al_2O_3	α(+θ+γ) reagent grade, washed	0.001–0.1 mol dm⁻³ NaCl	20	cip / pH	8.1 / 7.7–8	M Kosmulski. J.Colloid Int.Sci. 135: 590–593 (1990).
Al_2O_3	γ	0.001 mol dm⁻³ KCl		iep	8.7	F.J. Gil Llambias, J. Salvatierra, L. Bouyssieres, M. Escudey, and R. Cid. Appl.Catal. 59: 185 195 (1990).
Al_2O_3	α, Buehler, washed	0.005–0.139 mol dm⁻³ NaNO₃	25	cip	8.9	K.F. Hayes, G. Redden, W. Ela, and J.O. Leckie. J.Colloid Int.Sci. 142: 448–469 (1991).
Al_2O_3	γ, Degussa C	10⁻⁴ 10⁻¹ mol dm⁻³ NaClO₄	25	iep	8 8	L Righetto, G Bidoglio, G Azimonti, and I. R. Bellobono. Env.Sci.Technol. 25: 1913 1919 (1991).

Material	Description	Electrolyte	T	Method	pH	Reference
Al_2O_3	Degussa C, γ, calcined at 600 C	none	25	mass titration	7.43	S. Subramanian, M.S. Chattha, and C.R. Peters. J.Molecul. Catalysis 69: 235–245 (1991).
Al_2O_3				mass titration	8.3	C.Contescu, C.Sivaraj, and J.A. Schwarz. Appl. Catalysis 74: 95 108 (1991).
Al_2O_3	from isopropoxide	10^{-3}–10^{-1} mol dm^{-3} NaNO$_3$		mass titration	8.2	J.A. Schwarz, C.T. Ugbor, and R. Zhang. J. Catalysis 138: 38 54 (1992).
Al_2O_3	α, ground membranes prepared by sol-gel processing	0.001–0.1 mol dm^{-3} KCl		cip iep	8.52 9.5 10	M.J. Gieselmann. Langmuir 8: 1342 1346 (1992).
Al_2O_3	γ, Houdry 415, calcined at 600 C	0.01 mol dm^{-3} NH$_4$NO$_3$	room	iep	8	L. Vordonis, N. Spanos, P.G. Koutsoukos, and A. Lycourghiotis. Langmuir 8: 1736 1743 (1992).
Al_2O_3	γ	HCl+KOH		iep	8.2	M.I. Zaki, S.A.A. Mansour, F. Taha, and G.A.H. Mekhemer. Langmuir 8: 727 732 (1992).
Al_2O_3	JRC-ALO-4 JRC-ALO-5	KNO$_3$		pH	9.1 9.3	P.K. Ahn, S. Nishiyama, S. Tsuruya, and M. Masai. Appl. Catalysis A 101: 207 219 (1993).
Al_2O_3	γ, C from Degussa	NH$_3$		acousto	9.6	M. de Boer, R.G. Leliveld, A.J. van Dillen, J.W. Geus, and H.G. Bruil. Appl. Catalysis A. 102: 35–51 (1993).
Al_2O_3	reagent grade, washed	0.001 mol dm^{-3} NaCl 0.1 mol dm^{-3} NaCl	25	iep pH	7 7.6	M. Kosmulski. J.Colloid Int.Sci. 156: 305–310 (1993).
Al_2O_3	δ			pH	8.6	E. Wieland, thesis, quoted after S.B. Haderlein, and R.P. Schwarzenbach. Environ.Sci.Technol. 27: 316–326 (1993).
Al_2O_3	γ, type C Degussa	NaNO$_3$	25	cip	8.9	A.T. Stone, A. Torrents, J. Smolen, D. Vasudevar, and J. Hadley. Environ.Sci.Technol. 27: 895 905 (1993).
Al_2O_3	AKP-15 AKP-10 AKP-30 Sumitomo	0.001 mol dm^{-3} KNO$_3$	25	iep	3.7 4.3 4.3	K. Furosawa, K. Nagashima, and C. Anzai. Kobunshi Ronbuhshu 50: 343 347 (1993).
Al_2O_3	δ, Degussa C	0.01 mol dm^{-3} NaClO$_4$	25	iep	9.2	D.C. Girvin, P.L. Gassman, and H. Bolton. Soil Sci.Soc.Am.J. 57: 47 57 (1993).
Al_2O_3	γ, high purity, BDH	KCl		iep	4.6	M.G. Cattania, S. Ardizzone, C.L. Bianchi, and S. Carella. Colloids Surf. A 76: 233–240 (1993).
Al_2O_3	α, high purity	KNO$_3$ or KCl		iep	8.3	M.G. Cattania, S. Ardizzone, C. L. Bianchi, and S. Carella. Colloids Surf. A 76: 233–240 (1993).

TABLE 3.1 (Continued)

Oxide	Description	Salt	T	Method	pH_0	Source
Al_2O_3	A160 Dia Showa	KOH/HCl		acousto	10	T.E. Petroff, M. Sayer, and S.A.M. Hesp. Colloids Surf. A 78: 235–243 (1993).
Al_2O_3	γ, Harshaw, calcined at 500 C	0.1 mol dm⁻³ NaCl		mass titration	7.8	M.A. Vuurman, F. D. Hardcastle, and I.E. Wachs. J.Mol. Catalysis 84: 193–205 (1993).
Al_2O_3	γ, Cyanamid, calcined	0.1 mol dm⁻³ NaNO₃	25	pH	6.8	C. Contescu, J. Hu, and J.A. Schwarz. J.Chem. Soc. Faraday Trans. 89: 4091–4099 (1993).
Al_2O_3	γ, Cyanamid	0.1–0.0001 mol dm⁻³ NaNO₃		cip	7.0	C. Contescu, J. Jagiello, and J.A. Schwarz. Langmuir 9: 1754–1765 (1993).
Al_2O_3	Mager			iep	9	J. Liu, S.M. Howard, and K.N. Han. Langmuir 9: 3635–3639 (1993).
Al_2O_3	Houdry 415, γ	0.01 mol dm⁻³ NH₄NO₃	room	iep	8	N. Spanos, and A. Lycourghiotis, Langmuir 10: 2351–2362 (1994).
Al_2O_3	α, AKP-HP40, Sumitomo	0-0.01 mol dm⁻³ KNO₃	25	cip iep coagulation	9.0	H.M. Jang, J.H. Moon, and B.H. Kim. J.Ceram.Soc.Jpn. 102: 119–127 (1994).
Al_2O_3	γ, Harshaw	NH₄NO₃		pH	7.6	F.M. Mulcahy, M. Houalla, and D.M. Hercules. J.Catal. 148: 654–659 (1994).
Al_2O_3	commercial, CCI	KNO₃		iep	8.8	G. Gonzalez, S.M. Saraiva, and W. Aliaga. Dispersion Sci.Technol. 15: 123–132 (1994).
Al_2O_3	γ, reagent grade, washed	KCl	25	salt titr.	8.43	M. Kosmulski, J. Matysiak, and J. Szczypa. J.Colloid. Int.Sci. 164: 280–284 (1994).
Al_2O_3	A-16 Alcoa	10⁻³ mol dm⁻³ NaNO₃	room	acousto	9.9	Pradip, R.S. Premachandran, and S.G. Malghan. Bull.Mater.Sci. 17: 911–920 (1994).
Al_2O_3	γ, Alcoa F-1, washed	10⁻³–10⁻¹ mol dm⁻³ NaNO₃	20	cip	8.4	C.D. Cox, and M.M. Ghosh. Water Res. 28: 1181–1188 (1994).
Al_2O_3	γ, Degussa C	1 mol dm⁻³ NaClO₄	25	pH	6.45	P.V. Brady. Geochim. Cosmochim. Acta 58: 1213–1217 (1994).
Al_2O_3	γ, Degussa C	NaCl	25	cip	8.4	M.A. Schlautman, and J.J. Morgan. Geochim. Cosmochim. Acta 58: 4293–4303 (1994).
Al_2O_3	AKP-30 Sumitomo, washed			mass titration (modified)	7.88	J. Wernet, and D.L. Feke. J.Am.Ceram.Soc. 77: 2693 2698 (1994).

Material	Description	Electrolyte	T/°C	Method	pH/iep	Reference
Al_2O_3	γ, Cyanamid, calcined	0.1–0.001 mol dm^{-3} NaNO$_3$	25	cip	7.4	C.Contescu, A. Contescu, and J.A. Schwarz. J.Phys.Chem. 98: 4327–4335. (1994).
Al_2O_3	γ, LaRoche	0.1 mol dm^{-3} NaNO$_3$		mass titration	8.0	J.Park, and J.R. Regalbuto. J.Colloid Int.Sci. 175: 239-252 (1995).
Al_2O_3	α, LaRoche, calcined	0.1 mol dm^{-3} NaNO$_3$		mass titration	8.3–8.5	J.Park, and J.R. Regalbuto. J.Colloid Int.Sci. 175: 239 252 (1995)
Al_2O_3	α, Asahi, high purity	NH$_4$Cl		iep	6.7	Y. Hirata, and K. Onoue. Eur. J.Solid State Inorg.Chem. 32: 663–672 (1995).
Al_2O_3	α, Queensland Alumina	0.001 mol dm^{-3} KNO$_3$	25	pH	8.6	K.M. Spark, B.B. Johnson, and J.D. Wells. Eur.J.Soil Sci. 46: 621–631 (1995).
Al_2O_3				iep	8.7	
Al_2O_3	Sumitomo AKP-50	0.01 mol dm^{-3} NH$_4$Cl		iep	8.7	E.P. Luther, J.A. Yanez, G.V. Franks F.E. Lange, and D.S. Pearson. J.Am.Ceram. Soc. 78: 1495-1500 (1995).
Al_2O_3	Harshaw			pH	8.9	H. Hu, I.E. Wachs, and S.R. Bare. J.Phys.Chem. 99: 10897 10910 (1995).
Al_2O_3	α, high purity, LAKP-50 Sumitomo	0.005 mol dm^{-3} KCl		iep	8.2	J.S. Jeon, R.P. Sperline, S. Raghavan, and J.B. Hiskey. Colloids Surf. A 111: 29 38 (1996).
Al_2O_3	Degussa C	0.05 mol dm^{-3} NaClO$_4$	25	iep	9.65	H.L. Yao, and H.H. Yeh. Langmuir 12: 2981 2988; 2989 2994 (1996).
Al_2O_3	sapphire, single crystal, α	10^{-3} mol dm^{-3} KCl		AFM	9.3	S. Veeramasuneni, M.R. Yalamanchili, and J.D. Miller. J.Colloid Interf.Sci. 184: 594-600 (1996).
Al_2O_3	α, Alfa Aesar			iep	9.1	S. Veeramasuneni, M.R. Yalamanchili, and J.D. Miller. J.Colloid Interf.Sci. 184: 594-600 (1996).
Al_2O_3	γ, JRC ALO-4 washed	0.1 mol dm^{-3} NaNO$_3$	25	pH	8.5	H. Tamura, N. Katayama, and R. Furuichi. Environ.Sci.Technol. 30: 1198–1204 (1996).
Al_2O_3	γ	0.01 mol dm^{-3} NaNO$_3$	22–24	pH	7.8	B. Nowack, J. Lützenkirchen, P. Behra, and L. Sigg. Environ.Sci.Technol. 30: 2397–2405 (1996).
Al_2O_3	α, Union Carbide	0.03 mol dm^{-3} NaCl	43	pH	8	N.P.Hankins, J.H. O'Haver, and J.H. Harwell. Ind.Eng.Chem.Res. 35: 2844 2855 (1996).
Al_2O_3	amorphous δ, Degussa C	0.01 mol dm^{-3} NaCl		iep	9.3 9.3	S. Goldberg, H.S.Forster, and C.L. Godfrey. Soil Sci. Soc.Am.J. 60: 425–432 (1996).

TABLE 3.1 (Continued)

Oxide	Description	Salt	T	Method	pH$_0$	Source
Al$_2$O$_3$	γ, Johnson–Matthey	0.005 mol dm^{-3} NaNO$_3$	25	iep	7.7	C.Quang, S.L. Petersen, G.R. Ducatte, and N.E. Ballou. J.Chromatogr. A. 732: 377–384 (1996).
Al$_2$O$_3$	α + γ, Johnson Matthey	0.005 mol dm^{-3} NaNO$_3$	25	iep	7	C.Quang, S.L. Petersen, G.R. Ducatte, and N.E.Ballou. J.Chromatogr. A. 732: 377–384 (1996).
Al$_2$O$_3$	α, Martinswerk, washed	KOH + HNO$_3$	25	acousto	9.3	P.C.Hidber, T.J. Graule, and L.J.Gauckler. J.Am.Ceramic Soc. 79: 1857–1867 (1996).
Al$_2$O$_3$	A-16 Alcoa	0.01 mol dm^{-3} NaNO$_3$	25	iep	8.8	V.Ramakrishnan, Pradip, and S.G. Malghan. J.Am.Ceram.Soc. 79: 2567–2576 (1996).
Al$_2$O$_3$	α, Showa Denko	0.001 mol dm^{-3} KNO$_3$		pH	9.0	S. Fukuzaki, H. Urano, and K.Nagata. J.Fermentation Bioeng. 81: 163–167 (1996).
Al$_2$O$_3$	α, UCAR	0.001 mol dm^{-3} KCl		iep	7.5	P.Gil, M.Almeida, and H.M.M. Diz. Brit.Ceram. Trans. 95: 254–257 (1996).
Al$_2$O$_3$	α, flat specimen	KCl		AFM	7	M.E.Karaman, R.M. Pashley, T.D. Waite, S.J. Hatch, and H.Bustamante. Colloids Surf. A 129 130: 239–255 (1997).
Al$_2$O$_3$	C 100 Degussa	NH$_4$Cl		acousto	8.3–8.6 (hysteresis)	R.A. Overbeek, E.J. Bosma, D.W.H. de Blauw, A.J. van Dillen, H.G. Bruil, and J.W.Geus. Applied Catal. A 163: 129–144 (1997).
Al$_2$O$_3$	A16 SG, Alcoa	10^{-3} mol dm^{-3} KCl		iep	8.1	G.Tari, J.M.F. Ferreira, and O.Lyckfeldt. J.Eur.Ceram.Soc. 17: 1341–1350. (1997).
Al$_2$O$_3$	γ, Degussa C	0.001 mol dm^{-3} NaCl		iep	9.6	I.Sondi, O.Milat, and V.Pravdic. J.Colloid Interf. Sci. 189: 66–73 (1997).
Al$_2$O$_3$	AlCl$_3$ + NH$_4$OH, then drying at 200 C	0.01 mol dm^{-3} NaCl	20	iep	8.5	L.E. Ermakova, M.P. Sidorova, and V.M. Smirnov. Kolloid.Zh. 59: 563–565 (1997).
Al$_2$O$_3$	α, A 11 Alcoa	0.001 mol dm^{-3} KNO$_3$		iep	9.2	D.Santhiya, G. Nandini, S. Subramanian, K.A.Natarajan, and S.G. Malghan. Colloids Surf.A 133: 157–163 (1998).
Al$_2$O$_3$	AKP-30 Sumitomo	0.01 mol dm^{-3} KNO$_3$		acousto	9.7	S.B.Johnson, A.S.Russel, and P.J.Scales. Colloids Surf.A. 141: 119–130 (1998).

Material	Description	Electrolyte	t/°C	Method	pH	Reference
Al$_2$O$_3$	γ, Merck, washed	0.001–0.1 mol dm^{-3} KNO$_3$	20	cip	8.2	S.Mustafa, B. Dilara, Z.Neelofer, A.Naeem, and S.Tasleem. J.Colloid Interf.Sci. 204: 284 293 (1998).
Al$_2$O$_3$	α, > 99.99%, AKP 20 Sumitomo	0–0.01 mol dm^{-3} NaCl		pH	6.5 7.7	D.J.Kim, H. Kim, and J.K.Lee. J.Mater.Sci. 33: 2931 2935 (1998).
Al$_2$O$_3$	α, Fisher	0.01,0.1 mol dm^{-3} NaCl	25	iep / intersection	8 / 9.0	D.B.Ward, and P.V. Brady. Clays Clay Min. 46: 453–465 (1998).
Al$_2$O$_3$	δ, Degussa C, washed	0–0.1 mol dm^{-3} NaCl	20	cip	8.3	C.P.Schulthess, K.Swanson, and H. Wijnja. Soil Sci. Soc. Am.J. 62: 136 141 (1998).
Al$_2$O$_3$	δ, type C Degussa			iep	9.3	S. Goldberg, C.Su, and H.S.Forster in Adsorption of Metals by Geomedia (E.A.Jenne, ed.), Acacemic Press, San Diego 1998. pp. 401–426.
Al$_2$O$_3$	γ, Degussa	0.1 mol dm^{-3} NaCl	25	mass titration	6.9	J.P.Reymond, and F. Kolenda. Powder Techn. 103: 30 36 (1999).
Al$_2$O$_3$	γ, Procatalyse	0.1 mol dm^{-3} NaCl	25	mass titration	8.4	J.P. Reymond, and F. Kolenda. Powder Techn. 103: 30 36 (1999).
Al$_2$O$_3$	α, Sumitomo			iep	8	Y. Hirata, and H. Wakita. J. Ceram. Soc. Jpn. 107: 303–307 (1999).
Al$_2$O$_3$	γ, Merck, washed	NaClO$_4$	25	cip	8.76	M. Kosmulski, Colloids Surf. A 149: 397–408 (1999).
Al$_2$O$_3$	γ, Merck, washed	NaCl	25	cip	8.76	M.Kosmulski, and A. Plak. Colloids Surf.A 149: 409–412 (1999).
Al$_2$O$_3$	α, AKP-30 Sumitomo	0.01 1 mol dm^{-3} KCl, KBr, KI, NaNO$_3$, KNO$_3$, CsNO$_3$	25	acousto	9.4–9.6	S.B. Johnson, P.J. Scales, and T.W. Healy. Langmuir 15: 2836 2843 (1999).
Al$_2$O$_3$	α	NaCl	25	iep / AFM	5.6 / 5	H.G Pedersen. Langmuir 15: 3015 3017 (1999).
Al$_2$O$_3$	Sumitomo AKP-50	0.01–0.1 mol dm^{-3} NaCl, NaClO$_4$, NaNO$_3$	25	acousto	9.5	G.V. Franks, S.B. Johnson, P.J. Scales, D.V. Boger, and T.W. Healy. Langmuir 15: 4411–4420 (1999).
Al$_2$O$_3$	Woel Pharma 200	0.001–0.1 mol dm^{-3} KNO$_3$	25	cip	8.3	K. Csoban, and P. Joo. Colloids Surf. A 151: 97–112 (1999).
Al$_2$O$_3$	Dupont	0.001 mol dm^{-3} KNO$_3$	25	iep	9.1	G.E. Morris, W.A. Skinner, P.G. Self, and R.S.C. Smart. Colloids Surf. A 155: 27–41 (1999).
Al$_2$O$_3$	9 samples $\alpha,\gamma,\gamma,\eta$			mass titration	8.4 9.4	J.R.Regalbuto, A.Navada, S.Shadid, M.L.Bricker, and Q.Chen. J.Catalysis 184: 335–348 (1999).

TABLE 3.1 (Continued)

Oxide	Description	Salt	T	Method	pH$_0$	Source
Al$_2$O$_3$	α	0.001 mol dm^{-3} KCl		iep	7.5	G.Tari, I.Bobos, C.S.F.Gomes, and J.M.F. Ferreira. J.Colloid Interf.Sci. 210: 360-366 (1999).
Al$_2$O$_3$	α, A 16SG Alcoa, different volume fractions	conductivity 0.4-0.6 S/m	25	acousto	8.2-9	A.L.Costa, C.Galassi, and R.Greenwood. J.Colloid Interf.Sci. 212: 350-356 (1999).
Al$_2$O$_3$	A16, Alcoa, α	10^{-3} mol dm^{-3} KNO$_3$		iep	9	D.Santhiya, S.Subramanian, K.A. Natarajan, S.G. Malghan. J.Colloid Interf. Sci. 216: 143 153 (1999).
Al$_2$O$_3$	γ, Johnson-Matthey	0.001 mol dm^{-3} NaNO$_3$	23	pH	4	K.Nagashima, and F.Blum. J.Colloid Interf. Sci. 217: 28-36 (1999).
Al$_2$O$_3$	Prolabo	0.05 mol dm^{-3} NaNO$_3$	20±2	pH	7.2	L. Benyahya, and J.M. Garnier. Environ. Sci. Technol. 33: 1398-1407 (1999).
Al$_2$O$_3$	γ, American Cyanamid	0.001 mol dm^{-3} KCl		iep	8.85	F.Gil-Llambias, N.Escalona, C.Pfaff, C.Scott, and J.Goldwasser. React.Kinet.Catal.Lett. 66: 225-229 (1999).
Al$_2$O$_3$	γ, Degussa C	0.001-0.1 mol dm^{-3} NaClO$_4$		cip	8.6	T.Rabung, T.Stumpf, H. Geckeis, R. Klenze, and J.I.Kim. Radiochim. Acta 88: 711 716 (2000).
Al$_2$O$_3$·n H$_2$O	pseudoboehmite	0.01 mol dm^{-3} NaNO$_3$	25	iep	9.2	R.S.Alwitt. J.Colloid Int.Sci. 40: 195-198 (1972).
Al$_2$O$_3$·n H$_2$O	monodispersed, spherical		25	iep	9.4	E.Matijevic, A.Bell, R.Brace, and P.McFadyen. J.Electrochem.Soc. 120: 893 899 (1973).
Al$_2$O$_3$·n H$_2$O	hydrolysis of isopropoxide, bayerite + pseudoboehmite			cip	9.3	S.S.S.Rajan, K.W.Perrott, and W.M.H.Saunders. J.Soil Sci. 25: 438-447 (1974).
Al$_2$O$_3$·n H$_2$O	bayerite + pseudoboehmite	NaCl		author's	9.2	K.W.Perrott. Clays Clay Min. 25: 417-421 (1977).
Al$_2$O$_3$·n H$_2$O	Carlo Erba	KNO$_3$		cip	6.9	N.Peinemann, and A.K.Helmy. J. Electroanal. Chem. 78: 325 330 (1977).
Al$_2$O$_3$·n	pseudoboehmite synthetic			from acidity constants	9.3	P.Cambier, and G.Sposito. Clays Clay Miner. 39: 369 374 (1991).
		0.001 mol dm^{-3} KNO$_3$		pH	7.5	A.K.Helmy, S.G.de Bussetti, and E.A. Ferreiro. Colloids Surf. 58: 9 16 (1991).

Material	Description	Electrolyte	T/°C	Method	pH	Reference
$Al_2O_3 \cdot n\, H_2O$	α, Condea	0.1 mol dm^{-3} NaCl	25	mass titration	7.9	J.P.Reymond, and F.Kolenda. Powder Techn. 103: 30–35 (1999).
$Al_2O_3 \cdot n\, H_2O$	from nitrate	0.06–1.2 mol dm^{-3} NaNO$_3$	25	cip	8.9	P.Trivedi, and L.Axe. J.Colloid Interf.Sci. 218: 554–563 (1999).
AlOOH	cf. Table 3.8					
AlOOH	boehmite, monodispersed from chloride from chlorate VII			iep	9.3 / 8	W.B Scott, and E.Matijevic. J.Colloid Int.Sci. 66: 447–454 (1978).
AlOOH	boehmite from chlorate VII	9×10^{-5}–9×10^{-3} mol dm^{-3} NaClO$_4$		iep	10.4	W.F.Bleam, and M.B.McBride. J.Colloid Int. Sci. 103: 124–132 (1985).
AlOOH	δ, from butoxide, calcined at 25 C, different aging times	10^{-3} mol dm^{-3} NaCl	25	iep	9–9.4	B.I.Lee, and L.L.Hench. Colloids Surf. 23: 211 229 (1987).
AlOOH	boehmite layer on Al formed by treatment with boiling water	0.0001 mol dm^{-3} KCl		iep	7	B.Lovrecek, Z.Bolanca, and O.Korelic. Surf. Coa. Technol. 31: 351 364. (1987).
AlOOH	boehmite	0.1 mol dm^{-3} KNO$_3$		from acidity constants	7.45	B.I.Lobov, L.A. Rubina, I.F.Mavrin, I.G. Vinogradova, and Yu.I.Ratkovskii. Zh. Neorg. Khim. 34: 2499–2504 (1989).
AlOOH	boehmite	10^{-3} 10^{-1} mol dm^{-3} KNO$_3$	25	iep / cip / salt titr.	9.1 / 8.5 / 8.6	R.Wood, D.Fornasiero, and J.Ralston. Colloids Surf. 51: 389–403 (1990).
AlOOH	boehmite	0.001 mol dm^{-3} KCl		iep / pH	9.4 / 7.2	W.F.Bleam, P.E.Pfeffer, S.Goldberg, R.W.Taylor, and R.Dudley. Langmuir 7: 1702–1712 (1991).
AlOOH	boehmite, Condea	0.1 mol dm^{-3} NaCl	25	pH	8.5	E.Laiti, and L.O. Ohman, J.Colloid Interf.Sci. 183: 441–452 (1996).
AlOOH	boehmite, NaAlO$_2$ + HNO$_3$	0.001 1 mol dm^{-3} NaCl	20	iep / merge	9.5	M.P.Sidorova, L.E.Ermakova, I.A.Savina, and I.A.Kavokina. Kolloid Zh. 59:533–537 (1997).
AlOOH	boehmite, Reanal	KNO$_3$		iep / cip	9 / 9	M.Szekeres, E.Tombacz, K Ferencz, and I. Dekany. Colloids Surf. A 141: 319–325 (1998).
AlOOH	boehmite	0.05 mol dm^{-3} NaClO$_4$		iep	9.2	A.J Fairhurst, and P. Warwick. Colloids Surf. A 145: 229–234 (1998).
AlOOH	boehmite from chloride, heated at 160 C for 2 h-1 d	HCl + NaOH		iep	2 h 9.3 / 1 d 9	S.Music, D.Dragcevic, S.Popovic, and N.Vdovic. Mater.Chem.Phys. 59: 12–19 (1999).
$Al(OH)_3$	gibbsite, from chloride, 2 samples			iep	7.8 / 9.5	F.J Hingston, A.M.Posner, and J.P.Quirk. J.Soil Sci. 23: 177 192 (1972).

TABLE 3.1 (Continued)

Oxide	Description	Salt	T	Method	pH_0	Source
$Al(OH)_3$	amorphous monodispersed, treated with base and deionized to remove sulfates or w/o base treatment			iep	9.3 w 7 w/o	R.Brace, and E.Matijevic. J. Inorg. Nucl. Chem. 35: 3691–3705 (1973).
$Al(OH)_3$	gibbsite + admixture of bayerite, dehydration of reagent grade $Al(OH)_3$ Mallinckrodt	10^{-2} mol dm^{-3} NaNO$_3$		salt titr.	8.3	D.G.Kinniburgh, J.K.Syers, and M.L.Jackson. Soil Sci.Soc.Amer.Proc. 39: 464–470 (1975).
$Al(OH)_3$	amorphous, hydrolysis of nitrate	10^{-2} mol dm^{-3} NaNO$_3$		salt titr.	9.4 (?)	D.G.Kinniburgh, J.K.Syers, and M.L.Jackson. Soil Sci.Soc.Amer.Proc. 39: 464–470 (1975).
$Al(OH)_3$	gibbsite, from chloride	0.01–1 mol dm^{-3} NaNO$_3$	20	iep cip	10.0 9.8	B.V.Kavanagh, A.M.Posner, and J.P.Quirk. Faraday Disc.Chem.Soc. 59: 242 249 (1975).
$Al(OH)_3$	amorphous, from sulfate	0.01 mol dm^{-3} NaClO$_4$		iep	8.5	M.A. Anderson, J.F.Ferguson, and J.Gavis.J. Colloid Int. Sci. 54: 391 399 (1976).
$Al(OH)_3$	gibbsite Martifin	0.004 mol dm^{-3} NaCl	25	iep	9.1	W.B. Jepson, D.G.Jeffs, and A.P.Ferris. J.Colloid Int. Sci. 55: 454–461. (1976).
$Al(OH)_3$	precipitated from sulfate and washed	0.01 mol dm^{-3} NaNO$_3$	25	iep	8.5	H.Schott. J. Pharmac. Sci. 66: 1548 1550 (1977).
$Al(OH)_3$	amorphous, from sulfate	0.01 mol dm^{-3} NaClO$_4$	25	iep	8.6	M.A.Anderson, and D.T.Malotky. J.Colloid Int. Sci. 72: 413–427 (1979).
$Al(OH)_3$	gel, chloride containing	0.0005–0.6 mol dm^{-3} KCl		cip	9.65	J.R.Feldkamp, D.N.Shah, S.L.Meyer, J.L. White, and S.L.Hem. J.Pharm.Sci. 70: 638–640 (1981).
$Al(OH)_3$	from chloride	0.01–0.2 mol dm^{-3} KCl		cip	9.7	C.A.Beyrouty, G.E. van Scoyoc, and J.R. Feldkamp. Soil Sci. Soc. Am.J. 48: 284–287 (1984).
$Al(OH)_3$	amorphous, from chloride			pH	9.45	P.Melis, S.Dixit, A.Premoli, and C. Gessa. Studi Sassar. XXX 137 (1983), cited by: G.Micera, R.Dallocchio, S.Deiana, C. Gessa, P. Melis, and A.Premoli. Colloids Surf. 17: 395–400 (1986).
$Al(OH)_3$	amorphous, from nitrate	0.001 mol dm^{-3} NaCl		iep	9.5	S.Goldberg, and R.A. Glaubig. Clays Clay Miner. 35: 220–227 (1987).

material	description	electrolyte	T	method	IEP/PZC	reference
Al(OH)$_3$	gibbsite from chloride	0.005–0.5 mol dm^{-3} NaCl	20	merge	10	T.Hiemstra, W.H.van Riemsdijk, and M.G.M.Bruggenwert. Neth.J.Agric.Sci. 35: 281–293 (1987).
Al(OH)$_3$	dilution of solutions of polybasic aluminum chloride, two samples	0–1 mol dm^{-3} NaCl	25	iep cip	8.2; 8.4	E.Rakotonarivo, J.Y.Bottero, F.Thomas, J.E.Poirer, and J.M.Cases. Colloids Surf. 33: 191–207 (1988).
Al(OH)$_3$	gibbsite	0.1 mol dm^{-3} KNO$_3$		from acidity constants	7.95	B.I.Lobov, L.A.Rubina, I.F.Mavrin, I.G.Vinogradova, and Yu.I.Ratkovskii. Zh. Neorg.Khim. 34: 2499–2504 (1989).
Al(OH)$_3$	from nitrate			iep	8.9	P.R.Anderson, and M.M.Benjamin. Environ.Sci. Techn. 24: 1586-1592 (1990).
Al(OH)$_3$	bayerite from chloride	0.001 mol dm^{-3} KNO$_3$		pH	6.5	A.K.Helmy, S.G.de Bussetti, and E.A.Ferreiro. Colloids Surf. 58: 9 16 (1991).
Al(OH)$_3$	gibbsite				6.5	S.H.R.Davis, thesis, quoted after S.B.Haderlein, and R.P. Schwarzenbach. Environ.Sci.Technol. 27: 316-326 (1993).
Al(OH)$_3$	from nitrate	0.306 mol dm^{-3} NH$_3$		iep	8.5	C.S.Luo, and S.D.Huang. Separ. Sci.Technol. 28: 1253–1271 (1993).
Al(OH)$_3$	gibbsite	0.001 mol dm^{-3} NaCl		iep	9.6	S.Goldberg, H.S.Forster, and E.L. Heick. Soil Sci.Soc.Am.J. 57: 704–708 (1993).
Al(OH)$_3$	gibbsite from chloride	0.001 mol dm^{-3} NaCl		iep	9.8	B.A. Manning, and S. Goldberg. Soil Sci.Soc.Am. J. 60: 121 131 (1996).
Al(OH)$_3$	Hydral 710 Alcoa	0.01–0.1 mol dm^{-3} NaCl	25	acousto	9.1	W.N.Rowlands, R.W.O'Brien, R.J.Hunter, and V.Patrick. J.Colloid Interf.Sci. 188: 325–335 (1997).
Al(OH)$_3$	gibbsite, Baker	0.001 mol dm^{-3} NaCl		iep	5	I.Sondi, O.Milat, and V.Pravdic. J.Colloid Interf. Sci. 189: 66–73 (1997).
Al(OH)$_3$	gibbsite	0.01 mol dm^{-3} NaCl		iep	9.0	C.Su, and D.L.Suarez. Clays Clay Min. 45: 814-825 (1997).
Al(OH)$_3$	amorphous, synthetic	0.01 mol dm^{-3} NaCl		iep	9.2	C.Su, and D.L.Suarez. Clays Clay Min. 45: 814-825 (1997).
Al(OH)$_3$	bayerite, or γ (?), synthetic	KNC$_3$		iep cip	9 8.3	M.Szekeres, E.Tombacz, K Ferencz, and I. Dekany. Colloids Surf. A 141: 319 325 (1998).
Al(OH)$_3$	gibbsite			iep	9.41	S.Goldberg, C.Su, and H.S.Forster in Adsorption of Metals by Geomedia (E.A.Jenne, ed.), Academic Press, San Diego 1998. pp. 401-426.

TABLE 3.1 (Continued)

Oxide	Description	Salt	T	Method	pH_0	Source
Al(OH)₃	gibbsite, Hydral 710 Alcoa gibbsite, C31 Alcoa	10^{-4} 10^{-2} mol dm^{-3} NaCl	20– 25	iep	10.4 9.2	J.Addai-Mensah, J.Dawe, R.Hayes, C.Prestidge, and J.Ralston. J.Colloid Interf.Sci. 203: 115 121 (1998).
Al(OH)₃	gibbsite from nitrate, 3 samples	0.005, 0.1 mol dm^{-3} NaNO₃	20	pH	10	T.Hiemstra, H.Yong, and W.H.van Riemsdijk. Langmuir 15: 5942–5955 (1999).
Al(OH)₃	bayerite from nitrate	0.005, 0.1 mol dm^{-3} NaNO₃	20	intersection	9.1	T.Hiemstra, H.Yong, and W.H.van Riemsdijk. Langmuir 15: 5942 5955 (1999).
Al(OH)₃	hydrolysis of AlOCl			iep	10.2 10.5	I.M.Solomentseva, N.G. Gerasimenko, and S. Barany. Colloids Surf.A 151: 113 126 (1999).
Al(OH)₃	bayerite (with or without admixture of gibbsite and boehmite) from chloride, heated at 50 C for 2 h–7d	HCl+NaOH		iep	2 h 8.5 1 d 9.1 7 d 9.3	S.Music, D.Dragcevic, S.Popovic, and N.Vdovic. Mater.Chem.Phys. 59: 12 19 (1999).
Al₁₃ colloid	hydrolysis of AlCl₃	0–1 mol dm^{-3} NaCl		cip iep	8.2 8.4	J.Y.Bottero, and J.M.Cases in Adsorption on New and Modified Inorganic Sorbents, A.Dabrowski, and V.A.Tetrykh, Eds. Elsevier 1996. pp. 319–331.
BeO	reagent, washed	none, 0.001 mol dm^{-3} KCl and CaCl₂	25	iep	7.1	Fouad Taha, A.M.El-Roudi, A.A.Abd El Gaber, and F.M.Zahran. Rev.Roum.Chim. 35: 503–509 (1990).
Be(OH)₂	hydrolysis of BeCl₂ at room T		25	iep	11	P.Benes, and V.Jiranek. Radiochim. Acta 21: 49 53 (1974).
Bi₂O₃ Bi₂O₃	cf. Table 3.8	0.01 mol dm^{-3} KCl		iep	9.3	E.A.Nechaev, and V.N.Sheyin. Kolloidn.Zh. 41: 361–363 (1979).
Bi₂O₃ CdO		0.01 mol dm^{-3} KOH(?) 0.01 mol dm^{-3} KCl		iep iep	9.35 10.5	E.A.Nechaev, and V.N.Sheyin. Kolloidn. Zh. 42: 371 373 (1980). E.A.Nechaev, and V.N.Sheyin. Kolloidn.Zh. 41: 361 363 (1979).
CdO CdO		0.01 mol dm^{-3} KOH (?) 0.01 mol dm^{-3} KCl		iep iep	10.6 10.6	E.A.Nechaev, and V.N.Sheyin. Zh. 42: 371 373 (1980). E.A.Nechaev, and G.V.Zvonareva. Kolloidn.Zh. 42: 511 516 (1980).

Material	Description	Electrolyte	T	Method	iep/cip/pH	Reference
$Cd(OH)_2$	hydrolysis of $Cd(NO_3)_2$ at room T	none	22	iep	11	P.Benes, and K.Kopicka. J.Inorg.Nucl.Chem. 38: 2043–2048 (1976).
$Cd(OH)_2$	hexagonal, hydration of oxide	0.001–0.1 mol dm^{-3} $NaClO_4$	25	pH	10.2 10.4 (no sharp cip)	W.Janusz. J.Colloid Interf.Sci. 145: 119 126 (1991).
CeO_2	hydrolysis of sulfate	10^{-3} 10^{-1} mol dm^{-3} NaCl, $NaNO_3$	35	cip iep	7.6	K.C.Ray, P.K.Sengupta, and S.K.Roy. Indian J.Chem. A 17: 348–351 (1979).
CeO_2		0.01 mol dm^{-3} KCl		iep	8.2	E.A.Nechaev, and V.N.Sheyin. Kolloidn.Zh. 41: 361–363 (1979).
CeO_2	hydrolysis of sulfate, monodispersed	0.001 mol dm^{-3} $NaNO_3$		iep	5.2 rods 6.2 spheres	W.P.Hsu, L.Rönnquist, and E.Matijevic. Langmuir 4: 31–37 (1988).
CeO_2	from nitrate	0.1–0.5 mol dm^{-3} $NaClO_4$	25	cip	8.6 (pure) 10	M.Nabavi, O.Spalla, and B.Cabane. J.Colloid Int.Sci. 160: 459–471 (1993).
CeO_2	decomposition of chloride	0.005–0.3 mol dm^{-3} KNO_3		cip	8.1	L.A.de Faria, and S.Trasatti. J.Colloid Int.Sci. 167: 352–357 (1994).
CeO_2	decomposition of nitrate at 800 C	0.1 mol dm^{-3} $NaNO_3$		mass titration	7.1	J.Park, and J.R.Regalbuto. J.Colloid Int. Sci. 175: 239 252 (1995).
CoO	Co_3O_4 treated at different T	10^{-3} mol dm^{-3} NaCl	25	iep	9.9–10.5 (>1273 C)	S.Kittaka, and T.Morimoto. J.Colloid Int. Sci. 75: 398–403 (1980).
$Co(OH)_2$	synthetic	0.001 mol dm^{-3} KNO_3	25	iep	11	R.O James, and T.W.Healy. J.Colloid Int.Sci. 40: 53–64 (1972).
$Co(OH)_2$		0.01 mol dm^{-3} KNO_3	30	iep	11	P.H.Tewari, and W.Lee.J.Colloid Int.Sci. 52: 77–88 (1975).
$Co(OH)_2$	high purity	KNO_3	25	pH cip iep	8.3 11.4 11.5	P.H. Tewari, and A.B. Campbell. J.Colloid Int.Sci. 55: 531–539 (1976).
Co_3O_4	high purity	KNO_3	25	pH cip iep iep	10.4 11.4 11.4	P.H Tewari, and A.B. Campbell. J.Colloid Int.Sci. 55: 531 539 (1976).
Co_3O_4	monodispersed, cubic, hydrolysis of acetate	10^{-2} mol dm^{-3} KNO_3	25	iep	5.4	T. Sugimoto, and E. Matijevic. J.Inorg.Nucl. Chem. 41: 165 172 (1979).
Co_3O_4	treated at different T	10^{-3} mol dm^{-3} NaCl	25	iep	7.9(600 C)	S. Kittaka, and T. Morimoto. J.Colloid Int.Sci. 75: 398–403 (1980).
Co_3O_4	powder, thermal decomposition of salt	0.001 1 mol dm^{-3} $NaNO_3$	25	pH, merge	5.0	G.A. Kokarev, V.A. Kolesnikov, A.F. Gubin, and A. Korobanov. Elektrokhimiya 18: 466–470 (1982).

TABLE 3.1 (Continued)

Oxide	Description	Salt	T	Method	pH_0	Source
Co_3O_4	nitrate decomposition at various temperatures	0.001–0.2 mol dm⁻³ KNO₃		cip iep	7.5 (200- 500 C) 8.2 (600- 700 C) 9.2 (800 C)	C. Pirovano, and S. Trasatti. J.Electroanal.Chem. 180: 171–184 (1984).
Co_3O_4	Ventron	0.001–0.2 mol dm⁻³ KNO₃		cip iep	8.8- 8.9 8.7	C. Pirovano, and S. Trasatti. J.Electroanal.Chem. 180: 171 184 (1984).
Co_3O_4	active mass of Co₃O₄ electrode	10⁻³–10⁻¹ mol dm⁻³ KCl		iep	7.3	G.A. Kokarev, A.F. Gubin, V.A. Kolesnikov, and S.A. Skobelev. Zh.Fiz.Khim. 59: 1660–1663 (1985).
Co_3O_4	thermal decomposition of basic carbonate, calcined at different T	0.005–0.12 mol dm⁻³ KNO₃		cip≈iep	6.7 600- 700 C 7 7.3 200- 500, and 800 C	S. Ardizzone, G. Spinolo, and S. Trasatti. Electrochem.Acta 40: 2683 2686 (1995).
Co_3O_4	spinel, thermal decomposition of nitrate at 400 C	0.005–0.1 mol dm⁻³ KNO₃		cip	7.6	L.A. de Faria, M. Prestat, J.F. Koenig, P. Chertier, and S. Trasatti. Electrochim. Acta 44: 1481 1489 (1998).
Co_2O_3		0.01 mol dm⁻³ KCl		iep	7	E.A. Nechaev, and V.N. Sheyin. Kolloidn.Zh. 41: 361 363 (1979).
Co_2O_3		0.01 mol dm⁻³ KOH(?)		iep	6.9	E.A. Nechaev. Kolloidn.Zh. 42: 371 373 (1980).
Co_2O_3		0.01 mol dm⁻³ KCl		iep	6.9	E.A. Nechaev, and G.V. Zvonareva. Kolloidn.Zh. 42: 511–516 (1980).
$HCoO_2$	oxidation of β Co(OH)₂	10⁻³ mol dm⁻³ KNO₃		iep	6.2	S. Kittaka, S. Yamanaka, N. Yanagawa, and T. Okabe. Bull.Chem.Soc.Jpn. 63: 1381 1388 (1990).
Cr_2O_3	synthetic, treated at different T α, from nitrate	10⁻³ mol dm⁻³ NaCl KNO₃	25	iep salt titration	9.2 6.35	S. Kittaka. J.Colloid Int.Sci. 48: 327–333 (1974). D.E. Yates, and T.W. Healy. J.Colloid Interf.Sci. 52: 222-228 (1975).
Cr_2O_3		0.01 mol dm⁻³ KCl		iep	7	E.A. Nechaev, and V.N. Sheyin. Kolloidn.Zh. 41: 361 363 (1979).
Cr_2O_3 Cr_2O_3	hexagonal, reagent grade	0.01 mol dm⁻³ KOH (?) 0.001 mol dm⁻³ KNO₃		iep pH	7.25 7.3 7.4	E.A. Nechaev. Kolloidn.Zh. 42: 371 373 (1980). N.I. Ampelogova. Radiokhimiya 25: 579 584 (1983).

Material	Description	Electrolyte	T	Method	pH$_0$	Reference
Cr_2O_3	reagent, washed	none, 0.001 mol dm^{-3} KCl and CaCl$_2$	25	iep	2.25	Fouad Taha, A.M. El-Roudi, A.A. Abd El Gaber, and F.M. Zahran. Rev.Roum.Chim. 35: 503–509 (1990).
Cr_2O_3	reduction of CrO_2	10^{-3} mol dm^{-3} KNO$_3$		iep	6.8	S. Kittaka, S. Yamanaka, N. Yanagawa, and T. Okabe. Bull.Chem.Soc.Jpn. 63: 1381–1388 (1990).
Cr_2O_3	thermal decomposition of α HCrO$_2$	10^{-3} mol dm^{-3} KNO$_3$		iep	6.8	S. Kittaka, S. Yamanaka, N. Yanagawa, and T. Okabe. Bull.Chem.Soc.Jpn. 63: 1381–1388 (1990).
Cr_2O_3	amorphous	10^{-3} mol dm^{-3} KNO$_3$	25	iep	9.2	R.J. Crawford, I.H. Harding, and D.E. Mainwaring. Langmuir 9: 3050–3056 (1993).
Cr_2O_3	α, commercial	0.01 mol dm^{-3} KCl	25	iep	7.9	M.A. Blesa, G. Magaz, J.A. Salfity, and A.D. Weisz. Solid State Ionics 101 103 1235–1241 (1997)
Cr_2O_3		0.001–0.1 mol dm^{-3} NaCl	25	merge iep	6.1	E.V. Golikova, O.M. Ioganson, L.V. Duda, M.G. Osmolovskii, A.I. Yanklovich, and Y.M. Chernoberezhskii. Kolloid.Zh. 60: 188 193 (1998).
$Cr_2O_3 \cdot n H_2O$	monodispersed, spherical		25	iep	7.9	E. Matijevic, A. Bell, R. Brace, and P. McFadyen. J.Elctrochem.Soc. 120: 893–899 (1973).
$Cr_2O_3 \cdot n H_2O$	hydrolysis of chrom alum, monodispersed spherical	0.01 mol dm^{-3}		iep	7.3	A. Garg, and E. Matijevic. Langmuir 4: 38–44 (1988).
$Cr_2O_3 \cdot n H_2O$	hydrolysis of nitrate	10^{-3} mol dm^{-3} KNO$_3$		iep	7.35	S. Kittaka, S. Yamanaka, N. Yanagawa, and T. Okabe. Bull.Chem.Soc.Jpn. 63: 1381–1388 (1990).
$Cr_2O_3 \cdot n H_2O$	synthetic		25	iep	9.2	R.J. Crawford, I.H. Harding, and D.E. Mainwaring. J.Colloid Interf.Sci. 181: 561–570 (1996).
$Cr_2O_3 \cdot n H_2O$	hydrolysis of nitrate	0.01 mol dm^{-3} KCl	25	iep	8.3	M.A. Blesa, G. Magaz, J.A. Salfity, and A.D. Weisz. Solid State Ionics 101 103: 1235–1241 (1997).
$Cr_2O_3 \cdot n H_2O$	monodispersed, spherical from nitrate 3 samples	10^{-2} mol dm^{-3} KCl	25	iep	8.3–8.6	G.E. Magaz, L.G. Rodenas, P.J. Morando, and M.A. Blesa. Croat.Chem.Acta 71: 917 927 (1998).
CrOOH	α, 4 samples: hydrothermal hydrolysis of nitrate at different conditions	10^{-3} mol dm^{-3} KNO$_3$	25	iep pH	6.85 7 7.3 7.55	S. Kittaka, S. Yamanaka, N. Yanagawa, and T. Okabe. Bull.Chem.Soc.Jpn. 63: 1381–1388 (1990).
CrOOH	α, hydrothermal treatment of nitrate	0.01 mol dm^{-3} KCl	25	iep	9.1	M.A. Blesa, G. Magaz, J.A. Salfity, and A.D. Weisz. Solid State Ionics 101–103: 1235–1241 (1997).

TABLE 3.1 (Continued)

Oxide	Description	Salt	T	Method	pH_0	Source
$Cr(OH)_3$	synthetic	10^{-4}–10^{-1} mol dm⁻³ NaClO₄	25	iep	8.4	R. Sprycha, and E. Matijevic. Langmuir 5: 479 485 (1989).
$Cr(OH)_3$	hydrolysis of alum, sulfate free, 2 samples	0.001–0.1 mol dm⁻³ KNO₃		cip coagulation	8.4 7.9 8	C.E. Giacomelli, M.J. Avena, O.R. Camara, and C.P. de Pauli. J.Colloid Interf.Sci. 169: 149 160 (1995).
CrO_2	commercial, DuPont	0.1 mol dm⁻³ KCl		pH	2.59	R. Manoharan, and J.B. Goodenough. Electrochim.Acta 40: 303–307 (1995).
Cu_2O	mineral, washed	none, 0.001 mol dm⁻³ KCl and CaCl₂	25	iep	11.5	Fouad Taha, A.M. El-Roudi, A.A. Abd El Gaber, and F.M. Zahran. Rev.Roum.Chim. 35: 503 509 (1990).
Cu_2O-n H_2O	monodispersed, cubic		25	iep	5	E. Matijevic, A. Bell, R. Brace, and P. McFadyen. J.Electrochem.Soc. 120: 893–899 (1973).
CuO		0.01 mol dm⁻³ KCl		iep	9.6	E.A. Nechaev, and V.N. Sheyin. Kolloidn.Zh. 41: 361–363 (1979).
CuO		0.01 mol dm⁻³ KOH (?)		iep	9.4	E.A. Nechaev. Kolloidn.Zh. 42: 371–373 (1980).
CuO	treated at different T	10^{-3} mol dm⁻³ NaCl	25	iep	9.7 (600 C)	S. Kittaka, and T. Morimoto. J.Colloid Int.Sci. 75: 398–403 (1980).
CuO		0.005–0.1 mol dm⁻³ NaClO₄		iep	9.2 9.6	J.C. Liu, and C.P. Huang. Langmuir 8: 1851–1856 (1992).
CuO	reagent grade, Merck	KNO₃		iep	8.5	G.Gonzalez, S.M. Saraiva, and W. Aliaga. Dispersion Sci.Technol. 15: 123–132 (1994).
$Cu(OH)_2$	from chloride	0.002 mol dm⁻³ NaCl		iep	9.8	R.J. Pugh, and K. Tjus. J.Colloid Interf.Sci. 117: 231 241 (1987).
$Cu(OH)_2$	from nitrate	0.001 mol dm⁻³ NaNO₃	25	iep	10.3	K.K. Das, Pradip, and K.A. Natarajan. J.Colloid Interf.Sci. 196: 1 11 (1997).
Fe_3O_4	cf. Table 3.9					
Fe_3O_4	natural	10^{-3} 1 mol dm⁻³ KNO₃		cip	6.4	S.M. Ahmed, and D. Maksimov. Can.J.Chem. 46: 3841 3846 (1968).
Fe_3O_4	from sulfate	0.005–0.5 mol dm⁻³ KNO₃	25	cip	6.6	P.H. Tewari, and A.W. McLean. J.Colloid Interf. Sci. 40: 267–272 (1972).

Material	Description	Electrolyte	T	Method	Value	Reference
Fe_3O_4	pure iron oxidized in air at 1400 C then heated in air at different T	10^{-3} mol dm^{-3} NaCl	25	iep	200 C 9, 400 C 8.8, 800 C 8.1, 1200 C 5.2, 1400 C 5.1	T. Morimoto, and S. Kittaka. Bull.Chem.Soc.Jpn. 46: 3040-3043 (1973).
Fe_3O_4	natural, 2.4% SiO_2, 0.9% Al_2O_3	0.1 mol dm^{-3} NaCl, KCl	23±1	pH	6.5	S.K. Milonjic, A.L. Ruvarac, and M.V. Susic. Thermochim.Acta 11: 261-266 (1975).
Fe_3O_4		0.01 mol dm^{-3} KNO$_3$	30	iep	7	P.H. Tewari, and W. Lee. J.Colloid Int.Sci. 52: 77-88 (1975).
Fe_3O_4	$FeCl_3 + FeSO_4 + NH_3$	NaCl		pH	6.5	B. Venkataramani, K.S. Venkateswarlu, and J. Shankar. J.Colloid Int.Sci. 67: 187-194 (1978).
Fe_3O_4	Toda, dialysed	$N(CH_3)_4ClO_4$	25	cip	7.1	R.D. Astumian, M. Sasaki, T. Yasunaga, and Z.A. Schelly. J.Phys.Chem. 85: 3832-3835 (1981).
Fe_3O_4	magnetite, from FeO	0.001 mol dm^{-3} KNO$_3$		pH	7.3-7.6	N.I. Ampelogova. Radiokhimiya 25: 579-584 (1983).
Fe_3O_4	heated in autoclave with water at 280 C	0.001 mol dm^{-3} KNO$_3$		pH	6.7	N.I. Ampelogova. Radiokhimiya 25: 579-584 (1983).
Fe_3O_4	synthetic	10^{-3} mol dm^{-3} KNO$_3$	25	iep	6.9	M.A. Blesa, R.M. Larotonda, A.J.G. Maroto, and A.E. Regazzoni. Colloids Surf. 5: 197-208 (1982).
Fe_3O_4	natural magnetite, 8.4% SiO_2	0.01-1 mol dm^{-3} KCl; 0.25 mol dm^{-3} LiCl, NaCl	25	pH	6.4	S.K. Milonjic, M.M. Kopecni, and Z.E. Ilic. J.Radioanal.Chem.Nucl.Chem. 78: 15-24 (1983).
Fe_3O_4				pH	9.96	E.A. Nechaev, and V.A. Volgina, Dep. VINITI 1975, cited after S.K. Milonjic, M.M. Kopecni, and Z.E. Ilic. J.Radioanal.Chem.Nucl.Chem. 78: 15-24 (1983).
Fe_3O_4	synthetic Fe(II)/Fe(III) 0.45	KNO$_3$		cip	6.6	R. Biagotti, Thesis. Cited by A. Daghetti, G. Lodi, and S. Trasatti. Mater.Chem.Phys. 8: 1-90 (1983).
Fe_3O_4	synthetic Fe(II)/Fe(III) 0.35	KNO$_3$		cip	7.2	R. Biagotti, Thesis. Cited by A. Daghetti, G. Lodi, and S. Trasatti. Mater.Chem.Phys. 8: 1-90 (1983).
Fe_3O_4	synthetic magnetite	10^{-3}-10^{-1} mol dm^{-3} KNO$_3$	25, 30, 30	iep, cip, pH	6.85, 6.8, 6.4	A.E. Regazzoni, M.A. Blesa, and A.J.G. Maroto. J.Colloid Int.Sci. 91: 560-570 (1983).

TABLE 3.1 (Continued)

Oxide	Description	Salt	T	Method	pH_0	Source
Fe_3O_4	synthetic magnetite	10^{-3}–10^{-1} mol dm^{-3} KNO$_3$	30	cip	6.8	M.A. Blesa, N.M. Figliolia, A.J.G. Maroto, and A.E. Regazzoni. J.Colloid Int. Sci. 101: 410–418 (1984).
Fe_3O_4	spherical, monodispersed	10^{-3} mol dm^{-3} KNO$_3$		iep	5.1	S. Kittaka, S. Sasaki, and T. Morimoto. J.Mater. Sci. 22: 557–564 (1987).
Fe_3O_4	synthetic	10^{-3} mol dm^{-3} KCl		iep	6.1	U. Künzelmann, H.J. Jacobasch, and G. Reinhard. Werkst. Korrosion 40: 723–728 (1989).
Fe_3O_4	Toda Kogyo, washed	0.05-1 mol dm^{-3} NaNO$_3$, NaCl, NaClO$_4$	25	cip	5.4	H. Moriwaki, Y. Yoshikawa, T. Morimoto. Langmuir 6: 847 850 (1990).
Fe_3O_4	Kanto	0.1 mol dm^{-3} NaNO$_3$	25	pH	6.25	H. Tamura, and R. Furuichi. Bunseki Kagaku. 40: 635–640 (1991).
Fe_3O_4	chemically pure	0.01 mol dm^{-3} NaCl		iep	6.7	Q.Y. Song, F. Xu, and S.C. Tsai. Int.J.Miner.Proc. 34: 219–229 (1992).
Fe_3O_4	Kanto, washed	0.1 mol dm^{-3} NaNO$_3$	25	pH	6.25	H. Tamura, N. Katayama, and R. Furuichi. Environ. Sci. Technol. 30: 1198–1204 (1996).
Fe_3O_4	from FeSO$_4$	HNO$_3$/NaOH		pH	7.2	B.S. Mathur, and B. Venkataramani. Colloids Surfaces A 140: 403–416 (1998).
Fe_3O_4	natural magnetite, 2.4% SiO$_2$	0.01-1 mol dm^{-3} NaNO$_3$		cip	5.5	H. Catalette, J. Dumonceau, and P. Ollar. J. Contamin.Hydrol. 35: 151 159 (1998).
Fe_3O_4	Alfa Aesar 97%	0.001-0.1 mol dm^{-3} NaNO$_3$		cip	8.2	J. Shen, A.D. Ebner, and J.A. Ritter. J.Colloid Interf.Sci. 214: 333 343 (1999).
$Fe_3O_4 \cdot n H_2O$	hydrous synthetic	10^{-1} mol dm^{-3} NaNO$_3$	25	pH	6.47	A.R. Gupta, and B. Venkataramani. Bull.Chem. Soc.Jpn. 61: 1357 1362 (1988).
Fe_2O_3	cf. Table 3.9					
Fe_2O_3	hematite from nitrate	0.001-0.1 mol dm^{-3} NaClO$_4$	22	cip	8.3	G.Y. Onoda, and P.L. de Bruyn. Surf.Sci. 4: 48 63 (1966).
Fe_2O_3	α, hydrolysis of nitrate, 2 samples	0.002-1 mol dm^{-3} KCl	20± 1.5	cip	9.27, 8.9	P.J. Atkinson, A.M. Posner, and J.P. Quirk. J.Phys.Chem. 71: 550 558 (1967).
Fe_2O_3	α, hydrolysis of nitrate, 2 samples	KNO$_3$	20± 1.5	cip	8.45, 8.6	P.J. Atkinson, A.M. Posner, and J.P. Quirk. J.Phys.Chem. 71: 550-558 (1967).

Fe_2O_3	α, natural	0.001–1 mol dm⁻³ KNO₃		merge	5.3	S.M. Ahmed, and D. Maksimov. Can.J.Chem. 46: 3841–3846 (1968).
Fe_2O_3	from nitrate, different T treatment, not purified	0–0.01 mol dm⁻³ KCl, KNO₃	30	iep	6.3–6.9	H. Fukuda, and M. Miura. J.Sci.Hiroshima Univ. Ser.A. 36: 77–86 (1972).
Fe_2O_3	from nitrate, electrodialysis	0–0.01 mol dm⁻³ KCl	30	iep	8.7–8.9	H. Fukuda, and M. Miura. J.Sci.Hiroshima Univ. Ser.A. 36: 77–86 (1972).
Fe_2O_3	hematite, from nitrate	10⁻³–1 mol dm⁻³ KCl; 10⁻³–10⁻² mol dm⁻³ LiCl	20	cip	8.5	A. Breeuwsma, and J. Lyklema. Disc. Faraday Soc. 52: 324–333 (1972).
Fe_2O_3	hematite, synthetic	KCl	20	cip	8.5	A. Breeuwsma, and J. Lyklema. J.Colloid Interf. Sci. 43: 437–448 (1973).
Fe_2O_3	reagent grade	0.002 mol dm⁻³ NaCl	23	iep	8.5	K.N. Han, T.W. Healy, and D.W. Fuerstenau. J.Cclloid Int.Sci. 44: 407–414 (1973).
Fe_2O_3	product of hydrolysis of nitrate dehydrated in air at 600 C then heated in air at different T	10⁻³ mol dm⁻³ NaCl	25	iep	600 C 9.5 800 C 9.5 1200 C 4–4.5 1400 C 3.3–3.9	T. Morimoto, and S. Kittaka. Bull.Chem.Soc.Jpn. 46: 3040–3043 (1973).
Fe_2O_3	synthetic, treated at different T	10⁻³ mol dm⁻³ NaCl		iep	4(>1000 C) 9.3 (<800 C)	S. Kittaka. J.Colloid Int.Sci. 48: 327–333 (1974).
Fe_2O_3	hematite, commercial pigment, after Soxhlet extraction	0.005 mol dm⁻³ NaCl, 0.005 mol dm⁻³ NaNO₃	25	iep	4	D. Balzer, and H. Lange. Colloid Polym. Sci. 255: 140–152 (1977).
Fe_2O_3	hematite, commercial pigment, after Soxhlet extraction	0.005 mol dm⁻³ NaCl	25	iep	6.5	D. Balzer, and H. Lange. Colloid Polym. Sci. 255: 140-152 (1977).
Fe_2O_3	hydrolysis of nitrate			iep	9.1	G.C. Bye, and G.T. Simpkin. J.Appl.Chem. Biotechnol. 28: 116–118 (1978).
Fe_2O_3	α, various shapes from chloride, nitrate, chlorate (VII)			iep	6–6.7	E. Matijevic, and P. Scheiner. J.Colloid Int.Sci. 63: 509 524 (1978).
Fe_2O_3		0.01 mol dm⁻³ KCl		iep	8.2	E.A. Nechaev, and V.N. Sheyin. Kolloidn.Zh. 41: 361–363 (1979).
Fe_2O_3	α, synthetic	0.01 mol dm⁻³ KOH (?)	10	iep	8.15	E.A. Nechaev. Kolloidn. Zh. 42: 371–373 (1980).
Fe_2O_3		0.002 mol dm⁻³ NaCl		iep	6.3	E. Tipping. Geochim. Cosmochim. Acta 45: 191 199 (1981).
Fe_2O_3	hematite, with traces of goethite hydrolysis of $Fe(NO_3)_3$	$N(CH_3)_4ClO_4$	25	cip	8.4	R.D. Astumian, M. Sasaki, T. Yasunaga, and Z.A. Schelly. J.Phys.Chem. 85: 3832 3835 (1981).

TABLE 3.1 (Continued)

Oxide	Description	Salt	T	Method	pH_0	Source
Fe_2O_3	hematite, reagent grade	0.001 mol dm^{-3} KNO_3		pH	7.5	N.I. Ampelogova. Radiokhimiya 25: 579 584 (1983).
Fe_2O_3	hematite, synthetic, 3 samples			salt addition	5.9–7.3	O.K. Borggaard. Clays Clay Min. 31: 230–232 (1983).
Fe_2O_3	maghemite, synthetic, 2 samples			salt addition	7.2; 7.3	O.K. Borggaard. Clays Clay Min. 31: 230–232 (1983).
Fe_2O_3	amorphous, from nitrate			salt addition	7.2	O.K. Borggaard. Clays Clay Min. 31: 230–232 (1983).
Fe_2O_3	hematite, synthetic	0.01 1 mol dm^{-3} NaCl		cip	6.7	L. Madrid, E. Diaz, F. Cabrera, and P. de Arambarri. J.Soil Sci. 34: 57–67 (1983).
Fe_2O_3	hematite synthetic	$NaNO_3$		iep; cip, salt titration	7.2; 7.5	C.K.D. Hsi, and D. Langmuir. Geochim. Cosmochim. Acta 49: 1931–1941 (1985).
Fe_2O_3	hematite natural	$NaNO_3$		iep; cip, salt titration	7.0; 7.8	C.K.D. Hsi, and D. Langmuir. Geochim. Cosmochim. Acta 49: 1931–1941 (1985).
Fe_2O_3	hematite from chloride			iep	7.6	C.H. Ho, and D.C. Doren. Can.J.Chem. 63: 1100–1104 (1985).
Fe_2O_3	hematite, monodispersed	10^{-2} 1 mol dm^{-3} KCl		iep	9.4	N.H.G. Penners, L.K. Koopal, and J. Lyklema. Colloids Surf. 21: 457–468 (1986).
Fe_2O_3	hematite, from nitrate	10^{-2}–1 mol dm^{-3} KCl		cip	9.5; 8.4; purified: 9.5	N.H.G. Penners, L.K. Koopal, and J. Lyklema. Colloids Surf. 21: 457–468 (1986).
Fe_2O_3	α, synthetic, alkali washed	0.01–1 mol dm^{-3} KCl	20	cip; iep	6.7	H. Watanabe, and J. Seto. Bull.Chem.Soc.Jpn. 59: 2683–2687 (1986).
Fe_2O_3	α, synthetic, water washed	0.01–1 mol dm^{-3} KCl	20	cip	3.2	H. Watanabe, and J. Seto. Bull.Chem.Soc.Jpn. 59: 2683–2687 (1986).
Fe_2O_3	α, synthetic, alkali washed, hydrated for 20 h	0.00004 mol dm^{-3} KCl	20	iep	7.9	H. Watanabe, and J. Seto. Bull.Chem.Soc.Jpn. 59: 2683–2687 (1986).
Fe_2O_3	γ, synthetic, water washed	0.01–1 mol dm^{-3} KCl	20	cip	3.3	H. Watanabe, and J. Seto. Bull.Chem.Soc.Jpn. 59: 2683–2687 (1986).
Fe_2O_3	γ, synthetic, alkali washed	0.01–1 mol dm^{-3} KCl	20	cip; iep	5.5	H. Watanabe, and J. Seto. Bull.Chem.Soc.Jpn. 59: 2683–2687 (1986).

	Description	Electrolyte	T	Method	pH	Reference
Fe₂O₃	γ, synthetic, alkali washed, hydrated for 20 h spherical, monodispersed	0.00004 mol dm⁻³ KCl	20	iep	6.8	H. Watanabe, and J. Seto. Bull.Chem.Soc.Jpn. 59: 2683–2687 (1986).
Fe₂O₃		10⁻³ mol dm⁻³ KNO₃		iep	5.1	S. Kittaka, S. Sasaki, and T. Morimoto. J.Mater. Sci. 22: 557 564 (1987).
Fe₂O₃	hematite monodispersed	0.001–0.1 mol dm⁻³ NaNO₃	25	cip / iep	7.2 /	P. Hesleitner, D. Babic, N. Kallay, and E. Matijevic. Langmuir 3: 815–820 (1987).
Fe₂O₃	α, hydrolysis of nitrate	0.01 mol dm⁻³ KCl	25	iep	8.5	A.E. Regazzoni, M.A. Blesa, and A.J.G. Maroto. J.Colloid Interf.Sci. 122: 315–325 (1988).
Fe₂O₃	hematite, reagent grade, Carlo Erba	10⁻³ 10⁻¹ mol dm⁻³ KNO₃, NaNO₃, NaClO₄		cip	8.5 / 8.5 / 8.15	S. Ardizzone. J.Electroanal.Chem. 239: 419–425 (1988).
Fe₂O₃	hematite, synthetic	10⁻³–10⁻¹ mol dm⁻³ KNO₃, NaNO₃, NaClO₄		cip	8.63 / 8.63 / 8.7	S. Ardizzone. J.Electroanal.Chem. 239: 419–425 (1988).
Fe₂O₃	hematite monodispersed from chloride	0.001 mol dm⁻³ NaNO₃		iep	7.9	I. Kobal, P. Hesleitner, and E. Matijevic. Colloids Surf. 33: 167 174 (1988).
Fe₂O₃	hydrolysis of FeCl₃ in the presence of phosphate, monodispersed spindle like	0.01 mol dm⁻³		iep	6.5	A. Garg, and E. Matijevic. Langmuir 4: 38–44 (1988).
Fe₂O₃	α	0.01 mol dm⁻³ KNO₃	25	iep	4.2	L. Vordonis, C. Kordulis, and A. Lycourghiotis. J.Chem.Soc.Faraday Trans.I 84: 1593–1601 (1988).
Fe₂O₃	α, from nitrate	0.005–1 mol dm⁻³ KNO₃	20	cip	8.6	L.G.J. Fokkink, A de Keizer, and J. Lyklema. J.Colloid Int.Sci. 127: 116–131 (1989).
Fe₂O₃	α, from chloride	10⁻³ mol dm⁻³ KCl		iep	7	U. Künzelmann, H.J. Jacobasch, and G. Reinhard. Werkst. Korrosion 40: 723–728 (1989).
Fe₂O₃	γ, from chloride	10⁻³ mol dm⁻³ KCl		iep	5.6	U. Künzelmann, H.J. Jacobasch, and G. Reinhard. Werkst. Korrosion 40: 723–728 (1989).
Fe₂O₃	hematite monodispersed	0.005–0.1 mol dm⁻³ KNO₃	20	iep / cip	9.1 / 8.9	A.W. Gibb and L.K. Koopal. J.Colloid Interf.Sci. 134: 122–138 (1990).
Fe₂O₃	α, calcination of α-FeOOH	0.05–1 mol dm⁻³ NaNO₃, NaCl, NaClO₄	25	cip	7.5	H. Moriwaki, Y. Yoshikawa, and T. Morimoto. Langmuir 6: 847–850 (1990).
Fe₂O₃	hematite, heat treatment of maghemite in air at 550 C, purified by electrodialysis	0.01–1 mol dm⁻³ KCl	25	cip	6.7	H. Watanabe, and J. Seto. Bull.Chem.Soc.Jpn. 63: 2916–2921 (1990).

TABLE 3.1 (Continued)

Oxide	Description	Salt	T	Method	pH$_0$	Source
Fe_2O_3	maghemite dehydration, reduction and oxidation of synthetic goethite, purified by electrodialysis	0.01–1 mol dm^3 KCl	25	cip	5.5	H. Watanabe, and J. Seto. Bull.Chem.Soc.Jpn. 63: 2916–2921 (1990).
Fe_2O_3	maghemite, commercial, purified by electrodialysis	0.01–1 mol dm^{-3} KCl	25	cip	5.5	H. Watanabe, and J. Seto. Bull.Chem.Soc.Jpn. 63: 2916–2921 (1990).
Fe_2O_3	hematite monodispersed, $FeCl_3$ hydrolysis, washed	10^{-2} mol dm^{-3} NaNO$_3$, NaCl, NaClO$_4$	25	cip / iep	7.5 / 7.5	M. Colic, D.W. Fuerstenau, N. Kallay, and E. Matijevic. Colloids Surf. 59: 169–185 (1991).
Fe_2O_3	hematite	0.001–0.01 mol dm^{-3} NaNO$_3$	25	cip / iep	9.1	P. Hesleitner, N. Kallay, and E. Matijevic. Langmuir 7: 178–184; 1554 (1991).
Fe_2O_3	monodispersed from chloride	NaNO$_3$				S. Zalac, and N. Kallay. J.Colloid Int.Sci 149: 233–240 (1992).
Fe_2O_3	hematite, Alfa	NaNO$_3$	25	mass titr. / inflection	5.95 / 5.9	
Fe_2O_3	hematite, monodispersed, spherical	0.001 mol dm^{-3} KCl	25	iep	9.2	K. Furusawa, Z. Shou, and N. Nagahashi. Colloid Polym. Sci. 270: 212–218 (1992).
Fe_2O_3	natural specularite (= hematite)	0.01 mol dm^{-3} NaCl		iep	5.4	Q.Y. Song, F. Xu, and S.C. Tsai. Int.J.Miner.Proc. 34: 219–229 (1992).
Fe_2O_3	from chloride, cubic, monodispersed	10^{-4} mol dm^{-3} KCl	20	iep	3–3.5	K. Kandori, Y. Kawashima, and T. Ishikawa. J.Mater.Sci.Letters 12: 288–290 (1993).
Fe_2O_3	quantum size, hydrolysis of $FeCl_3$, dialysis	NaOH + HCl	25	inflection	7.5	D.W. Bahnemann. Isr.J.Chem. 33: 115–136 (1993).
Fe_2O_3	monodispersed spheres, hydrolysis of $FeCl_3$	KCl / KNO$_3$	25	iep / salt titr. / salt titr.	6 / 8.8 / 8.0	M. Kosmulski, J. Matysiak, and J. Szczypa. Bull.Pol.Acad.Sci.Chem. 41: 333–337 (1993).
Fe_2O_3	maghemite synthetic	0.01 mol dm^{-3} KCl	25	pH	5.5	H. Watanabe, J. Seto and Y. Nishiyama. Bull. Chem.Soc.Jpn. 66: 2751–2753 (1993).
Fe_2O_3	hematite, from chlorate VII				7.4	Z. Zhang, C. Boxall, and G.H. Kelsall. Colloids Surf.A 73: 145 163 (1993).
Fe_2O_3	3 samples: Ventron, Carlo Erba, synthetic	KNO$_3$		cip / iep	8.2 8.6 / 8.2–8.5	M.G. Cattania, S. Ardizzone, C.L. Bianchi, and S. Carella. Colloids Surf. A 76: 233 240 (1993).
Fe_2O_3	Gregory, Bottley & Lloyd	0.01 mol dm^{-3} NaClO$_4$		iep	6.3	G.H. Kelsall, Y. Zhu, and H.A. Spikes. J.Chem. Soc.Faraday Trans. 89: 267 272 (1993).

	Description	Electrolyte	T	Method	pH	Reference
Fe_2O_3	amorphous	10^{-3} mol dm^{-3} KNO$_3$	25	iep	8.2	R.J. Crawford, I.H. Harding, and D.E. Mainwaring. Langmuir 9: 3050–3056 (1993).
Fe_2O_3	natural	KNO$_3$		iep	6.8	G. Gonzalez, S.M. Saraiva, and W. Aliaga. J. Dispersion Sci.Technol. 15: 123–132 (1994).
Fe_2O_3	hematite, monodispersed	10^{-2} mol dm^{-3}		iep	7.5	A. Ben-Taleb, P. Vera, A.V. Delgado, and V. Gallardo. Mater.Chem.Phys. 37: 68–75 (1994).
Fe_2O_3	α		23	iep	3.7, 5.2	P. Jayaweera, S. Hettiarachchi, and H. Ocken. Colloids Surf. A 85: 19–27 (1994).
Fe_2O_3	hematite natural, purified			iep	5.3	Y. Wang, R.J. Pugh, and E. Forssberg. Colloids Surf.A 90: 117 133 (1994).
Fe_2O_3	hematite, Alfa	10^{-4} mol dm^{-3} NaNO$_3$	25	mass titration	6.1	N. Kallay, S. Zalac, J. Culin, U. Bieger, A. Pohlmeier, and H.D. Narres. Progr.Coll.Polym. Sci. 95: 108–112 (1994).
Fe_2O_3	amorphous	0.01 mol dm^{-3} NaCl		iep	7.2	S. Goldberg, H.S. Forster, and C.L. Godfrey. Soil Sci.Soc.Am.J. 60: 425–432 (1996).
Fe_2O_3	Polysciences	0.005 mol dm^{-3} NaNO$_3$	25	iep	4	C. Quang, S.L. Petersen, G.R. Ducatte, and N.E. Ballou. J.Chromatogr.A. 732: 377 384 (1996).
Fe_2O_3	hematite from chloride			iep	8.8	I. ul Haq, and E. Matijevic. J.Colloid Interf.Sci. 192 104-113 (1997).
Fe_2O_3	from chloride, 2 samples	0.005-0.5 mol dm^{-3} NaNO$_3$	25	iep cip coagulation	9.2	M. Schudel, S.H. Behrens, H. Holthoff, R. Kretzschmar, and M. Borkovec. J.Colloid Interf. Sci. 196: 241–253 (1997).
Fe_2O_3	hematite hydrolysis of chloride in presence of phosphate, monodispersed ellipsoidal		25	iep	7	T. Radeva, and I. Petkanchin. J.Colloid Interf.Sci. 196: 87–91 (1997).
Fe_2O_3	hematite	0.0025-0.1 mol dm^{-3} NaCl		cip	8.5	L. Liang. Thesis. quoted after N. Sahai, and D.A. Sverjensky. Geochim. Cosmochim. Acta 61: 2801-2826 (1997).
Fe_2O_3	hematite, synthetic	0.001-0.1 mol dm^{-3} KNO$_3$	21	cip	8.7	A.W.P. Vermeer, W.H. van Riemsdijk, and L.K. Koopal. Langmuir 14: 2810–2819 (1998).
Fe_2O_3	hematite synthetic	0.005-0.1 mol dm^{-3} NaNO$_3$		cip	9.4	P. Venema, T. Hiemstra, P.G. Weidler, and W.H. van Riemsdijk. J.Colloid Interf.Sci. 198: 282–295 (1998).
Fe_2O_3	maghemite, from chloride	0.001-0.1 mol dm^{-3} KNO$_3$	25	cip iep	6.6	L. Garcell, M.P. Morales, M. Andres-Verges, P. Tartaj, and C.J. Serna. J.Colloid Interf.Sci. 205: 470–475 (1998).

TABLE 3.1 (Continued)

Oxide	Description	Salt	T	Method	pH$_0$	Source
Fe$_2$O$_3$	hematite, Reachim, washed	0–1 mol dm^{-3} NaCl	25	cip / pH	8.1 / 8.2–8.7	S. Pivovarov. J.Colloid Interf.Sci. 206: 122 130 (1998).
Fe$_2$O$_3$	natural hematite, 1.8% SiO$_2$	0.001–0.1 mol dm^{-3} NaClO$_4$		cip	6.1	T. Rabung, H. Geckeis, J.I. Kim, and H.P. Beck. J.Colloid Interf.Sci. 208: 153 161 (1998).
Fe$_2$O$_3$	α, calcination of goethite at 415 C for 1 h	0.001–0.1 mol dm^3 NaCl		cip / iep	7.8	E.V. Golikova, O.M. Ioganson, L.V. Duda, M.G. Osmolovskii, A.I. Yanklovich, and Y.M. Chernoberezhskii. Kolloid.Zh. 60: 188 193 (1998).
Fe$_2$O$_3$	spherical, pseudocubic hematite, from chloride	10^{-2} mol dm^{-3} NaNO$_3$	25	iep	7.6	A.V. Delgado, and F. Gonzalez-Caballero. Croat.Chem.Acta 71: 1087 1104 (1998).
Fe$_2$O$_3$	hematite, Alfa	0.001 mol dm^3 KNO$_3$	25	mass titration / iep	6.2	T. Preocanin, and N. Kallay. Croat.Chem.Acta 71: 1117 1125 (1998).
Fe$_2$O$_3$	hematite monodispersed, FeCl$_3$ hydrolysis, washed	10^3 mol dm^{-3} NaNO$_3$	20	iep	9.5	M. Colic, and D.W. Fuerstenau. Powder Technol. 97: 129 138 (1998).
Fe$_2$O$_3$	hematite, ellipsidal, uniform, hydrolysis of chlorate VII in the presence of phosphate, NaOH washed	0.01 mol dm^3 KNO$_3$	(?)	iep	6	M. Stachen, M.P. Morales, M. Ocana, and C.J. Serna. Phys.Chem.Chem.Phys. 1: 4465–4471 (1999).
Fe$_2$O$_3$	α, hydrolysis of FeCl$_3$ at 100 C, washed	10^{-4} 10^{-1} mol dm^{-3} NaCl		iep / cip / pH	6 / 8.3 / 5.5–8	W. Janusz, A. Sworska, and J. Szczypa. Colloids Surf. A 149: 421–426 (1999).
Fe$_2$O$_3$	hematite, synthetic monodispersed	0.006–0.5 mol dm^3 NaClO$_4$	25	cip	9.25	R.J. Murphy, J.J. Lenhart, and B.D. Honeyman. Colloids Surf.A 157: 47–62 (1999).
Fe$_2$O$_3$	hematite, Alfa	0.001 mol dm^{-3} KNO$_3$	25	mass titration	6.2	N. Kallay, T. Preocanin, S. Zalac, H. Lewandowski, and H.D. Narres. J.Colloid Interf.Sci. 211: 401–407 (1999).
Fe$_2$O$_3$	hematite, synthetic monodispersed	0.01–0.27 mol dm^{-3} NaClO$_4$	25	cip	9.4	M. Kohler, B.D. Honeyman, and J.O. Leckie. Radiochim.Acta 85: 33–48 (1999).
Fe$_2$O$_3$·H$_2$O hydrolysis of nitrate		10^{-2} mol dm^3 NaNO$_3$		salt titr.	8.1	D.G. Kinniburgh, J.K. Syers, and M.L. Jackson. Soil Sci.Soc.Amer.Proc. 39: 464–470 (1975).
Fe$_2$O$_3$·H$_2$O from nitrate		NaClO$_4$		cip	8.1	J.C. Ryden, J.R. McLaughlin, and J.K. Syers. J.Soil Sci. 28: 72 92 (1977).

Sample	Electrolyte	T	Method	pH	Reference
$Fe_2O_3 \cdot H_2O$ from chloride	KNO_3		cip	7.8	N. Peinemann, and A.K. Helmy. J.Electroanal. Chem 78: 325- 330 (1977).
$Fe_2O_3 \cdot H_2O$ synthetic	0.0115, 0.112 mol dm^{-3} NaNO$_3$		salt titration	7.9	J.A. Davis, and J.O. Leckie. J.Colloid Int.Sci. 67: 90-107 (1978).
$Fe_2O_3 \cdot H_2O$ from nitrate or chloride, fresh and aged	0.002 mol dm^{-3} NaCl	10	iep	a 8.1 f 8.2	E. Tipping. Geochim. Cosmochim. Acta 45: 191 199 (1981).
$Fe_2O_3 \cdot H_2O$ from nitrate	0.01-1 mol dm^{-3} KNO$_3$	25	cip	8.0	U. Schwertmann, and H. Fechter. Clay Min. 17: 471–476 (1982).
$Fe_2O_3 \cdot H_2O$ from nitrate, chloride and sulfate, 8 different samples	0.01 1 mol dm^{-3} NaCl, NaClO$_4$, KNO$_3$	25	pH (no cip)	4.1 9.9	S.I. Pechenyuk, and E.V. Kalinkina. Kolloidn.Zh. 52: 716-721 (1990).
$Fe_2O_3 \cdot H_2O$ amorphous synthetic	0.001–0.1 mol dm^{-3} NaNO$_3$	25	cip	8.0	D.C. Girvin, L.L. Ames, A.P. Schwab, and J.E. Mc Garrah. J.Colloid Int.Sci. 141: 67 78 (1991).
$Fe_2O_3 \cdot H_2O$ natural limonite	0.01 mol dm^{-3} NaCl		iep	6.6	Q.Y. Song, F. Xu, and S.C. Tsai. Int.J.Miner.Proc. 34: 219-229 (1992).
$Fe_2O_3 \cdot n$ H$_2$O precipitated from nitrate	0.015-0.1 mol dm^{-3} NaNO$_3$	20	cip	8.0	C.D Cox, and and M.M. Ghosh. Water Res. 28: 1181 1188 (1994).
$Fe_2O_3 \cdot H_2O$ hydrolysis of chloride	10^{-3}–10^{-2} mol dm^{-3} NaClO$_4$	20 (?)	cip	8.1	H.C B. Hansen, T.P. Wetche, K. Raulund-Rasmussen, and O.K. Borggaard. Clay Miner. 29: 341 350 (1994).
$Fe_2O_3 \cdot H_2O$ from nitrate	0.1 mol dm^{-1} NaNO$_3$		iep	8.0	T.H Hsia, S.L. Lo, C.F. Lin, and D.Y. Lee. Colloid Surf. A 85: 1 7 (1994).
$Fe_2O_3 \cdot H_2O$ hydrolysis of nitrate	0.001-0.1 mol dm^{-3} NaNO$_3$		cip	9.2	J. Xue, and P.M. Huang. Geoderma 64: 343 356 (1995).
$Fe_2O_3 \cdot H_2O$ synthetic	0.0175 mol dm^{-3} KNO$_3$	25	iep	8.2	R.J. Crawford, I.H. Harding, and D.E. Mainwaring. J.Colloid Interf.Sci. 181: 561 570 (1996).
$Fe_2O_3 \cdot H_2O$ from nitrate	0.01 mol dm^{-3} NaNO$_3$	22 24	pH	6.6-6.7	B. Nowack, J. Lützenkirchen, P. Behra, and L. Sigg. Environ. Sci. Technol. 30: 2397-2405 (1996).
$Fe_2O_3 \cdot H_2O$ hydrolysis of chloride	10^{-3} mol dm^{-3} NaCl	25	pH	6.8	H.F. Ghoneimy, T.N. Morcos, and N.Z. Misak. Colloids Surf. A 122: 13-26 (1997).
$Fe_2O_3 \cdot H_2O$ from nitrate		25		7.1	A. Amirbahman, L. Sigg, and U. von Gunten. J.Colloid Interf.Sci. 194: 194-206 (1997).
$Fe_2O_3 \cdot H_2O$ from nitrate	0.3 mol dm^{-3} NaNO$_3$	25	pH	7.9	P.J. Pretorius, and P.W. Linder. Chem.Spec. Bioavail. 10: 115-119 (1998).
$Fe_2O_3 \cdot H_2O$ from nitrate			cip	8.5	K.P. Raven, A. Jain, and R.H. Loeppert. Environ. Sci.Technol. 32: 344-349 (1998).

TABLE 3.1 (Continued)

Oxide	Description	Salt	T	Method	pH_0	Source
Fe₂O₃ + FeOOH FeOOH	maghemite + goethite, oxidation of Fe(OH)₂ from chlorate VII with air cf. Table 3.8	0.01 1 mol dm⁻³ NaCl		cip	7	C. Liu, and P.M. Huang, Soil Sci.Soc.Am.J. 63: 65–72 (1999).
FeOOH	α, hydrolysis of nitrate	0.002–1 mol dm⁻³ KCl	20	cip	7.55	R.J. Atkinson, A.M. Posner, and J.P. Quirk. J.Phys.Chem. 71: 550–558 (1967).
FeOOH	goethite, 4 samples	0.01 1 mol dm⁻³ NaCl		cip	7.8 8.0 8.3	F.J. Hingston, A.M. Posner, and J.P. Quirk. J.Soil Sci. 23: 177 192 (1972).
FeOOH	goethite, from nitrate	10⁻³–10⁻¹ mol dm⁻³ KNO₃	25	cip	7.5	D.E. Yates, and T.W. Healy. J.Colloid Interf.Sci. 52: 222–228 (1975).
FeOOH	goethite precipitated from nitrate	0.1 mol dm⁻³ NaCl	20	pH	8.7	L. Madrid, and P. De Arambarri. Geoderma 21: 199–208 (1978).
FeOOH	β, rods from FeCl₃			iep	7.3	E. Matijevic, and P. Scheiner. J.Colloid Int.Sci. 63: 509–524 (1978).
FeOOH	goethite synthetic	0.001–0.1 mol dm⁻³ NaI	25	cip pH	9.7 9.3–9.5	T.D. Evans, J.R. Leal, and P.W. Arnold. J. Electroanal.Chem. 105: 161 167 (1979).
FeOOH	goethite, synthetic	10⁻¹ mol dm⁻³ NaClO₄		pH	7	L. Sigg, and W. Stumm. Colloids Surf. 2: 101–117 (1980).
FeOOH	goethite, synthetic, 3 samples	0.002 mol dm⁻³ NaCl	10	iep	8.4 8.2 7	E. Tipping. Geochim. Cosmochim. Acta 45: 191 199 (1981).
FeOOH	goethite, Toda, dialysed	N(CH₃)₄ClO₄	25	cip	8.2	R.D. Astumian, M. Sasaki, T. Yasunaga, and Z.A. Schelly. J.Phys.Chem. 85: 3832 3835 (1981).
FeOOH	goethite, synthetic, 2 samples			salt addition	7.2; 7.6	O.K. Borggaard. Clays Clay Min. 31: 230–232 (1983).
FeOOH	akaganeite synthetic			salt addition	7.2	O.K. Borggaard. Clays Clay Min. 31: 230–232 (1983).
FeOOH	lepidocrocite, synthetic, 2 samples			salt addition	5.8, 7.1	O.K. Borggaard Clays Clay Min. 31: 230–232 (1983).
FeOOH	feroxyhite = δ, synthetic			salt addition	7.5	O.K. Borggaard Clays Clay Min. 31: 230–232 (1983).

		Electrolyte	t	Method	pH	Reference
FeOOH	goethite, synthetic	0.01–1 mol dm⁻³ NaCl		cip	8.5	L. Madrid, E. Diaz, F. Cabrera, and P. de Arambarri. J.Soil Sci. 34: 57–67 (1983).
FeOOH	lepidocrocite, synthetic	0.01 1 mol dm⁻³ NaCl		cip	7.4	L. Madrid, E. Diaz, F. Cabrera, and P. de Arambarri. J.Soil Sci. 34: 57–67 (1983).
FeOOH	synthetic amorphous	0.001 mol dm⁻³	22.5	iep	9.2	M. Escudey, and G. Galindo. J.Colloid Int.Sci. 93: 78–83 (1983).
FeOOH	α, from nitrate	0–0.1 mol dm⁻³ NaCl		cip	8.7	R. Paterson, and H. Rahman. J.Colloid Interf.Sci. 98: 494–499 (1984).
FeOOH	goethite from chlorate VII	9×10^{-4}–9×10^{-3} mol dm⁻³ NaClO₄		iep	9	W.F. Bleam, and M.S. McBride. J.Colloid Interf. Sci. 103: 124–132 (1985).
FeOOH	goethite synthetic	10⁻³ mol dm⁻³ KNO₃	25	iep	8.4	E.H Rueda, R.L. Grassi, and M.A. Blesa. J.Colloid Int.Sci. 106: 243–246 (1985).
FeOOH	goethite synthetic	0.001–0.7 mol dm⁻³ NaNO₃	25	iep; cip, salt titration	8.9; 8.5	C.K.D. Hsi, and D. Langmuir. Geochim. Cosmochim. Acta. 49: 1931–1941 (1985).
FeOOH	goethite synthetic			iep	9.5	M.I Tejedor-Tejedor, and M.A. Anderson. Langmuir 2: 302–210 (1986).
FeOOH	goethite, from nitrate	NaNO₃	25	cip	8	M.L. Machesky, and M.A. Anderson. Langmuir 2: 582 587 (1986).
FeOOH	goethite	NaClO₄	25	cip	8.1	W.A. Zeltner, E.C. Yost, M.L. Machesky, M.I. Tejedor-Tejedor, and M.A. Anderson. Geochemical Processes at Mineral Surfaces, J.A. Davis, K.F. Hayes, Eds. ACS Symp. Series 323: 142–161 (1986).
FeOOH	goethite synthetic	0.01 1 mol dm⁻³ NaCl		cip	8.6 (text); 9.5 (Fig. 1)	P.M. Bloesch, L.C. Bell, and J.D. Hughes. Aust.J.Soil Res. 25: 377–390 (1987).
FeOOH	β, from chloride and urea	0.001–0.1 mol dm⁻³ KNO₃, NaNO₃, LiNO₃ KCl		cip; iep; cip	7.6; 7.5; 8.2	K.M. Parida. Ads. Sci. Technol. 3: 89–94 (1987).
FeOOH	goethite from nitrate	NaClO₄	25	iep; cip	9.7; 9	W.A. Zeltner, and M.A. Anderson. Langmuir 4: 469–474 (1988)
FeOOH	goethite, hydrolysis of nitrate	0.005–0.1 mol dm⁻³ NaNO₃	21	merge	10.2	T. Hiemstra, J.C.M. deWit, and W.H. van Riemsdijk. J.Colloid Interf.Sci. 133: 105–117 (1989).

TABLE 3.1 (Continued)

Oxide	Description	Salt	T	Method	pH_0	Source
FeOOH	goethite from sulfate	10^{-3} mol dm^{-3} KCl		iep	9.1	U. Künzelmann, H.J. Jacobasch, and G. Reinhard. Werkst. Korrosion 40: 723–728 (1989).
FeOOH	goethite synthetic			pH	8	M.L. Machesky, B.L. Bischoff, and M.A. Anderson. Env. Sci. Technol. 23: 580–587 (1989).
FeOOH	β, from FeCl$_3$	0.001–1 mol dm^{-3} KNO$_3$	25 27	iep cip salt titration	7.2 7.2 7.0	S.B. Kanungo, and D.M. Mahapatra. Colloids Surf. 42: 173–189 (1989).
FeOOH	goethite, hydrolysis of nitrate		25	iep	9.1	B.B. Johnson. Environ. Sci. Technol. 24: 112–118 (1990).
FeOOH	goethite from chloride			iep	10	I. Nirdosh, W.B. Trembley, and C.R. Johnson. Hydrometallurgy 24: 237–248 (1990).
FeOOH	natural, washed	none, 0.001 mol dm^{-3} KCl and CaCl$_2$	25	iep	6.6	Fouad Taha, A.M. El-Roudi, A.A. Abd El Gaber, and F.M. Zahran. Rev. Roum. Chim. 35: 503–509 (1990).
FeOOH	α, Fe(CO)$_5$ + H$_2$O$_2$			pH	6.8	M. Parkanyi-Berka, and P. Joo. Colloids Surf. 49: 165–182 (1990).
FeOOH	α, Toda Kogyo, washed	0.05–1 mol dm^{-3} NaNO$_3$, NaCl, NaClO$_4$	25	cip	5.6	H. Moriwaki, Y. Yoshikawa, and T. Morimoto. Langmuir 6: 847–850 (1990).
FeOOH	synthetic goethite	0.005–0.087 mol dm^{-3} NaNO$_3$	25	cip	8.6	K.F. Hayes, G. Redden, W. Ela, and J.O. Leckie. J.Colloid Int.Sci. 142: 448–469 (1991).
FeOOH	goethite from nitrate	0.001–0.01 mol dm^{-3} NaNO$_3$ NaClO$_4$		cip	7.9	A.B. Ankomah. Clays Clay Min. 39: 100–102 (1991).
FeOOH	α, synthetic			iep	5.6	T.F. Barton, T. Price, K. Becker, and J.G. Dillard. Colloids Surf. 53: 209–222 (1991).
FeOOH	goethite synthetic	10^{-3} mol dm^{-3} KNO$_3$	20	iep	8.4	M. Djafer, R.K. Khandal, and M. Terce Colloids Surf. 54: 209–218 (1991).
FeOOH	goethite from nitrate	KNO$_3$	25	pH iep	8.6	I. Lamy, M. Djafer, and M. Terce Water, Air Soil Pol. 57–58: 457–465 (1991).
FeOOH	goethite synthetic	0.1 mol dm^{-3} NaNO$_3$		pH	8.6	T. Hiemstra, and W.H. van Riemsdijk. Colloids Surf. 59: 7–25 (1991).

Formula	Description	Electrolyte	T	Method	pH	Reference
FeOOH	α, synthetic		25	cip	9.3	K. Mesuere, and W. Fish. Environ. Sci. Technol. 26: 2357-2364 (1992).
FeOOH	goethite from nitrate	$0.001-0.1$ mol dm^{-3} NaNO$_3$	25	pH, merge	7.8	N.J. Barrow, and V.C. Cox. J.Soil Sci. 43: 295-304 (1992).
FeOOH	goethite, from nitrate		25	cip	8.3	D.F. Rodda, B.B. Johnson, and J.D. Wells. J.Colloid Interf.Sci. 161: 57-62 (1993).
FeOOH	amorphous synthetic	$10^{-3}-10^{-2}$ mol dm^{-3} KNO$_3$ $10^{-3}-0.5$ mol dm^{-3} NaCl	25	iep cip iep	7 7.3 6.5	S.B Kanungo. J.Colloid Interf.Sci. 162: 86-92 (1993).
FeOOH	akaganeite synthetic	$0.001-0.5$ mol dm^{-3} NaCl KNO$_3$	25	cip iep cip iep	7 7.2 7.2 7.2	S.B. Kanungo. J.Colloid Interf.Sci. 162: 86-92 (1993).
FeOOH	goethite, hydrolysis of nitrate	NaNO$_3$	25	cip	7.9	A.T. Stone, A. Torrents, J. Smolen, D. Vasudevan, and J. Hadley. Environ.Sci.Technol. 27: 895-909 (1993).
FeOOH	goethite, synthetic	0.001 mol dm^{-3} NaCl		iep	8.5	S. Goldberg, H.S. Forster, and E.L. Heick. Soil Sci.Soc.Am.J. 57: 704-708 (1993).
FeOOH	goethite	NaNO$_3$		cip	7.9	U. Hoins, thesis, quoted after A. Scheidegger, M. Borkovec, and H. Sticher. Geoderma 58: 43-65 (1993).
FeOOH	goethite, Research Organic/Inorganic Chemical Corp., washed	$0.01-1$ mol dm^{-3} NaNO$_3$		cip	7.8	U. Hoins, L. Charlet, and H. Sticher. Water Air Soil Poll. 68: 241 255 (1993).
FeOOH	goethite, from nitrate	0.001 1 mol dm^{-3} NaCl	25	cip	9.1	D.G. Lumsdon, and L.J. Evans. J.Colloid Interf. Sci. 164: 119-125 (1994).
FeOOH	goethite from nitrate	$0.01-0.1$ mol dm^{-3} NaClO$_4$	25	cip	8.9	A. van Geen, A.P. Robertson, and J.O. Leckie. Geochim. Cosmochim. Acta 58: 2073-2086 (1994).
FeOOH	goethite, from nitrate	$0.005-0.05$ mol dm^{-3} NaClO$_4$		cip	8.4	R.S. Rundberg, Y. Albinsson, and K. Vannerberg. Radiochim. Acta 66/67: 333-339 (1994).
FeOOH	goethite from nitrate	0.001 mol dm^{-3} KNO$_3$	25	pH iep	9.3 9.1	K.M. Spark, B.B. Johnson, and J.D. Wells. Eur.J.Soil.Sci. 46: 621-631 (1995).
FeOOH	goethite, hydrolysis of nitrate	0.005 mol dm^{-3} NaClO$_4$	22	iep	8.6	V.A. Hackley, R.S. Premachandran, S.G. Malghan, and S.B. Schiller. Colloids Surf. A 98: 209-224 (1995).
FeOOH	goethite, hydrolysis of nitrate	$0.005-0.1$ mol dm^{-3} NaNO$_3$		merge iep	9.5 9.4	T. Hiemstra, and W.H. van Riemsdijk. J.Colloid Interf.Sci. 179: 488-508 (1996).

TABLE 3.1 (Continued)

Oxide	Description	Salt	T	Method	pH$_0$	Source
FeOOH	goethite from nitrate	0.005–0.1 mol dm^{-3} NaNO$_3$		merge	9.3	P. Venema, T. Hiemstra, and W.H. van Riemsdijk. J.Colloid Interf.Sci. 183: 515–527 (1996).
FeOOH	goethite synthetic		25	cip	8.3	D.P. Rodda, J.D. Wells, and B.B. Johnson. J.Colloid Interf.Sci. 184: 564–569 (1996).
FeOOH	goethite, from nitrate	0.005–0.1 mol dm^{-3} NaCl	20	cip	8	M.A. Ali, and D.A. Dzombak. Environ. Sci. Technol. 30: 1061–1071 (1996).
FeOOH	lepidocrocite synthetic	0.01 mol dm^{-3} NaNO$_3$	22–24 pH		6.5	B. Nowack, J. Lützenkirchen, P. Behra, and L. Sigg. Environ. Sci. Technol. 30: 2397–2405 (1996).
FeOOH	goethite from nitrate	0.001 mol dm^{-3} NaCl		iep	8.7	B.A. Manning, and S. Goldberg. Soil Sci.Soc. Am.J. 60: 121–131 (1996).
FeOOH	goethite poorly crystalline	0.01 mol dm^{-3} NaCl		iep	7.8	S. Goldberg, H.S. Forster, and C.L. Godfrey. Soil.Sci.Soc.Am.J. 60: 425–432. (1996).
FeOOH	α, aging of Fe(OH)$_3$ suspension	0.01 mol dm^{-3} KCl	25	iep	8.4	M.A. Blesa, G. Magaz, and J.A. Salfity, A.D. Weisz. Solid State Ionics 101–103: 1235–1241 (1997).
FeOOH	goethite from nitrate	0.01 mol dm^{-3} NaCl		iep	9.3	C. Su, and D.L. Suarez. Clays Clay Min. 45: 814–825 (1997).
FeOOH	goethite (or β') from nitrate	0.05 mol dm^{-3} NaClO$_4$		iep	6.7	A.J. Fairhurst, and P. Warwick. Colloids Surf. A 145: 229–234 (1998).
FeOOH	goethite synthetic	0.005–0.1 mol dm^{-3} NaNO$_3$		merge	9.3	P. Venema, T. Hiemstra, P.G. Weidler, and W.H. van Riemsdijk. J.Colloid Interf.Sci. 198: 282–295 (1998).
FeOOH	lepidocrocite, synthetic	0.005–0.1 mol dm^{-3} NaNO$_3$		cip	8.0	P. Venema, T. Hiemstra, P.G. Weidler, and W.H. van Riemsdijk. J.Colloid Interf.Sci. 198: 282–295 (1998).
FeOOH	goethite, synthetic			iep	8.82	S. Goldberg, C. Su, and H.S. Forster in Adsorption of Metals by Geomedia (E.A. Jenne, ed.), Academic Press, San Diego 1998. pp. 401–426.
FeOOH	α	0.001 0.1 mol dm^{-3} NaCl		cip iep	7.6	E.V. Golikova, O.M. Ioganson, L.V. Duda, M.G. Osmolovskii, A.I. Yanklovich, and Y.M. Chernoberezhskii. Kolloid.Zh. 60: 188–193. (1998)

Material	Description	T (°C)	Method	Electrolyte	pH	Reference
FeOOH	goethite from nitrate	27	iep		7.5	K.A. Matis, A.I. Zouboulis, D. Zamboulis, and A.V. Valtadorou. Water, Air Soil Poll. 111: 297–316 (1999).
FeOOH	goethite hydrolysis of nitrate	25	iep	NaCl	8.0	B.C. Barja, M.I. Tejedor-Tejedor, and M.A. Anderson. Langmuir 15: 2316–2321 (1999).
FeOOH	goethite from nitrate		cip	0.005–0.1 mol dm^{-3} NaClO$_4$	8.4	D. Peak, R.G. Ford, and D.L. Sparks. J.Colloid Interf.Sci. 218: 289–299 (1999).
FeOOH	from nitrate, amorphous $\cdot n\,H_2O$	25	iep / cip / salt titration	0.001–1 mol dm^{-3} KNO$_3$	6.5 (7.1?) / 7.15 / 6.95	S.B. Kanungo, and D.M. Mahapatra. Colloids Surf. 42: 173–189 (1989).
Fe(OH)$_3$	cf. Table 3.9					
Fe(OH)$_3$	amorphous from sulfate	25	iep	0.01 mol dm^{-3} NaClO$_4$	9.8	M.A. Anderson, and D.T. Malotky. J.Colloid Int.Sci. 72: 413–427 (1979).
Fe(OH)$_3$	amorphous, synthetic	25	cip, salt titration	0.001–0.1 mol dm^{-3} NaNO$_3$	7.9	C.K.D. Hsi, and D. Langmuir. Geochim. Cosmochim. Acta 49: 1931–1941 (1985).
Fe(OH)$_3$	amorphous from nitrate		iep	0.001 mol dm^{-3} NaCl	7.2	S. Goldberg, and R.A. Glaubig. Clays Clay Miner. 35: 220–227 (1987).
Fe(OH)$_3$			salt addition		7.2	P.R. Anderson, and M.M. Benjamin. Environ. Sci. Techn. 24: 1586–1592 (1990).
Fe(OH)$_3$	precipitated	25	iep	0.001 mol dm^{-3} NaNO$_3$	8	K.K. Das, Pradip, and K.A. Natarajan. J.Colloid Interf.Sci. 196: 1–11 (1997).
Fe(OH)$_3$	amorphous, synthetic		iep	0.01 mol dm^{-3} NaCl	8.5	C. Su, and D.L. Suarez. Clays Clay. Min. 45: 814–825 (1997).
Fe(OH)$_3 \cdot H_2O$			merge	0.01–1 mol dm^{-3} NaNO$_3$	8	M.A.F. Pyman, J.W. Bowden, and A.M. Posner. Clay Min. 14: 87–92 (1979).
Ga$_2$O$_3$			iep	0.01 mol dm^{-3} KOH (?)	7.5	E.A. Nechaev. Kolloidn. Zh. 42: 371–373 (1980).
Ga$_2$O$_3$			from acidity constants	0.1 mol dm^{-3} KNO$_3$	6.95	B.I. Lobov, L.A. Rubina, I.G. Vinogradova, and I.F. Mavrin. Zh. Neorg. Khim. 34: 2495–2498 (1989).
HfO$_2$	from chloride		iep		5.02	L.G. Maidanovskaya, and T.V. Skripko. Zh. Fiz. Khim. 46: 115–118 (1972).
HfO$_2$	reagent grade, washed	25	salt titr. / salt titr. / cip / iep	KCl / NaClO$_4$ / 10^{-4}–10^{-1} mol dm^{-3} NaCl / 2×10^{-3} mol dm^{-3} NaClO$_4$	7.6 / 7.4 / 7.6 / 7.1	M. Kosmulski. Langmuir 13: 6315–6320 (1997).

TABLE 3.1 (Continued)

Oxide	Description	Salt	T	Method	pH_0	Source
HfO_2	Aldrich	10^{-3}–10^{-1} mol dm^{-3} NaCl		cip	7.8	W. Janusz. Ads. Sci. Technol. 18: 117-134 (2000).
HgO	cf. Table 3.9					
HgO	yellow synthetic	10^{-4}–10^{-2} mol dm^{-3} KNO$_3$		iep	5.8	M.G. MacNaughton, and R.O. James. J.Colloid Interf.Sci. 47: 431-440 (1974).
In_2O_3		0.01 mol dm^{-3} KOH (?)		iep	7.25	E.A. Nechaev. Kolloidn.Zh. 42: 371-373 (1980).
In_2O_3		0.01 mol dm^{-3} KCl		iep	7.25	E.A. Nechaev, and G.V. Zvonareva. Kolloidn.Zh. 42: 511-516 (1980).
In_2O_3		0.1 mol dm^{-3} KNO$_3$		from acidity constants	7.00	B.I. Lobov, L.A. Rubina, I.G. Vinogradova, and I.F. Mavrin.Zh.Neorg.Khim. 34: 2495-2498 (1989).
In_2O_3	calcination of In(OH)$_3$ at 320 C	0.1 mol dm^{-3} NaNO$_3$	25	pH	7.7	S. Hamada, Y. Kudo, and K. Minagawa. Bull. Chem.Soc.Jpn. 63: 102-107 (1990).
In_2O_3	calcination of In(OH)$_3$	0.1 mol dm^{-3} NaNO$_3$	25	pH	7	S. Hamada, Y. Kudo, and T. Kobayashi. Colloids Surf. A 79: 227-232 (1993).
In(OH)$_3$	monodispersed, hydrolysis of nitrate	0.1 mol dm^{-3} NaNO$_3$	25	pH	7.7	S. Hamada, Y. Kudo, and K. Minagawa. Bull. Chem.Soc.Jpn. 63: 102-107 (1990).
In(OH)$_3$	monodispersed, hydrolysis of 2-aminobutyrate	0.1 mol dm^{-3} NaNO$_3$	25	pH	7.6	S. Hamada, Y. Kudo, and T. Kobayashi. Colloids Surf. A 79: 227-232 (1993).
IrO_2	cf. Tables 3.8 and 3.9					
IrO_2	electrode, thermal decomposition of salt, washed	0.001-1 mol dm^{-3} NaNO$_3$	25	pH	5.6	G.A. Kokarev, V.A. Kolesnikov, A.F. Gubin, and A. Korobanov. Elektrokhimiya 18: 466-470 (1982).
IrO_2		0.4 mol dm^{-3} KCl		pH	3.96	J.B. Goodenough, R. Manoharan, and M. Paranthaman. J.Am.Chem.Soc. 112: 2076-2082 (1990).
IrO_2	synthetic	0.01 mol dm^{-3} KNO$_3$		iep	3.4	A. Vitins, O.A. Petri, B.B. Damaskin, Y.A. Ermakov, A. Grzejdziak, E.L. Kolomnikova, S.A. Sukhishvili, and A.A. Yaroslavov. Elektrokhimiya 28: 404-413 (1992).
La_2O_3		0.01 mol dm^{-3} KCl		iep	9.4	E.A. Nechaev, and V.N. Sheyin. Kolloidn.Zh. 41: 361-363 (1979).

Adsorbent	Description	Electrolyte	T	Method	pH	Reference
La₂O₃		0.01 mol dm⁻³ KOH (?)		iep	9.5	E.A. Nechaev. Kolloidn.Zh. 42: 371-373 (1980).
La₂O₃		0.01 mol dm⁻³ KCl		iep	9.5	E.A. Nechaev, and G.V. Zvonareva. Kolloidn.Zh. 42: 511-516 (1980).
La₂O₃	from nitrate	10⁻³–10⁻¹ mol dm⁻³ NaCl	35	cip, iep	9.6	S.K. Roy, and P.K. Sengupta. J.Colloid Interf.Sci. 125: 340-343 (1988).
La₂O₃	reagent, washed	none, 0.001 mol dm⁻³ KCl and CaCl₂	25	iep	6.7	Fouad Taha, A.M. El-Roudi, A.A. Abd El Gaber, and F.M. Zahran. Rev. Roum. Chim. 35: 503-509 (1990).
La₂O₃	American Chemicals, calcined at 600 C	none	25	mass titration	10.34	S. Subramanian, M.S. Chattha, and C.R. Peters. J.Molecul.Catalysis 69: 235-245 (1991).
MgO	precipitated MgSO₄ + NaOH, washed, but very likely containing sulfates	0.02 mol dm⁻³ NaNO₃		iep	10.8	H. Schott. J.Pharmac.Sci. 70: 486-488 (1981).
MgO	Baker Baikowski				12.0 / 12.0	W.C. Hasz. Thesis. quoted after W.M. Mullins, and B.L. Averbach. Surface Sci. 206: 41-51 (1988).
MgO	reagent, washed	none, 0.001 mol dm⁻³ KCl and CaCl₂	25	iep	9.8	Fouad Taha, A.M. El-Roudi, A.A. Abd El Gaber, and F.M. Zahran. Rev. Roum. Chim. 35: 503 509 (1990).
MgO	Mg(OH)₂ calcined at 700 C				11	G. Deo, and I.E. Wachs. J.Phys.Chem. 95: 5889-5895 (1991).
MgO	Nacalai Tesque	KNO₃		pH	12.7	P.K. Ahn, S. Nishiyama, S. Tsuruya, and M. Masai. Appl. Catalysis A 101: 207-219 (1993).
MgO	Merck, reagent grade	10⁻³ mol dm⁻³ NaCl	25	iep	10.8	G. Tari, J.M.F. Ferreira, and O. Lyckfeldt. J.Eur.Ceram Soc. 17: 1341-1350 (1997).
MgO	Rhone Poulenc	0.1 mol dm⁻³ NaCl	25	mass titration	11.5	J.P. Reymond, and F. Kolenda. Powder Techn. 103: 30-36 (1999).
Mg(OH)₂	cf. Table 3.9					
Mg(OH)₂	natural brucite	0.01 mol dm⁻³ KCl	25	iep	14.1	W.J. McLaughlin, J.L. White, and S.L. Hem. J.Colloid Interf.Sci. 157: 113-123 (1993).
Mg(OH)₂	commercial	0.01 mol dm⁻³ KCl	25	iep	13.2	W.J. McLaughlin, J.L. White, and S.L. Hem. J.Colloid Interf.Sci. 157: 113-123 (1993).
MnOₓ	cf. Table 3.9					
MnO	manganosite heating of MnCO₃ for 3 h at 900-1000 C in hydrogen	0.001-0.01 mol dm⁻³ NaCl		iep	5.5	M.A. Arafa, S.M.R. El-Nozahi, and A.A. Youssef. Neue Hutte 37: 451-457 (1992).

TABLE 3.1 (Continued)

Oxide	Description	Salt	T	Method	pH_0	Source
Mn_3O_4	spherical, monodispersed	10^{-3} mol dm^{-3} KNO$_3$		iep	3.9–4.3	S. Kittaka, S. Sasaki, and T. Morimoto. J.Mater. Sci. 22: 557–564 (1987).
MnO_x	x = 1.03–1.19, nominally Mn$_3$O$_4$, 3 samples			iep	3.2 3.3 4.5	T. Morimoto, H. Nakahata, and S. Kittaka. Bull.Chem.Soc.Jpn. 51: 3387–3388 (1978).
Mn_3O_4 + Mn_2O_3	hydrolysis of Mn (II) pentanedionate			iep	6	I. ul Haq, and E. Matijevic. J.Colloid Interf.Sci. 192: 104–113 (1997).
MnO_x	x = 1.21–1.27, nominally α-Mn$_2$O$_3$			iep	8.17	T. Morimoto, H. Nakahata, and S. Kittaka. Bull.Chem.Soc.Jpn. 51: 3387–3388 (1978).
Mn_2O_3	α, bixbyite heating of MnO for 2 h at 700–850 C in oxygen	0.001–0.01 mol dm^{-3} NaCl		iep	4.7	M.A. Arafa, S.M.R. El-Nozahi, and A.A. Youssef. Neue Hutte 37: 451–457 (1992).
$MnO_{1.75}$	initially amorphous, different T treatment	10^{-3} mol dm^{-3} NaCl	25	iep	3.8 (200 C) 4.8 (400 C) 8.8 (600–800 C, conv. to α) 4 (1000–1400 C, conv. to (Mn$_3$O$_4$))	T. Morimoto, and S. Kittaka. Bull.Chem.Soc.Jpn. 47: 1586–1588 (1974).
MnO_2 + MnO	hydrolysis of Mn (II) methoxide			iep	5	I. ul Haq, and E. Matijevic. J.Colloid Interf.Sci. 192: 104–113 (1997).
MnO_2 + Mn_2O_3	hydrolysis of Mn (II) pentanedionate, then treatment with H$_2$O$_2$			iep	6	I. ul Haq, and E. Matijevic. J.Colloid Interf.Sci. 192: 104–113 (1997).
$MnO_{1.8–1.9}$	δ, nearly amorphous, 2 samples containing 2–4% K$_2$O	10^{-4}–1 mol dm^{-3} NaCl, KNO$_3$	30	pH iep	≈3	S.B. Kanungo, and D.M. Mahapatra. J.Colloid Interf.Sci. 131: 103–111 (1989).
$3MnO_2$· Mn(OH)$_2$ ·nH$_2$O	synthetic	none, shift in the iep in the presence of NaNO$_3$		iep (extrapolated) coagulation	2	T.W. Healy, A.P. Herring, and D.W. Fuerstenau. J.Colloid Int.Sci. 21: 435–444 (1966).
MnO_2	cf. Table 3.8					
MnO_2	δ, synthetic	none (shift in the iep in the presence of NaNO$_3$)		coagulation	1.5±0.5	T.W. Healy, A.P. Herring, and D.W. Fuerstenau. J.Colloid Int.Sci. 21: 435–444 (1966).

			T	method	value	reference
MnO_2	α, synthetic	none, shift in the iep in the presence of $NaNO_3$		iep	4.6±0.2	T.W. Healy, A.P. Herring, and D.W. Fuerstenau. J.Colloid Int.Sci. 21: 435–444 (1966).
MnO_2	γ, Union Carbide	up to 1 mol dm^{-3} $NaNO_3$		coagulation, iep	5.6±0.2	T.W. Healy, A.P. Herring, and D.W. Fuerstenau. J.Colloid Int.Sci. 21: 435–444 (1966).
MnO_2	β, Baker	up to 1 mol dm^{-3} $NaNO_3$		coagulation, iep	7.3±0.2	T.W. Healy, A.P. Herring, and D.W. Fuerstenau. J.Colloid Int.Sci. 21: 435–444 (1966).
MnO_2	β, natural	0.001 mol dm^{-3} NaCl		iep	4.4 (abstract) 6.4 (Fig. 4)	A.A. Yousef, M.A. Arafa, and M.A. Malati. J.Appl.Chem.Biotech. 21: 200–207 (1971).
MnO_2	β, thermal decomposition of $Mn(NO_3)_2$	0.001, 0.01 mol dm^{-3} NaCl		iep	4.6	A.A. Yousef, M.A. Arafa, and M.A. Malati. J.Appl.Chem.Biotech. 21: 200–207 (1971).
MnO_2	initially β, different T treatment	10^{-3} mol dm^{-3} NaCl	25	iep	4 (200 C), 5.2 (400 C), 7.8 (600-800 C, conv. to α) 5.2 (1000-1400 C, conv. to Mn_3O_4)	T. Morimoto, and S. Kittaka. Bull.Chem.Soc.Jpn. 47: 1586–1588 (1974).
MnO_2		0.1 mol dm^{-3} KCl	room	pH	3.85	E.A. Nechaev, and L.M. Smirnova. Kolloid.Zh. 39: 186–190 (1977).
MnO_2	δ, synthetic	10^{-3}–10^{-1} mol dm^{-3} KCl	25	pH, coagulation, cip	3.3±0.5(?), 3.6±0.5 (?) 5.5	M.J. Gray, M.A. Malati, and M.W. Rophael. J.Electroanal.Chem. 89: 135–140 (1978).
MnO_2	γ, International common sample No. 5	10^{-3}–10^{-1} mol dm^{-3} KCl	25	cip, pH	5.9±0.3	M.J. Gray, M.A. Malati, and M.W. Rophael. J.Electroanal.Chem. 89: 135 140 (1978).
MnO_2	chiefly β	0.001 mol dm^{-3} KCl	29	iep	5.8	M.W. Rophael, T.A. Bibawy, L.B. Khalil, and M.A. Malati. Chemistry & Industry (1) 27–28 (1979).
			30	iep	6.0	
MnO_2		0.01 mol dm^{-3} KCl		iep	2.5	E.A. Nechaev, and V.N. Sheyin. Kolloidn.Zh. 41: 361–363 (1979).
MnO_2	electrode, electrodeposition	0.001–1 mol dm^{-3} $NaNO_3$	25	pH	3.7	G.A. Kokarev, V.A. Kolesnikov, A.F. Gubin, and A. Korobanov. Elektrokhimiya 18: 466-470 (1982).
MnO_2	powder	0.001–1 mol dm^{-3} $NaNO_3$	25	pH	5.4	G.A. Kokarev, V.A. Kolesnikov, A.F. Gubin, and A. Korobanov. Elektrokhimiya 18: 466-470 (1982).

TABLE 3.1 (Continued)

Oxide	Description	Salt	T	Method	pH_0	Source
MnO_2	α, synthetic, different samples	$LiNO_3$	27	pH	3–5.4	S.B. Kanungo, and K.M. Parida. J.Colloid Interf. Sci. 98: 252–260 (1984).
MnO_2	δ, synthetic, different samples	$LiNO_3$	27	pH	1.5 (extrapol)– 2.1	S.B. Kanungo, and K.M. Parida. J.Colloid Interf. Sci. 98: 252–260 (1984).
MnO_2	γ, synthetic, different samples	$LiNO_3$	27	pH	3.6–4.1	S.B. Kanungo, and K.M. Parida. J.Colloid Interf. Sci. 98: 252–260 (1984).
MnO_2	β, synthetic: BDH; Merck	$LiNO_3$	27	pH	7.1	S.B. Kanungo, and K.M. Parida. J.Colloid Interf. Sci. 98: 252–260 (1984).
MnO_2	γ, IC22	0.05–1 mol dm^{-3} $NaNO_3$	25	cip	4.8	H. Tamura, T. Oda, M. Nagayama, and R. Furuichi. J.Electrochem.Soc. 136: 2782–2786 (1989).
MnO_2	Fisher			iep	5	I. Nirdosh, W.B. Trembley, and C.R. Johnson. Hydrometallurgy 24: 237–248 (1990).
MnO_2	δ, reagent, washed	none, 0.001 mol dm^{-3} KCl and $CaCl_2$	25	iep	1.9	Fouad Taha, A.M. El-Roudi, A.A. Abd El Gaber, and F.M. Zahran. Rev. Roum. Chim. 35: 503–509 (1990).
MnO_2	IC1 IC12 IC22	0.1 mol dm^{-3} $NaNO_3$	25	pH	4.15 3.76 4.72	H. Tamura, and R. Furuichi. Bunseki Kagaku. 40: 635–640 (1991).
MnO_2	pyrolusite, heating of $Mn(NO_3)_2$	0.001–0.01 mol dm^{-3} NaCl		iep	4.4	M.A. Arafa, S.M.R. El-Nozahi, and A.A. Youssef. Neue Hutte 37: 451–457 (1992).
MnO_2	γ, IC1, IC12, IC22, washed	0.1 mol dm^{-3} $NaNO_3$	25	pH	4.2 3.7 4.6	H. Tamura, N. Katayama, and R. Furuichi. Environ. Sci. Technol. 30: 1198–1204 (1996).
$MnO_2 \cdot n$ H_2O	from $Mn(NO_3)_2$ and $NaMnO_4$, 3:2 molar ratio, 16 h aged at pH 7	0.015–1.5 mol dm^{-3} $NaNO_3$	25	pH, merge	2.6	P. Trivedi, and L. Axe. J.Colloid Interf.Sci. 218: 554–563 (1999).
NbO_2		0.4 mol dm^{-3} KCl		pH	7.3	J.B. Goodenough, R. Manoharan, and M. Paranthamam. J.Am.Chem.Soc. 112: 2076–2082 (1990).
Nb_2O_5		0.01 mol dm^{-3} KCl		iep	3.8	E.A. Nechaev, and V.N. Sheyin. Kolloidn.Zh. 41: 361 363 (1979).

Material	Description	Electrolyte	T	Method	pH	Reference
Nb_2O_5		0.01 mol dm^{-3} KOH (?)		iep	3.55	E.A. Nechaev. Kolloidn.Zh. 42: 371–373 (1980).
Nb_2O_5		0.01 mol dm^{-3} KCl		iep	3.55	E.A. Nechaev, and G.V. Zvonareva. Kolloidn.Zh. 42: 511–516 (1980).
Nb_2O_5		0.4 mol dm^{-3} KCl		pH	7.6	J.B. Goodenough, R. Manoharan, and M. Paranthamam. J.Am.Chem.Soc. 112: 2076–2082 (1990).
Nb_2O_5	commercial, orthorhombic and synthetic, monoclinic	5×10^{-4}–5×10^{-2} mol dm^{-3} KNO_3		iep	4	G. Gonzalez, S.M. Saraiva, and W. Aliaga. Dispersion Sci.Technol. 15: 123–132 (1994).
Nb_2O_5			23	iep	4.4	P. Jayaweera, S. Hettiarachchi, and H. Ocken. Colloids Surf.A. 85: 19 (1994).
Nb_2O_5	$Nb_2O_5 \cdot 4\,H_2O$ calcined at 500 C			pH	4	H. Hu, I.E. Wachs, and S.R. Bare. J.Phys.Chem. 99: 10897–10910 (1995).
Nb_2O_5	reagent grade, washed	KCl	25	salt titr.	4.1	M. Kosmulski. Langmuir 13: 6315–6320 (1997).
$Nb_2O_5 \cdot n\,H_2O$		2×10^{-3} mol dm^{-3} $NaClO_4$		iep	4.1	
NiO	cf. Table 3.9			pH	9.85	P.H. Tewari, and A.B. Campbell. J.Colloid Int.Sci. 55: 531–539 (1976).
NiO	cf. Table 3.9			cip	11.3	
NiO	high purity	KNO_3	25	iep	11.3	
NiO		0.01 mol dm^{-3} KCl		iep	10.2	E.A. Nechaev, and V.N. Sheyin. Kolloidn.Zh. 41: 361–363 (1979).
NiO	CP grade	0–0.001 mol dm^{-3} $NaCl$		iep	7.5	F.F. Aplan, E.Y. Spearin, and G. Simkovich. Colloids Surf. 1: 361 371. (1980).
NiO		0.01 mol dm^{-3} KOH (?)		iep	10.3	E.A. Nechaev. Kolloidn.Zh. 42: 371 373 (1980).
NiO	treated at different T	10^{-3} mol dm^{-3} $NaCl$	25	iep	10.5 (600 C)	S. Kittaka, and T. Morimoto. J.Colloid Int.Sci. 75: 398–403 (1980).
NiO	powder	0.001–1 mol dm^{-3} $NaNO_3$	25	pH	8.4	G.A. Kokarev, V.A. Kolesnikov, A.F. Gubin, and A. Korobanov. Elektrokhimiya 18: 466–470 (1982).
NiO	electrode	0.001–1 mol dm^{-3} $NaNO_3$	25	pH	8.2	G.A. Kokarev, V.A. Kolesnikov, A.F. Gubin, and A. Korobanov. Elektrokhimiya 18: 466–470 (1982).
NiO	reagent, washed	none, 0.001 mol dm^{-3} KCl and $CaCl_2$	25	iep	6.4	Fouad Taha, A.M. El-Roudi, A.A. Abd El Gaber, and F.M. Zahran. Rev.Roum.Chim. 35: 503 509 (1990).

TABLE 3.1 (Continued)

Oxide	Description	Salt	T	Method	pH₀	Source
NiO	thermal decomposition of Ni(OH)$_2$	0.05 1 mol dm^{-3} NaNO$_3$, NaCl, NaClO$_4$	25	cip	7.3	H. Moriwaki, Y. Yoshikawa, and T. Morimoto. Langmuir 6: 847–850 (1990).
NiO	Ni(OH)$_2$ calcined in air at 300 C	0.01 mol dm^{-3} NaClO$_4$		iep	12.7	L. Durand-Keklikian, I. Haq, and E. Matijevic. Colloids Surf. A 92: 267–275 (1994).
NiO	bunsenite, Fisher	0.1 mol dm^{-3} NaClO$_4$	25	pH	8.8	C. Ludwig and W.H. Casey. J.Colloid Interf.Sci. 178: 176–185 (1996).
NiO	cubic, thermal decomposition of nitrate at 400 C	0.005–0.1 mol dm^{-3} KNO$_3$		cip	9.5	L.A. de Faria, M. Prestat, J.F. Koenig, P. Chartier, and S. Trasatti. Electrochim. Acta 44: 1481–1489 (1998).
NiO$_x$	oxidation of Ni-NiO$_x$ nanoparticles	1 mol dm^{-3} KNO$_3$(?)		iep	9 (in air) 8 (in oxygen)	K.C. Liu, and M.A. Anderson. Mater.Res.Soc. Symp.Proc. 432: 221–229 (1997).
NiO$_x$	Aldrich	1 mol dm^{-3} KNO$_3$(?)		iep	6.8	K.C. Liu, and M.A. Anderson. Mater. Res. Soc. Symp. Proc. 432: 221–229 (1997).
Ni(OH)$_2$	cf. Table 3.9					
Ni(OH)$_2$	high purity	KNO$_3$	25	pH cip iep	8.1 11.2 11.1	P.H. Tewari, and A.B. Campbell. J.Colloid Int.Sci. 55: 531–539 (1976).
Ni(OH)$_2$	Ni(NO$_3$)$_2$ + NH$_3$	0.05–1 mol dm^{-3} NaNO$_3$, NaCl, NaClO$_4$	25	cip	8.7	H. Moriwaki, Y. Yoshikawa, and T. Morimoto. Langmuir 6: 847–850 (1990).
Ni(OH)$_2$	Ni(NO$_3$)$_2$ + NH$_3$, uniform platelets	0.01 mol dm^{-3} NaClO$_4$		iep	9.3	L. Durand-Keklikian, L.Haq, and E. Matijevic. Colloids Surf. A 92: 267 275 (1994).
Ni(OH)$_2$	β, hydrolysis of sulfate	0.003–0.13 mol dm^{-3} KNO$_3$	28± 2	merge iep	10.5	M.J. Avena, and C.P. dePauli. Colloids Surf.A 108: 181–189 (1996).
Ni(OH)$_2$	from chloride	0.1 mol dm^{-3} NaCl		mass titration	11.3	G.M.S. El Shafei. J.Colloid Interf.Sci. 182: 249–253 (1996).
PbO	reagent, washed	0.01 mol dm^{-3} KOH (?)		iep	11.6	E.A. Nechaev. Kolloidn.Zh. 42: 371–373 (1980).
PbO		none, 0.001 mol dm^{-3} KCl and CaCl$_2$	25	iep	10.7	Fouad Taha, A.M. El-Roudi, A.A. Abd El Gaber, and F.M. Zahran. Rev. Roum. Chim. 35: 503–509 (1990).
PbO	Ventron, litharge + massicot	10^{-4} mol dm^{-3} NaCl		iep	11.3	J. Biscan, M. Kosec, and N. Kallay. Colloids Surf. A 79: 217 226 (1993).

Material	Description	Electrolyte	T			Reference
Pb(OH)$_2$	cf. Table 3.9					
PbO$_2$	electrodeposition	0.1, 1 mol dm^{-3} KCl		intersection	7.4	E.A. Nechaev, and V.A. Volgina. Elektrokhimiya 13: 177 181 (1977).
PbO$_2$	electrodeposition	0.01 mol dm^{-3} KCl		iep	8.2	E.A. Nechaev, and V.A. Volgina. Elektrokhimiya 15: 1564-1568 (1979).
PbO$_2$		0.01 mol dm^{-3} KCl		iep	8.3	E.A. Nechaev, and V.N. Sheyin. Kolloidn.Zh. 41: 361-363 (1979).
PbO$_2$		0.01 mol dm^{-3} KOH (?)		iep	8.4	E.A. Nechaev. Kolloidn. Zh. 42: 371-373 (1980).
PbO$_2$	α, powder, washed	0.001-1 mol dm^{-3} NaNO$_3$	25	pH	7.3	G.A. Kokarev, V.A. Kolesnikov, A.F. Gubin, and A. Korobanov. Elektrokhimiya 18: 466-470 (1982).
PbO$_2$	β, powder, washed	0.001-1 mol dm^{-3} NaNO$_3$	25	pH	5.4	G.A. Kokarev, V.A. Kolesnikov, A.F. Gubin, and A. Korobanov. Elektrokhimiya 18: 466-470 (1982).
PbO$_2$	β, electrode, electrodeposition	0.001-1 mol dm^{-3} NaNO$_3$	25	pH	4.5-5.4	G.A. Kokarev, V.A. Kolesnikov, A.F. Gubin, and A. Korobanov. Elektrokhimiya 18: 466-470 (1982).
PbO$_2$	β, electrodeposition	10^{-4}-10^{-2} mol dm^{-3} KNO$_3$	25	cip	9.2	N. Munichandraiah. J.Electroanal.Chem.266: 179-184 (1989).
PbO$_2$	reagent, washed	none, 0.001 mol dm^{-3} KCl and CaCl$_2$	25	iep	1.75	Fouad Taha, A.M. El-Roudi, A.A. Abd El Gaber, and F.M. Zahran. Rev. Roum. Chim. 35: 503-509 (1990).
PrO$_2$		0.01 mol dm^{-3} KCl		iep	8	E.A. Nechaev, and V.N. Sheyin. Kolloidn.Zh. 41: 361-363 (1979).
PrO$_2$		0.01 mol dm^{-3} KOH (?)		iep	8	E.A. Nechaev. Kolloidn.Zh. 42: 371-373 (1980).
PtO$_2$				?	5	A. Sekki, G.A. Kokarev, and Yu. I. Kapustin. Elektrokhimiya 21: 1277 (1985).
RuO$_2$	thermal decomposition of RuCl$_3$ at 400 C	0.005-0.15 mol dm^{-3} KNO$_3$; KCl		cip,iep	5.1	S. Ardizzone, P. Siviglia, and S. Trasatti. J.Electroanal.Chem. 122: 395-401 (1981).
RuO$_2$	thermal decomposition of RuCl$_3$ at 700 C	0.005-0.15 mol dm^{-3} KNO$_3$		cip	5.7	S. Ardizzone, P. Siviglia, and S. Trasatti. J.Electroanal.Chem. 122: 395-401 (1981).
				cip	6.1	
RuO$_2$	electrode, thermal decomposition of salt, washed	0.001-1 mol dm^{-3} NaNO$_3$	25	pH	5.5	G.A. Kokarev, V.A. Kolesnikov, A.F. Gubin, and A. Korobanov. Elektrokhimiya 18: 466-470 (1932).

TABLE 3.1 (Continued)

Oxide	Description	Salt	T	Method	pH_0	Source
RuO_2	powder, washed	$0.001\ 1$ mol dm^{-3} $NaNO_3$	25	cip	4	G.A. Kokarev, V.A. Kolesnikov, A.F. Gubin, and A. Korobanov. Elektrokhimiya 18: 466–470 (1982).
RuO_2	From $RuCl_3$ prepared at different T	0.005–0.15 mol dm^{-3} KNO_3	25	cip iep	4 (300 C) 6.1 (700 C) 5 (400 C) 6 (700 C)	P. Siviglia, A. Daghetti, and S. Trasatti. Colloids and Surfaces 7: 15–27 (1983).
RuO_2	anhydrous Ventron	0.005–0.15 mol dm^{-3} KNO_3	25	cip	4.8	P. Siviglia, A. Daghetti, and S. Trasatti. Colloids and Surfaces 7: 15–27 (1983).
RuO_2	stoichiometric (chemical vapor transport)	0.005–0.15 mol dm^{-3} KNO_3	25	cip	7.3	P. Siviglia, A. Daghetti, and S. Trasatti. Colloids and Surfaces 7: 15–27 (1983).
RuO_2	thermal decomposition of $RuCl_3$ at 420 C	0.01–0.25 mol dm^{-3} KNO_3; up to 0.002 mol dm^{-3} KCl	20	cip iep	5.75	J.M. Kleijn, and J. Lyklema. J.Colloid Interf.Sci. 120: 511–522 (1987).
RuO_2	thermal decomposition of $RuCl_3$ at 420 C	0.01–0.25 mol dm^{-3} KNO_3	20	cip iep	5.75	J.M. Kleijn, and J. Lyklema. Colloid Polym.Sci. 265: 1105–1113 (1987).
RuO_2		0.4 mol dm^{-3} KCl		pH	3.88	J.B. Goodenough, R. Manoharan, and M. Paranthamam. J.Am.Chem.Soc. 112: 2076–2082 (1990).
RuO_2	from chloride	0.005–0.05 mol dm^{-3} KNO_3	25	cip	5.3	L.A. de Faria, and S. Trasatti. J.Electronal.Chem. 340: 145 152 (1992).
$RuO_2 \cdot 1\ H_2O$	cf. Table 3.9					
$RuO_2 \cdot 1\ H_2O$	thermal decomposition of $RuCl_3$ or commercial hydrate at different T	0.001–0.01 mol dm^{-3} KNO_3	25	cip	4 (low T) 5.6 (500 C)	S. Ardizzone, A. Daghetti, L. Franceschi, and S. Trasatti. Colloids Surf. 35: 85–96 (1989).
$RuO_2 \cdot n\ H_2O$	Alfa	none		iep	2.8	J. Kiwi, and M.Gratzel. Chem.Phys.Letters 78: 241–245 (1981).
$RuO_2 \cdot n\ H_2O$				iep	3.25	D.N. Furlong personal comm. cited by A. Daghetti, G. Lodi, and S. Trasatti. Mater.Chem.Phys. 8: 1–90 (1983).

Material	Description	Electrolyte	T (°C)	Method	pH	Reference
$Sb_2O_5 \cdot n\ H_2O$	cf. Table 3.9					
Sc_2O_3		0.01 mol dm^{-3} KCl		iep	7	E.A. Nechaev, and V.N. Sheyin. Kolloidn.Zh. 41: 361 363 (1979).
Sc_2O_3		0.01 mol dm^{-3} KOH (?)		iep	7.2	E.A. Nechaev. Kolloidn.Zh. 42: 371–373 (1980).
Sc_2O_3		0.01 mol dm^{-3} KCl		iep	7.2	E.A. Nechaev, and G.V. Zvonareva. Kolloidn.Zh. 42: 511–516 (1980).
SiO_2	cf. Tables 3.8 and 3.9					
SiO_2	precipitated BDH	10^{-3}–10^{-1} mol dm^{-3} KCl	20	cip	2.5–3	T.F. Tadros, and J. Lyklema. J.Electroanal.Chem. 17: 267–275 (1968).
SiO_2	Degussa	0.001–1 mol dm^{-3} KCl		cip	3–3.5	M. Tschapek, L. Tcheichvili, and C. Wasowski. Clay Miner. 10: 219–229 (1974).
SiO_2	quartz	10^{-3}–10^{-2} mol dm^{-3} NaCl		iep	3	J. Jednacak, V. Pravdic, and W. Haller. J.Colloid Int.Sci. 49: 16–23 (1974).
SiO_2	Aerosil 380 Degussa	0.01–1 mol dm^{-3} KCl		cip / EMF	3–3.6 / 3.3 (extrapol.)	M. Tschapek, R.M. Torres Sanchez, and C. Wasowski. Colloid Polym. Sci. 254: 516-521 (1976).
SiO_2	quartz, purified, iron free, Soxhlet extraction	NaCl	25	iep	1.3	D. Balzer, and H. Lange. Colloid Polym. Sci. 255: 140–152 (1977).
SiO_2		0.1 mol dm^{-3} KCl	room	pH	1.8	E.A. Nechaev, and L.M. Smirnova. Kolloid.Zh. 39: 186–190 (1977).
SiO_2	Cabosil			iep	2	H.A. Elliott, and C.P. Huang. J.Colloid Int.Sci. 70: 29–45 (1979).
SiO_2		0.01 mol dm^{-3} KCl		iep	1.8	E.A. Nechaev, and G.V. Zvonareva. Kolloid.Zh. 511 516 (1980).
SiO_2	BDH, Degussa TK900, Crosfield A,B,C	0.01 mol dm^{-3} KNO$_3$	25	pH	2.8–2.9	G.C.Bye, M.McEvoy, and M.A. Malati. J.Chem. Tech.Biotechnol. 32: 781–789 (1982).
SiO_2	vitreous (aged and steamed)	0.001–0.01 mol dm^{-3} KNO$_3$		iep	4 (< 3)	F. Grieser, R.N. Lamb, G.R. Wiese, D.E. Yates, R. Cooper, and T.W. Healy. Radiat.Phys.Chem. 23: 43–48 (1984).
SiO_2	from ethoxide	10^{-3}–10^{-1} mol dm^{-3} NaNO$_3$		merge	4.1	J.A. Schwarz, C.T. Driscoll, and A.K. Bhanot. J.Colloid Int.Sci. 97: 55–61 (1984).
SiO_2	quartz natural, washed	0.1 mol dm^{-3} NaCl	25	iep	2.5	S.N. Omenyi, B.J. Herren, R.S. Synder, and G.V.F. Seaman. J.Colloid Int.Sci. 110: 130–136 (1986).

TABLE 3.1 (Continued)

Oxide	Description	Salt	T	Method	pH$_0$	Source
SiO$_2$	Aerosil OX-50	10^{-4}–10^{-3} mol dm^{-3} KCl		iep	2.5	H. Sonntag, V. Itschenskij, and R. Koleznikova. Croat. Chim. Acta 60: 383 393 (1987).
SiO$_2$	Thiokol, washed	0.1 mol dm^{-3} NaCl		pH	3.3	J.F. Kuo, and T.F. Yen. J.Colloid Int.Sci. 121: 220–225 (1988).
SiO$_2$	two samples			mass titration	2.0	W.C. Hasz. Thesis. quoted after W.M. Mullins, and B.L. Averbach. Surface Sci. 206: 41–51 (1988).
SiO$_2$	Cabot L 90			iep	3.2	J.S. Noh, and J.A. Schwarz. J.Colloid Interf.Sci. 130: 157–164 (1989).
SiO$_2$	quartz glass				2	G.P. Golub, M.P. Sidorova, and D.A. Fridrikhsberg. Kolloid.Zh. 51: 987 989 (1989).
SiO$_2$	natural, washed	none, 0.001 mol dm^{-3} KCl and CaCl$_2$	25	iep	2.2	Fouad Taha, A.M. El-Roudi, A.A. Abd El Gaber, and F.M. Zahran. Rev.Roum.Chim. 35: 503–509 (1990).
SiO$_2$	Chromosorb W, and another sample			pH	2.3 2.9	M. Parkanyi-Berka, and P. Joo. Colloids Surf. 49: 165–182 (1990).
SiO$_2$	films, oxidation of Si	0.001–0.1 mol dm^{-3} NaCl	24	iep	2.6–3.2	L. Bousse, S. Mostarshed, B. Van der Shoot, N.F. de Rooij, P. Gimmel, and W. Gopel. J.Colloid Interface Sci. 147: 22–32 (1991).
SiO$_2$	Stöber	0.001–0.1 mol dm^{-3} KCl	25	iep	3.8	M. Kosmulski, and E. Matijevic. Langmuir 7: 2066-2071 (1991).
SiO$_2$	gel, hydrolysis of ethoxide			mass titration	3.8	C. Contescu, C. Sivaraj, and J.A. Schwarz. Appl. Catalysis 74: 95 108 (1991).
SiO$_2$	from ethoxide	10^{-3}–10^{-1} mol dm^{-3} NaNO$_3$		mass titration cip	3.52 3.2	J.A. Schwarz, C.T. Ugbor, and R. Zhang. J. Catalysis 138: 38 54 (1992).
SiO$_2$	Stöber	0.001–0.1 mol dm^{-3} KCl	25	iep	3.4	M. Kosmulski, and E. Matijevic. Langmuir 8: 1060–1064 (1992).
SiO$_2$	Aerosil 200			iep	1.9	M.I. Zaki, S.A.A. Mansour, F. Taha, and G.A.H. Mekhemer. Langmuir 8: 727 732 (1992).
SiO$_2$	JRC, gel	KNO$_3$		pH	1.8	P.K. Ahn, S. Nishiyama, S. Tsuruya, and M. Masai. Appl.Catalysis A 101: 207 219 (1993).

Material	Sample	Electrolyte	T	Method	Value	Reference
SiO$_2$	gel, Merck, washed	0.001 mol dm^{-3} NaCl	25	iep	3.8	M. Kosmulski. J.Colloid Int.Sci. 156: 305 310 (1993).
SiO$_2$	Aerosil 200 Degussa	10^{-3}, 10^{-2} mol dm^{-3} KNO$_3$ 10^{-3} mol dm^{-3} KBr 10^{-2} mol dm^{-3} KBr	25	iep iep pH iep pH	3.2 <3 2.8 3.4 2.8	J.C.J. van der Donck, G.E.J. Vaessen, and H.N. Stein. Langmuir 9: 3553-3557 (1993).
SiO$_2$	Custer			iep	2	J. Liu, S.M. Howard, and K.N. Han. Langmuir 9: 3635-3639 (1993).
SiO$_2$		10^{-4} mol dm^{-3} NaNO$_3$	25	inflection	4.37	N. Kallay, and S. Zalac. Croat.Chem.Acta. 67: 467-479 (1994).
SiO$_2$	Cab-O-Sil	0.1 mol dm^{-3} NaNO$_3$		mass titration	L-90 3.4 EH-5 3.0	J. Park, and J.R. Regalbuto. J.Colloid Int.Sci. 175: 239 252 (1995).
SiO$_2$	Cabosil EH-5 wetted and calcined at 500 C			pH	3.7-4.3	H. Hu, I.E. Wachs, and S.R. Bare. J.Phys.Chem. 99: 10897-10910 (1995).
SiO$_2$	Lichrospher Si 100 and 1000	0.001-0.1 mol dm^{-3} NaCl	25	pH, merge	4 5.7	W. Janusz. Ads. Sci. Technol. 14: 151-161 (1996).
SiO$_2$	quartz, Showa Denko	0.001 mol dm^{-3} KNO$_3$		pH	4.2	S. Fukuzaki, H. Urano, and K. Nagata. J. Fermentation Bioeng. 81: 163-167 (1996).
SiO$_2$	OX 50 Degussa	NH$_4$Cl		acousto	2-3 (hysteresis)	R.A. Overbeek, E.J. Bosma, D.W. H. de Blauw, A.J. van Dillen, H.G. Bruil, and J.W. Geus. Applied Catal.A 163: 129-144 (1997).
SiO$_2$	Hypersil Purospher ES-gel YMC Kromasil Nucleosil	0.001 mol dm^{-3} NaCl		iep	2.1 2.32 2.34 2.44 2.86 3.5 2.9	P.G. Dietrich, K.H. Lerche, J. Reusch, and R. Nitzsche. Chromatographia 44: 362-366 (1997).
SiO$_2$	ES-gel, treated natural, washed	10^{-3} mol dm^{-3} KNO$_3$		iep	1.8	F. Rashchi, Z. Xu, and J.A. Finch. Colloids Surf.A 132: 159-171 (1998).
SiO$_2$	Stöber	0.001-0.1 mol dm^{-3} KCl	24	pH iep	4.3-4.5 4.1	M. Szekeres, I. Dekany, and A. de Keizer. Colloids Surf. A 141: 327-336 (1998).
SiO$_2$	Aerosil "Chlorovinyl"	0-0.02 mol dm^{-3} NaCl		iep	2.2	V.M. Gunko, V.I. Zarko, V.V. Turov, R. Leboda, E. Chibowski, and V.V. Gunko, J.Colloid Interf. Sci. 205: 106-120 (1998).

TABLE 3.1 (Continued)

Oxide	Description	Salt	T	Method	pH_0	Source
SiO_2	Sigma	HCl HNO_3 $HClO_4$		iep	2 none none	M. Kosmulski. J.Colloid Interf.Sci. 208: 543-545 (1998).
SiO_2	Aerosil 130-380, Degussa	0.1 mol dm⁻³ NaCl	25	mass titration	3.5-3.7	J.P. Reymond, and F. Kolenda. Powder Techn. 103: 30-36 (1999).
SiO_2	Stober	10^{-3} mol dm⁻³ KCl		iep	3.17	X.C. Guo, and P. Dong. Langmuir 15: 5535 5540 (1999).
SiO_2	Aerosil 380, Degussa	0.01, 0.1 mol dm⁻³ NaCl		iep	3.8	W. Rudzinski, R. Charmas, W. Piasecki, B. Prelot, F. Thomas, F. Villieras, and J.M. Cases. Langmuir 15: 5977-5983 (1999).
SiO_2	Geltech	0.001 mol dm⁻³ KNO_3	25	iep	3	G.E. Morris, W.A. Skinner, P.G. Self, and R.S.C. Smart. Colloids Surf. A 155: 27-41 (1999).
SiO_2	quartz Sikron 6000 SF, Quarzwerke GmbH.	0.001 mol dm⁻³ KCl		iep	4.2	D. Bauer, H. Buchhammer, A. Fuchs, W. Jaeger, E. Killmann, K. Lunkwitz, R. Rehmet, and S. Schwarz. Colloids Surf.A 156: 291-305 (1999).
SiO_2	Stober	0.01 mol dm⁻³ NaCl		iep	3	M. Ocana, and A.R. Gonzalez-Elipe. Colloids Surf.A 157: 315-324 (1999).
SiO_2	Stöber	10^{-3} mol dm⁻³ KCl	25	iep	3	R. Lindberg, G. Sundholm, J. Sjöblom, P. Ahonen and E.I. Kauppinen. J.Disp.Sci. Technol. 20: 715-722 (1999).
SiO_2	quartz, washed	HCl	25	iep	1.5	P.R. Johnson. J.Colloid Interf.Sci. 209: 264-267 (1999).
SiO_2	Geltech	10^{-2} mol dm⁻³ KNO_3		iep	2.9	M.P. Sidorova, H. Zastrow, L.E. Ermakova, N.F. Bogdanova, and V.M. Smirnov. Kolloid.Zh. 61: 113-117 (1999).
Sm_2O_3	commercial, containing carbonate, monoclinic	0.01-1 mol dm⁻³ NaCl, KNO_3	25	cip	7.45 7.8	S.I. Pechenyuk.Zh.Fiz.Khim. 61: 165-169 (1987).
SnO		0.01 mol dm⁻³ KCl		iep	5.2	E.A. Nechaev, and V.N. Sheyin. Kolloidn.Zh. 41: 361 363 (1979).
SnO		0.01 mol dm⁻³ KOH (?)		iep.	5.2	E.A. Nechaev. Kolloidn.Zh. 42: 371-373 (1980).
SnO_2	natural	0.001 1 mol dm⁻³ KNO_3		pH, merge	5.6	S.M. Ahmed, and D. Maksimov. J.Colloid Interf. Sci. 29: 97-104 (1969).

Material	Description	Electrolyte	T	Method	pH/iep	Reference
SnO$_2$				iep	4.5	T.W. Healy, G.R. Wiese, D.E. Yates, and B.V. Kavanagh. J.Colloid Int.Sci. 42: 647–649 (1973).
SnO$_2$		0.1 mol dm^{-3} KCl	room	coagulation pH	4	E.A. Nechaev, and L.M. Smirnova. Kolloid.Zh. 39: 186–190 (1977).
SnO$_2$	hydrolysis of chloride, washed	0.001–0.5 mol dm^{-3} NaCl, NaNO$_3$	35	iep / cip	4.5	K.C. Ray, and S. Khan. Indian J.Chem. 16A: 12–15 (1978).
SnO$_2$		0.01 mol dm^{-3} KCl		iep	4.2	E.A. Nechaev, and V.N. Sheyin. Kolloidn.Zh. 41: 361–363 (1979).
SnO$_2$	powder, washed	0.001–1 mol dm^{-3} NaNO$_3$	25	pH	5.7	G.A. Kokarev, V.A. Kolesnikov, A.F. Gubin, and A. Korobanov. Elektrokhimiya 18: 466–470 (1982).
SnO$_2$	natural, upgraded by acid leaching, 94% SnO$_2$, 0.6% Fe Aldrich 99.9999%	0.001 mol dm^{-3} KNO$_3$		iep	4.3	L.J. Warren. Colloids Surf. 5: 301–319 (1982).
SnO$_2$		10^{-4}–1 mol dm^{-3} KNO$_3$	25 / 22	iep / cip	4.5 / 4.3	M.R. Houchin, and L.J. Warren. J.Colloid Int.Sci. 100: 278–286 (1984).
SnO$_2$		KNO$_3$			5.5	C.P. Huang, thesis, quoted after M.R. Houchin, and L.J. Warren. J.Colloid Int.Sci. 100: 278 286 (1984).
SnO$_2$	natural	HCl / HNO$_3$		iep	3.1 / 3.7	G. Zambrana, L.W. Pommier, and R. Medina, AIME Centennial Annual Meeting, NY 1971, 71-B-30, quoted after M.R. Houchin, and L.J. Warren. J.Colloid Int.Sci. 100: 278 286 (1984).
SnO$_2$	natural	KCl		iep	3.4–4.5	P. Blazy, P. Degoul, and R. Houot in 2nd Technical Conference on Tin, Bangkok 1969 (W. Fox, ed.) vol. 3 p. 937, quoted after M.R. Houchin, and L.J. Warren. J.Colloid Int.Sci. 100: 278–286 (1984).
SnO$_2$	natural	KCl			4	G. Gutierrez, and L.W. Pommier, in 2nd Technical Conference on Tin, Bangkok 1969 (W. Fox, ed.) vol. 3 p. 915, quoted after M.R. Houchin, and L.J. Warren. J.Colloid Int.Sci. 100: 278–286 (1984).
SnO$_2$	natural, washed, 4 samples	10^{-4}–1 mol dm^{-3} KNO$_3$	25 / 22	iep / cip / salt titration	4.1–4.5 / 4.1–5.2 / 4.2–4.6	M.R. Houchin, and L.J. Warren. Colloids Surf. 16: 117–126 (1985).

TABLE 3.1 (Continued)

Oxide	Description	Salt	T	Method	pH_0	Source
SnO_2	Aldrich gold label	10^{-4}–1 mol dm^{-3} KNO$_3$	25 22	iep cip salt titration	4.5 4.3 4.2–4.3	M.R. Houchin, and L.J. Warren. Colloids Surf. 16: 117–126 (1985).
SnO_2	Aldrich 99.9999%	10^{-3} mol dm^{-3} KNO$_3$, Na$_2$CO$_3$		iep	4.5 4.3	C. Biegler, and M.R. Houchin. Colloids Surf. 21: 267–278 (1986).
SnO_2	reagent grade, sintered at different T	NaOH + HCl		iep	6 (5–6.2 claimed by the authors)	S.B. Balachandran, G. Simkovich, and F.F. Aplan. Int.J.Miner.Proc. 21: 157–171 (1987).
SnO_2	natural, 2 samples	NaOH + HCl		iep	5 6.5	S.B. Balachandran, G. Simkovich, and F.F. Aplan. Int.J.Miner.Proc. 21: 157–171 (1987).
SnO_2	reagent, washed	none, 0.001 mol dm^{-3} KCl and CaCl$_2$	25	iep	2.0	Fouad Taha, A.M. El-Roudi, A.A. Abd El Gaber, and F.M. Zahran. Rev. Roum. Chim. 35: 503–509 (1990).
SnO_2	Aldrich	10^{-4} mol dm^{-3} KCl		iep	4.7	M.G. Cattania, S. Ardizzone, C.L. Bianchi, and S. Carella. Colloids Surf.A 76: 233–240 (1993).
SnO_2	from SnCl$_4$	0.1 mol dm^{-3} LiNO$_3$, KNO$_3$, CsNO$_3$		iep	4	R. Rautiu, and D.A. White, Solv.Extr.Ion Exch. 14: 721–738 (1996).
SnO_2	hydrous, α, hydrolysis of chloride	10^{-3} mol dm^{-3} NaCl	25	pH	4.4	H.F. Ghoneimy, T.N. Morcos, and N.Z. Misak. Colloids Surf.A 122: 13–26 (1997).
SnO_2				authors'	5.6,6	H.K. Liao, L.L. Chi, J.C. Chou, W.Y. Chung, T.P. Sun, and S.K. Hsiung. Mater.Chem.Phys. 59: 6–11 (1999).
$Sn(OH)_4$	from chloride	0.306 mol dm^{-3} NH$_3$		iep	7	C.S. Luo, and S.D. Huang. Separ.Sci.Technol. 28: 1253–1271 (1993).
Ta_2O_5	cf. Table 3.8					
Ta_2O_5	Johnson Matthey			authors'	2.9	M.A. Butler, and D.S. Ginley. J.Electrochem.Soc. 125: 228–232 (1978).
Ta_2O_5	powder	0.001–1 mol dm^{-3} NaNO$_3$	25	pH	2.8	G.A. Kokarev, V.A. Kolesnikov, A.F. Gubin, and A.A. Korobanov. Elektrokhimiya 18: 466–470 (1982).
Ta_2O_5	β, oxidation of metal at 1350 C	KCl		pH	5.2–8.2	E. Yu. Chardymskaya, M.P. Sidorova, O.S. Semenova, and M.D. Khanin. Kolloidn. Zh. 51: 607–610 (1989).

Material	Description	Electrolyte	T	Method	pH	Reference
Ta$_2$O$_5$	films, oxidation of Ta	0.001–0.1 mol dm^{-3} NaCl	24	iep	2.7–3	L. Bousse, S. Mostarshed, B. van der Shoot, N.F. de Rooij F, P. Gimmel and W. Gopel J.Colloid Interface Sci. 147: 22–32 (1991).
Ta$_2$O$_5$			23	iep	4.6	P. Jayaweera, S. Hettiarachchi, and H. Ocken. Colloids Surf. A 85: 19–27 (1994).
Ta$_2$O$_5$	Merck, washed	KCl; NaClO$_4$; 10^{-4}–10^{-1} mol dm^{-3} NaCl; 2×10^{-3} mol dm^{-3} NaClO$_4$	25	salt titr.; salt titr.; cip; iep	5.0; 5.2; 5.2; 5.3	M. Kosmulski. Langmuir 13: 6315–6320 (1997).
Ta$_2$O$_5$	oxidation of Ta in dry oxygen at 500 C for 2 h	0.001–0.01 mol dm^{-3} NaCl		iep	4.4	M.P. Sidorova, N.F. Bogdanova, I.E. Ermakova, P.V. Bobrov. Kolloid.Zh. 59: 568–571 (1997).
Ta$_2$O$_5$.n H$_2$O	monodispersed spheres (hydrolysis of ethoxide)	0.001–0.1 mol dm^{-3} LiCl HCl, NaOH		iep	3.6; 4–5.5 (no data points in this range)	K. Nakanishi, Y. Takamiya, and T. Shimohira. Yogyo Kyokaishi 94: 1023–1028 (1986).
ThO$_2$	thorianite	10^{-3}, mol dm^{-3} KNO$_3$	25	pH intersection	5.9–6.8; 7.3	S.M. Ahmed. Can.J.Chem. 44: 1663–1670 (1966).
ThO$_2$	cubic Norton	0.001–1 mol dm^{-3} KCl, NaClO$_4$	25	pH	6.3–7	S.M. Ahmed. J.Phys.Chem. 73: 3546–3555 (1969).
ThO$_2$	from nitrate	0.01 mol dm^{-3} KOH (?)		iep	8.4	E.A. Nechaev. Kolloidn.Zh. 42: 371–373 (1980).
ThO$_2$.n H$_2$O	from nitrate	0.1 mol dm^{-3} NaNO$_3$	25	pH	6.79	A.R. Gupta, and B. Venkataramani. Bull.Chem. Soc.Jpn. 61: 1357–1362 (1988).
ThO$_2$.n H$_2$O	from nitrate	0.1 mol dm^{-3} LiCl	25	pH	7	H.S. Mahal, and B. Venkataramani. Indian J.Chem. 37A: 993–1001 (1998).
Th(OH)$_4$	synthetic	0.01 mol dm^{-3} KNO$_3$	25	iep	10	R.O. James, and T.W. Healy. J.Colloid Int.Sci. 40: 53–64 (1972).
Th(OH)$_4$		0.01 mol dm^{-3} KCl		iep	8.5	E.A. Nechaev, and V.N. Sheyin. Kolloidn.Zh. 41: 361–363 (1979).
TiO$_2$	cf. Tables 3.8 and 3.9					
TiO$_2$	anatase from sulfate		25	iep	6.2	M. Miura, H. Naono, and T. Iwaki. J.Sci. Hiroshima Univ. 30: 57–63 (1966).
TiO$_2$	rutile from chloride; anatase from TiOSO$_4$	0.001–0.1 mol dm^{-3} NaClO$_4$, NaCl		cip	6.0; 5.8	Y.G. Berube, and P.L. de Bruyn. J.Colloid Int.Sci. 27: 305–318 (1968).
TiO$_2$	natural rutile	0.001–1 mol dm^{-3} KNO$_3$		pH, merge	5.3	S.M. Ahmed, and D. Maksimov. J.Colloid Interf. Sci. 29: 97–104 (1969).

TABLE 3.1 (Continued)

Oxide	Description	Salt	T	Method	pH$_0$	Source
TiO$_2$	Degussa P 25	HCl/NaOH		iep	6.6	M. Herrmann, and H.P. Boehm. Z. Anorg.Allg. Chemie 368: 73–86 (1969).
TiO$_2$	anatase, Bayer	HCl/NaOH		iep	6.3	M. Herrmann, and H.P. Boehm. Z. Anorg.Allg. Chemie 368: 73–86 (1969).
TiO$_2$	anatase "A", Bayer, washed	HCl/NaOH		iep	6.4	M. Herrmann, and H.P. Boehm. Z. Anorg. Allg.Chemie 368: 73–86 (1969).
TiO$_2$	rutile "R-U", Bayer, washed	HCl/NaOH		iep	4	M. Herrmann, and H.P. Boehm. Z. Anorg. Allg. Chemie 368: 73–86 (1969).
TiO$_2$	anatase P 25	3 mol dm^{-3} NaClO$_4$	25	pH	6.4	P.W. Schindler, and H. Gamsjäger. Kolloid. Z.Z. Polymer. 250: 759-763 (1972).
TiO$_2$	from sulfate different T treatment	0.005 mol dm^{-3} KNO$_3$	30	iep	5.8 100 C 5.3 300 C 5.2 500 C 3.7 700°C 3.5 1000 C	H. Fukuda, and M. Miura. J.Sci.Hiroshima Univ. Ser.A. 36: 77–86 (1972).
TiO$_2$	from chloride different T treatment	0.005 mol dm^{-3} KNO$_3$	30	iep	6.6 100 C 6 600 C 4.8 1000 C	H. Fukuda, and M. Miura. J.Sci.Hiroshima Univ. Ser.A. 36: 77–86 (1972).
TiO$_2$	rutile from chloride	0.001 mol dm^{-3} KNO$_3$	25	iep	5.6	R.O. James, and T.W. Healy. J.Colloid Int.Sci. 40: 53–64 (1972).
TiO$_2$	rutile from chloride	0.01–0.07 mol dm^{-3} KNO$_3$	25	iep	4.6	G.D. Parfitt, J. Ramsbotham, and C.H. Rochester. J.Colloid Int.Sci. 41: 437-444 (1972).
TiO$_2$	rutile from chloride			iep	3.8	L.G. Maidanovskaya, and T.V. Skripko. Zh.Fiz. Khim. 46: 115–118 (1972).
TiO$_2$	anatase from chloride			iep	5.2	L.G. Maidanovskaya, and T.V. Skripko. Zh.Fiz. Khim. 46: 115–118 (1972).
TiO$_2$				iep	5.9	T.W. Healy, G.R. Wiese, D.E. Yates, and B.E. Kavanagh. J.Colloid Int.Sci. 42: 647–649 (1973).
TiO$_2$	rutile from chloride	0.0001–0.01 mol dm^{-3} KNO$_3$	25	iep coagulation	5.9	G.R. Wiese, and T.W. Healy. J.Colloid Int.Sci. 51: 427–433 (1975).

			t			reference
TiO_2	99.5%, reagent	0.01 mol dm^{-3} KNO$_3$	30	iep	5.5	P.H. Tewari, and W. Lee. J.Colloid Int.Sci. 52: 77–88 (1975).
TiO_2	rutile, Laporte	KNO$_3$		iep / cip / pH	3.4 / 7.5 / 7.0	R.M. Cornell, A.M. Posner, and J.P. Quirk. J.Colloid Int.Sci. 53: 6–13 (1975).
TiO_2	rutile, Laporte, washed	KNO$_3$		iep / cip / pH	8.0 / 3.7 / 3.6	R.M. Cornell, A.M. Posner, and J.P. Quirk. J.Colloid Int.Sci. 53: 6–13 (1975).
TiO_2	rutile, Laporte, washed	KNO$_3$		iep / cip / pH	5.5 / 5.5 / 5.6	R.M. Cornell, A.M. Posner, and J.P. Quirk. J.Colloid Int.Sci. 53: 6–13 (1975).
TiO_2	rutile, Laporte	KNO$_3$		iep / cip / pH	7.3 / 4.7 / 4.6	R.M. Cornell, A.M. Posner, and J.P. Quirk. J.Colloid Int.Sci. 53: 6–13 (1975).
TiO_2	rutile, Laporte, washed	KNO$_3$		iep / cip / pH	5.8 / 5.2 / 5.0	R.M. Cornell, A.M. Posner, and J.P. Quirk. J.Colloid Int.Sci. 53: 6–13 (1975).
TiO_2	rutile, hydrous, hydrolysis of TiCl$_4$	KNO$_3$		iep / cip	3.2 / 7.1	R.M. Cornell, A.M. Posner, and J.P. Quirk. J.Colloid Int.Sci. 53: 6–13 (1975).
TiO_2	rutile, hydrous, hydrolysis of TiCl$_4$ washed	KNO$_3$		iep / cip / pH	6.0 / 5.8 / 5.7	R.M. Cornell, A.M. Posner, and J.P. Quirk. J.Colloid Int.Sci. 53: 6–13 (1975).
TiO_2	rutile from chloride			iep	5.6	D.N. Furlong, quoted after G.D. Parfitt in Progress in Surface and Membrane Sci. 11: 182 (1976).
TiO_2	P25 Degussa	0.01 1 mol dm^{-3} mol dm^{-3} KCl		cip / EMF	6.6 / 6.6	M. Tschapek, R.M. Torres Sanchez, and C. Wasowski. Colloid Polym.Sci. 254: 516–521 (1976).
TiO_2	Degussa P 25 anatase	10^{-2} 1 mol dm^{-3} LiCl KCl		cip	5.5 / 5.3	M. Tschapek, C. Wasowski, and R.M. Torres Sanchez. J.Electroanal.Chem. 74: 167 176 (1976).
TiO_2		10^{-3} mol dm^{-3} KCl, LiCl 0.1 mol dm^{-3} KCl	room	iep / pH	6.6 / 3.6	E.A. Nechaev, and L.M. Smirnova. Kolloid.Zh. 39: 186–190 (1977).
TiO_2	high purity reagent			authors'	5.8	M.A. Butler, and D.S. Ginley. J.Electrochem.Soc. 125: 228 232 (1978).

TABLE 3.1 (Continued)

Oxide	Description	Salt	T	Method	pH_0	Source
TiO_2	rutile	10^{-3}–10^{-1} mol dm^{-3} KNO$_3$	25	cip	5.7	D.E. Yates, thesis, quoted after J.A. Davis, R.O. James, and J.O. Leckie. J.Colloid Interf.Sci. 63: 480–499 (1978).
TiO_2	rutile, different samples	0.01 mol dm^{-3} KNO$_3$	25	iep	3.4–5.5	D.N. Furlong, and G.D. Parfitt, J.Colloid Int.Sci. 65: 548 554 (1978).
TiO_2	anatase, different samples	0.01 mol dm^{-3} KNO$_3$	25	iep	2.7–6.0	D.N. Furlong, and G.D. Parfitt, J.Colloid Int.Sci. 65: 548–554 (1978).
TiO_2	anatase, washed				5.8	M. Ashida, M. Sasaki, H. Kan, T. Yasunaga, K. Hachiya, and T. Inoue. J.Colloid Int.Sci. 67: 219–225 (1978).
TiO_2		0.01 mol dm^{-3} KCl		iep	5	E.A. Nechaev, and V.N. Sheyin. Kolloidn.Zh. 41: 361 363 (1979).
TiO_2	monodispersed, hydrolysis of ethoxide and TiCl$_4$ aerosols	10^{-3} mol dm^{-3} KNO$_3$		iep	shells 4.5 spheres 5.2	M. Visca, and E. Matijevic. J.Colloid Interf.Sci. 68: 308–319 (1979).
TiO_2	rutile, from TiCl$_4$	0.01 mol dm^{-3} KNO$_3$		iep	5.5	D.N. Furlong, K.S.W. Sing, and G.D. Parfitt. J.Colloid Int.Sci. 69: 409–419 (1979).
TiO_2	Anatase HR, Titan			iep	3	H.A. Elliott, and C.P. Huang, J.Colloid Int.Sci. 70: 29–45 (1979).
TiO_2	anatase, Pfalz and Bauer	0.01 mol dm^{-3} NaClO$_4$	25	iep	8.3	M.A. Anderson, and D.T. Malotky. J.Colloid.Int. Sci. 72: 413–427 (1979).
TiO_2		0.01 mol dm^{-3} KCl		iep	5	E.A. Nechaev, and G.V. Zvonareva. Kolloidn.Zh. 42: 511–516 (1980).
TiO_2	gel 27% water, from hydrolysis of TiCl$_4$			iep	3.9	N. Jaffrezic-Renault, and H. Andrade-Martins. J.Radioanal.Chem. 55: 307 316 (1980).
TiO_2	rutile, hydrolysis of TiCl$_4$	10^{-2}–2.9 mol dm^{-3} KNO$_3$ 10^{-3}–0.1 mol dm^{-3} LiNO$_3$ 10^{-3}–0.1 mol dm^{-3} (CH$_4$)$_4$NCl	25	cip	5.8–6	D.E. Yates, and T.W. Healy. J.Chem.Soc. Faraday I 176: 9–18 (1980).
TiO_2	from chloride, different calcination	0.01 mol dm^{-3} KCl	40	pH	8.5 (300 C) 7.2 (500 C, anatase)	H. Kita, N. Henmi, K. Shimazu, H. Hattori, and K. Tanabe. J.Chem.Soc.Faraday Trans.I 77: 2451–2463 (1981).

Material	Description	Electrolyte	t/°C	Method	pH	Reference
TiO_2	rutile, Tioxide	10^{-3} mol dm^{-3} KNO$_3$		iep	5.9	D.N. Furlong, P.A. Freeman, and A.C.M. Lau. J.Colloid Int.Sci. 80: 20–31 (1981).
TiO_2	Degussa P 25, washed with boiling water	10^{-3} 10^{-1} NaCl		cip iep	6.2	A. Foissy, A.M. Pandou, J.M. Lamarche, and N. Jaffrezic-Renault. Colloids Surf. 5: 363 368 (1982).
TiO_2	anatase, Merck	0.001–0.1 mol dm^{-3} LiCl,NaCl,NaI	25	cip	6	R. Sprycha. J.Colloid Interf.Sci. 102: 173–185 (1984).
TiO_2	monodispersed, from ethoxide	10^{-3} 10^{-1} mol dm^{-3} KCl	25	cip iep	5.2 5.5	E. Barringer, and H.K. Bowen. Langmuir 1: 420–428 (1985).
TiO_2	Degussa 70% anatase	0.001 mol dm^{-3} KCl	23	iep	6.3	F.J. Gil-Llambias, A.M. Escudey, J.L.G. Fierro, and A. Lopez Agudo. J. Catalysis 95: 520–526 (1985).
TiO_2	Degussa P 25	10^{-3} 10^{-1} mol dm^{-3} KCl, KNO$_3$	20 25	cip iep coagulation	6.6	M.J.G. Janssen, and H.N. Stein. J.Colloid Interf. Sci 109: 508–515 (1986).
TiO_2	Merck 808 washed	10^{-2} mol dm^{-3} KNO$_3$	20 25	pH iep	7 <5	M.J.G. Janssen, and H.N. Stein. J.Colloid Interf. Sci. 109: 508–515 (1986).
TiO_2	rutile natural, washed	0.1 mol dm^{-3} NaCl	25	iep	4.1	S.N. Omenyi, B.J. Herren, R.S. Snyder, and G.V.F. Seaman. J.Colloid Int.Sci. 110: 130–136 (1986).
TiO_2	anatase	10^{-4}–10^{-2} mol dm^{-3} NaI, NaCl, CsCl		iep	6	R. Sprycha. J.Colloid Interf.Sci. 110: 278–281 (1986).
TiO_2	from chloride	0.009 mol dm^{-3} NaClO$_4$		iep	6.2	W.F. Bleam, and M.B. McBride. J.Colloid Interf. Sci 110: 335–346 (1986).
TiO_2	anatase, synthetic, aerosol procedure	0.001–0.01 mol dm^{-3} NaNO$_3$	25	intersection iep	6.5 4.4	N. Kallay, D. Babic, and E. Matijevic. Colloids Surf. 19: 375 386 (1986).
TiO_2	rutile, Tioxide	NaNO$_3$	25	cip	6	M.L. Machesky, and M.A. Anderson, Langmuir 2: 582 587 (1986).
TiO_2	rutile, Tioxide	10^{-3} 10^{-1} mol dm^{-3} KNO$_3$	21	cip iep	5.9	H.M. Jang, and D.W. Fuerstenau. Colloids Surf. 21: 235 257 (1986).
TiO_2	rutile, Tioxide	0.02 mol dm^{-3} NH$_4$NO$_3$		iep	6.2	D.W. Fuerstenau, and K. Osseo-Asare. J.Colloid Interf.Sci. 118: 524–542 (1987).
TiO_2	commercial, Tioxide	0.001 mol dm^{-3} KNO$_3$		iep	5.4–5.5	A.S. Rao. J.Disp.Sci.Technol. 8: 457–476 (1987).
TiO_2	Degussa P 25			mass titration	6.2	S. Subramanian, J.S. Noh, and J.A. Schwarz. J. Catalysis 114: 433–439 (1988).

TABLE 3.1 (Continued)

Oxide	Description	Salt	T	Method	pH$_0$	Source
TiO$_2$	rutile, Ventron, washed	10^{-3} 10^{-1} mol dm^{-3} NaCl, NaClO$_4$	20	cip	5.3	J. Szczypa, M. Kosmulski, and L. Wasowska, J.Colloid.Int.Sci. 126: 592 595 (1988).
TiO$_2$	quantum size, hydrolysis of TiCl$_4$, dialysis	0.01 mol dm^{-3}	25	inflection, coagulation	5.1	C. Kormann, D.W. Bahnemann, and M.R. Hoffmann. J.Phys.Chem. 92: 5196–5201 (1988).
TiO$_2$	Degussa P-25		23	mass titration pH	4 4.4	S. Subramanian, J.A. Schwarz, and Z. Hejase. J.Catalysis. 117: 512–518 (1989).
TiO$_2$	rutile from chloride	0.005–0.2 mol dm^{-3} KNO$_3$	20	cip	5.6	L.G.J. Fokkink, A de Keizer, and J. Lyklema. J.Colloid.Int.Sci. 127: 116–131 (1989).
TiO$_2$	Degussa P 25	10^{-3}–10^{-1} mol dm^{-3} NaCl		cip mass titration	4.9 4.0	J.S. Noh, and J.A. Schwarz. J.Colloid.Interf.Sci. 130: 157 164 (1989).
TiO$_2$	Degussa P 25	0.01–0.5 mol dm^{-3} NaCl	25	cip	6.3	S.R. Mehr, D.J. Eatough, L.D. Hansen, E.A. Lewis, and J.A. Davis. Thermochim.Acta. 154: 129–143 (1989).
TiO$_2$	rutile	0.001–0.1 mol dm^{-3} NaCl		pH, merge	5.1	W. Janusz. Mater.Chem.Phys. 24: 39–50 (1989).
TiO$_2$	rutile	0.002–0.1 mol dm^{-3} KNO$_3$	20	cip iep	5.8 5.9	A.W. Gibb, and L.K. Koopal. J.Colloid Interf.Sci. 134: 122–138 (1990).
TiO$_2$	rutile	0.02 mol dm^{-3} KNO$_3$	20	iep	5.9	L.G.J. Fokkink, A de Keizer, and J. Lyklema. J.Colloid Int.Sci. 135: 118–131 (1990).
TiO$_2$	rutile, from TiCl$_4$, 4 different samples	different 1 1 salts		coagulation	2.8 5.9	F. Dumont, J. Warlus, and A. Watillon. J.Colloid Int.Sci. 138: 543 554 (1990).
TiO$_2$	reagent, washed	none, 0.001 mol dm^{-3} KCl and CaCl$_2$	25	iep	2.0	Fouad Taha, A.M. El-Roudi, A.A. Abd El Gaber, and F.M. Zahran. Rev.Roum.Chim. 35: 503–509 (1990).
TiO$_2$	Degussa P 25	0.01(?)–0.25 mol dm^{-3} KNO$_3$	25	cip	6.3	K.C. Akratopulu, C. Kordulis, and A. Lycourghiotis. J.Chem.Soc.Faraday Trans. 86: 3437 3440 (1990).
TiO$_2$	JRC TiO$_2$ 5	0.1 mol dm^{-3} NaNO$_3$	25	pH	5.36	H. Tamura, and R. Furuichi. Bunseki Kagaku. 40: 635–640 (1991).

Material	Sample	Electrolyte	T	Method	pH	Reference
TiO_2	anatase, British Drug Houses	$NaNO_3$	25	mass titr. / inflection	6.55 / 6.6	S. Zalac, and N. Kallay. J.Colloid Int.Sci. 149: 233-240 (1992).
TiO_2	anatase from chloride	0.005 0.05 mol dm^{-3} KNO_2	25	cip	6.2	L.A. de Faria, and S. Trasatti. J.Electroanal.Chem. 340: 145-152 (1992).
TiO_2	anatase, Aldrich, washed	10^{-2} mol dm^{-3} KCl 0.001–0.1 KCl	25	iep	6.8	M. Kosmulski, and E. Matijevic. Colloids Surf. 64: 57-65 (1992).
TiO_2	anatase, spherical aerosol technique	0.001 mol dm^{-3} KCl	25	cip / iep	6 / 6.8	M. Kosmulski, and E. Matijevic. Colloids Surf. 64: 57-65 (1992).
TiO_2	rutile, hydrolysis of $TiCl_4$, soxhlet extraction	0.0001, 0.001 mol dm^{-3} KNO_3	25	iep	5.6	I. Larson, C.J. Drummond, D.Y.C. Chan, and F. Grieser. J.Am.Chem.Soc. 115: 11885–11890 (1993).
TiO_2	P 25 Degussa	$NaNO_3$	25	cip	6.5	A.T. Stone, A. Torrents, J. Smolen, D. Vasudevan, and J. Hadley. Environ. Sci. Technol. 27: 895–909 (1993).
TiO_2	rutile synthetic	0.001 mol dm^{-3} KNO_3	20	iep	5.7	A. Giatti, and L.K. Koopal. J.Electroanal.Chem. 352: 107–118 (1993).
TiO_2	Degussa P 25	0.002–1 mol dm^{-3} KNO_3	25	cip	5.8	M.J. Avena, O.R. Camara, and C.P. de Pauli. Colloids Surf. 69: 217 228 (1993).
TiO_2	Tioxide	KNO_3		iep / cip	5.0	M.G. Cattania, S. Ardizzone, C.L. Bianchi, and S. Carella. Colloids Surf.A 76: 233–240 (1993).
TiO_2	rutile, Tioxide	KOH+HCl		acousto	3	T.E. Petroff, M. Sayer, and S.A.M. Hesp. Colloids Surf.A. 78: 235–243 (1993).
TiO_2	Fluka anatase	10^{-4} mol dm^{-3} NaCl		iep	3.5	J. Eiscan, M. Kosec, and N. Kallay. Colloids Surf. A 79: 217–226 (1993).
TiO_2	anatase, BDH	10^{-4} mol dm^{-3} $NaNO_3$	20	mass titration	5.96	N. Kallay, S. Zalac, and G. Stefanic. Langmuir 9: 3457 3460 (1993).
TiO_2	rutile, Aldrich	KCl	25	salt titr.	5.97	M. Kosmulski, J. Matysiak, and J. Szczypa. J.Colloid.Int.Sci. 164: 280–284 (1994).
TiO_2	Amend, washed	KNO_3		iep	6.0	A.S. Desai, G.E. Peck, J.E. Lovell, J.L. White, and S.L. Hem. J.Colloid Interf.Sci. 166: 29–34 (1994).
TiO_2	rutile, Tioxide, washed	0.01, 0.1 mol dm^{-3} KCl, LiCl, CsCl, KNO_3, $KClO_4$	25	iep / cip	6.0 / 6.0	N. Kallay, M. Colic, D.W. Fuerstenau, H.M. Jang, and E. Matijevic. Colloid Polym.Sci. 272: 554–561 (1994).
TiO_2	anatase Fisher	10^{-3}–10^{-1} mol dm^{-3} NaCl	25	cip	6	W. Janusz. Pol.J.Chem. 68: 1871–1880 (1994).

TABLE 3.1 (Continued)

Oxide	Description	Salt	T	Method	pH$_0$	Source
TiO$_2$			23	iep	8.9	P. Jayaweera, S. Hettiarachchi, and H. Ocken. Colloids. Surf. A 85: 19–27 (1994).
TiO$_2$	anatase	0.1 mol dm^{-3} KNO$_3$	25	pH	$-$log [H] = 6.58	C. Ludwig, and P.W. Schindler. J.Colloid Interf. Sci. 169: 284–290 (1995).
TiO$_2$	anatase, hydrolysis of isopropoxide	0.001–0.1 mol dm^{-3} NH$_4$NO$_3$	25	cip iep	6.3	N. Spanos, I. Georgiadou, and A. Lycourghiotos. J.Colloid Int.Sci. 172: 374–382 (1995).
TiO$_2$	rutile, calcination of anatase	0.001–0.1 mol dm^{-3} NH$_4$NO$_3$	25	cip iep	5.5	N. Spanos, I. Georgiadou, and A. Lycourghiotos. J.Colloid Int.Sci. 172: 374–382 (1995).
TiO$_2$	Degussa P 25	10^{-4}–10^{-2} mol dm^{-3} NaCl		iep	5.85	K.W. Bush, M.A. Bush, S. Gopalakrishnan, and E. Chibowski. Coll.Polym.Sci. 273: 1186–1192 (1995).
TiO$_2$	rutile	0.001–0.1 mol dm^{-3} KCl		cip cip iep	5.75 5.6	E.V. Golikova, O.M. Rogoza, D.M. Shelkunov, and Y.M. Chernoberezhskii. Kolloidn.Zh. 57: 25–29 (1995).
TiO$_2$	Degussa P-25			pH	6.0–6.4	H. Hu, I.E. Wachs, and S.R. Bare J.Phys.Chem. 99: 10897–10910 (1995).
TiO$_2$	rutile	0.001 mol dm^{-3} KNO$_3$		iep AFM	5.6	I. Larson, C.J. Drummond, D.Y.C. Chan, and F. Grieser, J.Phys.Chem. 99: 2114–2118 (1995).
TiO$_2$	anatase, Aldrich, washed	0.01 mol dm^{-3} NaNO$_3$	25	acousto	5.85	M. Kosmulski, and J.B. Rosenholm. J.Phys.Chem. 100: 11681–11687 (1996).
TiO$_2$	Degussa P-25	0.001–0.1 mol dm^{-3} KCl	25	cip iep	6.5	R. Rodriguez, M.A. Blesa, and A.E. Regazzoni. J.Colloid Interf.Sci. 177: 122 131 (1996).
TiO$_2$	rutile synthetic	0.001–0.1 mol dm^{-3} NaCl	20	cip	5.8	E.M. Lee, and L.K. Koopal. J.Colloid Int.Sci. 177: 478–489 (1996).
TiO$_2$	rutile, Alfa Aesar, different treatment	0.01 mol dm^{-3} NaNO$_3$		pH	4.2 7	C. Contescu, V.T. Popa, and J.A. Schwarz. J.Colloid Interf.Sci. 180: 149–161 (1996).
TiO$_2$	Degussa P 25: 60% anatase 40% rutile	0.01 mol dm^{-3} NaNO$_3$		pH	4.8	C. Contescu, V.T. Popa, and J.A. Schwarz. J.Colloid Interf.Sci. 180: 149–161 (1996).
TiO$_2$	anatase, Alfa Aesar, different treatment	0.01 mol dm^{-3} NaNO$_3$		pH	< 3–6.8	C. Contescu, V.T. Popa, and J.A. Schwarz. J.Colloid Interf.Sci. 180: 149–161 (1996).
TiO$_2$	rutile, JRC TIO-5, washed	0.1 mol dm^{-3} NaNO$_3$	25	pH	5.5	H. Tamura, N. Katayama, and R. Furuichi. Environ. Sci. Technol. 30: 1198- 1204 (1996).

		electrolyte	T	method	pzc/iep	reference
TiO_2	xerogel; sintered			iep	x 6.2 / s 6.2	X. Fu, L.A. Clark, Q. Yang, and M.A. Anderson. Environ.Sci.Technol. 30: 647–653 (1996).
TiO_2	rutile, Polysciences	0.005 mol dm^{-3} NaNO$_3$	25	iep	5	C. Quang, S.L. Petersen, G.R. Ducatte, and N.E. Ballou. J.Chromatogr.A. 732: 377–384 (1996).
TiO_2	rutile, Ishihara Sangyo	0.001 mol dm^{-3} KNO$_3$		pH	6.1	S. Fukuzaki, H. Urano, and K. Nagata. J. Fermentation Bioeng. 81: 163–167 (1996).
TiO_2	P 25 Degussa	NH$_4$Cl		acousto	6.3–7 (hysteresis)	R.A. Overbeek, E.J. Bosma, D.W.H. de Blauw, A.J van Diillen, H.G. Bruil, and J.W. Geus. Applied Catal.A 163: 129–144 (1997).
TiO_2	anatase, Zaklady Chemiczne Police, washed	10^{-4}–10^{-1} mol dm^{-3} NaCl CsCl	25	cip / pH / cip / pH	6.4 / 6.4 / 5.3 / 5.8–6.9	W. Janusz, I. Kobal, A. Sworska, and J. Szczypa. J.Colloid Interf.Sci. 187: 381–387 (1997).
TiO_2	TiCl$_4$ + NH$_4$OH, then drying at 200 C	0.01 mol dm^{-3} NaCl	20	iep	4.7	L.E. Ermakova, M.P. Sidorova, and V.M. Smirnov. Kolloid.Zh. 59: 563–565 (1997).
TiO_2	rutile, Tioxide	HCl/NaOH		iep	4.2	P. Mikulasek, R.J. Wakeman, and J.Q. Marchant. Chem.Eng.J. 67: 97–102 (1997).
TiO_2	Degussa P25	NaClO$_4$	25	cip	6.2	A.M. Jakobsson, Y. Albinsson, and R.S. Rundberg. Technical Report TR-98-15, Swedish Nuclear Fuel and Waste Management Co. (1998).
TiO_2	rutile, Tioxide	0.03–1 mol dm^{-3} NaCl	25	cip	5.4	M.L. Machesky, D.J. Wesolowski, D.A. Palmer, and K. Ichiro-Hayashi. J.Colloid Interf.Sci. 200: 298–309 (1998).
TiO_2	Degussa P 25 washed	0–0.1 mol dm^{-3} NaCl	20	cip	5.73	C.P. Schulthess, and J.Z. Belek. Soil.Sci.Soc.Am.J. 62: 348–353 (1998).
TiO_2	reagent grade, Degussa, P 25 washed (different solid to liquid ratios)	0.005, 0.1 mol dm^{-3} NaCl	25	intersection / pH	6.3 / 5.6–5.9	A.M. Jakobsson, and Y. Albinsson, Radiochim. Acta 82: 257–262 (1998).
TiO_2	anatase, Sakai, washed	10^{-3} mol dm^{-3} NaNO$_3$		iep	6.2	M. Colic, and D.W. Fuerstenau. Powder Technol. 97: 129–138 (1998).
TiO_2	from butoxide, calcined at 450 C	0.0001–0.01 mol dm^{-3}		iep	KCl 5.7 / KNO$_3$ 5.9	G.A. Gomes, and J.F.C. Boodts. J.Braz.Chem.Soc. 10: 92–96 (1999).
TiO_2	from chloride, purified	0.001 mol dm^{-3} KNO$_3$		iep	3.5	G.A. Gomes, and J.F.C. Boodts. J.Braz.Chem.Soc. 10: 92–96 (1999).
TiO_2	anatase, Aldrich, washed	NaCl	25	cip	6.1	M. Kosmulski, and A. Plak. Colloids Surf. A 149: 409–412 (1999).

TABLE 3.1 (Continued)

Oxide	Description	Salt	T	Method	pH$_0$	Source
TiO$_2$	from n-butoxide	10^{-3} mol dm^{-3} KCl		iep	3.54	X.C. Guo, and P. Dong. Langmuir 15: 5535–5540 (1999).
TiO$_2$	anatase, from ethoxide	10^{-4}–10^{-1} mol dm^{-3} NaCl		cip iep	6.2 6.5	W. Janusz, A. Sworska, and J. Szczypa. Colloids Surf. A 152: 223–233 (1999).
TiO$_2$	rutile, Aldrich	0.001 mol dm^{-3} KNO$_3$	25	iep	5	G.E. Morris, W.A. Skinner, P.G. Self, and R.S.C. Smart. Colloids Surf.A 155: 27–41 (1999).
TiO$_2$	Merck, anatase + 15% rutile Aldrich, anatase	0.01 mol dm^{-3} NaNO$_3$	20	iep	5.2 5.9	M. Kosmulski, S. Durand-Vidal, J. Gustafsson, and J.B. Rosenholm. Colloids Surf.A 157: 245–259 (1999).
TiO$_2$	anatase, Aldrich, washed	0.005–0.3 mol dm^{-3} NaNO$_3$	25	cip iep	5.7 5.8	M. Kosmulski, J. Gustafsson, and J.B. Rosenholm. J.Colloid Int.Sci. 209: 200–206 (1999).
TiO$_2$	rutile, Aldrich, washed	0.001–0.1 mol dm^{-3} NaNO$_3$	25	cip iep	6.2	M. Kosmulski, J. Gustafsson, and J.B. Rosenholm. Colloid Polym.Sci. 277: 550–556 (1999).
TiO$_2$	hydrolysis of TiCl$_4$	10^{-3}–10^{-2} mol dm^{-3} NaCl	20	iep	5.5 (no clear cip)	M.P. Sidorova, L.E. Ermakova, N.F. Bogdanova, and M.A. Proner. Kolloid. Zh. 61: 108–112 (1999).
TiO$_2$		10^{-2} mol dm^{-3} KNO$_3$		iep	7.3	M.P. Sidorova, H. Zastrow, L.E. Ermakova, N.F. Bogdanova, and V.M. Smirnov. Kolloid.Zh. 61: 113–117 (1999).
TiO$_2$·1.14 H$_2$O	amorphous; TiCl$_4$ + water + NaOH, hydrolysis at pH 7–8, 20 C, dried at 20 C	0.1 mol dm^{-3} NaCl	30	pH	4.4	M. Sugita, M. Tsuji, and M. Abe. Bull.Chem. Soc.Jpn. 63: 559–564 (1990).
TiO$_2$·1.25 H$_2$O	amorphous; TiCl$_4$ + water + NaOH, hydrolysis at pH >13, 20 C, dried at 20 C	10^{-1} mol dm^{-3} NaCl KCl	30	pH	3.6 3.3 2.7	M. Sugita, M. Tsuji, and M. Abe. Bull.Chem. Soc.Jpn. 63: 559–564 (1990).
TiO$_2$·1.36 H$_2$O	anatase-type; Ti isopropoxide + cyclohexane + water, hydrolysis at pH 7, 10 C, dried at 85 C	0.1 mol dm^{-3} NaCl	30	pH	4.4	M. Sugita, M. Tsuji, and M. Abe. Bull.Chem. Soc.Jpn. 63: 559–564 (1990).
TiO$_2$·1.68 H$_2$O	amorphous; Ti isopropoxide + water, hydrolysis at pH 7, 5 C, dried at 20 C	10^{-1} mol dm^{-3} LiCl NaCl KCl	30	pH	4.5 4.4 4.0	M. Sugita, M. Tsuji, and M. Abe. Bull.Chem. Soc.Jpn. 63: 559 564 (1990).

Adsorbent	Description	Electrolyte	T	Method	pH	Reference
$TiO_2 \cdot 1.68$ amorphous; H_2O	Ti isopropoxide + water + NaOH, hydrolysis at pH > 13;5 C, dried at 20 C	0.1 mol dm^{-3} LiCl	30	pH	3.6	M. Sugita, M. Tsuji, and M. Abe. Bull.Chem.Soc.Jpn. 63: 559–564 (1990).
$TiO_2 \cdot n$ H_2O	from sulfate	0.1 mol dm^{-3} NaNO$_3$	25	pH	6.55	A.R. Gupta, and B. Venkataramani. Bull.Chem.Soc.Jpn. 61: 1357–1362 (1988).
$TiO_2 \cdot n$ H_2O	from sulfate	0.1 mol dm^{-3} NaNO$_3$	25	pH	6.7	A. Suzuki, H. Seki, and H. Maruyama. J.Chem.Eng.Jpn. 27: 505–511 (1994).
UO_2	from hexafluoride	0–0.001 mol dm^{-3} KClO$_4$, KCl, KNO$_3$	25	iep, coagulation	4.5	A.J.G. Maroto. Anales.Asoc.Quim.Argent. 58: 187–190 (1970).
UO_2	from nitrate, contains $UO_{2.9}$ and U_3O_8	0–0.05 mol dm^{-3} LiNO$_3$, KNO$_3$	25	cip / pH / pH	5.8 / 6.4–6.6 / 4.9	A. Fusain. J.Colloid.Int.Sci. 102: 389–399 (1984).
V_2O_5	cf. Table 3.9					
V_2O_5		0.1 mol dm^{-3} KCl	room	pH	3.1	E.A. Nechaev, and L.M. Smirnova. Kolloid.Zh. 39: 186–190 (1977).
V_2O_5		HCl	23	iep	1.4	F.J. Gil-Llambias, A.M. Escudey, J.L.G. Fierro, and A. Lopez Agudo. J. Catalysis 95: 520–526 (1985).
$WO_{2.87}$	thermal decomposition of ammonium tungstate	0.1 mol dm^{-3} NaCl, KCl		salt titration	< 2.5	T. Szalay, T. Nemet, and L. Bartha. Z.Phys.Chem. (Leipzig) 259: 641–652 (1978).
WO_3	cf. Table 3.9					
WO_3		0.01 mol dm^{-3} KOH (?)		iep	1.5	E.A. Nechaev. Kolloidn.Zh. 42: 371–373 (1983).
WO_3	reagent, washed	none, 0.01 mol dm^{-3} KCl	25	iep	1.95	Fouad Taha, A.M. El-Roudi, A.A. Abd El Gaber, and F.M. Zahran. Rev. Roum. Chim. 35: 503–509 (1990).
WO_3		0.001 mol dm^{-3} KCl		iep	0.3	F.J. Gil Llambias, J. Salvatierra, L. Bouyssieres, M. Escudey, and R. Cid. Appl. Catal. 59: 185–195 (1990).
WO_3			23	iep	4.9	P. Jayaweera, S. Hettiarachchi, H. Ocken. Colloids Surf. A 85: 19–27 (1994).
Y_2O_3		0.01 mol dm^{-3} KOH (?)		iep	7.35	E.A. Nechaev. Kolloidn.Zh. 42: 371–373 (1980).
Y_2O_3	commercial, low in carbonate cubic	10^{-2}–1 mol dm^{-3} NaClO$_4$, NaCl, KNO$_3$	25	cip	7.45 / 7.6 / 7.8	S.I. Pechenyuk. Zh.Fiz.Khim. 61: 165–169 (1987).

TABLE 3.1 (Continued)

Oxide	Description	Salt	T	Method	pH$_0$	Source
Y$_2$O$_3$	plasma prepared	0.001 mol dm^{-3} NaCl		iep	8.8	M. Kagawa, M. Omori, Y. Syono, Y. Imamura, and S. Usui. J.Am.Ceram.Soc. 70: c212–c213 (1987).
Y$_2$O$_3$	spherical, monodispersed (thermal decomposition of basic carbonate)	10^{-4}–10^{-2} mol dm^{-3} NaClO$_4$		cip iep	9.0	R. Sprycha, J. Jablonski, and E. Matijevic. J.Colloid Interf.Sci. 149: 561 568 (1992).
Y$_2$O$_3$	spherical, monodispersed (thermal decomposition of basic carbonate)	0.1 mol dm^{-3} KCl	25	iep	10.2	M. Kosmulski, and E. Matijevic. Colloids Surf. 64: 57–65 (1992).
Y$_2$O$_3$	99.99%, Rhone Poulenc	10^{-3} mol dm^{-3} NaNO$_3$	25	iep	8.5	M. Yasrebi, M. Ziomek-Moroz, W. Kemp, and D.H. Sturgis. J.Am.Ceram.Soc. 79: 1223–1227 (1996).
Y$_2$O$_3$	thermal decomposition of basic carbonate		25	iep	8.6	R.C. Plaza, J.D.G. Duran, A. Quirantes, M.J. Ariza, and A.V. Delgado. J.Colloid Interf.Sci. 194: 398–407 (1997).
Y$_2$O$_3$	Starck	10^{-3} mol dm^{-3} KCl	25	iep	8 / 7.6 (aged dispersion)	J.B. Rosenholm, F. Manelius, J. Stranden, M. Kosmulski, H. Fagerholm, H. Byman-Fagerholm, and A.B.A. Pettersson in Ceramic Interfaces: Properties and Applications III, R. St. C. Smart and J. Nowotny, eds., Institute of Materials Press, London, 1998, pp. 433–60
Yb$_2$O$_3$		0.01 mol dm^{-3} KOH (?)		iep	6.8	E.A. Nechaev. Kolloidn.Zh. 42: 371 373 (1980).
Yb$_2$O$_3$	commercial, low in carbonate, cubic	0.01–1 mol dm^{-3} NaCl, KNO$_3$	25	cip	7.15 / 7.2	S.I. Pechenyuk. Zh.Fiz.Khim. 61: 165–169 (1987).
ZnO	cf. Table 3.9					
ZnO	synthetic from chloride	0.0002–0.01 mol dm^{-3} NaCl	25	cip	8.8	L. Blok, and P.L. de Bruyn. J.Colloid Interf.Sci. 32: 518–526 (1970).
ZnO	synthetic 5 samples	NaBr, NaI, NaNO$_3$ 0.0002–0.01 mol dm^{-3} NaNO$_3$	25	cip	8.6 / <8–10	L. Blok, and P.L. de Bruyn. J.Colloid Interf.Sci. 32: 518–526 (1970).
ZnO		0.01 mol dm^{3} KCl		iep	9.3	E.A. Nechaev, and V.N. Sheyin. Kolloidn.Zh. 41: 361 363 (1979).
ZnO		0.01 mol dm^{-3} KOH (?)		iep	9.3	E.A. Nechaev. Kolloidn. Zh. 42: 371–373 (1980).

		Electrolyte	T			Reference
ZnO	treated at different T	10^{-3} mol dm^{-3} NaCl	25	iep	9.8 (600 C)	S. Kittaka, and T. Morimoto. J.Colloid.Int.Sci. 75: 398–403 (1980).
ZnO	Highways	0.1–0.001 mol dm^{-3} NaCl	20	?	8.7	H.F.A. Trimbos, and H.N. Stein. J.Colloid Interf. Sci. 77: 386–396 (1980).
ZnO	quantum size, hydrolysis of acetate	0.002 mol dm^{-3} acetate	25	inflection, coagulation	9.2	D.W. Bahnemann, C. Kormann, and M.R. Hoffmann. J.Phys.Chem. 91: 3789–3798 (1987).
ZnO	reagent, washed	none, 0.001 mol dm^{-3} KCl and CaCl$_2$	25	iep	6.9	Fouad Taha, A.M. El-Roudi, A.A. Abd El Gaber, and F.M. Zahran. Rev. Roum. Chim. 35: 503–509 (1990).
ZnO				pH	9.2	W. Janusz, quoted after R. Sprycha, J. Jablonski, and E. Matijevic. J.Colloid Interf.Sci. 149: 561–568 (1992).
ZnO		0.05 mol dm^{-3} NaClO$_4$		iep	9.8	J.C. Liu, and C.P. Huang. Langmuir 8: 1851 1856 (1992).
ZnO	amorphous	10^{-3} mol dm^{-3} KNO$_3$	25	iep	9.8	R.J. Crawford, I.H. Harding, and D.E. Mainwaring. Langmuir 9: 3050–3056 (1993).
ZnO	reagent washed	0.001–0.1 mol dm^{-3} KNO$_3$	20	cip	8.1	S. Mustafa, S. Parveen, A. Ahmad, and D. Begum. Ads. Sci. Technol. 15: 789–802 (1997).
ZnO		NaOH + HCl		iep	9.3	M. Colic, A. Chien, and D. Morse. Croat.Chem. Acta 71: 905–916 (1998).
ZnO · n H$_2$O	synthetic	0.0175 mol dm^{-3} KNO$_3$	25	iep	9.8	R.J. Crawford, I.H. Harding, and D.E. Mainwaring. J.Colloid Interf.Sci. 181: 561–570 (1996).
ZrO$_2$ / ZrO$_2$	cf. Tables 3.8 and 3.9 monoclinic	10^{-3}, 1 mol dm^{-3} KNO$_3$	25	pH intersection	5.5–6.2 / 6.5	S.M. Ahmed. Can.J.Chem 44: 1663 1670 (1966).
ZrO$_2$	monoclinic Norton	0.001–1 mol dm^{-3} KCl, NaClO$_4$	25	pH	5.7–6.6	S.M. Ahmed. J.Phys.Chem. 73: 3546–3555 (1969).
ZrO$_2$	from sulfate			iep	4.8	L.G. Maidanovskaya, and T.V. Skripko. Zh.Fiz.Khim. 46: 115–118 (1972).
ZrO$_2$	hydrolysis of nitrate	10^{-3}–0.5 mol dm^{-3} NaCl	35	cip / iep	6.7	K.C. Ray, and S. Khan. Indian J.Chem. 13: 577–580 (1975).
ZrO$_2$	99.9% reactor grade	0.01 mol dm^{-3} KNO$_3$	30	iep	6.8	P.H. Tewari, and W. Lee. J.Colloid Int.Sci. 52: 77–88 (1975).
ZrO$_2$		0.01 mol dm^{-3} KCl		iep	6.7	E.A. Nechaev, and V.N. Sheyin. Kolloid.Zh. 41: 361–363 (1979).

TABLE 3.1 (Continued)

Oxide	Description	Salt	T	Method	pH_0	Source
ZrO_2	reagent grade, Johnson-Matthey monoclinic + tetragonal	0.01 mol dm^{-3} $NaNO_3$		iep	4.9	E. Crucean, and B. Rand, T.J. Brit. Ceram.Soc. 78: 96–98 (1979).
ZrO_2	synthetic, from chloride and nitrate, 7 different samples, monoclinic and amorphous	0.01 mol dm^{-3} $NaNO_3$		iep	am 4–4.7 m 4.8	E. Crucean, and B. Rand. T.J. Brit. Ceram.Soc. 78: 96–98 (1979).
ZrO_2		0.01 mol dm^{-3} KOH (?)		iep	6.75	E.A. Nechaev. Kolloidn.Zh. 42: 371–373 (1980).
ZrO_2		0.01 mol dm^3 KCl		iep	6.75	E.A. Nechaev, and G.V. Zvonareva. Kolloid.Zh. 42: 511–516 (1980).
ZrO_2	monoclinic, reagent grade	0.001 mol dm^{-3} KNO_3		pH	6.4–6.5	N.I. Ampelogova. Radiokhimiya 25: 579–584 (1983).
ZrO_2	high purity, monoclinic	3×10^{-4}–10^{-1} mol dm^{-3} KCl		cip iep	4.0	A.A. Baran, N.S. Mitina, and B.E. Platonov. Kolloid Zh. 44: 964–968 (1982).
ZrO_2	baddelyite high purity from INVAP	10^{-3}–10^{-1} mol dm^{-3} KNO_3	25 30 30	iep cip pH	6.5 6.4 6.0	A.E. Regazzoni, M.A. Blesa, and A.J.G. Maroto. J.Colloid Int.Sci. 91: 560–570 (1983).
ZrO_2	monoclinic, hydrolysis of $ZrOCl_2$	NaCl		cip	8.1	R. Paterson, and H. Rahman. J.Colloid. Int.Sci. 103: 106–111 (1985).
ZrO_2	amorphous, hydrolysis of $ZrOCl_2$	0.001–0.1 mol dm^{-3} NaCl	21	pH cip	6.6 6.5–8	J.B. Stankovic, S.K. Milonjic, M.M. Kopecni, and T.S. Ceranic. J.Serb.Chem.Soc. 51: 95–104 (1986).
ZrO_2	TZO Toyo Soda	0.001 mol dm^{-3} KCl	25	iep	6.5	E.M. de Liso, W.R. Cannon, and A.S. Rao. Mat.Res.Soc.Symp.Proc. 60: 43–50 (1986).
ZrO_2	SC30 Magnesium Elektron	0.001 mol dm^{-3} KCl	25	iep	6.2	E.M. de Liso, W.R. Cannon, and A.S. Rao. Mat.Res.Soc.Symp.Proc. 60: 43–50 (1986).
ZrO_2				iep	≤ 5.5	F. Manelius. Finn.Chem.Lett. 14: 167–168 (1987).
ZrO_2	plasma prepared	0.001 mol dm^{-3} NaCl		iep	6.5	M. Kagawa, M. Omori. Y. Syono, Y. Imamura, and S. Usui, J. Am. Ceram. Soc. 70: c212–c213 (1987).
ZrO_2	dehydrated, then refluxed in water for 170 h	0.001 mol dm^3 NaCl		iep	5	M. Kagawa, M. Omori, Y.Syono, Y. Imamura, and S. Usui. J.Am.Ceram.Soc. 70: c212–c213 (1987).

Material	Description	Electrolyte	T	Method	pH/value	Reference
ZrO_2	monoclinic, Toyo Soda, TZ-0	NH_4Cl		iep	5.0	M. Hashiba, H. Okamoto, Y. Nurishi, and K. Hiramatsu, J. Mater. Sci. 23: 2893–2896 (1988).
ZrO_2				iep	5.8	N. Grine, M. Lindheimer, A. Larbot, and S. Partyka. Proc.Int.Conf.Inorg.Membr. ENSC Montpellier 465 (1989).
ZrO_2	Merck, washed	$0.001–0.1$ mol dm^{-3} NaCl		pH	4.3–4.5	W. Janusz. J.Radioanal.Nucl.Chem. 125: 393–401 (1989).
ZrO_2	monoclinic, hydrolysis of $ZrCl_4$, prepared at different T	$0.001–0.1$ mol dm^{-3} KNO_3		cip	8.0	S. Ardizzone, G. Bassi, and G. Liborio. Colloids Surf. 51: 207–217 (1990).
ZrO_2			25	cip	6.5	N.M. Figliolia, thesis, quoted after, M.A. Blesa, A.J.G. Maroto, and A.E. Regazzoni. J.Colloid Interf.Sci. 140: 287–290 (1990).
ZrO_2	reagent grade baddeleyite	$10^{-4}–1$ mol dm^{-3} LiCl, NaCl,KCl,CsCl		cip iep	5.2	E. Yu. Chardymskaya, M.P. Sidorova. O.V. Semenova, and V.E. Shubin. Kolloidn. Zh. 32: 547–552 (1990).
ZrO_2	Aldrich; Ventron, monoclinic, conditioned for 16 h at pH 4.5	$10^{-3}–10^{-1}$ mol dm^{-3} KNO_3	20	merge	> 9.4 > 8.8 9.1	S. Ardizzone, G. Chidichimo, A. Golemme, and M. Radaelli. Croat.Chem.Acta 63: 545–552 (1990).
ZrO_2	reagent grade, monoclinic, conditioned for 53 h, Aldrich	$10^{-3}–10^{-1}$ mol dm^{-3} KNO_3	20	iep cip	7.0 (fast) 5.6(slow)	S. Ardizzone, G. Chidichimo, A. Golemme, and M. Radaelli. Croat.Chem.Acta 63: 545–552 (1990).
ZrO_2	synthetic, monoclinic, different aging	$0.001–0.01$ mol dm^{-3} KNO_3		cip	6.8–8.2	S. Ardizzone, and M. Sarti. Mater.Chem.Phys. 28: 191–198 (1991).
ZrO_2	synthetic, monoclinic, 3 different preparation routes	$0.001–0.1$ mol dm^{-3} KNO_3		cip	7.4–7.7	S. Ardizzone, P. Lazzari, M. Sarti, and M.G. Cattania. Mater.Chem.Phys. 28: 399–412 (1991).
ZrO_2	synthetic, sol-gel process	10^{-4}, 10^{-2} mol dm^{-3} NaCl	25	iep pH	6.7 3.9	J. Randon, A. Larbot, C. Guizard, L. Cot, M. Lindheimer, and S. Partyka, Colloids Surf. 52: 241–255 (1991).
ZrO_2	calcination in O_2	$0.001–0.1$ mol dm^{-3} KNO_3		cip, inflection	7.7	S. Ardizzone, M.G. Cattania, and M. Sarti. Colloids Surf. 68: 25–35 (1992).
ZrO_2	monoclinic, from n-propoxide, calcined at 500°C for 3 h	NaOH + HCl		iep	7.3	S. Soled, and G.B. Mc Vicker. Catalysis Today 14: 189–194 (1992).
ZrO_2	synthetic from chloride	$0.001–0.1$ mol dm^{-3} KNO_3		cip	8 (aged at pH 4.4) 7.3 (aged at pH 9.4)	S. Ardizzone, and S. Carella. Mater.Chem.Phys. 31: 351–354 (1992).

TABLE 3.1 (Continued)

Oxide	Description	Salt	T	Method	pH_0	Source
ZrO_2	calcination in O_2, then H_2 treatment, 13 samples	0.001–0.1 mol dm^{-3} KNO_3		cip, inflection	7.1–8.1	S. Ardizzone, M.G. Cattania, and M. Sarti. Colloids Surf. 68: 25–35 (1992).
ZrO_2	flame pyrolysis of isopropoxide	KCl		iep	7.65–8.5 7.9	M. Schultz, S. Grimm, and W. Burckhardt. Solid State Ionics 63–65: 18–24 (1993).
ZrO_2	$ZrOCl_2$ + NH_3 pretreatment at 140°C	KCl		iep	7.2	M. Schultz, S. Grimm, and W. Burckhardt. Solid State Ionics 63–65: 18–24 (1993).
ZrO_2	$ZrOCl_2$ + urotropine, iep dependent on pretreatment	KCl		iep	140°C 5.3 300°C 3.8 600°C 5.5 1200°C 3.2	M. Schultz, S. Grimm, and W. Burckhardt. Solid State Ionics 63–65: 18–24 (1993).
ZrO_2	3 samples, monoclinic and monocl. + tetragonal	KNO_3		cip	7.65–8.2	M.G. Cattania, S. Ardizzone, C.L. Bianchi, and S. Carella. Colloids Surf. A 76: 233–240 (1993)
ZrO_2	Z-Tech, monoclinic	KOH/HCl		acousto	7	T.E. Petroff, M. Sayer, and S.A.M. Hesp. Colloids Surf.A. 78: 235–243 (1993).
ZrO_2	Ventron, baddeleyite	10^{-4} mol dm^{-3} NaCl		iep	7	J. Biscan, M. Kosec, and N. Kallay. Colloids Surf.A. 79: 217–226 (1993).
ZrO_2	Z-Tech	0.01 mol dm^{-3} KNO_3	25	acousto	7.3	Y.K. Leong, P.J. Scales, T.W. Healy, D.V. Boger and R. Buscall. J.Chem.Soc.Faraday Trans. 89: 2473–2478 (1993).
ZrO_2			23	iep	7.4	P. Jayaweera, S. Hettiarachchi, and H. Ocken. Colloids Surf. A 85: 19–27 (1994)
ZrO_2	synthetic, from $ZrCl_4$, monoclinic M, tetragonal T, and their mixtures	0.001–0.1 mol dm^{-3} KNO_3		cip	M 7.9 T 6.2 ½ M + ¼ T + ¼ C 7.8 1/3 M + 1/3 T + 1/3 C 7.8	S. Ardizzone, M.G. Cattania, and P. Lugo. Electrochim. Acta 39: 1509–1517 (1994).
ZrO_2	SC105 Magnesiun Elektron, washed			mass titration (modified)	6.49	J. Wernet, and D.L. Feke. J.Am.Ceram.Soc. 77: 2693–2698 (1994).

Material	Description	Electrolyte	T	Method	pH_0	Reference
ZrO_2	synthetic, from $ZrCl_4$	0.001–0.03 mol dm^{-3} KNO$_3$		cip	7.56	S. Ardizzone, M.G. Cattania, P. Lazzari, and P. Lugo. Colloids Surf. A 90: 45–54 (1994).
ZrO_2	monoclinic	0.001–0.1 mol dm^{-3} KCl		iep merge	6.5	E.V. Golikova, O.M. Rogoza, D.M. Shelkunov, and Y.M. Chernoberezhskii. Kolloidn.Zh. 57: 25–29 (1995).
ZrO_2	Z-Tech	0.001–0.1 mol dm^{-3} KNO$_3$		acousto cip	7.0	Y.K. Leong, P.J. Scales, T.W. Healy, and D.V. Boger Colloids Surf. A 95: 43–52 (1995).
ZrO_2	synthetic different calcination T	KNO$_3$	25	cip iep	6 (450 C) 5.1 (650°C)	M. Prica, thesis, quoted after J. Lyklema Fundamentals of Interface and Colloid Science, Academic Press, New York 1995, vol 2. pp. A3.1–A3.6.
ZrO_2				pH	7.6–8.1 (850 C) 5.9–6.1	H. Hu, I.E. Wachs, and S.R. Bare. J.Phys.Chem. 99: 10897–10910 (1995).
ZrO_2	Merck, washed	0.01 mol dm^{-3} NaNO$_3$		acousto	7.6	M. Kosmulski, and J.B. Rosenholm. J.Phys.Chem. 100: 11681–11687 (1996).
ZrO_2	monoclinic, Ishihara Sangyo	0.001 mol dm^{-3} KNO$_3$		pH	7.3	S. Fukuzaki, H. Urano, and K. Nagata. J. Fermentation Bioeng. 81: 163–167 (1996).
ZrO_2	hydrous, hydrolysis of $ZrCl_4$	0.01 mol dm^{-3} NaCl		iep	8.2	L.A. Perez-Maqueda, and E. Matijevic. J. Mater.Res. 12: 3286–3292 (1997).
ZrO_2	from Magneson Electron	0.01 mol dm^{-3} KCl	20	iep	6.1	B. Futman, P. van der Meeren, and D. Thierens. Colloids Surf.A. 121: 81–88 (1997).
ZrO_2	Z-Tech	0.1 mol dm^{-3} KCl	25	acousto	7.8	M.J. Solomon, T. Saeki, M. Wan, P.J. Scales, D.V. Boger, and H. Usui. Langmuir 15: 20–26 (1999).
ZrO_2	Tosoh Ceramics TZ-0	0.01 mol dm^{-3} NaCl, NaClO$_4$, NaNO$_3$	25	acousto	8.2	G.V. Franks, S.B. Johnson, P.J. Scales, D.V. Boger, and T.W. Healy. Langmuir 15: 4411–4420 (1999).
ZrO_2	from chloride, different pH of precipitation, monocl., tetragon., and mixtures (linear change of PZC with composition)	10^{-3}–10^{-1} mol dm^{-3} KNO$_3$		cip	M 8.6 T 6.5–6.8 M+T 7.9–8.3	S. Ardizzone, and C.L. Bianchi. J.Electroanal. Chem. 465: 136–141 (1999).
ZrO_2	Ventron	10^{-3}–10^{-1} mol dm^{-3} NaCl		cip	7.8	W. Janusz. Ads. Sci. Technol. 18: 117–134 (2000).
ZrO_2·nH_2O	from ZrO (NO$_3$)$_2$, 5 samples, amorphous	10^{-3}–10^{-1} mol dm^{-3} KNO$_3$	31	cip	6–6.6	F.S. Mandel, and H.G. Spencer. J.Colloid Int.Sci. 77: 577–579 (1980).

TABLE 3.1 (Continued)

Oxide	Description	Salt	T	Method	pH_0	Source
ZrO_2 $\cdot nH_2O$	powder, washed	0.001–1 mol dm^{-3} NaNO$_3$	25	pH	4.6	G.A. Kokarev, V.A. Kolesnikov, A.F. Gubin, and A. Korobanov. Elektrokhimiya 18: 466–470 (1982).
ZrO_2 $\cdot nH_2O$	amorphous, hydrolysis of ZrOCl$_2$	0.1–1 mol dm^{-3} NaCl	25	pH	4.0	S.K. Milonjic, Z.E. Ilic, and M.M. Kopecni. Colloids Surf. 6: 167–174 (1983).
$ZrO_2 \cdot n$ H_2O	monodispersed, hydrolysis of ZrOCl$_2$ in the presence of sulfate	10^{-3} mol dm^{-3} KNO$_3$	25	iep	5.7	M.A. Blesa, A.J.G. Maroto, S.I. Passaggio, N.E. Figliola, and G. Rigotti. J. Mater. Sci. 20: 4601–4609 (1985).
$Zr(OH)_4$	membrane	10^{-3} mol dm^{-3} KCl	31	iep	5.8	Y. Sun, and H.G. Spencer. J.Colloid Interf.Sci. 126: 361–366 (1988).

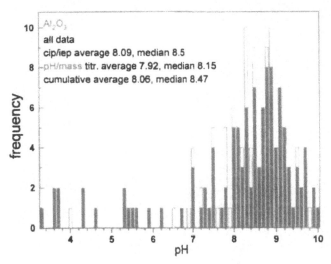

FIG. 3.11 Distribution of PZC values reported for all types of Al_2O_3 in the literature. Dark gray columns: "cip", "iep", and salt titration. White columns: "pH" and mass titration.

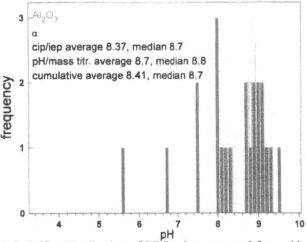

FIG. 3.12 Distribution of PZC values reported for α Al_2O_3 in the literature.

cited in Table 3.1 very likely report the same data. Thus, it is not surprising that the distributions shown in Figs. 3.11–3.37 do not resemble normal distribution. The frequency of PZC in the range between x-0.09 and x obtained by means of "cip", "iep", and "acousto" methods is represented by a gray column and a white column (on the top of the gray column if necessary) represents the "pH" and mass titration methods. The total height of a column (gray + white) represents the total frequency of PZC in a 0.1 pH unit wide range. When a sample was studied by different methods only one PZC value is used. In case of discrepancies the "cip" results are considered as the most significant, then "iep", then "acousto", then "pH", and then mass titration, but usually different methods produced identical or similar results so any other ranking in significance of the methods would not dramatically change the

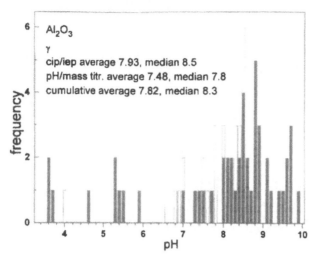

FIG. 3.13   Distribution of PZC values reported for γ Al₂O₃ in the literature.

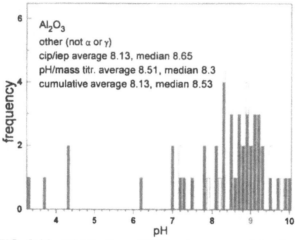

FIG. 3.14   Distribution of PZC values reported for Al₂O₃ (other than α or γ or structure unknown) in the literature.

results. When different salts gave different zero points nitrates V and chlorates VII were favored over halides, and Na and K salts over those of Li, Rb, or Cs. The studies reporting ranges of PZC rather than single values are neglected unless the range is less than 0.2 pH unit wide.

The results reported for Al_2O_3 are presented in Fig. 3.11 (cumulative), 3.12 (the α form), 3.13 (the γ form), and 3.14 (other forms, mixtures of different forms, or the crystalline form not reported). The median value of PZC is higher that the average because of a few studies reporting outstanding, very low PZC. The average PZC obtained by "cip", "iep", and "acousto" methods is higher than that obtained by the "pH" method and mass titration. This suggests presence of acid associated with dry alumina samples. Such a result is not surprising in view of the PZC at pH > 7 and thus tendency to adsorb protons from neutral solutions (cf. Section A). Taking into

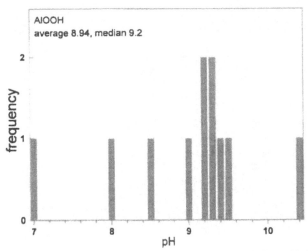

FIG. 3.15 Distribution of PZC values reported for AlOOH in the literature.

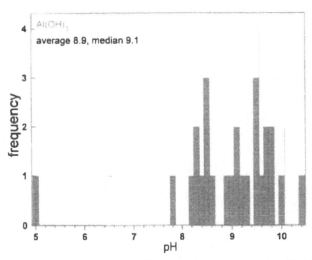

FIG. 3.16 Distribution of PZC values reported for Al(OH)₃ in the literature.

account all methods, the PZC of the α form seems to be higher than that of the γ form, but when only "cip", "iep", and "acousto" data is considered the difference between the α and γ forms is rather insignificant. The apparent low PZC of γ Al_2O_3 is due to many "pH" and mass titration results reported for this modification.

The PZC values for AlOOH (Fig. 3.15) and Al(OH)₃ (Fig. 3.16) are very scattered. The average and median PZC for AlOOH and Al(OH)₃ are equal but this match and the difference between the average PZC of AlOOH and Al(OH)₃ on the one hand and Al_2O_3 on the other are not positively significant.

The number of PZC values reported for Cr hydr(oxides) is modest, and they were collected in one graph (Fig. 3.17). Considering the discussed above results for analogous Al compounds no dramatic difference between Cr oxide, oxohydroxide

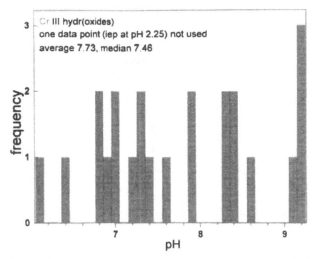

FIG. 3.17 Distribution of PZC values reported Cr (III) hydr(oxides) in the literature.

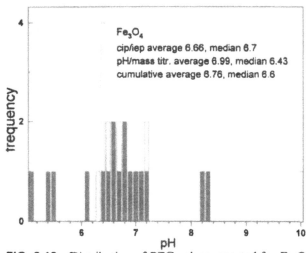

FIG. 3.18 Distribution of PZC values reported for Fe_3O_4 in the literature.

and hydroxide is expected. The results are almost evenly distributed over a 3-pH-units-wide range. In contrast, the distribution of PZC of magnetite (Fig. 3.18) is only one-pH-unit wide with rather few outstanding results.

   In view of the well known difference in the PZC between synthetic (Fig. 3.19) and natural (Fig. 3.20) hematite on the one hand and between hematite and maghemite (Fig. 3.21) on the other, these results are presented separately. Figure 3.22 shows the remaining Fe_2O_3 data (other crystalline forms, mixtures, or unknown form). These figures confirm that the average PZC of synthetic hematite is higher by 2 pH units than that of natural hematite (probably due to presence of silica in natural samples) or that of maghemite, and because of this substantial difference there is no cumulative graph for Fe_2O_3. The hematite-maghemite case delivers

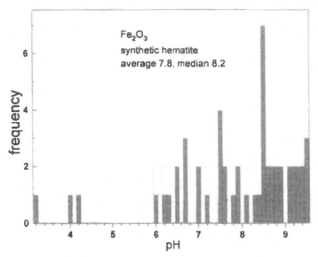

FIG. 3.19 Distribution of PZC values reported for synthetic hematite in the literature.

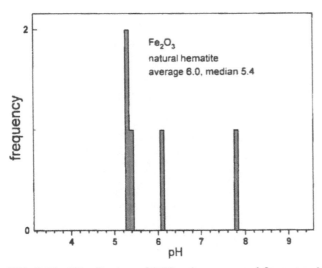

FIG. 3.20 Distribution of PZC values reported for natural hematite in the literature.

a strong evidence that the chemical formula alone is not sufficient to predict the PZC.

Figure 3.23 presents the zero points for hydrous iron oxide and hydroxide and Figs. 3.24–3.26 for FeOOH (synthetic goethite, other than synthetic goethite or crystallographic form unknown, and cumulative, respectively). The average PZC of synthetic goethite is higher than that of FeOOH other than synthetic goethite, but the difference is less significant than that between synthetic and natural hematite. On the other hand the average PZC of synthetic hematite, synthetic goethite, and iron hydrous oxide and hydroxide are practically equal. It is also noteworthy that the median and average values for iron hydr(oxides) are in much better harmony than analogous results for Al_2O_3, and that the error in the "pH" and mass titration data

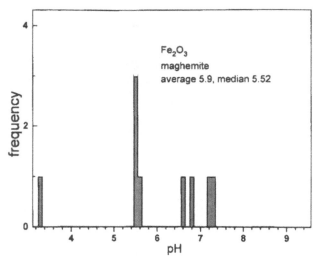

FIG. 3.21 Distribution of PZC values reported for maghemite in the literature.

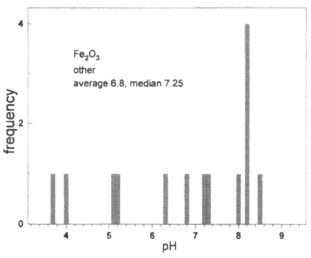

FIG. 3.22 Distribution of PZC values reported for Fe_2O_3 (other than hematite or maghemite or structure unknown) in the literature.

[caused by the acid/base associated with the dry samples of iron hydr(oxides)] is rather insignificant.

Relatively many PZC values are reported for MnO_2, but the results are very scattered probably due to existence of many crystalline forms and differences in stoichiometry from one sample to another.

Some studies with SiO_2 resulted in no specific PZC value (PZC beyond the experimental pH range if any), and in this respect the significance of analysis of the SiO_2 results from Table 3.1 is different than for other materials. The meaning of the average values for quartz (Fig. 3.27) on the one hand and amorphous materials (Fig. 3.28) on the other is confined, but the average PZC of the former is lower by one pH unit. Cumulative results are also presented (Fig. 3.29). The problem with

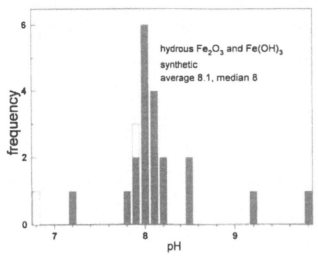

FIG. 3.23 Distribution of PZC values reported for hydrous Fe_2O_3 and $Fe(OH)_3$ in the literature.

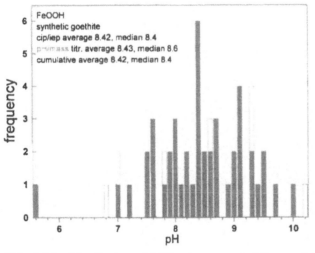

FIG. 3.24 Distribution of PZC values reported for synthetic goethite in the literature.

utilizing the data from Tables 3.8 and 3.9 in this analysis is in very different pH ranges covered in particular studies. For example when only negative σ_0 and ζ potentials are reported at pH > 3, the zero point can be anywhere below this pH. Thus, specific PZC cannot be assigned, but such a result in some sense discredits the studies reporting PZC of silica at pH > 3.

The PZC reported for SnO_2 (Fig. 3.30) are relatively consistent, with a few outstanding values.

Some results for TiO_2 are presented in Tables 3.8 and 3.9, but the number of relevant entries in these tables is rather insignificant as compared with the number of PZC values cited in Table 3.1. The PZC of rutile (Fig. 3.31), anatase (Fig. 3.32), and

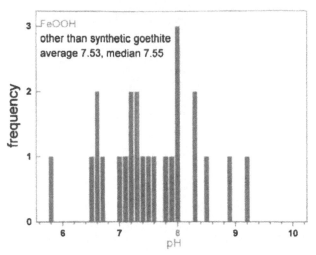

FIG. 3.25 Distribution of PZC values reported for FeOOH (other than synthetic goethite or structure unknown) in the literature.

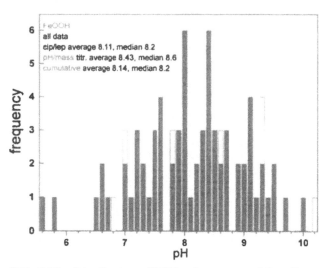

FIG. 3.26 Distribution of PZC values reported for all types of FeOOH in the literature.

mixtures thereof or materials of unknown structure (Fig. 3.33) are presented separately, and in a cumulative graph (Fig. 3.34). In contrast with most materials discussed in this section the distributions of PZC for rutile and anatase are nearly normal, and the average PZC of the former is lower by about 0.5 pH unit. Interestingly the average PZC of "other" TiO_2 materials which are chiefly mixtures of rutile and anatase is half way in between.

The PZC reported for ZrO_2 are presented in Figs. 3.35 (monoclinic), 3.36 (other than monoclinic or the form is unknown) and 3.37 (cumulative). Unlike TiO_2 whose PZC are amassed around the average value, the distribution for ZrO_2 is very broad. There is also no systematic difference in average PZC between monoclinic ZrO_2 and other forms of this material. The average PZC shown in Figs. 3.11–3.37 are close to

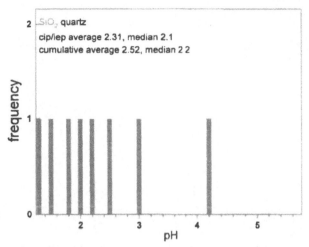

FIG. 3.27   Distribution of PZC values reported for quartz in the literature.

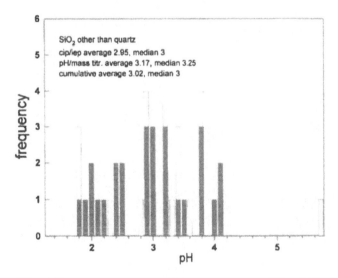

FIG. 3.28   Distribution of PZC values reported for SiO_2 (other than quartz or structure unknown) in the literature.

"recommended" PZC values discussed in the next section. However, the presented distributions are rather broad, and they support the approach that each sample of a material has its characteristic PZC even within a series of relatively pure samples having the same crystallographic structure. Also the results presented in Table 3.1 and not used in Figs. 3.11–3.37 favor this hypothesis. On the other hand, many outstanding PZC values can be a result of impurities and/or erroneous procedures.

D.   Collections of Zero Points

The number of citations is often considered as the quantitative measure of success of a scientific publication. In this respect the compilation of points of zero charge and

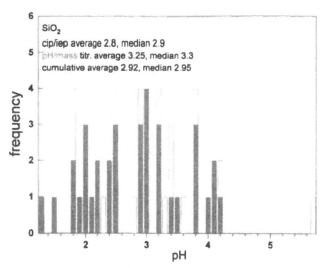

FIG. 3.29   Distribution of PZC values reported for all types of SiO_2 in the literature.

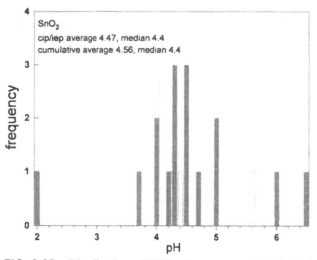

FIG. 3.30   Distribution of PZC values reported for SnO_2 in the literature.

isoelectric points of oxides, oxohydroxides, and hydroxides published by Parks [92] has been one of the most successful publications in colloid chemistry ever. Moreover, although the publication is already 35-years-old, the rate of citations shows a clearly increasing trend as shown in Fig. 3.38. A number of updates has been published afterwards (although they covered only a limited number of oxides), but the Parks' paper is still the most frequently used source of information about the zero points. The results quoted by Parks are not used in Table 3.1, but they are briefly summarized in Table 3.2. A necessity to update the Parks' review has been postulated [93], but only a few large collections of PZC were published after Parks and neither of them was so comprehensive. A collection of collections of PZC values can be found in Ref. [94]. An updated collection of collections together with the

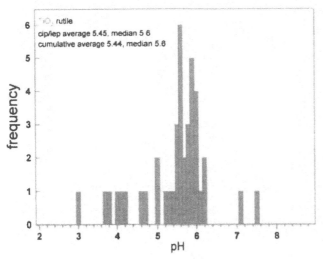

FIG. 3.31 Distribution of PZC values reported for rutile in the literature.

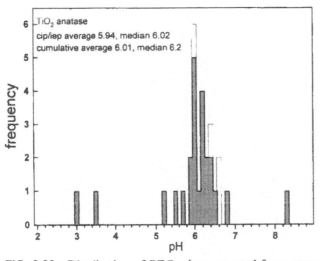

FIG. 3.32 Distribution of PZC values reported for anatase in the literature.

average and median PZC from Figs. 3.11–3.37 is presented in Table 3.2. Not always did the inspection of original source confirm the zero points reported in the collections but no comments on such discrepancies are made in Table 3.2, and the zero points are reported as they are in the collection. The danger of using data from secondary sources can be illustrated by the following example. In one of the collections pH 2.8 is quoted as PZC of Nb_2O_5. In the cited paper the PZC of Nb_2O_5 is reported as taken from another publication. This publication still reports 2.8 as PZC of Nb_2O_5 but with the footnote: "*the PZC of Nb_2O_5 is similar to that of Ta_2O_5*", and the original source is finally cited in which PZC of Ta_2O_5 was experimentally determined and Nb_2O_5 was not mentioned. Therefore, the readers

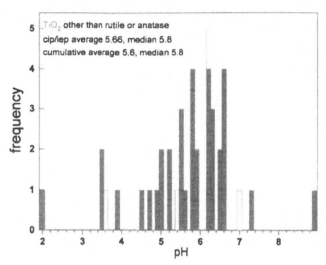

FIG. 3.33 Distribution of PZC values reported for TiO_2 (other than rutile or anatase or structure unknown) in the literature.

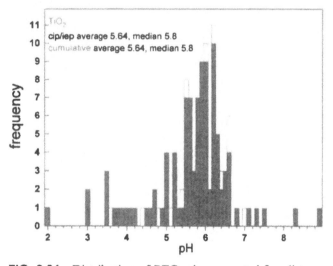

FIG. 3.34 Distribution of PZC values reported for all types of TiO_2 in the literature.

who want to use the data presented in Table 3.1 in their work are advised to refer to primary sources listed in the last column.

In spite of the evidence presented in Table 3.1 and Figs. 3.11–3.37 and of general awareness of inevitable scatter of PZC data, "recommended" PZC values for particular materials (in terms of chemical formula and crystalline structure) are very useful, e.g. to test purity of materials and correctness of experimental procedures. Moreover, a set of unique PZC values is necessary to verify attempts to derive PZC from first principles or to find correlations of the PZC with other physical properties. The recent developments in theory (cf. Chapter 5) are rather not in favor of the concept of unique PZC, unless crystals of the same morphology are compared, but

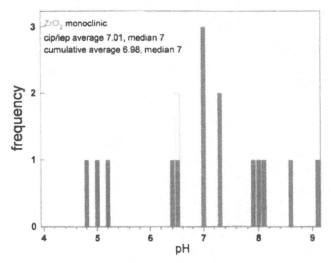

FIG. 3.35   Distribution of PZC values reported for monoclinic ZrO_2 in the literature.

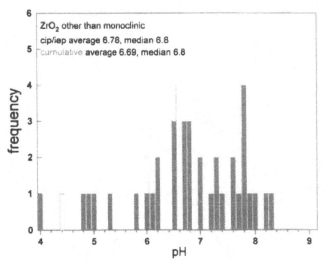

FIG. 3.36   Distribution of PZC values reported for ZrO_2 (other than monoclinic or structure unknown) in the literature.

some correlations (discussed in more detail in Section II) involving these unique PZC are very convincing. One danger with such correlations is that the choice of unique PZC can be biased, and with rather broad distributions reported in Figs. 3.11–3.37 it is always possible to find a value that supports this or another theory. The SiO_2/SnO_2 case is a classical example. Both materials show rather broad ranges of reported PZC which partially overlap. Using selected values it can be "proved" that their PZC are almost equal on the one hand or that they differ by 3 or more pH units on the other. A small difference in PZC between SiO_2 and SnO_2 favors the correlation between PZC and electronegativity (which is similar for Si and Sn), and a large difference supports correlation with the ionic radius (Sn^{4+} is much larger than Si^{4+}).

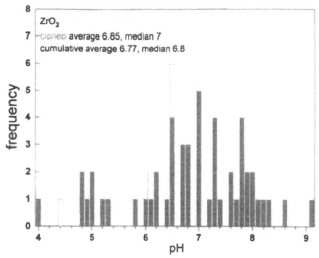

FIG. 3.37   Distribution of PZC values reported for all types of ZrO$_2$ in the literature.

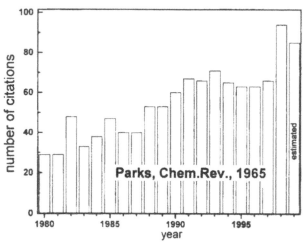

FIG. 3.38   Citations of the review on PZC of hydr(oxides) published by Parks in *Chemical Reviews* in 1965.

This example is not meant to suggest that any collection presented in Table 3.2 or correlation discussed in section II was biased, e.g. some unusual PZC values used to prove such correlations can very well be a result of incomplete literature searches. It would be very difficult to find all PZC published for given material and to properly assess their significance, and even Table 3.1, which is the largest compilation ever published is probably very incomplete.

E.   Mixed Oxides, Complex, and Ill-Defined Materials

Complex materials are more abundant and have more applications than well-defined pure oxides, but information about their zero points is scarce. Many publications

(*Text continues on pg. 179*)

TABLE 3.2 Collections of PZC

Ref.	92	95	96	97	98	99	100	93	101	102	103	104	105
Ag_2O			11.2	11.2									
Al_2O_3		9.2	5–9	9.2			9			8.8	9–9.4	8.5–9.02	
amorphous Al_2O_3	7.5–8												
γ-Al_2O_3	7.4–8.6							4.6–5.3				8.1–9.7	
α-Al_2O_3	2.3–10				9.1				9.1			8.8–9.1	
pseudoboehmite												8.8–9.2	
AlOOH					7.5							8.5	
boehmite γ-AlOOH												10.4	
boehmite α-AlOOH (?)	<2–9.4												
diaspore γ-AlOOH (?)	<2–7.5												
$Al(OH)_3$	3.8–5.2				5–5.2			9.5					7.7
gibbsite α-$Al(OH)_3$ (?)					7.7								
α-$Al(OH)_3$									10				
γ-$Al(OH)_3$ bayerite (?)	5.4–9.3												
$Am(OH)_4$			6.8										
BeO	10.2				10.2				10.2				
BeO hydrous		10.2											
Bi_2O_3													
$Bi(OH)_3$													9.4
CdO	10.4	10.4	10.4–11.6	11.6									
$Cd(OH)_2$	>10.5												
CeO_2	6.75		6.7					7				8.1	
CoO	11.4	11.4	11.4		11.4								9.9–10.5
$Co(OH)_2$												11.42	11.4
Co_3O_4						5.4–11.4						11.35	
Cr_2O_3	7		7–8	8.1	7						6.2–6.3		
α-Cr_2O_3								6.35				6.35	7.0

TABLE 3.2 (Continued)

Ref.	92	95	96	97	98	99	100	93	101	102	103	104	105
$Cr(OH)_3$ sol												8.4	
CuO	9.5		9.5	9.5	9.4		9.5		9.5	8.5			
$Cu(OH)_2$	9.4	9			7.7								
$Cu(OH)_2$ hydrous	7.6												
$Fe(OH)_2$	12	12	12		12								
Fe_3O_4	6.5			6.5		6.5–8.8		3.8–6.5	6.6			6.9	
Fe_2O_3		5.7	5.5–7.2	8.6			7			6.8	8.5		
α-Fe_2O_3	1.9–9.04				9.04			3.2–9.6	8.5			8.5–9	
γ-Fe_2O_3	6.7							3.3–5.4					
Fe_2O_3 homodisperse												7.2–9.5	
$FeOOH$									9–9.7				
α-$FeOOH$	3.2–6.8				6.7			7.5–10.2				7.5–9.3	
β-$FeOOH$								7.6–8.5				7.1	
γ-$FeOOH$	5.4–7.4				7.4								
$FeOOH$ amorphous, hydrous												7.05–7.25	
ferrihydrite, $Fe(OH)_3$ amorphous	4.3–8.8	9						5.1–8.7					
HfO_2													
HgO	7.3	7.3	7.3	7.3	7.3								
In_2O_3			7.8										
IrO_2													
La_2O_3	10.4		10.4		10.4								
MgO	12.4		12	12.4	12.4		12.4		12.4	12.4	12.4	9.6	10.4
MgO hydrous		12.7											
$Mg(OH)_2$					12								
$Mn(OH)_2$	12		11.4										
$Mn(II)$ manganite	7											1.8	

Material										
MnO$_2$	4–4.5		2.4–4.4					5.6		4–4.5
β-MnO$_2$					<3–7.3		4.6–7.3				7.3
δ-MnO$_2$					1.5–3.6						1.5–3.0
α-MnO$_2$											4.5
γ-MnO$_2$					5–5.9						4.8–5.6
MoO$_3$									2.8 (?)	1.8–2.1	
Nb$_2$O$_5$											
Nb(OH)$_5$			10.3	11.1							11.4
NiO	10.3	9.9	1.9	10.2	10.3			9.85–11.3	9.8		11.3
Ni(OH)$_2$	11.1–12	12									11.2
NiO$_x$											
PbO			11								
Pb(OH)$_2$	9.8–11	10.3	7.2								11
PbO$_2$											
α-PbO$_2$											
β-PbO$_2$					7.4–8.4						9.2
PtO	≥14										
PtO$_x$	8.6–9			9			9				9
PuO$_2$			8.7		2.8–7.3	5.4	5.4		5.1		
RuO$_2$	6.1		6.1								5.75
Ru(OH)$_4$	<0.4	0.4						1.9			
Sb$_2$O$_5$				1.8					1.8–2.2	1.8–2.2	<1.8–3.5
SiO$_2$	1.7	2	2.2								
SiO$_2$ quartz	1–3.7	3.7					2.9				
SiO$_2$ sols and gels	0.5–3						3.5				
Sm$_2$O$_3$	6.6					7.45–8.8					
SnO	6.6		4.3–7.3	4.3			7.3				6.6
SnO$_2$	4.5–9.5	4.5						4.7			5.4–5.5
SnO$_2$ hydrous	3.9		2.9								2.8
Ta$_2$O$_5$											
ThO$_2$	4.2–11		7.5	9–9.3	9–9.3	10–11		5.8	5.3		9–9.4
TiO$_2$	4.7–6	6.7	4.5–6.2	6.7			5.8				6.7
TiO$_2$ rutile	3.5–6.7				5	5.1–5.8		5.1–5.8			2.9–6.0

TABLE 3.2 (Continued)

Ref.	92	95	96	97	98	99	100	93	101	102	103	104	105
TiO$_2$ anatase								6.0				5.3–6.6	
TiO$_2$ homodisperse												5.2	
UO$_2$	4.7–5.8		6.6										
UO$_{2.08}$	6.6												
UO$_{2.33}$	6												
UO$_{2.45}$	3.5												
UO$_{2.3}$	3–4.5												
U$_3$O$_8$	4												
UO$_2$(OH)$_2$			4.3										
V$_2$O$_5$			3										
WO$_3$	<0.3–0.5		0.3	0.43	0.5					0.3			
Y$_2$O$_3$	8.95		8.9		9			7.45–8.75		8.6			9.0
Y$_2$O$_3$ hydrous													
Yb$_2$O$_3$							9	7.15–8.25					
ZnO	8.7	8.7	8.8	8.8	9.3				8.7 9.3		9.2–10.3	8.5–9.5	
Zn(OH)$_2$	10.3	10			7.8								
ZrO$_2$	4–11		4.9–6.8	6.7	10–11			9–10		6.8	5.5–6.3	5.1–8.1	10.3
ZrO$_2$ hydrous	6.7											6–6.4	
Zr(OH)$_4$													
ZrO$_2$·4H$_2$O								4.3–6.5					

Ref.	15	106	107	108	Table 3.1 average	Table 3.1 median
Ag$_2$O			11.2			
Al$_2$O$_3$			9.2	7.7	8.1	8.5
amorphous Al$_2$O$_3$				7.7		

γ-Al₂O₃	8.2–8.7			5.6–7.3	7.8	8.3
α-Al₂O₃	9.1				8.4	8.7
pseudoboehmite						
AlOOH	8.2			7.2	8.9	9.2
boehmite γ-AlOOH						
boehmite α-AlOOH (?)						
diaspore γ-AlOOH (?)						
Al(OH)₃	8.5			6.8 (am)	8.9	9.1
gibbsite α-Al(OH)₃ (?)						
α-Al(OH)₃						
γ-Al(OH)₃ bayerite (?)						
Am(OH)₄						
BeO			7.4			
BeO hydrous						
Bi₂O₃						
Bi(OH)₃				9.4		
CdO			10.3			
Cd(OH)₂		7.6–8.6				
CeO₂				8.1		
CoO						
Co(OH)₂		5–7.5				
Co₃O₄						
Cr₂O₃			6.4		7.7	7.5
α-Cr₂O₃				7.9–8.4		
Cr(OH)₃ sol						
CuO			9.5			
Cu(OH)₂						
Cu(OH)₂ hydrous						
Fe(OH)₂	6.5					
Fe₃O₄					6.8	6.6
Fe₂O₃			8.6		7.8 synt	8.2 synt
					6 nat	5.4 nat
α-Fe₂O₃	8.4–8.5					

TABLE 3.2 (Continued)

Ref.	15	106	107	108	Table 3.1 average	Table 3.1 median
γ-Fe$_2$O$_3$					5.9	5.5
Fe$_2$O$_3$ homodisperse						
FeOOH					8.1	8.2
α-FeOOH	6.1–7.7			8.3–8.5	8.4 synt	8.4 synt
β-FeOOH				7.25		
γ-FeOOH						
FeOOH amorphous, hydrous						
ferrihydrite, Fe(OH)$_3$ amorphous	8–8.1			7.1	8.1	8
HfO$_2$			7.1			
HgO			7.3			
In$_2$O$_3$				7.3		
IrO$_2$		0.5–5.6	0.5			
La$_2$O$_3$			12.4			
MgO						
MgO hydrous						
Mg(OH)$_2$						
Mn(OH)$_2$						
Mn(II) manganite						
MnO$_2$			4.8			
β-MnO$_2$	7.2	7.1		6.8		
δ-MnO$_2$	1.7 2.3	1.5 2.15				
α-MnO$_2$		3–5.4				
γ-MnO$_2$		3.6–4.82				
MoO$_3$			2			
Nb$_2$O$_5$			4.1			
Nb(OH)$_5$						
NiO			10.3			
Ni(OH)$_2$				11.3		

NiO$_x$		8.2–10.5				
PbO			11.3			
Pb(OH)$_2$						
PbO$_2$						
α-PbO$_2$		7.3				
β-PbO$_2$		4.5–9.2				
PtO		5				
PtO$_x$						
PuO$_2$						
RuO$_2$		3.88–7.3		5.5		
Ru(OH)$_4$						
Sb$_2$O$_5$	2–2.9					
SiO$_2$			0.4		2.9	3
SiO$_2$ quartz			2		2.5	2.2
SiO$_2$ sols and gels				2	3	3
Sm$_2$O$_3$						
SnO						
SnO$_2$		2–5.7	4.3	3	4.6	4.4
SnO$_2$ hydrous		2.7–3	2.9			
Ta$_2$O$_5$						
ThO$_2$						
TiO$_2$		4–5.9	5.8	6.7 (am)	5.6	5.8
TiO$_2$ rutile		3.9–6.5		5.6–8.2	5.4	5.6
TiO$_2$ anatase				6.6	6	6.2
TiO$_2$ homodisperse						
UO$_2$						
UO$_{2.08}$						
UO$_{2.33}$						
UO$_{2.45}$						
UO$_{2.3}$						
U$_3$O$_8$						
UO$_2$(OH)$_2$						
V$_2$O$_5$						

TABLE 3.2 (Continued)

Ref.	15	106	107	108	Table 3.1 average	Table 3.1 median
WO$_3$						
Y$_2$O$_3$			0.43			
Y$_2$O$_3$ hydrous						
Yb$_2$O$_3$						
ZnO			9.8			
Zn(OH)$_2$						
ZrO$_2$		4–8	5.9	6.5 (m)	6.8	6.8
					7 (m)	7 (m)
ZrO$_2$ hydrous		<2.7–10.4		4–6.6		
Zr(OH)$_4$						
ZrO$_2$·4H$_2$O						

report "PZC" of physical mixtures of different powders [63], streaming potentials measured in plugs packed with mixtures of oxides [109], etc. This kind of data is not discussed in this section. Charging behavior in mechanical mixtures is affected by sorption of products of dissolution of one component on other component(s) of the mixture, and this chapter is confined to surface charging in absence of specific adsorption.

The interest in mixed oxides dates back to the sixties, and Parks [110] proposed the following equation for their PZC:

$$PZC = \sigma_i F/(RT\varepsilon\kappa) + \Sigma f_i PZC_i \tag{3.9}$$

where σ_i is the intrinsic structural charge and f_i is the atomic fraction of the component i on the surface. The first term on rhs applies for clay minerals, while for crystalline oxides it equals zero. Experimental or calculated PZC can be used in Eq. (3.9). The idea that the PZC and/or σ_0 of mixed oxide can be calculated as linear combination of the corresponding quantities of the components was utilized in many subsequent publications, e.g. [111] This is true for some mechanical mixtures, but not necessarily for composite materials obtained, e.g. by coprecipitation. A few examples of linear dependence of PZC of mixtures of different crystallographic forms of oxides corresponding to the same chemical formula on their composition are reported in Table 3.1. The difference in the zero point was even proposed as a method to determine anatase to rutile ratio in samples containing both polymorphs. Other publications emphasize that the ratios obtained from charging data relate to surface composition rather than the bulk one. Some data apparently contradict Eq. (3.9), namely the IEP reported for some double oxides are much lower than the IEP of either component. Not necessarily do this data represent pristine values, and the effect of impurities cannot be excluded (cf. Sections A and B). It would be interesting to check if the low IEP of these mixed oxides is accompanied by low CIP.

Table 3.3 in this section is probably the first ever published compilation of PZC of mixed oxides and other complex materials. A few results have been presented in compilations of PZC discussed in Section D in addition to data on hydr(oxides). Table 3.3 presents data related to PZC of all inorganic materials (selected organic materials are discussed in Chapter 6) other than simple oxides whose PZC are collected in Table 3.1, composite materials having a layer structure (coatings, Section F), and salts of water soluble acids or bases (Section G). Most materials whose PZC are listed in Table 3.3 belong to one of the following groups:

- Salt type mixed oxides involving two or more insoluble oxides.
- Non stoichiometric mixed oxides, e.g., obtained by coprecipitation and wet impregnation.
- Clay minerals.
- Clays.
- Glasses.
- Soils.

PZC data on oxides containing small amounts of impurities, e.g. natural minerals are discussed in Section C, but data on materials with purposely added admixtures (e.g. yttria-stabilized tetragonal zirconia) are presented in Table 3.3. However, Table 3.1 may also contain some data on modified materials, namely the information about sample composition in original publications was not always complete. The bulk composition of complex materials is often very different from the surface composition in dry state, and even for originally homogeneous materials selective leaching can lead to difference between bulk and surface composition in wet systems. Therefore the PZC of mixed oxides and other complex materials is more sensitive to pretreatment, e.g. aging and grinding than that of simple oxides. The method of presentation of data in Table 3.3 and abbreviations are the same as in Table 3.1 and the organization of all tables is explained in the Introduction. The definitions of zero points discussed in section A for pure oxides are also applicable for mixed oxides. The other materials often carry permanent charge, and their PZC must not be identified with the CIP. This is why in contrast with simple oxides (Table 3.1) rather few "cip" data are reported for clays and clay minerals, and some results suggest, that the charging curves of these materials do not have a CIP at all. The σ_0 of clay minerals calculated from Eq. (3.1) may need correction, and the concerns about the meaning of the "pH" data discussed in section A apply also to clay minerals.

F. Composite Materials (Coatings)

The distribution of components in the composite materials discussed in Section E is more or less homogeneous, and their surface properties are expected to be a weighted average of the properties of their components. With layer structure, i.e. core + coating, the effect of the chemical nature of the core material on the surface charging of the particles is often insignificant, and their PZC reported in Table 3.4 are defined exclusively by the coating when it is sufficiently thick. The data in Table 3.4 is sorted by the coating material (according to the rules outlined in the Introduction) and then by the core material. The core serves as a template to receive particles of desired size, shape, specific density, magnetic properties, etc. Multiple publications reporting PZC of coating materials can be found in Table 3.1, although IrO_2 data are rather scarce.

There is no sharp difference between, e.g. silica coating on alumina (this section) and silicate sorbed on alumina (Chapter 4) or silica-alumina mixed oxide (Section E). No special convention was introduced to distinguish between these three cases (e.g. the coating usually contains some admixture of the core material) and the viewpoint of the authors of original publications was usually respected.

A few coatings belong to salts, a separate group of materials whose surface charging is discussed in Section G.

G. Salts

AgI has been extensively studied as a model colloid and historically the studies of surface charging of AgI precede the studies performed with metal oxides. The surface charge and PZC of AgI are defined by the activities of Ag^+ and I^- ions in the solution [104]. The solubility product of AgI is constant, so only one of these activities can be freely adjusted. The ionic strength effects on the ζ potential and σ_0 of AgI (plotted as a function of pAg) are similar to those shown in Fig. 3.4.

(*Text continues on pg. 202*)

TABLE 3.3 PZC of Mixed Oxides, Complex, and Ill-Defined Systems

Material	Description	Salt	T	Method	pH$_0$	Ref.
aluminates and materials containing Al$_2$O$_3$						
90% Al$_2$O$_3$-10% CeO$_2$	from citrate precursor, 2 different routes			pH	9.5 9.65	112
CoO + Al$_2$O$_3$	synthetic, different proportions	10^{-3} mol dm^{-3} NaCl	25	iep	9.5–10	41
Cr$_2$O$_3$ + Al$_2$O$_3$	synthetic, different proportions	10^{-3} mol dm^{-3} NaCl	25	iep	9.5	113
1.9% Cr$_2$O$_3$ + Al$_2$O$_3$	fired at 600°C, γ			iep	9.5	114
Fe$_2$O$_3$ + Al$_2$O$_3$	synthetic, different proportions	10^{-3} mol dm^{-3} NaCl	25	iep	8.3 (10 mol% Al)–9.5 (40 + mol% Al)	113
1.9% Fe$_2$O$_3$ + Al$_2$O$_3$	fired at 600 C			iep	9.7	114
Al(OH)$_3$ + Fe(OH)$_3$	coprecipitated			iep	9.2	115
La-Al composite oxide	calcined at 600°C	none	25	mass titration	2% La 8.21 5% La 9.61 8% La 9.9 20% La 10.15	116
LaAlO$_3$ calcined at 600 °C		none	25	mass titration	5.69	116
γ-Al$_2$O$_3$ + MgO	wet impregnation, calcination	NH$_4$NO$_3$		pH	8,8.3,8.6,8.7,8,8.9 for 1,2,5,7,9 and 11 atom% Mg/Al	117
magnesium aluminum hydroxide	from chlorides, different Al/(Al + Mg)	NaCl	25	cip iep	12.02–12.3 10.95–11.93	7,118 119 120

TABLE 3.3 (Continued)

Material	Description	Salt	T	Method	pH$_0$	Ref.
MoO$_3$/γ-Al$_2$O$_3$	wet impregnation, up to 13.8% MoO$_3$ by weight	0.001 mol dm^{-3} NaCl, KCl, KNO$_3$, KClO$_4$, NH$_4$Cl	22.5	iep	7.45–7.95 for 13.8% MoO$_3$ by weight	121 122
24% Al$_2$O$_3$, 56% SiO$_2$, 20% water	cracking catalyst	0.001–0.1 mol dm^{-3} KNO$_3$		cip, iep, pH	4.8 4.8 6	123
hydrous alumina-silica	synthetic, different Al/(Al+Si) molar ratios	0.1 mol dm^{-3} NaCl		author's	Al/(Al+Si) 0.87 8.7 0.64 7.3 0.56 6.2 0.45 4	22
silica alumina	N-631(L) Nikki	0.01 mol dm^{-3} KCl	40	pH	4.6	124
Al$_2$O$_3$ + SiO$_2$	from AlCl$_3$ + Si(EtO)$_4$	10^{-3}–10^{-1} mol dm^{-3} NaNO$_3$		cip	% Al$_2$O$_3$ 10 4.4 25 4.85 66 5.85 75 6.28 90 6.73	125
silica on alumina	deposited from ethoxide	10^{-3} 10^{-1} mol dm^{-3} NaNO$_3$		cip mass-titration	SiO$_2$ w/w 3% 7.29 7% 6.39 10% 5.37 1% 7.81 5% 7.07	126

adsorbent	description	electrolyte	T	method	pH_0	ref.
alumina on silica	deposited from isopropoxide	10^{-3}–10^{-1} mol dm^{-3} NaNO$_3$		cip	7% 6.5 10% 5.75	126
				mass titration	Al$_2$O$_3$ w/w 3% 3.79 7% 3.94 1% 5.37 3% 3.64 7% 3.87 20% 4.75 40% 5.16 60% 5.71 80% 6.52	
Al$_2$O$_3$ + 3% SiO$_2$	Akzo	HCl + KOH	20	iep	4.6	127
64% Al$_2$O$_3$ + 27% TiO$_2$ + 9% SiO$_2$	membrane, commercial	0.001–0.05 mol dm^{-3} NaCl	25	iep cip pH pH salt titr.	4.7 8.2 8.2 (pulverized membr.) 5.7 8.2	128
70% Al$_2$O$_3$ + 30% TiO$_2$	membrane	0.0001–0.01 mol dm^{-3} KCl, NaCl	20	iep	4.5	129
Al$_2$O$_3$ + TiO$_2$	membrane support, TAMI	0.001 mol dm^{-3} NaCl	25	iep	6.4	130,55
64% Al$_2$O$_3$ + 27% TiO$_2$ + 9% SiO$_2$	membrane	0.01,0.001 mol dm^{-3} NaCl	25	iep	4.7	131
γ-Al$_2$O$_3$ + V$_2$O$_5$	wet impregnation (ammonium vanadyl oxalate) calcination at 500°C	0.001 mol dm^{-3} KCl	23	iep	8.25 (1.1×10^{13} V$_2$O$_5$ molecules cm^{-2})–6.8 (2.29×10^{14} V$_2$O$_5$ molecules cm^{-2})	132

TABLE 3.3 (Continued)

Material	Description	Salt	T	Method	pH_0	Ref.
WO_3/Al_2O_3	WO_3 deposited on γ Al_2O_3 from ammonium tungstate at pH 3–9, calcined at 550°C extracted in ammonia	0.001 mol dm^{-3} KCl		iep	before extraction 2.4 (deposition at pH 3)–7.4 (deposition at pH 9) extracted 5.75 (deposition at pH 3)– 7.45 (deposition at pH 9)	133
WO_3/Al_2O_3	Al_2O_3 (Harshaw) impregnated with soln. of ammonium metatungstate, calcined at 500°C	0.1 mol dm^{-3} NaCl		mass titration	1% WO_3 7.45; 5% WO_3 7.15; 10% WO_3 6.06; 15% WO_3 5.72; 25% WO_3 = 90% monolayer 4.38	134
10% WO_3/various metal oxides/Al_2O_3 (70% monolayer)	Al_2O_3 (Harshaw) impregnated with soln. of ammonium metatungstate, calcined at 500°C	0.1 mol dm^{-3} NaCl		mass titration	6.5% SnO_2 6.05; 3.4% Fe_2O_3 6.45; 3.2% NiO 6.54; 3.5% ZnO 6.74; 7.7% CeO_2 6.8; 3.2% CoO 6.93; 7% La_2O_3 7.47	134
WO_3/Al_2O_3	catalyst, adsorption equilibrium method, γ-alumina + ammonium metatungstate	0.001 mol dm^{-3} KCl		iep	5.9% W 8; 10.1% W 7.1; 13.1% W 6.3	135
ZnO + Al_2O_3	Synthetic, different proportions	10^{-3} mol dm^{-3} NaCl	25	iep	9.5–10	41
ZrO_2 + Al_2O_3	8 mol% ZrO_2, hydrolysis of alkoxides, monodispersed spherical particles	NaOH		iep	10	136

aluminosilicates

			T			
chlorite	natural	10^{-4}–0.56 mol dm^{-3} NaCl		iep	5	137

clay minerals

			T			
beidellite	SBCa-1 milled	10^{-3} mol dm^{-3} NaCl		iep	3	138
illite	Clay Minerals Society's Source Clays Repository	10^{-2} mol dm^{-3} NaCl	25	pH	9.6	139
Fe (III) illite	10.3% Fe	0.01 mol dm^{-3} NaNO$_3$		iep	6.5	140
kaolinite	cf. Tables 3.8 and 3.9			intersection	2.5–4.2	141
kaolinite	9 different samples modified by grinding and chemical treatment	10^{-3}, 10^{-1} mol dm^{-3} KCl				
kaolinite	BDH	10^{-2} mol dm^{-3} NaCl	25	pH	4.5	139
kaolinite	Georgia gibbsite plane edge aluminol			intersection	3.88 5.5 7.5	142 143
kaolinite	KGa-2	0.01 mol dm^{-3} NaCl		iep	3.3	144
kaolinite	Clay Mineral Society Repository	0.01 mol dm^{-3} KNO$_3$		pH	4.5	145
kaolinite	KGa-2	0.1 mol dm^{-3} NaCl		iep	2.93	146
kaolinite	KGa-1, 1–16 h aged	0.01 mol dm^{-3} LiCl		pH	4.6–5	147
kaolinite	KGa-2, 1–16 h aged	0.01 mol dm^{-3} LiCl		pH	5.2–5.4	147
kaolinite	KGa-2 washed in 1 mol dm^{-3} NaCl and water	0.01 mol dm^{-3} NaClO$_4$	25	iep, coagulation	4.8	148 149
kaolinite	KGa-1			iep	2.88	150

TABLE 3.3 (Continued)

Material	Description	Salt	T	Method	pH$_0$	Ref.
kaolinite	Ajax		10, 40	pH	4.7	151
kaolinite	DBK, 6 GTile, original sample calcined at 980°C ground	KCl		EMF	4.5 5.7 6	152
kaolinite	English Clay Lovering Pochin	0.05 mol dm^{-3} NaNO$_3$	20± 2	pH	4.5	153
Fe (III) kaolinite	8.1% Fe	0.01 mol dm^{-3} NaNO$_3$		iep	6.5	140
mica	Clay Mineral Society Repository	0.01 mol dm^{-3} KNO$_3$		pH	5.5	145
montmorillonite	cf. Tables 3.8 and 3.9					
montmorillonite	Fluka	10^{-2} mol dm^{-3} NaCl	25	pH	3.8	139
montmorillonite K$^+$ montmorillonite	commercial bentonite (Volclay) treated with 1 mol dm^{-3} KCl and then washed with water	0.001 mol dm^{-3} KCl		intersection EMF	3.04 1 (extrapolated)	142 69
montmorillonite	from commercial bentonite	0.005–0.5 mol dm^{-3} NaNO$_3$		pH	7	154
ripidolite	CCa-1 milled	10^{-3} mol dm^{-3} NaCl		iep	6	138
smectite	different mechanical and thermal treatment	0.001–0.005 mol dm^{-3} KCl		EMF	2.5 (extrapolated)–7.5	155
smectite	Clay Mineral Society Repository	0.01 mol dm^{-3} KNO$_3$		pH	9.4	145

Material	Description	Electrolyte	T (°C)	Method	Value	Ref.
vermiculite	Clay Mineral Society Repository	0.01 mol dm^{-3} KNO$_3$		pH	9.4	145
materials containing cobalt oxides						
Ni+Co oxide spinel at 0–80 mol% Ni, cubic NiO at 60–100 mol% Ni (bulk composition)	thermal decomposition of nitrates at 400°C	0.005–0.1 mol dm^{-3} KNO$_3$		cip	10 mol% Ni 8.1 / 20 mol% Ni 9.0 / 30+ mol% Ni 9.5	156
chromates III and materials containing Cr$_2$O$_3$						
corrosion products of stainless steel	316 L, heated in air 15 min at 900°C / 2 h at 1000°C / treated with 1% NaCl solution at room temp.	NaOH, HNO$_3$		adhesion	3.3 / 2.8 / 3.7–5.2	71
chromite	natural, purified, FeCr$_2$O$_4$ + MgCr$_2$O$_4$	0.01 mol dm^{-3}		iep	6	157
corrosion products of stainless steel	AISI 304, different treatments			iep	≈4	158
LaCr$_{0.9}$Ni$_{0.1}$O$_{2.95}$	combustion spray pyrolysis, 650°C	0.01 mol dm^{-3} KNO$_3$	24	iep	7.6	159
LaCr$_{0.9}$Zn$_{0.1}$O$_{2.95}$	combustion spray pyrolysis, 650°C	0.01 mol dm^{-3} NaCl	24	iep	7.6	159
LaCrO$_3$ doped with Ca and Sr	combustion spray pyrolysis, 650°C	0.01 mol dm^{-3} NaCl	24	iep	3.7–6.8	159
materials containing iron hydr(oxides)						
CoO + Fe$_2$O$_3$	synthetic, different proportions	10^{-3} mol dm^{-3} NaCl	25	iep	Co/(Fe+Co) 0.1 3.9 / 0.2 3.7 / 0.3 3.4 / 0.4 6.2 / 0.6 9.6 / 0.8 10	41

TABLE 3.3 (Continued)

Material	Description	Salt	T	Method	pH$_0$	Ref.
Co$_x$Fe$_{3-x}$O$_4$	synthetic, hydrolysis of Co(NO$_3$)$_2$ + FeSO$_4$	10^{-3}–10^{-1} mol dm^{-3} KNO$_3$		cip, iep	Co/Fe 0.1 6.5 0.26 7.0 0.33 7.3 0.5 8.0	160
Fe$_3$O$_4$ substituted by Co, Cr, or Ni	from mixture of sulfates	HNO$_3$/NaOH		pH	6 (20–40% Ni) 6.5 (50% Ni) 7 (70% Ni) 7 (15–70% Co) 7.8 (85% Co) 7 (10–30% Cr) 8.5 (60–90% Cr)	161
Fe$_2$O$_3$-Cr$_2$O$_3$	synthetic, different proportions	10^{-3} mol dm^{-3} NaCl	25	iep	7(10–20 mol% Fe) - 4.5(60 + mol% Fe)	113
α-Fe$_{1-x}$Cr$_x$OOH	nitrates + KOH, 70°C, 0.04 < x < 0.13	0.01 mol dm^{-3} KCl	25	iep	8.5	19
CuO + Fe$_2$O$_3$	synthetic, different proportions	10^{-3} mol dm^{-3} NaCl	25	iep	Cu/(Fe + Cu) 0.2 8.4 0.3 6.3 0.4 8.6 0.6 8.8 0.8 9.1	41
NiFe$_2$O$_4$	decomposition of mixture of oxalates	0.01 mol dm^{-3} KNO$_3$	30	iep	3.5	162
NiFe$_2$O$_4$	autoclaving suspension of NiO and Fe$_3$O$_4$	0.01 mol dm^{-3} KNO$_3$	30	iep	7	162
NiO + Fe$_2$O$_3$	synthetic, different proportions	10^{-3} mol dm^{-3} NaCl	25	iep	Ni/(Fe + Ni) 0.1 5.2	41

Material	Description	Electrolyte	t	Method	Results	Ref
$Ni_xFe_{3-x}O_4$	$x = 0.18–0.79$ monodispersed, spherical particles	10^{-2} mol dm^{-3} KNO$_3$	25	iep	0.2 4.2 0.3 4.6 0.4 4.8 0.6 8.3 0.8 9.5 6.7	163
ferrihydrite (containing silica)	natural	0.01–1 mol dm^{-3} KNO$_3$	25	cip	Si/(Si + Fe) 0.1 7.5 0.11 6.3 0.27 5.3 0.17 6.8 0.29 6	164
goethite, 0.86% Si	synthetic from chloride			pH cip	6 6.4	165
hydrous Fe(III)-Sn oxide	hydrolysis of chlorides, 1:1 molar ratio	10^{-3} mol dm^{-3} NaCl	25	pH	6.2	166
$ZnO + Fe_2O_3$	synthetic, different proportions	10^{-3} mol dm^{-3} NaCl	25	iep	Zn/(Fe + Zn) 0.2 7.3 0.3 3.5 0.4 6.6 0.6 8.7 0.8 9.2	41
α-FeOOH	synthetic, grown in the presence of Zn and organic acids	NaClO$_4$		iep	4.4–7	167
materials containing IrO$_2$, MnO$_2$, and NiO						
$Pb_2Ir_2O_{7-y}$		0.4 mol dm^{-3} KCl		pH	3.3	168

TABLE 3.3 (Continued)

Material	Description	Salt	T	Method	pH$_0$	Ref.
MnO$_2$	chiefly β, doped (at%)	0.001 mol dm^{-3} KCl	29–30	iep	0.1% Cr(III) 6.4 0.1% Th(IV) 5.8 0.1% Mo(VI) 5.1 0.05% V(V) 6.1 0.1% V(V) 3.3 0.2% V(V) 2.0	169
NiO	doped with 1% of Cr$_2$O$_3$ or Al$_2$O$_3$	0.001 mol dm^{-3} NaCl		iep	9.5	170
materials containing RuO$_2$						
Pb$_2$Ru$_2$O$_{7-y}$		0.4 mol dm^{-3} KCl		pH	10.74	168
Ru$_{0.3}$Ti$_{0.7-x}$Ce$_x$O$_2$	from chlorides, calcined at 550°C in oxygen	0.005–0.3 mol dm^{-3} KNO$_3$		cip	5.7 $x = 0$ 6.5 $x = 0.1$ 7.3 $x = 0.3$ 8.2 $x = 0.5$ 8.6 $x = 0.7$	171
Ru$_{0.7}$Rh$_{0.3}$O$_2$		0.4 mol dm^{-3} KCl		pH	7.93	168
RuO$_2$ + TiO$_2$	from mixture of chlorides, different T treatment	0.005–0.05 mol dm^{-3} KNO$_3$	25	cip	5.8 20%RuO$_2$ 5.3 40 + %RuO$_2$	172
silicates and materials containing SiO$_2$						
sand, sandstone	baked, crushed, different treatment	0.001–0.01 mol dm^{-3} NaCl	25	iep	3–5	173

Material	Description	Electrolyte	Method	Value	Ref.
SiO$_2$ + Al$_2$O$_3$	coprecipitated Na$_2$SiO$_3$ + Al(NO$_3$)$_3$	NaNO$_3$	salt titration	4.2 (50–60% mol Al) 5 (70% mol Al) 7 (80% mol Al)	174
mullite 10% Al$_2$O$_3$ + silica gel	Al$_2$O$_3$ from propoxide		mass titration	8.5 5.09	175 176
silica-alumina silica-alumina	JCR-SAL-2 hydrolysis of Si(EtO)$_4$ + Al(t-BuO)$_3$ in the presence of sulfate	KNO$_3$	pH iep	4.2 9.0 (Si/Al = 4) 7.5 (Si/Al = 6) <3 if any (Si/Al = 8)	177 178
quartz + Al$_2$O$_3$, 6.48% Al	Al$_2$O$_3$ precipitated from chloride, not aged aged	0, 0.0488 mol dm^{-3} NaCl	intersection	8 7.1	179
1.67 SiO$_2$·Al$_2$O$_3$·0.04 Na$_2$O·0.05 H$_2$O	amorphous, hydrolysis of Na$_2$SiO$_3$ + AlCl$_3$	10^{-3} 10^{-2} mol dm^{-3} NaCl	iep	5.5	180
SiO$_2$ + Al$_2$O$_3$	10% v/v excess alumina over stoichiometric mullite	10^{-3} mol dm^{-3} NaCl	iep	8.2 not aged 5.8–6.6 aged	181
SiO$_2$ + Fe$_2$O$_3$	coprecipitated Na$_2$SiO$_3$ + Fe(NO$_3$)$_3$	NaNO$_3$	salt titration	3.5 (60% mol Fe) 4 (70% mol Fe) 5 (80% mol Fe) 6.5 (90% mol Fe)	174
quartz + Fe$_2$O$_3$, 6.68% Fe	Fe$_2$O$_3$ precipitated from nitrate	0, 0.0488 mol dm^{-3} NaCl	intersection	8	179
SiO$_2$ + magnetite	precipitation of silica from ethoxide in the presence of magnetite	0.001–0.1 mol dm^{-3} NaNO$_3$	cip	6.2 (20% Fe$_3$O$_4$ w/w) 6.8 (40%, Fe$_3$O$_4$) 7.2 (60% Fe$_3$O$_4$) 8 (80% Fe$_3$O$_4$)	182
silica-germania	chemical vapor deposition, different% GeO$_2$ w/w	none	iep	1% 3 3% 3.2 6% 3.4 11% 2.6 20% 2.6	183

TABLE 3.3 (Continued)

Material	Description	Salt	T	Method	pH$_0$		Ref.
fosterite	natural, artificially weathered		25± 1	iep	8.4 (1h), 8.0 (4h)		184
chrysotile	cf. Table 3.8						
chrysotile asbestos	synthetic	none	22,30	iep	12.3		185
silica gel + MgO	impregnation of silica with Mg acetate, calcination at 500°C	KNO$_3$		pH	9 10.5	(1% MgO) (40% MgO)	177
Ti-Si gel	from TiCl$_4$ and Na$_2$SiO$_3$, 1:1 molar ratio	0.05 mol dm^{-3} NaNO$_3$		pH	4.2		186
Zn$_2$SiO$_4$	from Zn(NO$_3$)$_2$ and Si ethoxide	10^{-3}–10^{-1} mol dm^{-3} KNO$_3$	25	pH	7.4		187
silica zirconia 7:3	from ZOCl$_2$ and silicon ethoxide	0.01 mol dm^{-3} KCl	40	merge pH	3.8		124
ZrSiO$_4$	natural	KNO$_3$	25	iep cip salt titr. iep cip salt titr.	5.7 5.9 5.9 5.5 6.1 6.1		188
ZrSiO$_4$	natural	KNO$_3$		iep cip	5.7 5.9		188
ZrSiO$_4$	natural	0.01 mol dm^{-3} KCl	25	iep	5.4		189
materials containing SnO$_2$ and Ta$_2$O$_5$							
SnO$_2$ + other oxides	reagent grade sintered with 1% of other oxides at 1050–1400 C	NaOH + HCl		iep	WO$_3$ Sb$_2$O$_5$	2.25 2.5–6.8	190

Material	Description	Electrolyte	T	Method	PZC	Ref
$FeTa_2O_6$	synthetic: $Fe + Fe_2O_3 + Ta_2O_5$	10^{-4}–10^{-1} mol dm^{-3} KNO_3		cip; iep	Fe_2O_3 7; Cr_2O_3 7; Nb_2O_5 5–5.5; 3.8; 3.9	191
titanates and materials containing TiO_2						
TiO_2 pigments	P1 0.91% Al_2O_3; P2 3.11% Al_2O_3 0.11% SiO_2; P3 2.99% Al_2O_3 2.56% SiO_2	0.001 mol dm^{-3} KNO_3	25	iep	7.8; 8.1; 6.6	192
$TiO_2 + Al_2O_3$	flame oxidation of $TiCl_4 + AlCl_3 + PCl_3$	10^{-3} mol dm^{-3} KNO_3	22–24	iep	6.5	193
$FeTiO_3$			25	author's	6.3	97
Fe(III)–Ti(IV) oxide	quantum size, hydrolysis of $TiCl_4 + FeCl_3$ in water			inflection, coagulation	5.5 (small iron content); 7.0 (50% Fe(III), pseudobrookite)	194
$FeTiO_3$	ilmenite, commercial, fresh and leached	0.01 mol dm^{-3} KCl	25	iep	3.1–3.6	19
$TiO_2 + SiO_2$	xerogel sintered	0.01 mol KCl	25	iep	5.2; 5.2	195
TiO_2 (Degussa, 70% anatase) + V_2O_5	wet impregnation (ammonium vanadyl oxalate) calcination at 500 C	0.001 mol dm^{-3} KCl	23	iep	4.5 (5.7×10^{13} V_2O_5 molecules cm^{-2}) 1.95 (7.2×10^{14} V_2O_5 molecules cm^{-2})	132
titania zirconia 63:37	from $ZrOCl_2$ and $TiCl_4$	0.01 mol dm^{-3} KCl	40	pH	5	124

TABLE 3.3 (Continued)

Material	Description	Salt	T	Method	pH$_0$	Ref.
79% TiO$_2$ + 21% ZrO$_2$	11 different samples prepared from chloride, isopropoxide, and reagent grade oxides.	10^{-3}–10^{-1} mol dm^{-3} KNO$_3$	40	cip	2.45 4.45 4.75 5.15 5.45 5.75 5.9 6.25 6.3 6.95 10.1	196
TiO$_2$ + ZrO$_2$	xerogel sintered			iep	5.9 6.3	195
tungstates						
FeWO$_4$	natural ferberite	10^{-3} mol dm^{-3} KNO$_3$		iep	2.1	197
FeWO$_4$	natural hubnerite	10^{-3} mol dm^{-3} KNO$_3$		iep	2.8	197
materials containing ZrO$_2$						
Pb$_{1-x}$Zr$_{0.52}$Ti$_{0.48}$O$_3$	x = 0, 0.025, or -0.01; solid state reaction at 900°C	10^{-4} mol dm^{-3} NaCl		iep	6.2–6.8	198
Zr$_{.75}$Si$_{.75}$O$_2$ Zr$_{.75}$Si$_{.25}$O$_2$	amorphous, from Zr n-propoxide + Si methoxide, calcined at 500 C for 3 h	NaOH + HCl		iep	3.3 4.8	199
ZrO$_2$ + 3.5% Y$_2$O$_3$	TZ2 Toyo Soda	0.001 mol dm^{-3} KCl	25	iep	7.4	200

Material	Description	Electrolyte	T	Method	pH	Ref.
ZrO_2 + 3 mol % Y_2O_3	plasma prepared	0.001 mol dm^{-3} NaCl		iep	7.7	201
ZrO_2 + 3 mol % Y_2O_3	dehydrated, then refluxed in water for 170 h	0.001 mol dm^{-3} NaCl		iep	5.5	201
ZrO_2 + 3% mol Y_2O_3	tetragonal, Toyo Soda, TZ-3Y	NH_4Cl		iep	6.5	202
ZrO_2 + 9 mol% Y_2O_3	$ZrOCl_2$ + Y_2O_3 + urotropine, iep dependent on pretreatment	0.001 mol dm^{-3} KCl		iep	140°C 5.7 300°C 4.4 600°C 4.9 1200°C 3.0	203
ZrO_2 + 9 mol% Y_2O_3	$ZrOCl_2$ + Y_2O_3 + KOH, iep dependent on pretreatment	0.001 mol dm^{-3} KCl		iep	140°C 5.5 300°C 5.5 600°C 5.4 1200°C 4.2	203
ZrO_2 + 9 mol% Y_2O_3	flame pyrolysis of Zr isopropoxide and Y(acac)$_3$	0.001 mol dm^{-3} KCl		iep	8.0	203
Y stabilized ZrO_2	SY 5.2 from Z-Tech	10^{-3} mol dm^{-3} NaNO$_3$	room	acousto	6.65	204
ZrO_2 + 5% Y_2O_3	SYP 5.2 from Z-Tech (without organic binder)	0.01 mol dm^{-3} NaNO$_3$	25	iep	7.8	205
ZrO_2 + 3% mol Y_2O_3	tetragonal, Toyo Soda, TZ-3 YS	10^{-3} mol dm^{-3} NH$_4$NO$_3$ KCl	25	iep	6.5 4 (aged dispersion)	206
ZrO_2 + 5% Y_2O_3	SY 5.2; SYP 5.2 from Z-Tech (without organic binder)	0.01 mol dm^{-3} NaNO$_3$		iep	6.9	207
ZrO_2 + 3% mol Y_2O_3	synthetic	0.001 mol dm^{-3} KCl	25?	iep	5.8	208
ZrO_2 + 3% mol Y_2O_3	nanoparticles, coprecipitation method	HCl/NaOH	25	iep	5.8	209
zeolites	cf. Table 3.9					

TABLE 3.3 Continued

Material	Description	Salt	T	Method	pH$_0$	Ref.
zeolite A		0.01 mol dm^{-3} NaClO$_4$		iep	7.2	210
blazer		0.01 mol dm^{-3} NaClO$_4$		iep	5.9	210
Na-Y zeolite	JRC	KNO$_3$		pH	9.8	177
Na-ZSM-5 zeolite	synthetic	KNO$_3$		pH	1.9	177
H-ZSM-5 zeolite	synthetic	KNO$_3$		pH	1.2	177
clays						
Fe (III) bentonite	16.9% Fe	0.01 mol dm^{-3} NaNO$_3$		iep	5.4	140
kaolin	3 different samples, calcined at different T	0.001–1 mol dm^{-3} KCl		cip	<200°C 3 400°C 4 600–900°C 4.6 1100°C 4.3 1300°C 5	211
kaolin	3 different samples	0.001–1 mol dm^{-3} KCl		cip	2.8–3	211
kaolin		0.001–0.1 mol dm^{-3} KNO$_3$		cip, iep, pH	4.0 4.0 3.3	123
kaolin, H$^+$	commercial (Georgia Kaolin Co.) treated with 0.05 dm^{-3} HCl and then washed with water	0.01–1 mol dm^{-3} KCl		cip EMF	3 3 (extrapolated)	68
kaolin kaolin + Al$_2$O$_3$	Al$_2$O$_3$ precipitated from chloride, aged and not aged	0, 0.0488 mol dm^{-3} NaCl		iep intersection	4.4 5.5 (1% Al)–7.6 (6% Al)	212 179

Material	Description	Electrolyte	Method	Value	Ref
kaolin modified by sorption of Al_{13} polycationic species	$1\ 7\times10^{-5}$ mol Al/g kaolin, prepared at different pH		iep	4.3–4.7	213
kaolin + Fe_2O_3	Fe_2O_3 precipitated from nitrate	0, 0.0488 mol dm^{-3} NaCl	intersection	5.2 (2% Fe)–7.8 (7% Fe)	179
3 clays	cf. Tables 3.8 and 3.9	0.01–1 mol dm^{-3} NaCl	cip	6.2–6.8	214
glasses					
Na and K responsive glass	Electrofact 66% SiO_2, 2% GeO_2, 2% Al_2O_3, 13% B_2O_3, 17% K_2O or Na_2O w/w	0.001–0.1 mol dm^{-3} LiCl, NaCl, KCl, CsCl, and $(C_2H_5)_4NCl$	pH	6	215
pyrex glass		0.0001–0.01 mol dm^{-3} KCl	iep	4 3 (aged) < 3.2 (aged and steamed) 3 (γ-irradiated)	216
13 controlled pore glasses	various heat treatment and leaching	0.1 mol dm^{-3} NaCl	pH	3–6.5	217
SiO_2 glass	mole %: Na_2O 25 Al_2O_3 4 Li_2O 24 Fe_2O_3 10 after thermal treatment		iep	0.8 2.3 4.4	218
5 controlled pore glasses	various heat treatment and leaching	0.01–0.001 mol dm^{-3} NaCl	pH	4–6	219
6 controlled pore glasses	various heat treatment and leaching	0.1 mol dm^{-3} NaCl	pH	4–6.2	220
glass	Ingold Pyrex	KNO₃	iep	3.85 4.4	221
other materials					
activated red mud	acid dissolution followed by ammonia treatment	0.001–0.1 mol dm^{-3} KNO₃	cip	8.5	222

TABLE 3.3 (Continued)

Material	Description	Salt	T	Method	pH$_0$	Ref.
soils						
3 soil samples	untreated	0.001 mol dm^{-3}	25	iep	8.7–9.3	223
	deferrated				4.3–4.9	
26 soils				intersection	4.7–6.2	224
2 soils				intersection	2.44, 2.8	142
red soil	treated at 373–1373 K	0.001–0.1 mol dm^{-3} KCl		iep, intersection	3.6–5	225

TABLE 3.4 PZC of Composite Materials (Coatings)

Coating	Core	Description	Salt	T	Method	pH_0	Remarks	Ref.
Al_2O_3	Fe_2O_3		0.001–0.1 mol dm^{-3} KNO$_3$	25	iep cip	6.5	0.000974 mol Fe/g Al$_2$O$_3$ IEP≈CIP, for lower surface coverages the shift in IEP of alumina is approximately proportional to the surface coverage	226
Al_2O_3	SiO_2	Positive sol 130 M from du Pont, purified	LiCl, NaCl, KCl	room	iep, coagulation	9 9.6	aged for 1 week–1 month after pH adjustment aged for 10 min–1 h after pH adjustment	227
Al_2O_3	TiO_2 rutile	from chloride, sulfate, and NaAlO$_2$	0.01 1 mol dm^{-3} KNO$_3$		cip	5–7.5	0.5–8% Al$_2$O$_3$	228
Al_2O_3 Al_2O_3	TiO_2 rutile	from isopropoxide hydrolysis of Al (III) at different pH			mass titration iep	6.9–8.06 8	0.5–7.8% alumina w/w for hydrolysis at pH > 3.8. Otherwise three IEPs.	229 230
$Al(OH)_3$	kaolinite, montmorillonite, 2 soils	2–10%			intersection	4–6.8		142
Co(II) (hydr)oxide	Stober silica	from acetate	0.01 mol dm^{-3} NaCl		iep	6	Co/Si atomic 0.69 from XPS	231
$Cr_2O_3 \cdot n\ H_2O$	hematite, spindle like, monodispersed	from chrom alum	0.01 mol dm^{-3}		iep	7.3		232
Fe_2O_3	sand	from nitrate	0.05–1 mol dm^{-3} NaNO$_3$	25	cip	8.2		233
FeOOH	quartz	FeCl$_3$ + NaOH		25	iep	5.1		234

TABLE 3.4 (Continued)

Coating	Core	Description	Salt	T	Method	pH$_0$	Remarks	Ref.
Fe(OH)$_3$	kaolinite, montmorillonite, 2 soils	2–10%			intersection	2.9–6.3		142
IrO$_2$	SiO$_2$	from chloride	10^{-3} mol dm^{-3} KNO$_3$	20	iep	2.6	titration curves (4×10^{-3}–10^{-1} mol dm^{-3} KNO$_3$) merge at pH 4.	235
Mn(OH)$_2$	hematite	hydrolysis of Mn(II) 2,4 pentanedionate			iep	4.5		236
MnO + MnO$_2$	hematite	hydrolysis of methoxide			iep	4.8		236
Ni(II) (hydr)oxide	Stober silica	from sulfate	0.01 mol dm^{-3} NaCl		iep	7.1	Ni/Si atomic 0.89 from XPS	231
RuO$_2$	silica, Spherosil	prepared at 300–600°C	0.005–0.15 mol dm^{-3} KNO$_3$	25	cip	5.6–6.3	compact layer, no cracks	237
SiO$_2$	ferrihydrite		0.01–1 mol dm^{-3} KNO$_3$	25	cip	6.65 5.7	17 mol% Si 29 mol% Si	164
SiO$_2$	rutile covered by alumina	hydrolysis of silicate at pH 6.1–8.5			iep	< 3 if any		230

SiO$_2$	rutile	hydrolysis of silicate	0.01 mol dm^{-3} KNO$_3$		iep	3	4.99% silica, higher IEP for lower coverages	238
TiO$_2$	Al$_2$O$_3$, SiO$_2$	from isopropoxide	10^{-3} mol dm^{-3} KCl		mass titration	8.14–7.29	2–11.4% titania w/w	229
TiO$_2$		from n-butoxide			iep	3.54	24 and 46 nm thick coating on 545 nm in diameter core (Stober silica). IEP equals to that of bulk titania reported in the same study	239
SiO$_2$	SiO$_2$	from Ti(diisopropoxide) bis-2,4,-pentadionate	10^{-3} mol dm^{-3} KNO$_3$	20	iep	3.9–4.1	titration curves (10^{-3}–10^{-1} mol dm^{-3} KNO$_3$) merge at pH 5.	235
V$_2$O$_5$; V$_2$O$_5$ + P$_2$O$_5$	Al$_2$O$_3$, SiO$_2$, TiO$_2$	hydrolysis of V species at different oxidation stages			acousto	2.3–8	Some materials have negative or positive ζ potentials over the entire pH range. Hysteresis.	240
Y$_2$O$_3$	hematite	thermal decomposition of YOHCO$_3$	0.005 mol dm^{-3} NaCl	25	iep	7.8–8.5	3 samples	241
MnCO$_3$	hematite	MnSO$_4$ + urea at 90°C	0.005 mol dm^{-3} NaCl	25	iep	pH 6		236
YOHCO$_3$	hematite	urea + Y(NO$_3$)$_3$			iep	pH 7.7–8.1	11 samples	241
YOHCO$_3$	polystyrene latex	hydrolysis of nitrate	10^{-3} mol dm^{-3} NaNO$_3$		iep	pH 7.7	ratio inner: outer diameter 0.65	242

Analogous dependence of surface charging of the other sparingly soluble salts on the activities of their ions can be expected, i.e. Ba^{2+} and SO^{2-}_4 are expected to be potential determining ions for $BaSO_4$, etc. but pH was often used as the independent variable. Many salts belong to the family of materials whose surface charging is pH dependent, and they will be discussed as a separate category of materials in this section. Surface charging of silver halides is not addressed here.

There is no contradiction between using the activity of the cation of the salt on the one hand, and the pH on the other as the independent variable. All sparingly soluble salts involve weak acids and/or hydrolyzable cations, thus these two quantities are interrelated, and the measurement of pH is certainly easier and more precise than measurements of activity of metal cations. It should be emphasized that the activity of the cation of the sparingly soluble salt on the one hand and analytical concentration of the metal in solution on the other are two different quantities, although they are often confused. The isoelectric points of salts expressed in terms of pH can be recalculated into the corresponding concentrations of ions—constituents of the salts and *vice versa*.

Example 1

Most IEP values reported for ZnS are at pH 3 or below (*vide infra*). In this pH range hydrolysis of Zn^{2+} ions is negligible and $[H_2S] >> [HS^-] + [S^{2-}]$. The equilibrium constant (cf. Eq. 2.23) of the reaction.

$$ZnS + 2H^+ = H_2S + Zn^{2+} \tag{3.10}$$

equals to $10^{-24.53}/10^{-19} = 10^{-5.53}$ since this reaction can be expressed as a difference of two reactions: dissolution of ZnS ($K = 10^{-24.53}$) and complete dissociation of H_2S ($K = 10^{-19}$). When dissociation of H_2S and hydrolysis of Zn^{2+} is negligible, and there is no H_2S or Zn^{2+} in solution other than that due to dissolution of ZnS, reaction (3.10) gives

$$[H_2S] = [Zn^{2+}] \tag{3.11}$$

At a low ionic strength, the expression for equilibrium constant of reaction (3.10) combined with Eq. (3.11) yields

$$K_{(3.10)} = 10^{-5.53} = [Zn^{2+}]^2/[H^+]^2 \tag{3.12}$$

thus, $[Zn^{2+}] = [H^+] \times 10^{-2.76}$. The equilibrium concentration of Zn^{2+} ions in contact with ZnS at low ionic strength and in the presence of 0.1 mol dm^{-3} of 1-1 electrolyte (the activity coefficient estimated using Davies formula) is plotted as a function of pH in Fig. 3.39. Certainly, the equilibrium activity of Zn^{2+} ions at given pH is independent of the ionic strength. Similar calculations can be performed for other sulfides.

Usually, more than one species involving the metal cation of interest (hydrolysis products) and/or more than one species involving the anion of interest (anions at different degree of deprotonation) are present at comparable concentrations in the pH range near the IEP of a sparingly soluble salt. In such general case the simplifications similar to those made in Example 1 are not justified, and exact analytical solution would be very tedious. Fortunately, specialized computer programs are available. For information about these speciation programs, and their

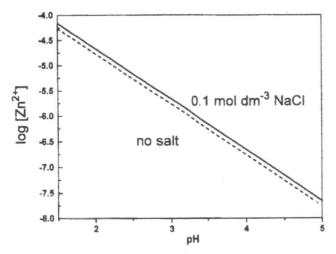

FIG. 3.39   Concentration of Zn^{2+} ions in contact with ZnS as a function of pH.

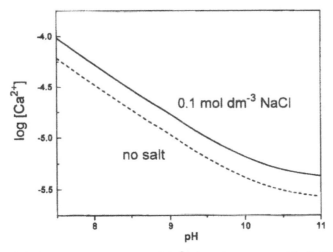

FIG. 3.40   Concentration of Ca^{2+} ions in contact with $CaCO_3$ as a function of pH.

terminology see Chapter 5. The most convenient representation for solubility problems is to set the sparingly soluble salt as one of the components with constant activity (concentration) equal to 1. Concentration of Ca^{2+} ions in equilibrium with $CaCO_3$ as a function of pH is shown in Fig. 3.40. The corresponding concentration of the CO_3^{2-} ions can be calculated from the solubility product. The calculations were performed using the following values of pK: 11.21 for the solubility product of $CaCO_3$, and 6.37 and 10.25 for the first and second dissociation constant of H_2CO_3. The results presented in Ref. [243] are significantly different from those shown in Fig. 3.40, probably because different pK values were used.

The effects of inert electrolyte on surface charging of salts other than AgI are not very well documented. The electrokinetic mobility of two samples of calcite was studied as a function of pH at different concentrations of NaCl [244]. The natural sample showed a typical ionic strength dependence (cf. Fig. 3.4), but the electrophoretic mobility of the synthetic sample was independent of the ionic strength. Addition of small amounts of alkaline earth metal chlorides to positively charged $CaCO_3$ dispersions reversed the sign to negative, but on further addition the sign was reversed back to positive. In another study the electrokinetic mobility of $CaCO_3$ was plotted as a function of $[Ca^{2+}]$. The natural material was negatively charged for $[Ca^{2+}] < 10^{-4}$ mol dm^{-3} and positively charged for $[Ca^{2+}] > 10^{-4}$ mol dm^{-3}. The sign reversal was observed at $[Ca^{2+}] = 5 \times 10^{-4}$ mol dm^{-3} for the synthetic sample. In the presence of about 10^{-2} mol dm^{-3} Ca the electrokinetic mobility of calcium phosphate and carbonate was pH independent over the range 8–12 [245].

Mishra studied effects of addition of calcium and carbonates and phosphates (concentrations on the order of 10^{-4} mol dm^{-3}) on the ζ(pH) curves for natural calcite and apatite [246]. Usually the presence of calcium induced a shift in ζ to more positive and presence of carbonates and phosphates to more negative value, i.e. according to expectations, but over certain pH ranges there was no effect or even the effect was opposite to the expected one. The electrokinetic behavior of apatite in calcite supernatant was similar to that of calcite in water, and the electrokinetic behavior of calcite in apatite supernatant was similar to that of apatite in water [247]. The ζ potentials of calcite and apatite were positive and rather pH insensitive in the presence of $\geq 10^{-3}$ mol dm^{-3} of calcium, and negative and rather pH insensitive in the presence of $\geq 10^{-3}$ mol dm^{-3} of phosphate, and the effect of carbonates on ζ potentials of apatite was negligible.

Hysteresis in electrokinetic behavior of calcite has been also reported [248]. The PZC of salts are summarized in Table 3.5. The "PZC" column specifies not only the value but also the choice of independent variable (pH or concentration of the cation of the salt). The salts are sorted by the chemical symbol of the anion, and then of the cation.

So far the redox potential as the significant variable affecting the surface charging has not been addressed in this book. In contrast with pH the redox potential is rather difficult to measure and control, and its possible effects were neglected in most cited publications. This is fully justified for many systems, e.g. the course of surface charging curves of titania and alumina measured in hydrogen atmosphere on the one hand and in oxygen atmosphere on the other is identical [249]. Surface charging of magnetite and manganese oxides is probably somewhat sensitive to the redox conditions, but systematic studies in this direction have not been reported, and considering rather consistent PZC values reported for the magnetite in different studies, the effect of redox conditions is not very dramatic. On the other hand the surface charging behavior of sparingly soluble sulfides appears to be extremely redox-sensitive. Sulfides will be discussed separately from other salts and their zero points are not listed in Table 3.5.

Many investigations, chiefly electrokinetic have been published on pH dependent surface charging of metal sulfides, and the redox conditions have been considered as a major variable in many papers. Dekkers and Schoonen [250] estimated dissolved oxygen concentration in water (nitrogen purged for 30 min) to be 0.25 mg dm^{-3}. This water was used to prepare 10^{-3}–10^{-2} mol dm^{-3} NaCl solutions in which synthetic greigite Fe_3S_4 and pyrrhotite $Fe_{1-x}S$ were dispersed, and their IEP were found at pH 3 and 2, respectively. The role of time of equilibration (short time is preferred to avoid oxidation) and of the nature of supporting electrolyte (chloride is preferred over nitrate or chlorate VII) was also emphasized.

An extensive electrokinetic study of ZnS and CdS at oxygen free conditions [251] suggests an IEP about pH 2. The results obtained at different Zn and Cd concentrations in solution resemble the electrokinetic behavior of silica in the presence of heavy metal cations (cf. Chapter 4): the sign of the ζ potential is reversed from negative to positive over a narrow pH range (near pH 6 for the ZnS-$ZnCl_2$ system) when a critical Zn concentration is exceeded, and this pH range broadens when Zn concentration increases. The pH range where the ζ potential is affected by a Zn salt can be shifted to lower or higher pH when chloride is replaced by sulfate or acetate, respectively. Similar results were reported for CdS except the effect of these anions was rather insignificant. Negative ζ potentials of ZnS and CdS were enhanced in the presence of $(NH_4)_2S$ but the IEP was not influenced. For monodispersed spherical CdS [252] an IEP at pH 3.5 was found, and no special effort was made to insulate the sample from air, but an IEP and CIP at pH as high as 7.5 has been reported [253] for a reagent grade CdS for three $NaClO_4$ concentrations, in solutions prepared with water saturated with N_2. Addition of Cd to the solution at constant pH (3–7.5) led to increase in ζ potential by about 6 m V per decade. The IEP of precipitated ZnS and commercial $Zn_{0.95}Cd_{0.05}S$ phosphor [254] was reported at pH 3, and 2.5, respectively, and no special effort was made to insulate the samples from air; in the basic region the electrokinetic curves of sulfides match that of ZnO. An IEP at pH 6.8 was reported [255] for ZnS under oxygen free conditions with 10^{-3}–10^{-2} KNO_3 mol dm^{-3} as the supporting electrolyte. Natural and synthetic ZnS [256] showed completely different electrokinetic behavior under oxygen free conditions (N_2 bubbling), namely the former had IEP at pH 3 (extrapolated) and the latter at pH 8, both in 0.002 mol dm^{-3} NaCl. Reagent grade HgS [257] has an IEP at pH 7 for three $NaClO_4$ concentrations, in solutions prepared with water saturated with N_2. Natural galena (PbS) containing 4–6% SiO_2 showed only negative ζ potentials at pH 2–12 [258]. No special precautions were taken to avoid oxidation. Addition of Pb(II) to the solution caused similar effect on electrokinetic curves of galena to those reported for silica in the presence of Pb(II) and other heavy metal cations (Chapter 4), and analogous to effects of Zn and Cd on the electrokinetic behavior of ZnS and CdS [251] discussed above. This result indicates a strong effect of SiO_2 impurities on the overall electrokinetic behavior. On the other hand synthetic reagent grade PbS containing 0.6–1% SiO_2 had a IEP at pH 6 under the same conditions. Pugh and Bergstrom cite a number of electrokinetic studies for galena which resulted in IEP at pH 2–4 or no IEP at all (only negative ζ potential). In another study with natural galena [259] no attempts were made to control oxygen, and IEP was at pH 2–7 and depended on the history of the sample. IEP at pH 2 for chalcopyrite and for sulfur (liberated from natural galena) is reported in the same publication. Elemental sulfur on the surface of oxidized galena was deemed responsible for its electrokinetic properties.

Electrokinetic behavior of reagent grade synthetic covellite CuS, pyrrhotite Fe_5S_6 and millerite NiS was studied over the pH range 2–11 [260]. Pyrrhotite showed IEP at pH 6. Covellite and millerite showed negative ζ potential at very low and very high pH and positive ζ potential over limited pH range 7–9 (CuS) and 8–10 (NiS). A series of $(CuS)_x(CdS)_{1-x}$ samples [261] had negative ζ potentials at pH 4 in 0.05

mol dm^{-3} KClO$_4$ and at pCu 1–8, while the sign of CuS was reversed to positive at pCu > 3. No special measures against oxidation were undertaken. Many IEP values for different sulfides are summarized in Ref. [262]. These results are very divergent, probably due to different degree of surface oxidation. The above discussion shows that the degree of oxidation of sulfides is difficult to control. For example, not necessarily is the N$_2$ bubbling a sufficient protection against oxygen and the sample can be already irreversibly oxidized to some degree before its contact with solution. A systematic study of the effect of oxidation of freshly ground natural pyrite (3% SiO$_2$) on its electrokinetic behavior suggests that "virgin" pyrite has an IEP at pH < 2 (if any) and the IEP shifts to pH as high as 7 when is it exposed to air for 1 day [263]. Shorter exposures lead to lower IEP values. The rate of oxidation reflected by the shift in the IEP is higher when air is replaced by oxygen, and it is also pH dependent (faster oxidation in acidic media). Freshly ground chalcopyrite has IEP at pH < 3 if any, but aged samples have positive ζ potentials at pH 4–8 [264]. The H$_2$O$_2$ treated samples behave like the aged ones. The sign of ζ potential of chalcopyrite is also reversed on addition of water soluble salts of Cu(II) and/or Fe(III).

Bebie at al. [265] studied electrokinetic properties of NiS$_2$, FeS$_2$, CoS$_2$, PbS, ZnS, FeCuS$_2$, FeAsS, and FeS at pH 2–11. The natural minerals (and a few synthetic samples) were washed with 0.5 mol dm^{-3} HCl and protected from oxygen. The traces of oxygen present in water were scavenged by addition of Na$_2$SO$_3$. With one exception of FeS (iep at pH 3.3) the sulfides showed only negative ζ potentials at pH > 2, and their IEP were estimated by extrapolation. The details of the extrapolation procedure were not given, e.g. the IEP of NiS$_2$ at pH 0.6 was claimed, while this material showed pH independent ζ potential of -45 mV over the pH range 2–8. According to Bebie et al. there is a similar linear correlation between the IEP of sulfides and the electronegativity as that discussed in Section II for oxides. However, the lines obtained for oxides on the one hand and sulfides on the other hand did not match. Sulfur is less electronegative than oxygen but the IEP reported for sulfides are lower than those of corresponding oxides, and this is a clear discrepancy.

II. CORRELATIONS BETWEEN ZERO POINTS AND OTHER PHYSICAL QUANTITIES

Two opposite approaches to zero points have been discussed in Sections I.C and I.D. The value of PZC can be attributed to specific sample on the one hand (Table 3.1) or to a chemical formula on the other (Table 3.2, recommended PZC values). Accordingly, two categories of correlations between zero points and other physical quantities can be distinguished: the correlations with quantities measured for specific samples (which can vary from one sample to another), and with quantities characterizing a chemical compound. Most sets of recommended PZC values (Section I.D) were complied in order to illustrate the latter type of correlations. Table 3.2 shows substantial discrepancies between particular sets. Even for TiO$_2$ whose PZC values reported in different sources are rather consistent (Section I.C) the recommended values range from 5 to 6.7 (paradoxically, the same source was quoted for both extreme values), and for other materials the differences are even more substantial. Although the range of recommended PZC values of SnO$_2$ is quite broad (pH 4.7–7.3), most results reported in Fig. 3.30 (PZC of SnO$_2$ taken from

(*Text continues on pg. 211*)

TABLE 3.5 PZC of Sparingly Soluble Salts

Salt	Description	Electrolyte	T	Method	PZC	Ref.
Ba-β-alumina	hydrolysis of nitrates	1 mol dm^{-3} KCl		pH	pH 9.5	64
SiC	α, Syowa Denko	NH$_4$Cl		iep	pH 4	202
SiC	β, Syowa Denko	NH$_4$Cl		iep	pH 5.5	202
Al hydroxycarbonate	Barcroft (IEP dependent on the solid to liquid ratio)	0.016 mol dm^{-3} KCl		iep	pH 7–8	267
Al hydroxycarbonate	amorphous, commercial dispersion, diluted	0.01 mol dm^{-3} KCl	25	iep	pH 7.5–(>9)	268
CaCO$_3$	Iceland Spar		25	iep	pH 9.5	269
CaCO$_3$	natural, 0.5% iron	0.002 mol dm^{-3} NaClO$_4$	25	iep	pH 8.2	246
CaCO$_3$	synthetic	0.01–0.15 mol dm^{-3}	25	iep	pCa 4.5 (extrapol.)	245
CaCO$_3$	synthetic	0–0.002 mol dm^{-3} KNO$_3$		iep	pH $\approx$ 11	247
CaCO$_3$	precipitated calcite	0.005 mol dm^{-3} NaCl		iep	pH <7 pCa 3.4–4	248
CaCO$_3$	natural	NaCl+H$_2$CO$_3$; NaCl+NaHCO$_3$+H$_2$CO$_3$; NaCl+NaHCO$_3$+Ca(OH)$_2$;NaCl+CaCl$_2$; NaCl; NaCl+NaHCO$_3$; NaCl+CaCl$_2$; NaCl		iep	pCa 2 extrapol.	244
CaCO$_3$	synthetic	NaCl		iep	pH 9.6 pCa 4	244
CaCO$_3$	precipitated Pfizer	CaCl$_2$		acousto	pH 9.6 pCa 3.5; pCa 4.34 (extrapol.)	270
CaCO$_3$		KCl	25	iep	pCa 2.7 pH <8.5	243
CaCO$_3$	Baker	0.001 mol dm^{-3} KNO$_3$		iep	pH <6	271
FeCO$_3$	natural	0.0001–0.01 mol dm^{-3} KNO$_3$		iep	pH 7.9	272

TABLE 3.5 (Continued)

Salt	Description	Electrolyte	T	Method	PZC	Ref.
$FeCO_3$	synthetic	0.1 mol dm^{-3} NaCl	25	pH	pH 5.3 at 0.5 atm CO_2	273
$FeCO_3$	natural siderite	0.01 mol dm^{-3} NaCl		iep	pH > 12	274
$MgCO_3$	natural	$0-0.01$ mol dm^{-3} KCl	25	iep	pH < 7	275
$(Ca, Mg)CO_3$	natural	$0-0.01$ mol dm^{-3} KCl	25	iep	pH < 9	275
$MnCO_3$		$0.003, 1$ mol dm^{-3} NaCl	25	inter-section	pCa 2.7 pH 5.5 at 0.5 atm CO_2	273
$MnCO_3$	$MnSO_4$ + urea at 90°C			iep	pH 6	236
$PbCO_3$	natural cerussite			iep	pH 4	276
$PbCO_3$	Aldrich	0.001 mol dm^{-3} KNO$_3$	22	iep	pH 12	277
$PbCO_3$	Koch Light	0.001 mol dm^{-3} KNO$_3$		iep	pH << 6	271
$Pb_3(OH)_2(CO_3)_2$	monodispersed, spherical	positive ζ at pH 5.5–7, negative beyond this pH range		iep	pH 8.6	278
$YOHCO_3$		$10^{-4}-10^{-2}$ mol dm^{-3} NaClO$_4$		cip	pH 7.7	279
$YOHCO_3$	urea + Y(NO$_3$)$_3$	NaCl	25	iep	pH 8.0	241
$Zn_5(OH)_6(CO_3)_2$	synthetic	$10^{-3}-10^{-1}$ mol dm^{-3} KClO$_4$		cip	pH 7.6	280
Zr molybdate	precipitated, 4 different samples	$0.05-0.5$ mol dm^{-3} LiNO$_3$, NaNO$_3$, KNO$_3$		cip	pH 2.9–3.6	281
Si_3N_4	Johnson Matthey	10^{-2} mol dm^{-3} NaNO$_3$		iep	4	282
Si_3N_4		10^{-3} mol dm^{-3} KNO$_3$		AFM	6	75
$AlPO_4$	anhydrous and hydrated			pzc	pH 4	175
$AlPO_4 \cdot n\ H_2O$	n = 2–3, Al(NO$_3$)$_3$ + Na$_3$PO$_4$ dialyzed, dried at 110°C	0.1 mol dm^{-3} KCl	30	pH	pH < 3.5	283
$Ca_3(PO_4)_2$	synthetic	$0.005-0.06$ mol dm^{-3}	25	iep	pCa 3.3	245
$Ca_5(PO_3)_3Cl$	natural, 1.5% silica	0.002 mol dm^{-3} NaClO$_4$	25	iep	pH 6.7	246
$Ca_5(PO_3)_3F$	natural, 2 samples 0.08% iron	0.002 mol dm^{-3} NaClO$_4$	25	iep	pH 3.5, 5	246
$Ca_{10}(OH)_2(PO_4)_6$	synthetic	NaCl			pH 7.6	284
$Ca_{10}(OH)_2(PO_4)_6$	synthetic	NaCl	20	cip	pH 7.6	285

Material	Description	Electrolyte	T/°C	Method	pH	Ref
Ca$_{10}$(OH)$_2$(PO$_4$)$_6$	synthetic hydroxyapatite	KCl, KNO$_3$, KClO$_4$, (CH$_3$)$_4$NCl		iep	pH 8.3–8.6, pH 8.5, pH 8.5, pH 8.4	247
Ca$_{10}$F$_2$(PO$_4$)$_6$	synthetic	0–0.002 mol dm^{-3} KNO$_3$, KCl, KClO$_4$		cip	pH 7.4, pH 6.7–6.9, pH 6.8	285
CrPO$_4$·n H$_2$O	n = 2–7, Cr(NO$_3$)$_3$ + Na$_3$PO$_4$ dialyzed, dried at 110°C	0.1 mol dm^{-3} KCl	30	pH	pH <4.5	283
FePO$_4$·n H$_2$O	n = 1 2, Fe(NO$_3$)$_3$ + Na$_3$PO$_4$ dialyzed, dried at 110°C	0.1 mol dm^{-3} KCl	30	pH	pH <4.5	283
Sn(HPO$_4$)$_2$·nH$_2$O	synthetic	0.01 mol dm^{-3} KNO$_3$		iep	pH 6	286
Th$_4$(PO$_4$)$_4$P$_2$O$_7$	from + NH$_4$H$_2$PO$_4$ + Th(NO$_3$)$_4$ solution, heated at 400°C	0.05 mol dm^{-3} KNO$_3$		iep	pH 6.8	287
Th$_3$(PO$_4$)$_4$	synthetic	0.005–0.2 mol dm^{-3} LiClO$_4$	25	iep	pH 2.5	288
Ti(HPO$_4$)$_2$·nH$_2$O	synthetic	0.01 mol dm^{-3} KNO$_3$		iep	pH 4	286
ZrP$_2$O$_7$	from ZrOCl$_2$ + NH$_4$H$_2$PO$_4$ solution, heated at 400°C	0.05 mol dm^{-3} KNO$_3$		iep	pH 3.6	287
Zr$_2$O(PO$_4$)$_2$	from ZrOCl$_2$ + NH$_4$H$_2$PO$_4$ solution, heated at 400°C	0.05 mol dm^{-3} KNO$_3$		iep	pH 4	287
Zr(HPO$_4$)$_2$·nH$_2$O	synthetic	0.01 mol dm^{-3} KNO$_3$		iep	pH 4	286
BaSO$_4$	Aldrich, 99%	none		iep	pH 5	289
CaSO$_4$	A & C	0.001 mol dm^{-3} KNO$_3$	25	iep	negative ζ between pH <2 and 10.5	271
PbSO$_4$	natural anglesite	(three IEP for one sample)		iep	pH 4; 6; and 11.5	276
PbSO$_4$	Aldrich	0.001 mol dm^{-3} KNO$_3$ (positive ζ at pH > IEP)		iep	pH 6	271
SrSO$_4$	natural celestine, washed			iep	pH 2.8	290
BaTiO$_3$	high purity reagent			author's	pH 9.9	97
BaTiO$_3$	TICON HPB, TAM			iep	pH >10.2	291
BaTiO$_3$				acousto	pH 9.6	292

TABLE 3.5 (Continued)

Salt	Description	Electrolyte	T	Method	PZC	Ref.
$SrTiO_3$	high purity reagent			author's	pH 8.6	97
$CaWO_4$	natural	$0.01–0.03$ mol dm^{-3} NaCl, pH 4–11	20–25	iep	pCa 1–2 pH < 3	293
$Cs_3PW_{12}O_{40}$	monodispersed spheres			iep	pH 1.5	294
Th salt of $H_3PW_{12}O_{40}$	monodispersed spheres			iep	pH 5.4	294
Th-tungstosilicic acid compound	monodispersed spheres			iep	pH 3.5	295
Zr-tungstosilicic acid compound	monodispersed spheres			iep	pH 3	295

original papers published after 1965) fall beyond this range. Also the choices of materials whose PZC were used to test particular correlations were very different. Most sets of data used to test correlations between PZC and other quantities, even in quite recent papers, were based on selected PZC from Parks' review [92], sometimes combined with authors' own results, and recently published results were usually ignored. The trustworthiness of parameters of many empirical equations presented in this section is confined by the choice of "recommended" PZC. Most correlations discussed in this section were derived for simple (hydr)oxides, but some correlations are applicable to more complex materials. There is not too much sense in discussing PZC of oxides which react with water and form strong, water soluble bases or acids but the correlations (if applicable) are expected to produce very high PZC for the former and very low PZC for the latter.

Many methods discussed in Section I.B are based on correlations between PZC and other quantities, and these correlations will not be discussed in this section.

A. Ionic Radii, Bond Valence, and Related Quantities

Parks [92] proposed the following equation:

$$PZC = 18.6 - 11.5(Z/R)_{eff} \tag{3.13}$$

where

$$(Z/R)_{eff} = Z/R + 0.0029\,CFSE\,(kcal) + a \tag{3.14}$$

where Z is the cation charge, R is the cation radius, CFSE is the crystal field stabilization energy and $a = 0$ for $CN = 6$ for hydroxides, oxohydroxides and hydrous oxides. For $CN = 4$–8 and for anhydrous oxides a ranges from -0.18 to $+0.39$.

The same equation can be written as

$$PZC = A_{eff} - 11.5\,(Z/R + 0.0029\,CFSE) \tag{3.15}$$

where A_{eff} has different values (14.1–20.7) for anhydrous oxides on the one hand or for hydroxides, oxohydroxides and hydrous oxides on the other, and it also depends on CN.

The coefficients in the following equation [98]

$$PZC = 18.43 - 53.12\,(\nu/L) - 1/2\,\log[(2 - \nu)/\nu] \tag{3.16}$$

were calculated using experimental PZC values for α-Al_2O_3 (9.1) and MgO (12.4). In Eq. (3.16) $\nu = Z/CN$ and L is a sum of center to center distances metal–O and O–H, and it was tested using data for 21 oxides, oxohydroxides and hydroxides (some PZC were beyond the range of results presented in Table 3.1, e.g. 10–11 for ZrO_2 and 5–5.2 for $Al(OH)_3$). With six other oxides and hydroxides with metal ions whose d-electrons are involved in hybrid orbitals (Cu, Cr, Fe(II), Co, Ni), a correction for crystal field stabilization energy was used, namely ν/L in Eq. (3.16) was replaced by

$$(\nu/L)_{eff} = \nu/L + 5.61 \times 10^{-4}C \tag{3.17}$$

where C represents the crystal field stabilization energy of the metal cation (nonelectrostatic contribution to the energy of hydration of the cation, in kcal/mol) and 5.61×10^{-4} is an empirical constant (calculated assuming PZC of $Co(OH)_2$ at pH 11.4). By defining the constant K representing the affinity of proton to oxide as

$$\log K = \log \left[(2 - \nu)/\nu \right] + 2PZC \tag{3.18}$$

Eq. (3.16) we can be rewritten as

$$\log K = A - B(\nu/L) \tag{3.19}$$

where A and B are empirical constants. Bleam [296] showed that the following equation

$$\log K = A - B\nu \tag{3.20}$$

gives an equally good correlation as Eq. (3.19), naturally with different values of empirical constants.

Sverjensky [101] proposed the following equation:

$$PZC = 21.1158 \, \varepsilon_r^{-1} - 42.9148 \, (\nu/L) + 14.686 \tag{3.21}$$

where $L = 0.101$ nm $+ r_{M-O}$. The values of empirical constants were calculated from experimental PZC for 6 oxides, $Al(OH)_3$ and $Al_2Si_2O_5(OH)_4$. This approach makes it possible to calculate PZC of composite materials by using an average $\nu/L = \Sigma \left[(\nu/L)N_i \right] / \Sigma \, N_i$ over all cation sites in Eq. (3.21). The first term on the rhs represents Born solvation energy and the second term represents electrostatic repulsion between surface protons and lattice cations (as in Parks' classical approach). The third term combines different attractive interactions. Many crystals show anisomorphism of dielectric constants, but this problem was not addressed.

Sverjensky and Sahai [297] found similar correlations:

$$PZC = 11.43 \, \varepsilon_r^{-1} - 33.72 \, (\nu/L) + 13.38 \tag{3.22}$$
$$PZC = 11.07 \, \varepsilon_r^{-1} - 33.48 \, (\nu/L) + 13.39 \tag{3.23}$$

using the PZC data from different sources.

Dependence of PZC on Z/R for 36 oxides, oxohydroxides and hydroxides (3 PZC values calculated assuming that PZC corresponds to the pH of minimum solubility, 6 self measured values, the other values taken from literature) is nearly linear [96], but the results (presented in from of graph) show substantial scatter.

The following correlation was proposed [105]

$$PZC = A - B[(Z/r^2)/CN_o] \tag{3.24}$$

where $A = 13$ for hydroxides and 12 for oxides and $B = 1.5$ for hydroxides and 2 for oxides and CN_o is coordination number of oxygen. Literature data for 7 hydroxides (including $Ca(OH)_2$) and 9 oxides (plus one calculated value) were used. It was admitted that Eq. (3.24) did not work for silica, and the calculated value for $Al(OH)_3$ (PZC at pH 5) is much lower than most results reported in Table 3.1.

Ray and Sen [95] report almost ideal correlation between partial charge of oxygen and PZC of 19 oxides and hydroxides taken from literature. The pH 10 used in this study as PZC for ZrO_2 is much higher than most results reported in Table 3.1.

B. Electronegativity

PZC was found to be a linear function of the electronegativity of the oxide [97] which is defined as

$$x(M_aO_b) = [x^a(M)x^b(O)]^{1/(a+b)} \tag{3.25}$$

where $x(M)$ and $x(O)$ are the electronegativities of the metal and of oxygen.

The coefficients in the equation

$$PZC = -8.48x \,(\text{in eV}) + 57.3 \tag{3.26}$$

were determined using PZC for 15 oxides taken from literature, 2 self measured values for oxides and 3 self measured values for titanates.

Similar correlation with somewhat different coefficients was found for 24 oxides [107]

$$PZC = -9.25x \,(\text{in eV}) + 61.4 \tag{3.27}$$

The following correlation

$$PZC = 14.43 - 0.58X_i \tag{3.28}$$

was found for 10 PZC values taken from literature and 3 self measured values (actually IEP was determined) [102], where X_i is defined as $(1 + 2\,Z)X_o$, where Z is the charge of metal ion and the electronegativity of the metal at given oxidation state X_o was taken from Ref. [298]. It should be emphasized that different definitions of electronegativity have been published, and each definition leads to different values.

C. The pH of Saturated Solution

The following correlations were proposed by Schott [299] for hydroxides ($T = 25°C$):

$$PZC = 3.82 + 0.764\,pH_{sp} \tag{3.29}$$

and

$$PZC = 15.41 - 0.355\,pK_{sp} + 0.00371\,pK_{sp}^2 \tag{3.30}$$

where K_{sp} is the solubility product of hydroxide and $pH_{sp} = 14 + (\log Z - pK_{sp})/(Z+1)$ represents the pH of hydroxide suspension in pure water.

A linear correlation between PZC and the pH of saturated solution of 10 hydroxides (slope ≈ 1) is reported in Ref. [300].

D. Heat of Immersion

Heat of immersion is proportional to PZC according to Ref. [301]. The following straight line

$$-\Delta H_{immersion} = 0.04 + 78\,PZC(Jm^{-2}) \tag{3.31}$$

was based on heats of immersion of five samples of silica, three samples of SnO_2, two samples of rutile, two samples of α-Al_2O_3, and Fe_2O_3, and Cr_2O_3 (both hexagonal). Points of zero charge were not measured with the same samples of oxides and the

PZC value used for hematite (pH 6.7) was rather low. Linear correlation between heat of immersion in hexane and PZC in water was also claimed.

A similar correlation was found in Ref. [302]

$$-\Delta H_{immersion} = 0.05 + 80 PZC (\text{J m}^{-2}) \tag{3.32}$$

using data for 2 samples of Fe_3O_4, 3 samples of Al_2O_3, 2 samples of SiO_2 (gel and quartz), SnO_2, anatase, rutile, hematite, ZrO_2, and Cr_2O_3.

More recent studies [303,304], suggest that the heat of immersion of oxides is pH dependent (minimum at PZC), thus a question arises which value should be taken into account in such a correlation. Most reported heats of immersion were measured without pH control. A few examples of heats of immersion from the literature are summarized in Table 3.6. These results indicate a substantial effect of outgasing conditions on the $\Delta H_{immersion}$. Thus, significance of results obtained by comparison of different materials outgased under different conditions is problematic. Interpretation of heats of immersion of oxides in the framework of 2-pK model (cf. Chapter 5) has been discussed by Rudzinski et al. [305].

E. Viscosity

The rheology of concentrated dispresions (volume fraction of solid of at least a few %) is governed by surface charging. A few examples of correlation between the viscosity at constant shear rate and other rheological properties, e.g. yield stress on the one hand and the IEP on the other are compiled in Table 3.7.

In the Bingham model the shear stress is a linear function of the shear rate

$$\tau = \tau_B + \eta_{pl}\gamma^\bullet \tag{3.33}$$

where η_{pl} is plastic viscosity, but unlike for Newtonian fluid the line does not go through the origin. In real systems the low shear rate portion of the flow curve shows deviations from Eq. (3.33), and the Bingham yield stress τ_B is obtained by extrapolation of the high shear rate portion to $\gamma^\bullet = 0$. Typical τ_B(pH) curves at different volume fractions of solid are shown in Fig. 3.41. The IEP is sensitive to the presence of strongly adsorbing species, and so is the maximum in a τ_B(pH) curve (Fig. 3.42). Thus, the maximum obtained with impure material does not represent the pristine value of the IEP.

F. Physical Properties of Dry Surfaces

The following correlations were reported by Ardizzone et al. [221,339]

$$PZC = 11.32 - 1.14(DO + DM) \tag{3.34}$$
$$PZC = 11.63 - 1.17(DO' + DM) \tag{3.35}$$

where DO is oxygen 1 s binding energy minus 530 eV, DO' is the DO corrected for a high energy shoulder attributed to surface OH groups and DM is cation binding energy minus metal binding energy (eV), from XPS spectra. The empirical coefficients in Eqs. (3.34) and (3.35) were obtained using 3 samples of hematite, 3 samples of zirconia, 2 samples of alumina, 2 samples of glass, one sample of titania and SnO_2. For some materials CIP of titration curves were taken as PZC while for other materials IEP were used. Some of the PZC values used to obtain this

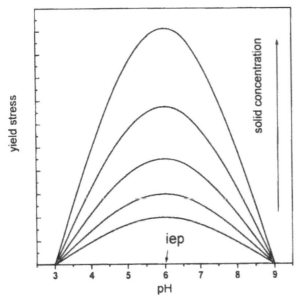

FIG. 3.41 Yield stress of concentrated dispersions of an oxide as a function of pH.

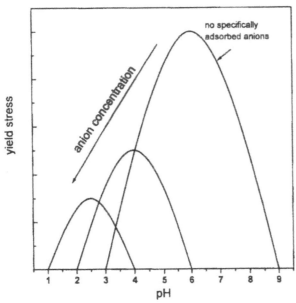

FIG. 3.42 Effect of specific adsorption of anions on the yield stress of concentrated dispersions.

correlation (e.g. 4.6 for γ-alumina) are far from results reported by other authors for the same compounds (Table 3.1).

Somewhat different coefficients, namely,

$$PZC = 12.2 - 1.35(DO + DM) \tag{3.36}$$

(*Text continues on pg. 222*)

TABLE 3.6 Heats of Immersion

Material	Description	Electrolyte	T	pH	$-\Delta H_{immersion}/$J m^{-2}	Ref.
Al$_2$O$_3$	α, Linde A evacuated at 400°C	0.002 mol dm^{-3} NaCl	25, 45	2–11	0.9 at PZC (minimum) >1 at pH <4	306
Fe$_3$O$_4$	natural	none	25		0.646±0.025	302
Fe$_2$O$_3$	hematite, commercial pigment, after Soxhlet extraction	none	25		0.466	307
Fe$_2$O$_3$	hematite, commercial pigment, after Soxhlet extraction	none	25	7	1±0.05	307
Fe$_2$O$_3$	hematite, from maghemite	none	25		0.317 (outgasing at 25°C) 0.602 (outgasing at 200°C)	308
Fe$_2$O$_3$	maghemite, from goethite	none	25		0.304 (outgasing at 25°C) 0.513 (outgasing at 200°C)	308
Fe$_2$O$_3$	maghemite, synthetic	0.01 mol dm^{-3} KCl	25	6.8 3–11	0.3 >0.3	304
FeOOH·0.5H$_2$O	from chloride	none			0.14	309
SiO$_2$	Aerosil Degussa	10^{-3}–10^{-2} mol dm^{-3} HCl, HNO$_3$			0.154 (outgasing at 110°C) 0.068 (outgasing at 800°C)	310
SiO$_2$	XL Nissan				0.691	311
SiO$_2$ coating on rutile					0.70 0% SiO$_2$ 0.37 1.27% SiO$_2$ 0.245 3.95% SiO$_2$ 0.18 8.57% SiO$_2$	230

TABLE 3.7 Effects of Surface Charging and Adsorption on Viscosity and Other Rheological Properties

Material	Description	Salt	Property	Shear rate	Solid concentration	pH	Result	Ref.
Al_2O_3	commercial, Tioxide	0.001 mol dm^{-3} KNO$_3$	yield stress, flow behavior index, flow consistency number	10–90 s^{-1}	10–60% w/w	3–10.5		312
Al_2O_3	α, commercial, Sumitomo, AKP-50, AKP-15	NH$_4$Cl, KI, NH$_4$NO$_3$	yield stress	0.01–100 s^{-1}	10–30% v/v	2–9	At pH 9 presence of 1 mol dm^{-3} NH$_4$Cl depresses the viscosity of 20% v/v dispersion (with respect to "no salt"). Viscosity reaches a maximum at pH 2 for 1.9 mol dm^{-3} NH$_4$Cl, at pH 4 at 1.7 mol dm^{-3} NH$_4$Cl, 1.9 mol dm^{-3} KI and 2.2 mol dm^{-3} NH$_4$NO$_3$ and at pH 6 at 13 mol dm^{-3} NH$_4$Cl. These concentrations are rather insensitive to the solid v/v fraction. Further increase of ionic strength does not significantly affect the viscosity.	313
Al_2O_3	α	NH$_4$Cl	shear modulus	2–400	10% v/v	5.9–7	Shear stress at pH 7 > > than at pH 6.5 > than at pH 5.9.	314
Al_2O_3	α, commercial, Sumitomo, AKP-50	ammonium citrate		0.1–1000 s^{-1}	20% v/v	3–9	At pH 3 maximum viscosity at 0.015 mol dm^{-3} citrate (this corresponds roughly to $\zeta = 0$ at pH 3). No further increase in viscosity when more citrate is added. At pH 8, the viscosity has a minimum at 0.03 mol dm^{-3} citrate (a Newtonian liquid). Further increase in citrate concentration up to 0.5 mol dm^{-3} causes an increase in viscosity.	315
Al_2O_3	commercial, A-16 Alcoa	0.01 mol dm^{-3} NaNO$_3$	yield stress		27–30% v/v	5–11	The yield stress varies as fourth power of solids loading and peaks at pH 8 (IEP at pH 8.8).	316
Al_2O_3	α, commercial, HRA 10 Martinswerk, washed	citrate	yield stress	51 s^{-1}	70% w/w	4–12	The pH of maximum viscosity and yield stress decreases as citric acid (0.05–0.5% w/w) is added, and it roughly reflects the shift of the IEP.	317
Al_2O_3	α, commercial, UCAR		yield stress		55% w/w	2–11	The yield stress and viscosity peak at pH 9 for milled powder. The yield stress and viscosity of as received alumina are rather insensitive to pH.	318

TABLE 3.7 (Continued)

Material	Description	Salt	Property	Shear rate	Solid concentration	pH	Result	Ref.
Al_2O_3	α, AKP-50 Sumitomo	0.5 mol dm^{-3} LiCl, NaCl, KCl, CsCl, (CH$_3$)$_4$ NCl	yield stress	1–600 s^{-1}	20–30% v/v	9–12	At pH 9 the difference between results obtained with different chlorides is insignificant, but at pH 12 the viscosity and yield stress decreases in the series LiCl > NaCl > KCl > CsCl > (CH$_3$)$_4$NCl.	319
Al_2O_3	commercial, A-16 Alcoa	0.01 mol dm^{-3} NaNO$_3$	yield stress	0.38–384 s^{-1}	30% v/v	4–12	The yield stress peaks at pH 8 (IEP at pH 8.8) but the viscosity is rather insensitive to pH over the range 6.5–11. Minimum stability at pH 9.	207
Al_2O_3	AKP-30 Sumitomo, washed	0.01 mol dm^{-3} KNO$_3$	yield stress	0.2 rpm	20–30% v/v	8–11	Yield stress peaks at pH 9.3, slightly below the IEP; linear plot yield stress vs. ζ^2.	320
Al_2O_3	α, AKP-50 Sumitomo	KOH		0.1–300 s^{-1}		9	The viscosity is unaffected by 0.002 mol dm^{-3} SDS, 0.002 mol dm^{-3} Triton TX-100, and 0.002 mol dm^{-3} SDS + 0.004 mol dm^{-3} TX-100. The viscosity is depressed by a factor of 3 by 0.002 mol dm^{-3} SDS + 0.0004 mol dm^{-3} TX-100, and by a factor of 10 by 0.005 mol dm^{-3} SDS + 0.002 mol dm^{-3} TX-100.	321
Al_2O_3	α, AKP 20 Sumitomo	HCl		10.2 s^{-1}	20,50,58 % v/v	1–7	The viscosity shows a minimum at pH 4.	322
Al_2O_3	α, AKP-50 Sumitomo	0.01–1 mol dm^{-3} different 1–1 salts	yield stress	0.2 rpm	25% v/v	5–12.5	The yield stress in KCl, KBr, KI, NaNO$_3$, KNO$_3$, and CsNO$_3$ peaks at pH 9–9.3 corresponding to the IEP, and the yield stress (pH) curves are symmetrical with respect to the IEP. At >0.1 mol dm^{-3} LiNO$_3$ the maximum in the yield stress is shifted to higher pH values and the curves are not symmetrical, namely there is very little decrease in the yield stress at pH above the value corresponding to the maximum. This reflects the shift in the IEP at high LiNO$_3$ concentrations. Yield stress is a linear function of ζ^2.	323
Al_2O_3	AKP-50 Sumitomo	0.01–1 mol dm^{-3} NaCl, NaNO$_3$,	yield stress	0.2 rpm	25% v/v	4–12	At low ionic strength the yield stress peaks at pH 9.5 which is close to the IEP except for NaIO$_3$ which	324

Material	Description	Electrolyte	Method	Shear rate	Concentration	pH range	Comments	Ref
Al_2O_3	α, A-16 SG Alcoa	$NaClO_4$, $NaBrO_3$, $NaIO_3$	yield stress	518.5 s^{-1}	25% v/v	8–10	shows a maximum in the yield stress at pH 7.5 for 0.01 mol dm^{-3}. For 0.01 mol dm^{-3} $NaIO_3$ the yield stress continuously increases when pH decreases and there is no IEP. For 1 mol dm^{-3} $NaBrO_3$ the yield stress has a broad maximum about pH 8.	325
$Al(OH)_3$	commercial samples containing carbonates	0.03 mol dm^{-3} KCl			9%	5–9	Yield stress and viscosity peak at pH 9 (IEP). The viscosity of a sample having CIP at pH 6.95 peaks at pH 7, the viscosity of a sample having CIP at pH 6.3 peaks at pH 6.5.	326
$Al(OH)_3$	gibbsite Alcoa Hydral 710	10^{-2} mol dm^{-3} NaCl	yield stress	0–300 s^{-1}	17% w/w	6–12	The extrapolated yield stress is a linear function of ζ^2 at pH<10, but not at pH>10.	327
Fe_2O_3	maghemite synthetic, 3 samples		yield stress		10% w/w	5.5–8	The yield stress peaks at pH 6.5–7 corresponding to the IEP.	328
SiO_2	Geltec, high purity	0.03 mol dm^{-3} $NaNO_3$		0.1–215 s^{-1}	54–58% v/v	3–13	The log (viscosity) decreases linearly when pH decreases from 3 to 10, very high viscosity at pH>11. The results obtained with polyethylene oxide at pH 9.5 do not show any clear trend.	329
TiO_2	commercial, Tioxide	0.001 mol dm^{-3} KNO_3	yield stress, flow behavior index, flow consistency number	10–90 s^{-1}	5–30% w/w	3–10.5		312
TiO_2	rutile, Tioxide			500 s^{-1}	1–50% v/v	2–13	A broad minimum: region of low and pH independent viscosity in the center of pH scale, pH 5–12 for 10%, and 6–12 for 30% v/v dispersion. With 50% v/v a sharp minimum at pH 8.5.	330
TiO_2	anatase Aldrich	0.005–0.5 mol dm^{-3} $NaNO_3$, NaI	yield stress	0.05–1160 s^{-1}	38% w/w	4–10	At >0.1 mol dm^{-3} the IEP shifts to high pH values and maximum in viscosity at 1.16 s^{-1} shifts accordingly, but no exact match was observed. Similar shifts of the maximum in viscosity to high pH are observed when water is replaced by mixed water + organic solvent, and in the presence of divalent cations.	331
TiO_2	rutile Aldrich	10^{-4}–10^{-2} mol dm^{-3} $Ca(NO_3)_2$, $Ba(NO_3)_2$, $Mg(NO_3)_2$	yield stress	0.05–1160 s^{-1}	38% w/w	4–11	The viscosity of dispersions at pH>6 is rather insensitive to pH in the presence of 10^{-3} mol dm^{-3} $Ca(NO_3)_2$, $Ba(NO_3)_2$, or $Mg(NO_3)_2$. The viscosity increases in the series Mg<Ca<Ba.	332

TABLE 3.7 (Continued)

Material	Description	Salt	Property	Shear rate	Solid concentration	pH	Result	Ref.
ZrO_2	Z-Tech (2% HfO_2)	citrate, phosphate up to 1.1% dry weight basis, sulfate, benzene 1,2,3, carboxylate	yield stress		45-65% w/w	1-10	In absence of specifically adsorbed anions the yield stress peaks at pH 7 (IEP) and the yield stress (ζ^2) plots are linear. In the presence of specifically adsorbed anions the maximum shifts to low pH values according to the shift in the IEP and yield stress at the maximum is depressed when the anion concentration increases.	333
ZrO_2	Z-Tech	0.05 mol dm^{-3} KNO_3	yield stress		18% v/v	2-10	The maximum yield stress corresponds to the IEP. Presence of polyacrylic acid (different MW) shifts the IEP to lower pH, and the maximum of yield stress is shifted accordingly.	334
ZrO_2	TZ-0 Tosoh Ceramics	0.1 mol dm^{-3} NaCl, $NaNO_3$, $NaClO_4$, $NaBrO_3$, $NaIO_3$	yield stress	0.2 rpm	20% v/v	2-12	The yield stress peaks at pH 8.5 which is close to the IEP except for $NaIO_3$ which does not show IEP or maximum in yield stress.	324
ZrO_2	Z-Tech	0.1 mol dm^{-3} KCl	yield stress		57% w/w	3-11	The yield stress peaks at pH 7 corresponding to IEP. On addition of cationic surfactant the of yield stress at high pH increases and the maximum shifts to higher pH values, according to the shifts in the IEP. On addition of anionic surfactant the of yield stress at low pH increases and the maximum shifts to lower pH values, according to the shift in the IEP.	335
kaolinite	washed	0.017–0.25 mol dm^{-3} NaCl	Bingham yield stress		9% w/w	4-10	At pH 7.3 the Bingham yield stress is almost independent of the NaCl concentration. This result is interpreted as IEP of the edge surface at this pH.	336

Material	Sample / composition	Electrolyte	Property	Shear rate	Solids loading	pH	Remarks	Ref.
TiO_2 pigments	P1 0.91% Al_2O_3 P2 3.11% Al_2O_3 0.11% SiO_2 P3 2.99% Al_2O_3 2.56% SiO_2 P4 0.97% Al_2O_3 1.28% SiO_2	0.01 mol dm^{-3} KNO_3		0–300 s^{-1}	10% v/v	2.5–10	The maximum yield value corresponds roughly to the IEP except sample P4 (pH of maximum yield value = 3, IEP at pH < 2 if any).	192
ZrO_2 + 5 % Y_2O_3	commercial, SYP 5.2 Z-Tech, Tech, 4% w/w of organic binder	0.01 mol dm^{-3} $NaNO_3$	yield stress		18–27% v/v	5–11	The yield stress varies as fourth power of solids loading and peaks at pH 8 (IEP at pH 7.8).	205
ZrO_2 + 5 % Y_2O_3	commercial, SY 5.2 Z-Tech	0.01 mol dm^{-3} $NaNO_3$		0.38–384 s^{-1}	30% v/v	4–12	The viscosity peaks at pH 6–8 (IEP at pH 6.9). Minimum stability at pH 7.	207
bentonite	washed, 3 samples	none	yield stress	0.1–500 s^{-1}	5.5–12.5% w/w	2.3–12	For Wyoming bentonite the highest viscosity and yield stress are observed at pH 2.3; for two other materials a maximum of viscosity and yield stress is observed at pH 5–6 (there is no IEP, only negative ζ potentials).	337
kaolin	acid washed, Ajax	0.01 mol dm^{-3} KNO_3	yield stress	0.2 rpm	20–26% v/v	4–10	Yield stress peaks at pH 5.3 (there is no IEP).	320
3 lateritic sediments	goethite + serpentinite + maghemite + gibbsite + quartz (different proportions), different size fractions		yield stress	200 s^{-1}	36% w/w	3–12	Maximum of viscosity matches the IEP (pH 5–8.2).	338

were found by Delamar [100] who used PZC data for 8 oxides selected from Parks' review. A possibility to estimate the PZC from XPS measurements was also discussed in Ref. [340].

G. Interaction with Organic Molecules

Acid-base properties of oxide surfaces are employed in many fields and their relationship with PZC has been often invoked. Adsorption and displacement of different organic molecules from gas phase was proposed as a tool to characterize acid-base properties of dry ZnO and MgO [341]. Hammet acidity functions were used as a measure of acid-base strength of oxides and some salts [342]. Acidity and basicity were determined by titration with 1-butylamine and trichloroacetic acid in benzene using indicators of different pK_a. There is no simple correlation between these results and the PZC. Acid-base properties of surfaces have been derived from IR spectra of vapors of probe acids or bases, e.g. pyridine [343] adsorbed on these surfaces. The correlation between Gibbs energy of adsorption of organic solvents on oxides calculated from results obtained by means of inverse gas chromatography and the acceptor and donor ability of these solvents was too poor to use this method to characterize the donor-acceptor properties of the solids [344].

Certain organic molecules, e.g. 7,7,8,8,-tetracyanoquinodimethane TCNQ can accept electrons and form free anion radicals stable at room temperature. Concentrations of TCNQ and TCNQ radical can be determined using ESR, or even assessed on the basis of coloration. Meguro and Esumi [345] proposed a method to determine acid base properties of solid surfaces from radical concentration in the surface layer for a series of electron acceptors having different electron affinities. In this method a tacit assumption is made that except for the studied absorbent and TCNQ/TCNQ radical there are no other electron acceptors or donors in the system, which is not necessarily correct. This problem is analogous to assessment of acid base properties of materials based on their electrokinetic potentials in allegedly pure organic solvents (Section V).

H. Other

A linear correlation was found [107] between zero points of oxides and passive films and $V_{FB} + 0.6442\ E_g$ where V_{FB} is flat-band potential and E_g is the band gap.

$$PZC = 16.33 - 7.23(V_{FB} + 0.6442 E_g) \qquad (3.37)$$

The same publication reports on nonlinear dependence between pitting potential of ion-implanted binary surface alloys of aluminum and zero points of oxides of implanting metals.

The following correlation was proposed [346]

$$PZC = -2.9E_F + 16.8 \qquad (3.38)$$

and the empirical coefficients were calculated using results obtained with 2 samples of MgO, 3 samples of Al_2O_3, mullite, two samples of $AlPO_4$, and 2 samples of SiO_2. Vijh [103] found the following correlation

$$PZC = 12.72 - 0.48\ \Delta H_s / CN \qquad (3.39)$$

where ΔH_s is the heat of sublimation per mole of the metal (in kcal) and CN is the coordination number of the metal, and the empirical coefficients in Eq. (3.39) were calculated using results for 11 oxides. Two other parameters characterizing acid-base properties of oxides were found to show less spectacular correlation with the PZC in the same study.

The following correlation:

$$\text{PZC} = 12.22 - 0.89 \times \text{ionization potential of metal in MJ/mol} \qquad (3.40)$$

was found for 12 oxides [347] whose PZC were taken from Parks' classical review. A correlation coefficient of 0.93 was reported, but for some oxides Eq. (3.40) gives PZC, e.g. 4.4 for titania, very different from most values reported in Table 3.1. A correlation between the PZC and the color of a series of pH indicators adsorbed on a powder from heptane was studied in the same publication.

Siviglia et al. [237] found a correlation between the CIP of a series of RuO_2 powders and position of certain peak (in terms of 2θ) in their x-ray diffraction spectra. Extrapolation of the obtained curve to the theoretical position of this peak in pure, stoichiometric RuO_2 gives the PZC of "ideal" oxide. The curve can be linearized when CIP is plotted versus reciprocal square of the cation-cation spacing in the lattice.

Wachs et al. [348,349] published a series of papers on speciation of V(V), Cr(VI), Mo(VI), W(VI) and Re(VII) supported on different metal oxides, studied by Raman spectroscopy. The speciation corresponds roughly to speciation in aqueous solutions whose pH is equal to the PZC of the support. Thus, CrO_4^{2-} is the major species detected on MgO, a mixture of CrO_4^{2-} and $Cr_2O_7^{2-}$ is found on TiO_2, and a mixture of $Cr_2O_7^{2-}$ with higher condensation products on SiO_2.

Maximum in friction coefficient (defined as ratio of frictional force to normal reaction force) near the PZC was found for alumina [350]. Pristine PZC was used in calculations and friction experiments were performed in 0.02 mol dm^{-3} Na_2SO_4, thus the results could be affected by specific sorption of sulfate. Resistivity of alumina dispersions is lower than that of solutions without alumina except in the closest vicinity of the IEP [351].

III. SURFACE CHARGING IN INERT ELECTROLYTES

A. Surface Charge Density

The studies of surface charging of different materials, chiefly silica and clay minerals, which did not result in a specific PZC value are summarized in Table 3.8. Some publications reporting $\sigma_0 = 0$, but no positive σ_0 for these materials are also presented in Table 3.8, and PZC values found for the same materials in other publications are shown in Tables 3.1 and 3.3. The organization of Table 3.8 is similar to Tables 3.1 and 3.3 except the method is not specified (only "pH" results are presented in Table 3.8). Additional information about the effect of the nature of the salt on σ_0 is given in the column "results" (the abbreviation "KCl > NaCl" means that σ_0 in the presence of KCl is higher than in NaCl). The plots of σ_0 (pH) for a few common materials taken from the literature cited in Tables 3.1 and 3.8 are compared in Figs. 3.43–3.73. These materials include Al_2O_3 (Figs. 3.43–3.46), AlOOH (Figs. 3.47–3.50), synthetic hematite (Figs. 3.51–3.54), FeOOH (Figs. 3.55–3.58), maghemite (Figs. 3.59–3.60), SiO_2 (Figs. 3.61–3.66), TiO_2 (Figs. 3.67–3.70), and ZrO_2

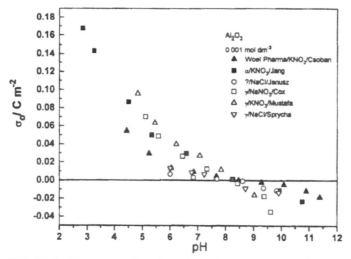

FIG. 3.43   The σ_0 as a function of pH in the $Al_2O_3 - 10^{-3}$ mol dm^{-3} inert electrolyte system.

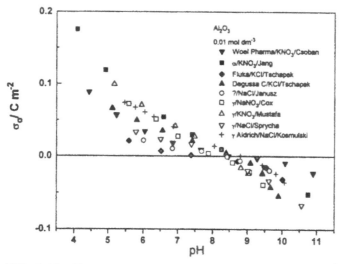

FIG. 3.44   The σ_0 as a function of pH in the $Al_2O_3 - 10^{-2}$ mol dm^{-3} inert electrolyte system.

(Figs. 3.71–3.73). The σ_0 (pH) curves for other (hydr) oxides are collected in Figs. 3.74–3.79. The σ_0 at given pH depends on the ionic strength (Fig. 3.4), although some publications, e.g. [352] suggest that the effect of ionic strength on σ_0 is rather insignificant. After all, all results presented in one graph were obtained at the some ionic strength equal to 10^{-3}, 10^{-2}, 10^{-1} or 1 mol dm^{-3}. A brief characterization of each sample, formula of the electrolyte, and the last name of the first author are given in

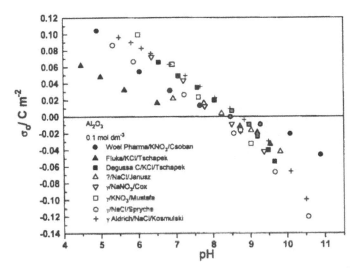

FIG. 3.45 The σ_0 as a function of pH in the $Al_2O_3 - 10^{-1}$ mol dm^{-3} inert electrolyte system.

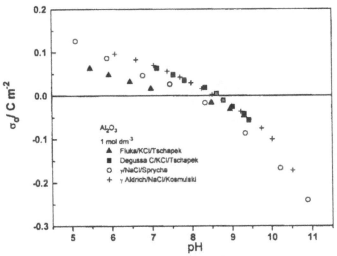

FIG. 3.46 The σ_0 as a function of pH in the Al_2O_3 1 mol dm^{-3} inert electrolyte system.

the figure, and more complete information on the sample and experimental conditions, and full bibliographical information can be found in Tables 3.1 and 3.8. Most results were presented in graphical form in original publications and the scanning procedure might have resulted in some errors, thus, the data points in Figs. 3.43–3.79 represent approximate values. To avoid overcrowding of symbols no more than a dozen of data points from one publication were used in the one graph. Data

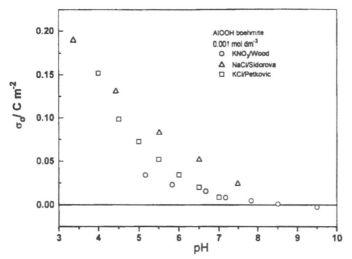

FIG. 3.47 The σ_0 as a function of pH in the boehmite – 10^{-3} mol dm^{-3} inert electrolyte system.

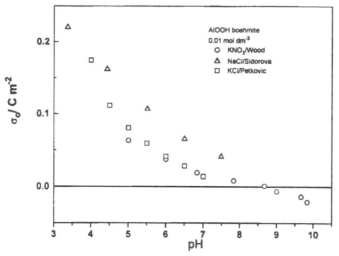

FIG. 3.48 The σ_0 as a function of pH in the boehmite – 10^{-2} mol dm^{-3} inert electrolyte system.

from publications which resulted in extreme pH$_0$ (far from average values, cf. Section I.B) were not used in Figs. 3.43–3.79. Many sets of data were originally presented in a different way, and sometimes recalculation was necessary, e.g. when σ_0 was expressed in C/g or mol/m^2. In many publications the σ_0 scale is reversed (positive σ_0 in the bottom) and in a few publications even the pH scale is reversed (high pH values to the left). Sometimes the pH and σ_0 axes are interchanged (pH is an ordinate). Positive values for σ_0 and ζ potential of silica at high pH are reported in a

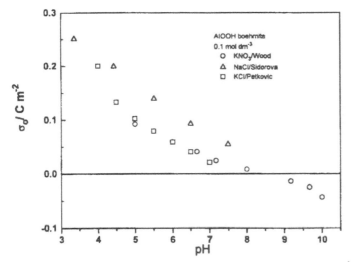

FIG. 3.49 The σ_0 as a function of pH in the boehmite – 10^{-1} mol dm^{-3} inert electrolyte system.

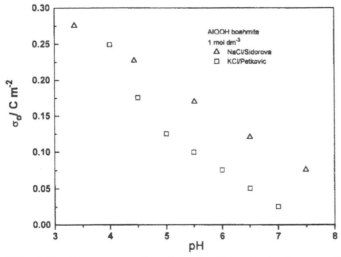

FIG. 3.50 The σ_0 as a function of pH in the boehmite – 1 mol dm^{-3} inert electrolyte system.

few publications. This is clearly a mistake rather than a real effect, i.e. these results are similar to normal behavior of silica when taken with minus sign. In some publications surface charge density is presented in form of Z(pH) plots where Z is the number of protons reacted per one surface hydroxyl group. This representation is more suitable for studies of acid base properties of well defined species, e.g. the $Al_{13}O_4(OH)_{24}(H_2O)_{12}^{7+}$ complex [353]. The σ_0 is obtained from such results by multiplying Z by the site density (cf. Chapter 5) expressed in C m^{-2}. Not necessarily

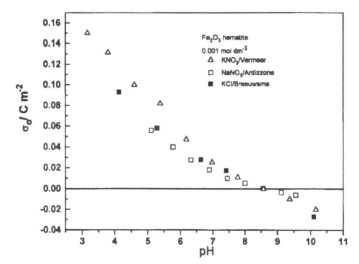

FIG. 3.51 The σ_0 as a function of pH in the hematite – 10^{-3} mol dm^{-3} inert electrolyte system.

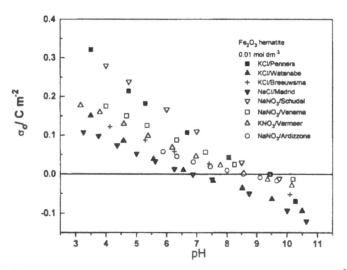

FIG. 3.52 The σ_0 as a function of pH in the hematite – 10^{-2} mol dm^{-3} inert electrolyte system.

do the results calculated from Eq. (3.1) represent the σ_0, namely high consumption of acid or base at extreme pH values can be chiefly due to dissolution of the material [354]. The surface area used in Eq.(3.1) to calculate σ_0 is most often referred to as BET in original publications, and this term is applied for different procedures that can give somewhat different surface areas for the same sample. On the other hand,

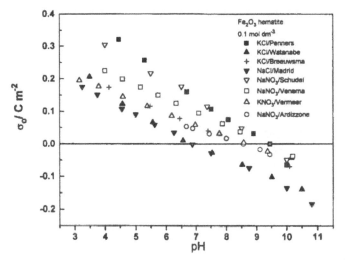

FIG. 3.53   The σ_0 as a function of pH in the hematite – 10^{-1} mol dm^{-3} inert electrolyte system.

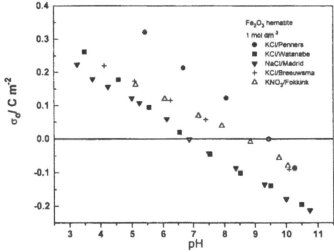

FIG. 3.54   The σ_0 as a function of pH in the hematite – 1 mol dm^{-3} inert electrolyte system.

the changes in surface area caused by aggregation of particles are probably negligible, e.g. the apparent σ_0 is not enhanced on ultrasonic treatment of dispersions [355].

   This is probably the largest compilation of the σ_0 from different sources ever published. Twenty-six sets of charging data for ten oxides in ten electrolytes (most of them for at least three ionic strengths) were compiled in tabular form by Sahai and

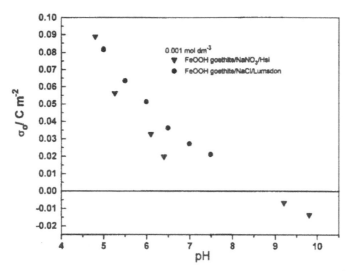

FIG. 3.55 The σ_0 as a function of pH in the goethite – 10^{-3} mol dm^{-3} inert electrolyte system.

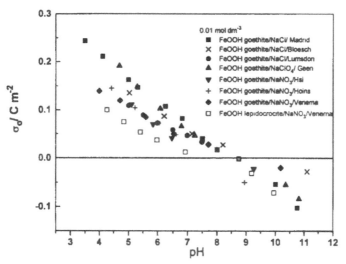

FIG. 3.56 The σ_0 as a function of pH in the FeOOH – 10^{-2} mol dm^{-3} inert electrolyte system.

Sverjensky [356]. Moreover a few compilations for specific materials have been published.

The σ_0 data reported for materials having the same chemical formula are often very divergent. These differences can be partially due to temperature and electrolyte effects, namely both T and the nature of supporting electrolyte affect the σ_0, although their effects (discussed in subsequent sections) are not sufficiently significant to

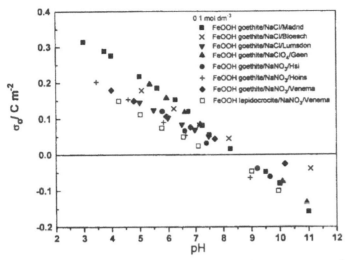

FIG. 3.57 The σ_0 as a function of pH in the FeOOH – 10^{-1} mol dm^{-3} inert electrolyte system.

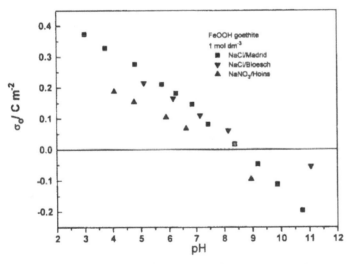

FIG. 3.58 The σ_0 as a function of pH in the goethite – 1 mol dm^{-3} inert electrolyte system.

explain these discrepancies. On the other hand, the range of σ_0 for different materials shown in Figs. 3.43–3.79 is rather narrow, and most values reported for pH$_0$ ±3 pH units are on the order of ±0.1 C m^{-2} with exception of PbO$_2$ and Cr(OH)$_3$ (Figs. 3.75 and 3.77), whose σ_0 on the order of 1 C m^{-2} has been reported. Different ideas and theories have been proposed regarding σ_0 of different materials on the one hand, and different samples having the same chemical formula on the other. Fokkink et al. [357]

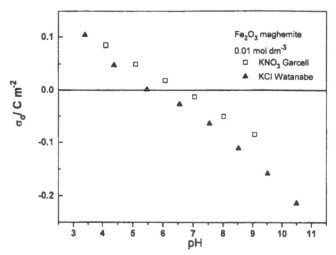

FIG. 3.59 The σ_0 as a function of pH in the maghemite – 10^{-2} mol dm^{-3} inert electrolyte system.

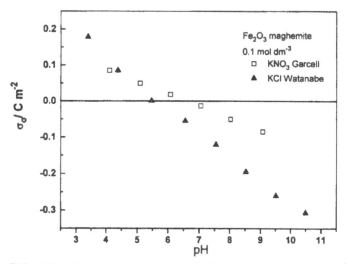

FIG. 3.60 The σ_0 as a function of pH in the maghemite – 10^{-1} mol dm^{-3} inert electrolyte system.

argue that charging curves for rutile, RuO_2 and hematite at three different KNO_3 concentrations plotted as a function of pH-pH$_0$ are practically identical (within experimental error). Lyklema [358] analyzed the surface charge densities of different types of silica for the same ionic strength (0.1 NaCl or KCl) and concluded that σ_0 decreases in the series precipitated silica > pyrogenic silica > quartz.

The σ_0 reported for given pH-pH$_0$ and ionic strength for different samples of alumina (Figs. 3.43–3.46) can vary by a factor of 5 on the positive and negative

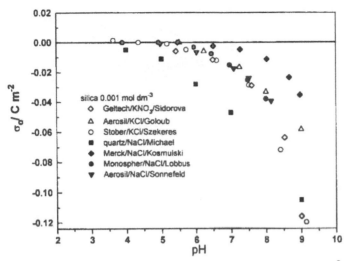

FIG. 3.61 The σ_0 as a function of pH in the $SiO_2 - 10^{-3}$ mol dm^{-3} inert electrolyte system.

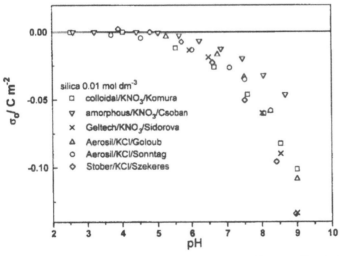

FIG. 3.62 The σ_0 as a function of pH in the $SiO_2 - 10^{-2}$ mol dm^{-3} inert electrolyte system (potassium salts).

branch of charging curves. The results for positive branch of charging curves of synthetic hematite (Figs. 3.51–3.54) are somewhat more consistent and the difference between the highest and lowest σ_0 is only by a factor of 3. Even more consistent results (factor 1.5) were reported for positive branch of charging curves of synthetic goethite, but the σ_0 reported for the negative branch are more divergent. The shape of charging curves of silica (Figs. 3.61–3.66) is very different from metal oxides,

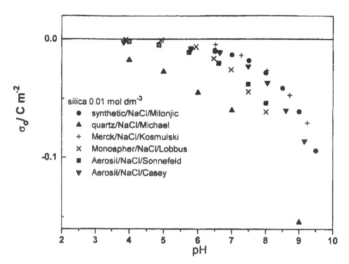

FIG. 3.63 The σ_0 as a function of pH in the SiO_2 – 10^{-2} mol dm^{-3} inert electrolyte system (sodium salts).

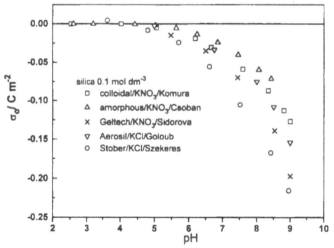

FIG. 3.64 The σ_0 as a function of pH in the SiO_2 – 10^{-1} mol dm^{-3} inert electrolyte system (potassium salts).

namely at pH < 5 the σ_0 is very low and rather insensitive to pH. One set of charging curves for quartz (Figs. 3.61, 3.63, and 3.65) does not confirm the mentioned above allegation, that its σ_0 is lower than that of other types of silica, and the results obtained at low ionic strengths suggest the opposite. The σ_0 of silica at the same conditions reported in different sources can differ by a factor of 5 (some even more divergent results were published but they are not used in the figures). The positive

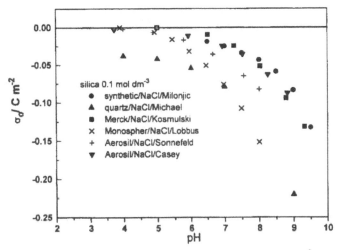

FIG. 3.65 The σ_0 as a function of pH in the $SiO_2 - 10^{-1}$ mol dm^{-3} inert electrolyte system (sodium salts).

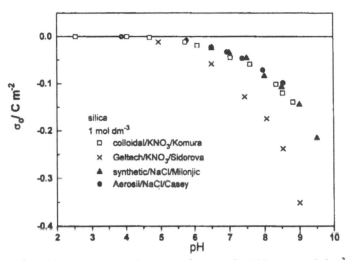

FIG. 3.66 The σ_0 as a function of pH in the $SiO_2 - 1$ mol dm^{-3} inert electrolyte system.

branches of charging curves of titania (Figs. 3.67–3.70) and zirconia (Figs. 3.71–3.73) are more consistent while on the negative branch the difference between the highest and lowest σ_0 at given pH and ionic strength is by factor > 5 for titania and > 10 for zirconia. The results shown in Figs. 3.67–3.70 do not indicate any substantial difference in the magnitude of σ_0 between anatase and rutile. The $\sigma_0(pH)$ curves at constant ionic strength for different oxides presented in Figs. 3.74, 3.76, and 3.78 are almost parallel (similar σ_0 at given pH-pH_0) with a few exceptions.

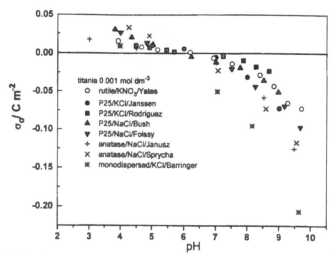

FIG. 3.67 The σ_0 as a function of pH in the $TiO_2 - 10^{-3}$ mol dm^{-3} inert electrolyte system.

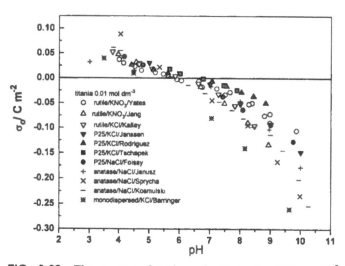

FIG. 3.68 The σ_0 as a function of pH in the $TiO_2 - 10^{-2}$ mol dm^{-3} inert electrolyte system.

B. Electrokinetic Potential

A precise definition of the ζ potential [407] is long and it can be hardly understood without deep knowledge in colloid chemistry. The present discussion is somewhat simplified. Studies of transport phenomena revealed that a kinetic unit moving with respect to aqueous medium consists of a solid particle and thin film of solution surrounding the particle. Consequently these phenomena are governed by the electric

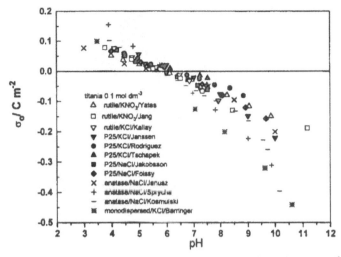

FIG. 3.69 The σ_0 as a function of pH in the $TiO_2 - 10^{-1}$ mol dm^{-3} inert electrolyte system.

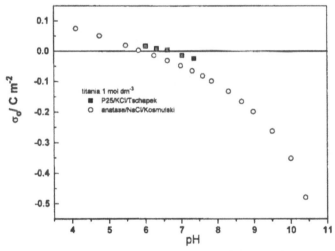

FIG. 3.70 The σ_0 as a function of pH in the $TiO_2 - 1$ mol dm^{-3} inert electrolyte system.

potential at the border between this film and bulk solution rather than by the surface potential.

Smoluchowski was the first to recognize that the interactions responsible for electrokinetic phenomena and colloid stability can be rationalized in terms of this potential (ζ potential), which must not be identified with the surface potential. In absence of strongly adsorbing species ζ potential has the same sign as the surface potential, but the absolute value of the ζ potential is lower, especially at high ionic

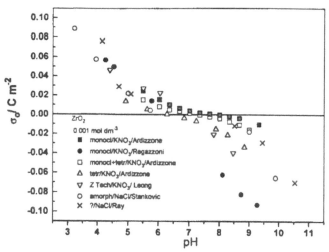

FIG. 3.71 The σ_0 as a function of pH in the ZrO_2 – 10^{-3} mol dm^{-3} inert electrolyte system.

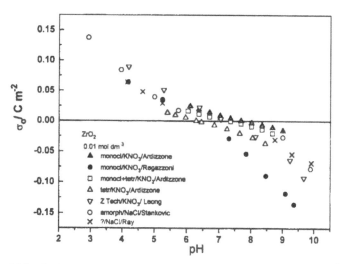

FIG. 3.72 The σ_0 as a function of pH in the ZrO_2 – 10^{-2} mol dm^{-3} inert electrolyte system.

strengths. Many attempts to determine the surface potential of semiconductor materials have been published. For instance, the measurements of open circuit potentials between TiO_2 films grown on Ti by different methods as a function of pH led to slopes of -15 to -40 mV/pH unit [408]. These slopes were rather insensitive to KNO_3 concentration. In case of thermally grown film the slope was about -15 mV/

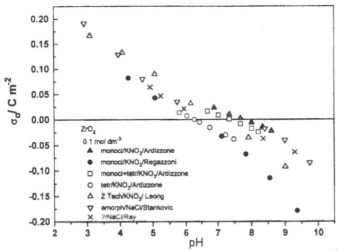

FIG. 3.73 The σ_0 as a function of pH in the ZrO_2 – 10^{-1} mol dm^{-3} inert electrolyte system.

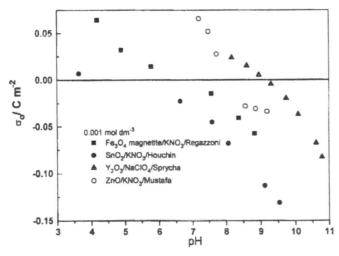

FIG. 3.74 The σ_0 as a function of pH in the oxide (Fe_3O_4, SnO_2, Y_2O_3, ZnO) – 10^{-3} mol dm^{-3} inert electrolyte system.

pH unit at pH < 5, and about -40 mV/pH unit at pH > 5. The slope of the surface potential of metal oxides as a function of pH is believed to be somewhat below the Nernstian slope (-59 mV/pH unit).

Smoluchowski proposed an equation allowing to calculate the ζ potential from experimentally determined electrokinetic mobility μ:

$$\mu = \zeta\varepsilon/\eta \qquad (3.41)$$

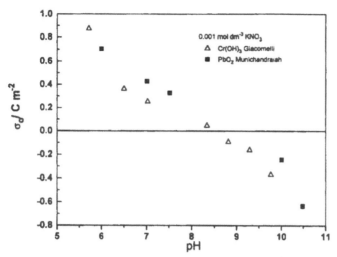

FIG. 3.75 The σ_0 as a function of pH in the (hydr)oxide [Cr(OH)$_3$, PbO$_2$] – 10^{-3} mol dm^{-3} inert electrolyte system.

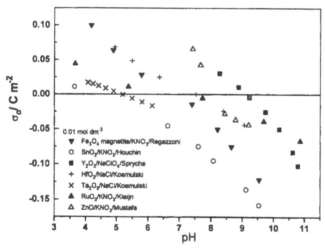

FIG. 3.76 The σ_0 as a function of pH in the oxide (Fe$_3$O$_4$, HfO$_2$, RuO$_2$, SnO$_2$, Ta$_2$O$_5$, Y$_2$O$_3$, ZnO) – 10^{-2} mol dm^{-3} inert electrolyte system.

where ϵ and η are the electric permittivity and viscosity of the medium. Unfortunately, this simple equation is only valid at high ionic strengths and for large particles (thin double layer). Many other approximated equations have been proposed, whose validity range is broader than that of Eq. (3.41) (cf. any handbook of colloid chemistry), but Smoluchowski equation is built in software of most commercial zetameters, and many values of ζ potential reported in literature were calculated using this equation. Some publications report the electrokinetic mobility

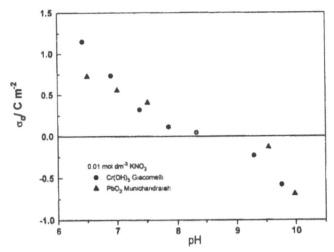

FIG. 3.77 The σ_0 as a function of pH in the (hydr)oxide [Cr(OH)$_3$, PbO$_2$] – 10^{-2} mol dm^{-3} inert electrolyte system.

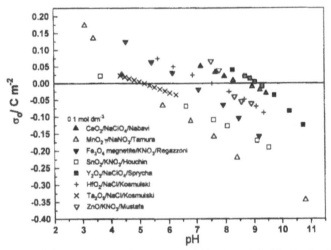

FIG. 3.78 The σ_0 as a function of pH in the oxide (CeO$_2$, Fe$_3$O$_4$, HfO$_2$, MnO$_2$, SnO$_2$, Ta$_2$O$_5$, Y$_2$O$_3$, ZnO) – 10^{-1} mol dm^{-3} inert electrolyte system.

rather than ζ potential. The ζ potentials calculated by means of Smoluchowski equation (straight line) are compared in Figs 3.80 and 3.81 with results obtained using exact theory [409] for typical ionic strengths at which electrokinetic measurements are carried out and for typical sizes of colloidal particles.

Figure 3.81 shows that a difference in the particle size can result in different mobilities at a constant ζ potential. Exact calculation of ζ potentials from experimental data is possible for spherical particles but not for irregularly shaped

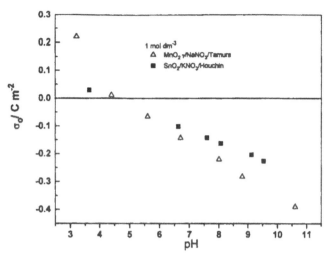

FIG. 3.79 The σ_0 as a function of pH in the oxide (MnO$_2$, SnO$_2$) – 1 mol dm^{-3} inert electrolyte system.

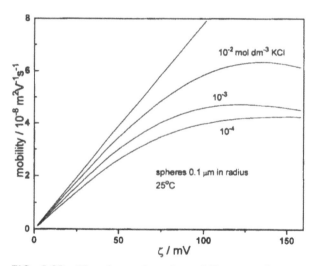

FIG. 3.80 The electrophoretic mobility as a function of ζ potential for different KCl concentrations.

particles, for which Eq. (3.41) is often used in hope that different factors that remain beyond control will cancel out. The theoretical curves in Figs. 3.80 and 3.81 were calculated assuming 100% dissociation of the electrolyte. The relationship between electrokinetic mobility and ζ potential in solutions of weak electrolytes was recently analyzed by Grosse and Shilov [410]. Their model calculation for thin double layer gave practically identical results for strong and weak electrolyte in the range of low ζ potentials. On the other hand the highest theoretically possible mobility (cf.

(*Text continues on pg. 247*)

TABLE 3.8 Surface Charging in Absence of Strongly Adsorbing Species

Material	Characterization	Electrolyte	Conc./ mol dm^{-3}	T	pH	Result	Ref.
AlOOH	boehmite colloidal from chloride	KNO$_3$, KCl, KBr, KI	0.001–1	20	4–7	KI > KBr > KNO$_3$ >> KCl	359
Bi$_2$O$_3$	crystalline, from nitrate	NaNO$_3$	0.1		9–11	Negative σ_0.	360
Fe$_2$O$_3$·H$_2$O	from chloride	KCl	1	22	2–10	The charge density is changing on aging for many days as a result of transformation into goethite.	361
Fe$_2$O$_3$·H$_2$O	hydrolysis of nitrate in the presence of citrate	NaNO$_3$	0.001–1		3.5–10	The charging curves merge at pH <4 but no CIP is observed.	362
FeOOH	β, from chloride	NaCl	0–0.002	25	4–11	Charging curves obtained at different ionic strengths merge at pH >9.5.	363
IrO$_2$	thermal decomposition of IrCl$_3$	KNO$_3$, K$_2$SO$_4$	0.005–0.1		4–8	No CIP in KNO$_3$, CIP at pH 4.9 in K$_2$SO$_4$. The ionic strength effect suggests PZC at pH <4.	364
IrO$_2$	thermal decomposition of H$_2$IrCl$_6$	KNO$_3$	0.1	25	3–8	Potential of IrO$_2$: 0.52–1.26 V—different charging curves, PZC at pH 6.5 (0.52 V)–3.5 (0.9 V), PZC at pH <3 if any for potentials >0.96 V.	365, 366
IrO$_2$	synthetic	KNO$_3$	0.0003–1	25	2–10	PZC at pH <2 if any.	367
MnO$_2$	δ, synthetic	NaCl	0.0004–0.7 0.7		2.5–9		368
MnO$_2$	δ, MnCl$_2$ + KMnO$_4$	NaCl, NaCl + KCl	0.016–1		2–8	Negative σ_0, the charging curves obtained at different ionic strengths merge at pH 2.	369
MnO$_2$	λ, leaching of LiMn$_2$O$_4$ with nitric acid	LiCl, NaCl, KCl, RbCl, CsCl	0.01–1	25	2–12	Negative σ_0; LiCl > NaCl > KCl = RbCl = CsCl, but at pH <6 the NaCl and KCl curves merge; zero charge at pH 2.8 (Li) and 4.2 (K).	370, 371
MnO$_2$	γ, IC12	LiNO$_3$, NaNO$_3$, KNO$_3$, CsNO$_3$	0.1	25	4–11	Li > Na > K – Cs, negative σ_0.	372
MnO$_2$	δ, KMnO$_4$ + conc. HCl	NaNO$_3$	0.01–0.5		2.5–10	Negative σ_0.	373

TABLE 3.8 (Continued)

Material	Characterization	Electrolyte	Conc./ mol dm^{-3}	T	pH	Result	Ref.
SiO$_2$	quartz	KNO$_3$	10^{-3}, 1	25	4–11	Negative σ_0.	374
SiO$_2$	pyrogenic Cab-O-Sil, dried and acid treated	KCl,	0.01–1	20	1.5–9	Cs > K > Li; at pH < 4 scattered values of $\sigma_0 \approx 0$ with no apparent trend with pH or ionic strength.	375
SiO$_2$	Aerosil	LiCl, CsCl HCl, HNO$_3$	10^{-3}–2.6		0–3	Positive σ_0 equal for HCl and HNO$_3$ at pH < 2. ΔH_{ads}(HCl, HNO$_3$) is negative.	310
SiO$_2$	colloidal	LiNO$_3$, KNO$_3$, CsNO$_3$	0.01–1	25	2.5–9	Cs > K > Li	376
SiO$_2$	from ethoxide	KCl	0.01	40	7–9.5	Negative σ_0.	124
SiO$_2$	quartz, Min-U-Sil	NaCl	0.001–0.1	25	4–9	Negative σ_0.	377
SiO$_2$	Ludox	NaCl, LiCl, KCl	0.01–0.7	25	7–9	K > Na > Li, negative σ_0.	378
SiO$_2$	hydrolysis of silicate	LiCl, NaCl, KCl, CsCl	0.001–1	25	6.5–9.5	K > Cs > Na > Li (0.001 mol dm^{-3}); Li > Na > Cs > K (0.5 mol dm^{-3}).	379
SiO$_2$	Aerosil OX-50	KCl, LiCl, RbCl	10^{-4}–10^{-2}		3–8	Rb > K > Li, negative σ_0.	380
SiO$_2$	Zeothix 265, washed	NaCl	0–0.35		2–10	Apparent PZC ranges from pH 2 (0.35 mol dm^{-3} NaCl) to 6 (no salt).	381
SiO$_2$	Nucleosil 100-30	NaCl	200 g/dm^3	30	4–10	Negative σ_0, also in D$_2$O, in which the σ_0 is lower by factor 2 than in H$_2$O.	382
SiO$_2$	quartz	LiCl	1		5–11		383
SiO$_2$	hydrolysis of silicate	LiCl,NaCl, KCl,CsCl	0.1–4	25	6.5–9.5	Cs > K = Na > Li (0.01 mol dm^{-3}); Li > Na > K > Cs (2.5 mol dm^{-3}).	384
SiO$_2$	sand, cristobalite	NaNO$_3$	0.01	25	7.5–10	Negative σ_0 (chromatographic method).	87
SiO$_2$	Aerosil 380	NaCl	0.01–1	25	3.5–9	Negative σ_0.	385
SiO$_2$	Aerosil 300	LiCl, NaCl, KCl, RbCl, CsCl	0.005–0.3	25	4–8	Cs > Rb > K > Na > Li, negative σ_0.	386
SiO$_2$	Ajax	KNO$_3$	10^{-3}	25	5–9	Negative σ_0.	387

Material	Description	Electrolyte	Concentration	Temp.	pH	Comments	Ref.
SiO_2	Aerosil OX 50	KCl (?)	0.001–0.1		4–10	Negative σ_0 (the curves are erroneously labeled).	388
SiO_2	Monospher 250 Merck	NaCl	0.001–0.1		4–8	Negative σ_0	389
SiO_2	quartz, Min-U-Sil 5	NaCl	0.0117, 0.094	25	4.5–9	Negative σ_0	390
SiO_2	amorphous, Machinery Nagel NH-R	KNO_3	0.01–0.1	25	2.5–9	Negative σ_0	391
SiO_2	chromatographic, Baker	NaCl	0.1		6–8	Negative σ_0	392
SiO_2	from ethoxide	$NaNO_3$	0.001–0.1		4–9	CIP at pH 5.7 is claimed. Actually at this pH the curves for three ionic strengths merge. (supporting information)	182
SiO_2	Aldrich	$NaNO_3$	0.05	20±2	4–10	Negative σ_0.	153
SiO_2	Geltech	KNO_3	0.001–1	20	4.5–9	Negative σ_0.	393
$SiO_2 \cdot H_2O$		$NaNO_3$	0.01–1		4–8	Negative σ_0.	174
Ta_2O_5	amorphous	KCl	0.0001–1		3–10	Negative σ_0.	394
Ta_2O_5	β, synthetic, calcined at 800°C	CsCl, KCl, NaCl, LiCl	0.0001–1		3–10	Cs > K > Na > Li, negative σ_0	394
TiO_2	anatase pigment grade	NaCl	0.0004–0.01		6.5–11	Negative σ_0.	395
ZrO_2	hydrolysis of $ZrOCl_2$	KCl	0.01	40	3–10	PZC at pH < 2.7.	124
ZrO_2	amorphous, hydrolysis of $ZrOCl_2$	LiCl, NaCl, KCl	0.1–1	25	7.5–10	At low pH and low concentration K > Na > Li. At high pH and high concentration Li > Na > K.	396
ZrO_2	commercial, 3 M	NaCl	200g/dm^3	30	5–11		382

materials containing Al_2O_3

Material	Description	Electrolyte	Concentration	Temp.	pH	Comments	Ref.
0.2% F doped Al_2O_3		$NaNO_3$	0.001–0.1		3–10	No clear CIP	397

clay minerals

Material	Description	Electrolyte	Concentration	Temp.	pH	Comments	Ref.
illite	5% chlorite; 5% quartz	$NaNO_3$	0.1, 0.5	25	3–9	Negative σ_0.	398
illite	3 samples	$NaNO_3$	0.1	25	3–10	$\sigma_0 = 0$ at pH < 3.2. Substantial dissolution of illite.	399
illite	Silver Hill Montana	$NaClO_4$	0.2, 0.5		4–10	Negative σ_0	400
kaolinite	KGa-1	NaCl	0.1	25	4–9	Negative σ_0	401, 402
kaolinite	KGa-1	NaCl	0.01, 0.1	25	3.5–9	Negative σ_0 except for pH < 4 with 0.01 mol dm^{-3} NaCl.	390

TABLE 3.8 (Continued)

Material	Characterization	Electrolyte	Conc./ mol dm^{-3}	T	pH	Result	Ref.
Na-montmorillonite	from commercial bentonite	NaNO$_3$ NaCl	0.005–0.5 0.5		3.5–10.5	No CIP, scattered results.	154
Na-montmorillonite	washed	NaClO$_4$	0.1, 0.5		4–10		403
silicates and materials containing SiO$_2$							
1.67 SiO$_2$·Al$_2$O$_3$·0.04 Na$_2$O·0.05 H$_2$O	amorphous, hydrolysis of Na$_2$SiO$_3$ + AlCl$_3$	NaCl	0.003–0.1		4–8	Charging curves obtained at different ionic strengths merge at pH 5, no CIP.	180
SiO$_2$ coated by goethite	sand, cristobalite	NaNO$_3$	0.01	25	4–9	The results suggest a PZC at pH 7.4 (chromatographic method).	87
sandstone	acid leached, washed	NaCl or KCl	0.01–0.1		2–12	At pH 2–5 $\sigma_0 = 0$.	173
chrysotile asbestos	synthetic	NaCl	0–1	30	9–12	No CIP.	185
clays							
kaolin	acid washed American standard				4.5–9		404
kaolin	commercial	NaCl	0, 0.0488		4–8	At pH 4 7 identical results in the presence and absence of salt.	179
glasses							
controlled pore glass		NaCl, CsCl	0.001–0.1	20	3–8	Charging curves for different ionic strengths merge at pH 3.	405
porous glasses		NaCl	10^{-3}–1		4–9	CIP at pH < 4 (if any).	406

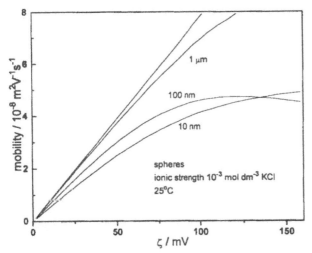

FIG. 3.81   The electrophoretic mobility as a function of ζ potential for different particle radii.

Figs. 3.80 and 3.81) for weak electrolyte is substantially higher than that for strong electrolyte. Dukhin et al. [411] discussed the errors that can be caused by interpretation of electroacoustic data obtained at high solid to liquid ratios by means of theories that are only valid for dilute systems. These theories give severely underestimated $|\zeta|$ at high solid to liquid ratios.

These few examples do not exhaust a long list of difficulties in calculation of exact values of ζ potential from electrophoretic or electroacoustic data, and most results reported in literature appear to be rough estimates rather than exact values, except for some results obtained with nearly spherical particles having very narrow size distributions, e.g. Stöber silica. Many publications tend to overestimate the accuracy of ζ potentials presented therein. For example in a recent publication a graph was presented showing the variations of the measured ζ potential as a function of time, and all data points showed in that graph ranged from 6 to 7 mV.

Matijevic [412, 413] reviewed on preparation and properties of "monodis-persed" colloidal particles. They are ideal model systems because mathematical equations regarding transport phenomena assume simple forms for particles having spherical symmetry. So called monodispersed colloids have in fact some particle size distribution and, e.g. the mass-average diameter and number-average diameter (their ratio is used as the measure of degree of polydispersity) differ by a factor of about 1.1 in typical inorganic "monodispersed" colloids. Different experimental methods lead to different types of "average" diameters. Only for ideally monodispersed colloids there would be no concern about application of experimentally determined effective particle size in interpretation of electrokinetic data; unfortunately, such colloids do not exist. Monodispersed colloids are usually prepared in the presence of complexing anions, and it is difficult to remove these anions from the final products. Slow release of these strongly interacting anions affects the IEP (Chapter 4). Thus, determination of exact values of ζ potential is not trivial, even with monodispersed colloids.

The relationship between the ζ potential and colloid stability is illustrated in Figs. 3.82 and 3.83. The repulsion of particles at very small distances is not shown.

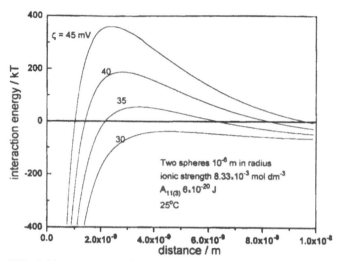

FIG. 3.82   The interaction energy between two spherical particles as a function of ζ potential.

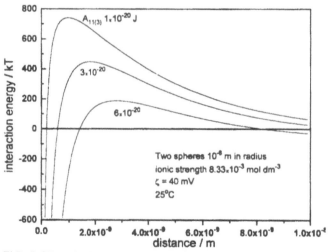

FIG. 3.83   The interaction energy between two spherical particles as a function of Hamaker constant.

The interaction energy between two identical particles depends on the ζ potential and retarded Hamaker constant (cf. any handbook of colloid chemistry), and Fig. 3.83 shows that a ζ potential of about (plus or minus) 40 mV assures an energy barrier that prevents fast coagulation even with relatively high Hamaker constant. When the absolute value of the ζ potential is lower this barrier disappears (Fig. 3.82), and the stability ratio (Fig. 3.6) approaches 1. The theory of colloid stability is discussed in detail in handbooks of colloid chemistry.

Results of electrokinetic studies of different materials, chiefly silica and clay minerals which produced only positive or only negative ζ potentials (no IEP) are shown in Table 3.9. A few sets of electrokinetic data with both negative and positive

ζ potentials, but no data points in the vicinity of the IEP are also presented. The IEP values for the same materials reported in other publications are presented in Tables 3.1 and 3.3. Organization of Table 3.9 is similar to the organization of Tables 3.1 and 3.3, and the "method" column is left empty when only classical electrokinetic methods were used. Table 3.9 substantiates the result obtained in Section I.C that the average of PZC values reported for quartz falls at lower pH than for silica gels, namely, in many studies only negative ζ potentials of quartz are reported.

The plots of ζ (pH) for a few common materials obtained using classical electrokinetic methods, taken from the literature cited in Tables 3.1 and 3.9 are compared in Figs 3.84–3.97. These materials include Al_2O_3 and $Al(OH)_3$ (Figs 3.84–3.86), Fe_2O_3 (synthetic hematite and maghemite, Figs. 3.87–3.89), SiO_2 (Figs. 3.90–3.93), and TiO_2 (Figs 3.94–3.97). The ζ(pH) curves for other (hydr) oxides are collected in Figs. 3.98–3.100. The ζ potential at given pH depends on the ionic strength (Fig. 3.4), and all results presented in one graph were obtained at the same ionic strength equal to 10^{-4}, 10^{-3}, 10^{-2}, or 10^{-1} mol dm^{-3}. A few results obtained at higher ionic strengths by means of electroacoustic method are discussed in Section C. The figure legends in this section, and concerns about comparison of ζ potentials taken from different sources are similar to those discussed in Section A for the σ_0 data. Some ζ potentials were estimated from mobilities taken from original publications using Smoluchowski equation. The results presented in Figs. 3.84–3.97 for different samples of the same material are much more consistent than corresponding σ_0 data, and the difference between ζ potentials of two samples of the same material at given pH-pH$_0$ and ionic strength does not exceed a factor of two. Also the magnitude of ζ potential at given pH-pH$_0$ and ionic strength for various materials is rather consistent. All results plotted together as a function of pH-pH$_0$ give master curves with the ζ potential changing nearly linearly around the IEP, and reaching a plateau at about (plus or minus) 80 mV for 10^{-4} or 10^{-3} mol dm^{-3} electrolyte, 60 mV for 10^{-2}, and 40 mV for 10^{-1} mol dm^{-3} electrolyte at pH < pH$_0$-3 or pH > pH$_0$ + 3. Such tendency to level

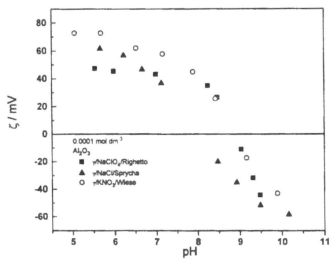

FIG. 3.84 The ζ potential as a function of pH in the Al_2O_3 – 10^{-4} mol dm^{-3} inert electrolyte system.

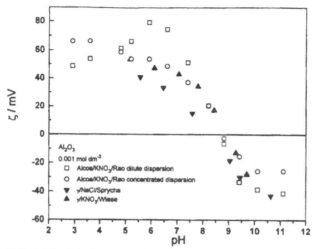

FIG. 3.85   The ζ potential as a function of pH in the $Al_2O_3 - 10^{-3}$ mol dm^{-3} inert electrolyte system.

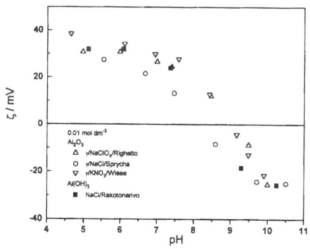

FIG. 3.86   The ζ potential as a function of pH in the Al_2O_3 (or AlOOH) $- 10^{-2}$ mol dm^{-3} inert electrolyte system.

out is not observed with σ_0(pH) curves; also the consistency in electrokinetic behavior as opposite to divergence in σ_0(pH) curves is somewhat paradoxical, especially considering that many sets of electrokinetic and charging curves presented in this section were taken from the same publications and they represent the same samples.

C.   Adsorption of Ions of Inert Electrolyte

The surface charge of materials showing pH dependent surface charging is balanced by counterions. Uptake of ions of inert electrolytes has been even used to define the

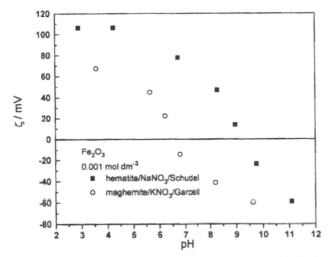

FIG. 3.87   The ζ potential as a function of pH in the $Fe_2O_3 - 10^{-3}$ mol dm^{-3} inert electrolyte system.

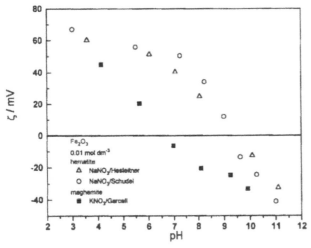

FIG. 3.88   The ζ potential as a function of pH in the $Fe_2O_3 - 10^{-2}$ mol dm^{-3} inert electrolyte system.

PZC (cf. Section I. A), but the number of publications reporting the uptake of counterions is small compared with the number of studies of uptake/release of protons. Table 3.10 presents the following types of results:

- Uptake of ions of inert electrolytes from solution.
- σ_0 measured in the presence of different inert electrolytes for the same sample at otherwise the same conditions.
- ζ determined in the presence of different inert electrolytes for the same sample at otherwise the same conditions.

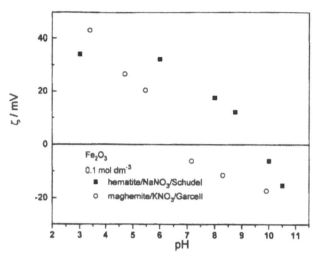

FIG. 3.89 The ζ potential as a function of pH in the $Fe_2O_3 - 10^{-1}$ mol dm^{-3} inert electrolyte system.

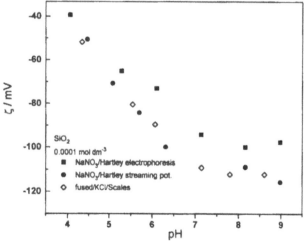

FIG. 3.90 The ζ potential as a function of pH in the $SiO_2 - 10^{-4}$ mol dm^{-3} inert electrolyte system.

- Effects of the nature of electrolyte on coagulation.

A few results obtains by other methods are also reported. The nature of the cation affects chiefly the negative branch of charging and electrokinetic curves, and coagulation of negatively charged particles, and the nature of the anion affects the positive branch, and in the vicinity of PZC both the anion and the cation may play a significant role at high salt concentrations. Some results showing the dependence of σ_0 on the nature of the electrolyte have been presented in Table 3.8. A few salts that

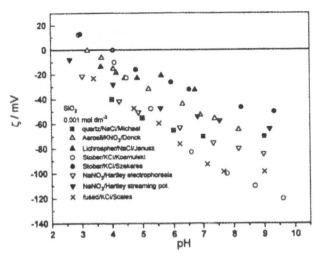

FIG. 3.91 The ζ potential as a function of pH in the SiO_2 – 10^{-3} mol dm^{-3} inert electrolyte system.

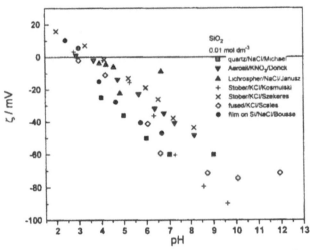

FIG. 3.92 The ζ potential as a function of pH in the SiO_2 – 10^{-2} mol dm^{-3} inert electrolyte system.

have been used in coagulation studies are not considered as inert electrolytes in this book (cf. Introduction), but the results are reported in Table 3.10 together with results obtained for inert electrolytes in the same study for sake of completeness. Sorption of anions from these salts (uptake, ζ potentials, etc.) is discussed in more detail in Chapter 4. The uptake of monovalent ions in the presence of alkaline earth cations has been extensively studied by soil scientists, but these results will not be discussed in this section. The kinetics of adsorption of ions from inert electrolytes

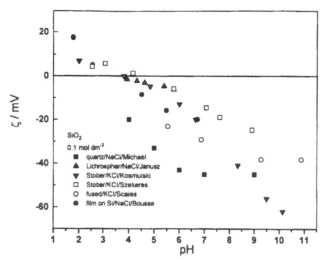

FIG. 3.93 The ζ potential as a function of pH in the $SiO_2 - 10^{-1}$ mol dm^{-3} inert electrolyte system.

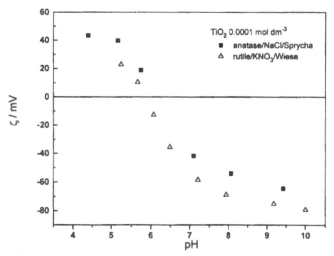

FIG. 3.94 The ζ potential as a function of pH in the $TiO_2 - 10^{-4}$ mol dm^{-3} inert electrolyte system.

and adsorption competition of these ions are discussed together with analogous results reported for strongly adsorbing species in Chapter 4.

The effects of the nature of the electrolyte on σ_0, ζ and coagulation rate are rather insignificant at low ionic strengths and near pH$_0$, but they can be substantial at high ionic strengths. Already in 1964 Yopps and Fuerstenau [462] noticed that the minimum in chain like curves representing stability of alumina dispersions (more

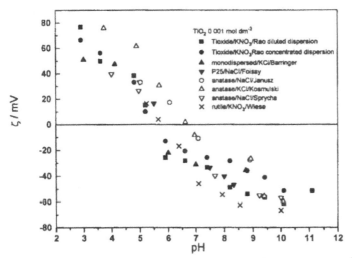

FIG. 3.95 The ζ potential as a function of pH in the $TiO_2 - 10^{-3}$ mol dm^{-3} inert electrolyte system.

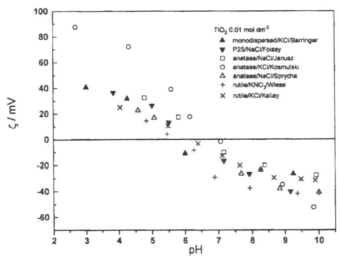

FIG. 3.96 The ζ potential as a function of pH in the $TiO_2 - 10^{-2}$ mol dm^{-3} inert electrolyte system.

precisely: relative time necessary to reach certain intensity of light passing through the dispersion) shifted to high pH values for KCl concentration of 1 mol dm^{-3}. Specific interactions at concentrations of "inert" electrolytes of > 0.1 mol dm^{-3} are commonplace. Gibb and Koopal [463] observed a shift in the IEP to high pH and PZC to low pH, which is typical for specific adsorption of cations (Chapter 4) for

256

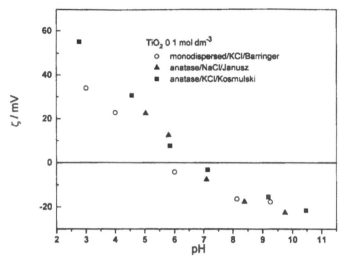

FIG. 3.97 The ζ potential as a function of pH in the TiO₂ – 10⁻¹ mol dm⁻³ inert electrolyte system.

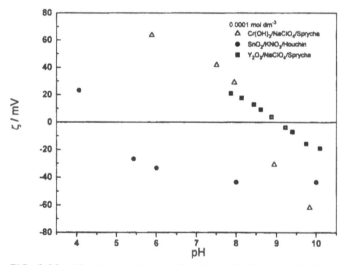

FIG. 3.98 The ζ potential as a function of pH in the (hydr)oxide [Cr(OH)₃, SnO₂, Y₂O₃] – 10⁻⁴ mol dm⁻³ inert electrolyte system.

hematite and rutile dispersions in KNO_3 at concentrations as low as 0.01 mol dm⁻³. This may suggest, that the IEP and CIP of charging curves obtained at ionic strengths of 0.01 or higher is not an appropriate method to obtain the pristine PZC, but these observations were not confirmed by other studies.

The " > " symbol used in Table 3.10 between chemical formulas of adsorbates denotes higher (" > > "= much higher) affinity to the surface, i.e. lower CCC (cf.

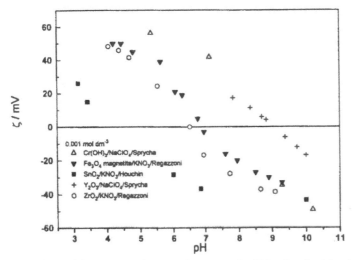

FIG. 3.99   The ζ potential as a function of pH in the (hydr)oxide [Cr(OH)$_3$, Fe$_3$O$_4$, SnO$_2$, Y$_2$O$_3$, ZrO$_2$] – 10^{-3} mol dm^{-3} inert electrolyte system.

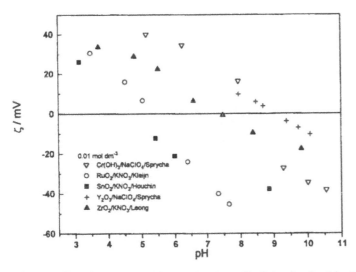

FIG. 3.100   The ζ potential as a function of pH in the (hydr)oxide [Cr(OH)$_3$, RuO$_2$, SnO$_2$, Y$_2$O$_3$, ZrO$_2$] – 10^{-2} mol dm^{-3} inert electrolyte system.

Fig. 3.6), higher uptake, and/or higher $|\sigma_0|$. When uptake of the anion and the cation was studied, the formula of the salt is given in the column "adsorbate". When only uptake of one of these ions was studied, the formula of the other ion is given in the column "electrolyte". Table 3.10 reports also experimental conditions, namely, the solid to liquid ratio and equilibration time. The abbreviation "ζ" in the column "method" is used for classical electrokinetic methods as opposite to the

(*Text continues on pg. 265*)

TABLE 3.9 Electrokinetic Potentials in Absence of Strongly Adsorbing Species

Adsorbent	Characterization	Electrolyte	Concentration/ mol dm^{-3}	T	pH	Method	Result	Ref.
Al$_2$O$_3$	sapphire crystal	NaCl	0.001	22	6.7–11	AFM	Absolute values of ζ potentials are obtained from force measurements.	79
Al$_2$O$_3$	γ, Houdry H 415	NH$_4$NO$_3$	0.01	25	4–10		IEP between pH 7.2 and 8.7 (no data in this pH range).	414
Fe$_3$O$_4$	magnetite, from nitrate	NaCl	0.0005	25	4–6.2		IEP at pH 6.4 (extrapolated).	415
Fe$_2$O$_3$	hematite, hydrolysis of nitrate	KClO$_4$	0.02		2–12		Positive ζ potentials at pH < 5; negative ζ potentials at pH > 11.	416
Fe$_2$O$_3$	from nitrate	NH$_3$	0.306		5–9		Negative ζ potentials over the entire pH range.	417
Fe$_2$O$_3$	hematite, synthetic, 2 samples				1–11		In addition to one sample which behaves normally, electrokinetic data for 2 other samples are presented and IEP at pH 1.5 is claimed for these samples. In fact one sample has negative ζ potentials over the entire pH range, for another a random mixture of positive and negative potentials was observed.	418
Fe$_2$O$_3$	Polysciences	NaNO$_3$	0.001		3–10		Negative ζ potentials over the entire pH range, dependent on the concentration of particles.	419
HgO	reagent, washed, yellow and red	KCl	0, 0.001	25	6–12		Negative ζ potentials over the entire pH range, dissolution at pH < 6.	25
IrO$_2$	thermal decomposition of IrCl$_3$	KNO$_3$	0.001		3–6		Mobility of -3×10^{-8} m^{2}V^{-1}s^{-1}, pH independent.	364
Mg(OH)$_2$					10.5–12		IEP at pH >12 if any.	420

		Electrolyte	Conc.	T	pH range	Comments	Ref.
MnO_x	nominally β MnO_2, x = 1.19–2.44 2 samples	NaCl				IEP at pH <3.	421
MnO_2	δ, synthetic	KNO_3	0.01		4–12	IEP at pH 2.4 (extrapolated).	368
$Nb_2O_5 \cdot n H_2O$	synthetic					IEP at pH <4 if any.	286
NiO	amorphous	KNO_3	0.001	25	8–11	Only positive ζ potentials.	422
$NiO \cdot n H_2O$	synthetic			25	8–11	Only positive ζ potentials.	423
$Ni(OH)_2$	Research Organic/Inorganic Chemical Corp.	NH_4NO_3	0.02	22	9–11	Only positive ζ potentials.	424
$Pb(OH)_2$	freshly precipitated				10–11.5	IEP at pH >11.5 if any.	276
PdO				23	2–8	Negative ζ potentials over the entire pH range.	425
Pr_6O_{11}	reagent, washed	KCl	0, 0.001	25	2–10	Negative ζ potentials over the entire pH range.	25
$RuO_2 \cdot 1.1 H_2O$	from commercial hydrate and from $RuCl_3$, different calcination temperatures					Scattered results, IEP < PZC, higher IEP for higher calcination T.	426
$Sb_2O_5 \cdot n H_2O$	synthetic	KNO_3	0.01		4–12	IEP at pH <4 if any.	286
SiO_2	precipitated, BDH	KCl	10^{-3}	20	2–5	IEP at pH <2 if any.	4
SiO_2	quartz	KNO_3	0.001	25	2–11	IEP at pH <2 if any.	427
SiO_2	quartz	HNO_3, HCl	0.01		2–9	$\zeta = 0$ at pH 2, but no positive values are reported.	428
SiO_2	vitreous	NaCl	10^{-3}–10^{-2}		2–7	IEP at pH <3 if any.	429
SiO_2	vitreous	NaCl	10^{-3}		2–6	IEP at pH <2 if any.	430
SiO_2	vitreous	NaCl	0.01		2–10	IEP at pH <2 if any.	431
SiO_2	fused	KCl	10^{-3}	21	3–6	IEP at pH <3 if any.	53
SiO_2	quartz, Min-U-Sil	NaCl	0.0001–0.1	25	4–9	Zero mobility at pH 2.3, but no positive ζ.	377
SiO_2	natural quartz				3–10	IEP at pH <3 if any.	420
SiO_2	quartz, acid washed	none		25	1.5–5	$\zeta = 0$ at pH 1.5, but no positive values are reported.	173

TABLE 3.9 (Continued)

Adsorbent	Characterization	Electrolyte	Concentration/ mol dm^{-3}	T	pH	Method	Result	Ref.
SiO$_2$	Ottawa sand	NaCl	0.001		2–11		IEP at pH <2 if any.	109
SiO$_2$	fused	KCl	10^{-4}–10^{-1}	20	5.8		-90 to -25 mV	432
SiO$_2$	flat specimen	KCl	10^{-3}	25	2–10		IEP at pH <2 if any.	433
SiO$_2$	quartz	NaCl	0.01		2–10		$\zeta = 0$ at pH 2.	274
SiO$_2$	Ludox	KNO$_3$	0.01	25		acousto	-78mV	434
SiO$_2$	fumed	none		20	2–10		IEP at pH <2 if any.	435
SiO$_2$	Stober	KCl or KNO$_3$	10^{-3}	25	4–9	acousto	IEP at pH <4 if any, also electrophoresis.	436
SiO$_2$	fused	KCl	10^{-4}–10^{-1}	20	3–11		Negative ζ potentials over the entire pH range.	437
SiO$_2$	quartz	NaCl	10^{-2}		2–10		$\zeta = 0$ at pH 3, but no positive values.	438
SiO$_2$	Stober	KCl	10^{-3}–10^{-1}		5.8		Negative ζ, but in 0.1 KCl $\zeta \approx 0$.	439
SiO$_2$	glass	KNO$_3$	0.001		3–9	AFM		440
SiO$_2$	quartz, washed			25	3–10	AFM	Negative ζ potentials over the entire pH range.	234
SiO$_2$	LiChrospher 100 and 1000	NaCl	0.001–0.1		3.5–6		IEP at pH <3.5 if any.	441
SiO$_2$	Ludox SM	NaNO$_3$	10^{-3}	25	3.5–10		IEP at pH <3.5 if any.	442
SiO$_2$	colloidal, Allied Signal, fused flats	NaNO$_3$	10^{-4}–10^{-3}	25	3–9	AFM	IEP at pH <3 if any (also streaming potential and electrophoresis).	443
SiO$_2$	amorphous Aldrich	NaCl	0.001		2–11		IEP at pH <2 if any.	444
SiO$_2$	quartz	NaNO$_3$	10^{-3}	25	3–9		IEP at pH <3 if any.	264
SiO$_2$	Eurospher	NaCl	0.001		1.7–10		IEP at pH <1.7 if any.	34
SiO$_2$	quartz	NaCl, CsCl	0.001–0.01	20	3–10		$\zeta = 0$ at pH 2, but no positive values are reported, also for 0.1 mol dm^{-3} NaCl.	445
SiO$_2$	Aerosil A-175	NaCl	0.01	20	3–7		$\zeta = 0$ at pH 3, but no positive values are reported.	446

Material	Description	Electrolyte	Concentration	T	pH	Method	Remarks	Ref
SiO$_2$	quartz capillary	NaCl	0.001		2–5		Zero mobility at pH 2, but no positive ζ.	447
SiO$_2$	fused, capillary	none		room			99 mV (minus sign missing?); 33–85 mV (?) in different buffers pH 4–10.9.	448
SiO$_2$	XL Nissan	NaCl	0–0.1		1.6–natural		Negative ζ.	311
SiO$_2$	quartz	NaCl	0.0005		1–11		IEP at pH < 1 if any.	449
SiO$_2$	Ludox; technical				2–10	acousto	IEP at pH < 2 if any.	56
SiO$_2$		KCl	10^{-3}		3–10		IEP at pH < 3 if any.	450
SiO$_2$	fumed Sigma	NaNO$_3$	10^{-3}	25	4–9	acousto	IEP at pH < 4 if any. (also electrophoresis).	57
TiO$_2$	anatase pigment grade	NaCl	0.0004–0.01		8–11		Negative ζ.	395
TiO$_2$		HNO$_3$, KOH		23	5–12			49
TiO$_2$	Degussa P 25	KCl	10^{-3}	20	3–11		|ζ| <35 mV, IEP between pH 6.6 and 8.5 (no data points in this pH range).	451
TiO$_2$	Degussa P 25 anatase:rutile 70:30	NaNO$_3$, NaCl, Na$_2$SO$_4$, CaCl$_2$	10^{-3}–10^{-2}		4, 5.5, 10.5		NaNO$_3$, NaCl, pH 5.5 the mobility is constant over the studied concentration range. NaNO$_3$, NaCl, CaCl$_2$ pH 4 the mobility increases when the ionic strength increases.	452
TiO$_2$	anatase, 3 samples	KCl	0.003–0.1		3.3–6.5		Negative ζ at pH 5.8, positive at pH 3.3	453
V$_2$O$_5$	Merck, orthorhombic >99.9%	KCl	0.01		1.5–3.5		Negative ζ potentials.	454
V$_2$O$_5$		KNO$_3$	5×10^{-4}–5×10^{-3}		1.5–5.5		Negative ζ potentials.	102
V$_2$O$_5$					2–8		Negative ζ potentials.	455
WO$_3$		KCl	0.01		1.5–4.5		Negative ζ potentials.	454
ZnO		NaCl	10^{-3}–10^{-2}		8–10		Negative ζ potentials (more negative at pH 8 than at pH 10).	456
ZnO	Puratronic grade I	KCl	0.005		7–12		Negative ζ potentials.	254

TABLE 3.9 (Continued)

Adsorbent	Characterization	Electrolyte	Concentration/ mol dm^{-3}	T	pH	Method	Result	Ref.
ZrO$_2$	spray-ICP technique alkaline treated				2–6 3–4		Positive ζ potentials.	457
ZrO$_2$	3 samples, pyrolysis of ZrO(NO$_3$)$_2$				2–6		Negative ζ potentials.	457
ZrO$_2$		HNO$_3$, KOH		23	2–11		Also ζ potentials at 235°C are reported.	49
clay minerals and related materials								
beidellite	SBCa-1	NaCl	10^{-3}		2–11		Negative ζ potentials.	138
halloysite	7A	KCl	10^{-3}		3–10		IEP at pH <3 if any.	450
Na-illite		NaCl	10^{-4}–0.56		2–10		IEP at pH <2 if any.	140
illite	natural				2 10		IEP at pH <2 if any. No significant effect of ionic strength (10^{-4}–10^{-2} mol dm^{-3}) on ζ.	137
illite	5% chlorite; 5% quartz	NaNO$_3$	0.001	25	2–11		Negative ζ.	398
Na-kaolinite					2–10		IEP at pH <2 if any.	140
kaolinite	natural, washed	NaCl	0.1	25	2–6		Negative ζ potentials.	458
Na-kaolinite	Sigma	NaCl	0.005	25	3 8		Negative ζ potentials.	459
kaolinite, well and poor ordered		KCl	10^{-3}		3–10		IEP at pH <3 if any.	450
Na-kaolinite	washed in H$_2$O$_2$ and then in 1 mol dm^{-3} NaCl	NaCl	0.001		3–8		Negative ζ potentials.	460

Material	Description	Electrolyte	Concentration	T	pH range	Comments	Ref.
montmorillonite	natural	NaCl	10^{-4}–0.56		2–10	IEP at pH <2 if any. ζ decreases when ionic strength increases.	137
muscovite mica	natural, green and ruby, flat disks	KNO_3, KCl	10^{-4}–10^{-2}		3–10	Different results are obtained dependent on time and conditions of aging.	461
muscovite mica	flat plate	LiCl, NaCl, KCl, CsCl, HCl	3×10^{-5}–3×10^{-2}	20	3–6	In the concentration range 10^{-3} 10^{-2} mol dm^{-3} the ζ potentials in presence of Cs are significantly less negative than for Li.	432
mica	flat specimen	KCl	10^{-3}	25	2–10	IEP at pH <2 if any, but mica particles have an IEP at pH 3.	433
ripidolite	CCa-1	NaCl	10^{-3}		3–11	Negative ζ potentials.	138
silicates and materials containing SiO_2							
sand, sandstone	untreated; acid-leached	NaCl	0.001–0.01	25	2–11	IEP at pH <3 if any.	173
cordierite	Baikowski	none		20	2–11	IEP at pH <2 if any. Negative ζ except at pH 4.5–6 where ζ is positive.	435
ZrO_2-SiO_2	spray-ICP technique; as prepared and heated to 1200 C				5.5; 3.7	Mobility of -2×10^{-8} m^2/V/s for >20% mol SiO_2.	457
materials containing TiO_2							
TiO_2 pigment	0.97% Al_2O_3, 1.28% SiO_2	KNO_3	0.001	25	2–10	Negative ζ potentials.	192
zeolites							
Mordenite Zeolex 23 Zeo 49 Zeosyl 100		$NaClO_4$	0.01		3–9	Negative ζ potentials.	210

TABLE 3.9 (Continued)

Adsorbent	Characterization	Electrolyte	Concentration/ mol dm^{-3}	T	pH	Method	Result	Ref.
clays								
Na-bentonite					2–10		IEP at pH < 2 if any.	140
bentonite	purified, 3 samples	supernatant			2–12		Negative ζ potentials.	337
kaolin	acid washed, Ajax	KNO$_3$	0.01		4–11	acousto	Negative ζ potentials.	320
glasses								
controlled pore glass		NaCl	0.001, 0.01	20	3–8		IEP at pH < 2.5 if any.	405
porous glasses		NaCl	10^{-3}–10^{-1}		2–7		IEP at pH < 2 if any.	406

electroacoustic results. Onset of uptake of certain ion at specific pH is often reported in the column "results". A high pH promotes uptake of cations and low pH—of anions. Thus, for an anion "onset of uptake at pH X" means no uptake above pH X and substantial uptake below pH X, while for cations "onset of uptake at pH Y" means substantial uptake above this pH. Two opposite types of behavior have been reported. In some studies the uptake of coions is negligible and the onset of uptake of counterions is observed at pH≈PZC. In other studies, however, substantial uptake of coions was observed even far from PZC for similar materials. The latter behavior can be hardly explained in terms of purely electrostatic interaction. Many materials contain sodium and chloride impurities. Application of radiotracer method to measure uptake at low concentrations of inert electrolyte might have resulted in overestimated uptake as an effect of isotope or ion exchange. This and other concerns about the experimental methods used to study the uptake of ions from solution are discussed in more detail in Chapter 4.

The affinity series of alkali metal cations reported for the same material in different sources and obtained by various methods are rather consistent. Metal oxides have a higher affinity to Li (higher negative charge, lower CCC) than to Cs. The pH dependence and ion specificity of coagulation [464] of silica is different from metal oxides. Some publications reporting the surface charging of silica suggest the usual sequence (Li > Cs) at high ionic strengths, and reversed sequence (Cs > Li) at low ionic strengths. Different sequences of alkali metal adsorption by silica at low and high electrolyte concentrations can be explained in terms of different activity coefficients, namely the activity of Cs in one molar CsCl is considerably lower than the activity of Li in one molar LiCl. Different selectivity sequences for samples having the same formal chemical formula can be due to their different porous structures. Aging of γ-MnO_2 with 1 mol dm^{-3} alkali hydroxides and heating the filtered absorbent at 600°C, and then leaching with nitric acid leads to ion sieve type adsorbents whose selectivity depends on the nature of the alkali metal hydroxide used in the preparation procedure [465]. The selectivity sequence is Li > K > Rb > Cs > Na for materials prepared with LiOH, but it is Rb > K > Cs > Na = Li when NaOH or KOH is used. The selectivity coefficients (cf. Chapter 4) of LiOH derived materials can be as high as > 2900 (Li over Na) [466] while the corresponding value for γ-MnO_2 is only 27. In view of high selectivity over Na, and also Ca these materials can be even used to recover Li from sea water (the selectivity Li/Mg was not reported).

Many studies resulted in common crossover points of titration curves even for electrolyte concentrations > 1 mol dm^{-3}. On the other hand, the IEP is substantially different form the pristine value at electrolyte concentration as low as 0.1 mol dm^{-3}. Typical electrokinetic behavior of metal oxides at very high ionic strengths is illustrated in Fig. 3.101. The IEP is gradually shifted to high pH values when the electrolyte concentration increases, and at very high ionic strengths the ζ potential is always positive and rather insensitive to pH. These results have been recently confirmed by rheological measurements (cf. Table 3.7).

The shifts in IEP were observed for all salts, but only some Na and Li salts have an ability to reverse the sign of ζ to positive over the entire pH range (Figs. 3.102 and 3.103). The critical salt concentration inducing such an effect depends on the nature of the anion, and hard-soft acid-base interactions have been invoked to explain these effects. Small cations (Na^+, Li^+) and large anions (I^-) show a

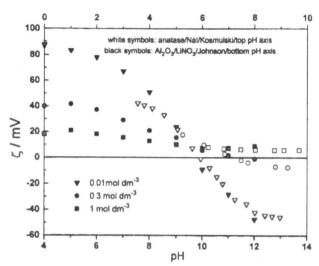

FIG. 3.101 The ζ potential in the Al_2O_3-$LiNO_3$ and anatase-NaI systems at very high ionic strengths.

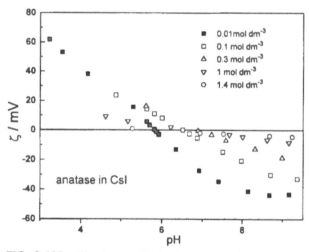

FIG. 3.102 The ζ potential in the anatase-CsI system at very high ionic strengths.

differentiating effect, and large cations (K^+) and small anions (Cl^-) do not, i.e. the high ionic strength electrokinetic behavior of titania in the presence of different chlorides is rather insensitive to the nature of the cation, and the electrokinetic behavior in the presence of different potassium salts is insensitive to the nature of the anion. Interpretation of the results presented in this section in terms of different models is discussed in Chapter 5.

IV. TEMPERATURE EFFECTS

Example 5 in Section 2.II shows that a change in the temperature by a few K does not induce a dramatic difference in ΔG of a chemical reaction. On the other hand in

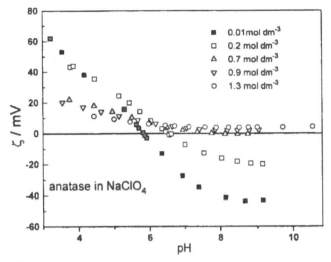

FIG. 3.103 The ζ potential in the anatase-NaClO$_4$ system at very high ionic strengths.

discussions of discrepancies in surface charging and electrokinetic data reported in different sources it was emphasized, that these results were obtained at various temperatures, and this could be an important factor. The temperature effects on PZC will be discussed briefly in this section. In principle this book is confined to sorption phenomena at room or nearly room temperature, but some studies of temperature effects go far behind the boiling point of water at atmospheric pressure. An instrument for electrokinetic measurements at extremely high temperatures and pressures was described by Jayaweera and Hettiarachchi [49]. The literature data on PZC at $T < 100°C$ were extrapolated to higher T using a third degree polynomial [540]. In this extrapolation the error of 0.1 pH unit in determination of the PZC at 80°C can result in a difference in excess of 5 pH units at 350°C. There are also no solid physical grounds that the PZC(T) should be a third degree polynomial. A few results obtained at extremely high T are shown in Table 3.11, but most studies of temperature effects on the PZC were carried out at 5–60°C.

It has been already discussed (e.g. Eq. (3.18)) that the PZC can be considered as a linear function of log K of a surface reaction responsible for proton adsorption and dissociation. Once this reaction is defined, the standard thermodynamic approach (Section 2.II) can be applied, and the temperature effect on the equilibrium constant can be calculated by combining Eqs. (2.20) and (2.24).

Modeling of surface charging in terms of different surface reactions is discussed in detail in Chapter 5, and the present discussion is confined to an empirical approach. Let us assume without proof that the surface reaction responsible for the proton adsorption can be defined in such way that

$$\log K = \text{pH}_0 \tag{3.42}$$

Combination of Eqs. (3.42), (2.20) and (2.24) gives

$$\Delta H_{ads} = RT^2 \ln(10) \times d\,\text{pH}_0/dT \tag{3.43}$$

(*Text continues on pg. 280*)

TABLE 3.10 Adsorption of Ions of Inert Electrolyte

Adsorbent	Adsorbate	Total concentration/ mol dm^{-3}	Solid to liquid ratio	Equilibration time	T	pH	Electrolyte	Method	Result	Ref.
Al$_2$O$_3$, γ, chromatographic, Fisher	NaBr	10^{-4}–10^{-1}		>7d		4–10	also Na uptake from NaCl	uptake	Crossover of log Na uptake and log Br uptake vs. pH curves at pH 7.6 (10^{-4} mol dm^{-3} NaBr) – 8.2 (0.1 mol dm^{-3} NaBr).	467
Al$_2$O$_3$, γ	NaBr	0.01–0.1	5 g (154 m^2/g)/50 cm^3			7–11		uptake	Substantial uptake of coions.	468
Al$_2$O$_3$ γ, Degussa C	NaClO$_4$	0.1						ζ	IEP shifts to 9.7 (9 is the low ionic strength value).	469
Al$_2$O$_3$ α, A 16 SG Alcoa	NaCl	10^{-3}–10^{-2}	100 g (8.4 m^2/g)/dm^3		25	3–11		uptake	Substantial uptake of coions, equal Na and Cl uptake at IEP.	470
Al$_2$O$_3$, Degussa C, washed	NaCl	0.01	0.4 g (83 m^2/g/35) cm^3			3–12		uptake	Onset of Na adsorption at pH 10, onset of Cl adsorption at pH 7.	471
Al$_2$O$_3$, POCh washed	NaCl	0.001				6–10		uptake	Onset of Cl uptake at pH 8, onset of Na uptake at pH 9.	472
Al$_2$O$_3$, γ	NaCl, NaI, CsCl	0.001–1	4000–30000 m^2/dm^3			6–10		uptake	Substantial uptake of coions.	473
Al$_2$O$_3$ Merck and Degussa C	KCl	0.01	1 g (105–117 m^2)/50 cm^3	30 min		5–10		uptake	Substantial uptake of coions.	474
Al$_2$O$_3$, reagent grade, 2 samples	NaCl	0.01	5 g (25 or 93 m^2/g)/50 cm^3		20	4–11		uptake	Onset of uptake at PZC.	475
Al$_2$O$_3$, Degussa C, washed	Cl	0–0.01	0.36 g (86 m^2/g)/35 cm^3	1 d		1–9	Na, 0–10$^{-3.48}$ mol dm^{-3} acetate or carbonate	uptake	Onset of uptake at pH 7.5. The effect of acetate and carbonate is insignificant.	476

Adsorbent	Electrolyte	Concentration	Solid/liquid ratio	Time	T (°C)	pH	Counterion	Method	Result	Ref
Al_2O_3 Merck, washed	Na	0.001–0.1	10 g (129 m^2/g)/50 cm^3		25	7–12	Cl	uptake	Negative adsorption at pH < 8, positive adsorption at pH > 8.	477
Al_2O_3 reagent grade	Na, Li, Cs	0.1	1 g/18 cm^3	1 d	25		Cl	uptake	The uptake measured radiometrically, refractometrically, and by AAS is equal: Li > Na > Cs.	478
Al_2O_3	Li, Cs, Cl	0.01				3–12		uptake	Li > Cs.	479
Al_2O_3 reagent grade, washed	$NaCl$, $CsCl$	0.1			15, 35	3–12		uptake	Na > Cs. The uptake of cations at constant pH increases with T but the uptake at constant σ_0 is independent of T. The uptake of Cl at constant pH slightly decreases with T.	480
Al_2O_3, γ, Merck	NO_3	0.1			20–60	3–6	K	uptake	Uptake decreases by factor 2 when T increases from 20 to 60°C.	481
Al_2O_3, α, Sumitomo	Li	0.1–1	2% v/v	20 min	25	4–12	NO_3	acousto	The IEP shifts to pH 11 at 0.3 mol dm^{-3} $LiNO_3$. With 1 mol dm^{-3} $LiNO_3$ ζ potentials are positive up to pH 12.	482
Al_2O_3, AKP-50 Sumitomo	Cl, NO_3, ClO_4, BrO_3, IO_3	0.01, 0.1	2% v/v	5 min	25	4–12	Na	acousto	Positive ζ potentials decrease in the series ClO_4 > Cl = NO_3 > BrO_3 > IO_3; or 0.1 mol dm^{-3} $NaIO_3$ only negative ζ potentials are observed.	324
Al_2O_3, γ, Johnson-Matthey	Na	0.001	7.5 g (55 m^2/g)/450 cm^3	1 d	23	2–5	NO_3	uptake	Onset of Na uptake at pH 3.8.	483
$AlOOH$ boehmite	NO_3, Cl, I	0.01–0.2		4 h	22	5.5–7.2	K	coagulation	Cl = NO_3 > I	484
$Al(OH)_3$ gel	$NaCl$					3–10		uptake	Onset of uptake at pH 9.	485
$Al(OH)_3$ Alcoa Hydral 710	$NaCl$	0.5–3	107 g/400 cm^3	10 min	25	8–13		acousto	The IEP shifts to pH > 10 at 0.5 mol dm^{-3} $NaCl$. With 3 mol dm^{-3} $NaCl$ ζ potentials are positive up to pH 13.	486
$Al(OH)_3$ gibbsite	$NaCl$	0.001				5–11		uptake	Onset of Cl uptake at pH 10; onset of Na uptake at pH 9.4.	487
Bi_2O_3	Cl	0.2	0.5 g/25 cm^3	1 d		natural, Na 1.3		uptake		488

TABLE 3.10 (Continued)

Adsorbent	Adsorbate	Total concentration/mol dm^{-3}	Solid to liquid ratio	Equilibration time	T	pH	Electrolyte	Method	Result	Ref.
Cd(OH)$_2$	NaCl, NaClO$_4$	10^{-3}–10^{-1}			25	8–11		uptake	Cd(OH)$_2$ can be transformed into basic chloride in the presence of NaCl.	489
CeO$_2$ hydrous, different samples	NO$_3$	0.1			25	2–6	Na	uptake		490
CeO$_2$ hydrous, different samples	Na, Cs	0.1			25	6.5–12.5	Cl	uptake		491
Co$_3$O$_4$	CsCl	0.001, 0.01				3–10		uptake	Equal uptake of Cs and Cl at pH 5.	492
Cr$_2$O$_3$	Cs	carrier free	0.1 g/20 cm^3	3h	25	8	none; 3 g H$_3$BO$_3$ dm^{-3}	uptake	18% uptake 3% uptake	493
Fe$_3$O$_4$	Cs	carrier free	0.1 g/20 cm^3	3h	25	3–11	none; 3 g H$_3$BO$_3$ dm^{-3}	uptake	Uptake peaks at pH 8 (50 and 30% uptake in absence and 26 and 11% in presence of borates for two different samples of magnetite) as a result of changes in surface charge and competition of alkali metal cations used to establish the pH. Uptake decreases with T (25–125 C range studied).	493
Fe$_3$O$_4$ synthetic	Li, Cs	0.01	0.4 g/25 cm^3	2 d		10–12	Cl	uptake	Uptake not affected by γ-irradiation.	494
Fe$_3$O$_4$ natural magnetite 2.4% SiO$_2$	Cs	2–8×10^{-5}	2 g (18.3 m^2/g)/dm^3		25	3–9		uptake	A pH independent uptake (40% with 2×10^{-5} mol dm^{-3} Cs).	495
Fe$_2$O$_3$ hematite, from nitrate	LiCl	10^{-3} 1	10 g (18–35 m^2/g)/ 500 cm^3	5 min	20	4–10		surface charge	PZC is shifted to lower pH at [LiCl] > 0.1 mol dm^{-3} (no such shift for KCl).	496
Fe$_2$O$_3$	many different 1–1							coagulation	Li > Na > K = Cs IO$_3$ > F > CH$_3$COO > CH$_2$ClCOO >	416

Adsorbent	salts		solid	time	T	pH		method		ref
Fe_2O_3	Cs	carrier free	0.1 g/20 cm³	3h	25	8	none; 3 g H_3BO_3 dm⁻³	uptake	$BrO_3 > SCN > CHCl_2COO > Br > NO_3 > ClO_3 > Cl > ClO_4 = I$ 30% uptake 4% uptake	493
Fe_2O_3 hematite	NaCl	0.01, 0.1				4,7, 10		uptake		214
Fe_2O_3 hematite, Fluka	Cs	10^{-3}–10^{-1}	1 g (4.2 m²)/20 cm³	45 min	20	4–10	chloride	uptake	Linear log–log adsorption isotherms at constant pH slope 0.8.	497
Fe_2O_3	many different anions	0.001–0.1			25	2–6	Na	coagulation	$IO_3 > SCN > F > NO_3 = I > Br = ClO_4 > Cl$	498
Fe_2O_3 hematite, hydrolysis of nitrate	many different 1–1 salts					2–12		coagulation	$Li > Na > K$ $IO_3 > F > BrO_3 > Br > NO_3 > ClO_3 > Cl > ClO_4$	499
Fe_2O_3	NaCl	10^{-2}–10^{-4}				5–10		uptake	Onset of Na uptake at pH 7 (CIP at pH 8). Onset of Cl uptake at pH 7 from 0.001 mol dm⁻³ NaCl, but substantial uptake from 0.01 mol dm⁻³ NaCl over the entire pH range. (figure captions are swapped in the original paper).	500
FeOOH, goethite	NaCl	0.005				3–9		uptake	Na uptake equals zero at pH <7, Cl uptake equals σ_0 at pH 4–7	501
FeOOH, goethite	NaCl	0.01, 0.1				4,7, 10		uptake		214
FeOOH, lepidocrocite	NaCl	0.01,0.1				4,7, 10		uptake		214
FeOOH, goethite	Cl, NO_3, ClO_4	10^{-2}–10^{-1}	100 g (80 m²/g)/dm³	1 d		pD 4–11	Na	ATR-cylindrical internal reflection spectroscopy		502

TABLE 3.10 (Continued)

Adsorbent	Adsorbate	Total concentration/ mol dm⁻³	Solid to liquid ratio	Equilibration time	T	pH	Electrolyte	Method	Result	Ref.
FeOOH, goethite	Na	0.005, 0.016	0.2 g (76 m²/g)/10 cm³	2 d		4-12	ClO₄	uptake	Positive adsorption at pH > 8.5; negative adsorption at pH < 8.5.	503
HfO₂ Aldrich	NaCl	0.001, 0.01				4-11		uptake	Onset of uptake at PZC.	504
MnO₂, δ	Na, K					1-7		uptake	No uptake at pH <2.2.	368
MnO₂ δ, MnCl₂ + KMnO₄	Na,K	0.001 K + 0.015 Na		1 d	25	2-8	Cl	uptake		369
MnO₂, λ, leaching of LiMn₂O₄ with nitric acid	Li,K,Cs	0.1	0.1 g/10 cm³		25	2-12	Cl	uptake	Negligible uptake of K and Cs, onset of Li uptake at pH 2.5.	370
MnO₂, λ, leaching of LiMn₂O₄ with nitric acid	Li,Na,K,Rb, Cs	0.01-1	0.1 g/10 cm³	21 d	25	2-12	Cl	uptake	Negligible uptake of Na,K,Rb and Cs, onset of Li uptake at pH 2 (1 mol dm⁻³), onset of Li uptake at pH 2 (1 mol dm⁻³)-3 (0.01 mol dm⁻³). Li uptake increases with T (10-40°C).	371
Nb₂O₅ hydrous, from NaCl chloride	NaCl	0.1	0.25 g/30 cm³	7 d	25	1-9		uptake	Onset of Cl uptake at pH 3.5; onset of Na uptake at pH 4.2.	505
RuO₂ thermal decomposition of RuCl₃ at 420 C	Cl	0.002-1	1-2 g (21.5 m²/g)/50 cm³		20	2-9	K	ζ, uptake, proton release	Shift of IEP to low and PZC to high pH at > 0.01 mol dm⁻³ Cl.	506
SiO₂ precipitated, BDH	Li,Na, K,Cs, (C₂H₅)₄N	0.001-0.1			20	3-10	Cl	pH, uptake	Cs>K > Li>(C₂H₅)₄N	4
SiO₂	Li,Na,K	0.1				6-12	ClO₄	uptake	Na uptake is slightly lower than σ₀. K > Na > Li at pH <10, at pH > 11 the sequence is reversed.	507
SiO₂, quartz sand, washed	KCl	0.01-0.1	1,25% moisture	>20 min	35	3-11		conductance	Mobility of ions in double layer is higher than that in solution.	508

Adsorbent	Salt	Concentration	Solid	Time	T	pH	Counterion	Method	Result	Ref
SiO$_2$, hydrous, amorphous	NaCl	0.1	0.2 g solid			3–9		uptake		22
SiO$_2$ vitreous, Herasil	Na, Br	0.006–0.015		3–21 h		10		uptake	Na penetration depth 0.3 nm.	431
SiO$_2$	Li,Na,K,Cs	0.01–1	1 g ($=310$ m^2)/250 cm^3	5 min	25	7–9.5	Cl	proton release, uptake	Parameters of Langmuir adsorption isotherm are listed for different pH values.	509
SiO$_2$ pyrogenic, Wacker	Cs,K,Li,Na, Rb	10^{-4}–10^{-1}			25	4–9	Cl	uptake	Cs > Rb > K > Na > Li	510
SiO$_2$ Stober	Li, Na, Rb, Cs	0.01, 0.1			25	3–10	Cl	ζ	At 0.01 mol dm^{-3} there is no cation specificity, at 0.1 mol dm^{-3} the IEP is shifted to pH 4.5 for RbCl and 5.2 for CsCl.	511
SiO$_2$	Li, Na, Cs	0.1–4	1 g (300 m^2)/250 cm^3		25	6.5–9.5	Cl, ClO$_4$, SO$_4$	uptake, coagulation	The effect of anion is insignificant. Cs > Na > Li (0.1 mol dm^{-3}) Li > Na > Cs (0.5 mol dm^{-3})	384
SiO$_2$ Merck, washed	Na	0.001–0.1	5 g (388 m^2/g)/50 cm^3		25	5–9	Cl	uptake	Uptake nearly matches the surface charge density.	477
SiO$_2$ reagent grade	Na, Li, Cs	0.1	1 g/18 cm^3		25	1 d	Cl	uptake	The uptake measured radiometrically is higher than that measured refractometrically: Cs > Na > Li.	478
SiO$_2$	Li, Cs	0.01				5–10	Cl	uptake	Cs > Li	479
SiO$_2$ Aerosil 200 Degussa	tetramethyla-mmonium, tetraethylam-monium, tetrapropyla-mmonium	0.001–0.1	5 g ($=1000$ m^2)/100–150 cm^3	1 d	25	2–8	Br	ζ, uptake	IEP: TMA 3.4,4.6, TEA 4,5.3, TPA 4,3,5.3 for 10^{-3} and 10^{-2} mol dm^{-3}, respectively. Adsorption isotherms at pH 3 and 5 are reported.	512
SiO$_2$ reagent grade, washed	NaCl, CsCl	0.1			15 – 35	4–10		uptake	Cs > Na. The uptake of cations at constant pH increases with T but the uptake of Na at constant σ_0 is independent of T.	480
SiO$_2$ Lichrospher Si 100 and 1000	NaCl	0.001–0.1			25	3–9		uptake	positive and pH independent uptake of Cl from 0.01 mol dm^{-3} NaCl	441

TABLE 3.10 (Continued)

Adsorbent	Adsorbate	Total concentration/ mol dm^{-3}	Solid to liquid ratio	Equilibration time	T	pH	Electrolyte	Method	Result	Ref.
SiO$_2$ quartz	Na, Cs	10^{-3}–1	10 g(0.51 m^2/g)/250 cm^3		24, 25	5–11	Cl	uptake, chromato graphy	log (Na uptake) is a linear function of pH, slope 0.34 (the figure caption is "surface charge density"). Stability constants of ≡SiO$^-$··Na$^+$ and ≡SiO$^-$··Cs$^+$ surface species are also reported (but their definition is problematic).	513
SiO$_2$ quartz	Cs	0.1			20	2–5.5 5.5		ζ	IEP at pH 3.3 (IEP at pH <2 if any at lower CsCl concentrations and in NaCl up to 0.1 mol dm^{-3}).	445
SiO$_2$ fumed, Sigma	CsCl, NaI, LiCl, CsNO$_3$	0.1–0.7	6% w/w		25	3–8		acousto	IEP is shifted to higher pH. The most significant shift to pH 5.5 was observed for 0.1 mol dm^{-3} Cs salts. Further increase in concentration of these salts shifts the IEP back to lower pH (or no IEP at all).	514
SiO$_2$ Ludox	Na, K	3				2–12	Cl	coagulation	Coagulation at pH 7, repeptization at pH 11 in KCl, no repeptization in NaCl	515
SiO$_2$ Geltech	K	10^{-3}–0.1				2–5.5	NO$_3$	ζ	The point at which ζ = 0 shifts from pH 2 to 4 when KNO$_3$ concentration increases from 10^{-3} to 0.1 mol dm^{-3}, but no positive ζ are reported except for 0.01 mol dm^{-3}.	393
SnO$_2$, α, hydrolysis of SnCl$_4$ in the presence of ammonia at pH 4	NaCl	0.8	500 mg/50 cm^3	3h		0–10		neutron activation of the solid	Cl uptake is zero at pH >5 and sodium uptake is zero at pH <2. Equal uptake of anion and cation at pH 3.8. Maximum uptake = 0.9	516

Solid	Electrolyte	Concentration	Solid/solution	Time	T/°C	pH	Method	Comments	Ref
SnO₂ hydrous from chloride	NaCl	0.1	0.25 g/30 cm³	4 d	25	1–9	uptake	mmol/g SnO₂ at pH 0 for Cl and pH 10 for Na, respectively.	505
SnO₂ hydrous, 3 samples	Na	0.1			25	7–11.5	uptake	Onset of Cl uptake at pH 3.5; onset of Na uptake at pH 4.	517
SnO₂ from SnCl₄	Cs	0.005–0.1	0.5 g/10 cm³	3 d	25	2–10	uptake	Cs uptake is enhanced by factor 20 at pH 2 in the presence of Co.	518
ThO₂ hydrous, synthetic, 4 samples	Li,Na,K,Cs, NH₄, Cl,Br,NO₃, CNS	0.04–0.1	0.5 g/25 cm³	1 d	25		uptake	Li>Na>K>Cs>NH₄ at constant concentration of hydroxide; Cl>NO₃>Br>CNS (natural pH).	519
ThO₂	Cl	0.2	0.5 g/25 cm³	1 d		natural, Na 1.3	uptake		488
ThO₂ hydrous	LiCl	0.1	0.3 g (77 m²/g)/dm³	4 h	25	3–11	uptake	Onset of Cl uptake at pH 6, onset of Li uptake at pH 7.2.	520
TiO₂, gel, 27% water, hydrolysis of TiCl₄	NaCl	0.8	500 mg/50 cm³	3h		1–12	neutron activation of the solid	Cl uptake is zero at pH>4.5 and sodium uptake is zero at pH<3.5.	521
TiO₂	Cl	0.2	0.5 g/25 cm³	1 d		natural, Na 1.3	uptake	Equal uptake of anion and cation at pH 4. Maximum uptake=0.5 mmol/g TiO₂ at pH 1.5 for Cl and 3 mmol/g TiO₂ at pH 10 for Na, respectively.	488
TiO₂	Cs	carrier free	0.1 g/20 cm³	3h	25	8; none; 3 g H₃BO₃ dm⁻³	uptake	18% uptake 2% uptake	493
TiO₂ Degussa P 25, washed with boiling water	NaCl	10⁻²				2–11	uptake	Onset of uptake at PZC; negligible uptake of coions.	522
TiO₂ anatase, Merck	Na, Cs, Cl	0.001–0.1			25	3–10	uptake	Substantial uptake of coions.	523
TiO₂ hydrous, from sulfate	Li, Cs	0.01	0.4 g/25 cm³	2 d		5–10, Cl	uptake	Uptake not affected by γ-irradiation, onset of Cs uptake at pH 5, and of Li at pH 6.	494

TABLE 3.10 (Continued)

Adsorbent	Adsorbate	Total concentration/ mol dm^{-3}	Solid to liquid ratio	Equilibration time	T	pH	Electrolyte	Method	Result	Ref.
TiO$_2$ rutile	NaCl	0.001-0.1				3-10		uptake	Uptake of Cl from 0.01 mol dm^{-3} NaCl is pH independent.	524
TiO$_2$ rutile, from TiCl$_4$, 4 different samples	many different 1-1 salts	10^{-2}-10^{-1}				1.5-12		coagulation	Various cation sequences, i.e., Li>Na>K>Cs or Cs>K>Na>Li for different samples. For anions always IO$_3$>BrO$_3$>NO$_3$>ClO$_3$>Cl>ClO$_4$=Br=I.	525
TiO$_2$ rutile Tioxide, washed	KCl, LiCl, CsCl, KNO$_3$, KClO$_4$	0.01, 0.1	1 g/100 cm^3		25	4-9		surface charge, ζ	Li>K>Cs Cl>NO$_3$>ClO$_4$ (but the CIP and IEP are not affected).	3
TiO$_2$ Fisher, anatase	NaCl	0.001-0.1			25	3-11		uptake	Appreciable uptake of coions.	526
TiO$_2$ anatase, Aldrich	15 different 1-1 electrolytes	up to 1.7	20% w/w		20	4-11		acousto	At >0.1 mol dm^{-3} the IEP shifts to high pH values. The effect is more pronounced for large anions and small cations. For some salts (only Li and Na) at sufficiently high concentration there is no IEP and the ζ is positive at pH up to 11.	527
TiO$_2$	(CH$_3$)$_4$NClO$_4$	0.005						uptake	Onset of (CH$_3$)$_4$N uptake at pH 6.5, onset of ClO$_4$ uptake at pH 3.5.	21
TiO$_2$	Na,Cs	10^{-4}-10^{-2}			25	5-10	Cl	uptake	Cs>Na	528
TiO$_2$ anatase, synthetic	NaCl	0.005-0.05			25	natural		uptake		529
TiO$_2$ anatase, from ethoxide	NaCl	10^{-4}-10^{-2}				4-10		uptake	Onset of uptake at PZC; negligible or negative uptake of coions.	530
TiO$_2$ Merck, anatase+15% rutile	NaCl, NaI, NaNO$_3$	0.1-0.9	5-6% w/w		20	4-10		acousto	At >0.1 mol dm^{-3} the IEP shifts to high pH values.	531

Adsorbent	Electrolyte	Concentration	Solid-to-liquid	Time	T	pH	Conditions	Method	Results	Ref
TiO$_2$ anatase, Aldrich	NaI, NaNO$_3$	0.1–0.5	20% w/w		20	4–9		acousto	At > 0.1 mol dm^{-3} the IEP shifts to high pH values.	331
WO$_3$ amorphous, precipitated	Li,Na,K,Rb, Cs	0.01–0.2				2–11		coagulation	Cs > Rb > K > Na > Li; extrapolated IEP at pH 1.8.	532
ZnO	NaBr	1.5×10^{-3}	single crystal	4 h		7.7, 10.6		uptake	Br > Na at pH 7.7; Na > Br at pH 10.6; results presented as profiles of Na and Br concentration as a function of depth.	456
ZrO$_2$ Ventron	NaCl	0.001–0.1				4–11		uptake	Substantial uptake of coions.	504
ZrO$_2$ hydrous	NaCl	0.05				2–12		uptake	Appreciable uptake of coions up to 1 pH unit from equal adsorption point (at pH 7).	533
ZrO$_2$	Cl	0.2	0.5 g/25 cm^3	1 d		1.3	natural, Na	uptake	20% uptake	488
ZrO$_2$	Cs	carrier free	0.1 g/20 cm^3	3h	25	8	none; 3 g H$_3$BO$_3$ dm^{-3}	uptake	3% uptake	493
ZrO$_2$ hydrous, hydrolysis of ZrOCl$_2$, various heat treatment	NaCl	0.1	0.2 g/20 cm^3	6 d	room	1–13		uptake	Onset of Na uptake at pH 6, onset of Cl uptake at pH 7.	534
ZrO$_2$ Merck	NaCl	0.001–0.1			20	2–9		uptake	Appreciable uptake of coions.	535
ZrO$_2$ Merck	NaNO$_3$, NaBr up to 1		20% w/w			4–11		ζ	At > 0.1 mol dm^{-3} the IEP shifts to high pH values. At sufficiently high concentration there is no IEP and the ζ is positive at pH up to 11.	527
ZrO$_2$ TZ-0 Tosoh Ceramics	Cl, NO$_3$, ClO$_4$, BrO$_3$, IO$_3$	0.01, 0.1	2% v/v	5 min	25	4–12	Na	acousto	Positive ζ potentials decrease in the series ClO$_4$ > Cl = NO$_3$ > BrO$_3$ > IO$_3$; for 0.01 and 0.1 mol dm^{-3} NaIO$_3$ only negative ζ potentials are observed, for 0.1 mol dm^{-3} NaClO$_4$, NaCl and NaNO$_3$ the IEP is slightly shifted to higher pH.	324

TABLE 3.10 (Continued)

Adsorbent	Adsorbate	Total concentration/ mol dm^{-3}	Solid to liquid ratio	Equilibration time	T	pH	Electrolyte	Method	Result	Ref.
materials containing Al$_2$O$_3$										
alumina + silica, hydrous, amorphous	NaCl	0.1	0.2 g solid			3–9		uptake		22
phyllosilicates										
muscovite mica	Li,Na,K,Rb, Cs	5×10^{-4}	10–150 mg/25 cm^3	2 d	room	natural	chloride	uptake	Cs = Rb = K > Na > Li	536
other mixes oxides										
Bi$_2$O$_3$ + ThO$_2$ (different ratios)	Cl, Br, I	0.2	0.5 g/25 cm^3	1 d	1.3–12	Na	uptake			488
Bi$_2$O$_3$ + TiO$_2$ (8:1)	Cl	0.2	0.5 g/25 cm^3	1 d	natural, Na 1.3		uptake			488
Bi$_2$O$_3$ + ZrO$_2$ (different ratios)	Cl	0.2	0.5 g/25 cm^3	1 d	1.3–12	Na	uptake		488	488
SnO$_2$ + SiO$_2$	Cs	0.005–0.1	0.5 g/10 cm^3	3 d	25	2–10		uptake	Cs uptake is enhanced by factor 30 at pH 2 in the presence of Co.	518
TiO$_2$ + Sb$_2$O$_5$ (8:1)	Cl	0.2	0.5 g/25 cm^3	1 d	natural, Na 1.3		uptake			488

Material	Ion	Concentration	Solid/liquid	Time	Temp	pH / electrolyte	Method	Comments	Ref
$Ce(OH)_2(HWO_4)_2 \cdot x H_2O$	Cs	0.01–0.1		7 d	25	0.01–0.1 mol dm^{-3} HCl	uptake	Cs uptake increases with T. Hysteresis. Also a study of uptake from aqueous methanol.	537
$ZrO_2 + Sb_2O_5$ (10:3)	Cl	0.2	0.5 g/25 cm^3	1 d		natural, Na 1.3	uptake		488
$ZrO_2 + TiO_2$ different proportions	NaCl	0.1	0.1 g/15 cm^3			1–13	uptake	The pH of the onset of Na and Cl adsorption match. They range from 3 (100% TiO_2) to 6.3 (100% ZrO_2).	538
clays									
3 clays	NaCl	0.01, 0.1				4,7,10	uptake		214
glasses									
porous glass	Li,Na,K	10^{-3}–10^{-2}				6–9	uptake	K > Na > Li The uptake is significant over the entire studied pH range and rather insensitive to heat treatment of CPG	539
4 controlled pore glasses	Na					3–8 chloride	uptake		219, 220

thus, the temperature effects on pH_0 can be conveniently expressed by the value of ΔH_{ads} of this reaction. The ΔH_{ads} calculated from the temperature effect on pH_0 by means of Eq. (3.43) is equal to the enthalpy (heat of proton adsorption) determined in a properly designed calorimetric experiment, but setup of such an experiment and interpretation of calorimetric measurements in complex hetero-geneous systems in anything but trivial [541]. Calorimetry gives the overall heat, which can be influenced by different reactions, e.g. the chemical dissolution of adsorbents. Temperature effect on chemical dissolution was studied in detail for silica [542].

Equation (3.43) was used in many original publications and in compilations of thermodynamic data (temperature effects on PZC and calorimetric) on surface charging of oxides [543–545]. Machesky [544] summarized also temperature effects on specific adsorption, which are discussed in detail in Chapter 4. Most ΔH_{ads} values reported in Refs. [543–545] are negative, i.e. proton adsorption is exothermic and the pH_0 decreases when T increases. Moreover, more negative ΔH_{ads} is observed for oxides whose PZC falls at high pH. However, different representations of the temperature effects on PZC than that based on Eqs. (3.42) and (3.43) have been also used, thus, special caution must be exercised by comparison of results reported in different sources. In some publications the surface reaction responsible for the proton adsorption is defined in such way that

$$\log K = pH_0 - 1/2\,pK_w \tag{3.44}$$

The rationale for such a representation is that the ionic product of water K_w is a function of the temperature and the equilibrium constant defined by Eq. (3.44) expresses the distance between PZC and neutral pH at given T. Indeed, pH 7 which is neutral at 25°C becomes basic at elevated T and acidic at $T < 25$°C. It can be easily shown that at 25°C

$$\Delta H\,(K\text{ defined by Eq.}(3.44)) = \Delta H(K\text{ defined by Eq. }(3.42)) + 1/2\Delta H_w \tag{3.45}$$

where the enthalpy of water autodissociation ΔH_w equals 55.81 kJ/mol. Interest-ingly, with K defined by Eq. (3.44) many ΔH values reported in literature become close to zero. The parameters $\Delta_{ch}H°$ (or $\Delta_d H° -\Delta_p H°$) used by Kallay, and ($Q_{a1} + Q_{a2}$) used by Rudzinski to characterize temperature dependence of PZC and calorimetric effects of surface charging are both equal to $-2\,\Delta H_{ads}$ (K defined by Eq. (3.42)). Rudzinski et al. [546] found the following correlation:

$$Q_{a1} + Q_{a2} = 13.04\,PZC - 28.9 \tag{3.46}$$

using selected data (temperature effects on PZC and calorimetric) from literature.

Sverjensky and Sahai [547] collected literature data on temperature effects on the PZC and proposed the following equation for proton adsorption on oxides

$$\Delta H_{ads} = -57.614[\varepsilon^{-1} + (T\varepsilon^{-2}\,d\varepsilon/dT)] + (125.208\,\nu/L) - 39.78 \tag{3.47}$$

(cf. Section II, ΔH_{ads} is expressed in kcal mol^{-1}, K defined by Eq. (3.42)). Similar correlations were obtained for enthalpies of reactions used in the TLM model (cf. Chapter 5). A few publications report the temperature effects on surface charging only in terms of ΔH for reactions defined by certain models, especially for materials that do not have a PZC.

For a constant (temperature independent) ΔH_{ads}, the PZC is a linear function of $1/T$ [Eq. (3.43)], and indeed, such linear dependence has been confirmed in many studies. In a few other studies the PZC is not a linear function of $1/T$. Thus, the slope depends on the choice of the temperature range, and in order to attribute the calculated ΔH_{ads} to specified temperature a relatively narrow T range should be chosen with such data sets. On the other hand, with a narrow T range (5 K or so) the difference in PZC is on the order of the experimental error, and the significance of the calculated ΔH_{ads} is problematic. Most likely, however, substantial deviations from linearity are due to insufficient purity of materials or some experimental errors. A few ΔH_{ads} values selected from Table 3.11 (K defined by Eq. (3.42), only "cip" and salt titration data, linear dependence between PZC and $1/T$) are shown in Fig. 3.104. Some ΔH_{ads} values in Table 3.11 and Fig. 3.104 were recalculated from raw data and they are different from the originally reported values. Figure 3.104 confirms that ΔH_{ads} is more negative for high pH_0, at least over the pH range 4–7.

The ΔH_{ads} values corresponding to PZC at pH $\approx$ 8 are very scattered. Similar trends are observed in analogous graphs presented in other compilations. It should be emphasized that ΔH_{ads} for given material is pH dependent (this is apparent from some calorimetric studies), and ΔH_{ads} from "cip" data can be attributed to pH = CIP, i.e. it is determined at different pH for different materials. Thus, very likely the ΔH_{ads} found for certain materials is more negative because it was determined at high pH rather than because this material has a high pH_0.

Studies of temperature effects on surface charging by methods other than calorimetry or potentiometric titration and related methods are rare. According to Revil et al. [548] the ζ potential of silica at given pH and ionic strength linearly increases with T. Two sets of experimental data were found in literature to support this hypothesis. Vlekkert et al. [549] studied temperature effects on the surface potential of Al_2O_3 by measuring the ISFET response at different pH. This method is not suitable to locate the PZC. The slope of the response-$(d\psi_o/dpH)$ linearly

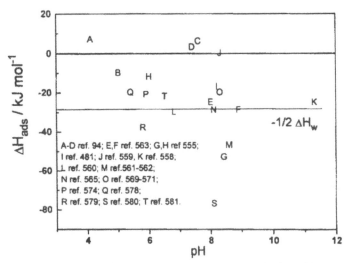

FIG. 3.104 Enthalpy of proton adsorption on (hydr)oxides as a function of their PZC.

increases with T. The entalpies of individual reactions of the 2-pK model (cf. Chapter 5) were calculated assuming a temperature independent PZC at pH 8, and this assumption contradicts the experimental evidence presented in Table 3.11.

V. NONAQUEOUS MEDIA

The temperature effects on surface charging discussed in Section IV are partially due to changes in the ionic product of water, which in turn is related to other physical quantities, e.g. electric permittivity. The same physical quantities can be changed at room temperature by admixture of organic cosolvents. In principle, the surface charging in the presence of organic species is discussed in Chapter 4 (devoted to specific adsorption), but lower alcohols, ethers, ketones, and amides, and a few other solvents miscible with water will be considered in this section as modifiers of the properties of the solvent water. It should be emphasized that physical properties of water at extremely high ionic strengths are also considerably different from those of pure water. Several up-to-date reviews on the surface charging in nonaqueous and mixed solvent systems are available [584–587].

Many studies in nonaqueous systems have been carried out in the presence of strongly adsorbing species, chiefly surfactants and polymers, but such systems are beyond the scope of this Chapter. Only surface charging in the presence of strong acids and bases and/or 1–1 salts will be discussed here. The experimental systems upon which the present discussion is based are summarized in Table 3.12.

A broad spectrum of physical and chemical properties can be found among nonaqueous solvents. Traditionally the electric permittivity was considered as the property responsible for dissociation of electrolytes, their solubility, and thus all related properties including surface charging. More recently the solvent properties have been also expressed in terms of empirical solvent scales. These scales are based on energies of the longest wavelength absorption peaks of specially selected organic molecules, which are different from one solvent to another (solvatochromic effect), heats of dissolution of Lewis acids (e.g. $SbCl_5$) and chemical shifts in NMR spectra of probe molecules (e.g. of ^{31}P in triethyloxophosphine) in these solvents, etc. These empirical solvent scales represent one of the following solvent properties: overall polarity/polarizability, ability to solvate cations or ability to solvate anions or combinations thereof and they are interrelated. Studies in water-organic mixtures allow to change these properties continuously from pure water to pure organic solvent.

The surface charging in mixed water-organic solvents rich in water has been extensively studied for a few common oxides but not for other materials. The σ_0(pH) and ζ(pH) curves in such mixtures are relatively insensitive to the concentration of organic cosolvent up to about 30% weight fraction at low ionic strengths, with a few exceptions. The $|\sigma_0|$ on the negative branch of charging curves of materials was substantially reduced in the presence of DMSO. The $|\sigma_0|$ of silica was also reduced in the presence of other organic solvents and this effect was more pronounced for cosolvents having a low value of $E_T(30)$, a solvatochromic polarity/polarizability parameter.

The electrokinetic curves observed in the presence of organic cosolvents are often similar to those shown in Figs. 3.101–3.103 (pure water, high ionic strengths).

(*Text continues on pg. 292*)

TABLE 3.11 Temperature Effects on the PZC and Related Studies

Material	Description	Salt	$T/°C$	Method	Result	Ref.
Al_2O_3	chromatographic, Fisher	$0.005-0.1$ mol dm^{-3} KNO_3	30–90	cip, pH	$\Delta H_{ads} = -28 kJ/mol$	550
Al_2O_3	sapphire	2×10^{-3} mol dm^{-3} KNO_3	25–80	ζ	The negative ζ at pH 9.4 increases as T increases (but does not return to initial value on cooling).	551
Al_2O_3	α, Linde A		23–43	pH	PZC at pH 8.8 (23°C) and 8.1 (43°C).	306
Al_2O_3	γ, Houdry 415	$0.001-0.1$ mol dm^{-3} KNO_3	10–50	cip	CIP at pH 4.45 at 10°C and 8.95 at 50°C.	552
Al_2O_3	γ, Cyanamid		23–82	mass titration	PZC at pH 7.2 at 23°C and 8.2 at 82°C. $\Delta H_{ads} = 67$ kJ/mol.	553
Al_2O_3	γ, Johnson Matthey and synthetic	$0.001-0.1$ mol dm^{-3} NaCl		calorimetry	$\Delta H_{ads} = -30$ kJ/mol at pH < 6; at pH > 6 ΔH_{ads} is more exothermic and ionic strength dependent.	554
Al_2O_3	γ, reagent grade, washed		15–35	salt titration	$\Delta H_{ads} = -53$ kJ/mol	555
Al_2O_3	γ, Degussa C	1 mol dm^{-3} $NaClO_4$	25–60	pH	PZC at pH 6.45 at 25°C and 6.32 at 60°C.	556
Al_2O_3	reagent grade, washed	0.1 mol dm^{-3} NaCl	15–35	pH	$\Delta H_{ads} = -49$ kJ/mol (pH 8–10)	480
Al_2O_3	γ, Cyanamid, calcined	$0.1-0.001$ mol dm^{-3} $NaNO_3$	10–50	cip	CIP at pH 6.3 at 10°C and 8.0 at 50°C.	557
Al_2O_3	γ, Merck, washed	$0.001-0.1$ mol dm^{-3} KNO_3	20–60	cip	CIP at pH 8.2 (20°C) and 7.8 (60°C). The authors used uncorrected σ_0 and obtained ionic strength dependent ΔH_{ads}.	481
Al_2O_3	α, reagent grade	$0.01-0.1$ mol dm^{-3} NaCl	25,60	pH		390
$Co(OH)_2$	high purity	KNO_3	25–80	pH cip	CIP at pH 11.4 at 25°C and 10.7 at 80°C.	558
Co_3O_4	high purity	KNO_3	25–80	pH cip	CIP at pH 11.4 at 25°C and 10.4 at 80°C.	558

TABLE 3.11 (Continued)

Material	Description	Salt	T/°C	Method	Result	Ref.
Cr_2O_3			235	iep	6.6	425
Fe_3O_4	synthetic	0.005–0.5 mol dm^{-3} KNO_3	25–90	cip, pH	5.4 (pH) at 90°C; 6.55 at 25°C (pH and cip).	550
Fe_3O_4	synthetic magnetite	0.001–0.1 mol dm^{-3} KNO_3	30–80	cip	CIP at pH 6.8 at 30°C and 6 at 80°C.	560
Fe_3O_4			235	iep	6.1–8.4	425
Fe_2O_3	synthetic	0.001–0.1 mol dm^{-3} $NaClO_4$	22–60	cip	No temperature effect on charging curves (fast titration).	559
Fe_2O_3	natural hematite	10^{-3} mol dm^{-3} KNO_3	25–80	iep	The negative ζ at pH 5 increases as T increases (but does not return to initial value on cooling).	551
Fe_2O_3	α, from nitrate	0.005–1 mol dm^{-3} KNO_3	5–60	cip, calorimetry	$\Delta H_{ads} = -47$ kJ/mol (recalculated) $\Delta H_{ads} = -36$ kJ/mol	561 562
Fe_2O_3	hematite, from chloride	KCl KNO_3	15–35	salt titration	$\Delta H_{ads} = -29$ kJ/mol (KCl); -25 kJ/mol (KNO_3)	563
Fe_2O_3 sintered			23,235	iep	IEP at pH 3.7 at 23°C and 3.4 (?) at 235°C.	425
Fe_2O_3	hematite, Alfa	10^{-4} mol dm^{-3} $NaNO_3$	5–35	mass titration, calorimetry	PZC at pH 6.3 at 5°C, and 6.0 at 35°C. The results calculated from T effect on PZC were confirmed by calorimetry.	564
Fe_2O_3	hematite, reagent grade, washed	0–1 mol dm^{-3} NaCl	25,60, 100	cip	CIP at pH 8.1 at 25°C, 7.5 at 60°C, and 6.9 at 100°C.	565
Fe_2O_3	hematite, reagent grade	0.001 mol dm^{-3} KNO_3	10–45	mass titration	$\Delta H_{ads} = -21$ kJ/mol	566
FeOOH	goethite, hydrolysis of nitrate	0.01–0.1 mol dm^{-3} $NaNO_3$	25	calorimetry		567
FeOOH	α			calorimetry		568 569 570
FeOOH	goethite, hydrolysis of nitrate		10–70	cip	PZC at pH 8.5 at 10 C and 7.8 at 70 C.	571

Material	Description	Electrolyte	T	Method	Notes	Ref
HfO_2	Aldrich 99.95%	KCl, NaClO$_4$	15–35	salt titration	$\Delta H_{ads} = 6$ kJ/mol, $\Delta H_{ads} = 3$ kJ/mol	94
MnO_2	γ	0.01 mol dm^{-3} KCl	25–73	pH	Positive and negative charge is depressed at high T.	572
MnO_2	λ, leaching of $LiMn_2O_4$ with nitric acid	0.1 mol dm^{-3} LiCl	10–40	pH	Negative charge increases with T.	371
Nb_2O_5	Schuchardt, 99.9%	KCl	15–35	salt titration	$\Delta H_{ads} = 7$ kJ/mol	94
NiO	high purity	KNO_3	25–80	pH cip	CIP at pH 11.3 at 25°C and 10.5 at 80°C.	558
$Ni(OH)_2$	high purity	KNO_3	25–80	pH cip	CIP at pH 11.2 at 25°C and 10.5 at 80°C.	558
SiO_2	quartz, Brazilian, washed	10^{-3}–10^{-2} mol dm^{-3} KNO_3	35,65	ζ	The negative ζ increases as T increases.	551
SiO_2		0.7 mol dm^{-3} NaCl	5,25	pH	The negative σ_0 increases as T increases.	378
SiO_2	Aerosil	0.01–1 mol dm^{-3} NaCl, KCl, $(CH_3)_4$NCl	25	calorimetry	$\Delta H_{ads} = -7$ kJ/mol in KCl and -8 kJ/mol in NaCl (limit for uncharged surface, the adsorption is increasingly exothermic for higher pH).	385
SiO_2		10^{-4} mol dm^{-3} $NaNO_3$	5–40	inflection, calorimetry	Inflection point at pH 4.57 at 5°C and 4.25 at 40°C, $\Delta H_{ads} = -17$ kJ/mol (calorimetry).	573
SiO_2	reagent grade, washed	0.1 mol dm^{-3} NaCl	15–35	pH	The negative σ_0 increases as T increases. Shift by about 0.013 pH unit/K.	480
SiO_2	quartz	0.0114–0.094 mol dm^{-3} NaCl	25,60	pH		390
Ta_2O_5	Merck	KCl	23,235	iep	IEP at pH 4.6 at 23°C and 6.3 at 235°C.	425
Ta_2O_5		0.1 mol dm^{-3} KCl	15–35	salt titration	$\Delta H_{ads} = -10$ kJ/mol	94
$ThO_2 \cdot n H_2O$	from nitrate	LiCl	25–70	pH	PZC at pH 7 at 25°C, 6.8 at 50°C and 6.75 at 70°C.	520

TABLE 3.11 (Continued)

Material	Description	Salt	T/°C	Method	Result	Ref.
TiO_2	rutile, from chloride	0.001–1 mol dm^{-3} NaCl	25–95	cip	$\Delta H_{ads}^{\circ} = -21$ kJ/mol	574
TiO_2	rutile, Tioxide	0.01–0.1 mol dm^{-3} NaNO$_3$	25	calorimetry		567
TiO_2	Degussa P 25		23–82	mass titration	PZC at pH 4 at 23°C and 6.2 at 82°C. $\Delta H_{ads} = 143$ kJ/mol	553
TiO_2	rutile, from chloride	0.02–1 mol dm^{-3}	5–50	cip	$\Delta H_{ads} = -18$ kJ/mol	561
				calorimetry	$\Delta H_{ads} = -21$ kJ/mol	562
TiO_2	Degussa P 25	0.01–0.5 mol dm^{-3} NaCl	25	calorimetry	$\Delta H_{ads} = -40$ kJ/mol at pH 10, at lower pH different results from acid and base titration (-10 to -20 kJ/mol at pH 4).	575
TiO_2	Degussa P 25	0.01 (?)–0.25 mol dm^{-3} KNO$_3$	10–45	cip	CIP at pH 3.8 at 10°C, 6.5 at 20°C, 6.3 at 25°C, 6.5 at 35°C, and 7.1 at 45°C.	576
TiO_2	anatase, reagent grade	10^{-4} mol dm^{-3} NaNO$_3$	5–45	mass titration	PZC at pH 6.05 at 5°C and 5.85 at 45°C. $\Delta H_{ads} = -8$ kJ/mol (confirmed by calorimetry)	577
TiO_2	rutile, Aldrich	KCl	15–35	salt titration	$\Delta H_{ads} = -12$ kJ/mol	555
TiO_2			23,235	iep	IEP at pH 8.9 at 23°C and 6.6 at 235°C.	425
TiO_2	rutile	0.03–1 mol dm^{-3} NaCl	25–250	cip	CIP at pH 5.4 at 25°C, 5.1 at 50°C, 4.7 at 100°C, 4.4 at 150°C, 4.3 at 200°C, and 4.2 at 250°C.	578
TiO_2	anatase, synthetic	0.005–0.05 mol dm^{-3} NaCl	25, 47	NaCl uptake, calorimetry	The NaCl uptake at natural pH increases with T.	529
TiO_2	Merck, anatase + 15% rutile	0.5 mol dm^{-3} NaCl	20–30	acousto	The IEP shifts to lower pH as T increases.	531
UO_2	from nitrate, contains $UO_{2.9}$ and U_3O_8	0–0.048 mol dm^{-3} LiNO$_3$	25 70	pH, merge	PZC at pH 5.8 at 25 C and 4.8 at 70°C.	579

Material	Description	Electrolyte	T	Method	Comments	Ref
ZnO	reagent, washed	$0.001–0.1$ mol dm^{-3} KNO$_3$	20–50	cip	$\Delta H_{ads} = -77$ kJ/mol	580
ZrO$_2$				cip	$\Delta H_{ads} = -22$ kJ/mol	581
ZrO$_2$			23,235	iep	IEP at pH 7.4 at 23°C and 7.2 (?) at 235 C.	425
ZrO$_2$	4 different membrane materials, different methods of measurement	10^{-3} mol dm^{-3} NaCl	50	iep	IEP at pH 4–7	582
ZrO$_2$	ceramic tube	10^{-1} mol kg^{-1}	200–400	iep	Negative ζ in 10^{-2} mol kg^{-1} HCl at T > 350 C.	583

Clay minerals

Material	Description	Electrolyte	T	Method	Comments	Ref
illite	Clay Minerals Society's Source Clays Repository	10^{-2} mol dm^{-3} NaCl	25, 40	pH	PZC at pH 9.6 at 25 C and 9.4 at 40 C.	139
kaolinite	BDH	10^{-2} mol dm^{-3} NaCl	25, 40	pH	PZC at pH 4.5 at 25°C and 4.2 at 40 C.	139
kaolinite	KGa-1	0.1 mol dm^{-3} NaCl	25,50, 70	pH		401, 402
kaolinite	Ajax		10,40, 70	pH	$XH + K^+ = XK + H^+$ ΔH 36 kJ/mol; $SOH = SO^- + H$ 32 kJ/mol; $SOH + H^+ = SOH_2^+$ 53 kJ/mol	151
kaolinite	KGa-1	$0.01–0.1$ mol dm^{-3} NaCl	25,60	pH		390
montmor-illonite	Fluka	10^{-2} mol dm^{-3} NaCl	25, 40	pH	PZC at pH 3.8 at 25°C and 3.7 at 40 C.	139

mixed oxides

Material	Description	Electrolyte	T	Method	Comments	Ref
79% TiO$_2$ + 21% ZrO$_2$	hydrolysis of mixture of isopropoxides	$10^{-3}–10^{-1}$ mol dm^{-3} KNO$_3$	30–60	cip	CIP decreases as T increases, $\Delta H_{ads} = -65$ kJ/mol.	196

TABLE 3.12 Surface Charging in Nonaqueous Media

Material	Characterization	Salt	T	Method	Solvent	Ref.
Al_2O_3	reagent grade, 2 samples, washed	$0.001–0.1$ mol dm^{-3} NaCl	20	pH uptake of Na and Cl	ethanol 12%, 1-propanol 10% 1-butanol 4% w/w	475
Al_2O_3	Merck, washed	$0.001–0.1$ mol dm^{-3} NaCl	25	pH uptake of Na, ζ	methanol 20%, ethanol 30% w/w	477
Al_2O_3	reagent grade	0.1 mol dm^{-3} NaCl, LiCl, CsCl	25	uptake of cations	50% methanol	478
Al_2O_3		0.1 mol dm^{-3} NaCl, 0.01 mol dm^{-3} LiCl, CsCl		pH, uptake of Li, Cl and Cs	methanol and ethanol 30% w/w	479
Al_2O_3	γ, reagent grade washed	KCl	25	salt titration	40% dioxane 60% methanol	555
Al_2O_3	reagent grade, washed	0.1 mol dm^{-3} NaCl, CsCl	15 35	pH, uptake of ions	30% methanol, 30% dioxane	480
Al_2O_3	AKP 50 Sumitomo	$0–10^{-3}$ mol dm^{-3} LiCl	25	ζ, coagulation	ethanol	588
Al_2O_3	Merck, washed	$10^{-2}–1$ mol dm^{-3} NaCl, CsCl, LiCl	25	pH	up to 20% dioxane, glycerol, DMSO, THF	589
CrO_2	DuPont	10^{-4} mol dm^{-3} NaClO$_4$		ζ	tetrahydrofuran, water up to 0.4%	590
Fe_2O_3	hematite, monodispersed, from chloride	$0.001–0.01$ mol dm^{-3} NaNO$_3$	25	pH iep	50% ethanol, 50% methanol	591
Fe_2O_3	hematite, from chloride	KCl KNO$_3$	25	salt titration	dioxane up to 50%	563
Fe_2O_3	α, hydrolysis	many 1–1		coagulation	urea up to 12 mol dm^{-3}	499

Material	Description	Salts	t/°C	Method	Solvent	Ref.
Fe_2O_3	of nitrate, α, hydrolysis of $FeCl_3$ at 100 C, washed	10^{-4}–10^{-1} mol dm^{-3} NaCl		iep	50% ethanol	500
Fe_2O_3	hematite synthetic cubic	0.01 mol dm^{-3} $NaClO_4$		pH; uptake of Na and Cl; iep	10%, 30% ethanol	592
$Fe_2O_3 \cdot 3.1 H_2O$	amorphous, from chloride	0.1 mol dm^{-3} NaBr, $NaNO_3$, 0.05 mol dm^{-3} Na_2SO_4	25	kinetic study of uptake	30–60% methanol and 1-propanol	593
FeOOH	goethite, hydrolysis of nitrate	0.1 mol dm^{-3} KCl, $(CH_3)_4NCl$	25	titration (no experimental data were published, only results of model calculations)	10–50% acetone, 20–80% methanol	594
SiO_2	pyrogenic, Wacker	10^{-4}–0.1 mol dm^{-3} LiCl, pH 9	25	Li uptake	0.2 mol dm^{-3} urea, 2 mol dm^{-3} N-methyl acetamide, methylurea	510
SiO_2	Aerosil	none	25	ζ	17 solvents (water free)	595
SiO_2	fused	0.01 mol dm^{-3} KCl + 0.001 mol dm^{-3} phosphate	25	ζ	up to 60 mol% methanol, ethanol, 1-propanol, acetonitrile, acetone, DMSO	596
SiO_2	Stober	10^{-6}, 10^{-4} mol dm^{-3} LiCl, NaCl, KCl, CsCl, HCl, KOH	25	ζ	80–95% w/w dioxane	597
SiO_2	Stober	0.01, 0.1 mol dm^{-3} LiCl, NaCl, RbCl, CsCl	25	ζ	30% methanol	511
SiO_2	Stober	10^{-3}–10^{-1} mol dm^{-3} KCl	25	ζ	30% ethanol, 30% methanol, 30% 1-propanol, 10% dioxane	598
SiO_2	Merck, washed	0.001–0.1 mol dm^{-3} NaCl	25	pH; uptake of Na, ζ	methanol 30%, ethanol 60% w/w	477
SiO_2	reagent grade	0.1 mol dm^{-3}	25	uptake of cations	50% methanol	478

TABLE 3.12 (Continued)

NaCl, LiCl, CsCl

Material	Characterization	Salt	T	Method	Solvent	Ref.
SiO_2		0.01 mol dm^{-3} LiCl, CsCl		pH, uptake of Li and Cs	30% methanol w/w	479
SiO_2	reagent grade, washed	0.1 mol dm^{-3} NaCl, CsCl	15 35	pH, uptake of ions	30% methanol, 30% dioxane	480
SiO_2	chromatographic, Merck	0.01–1 mol dm^{-3} NaCl, LiCl, KCl, CsCl	25	pH	> 20 organic co-solvents up to 20% w/w	599
SiO_2	Stober	0.001 mol dm^{-3} CaI$_2$, 0.0015 mol dm^{-3} NaI	25	ζ	acetone-water, DMF-water	600
SiO_2	Merck, washed	0.1 mol dm^{-3} NaClO$_4$	25	uptake of Ni	up to 2% DMSO, THF, methanol, glycerol	601
SiO_2	fused, capillary	none	room	ζ	acetonitrile, methanol (20–100%), DMF, formamide, DMSO	448
SiO_2		none		AFM	methanol, ethanol, 1-propanol, 30–99%	602
SiO_2	XL Nissan	0–10^{-3} mol dm^{-3} NaCl		ζ	methanol (up to 4.6% water), acetonitrile	311
SiO_2		none		AFM	10–99% w/w dioxane	603
SiO_2	chromatographic, Baker	0.001–0.1 mol dm^{-3} NaCl, LiCl	25	pH	4–8% w/w acetone	392
SiO_2		none		AFM	80–99% w/w n-alcohols (up to hexanol)	604
SiO_2	Stober	10^{-6}–0.01 mol dm^{-3} CsCl	25	ζ	25 solvents	605
TiO_2	Degussa	none	25	ζ	1-pentanol	606
TiO_2	rutile, from TiCl$_4$	none	25	ζ	n-alcohols up to C$_{10}$	607

Material	Description	Electrolyte	T	Method	Solvent	Ref.
TiO_2	Ventron, washed	0.001–0.1 mol dm⁻³ NaCl	20	pH	10% ethanol, 4% 1-butanol	608
TiO_2	anatase, Aldrich, washed	10^{-3} 10^{-1} mol dm⁻³ KCl, LiCl, CsCl, KI	25	ζ, pH	30% methanol, 30% ethanol w/w	609
TiO_2	reagent grade, rutile, Aldrich	KCl	25	salt titration	40% dioxane	555
TiO_2	anatase, Aldrich	10^{-6}–10^{-2} mol dm⁻³ CsCl; CsClO₄		zeta	95–99% dioxane, DMSO, acetone, alcohols up to C_3	610
TiO_2	rutile	none		particle size, sedimentation	dioxane, 2-propanol	611
TiO_2	anatase, reagent grade, washed	10^{-2}–1 mol dm⁻³ NaCl	25	pH	up to 20% dioxane, glycerol, DMSO, THF	589
TiO_2	anatase, from ethoxide	10^{-4}–10^{-1} mol dm⁻³ NaCl		ζ, uptake of ions, pH	ethanol 50% and 100%	530
TiO_2	Merck anatase + 15% rutile	0.16 mol dm⁻³ NaNO₃	20	acousto	methanol, ethanediol up to 30% 2-propanol 10%	531
TiO_2	anatase, Aldrich, washed	10^{-1} mol dm⁻³ NaNO₃	25	pH, acousto	20% 1-propanol	331
TiO_2	anatase, Aldrich, washed	up to 3×10^{-4} mol dm⁻³ HClO₄, CsOH	25	ζ	1-propanol, 2-butanol, 1-pentanol, dioxane	612
Y_2O_3	decomposition of basic carbonate, spherical	10^{-1} mol dm⁻³ KCl	25	ζ	30% ethanol w/w	609
ZnO	Merck	HCl, KOH		ζ	methanol, ethanol, 1 propanol, water up to 0.3%	613
Ca-montmorillonite	converted from natural mineral		25	ζ	2-propanol	614
controlled pore glass		0.001–0.1 mol dm⁻³ NaCl, CsCl	20	pH	12% ethanol, 9% 1-propanol, 4% 1-butanol	405

In water, substantial shifts of the IEP to high pH values (with respect to the pristine PZC) are observed at ionic strengths on the order of 1 mol dm^{-3}, and with aqueous ethanol and methanol (30% weight fraction) sign reversal of the ζ potential to positive over the entire pH range was observed at an ionic strength as low as 0.1 mol dm^{-3}. Another important difference between the effects presented in Figs. 3.101–3.103 for water and those observed in mixed water–organic solvents is that the latter are insensitive to the nature of the salt (Cs versus Li, or Cl versus I). Surface charging studies in water–organic mixtures rich in water have been carried out with most common organic solvents miscible with water, but the electrokinetic studies are confined to a few lower alcohols.

The possibility to use the pH as the major variable defining the surface charging of the materials in nonaqueous organic liquids and their mixtures with small amounts of water is limited by the difficulties in pH measurements. This does not mean, however, that the pH (or proton activity) in such systems is less important than in water or water-rich organic mixtures. In view of the difficulties with pH measurements, the procedure illustrated in Fig. 3.3 is not applicable, and σ_0 cannot be obtained from titration data, but electrokinetic methods can be applied in such systems. Most commercial zetameters are designed for measurements in aqueous dispersions, and not necessarily are they all suitable for systems whose conductance and other physical properties are considerably different from the properties of water. Also the interpretation of electrokinetic results in nonaqueous media can be more difficult than in water, e.g. due to ion pairing. Analytical concentration of strong acid or strong base can be used to express proton activity in such systems, keeping in mind that not necessarily are they fully dissociated in the organic solvent under study. The ζ potential of oxides and related materials in analytical reagent grade organic liquids without addition of any other chemicals is chiefly defined by acidic and basic impurities, which are present at typical concentrations on the order of 10^{-5} mol dm^{-3} even in the purest commercially available reagents. They cannot be removed from the solvent unless a purification procedure targeted at a specific impurity is applied. Amines, which are present in most organic reagents create a strongly basic environment (high pH) that leads to negative surface charge and ζ potential. They can be neutralized by mineral acids, and then the sign of the ζ potential of certain materials reverses to positive. Obviously the nature and concentration of the impurities changes from one lot of the solvent to another, and this results in different values of ζ potential in "pure" organic solvent from different sources and in different amounts of mineral acid necessary to neutralize the amines. Also traces of water may play a critical role, namely they act as a pH buffer, and this may even lead to sign reversal of the ζ potential of some materials in "pure" organic solvent on addition of water. A few irrational ideas related to ζ potentials in "pure" organic solvents have been published, and they are commented on elsewhere [585–587].

It was discussed in Section III.C that salts, which are inert at low concentrations and close to PZC can induce effects characteristic for specific adsorption at sufficiently high concentrations. In organic solvents (water free or containing small amounts of water) the specific interactions of electrolytes that are inert in water are commonplace, even at concentrations below 10^{-3} mol dm^{-3} (but the activities of alkali metal cations in such solutions can be higher that in 1 molar aqueous solution). Reversal of sign ζ potential of inorganic materials from negative (in "pure" organic solvents) to positive on addition of CsCl has been observed. The concentration of salt necessary to induce such a sign reversal depends on the nature of the salt (e.g. LiCl is less efficient than CsCl) and on the nature of the solvent. Very likely it also changes from one lot of solvent to another due to different concentrations of basic impurities therein, but systematic studies in this direction have not been carried out. Sign reversal of silica on addition of CsCl was only observed for solvents or solvent mixtures whose relative electric permittivity was < 24, while for more polar solvents the negative ζ potential as a function of [CsCl] asymptotically tended to zero.

These results indicate, that the mechanism of surface charging of mineral oxides and related materials in polar organic solvents is similar to that in water, and the main difference is due to technical and theoretical difficulties related to measurement and interpretation of activities of proton and other inorganic ions in such media.

REFERENCES

1. G. A. Kokarev, V. A. Kolesnikov, A. F. Gubin, and A. Korobanov. Elektrokhimiya 18: 466–470 (1982).
2. S. Ardizzone. J. Electroanal. Chem. 239: 419–425 (1988).
3. N. Kallay, M. Colic, D. W. Fuerstenau, H. M. Jang, and E. Matijevic. Colloid Polym. Sci. 272: 554–561 (1994).
4. T. F. Tadros, and J. Lyklema. J. Electroanal. Chem. 17: 267–275 (1968).
5. G. A. Gomes, and J. F. C. Boodts. J. Braz. Chem. Soc. 10: 92–96 (1999).
6. S. K. Roy, and P. K. Sengupta. J. Colloid Interf. Sci. 125: 340–343 (1988).
7. S. H. Han, W. G. Hou, C. G. Zhang, D. J. Sun, X. Huang, and G. T. Wang. J. Chem. Soc. Faraday Trans. 94: 915–918 (1998).
8. M. Herrmann, and H. P. Boehm. Z Anorg Allg Chemie 368: 73–86 (1969).
9. L. Madrid, and P. De Arambarri. Geoderma 21: 199–208 (1978).
10. J. Lützenkirchen, and P. Magnico. Colloids Surfaces A 137: 345–354 (1998).
11. A. N. Zhukov. Kolloid. Zh. 58: 280–282 (1996).
12. J. Lützenkirchen. J. Colloid Interf. Sci. 204: 119–127 (1998).
13. R. Charmas, W. Piasecki, and W. Rudzinski. Langmuir 11: 3199–3210 (1995).
14. L. Blok, and P. L. de Bruyn. J. Colloid Interf. Sci. 32: 518–526 (1970).
15. L. W. Zelazny, H. E. Liming, and A. N. Vanwormhoudt. In Methods of Soil Analysis. Part 3. Chemical Methods, SSSA, Madison 1996, pp. 1231 1253.
16. G. Sposito. Environ. Sci. Technol. 32: 2815–2819 (1998).
17. M. Sugita, M. Tsuji, and M. Abe. Bull. Chem. Soc. Jpn. 63: 559–564 (1990).
18. K. Jiratova. Appl. Catal. 1: 165–167 (1981).
19. M. A. Blesa, G. Magaz, J. A. Salfity, and A. D. Weisz. Solid State Ionics 101–103: 1235–1241 (1997).
20. J. F. Watts, and E. M. Gibson. Int. J. Adhesion Adhesives 11: 105–108 (1991).
21. K. D. Dobson, P. A. Connor, and A. J. McQuillan. Langmuir 13: 2614–2616 (1997).
22. K. W. Perrott. Clays Clay Min. 25: 417–421 (1977).
23. N. S. Petro, N. Z. Misak, and I. M. El-Naggar. Colloids Surf. 49: 211–218 (1990).
24. N. S. Petro, I. M. El-Naggar, E. S. I. Shabana, and N. Z. Misak. Colloids Surf. 49: 219–227 (1990).
25. F. Taha, A. M. El-Roudi, A. A. Abd El Gaber, and F. M. Zahran. Rev. Roum. Chim. 35: 503–509 (1990).
26. J. S. Moya, J. Rubio, and J. A. Pask. Ceramic Bull. 59: 1198–1200 (1980).
27. P. E. K. Donaldson. Med. Biol. Eng. Comput. 31: 75–78 (1993).

28. D. N. Furlong, P. A. Freeman, and A. C. M. Lau. J. Colloid Int. Sci. 80: 20–31 (1981).
29. W. A. Zeltner, and M. A. Anderson. Langmuir 4: 469–474 (1988).
30. T. D. Evans, J. R. Leal, and P. W. Arnold. J. Electroanal. Chem. 105: 161–167 (1979).
31. S. Hamada, Y. Kudo, and K. Minagawa. Bull. Chem. Soc. Jpn. 63: 102–107 (1990).
32. A. S. Desai, G. E. Peck, J. E. Lovell, J. L. White, and S. L. Hem. J. Colloid Interf. Sci. 166: 29–34 (1994).
33. R. M. Cornell, A. M. Posner, and J. P. Quirk. J. Colloid Int. Sci. 53: 6–13 (1975).
34. P. G. Dietrich, K. H. Lerche, J. Reusch, and R. Nitzsche. Chromatographia 44: 362–366 (1997).
35. L. Vordonis, P. G. Koutsoukos, and A. Lycourghiotis. J. Catalysis 98: 296–307 (1986).
36. J. P. Jolivet, and E. Tronc. J. Colloid Interf. Sci. 125: 688–701 (1988).
37. J. Biscan, M. Kosec, and N. Kallay. Colloids Surf. A 79: 217–226 (1993).
38. J. R. Regalbuto, A. Navada, S. Shadid, M. L. Bricker, and Q. Chen. J. Catalysis. 184: 335–348 (1999).
39. H. Watanabe, and J. Seto. Bull. Chem. Soc. Jpn. 59: 2683–2687 (1986).
40. S. Kittaka. J. Colloid Int. Sci. 48: 327–333 (1974).
41. S. Kittaka, and T. Morimoto. J. Colloid Int. Sci. 75: 398–403 (1980).
42. M. J. G. Janssen, and H. N. Stein. J. Colloid Interf. Sci. 109: 508–515 (1986).
43. K. W. Bush, M. A. Bush, S. Gopalakrishnan, and E. Chibowski. Coll. Polym. Sci. 273: 1186–1192 (1995).
44. J. Lyklema. J. Coll. Int. Sci. 99: 109–118 (1984).
45. C. P. Schulthess, and D. L. Sparks. Soil Sci. Soc. Am. J. 50: 1406–1411 (1986).
46. S. I. Pechenyuk. Izv. Akad. Nauk. Ser. Khim. 48: 228–238 (1999).
47. S. M. Ahmed, and D. Maksimov. Can. J. Chem. 46: 3841–3846 (1968).
48. T. W. Healy, A. P. Herring, and D. W. Fuerstenau. J. Colloid Int. Sci. 21: 435–444 (1966).
49. P. Jayaweera, and S. Hettiarachchi. Rev. Sci. Instrum. 64: 524–528 (1993).
50. R. Pelton, P. Miller, W. McPhee, and S. Rajaram. Colloids Surf. A. 80: 181–189 (1993).
51. A. Doren, J. Lemaitre, and P. G. Rouxhet. J. Colloid Int. Sci. 130: 146–156 (1989).
52. P. Bouriat, P. Saulnier, P. Brochette, A. Graciaa, and J. Lachaise. J. Colloid Interf. Sci. 209: 445–448 (1999).
53. J. S. Lyons, D. N. Furlong, A. Homola, and T. W. Healy. Aust. J. Chem. 34: 1167–1175 (1981).
54. S. Ross, and R. P. Long. Ind. Eng. Chem. 61: 58 (1969).
55. A. Szymczyk, P. Fievet, J. C. Reggiani, and J. Pagetti. Desalination 116: 81–88 (1998).
56. A. S. Dukhin, H. Ohshima, V. N. Shilov, and P. J. Goetz. Langmuir 15: 3445–3451 (1999).
57. M. Kosmulski, P. Eriksson, J. Gustafsson, and J. B. Rosenholm. J. Colloid Interf. Sci. 220: 128–132 (1999).
58. F. Dumont and A. Watillon. Disc. Faraday. Soc. 52: 352–360 (1972).
59. J. S. Noh, and J. A. Schwarz. J. Colloid Interf. Sci. 130: 157–164 (1989).
60. T. Preocanin, and N. Kallay. Croat. Chem. Acta 71: 1117–1125 (1998).
61. J. P. Reymond, and F. Kolenda. Powder Techn. 103: 30–36 (1999).
62. J. Park, and J. R. Regalbuto. J. Colloid Int. Sci. 175: 239–252 (1995).
63. J. Wernet, and D. L. Feke. J. Am. Ceram. Soc. 77: 2693–2698 (1994).
64. G. Busca, C. Cristiani, P. Forzatti, and G. Groppi. Catalysis Lett. 31: 65–74 (1995).
65. S. Zalac, and N. Kallay. J. Colloid Int. Sci. 149: 233–240 (1992).
66. A. K. Helmy, E. A. Ferreiro, and S. G. de Bussetti. Z. Phys. Chem. (Leipzig) 261: 1065–1073 (1980).
67. A. Scheidegger, M. Borkovec, and H. Sticher. Geoderma 58: 43–65 (1993).
68. M. Tschapek, R. M. Torres Sanchez, and C. Wasowski. Colloid Polym. Sci. 254: 516–521 (1976).

69. M. Tschapek, C. Wasowski, and S. Falasca. Colloid Polym. Sci. 269: 1190–1195 (1991).

70. N. Kallay, Z. Torbic, E. Barouch, and J. Jednacak-Biscan. J. Colloid Interf. Sci. 118: 431–435 (1987).

71. N. Kallay, D. Kovacevic, I. Dedic, and V. Tomasic. Corrosion 50: 598–602 (1994).

72. N. Kallay, Z. Torbic, M. Golic, and E. Matijevic. J. Phys. Chem. 95: 7028–7032 (1991).

73. H. Moriwaki, Y. Yoshikawa, and T. Morimoto. Langmuir 6: 847–850 (1990).

74. R. Lindberg, G. Sundholm, J. Sjöblom. P. Ahonen and E.I. Kauppinen. J. Disp. Sci. Technol. 20: 715–722 (1999).

75. X. Y. Lin, F. Creuzet, and H. Arribart. J. Phys. Chem. 97: 7272–7276 (1993).

76. T. Arai, D. Aoki, Y. Okabe, and M. Fujihira. Thin Solid Films 273: 322–326 (1996).

77. W. A. Ducker, Z. Xu, D. R. Clarke, and J. N. Israelachvili. J. Am. Chem. Soc. 77: 437–443 (1994).

78. K. Hu, F. R. F. Fan, A. J. Bard, and A. C. Hillier. J. Phys. Chem. B 101: 8298–8303 (1997).

79. R. G. Horn, D. R. Clarke, and M. T. Clarkson. J. Mater. Res. 3: 413–416 (1988).

80. T. Hiemstra, W. H. van Riemsdijk. Langmuir 15: 8045–8051 (1999).

81. C. M. Eggleston, and G. Jordan. Geochim. Cosmochim. Acta 62: 1919–1923 (1998).

82. B. Lovrecek, Z. Bolanca, and O. Korelic. Surf. Coat. Technol. 31: 351–364 (1987).

83. E. McCafferty, and J. P. Wightman. J. Colloid Interf. Sci. 194: 344–355 (1997).

84. G. T. Brown, and J. R. Darwent. J. Chem. Soc. Chem. Commun. 98–100 (1985).

85. A. B. Ankomah. Clays Clay Min. 39: 100–102 (1991).

86. V. I. E. Bruyere, P. J. Morando, and M. A. Blesa. J. Colloid Interf. Sci. 209: 207–214 (1999).

87. C. S. Bürgisser, A. M. Scheidegger, M. Borkovec, and H. Sticher. Langmuir 10: 855–860 (1994).

88. H. H. van den Vlekkert, and N. F. de Rooij. Analusis 16: 110–119 (1988).

89. A. A. Poghossian. Sensors Actuators B 44: 551–553 (1997).

90. C. Quang, S. L. Petersen, G. R. Ducatte, and N. E. Ballou, J. Chromatogr. A. 732: 377–384 (1996).

91. C. P. Schulthess, and D. L. Sparks. Soil Sci. Soc. Am. J. 52: 92–97 (1988).

92. G. A. Parks. Chem. Rev. 65: 177–198 (1965).

93. S. I. Pechenyuk. Usp. Khimii 61: 711–733 (1992).

94. M. Kosmulski. Langmuir. 13: 6315–6320 (1997).

95. K. C. Ray, and P. K. Sen. Indian J. Chem. 12: 170–173 (1974).

96. E. A. Nechaev, and L. M. Smirnova. Kolloid. Zh. 39: 186–190 (1977).

97. M. A. Butler, and D. S. Ginley. J. Electrochem. Soc. 125: 228–232 (1978).

98. R. H. Yoon, T. Salman, and G. Donnay. J. Colloid Interf. Sci. 70: 483–493 (1979).

99. A. Daghetti, G. Lodi, and S. Trasatti. Mater. Chem. Phys. 8: 1–90 (1983).

100. M. Delamar. J. Electron Spectrosc. 53: C11–C14 (1990).

101. D. A. Sverjensky. Geochim. Cosmochim. Acta 58: 3123–3129 (1994).

102. G. Gonzalez, S. M. Saraiva, and W. Aliaga. J. Dispersion Sci. Technol. 15: 123–132 (1994); G. Gonzalez, S. M. Saraiva, and W. Aliaga. South. Braz. J. Chem. 2: 5–20 (1994).

103. A. K. Vijh. Appl. Phys. Commun. 13: 275–281 (1994), (formulas of oxides are not given).

104. J. Lyklema. Fundamentals of Interface and Colloid Science. vol. 2 Academic Press London 1995. pp. A3.2–A3.6.

105. G. M. S. El Shafei. J. Colloid Interf. Sci. 182: 249–253 (1996).

106. S. Ardizzone, and S. Trasatti. Adv. Colloid Interf. Sci. 64: 173–251 (1996).

107. E. McCafferty. J. Electrochem. Soc. 146: 2863–2869 (1999).

108. S. I. Pechenyuk. Izv. Akad. Nauk Ser. Khim. 48: 1029–1035 (1999).

109. J. F. Kuo, M. M. Sharma, and T. F. Yen. J. Colloid Int. Sci. 126: 537–546 (1988).

110. G. A. Parks. Advan. Chem. Ser. 67: 121 (1967).
111. J. F. Kuo, and T. F. Yen. J. Colloid Int. Sci. 121: 220–225 (1988).
112. N. Balagopal, H. K. Varma, K. G. K. Warrier, and A. D. Damodaran. Ceramics Int. 18: 107–111 (1992).
113. S. Kittaka. J. Colloid Int. Sci. 48: 334–338 (1974).
114. G. C. Bye, and G. T. Simpkin. J. Appl. Chem. Biotechnol. 28: 116–118 (1978).
115. P. R. Anderson, and M. M. Benjamin. Environ. Sci. Techn. 24: 1586–1592 (1990).
116. S. Subramanian, M. S. Chatta, and C. R. Peters. J. Molecul. Catalysis 69: 235–245 (1991).
117. F. M. Mulcahy, M. Houalla, and D. M. Hercules. J. Catal. 148: 654–659 (1994).
118. S. H. Han, S. P. Xu, W. G. Hou, D. J. Sun, C. G. Zhang, and G. T. Wang. Chinese J. Chem. 15: 304–312 (1997).
119. S. H. Han, D. J. Sun, W. G. Hou, C. G. Zhang, and G.T. Wang. Chinese Chem. Lett 8: 87–90 (1997).
120. S. H. Han, W. G. Hou, Q. Dong, D. J. Sun, X. R. Huang, and C. G. Zhang. Chem. Res. Chinese Univ. 15: 58–62 (1998).
121. F. J. Gil-Llambias, and A. M. Escudey-Castro. J. Chem. Soc. Chem. Commun. 478–479 (1982).
122. M. Escudey, and F. Gil-Llambias. J. Colloid Int. Sci. 107: 272–275 (1985).
123. R. Vilcu, and M. Olteanu. Rev. Roum. Chim. 20: 901–910 (1975).
124. H. Kita, N. Henmi, K. Shimazu, H. Hattori and K. Tanabe. J. Chem. Soc. Faraday Trans. 1 77: 2451–2463 (1981).
125. J. A. Schwarz, C. T. Driscoll, and A. K. Bhanot. J. Colloid Int. Sci. 97: 55–61 (1984).
126. J. A. Schwarz, C. T. Ugbor, and R. Zhang. J. Catalysis 138: 38–54 (1992).
127. M. I. Zaki, S. A. A. Mansour, F. Taha, and G. A. H. Mekhemer. Langmuir 8: 727–732 (1992).
128. M. Mullet, P. Fievet, J. C. Reggiani, and J. Pagetti. J. Membr. Sci. 123: 255–265 (1997).
129. A. Szymczyk, A. Pierre, J. C. Reggiani, and J. Pagetti. J. Membrane Sci. 134: 59–66 (1997).
130. A. Szymczyk, P. Fievet, J. C. Reggiani, and J. Pagetti. Desalination 119: 303–308 (1998).
131. A. Szymczyk, P. Fievet, M. Mullet, J. C. Reggiani, and J. Pagetti. Desalination 119: 309–314 (1998).
132. F. J. Gil-Llambias, A. M. Escudey, J. L. G. Fierro, and A. Lopez Agudo. J. Catalysis 95: 520–526 (1985).
133. F. J. Gil Llambias, J. Salvatierra, L. Bouyssieres, M. Escudey, and R. Cid. Appl. Catal. 59: 185–195 (1990).
134. M. M. Ostromecki, L. J. Burcham, I. E. Wachs, N. Ramani, and J. G. Ekerdt. J. Mol. Catal. A 132: 43–57 (1998).
135. F. Gil-Llambias, N. Escalona, C. Pfaff, C. Scott, and J. Goldwasser. React. Kinet. Catal. Lett. 66: 225–229 (1999).
136. T. Ogihara, K. Wada, T. Yoshida, T. Yanagawa, N. Ogata, and K. Yoshida. Ceramics Int. 19: 159–168 (1993).
137. I. Sondi, J. Biscan, and V. Pravdic. J. Colloid Interf. Sci. 178: 514–522 (1996).
138. I. Sondi, and V. Pravdic. J. Colloid Int. Sci. 181: 463–469 (1996).
139. M. M. Motta, and C. F. Miranda. Soil Sci. Soc. Am. J. 53: 380–385 (1989).
140. P. Rengasamy, and J. M. Oades. Aust. J. Soil Res. 15: 235–242 (1977).
141. R. M. Torres Sanchez, E. F. Aglietti, and J. M. Porto-Lopez. Mater. Chem. Phys. 20: 27–38 (1988).
142. K. Sakurai, A. Teshima, K. Kyuma. Soil Sci. Plant Nutr. 36: 73–81 (1990).
143. E. Wieland, thesis, quoted after S. B. Haderlein, and R. P. Schwarzenbach. Environ. Sci. Technol. 27: 316–326 (1993).

144. S. Goldberg, H. S. Forster, and E. L. Heick. Soil Sci. Soc. Am. J. 57: 704–708 (1993).
145. I. Fox, and M. A. Malati. J. Chem. Tech. Biotechnol. 57: 97–107 (1993).
146. S. Goldberg, H. S. Forster, and C. L. Godfrey. Soil Sci. Soc. Am. J. 60: 425–432 (1996).
147. B. K. Schroth, and G. Sposito. Mater. Res. Soc. Symp. Proc. 432: 87–92 (1997).
148. R. Kretzschmar, D. Hesterberg, and H. Sticher. Soil Sci. Soc. Am. J. 61: 101–108 (1997).
149. R. Kretzchmar, H. Holthoff, and H. Sticher. J. Colloid Interf. Sci. 202: 95–103 (1998).
150. S. Goldberg. C. Su, and H. S. Forster. In Adsorption of Metals by Geomedia (E. A. Jenne, ed.), Academic Press, San Diego 1998. pp. 401–426.
151. M. J. Angove, B. B. Johnson, and J. D. Wells. J. Colloid Interf. Sci. 204: 93–103 (1998).
152. R. M. Torres Sanchez, E. I. Basaldella, and J. F. Marco. J. Colloid Interf. Sci. 215: 339–344 (1999).
153. L. Benyahya, and J. M. Garnier. Environ. Sci. Technol. 33: 1398–1407 (1999).
154. H. Wanner, Y. Albinsson, O. Karnland, E. Wieland, P. Wersin, and L. Charlet. Radiochim. Acta 66/67: 157–162 (1994).
155. C. Volzone, and R. M. Torres Sanchez. Colloids Surf. A 81: 211–216 (1993).
156. L. A. de Faria, M. Prestat, J. F. Koenig, P. Chertier, and S. Trasatti. Electrochim. Acta 44: 1481–1489 (1998).
157. Y. Wang, R. J. Pugh, and E. Forssberg. Colloids Surf. A 90: 117–133 (1994).
158. L. Boulange-Petermann, A. Doren, B. Baroux, and M. N. Bellon-Fontaine. J. Colloid Int. Sci. 171: 179–186 (1995).
159. G. Stakkestad, B. Grung, J. Sjoblom, and T. Sigvartsen. Colloid Polym. Sci. 277: 627–636 (1999).
160. S. Ardizzone. A. Chittofrati, and L. Formaro. J. Chem. Soc. Faraday Trans. I 83: 1159–1168 (1987).
161. B. S. Mathur, and B. Venkataramani. Colloids Surfaces A 140: 403–416 (1998).
162. P. H. Tewari, and W. Lee. J. Colloid Int. Sci. 52: 77–88 (1975).
163. A. E. Regazzoni, and E. Matijevic. Corrosion 38: 212–218 (1982).
164. U. Schwertmann, and H. Fechter. Clay Min. 17: 471–476 (1982).
165. G. W. Bruemmer, J. Gerth, and K. G. Tiller. J. Soil Sci. 39: 37–52 (1988).
166. H. F. Ghoneimy, T. N. Morcos, and N. Z. Misak. Colloids Surf. A 122: 13–26 (1997).
167. T. F. Barton, T. Price, K. Becker, and J. G. Dillard. Colloids Surf. 53: 209–222 (1991).
168. J. B. Goodenough, R. Manoharan, and M. Paranthamam. J. Am. Chem. Soc. 112: 2076–2082 (1990).
169. M. W. Rophael, T. A. Bibawy, L. B. Khalil, and M. A. Malati. Chemistry & Industry (1) 27–28 (1979).
170. F. F. Aplan, E. Y. Spearin, and G. Simkovich. Colloids Surf. 1: 361–371 (1980).
171. L. A. De Faria, J. F. C. Boodts, and S. Trasatti. Colloids Surfaces A 132: 53–59 (1998).
172. L. A. de Faria, and S. Trasatti. J. Electroanal. Chem. 340: 145–152 (1992).
173. M. M. Sharma, J. F. Kuo, and T. F. Yen. J. Colloid Int. Sci. 115: 9–16 (1987).
174. M. A. F. Pyman, J. W. Bowden, and A. M. Posner. Clay Min. 14: 87–92 (1979).
175. W. C. Hasz. Thesis. quoted after W. M. Mullins, and B. L. Averbach. Surface Sci. 206: 41–51 (1988).
176. C. Contescu, C. Sivaraj, and J. A. Schwarz. Appl. Catalysis 74: 95–108 (1991).
177. P. K. Ahn, S. Nishiyama, S. Tsuruya, and M. Masai. Appl. Catalysis A 101: 207–219 (1993).
178. S. Nishikawa, and E. Matijevic. J. Colloid Interf. Sci. 165: 141–147 (1994).
179. M. Arias, T. Barral, and F. Diaz-Fierros. Clays Clay Min. 43: 406–416 (1995).
180. M. J. Avena, and C. P. de Pauli. Colloids Surf. A 118: 75–87 (1996).
181. M. Zhou, J. M. F. Ferreira, A. T. Fonseca, and J. L. Baptista. J. Eur. Cer. Soc. 17: 1539–1544 (1997).
182. J. Shen, A. D. Ebner, and J. A. Ritter. J. Colloid Interf. Sci. 214: 333–343 (1999).

183. V. M. Gunko, V. I. Zarko, V. V. Turov, R. Leboda, E. Chibowski, and V. V. Gunko. J. Colloid Interf. Sci. 205: 106–120 (1998).
184. R. W. Luce, and G. A. Parks. Chem. Geology. 12: 147–153 (1973).
185. S. Ozeki, I. Takano, M. Shimizu, and K. Kaneko. J. Colloid Int. Sci. 132: 523–531 (1989).
186. K. A. Venkatesan, N. Sathi Sasidharan, and P. K. Wattal. J. Radioanal. Nucl. Chem. 220: 55–58 (1997).
187. R. Sprycha. Colloids Surf. 5: 147–157 (1982).
188. M. Mao, D. Fornasiero, J. Ralston, R. S. C. Smart, and S. Sobieraj. Colloids Surf. 85: 37–49 (1994).
189. M. Bjelopavlic, J. Ralston, and G. Reynolds. J. Colloid Interf. Sci. 208: 183–190 (1998).
190. S. B. Balachandran, G. Simkovich, and F. F. Aplan. Int. J. Miner. Proc. 21: 157–171 (1987).
191. M. R. Houchin. Colloids Surf. 13: 125–136 (1985).
192. G. E. Morris, W. A. Skinner, P. G. Self, and R. S. C. Smart. Colloids Surf. A 155: 27–41 (1999).
193. W. H. Morrison. J. Colloid Int. Sci. 100: 121–127 (1984).
194. D. W. Bahnemann. Isr. J. Chem. 33: 115–136 (1993).
195. X. Fu, L. A. Clark, Q. Yang, and M. A. Anderson. Environ. Sci. Technol. 30: 647–653 (1996).
196. B. Karmakar, and D. Ganguli. Bull. Chem. Soc. Jpn. 62: 1373–1375 (1989).
197. Z. Xu, Y. Hu, and Y. Li. J. Colloid Interf. Sci. 198: 209–215 (1998).
198. J. Biscan, M. Kosec, and N. Kallay. Colloids Surf. A 79: 217–226 (1993).
199. S. Soled, and G. B. McVicker. Catalysis Today 14: 189–194 (1992).
200. E. M. de Liso, W. R. Cannon, and A. S. Rao. Mat. Res. Soc. Symp. Proc. 60: 43–50 (1986).
201. M. Kagawa, M. Omori, Y. Syono, Y. Imamura, and S. Usui. J. Am. Ceram. Soc. 70: c212–c213 (1987).
202. M. Hashiba, H. Okamoto, Y. Nurishi, and K. Hiramatsu. J. Mater. Sci. 23: 2893–2896 (1988).
203. M. Schultz, S. Grimm, and W. Burckhardt. Solid State Ionics 63–65: 18–24 (1993).
204. Pradip, R. S. Premachandran, and S. G. Malghan. Bull. Mater. Sci. 17: 911–920 (1994).
205. V. Ramakrishnan, Pradip, and S. G. Malghan. J. Am. Ceram. Soc. 79: 2567–2576 (1996).
206. J. B. Rosenholm, F. Manelius, J. Stranden, M. Kosmulski, H. Fagerholm, H. Byman-Fagerholm, and A. B. A. Pettersson. In Ceramic Interfaces: Properties and Applications III, (R. St. C. Smart and J. Nowotny, eds.), Institute of Materials Press, London, 1998, pp. 433–60.
207. V. Ramakrishnan, Pradip, and S. G. Malghan. Colloids Surf. A 133: 135–142 (1998).
208. J. Wang, L. Gao, J. Sun, and Q. Li. J. Colloid Interf. Sci. 213: 552–556 (1999).
209. J. Wang, and L. Gao. J. Colloid Int. Sci. 216: 436–439 (1999).
210. C. P. Huang, and E. A. Rhoads. J. Colloid Int. Sci. 131: 289–306 (1989).
211. M. Tschapek, L. Tcheichvili, and C. Wasowski. Clay Miner. 10: 219–229 (1974).
212. J. Hong, and P. N. Pintauro Water Air Soil Pol. 86: 35–50 (1994).
213. L. B. Garrido, C. Volzone, and R. M. Torres-Sanchez. Colloids Surf. A 121: 163–171 (1997).
214. L. Madrid, E. Diaz, F. Cabrera, and P. de Arambarri. J. Soil Sci. 34: 57–67 (1983).
215. T. F. Tadros, and J. Lyklema. J. Electroanal. Chem. 22: 9–17 (1969).
216. F. Grieser, R. N. Lamb, G. R. Wiese, D. E. Yates, R. Cooper, and T. W. Healy. Radiat. Phys. Chem. 23: 43–48 (1984).
217. A. L. Dawidowicz, W. Janusz, J. Szczypa, and A. Waksmundzki. J. Colloid Interf. Sci. 115: 555–558 (1987).

218. G. P. Golub, M. P. Sidorova, and D. A. Fridrikhsberg. Kolloid. Zh. 51: 987–989 (1989).
219. J. Szczypa, A. L. Dawidowicz, R. Sprycha, and P. Golkiewicz. J. Radioanal. Nucl. Chem. 129: 171–180 (1989).
220. W. Janusz, A. L. Dawidowicz, and J. Szczypa. J. Mater. Sci. 26: 4865–4868 (1991).
221. M. G. Cattania, S. Ardizzone, C. L. Bianchi, and S. Carella. Colloids Surf. A 76: 233–240 (1993).
222. J. Pradhan, S. Das, and R. S. Thakur. J. Colloid Interf. Sci. 217: 137–141 (1999).
223. M. Escudey, and G. Galindo. J. Colloid Int. Sci. 93: 78–83 (1983).
224. K. Sakurai, Y. Ohdate, and K. Kyuma. Soil Sci. Plant Nutr. 35: 21–31 (1989).
225. R. M. Torres-Sanchez, and E. L. Tavani. J. Thermal Anal. 41: 1129–1139 (1994).
226. L. Vordonis, C. Kordulis, and A. Lycourghiotis. J. Chem. Soc. Faraday Trans. I 84: 1593–1601 (1988).
227. E. P. Katsanis, and E. Matijevic. Colloid Polym. Sci. 261: 255–264 (1983).
228. R. M. Cornell, A. M. Posner, and J. P. Quirk. Coll. Polym. Sci. 261: 137–142 (1983).
229. S. Subramanian, J. S. Noh, and J. A. Schwarz. J. Catalysis 114: 433–439 (1988).
230. G. D. Parfitt. Croat. Chem. Acta. 45: 189–194 (1973).
231. M. Ocana, and A. R. Gonzalez-Elipe. Colloids Surf. A 157: 315–324 (1999).
232. A. Garg, and E. Matijevic. Langmuir 4: 38–44 (1988).
233. M. F. Azizian, and P. O. Nelson. In Adsorption of Metals by Geomedia (E. Jenne, ed.), Academic Press, NY., 1998 pp. 165–180.
234. J. P. Loveland, J. N. Ryan, G. L. Amy, and R. W. Harvey. Colloids Surf. A 107: 205–221 (1996).
235. A. Giatti, and L. K. Koopal. J. Electroanal. Chem. 352: 107–118 (1993).
236. I. ul Haq, and E. Matijevic. J. Colloid Interf. Sci. 192: 104–113 (1997).
237. P. Siviglia, A. Daghetti, and S. Trasatti. Colloids Surfaces 7: 15–27 (1983).
238. D. N. Furlong, K. S. W. Sing, and G. D. Parfitt. J. Colloid Int. Sci. 69: 409–419 (1979).
239. X. C. Guo, and P. Dong. Langmuir 15: 5535–5540 (1999).
240. R. A. Overbeek, E. J. Bosma, D. W. H. de Blauw, A. J. van Dillen, H. G. Bruil, and J. W. Geus. Applied Catal. A 163: 129–144 (1997).
241. R. C. Plaza, J. D. G. Duran, A. Quirantes, M. J. Ariza, and A. V. Delgado. J. Colloid Interf. Sci. 194: 398–407 (1997).
242. N. Kawahashi, and E. Matijevic. J. Colloid Int. Sci. 138: 534–542 (1990).
243. D. S. Cicerone, A. E. Regazzoni, and M. A. Blesa. J. Colloid Int. Sci. 154: 423–433 (1992).
244. A. Pierre, J. M. Lamarche, R. Mercier, A. Foissy and J. Persello. J. Dispersion Sci. Technol. 11: 611–635 (1990).
245. T. Foxall, G. C. Peterson, H. M. Rendall, and A. L. Smith. J. Chem. Soc. Faraday Trans. 1 75: 1034–1039 (1979).
246. S. K. Mishra. Int. J. Miner. Proc. 5: 69–83 (1978).
247. J. O. Amankonah, and P. Somasundaran. Colloids Surf. 15: 335–353 (1985).
248. D. W. Thompson, and P. G. Pownall. J. Colloid Interf. Sci. 131: 74–82 (1989).
249. M. Kosmulski, and A. Plak, unpublished data.
250. M. J. Dekkers, and M. A. A. Schoonen. Geochim. Cosmochim. Acta 58: 4147–4153 (1994).
251. Y. F. Nicolau, and J. C. Menard. J. Colloid Interf. Sci. 148: 551–570 (1992).
252. E. Matijevic, and D. M. Wilhelmy. J. Colloid Int. Sci. 86: 476–484 (1982).
253. S. W. Park, and C. P. Huang. J. Colloid Int. Sci. 117: 431–441 (1987).
254. R. Williams, and M. E. Labib. J. Colloid Int. Sci. 106: 251–254 (1985).
255. M. S. Moignard, R. O. James, and T. W. Healy. Aust. J. Chem. 30: 733–740 (1977).
256. R. J. Pugh, and K. Tjus. J. Colloid Int. Sci. 117: 231–241 (1987).
257. Y. H. Hsieh, and S. Tokunaga. C. P. Huang. Colloids Surf. 53: 257–274 (1991).

258. R. J. Pugh, and L. Bergstrom. Colloids Surf. 19: 1–20 (1986).
259. S. Kelebek, and G. W. Smith. Colloids Surf. 40: 137–143 (1989).
260. S. Acar, and P. Somasundaran. Miner. Eng. 5: 27–40 (1992).
261. D. Tsamouras, E. Dalas, S. Sakkopoulos, and P. G. Koutsoukos. Langmuir 15: 8018–8024 (1999).
262. J. C. Liu, and C. P. Huang. Langmuir 8: 1851–1856 (1992).
263. D. Fornasiero, V. Eijt, and J. Ralston. Colloids Surf. 62: 63–73 (1992).
264. K. K. Das, Pradip, and K. A. Natarajan. J. Colloid Interf. Sci. 196: 1–11 (1997).
265. J. Bebie, M. A. A. Schoonen, M. Fuhrmann, and D. R. Strongin. Geochim. Cosmochim. Acta 62: 633–642 (1998).
266. A. K. Helmy, and E. A. Ferreiro. Z. phys. Chem. (Leipzig) 257: 881–892 (1976).
267. W. J. McLaughlin, J. L. White, and S. L. Hem. Drug Dev. Ind. Pharm. 18: 2081–2094 (1992).
268. W. J. McLaughlin, J. L. White, and S. L. Hem. J. Colloid Interf. Sci. 157: 113–123 (1993).
269. P. Somasundaran, and G. E. Agar. J. Colloid Interf. Sci. 24: 433–440 (1967).
270. Y. C. Huang, F. M. Fowkes, T. B. Lloyd, and N. D. Sanders. Langmuir 7: 1742–1748 (1991).
271. F. Rashchi, Z. Xu, and J. A. Finch. Colloids Surf. A 132: 159–171 (1998).
272. C. Biegler, and M. R. Houchin. Colloids Surf. 21: 267–278 (1986).
273. L. Charlet, P. Wersin, and W. Stumm. Geochim. Cosmochim. Acta 54: 2329–2336 (1990).
274. Q. Y. Song, F. Xu, and S. C. Tsai. Int. J. Miner. Proc. 34: 219–229 (1992).
275. J. J. Predali, and J. M. Cases. J. Colloid Int. Sci. 45: 449–458 (1973).
276. M. C. Fuerstenau, S. A. Olivas, R. Herrera-Urbina, and K. N. Han. Int. J. Miner. Proc. 20: 73–85 (1987).
277. R. Herrera-Urbina, and D. W. Fuerstenau. Colloids Surf. A 98: 25–33 (1995).
278. D. Fornasiero, F. Li, and J. Ralston. J. Colloid Interf. Sci. 164: 345–354 (1994).
279. R. Sprycha, J. Jablonski, and E. Matijevic. J. Colloid Interf. Sci. 149: 561–568 (1992).
280. J. Szczypa, W. Janusz, and M. Szymula. Fizykoch. Probl. Miner. 12: 101–114 (1980).
281. P. L. Brown, G. R. Erickson, and J. V. Evans. Colloids Surf. 62: 11–21 (1992).
282. N. Jaffrezic Renault, A. De, P. Clechet, and A. Maaref. Colloids Surf. 36: 59–68 (1989).
283. S. Mustafa, A. Naeem, S. Murtaza, N. Rehana, and S. Y. Samad. J. Colloid Interf. Sci. 220: 63–74 (1999).
284. S. K. Nicol, and A. J. Clarke. In Proc. 2nd Int. Symp. on Composition Properties and Fundamental Structure of Tooth Enamel, London 1969, quoted after L. C. Bell, A. M. Posner, and J. P. Quirk. J. Colloid Int. Sci. 42: 250–261 (1973).
285. L. C. Bell, A. M. Posner, and J. P. Quirk. J. Colloid Int. Sci. 42: 250–261 (1973).
286. R. Rautiu, D. A. White, S. A. Adeleye, and L. Adkins. Hydrometallurgy 35: 361–374 (1994).
287. R. Drot, C. Lindecker, B. Fourest, and E. Simoni. New J. Chem. 22: 1105–1109 (1998).
288. B. Fourest, N. Hakem, and R. Guillaumont. Radiochim. Acta 66/67: 173–179 (1994).
289. M. Balastre, J. Persello, A. Foissy, and J. F. Argiller. J. Colloid Interf. Sci. 219: 155–162 (1999).
290. F. Gonzalez-Caballero, M. A. Cabrerizo, J. M. Bruque, and A. Delgado. J. Colloid Interf. Sci. 126: 367–370 (1988).
291. A. W. M. de Laat, G. L. T. van den Heuvel, and M. R. Böhmer. Colloids Surf. A 98: 61–71 (1995).
292. U. Paik, V. A. Hackley, S. C. Choi, and Y. G. Jung. Colloids Surf. A 135: 77–88 (1998).
293. R. Arnold, and L. J. Warren. J. Colloid Int. Sci. 47: 134–144 (1974).
294. L. A. Perez-Maqueda, and E. Matijevic. Chem. Mater. 10: 1430–1435 (1998).

295. A. Koliadima, L. A. Perez-Maqueda, and E. Matijevic. Langmuir 13: 3733–3736 (1997).
296. W. F. Bleam. J. Colloid Interf. Sci. 159: 312–318 (1993).
297. D. A. Sverjensky, and N. Sahai. Geochim. Cosmochim. Acta 60: 3773–3797 (1996).
298. W. Gordy and W. J. Thomas. J. Chem. Phys. 24: 439–444 (1956).
299. H. Schott. J. Pharmac. Sci. 66: 1548–1550 (1977).
300. S. I. Pechenyuk. Thesis, quoted after S. I. Pechenyuk. Usp. Khimii 61: 711–733 (1992).
301. T. W. Healy, and D. W. Fuerstenau. J. Colloid Sci. 20: 376–386 (1965).
302. S. K. Milonjic, A. L. Ruvarac, and M. V. Susic. Thermochim. Acta 11: 261–266 (1975).
303. W. Rudzinski, R. Charmas, and S. Partyka. Langmuir 7: 354–362 (1991).
304. H. Watanabe, J. Seto and Y. Nishiyama. Bull. Chem. Soc. Jpn. 66: 2751–2753 (1993).
305. W. Rudzinski, R. Charmas, and S. Partyka. Colloids Surf. 70: 111–130 (1993).
306. D. A. Griffiths, and D. W. Fuerstenau. J. Colloid Interf. Sci. 80: 271–283 (1981).
307. D. Balzer, and H. Lange. Colloid Polym. Sci. 255: 140–152 (1977).
308. H. Watanabe, and J. Seto. Bull. Chem. Soc. Jpn. 61: 3067–3072 (1988).
309. S. Rohrsetzer, I. Paszli, F. Csempesz, and S. Ban. Coll. Polym. Sci. 270: 1243–1251 (1992).
310. M. Tschapek, S. G. DeBussetti, and G. Pozzo Ardizzi. Electroanal. Chem. Int. Electrochem. 52: 304–309 (1974).
311. A. Kasseh, and E. Keh. J. Colloid Interf. Sci. 197: 360–369 (1998).
312. A. S. Rao. J. Disp. Sci. Technol. 8: 457–476 (1987).
313. J. C. Chang, F. E. Lange, and D. S. Pearson. J. Am. Ceram. Soc. 77: 19–26 (1994).
314. Y. Hirata, and K. Onoue. Eur. J. Solid State Inorg. Chem. 32: 663–672 (1995).
315. E. P. Luther, J. A. Yanez, G.V. Franks F. F. Lange, and D. S. Pearson. J. Am. Ceram. Soc. 78: 1495–1500 (1995).
316. V. Ramakrishnan, Pradip, and S. G. Malghan. J. Am. Ceram. Soc. 79: 2567–2576 (1996).
317. P. C. Hidber, T. J. Graule, and L. J. Gauckler. J. Am. Ceramic Soc. 79: 1857–1867 (1996).
318. P. Gil, M. Almeida, and H. M. M. Diz. Brit. Ceram. Trans. 95: 254–257 (1996).
319. M. Colic, G. V. Franks, M. L. Fisher, and F. F. Lange. Langmuir 13: 3129–3135 (1997).
320. S. B. Johnson, A. S. Russel, and P. J. Scales. Colloids Surf. A 141: 119–130 (1998).
321. M. Colic, M. L. Fisher, and D. W. Fuerstenau. Colloid Polym. Sci. 276: 72–80 (1998).
322. D. J. Kim, H. Kim, and J. K. Lee. J. Mater. Sci. 33: 2931–2935 (1998).
323. S. B. Johnson, G. V. Franks, P. J. Scales, and T. W. Healy. Langmuir 15: 2844–2853 (1999).
324. G. V. Franks, S. B. Johnson, P. J. Scales, D. V. Boger, and T. W. Healy. Langmuir 15: 4411–4420 (1999).
325. A. L. Costa, C. Galassi, and R. Greenwood, J. Colloid Interf. Sci. 212: 350–356 (1999).
326. J. R. Feldkamp, D. N. Shah, S. L. Meyer, J. L. White, and S. L. Hem. J. Pharm. Sci. 70: 638–640 (1981).
327. J. Addai-Mensah, J. Dawe, R. Hayes, C. Prestidge, and Ralston. J. Colloid Interf. Sci. 203: 115–121 (1998).
328. L. Garcell, M. P. Morales, M. Andres-Verges, P. Tartaj, and C. J. Serna. J. Colloid Interf. Sci. 205: 470–475 (1998).
329. A. A. Zaman, B. M. Moudgil, A. L. Fricke, and H. El-Shall. J. Rheol. 40: 1191–1210 (1996).
330. P. Mikulasek, R. J. Wakeman, and J. Q. Marchant. Chem. Eng. J. 67: 97–102 (1997).
331. M. Kosmulski, J. Gustafsson, and J. B. Rosenholm. J. Colloid Int. Sci. 209: 200–206 (1999).

332. M. Kosmulski, J. Gustafsson, and J. B. Rosenholm. Colloid Polym. Sci. 277: 550–556 (1999).
333. Y. K. Leong, P. J. Scales, T. W. Healy, D. V. Boger and R. Buscall. J. Chem. Soc. Faraday Trans. 89: 2473–2478 (1993).
334. Y. K. Leong, P. J. Scales, T. W. Healy, and D. V. Boger. Colloids Surf. A 95: 43–52 (1995).
335. M. J. Solomon, T. Saeki, M. Wan, P. J. Scales, D. V. Boger, and H. Usui. Langmuir 15: 20–26 (1999).
336. B. Rand, and I. E. Melton. J. Colloid Interf. Sci. 60: 308–320 (1977).
337. M. Benna, N. Kbir-Ariguib, A. Magnin, and F. Bergaya. J. Colloid Interf. Sci. 218: 442–455 (1999).
338. A. Cerpa, M. T. Garcia-Gonzalez, P. Tartaj, J. Requena, L. Garcell, and C. J. Cerna. Clays Clay Min. 47: 515–521 (1999).
339. S. Ardizzone, M. G. Cattania, and P. Lugo. Electrochim. Acta 39: 1509–1517 (1994).
340. M. Casamassima, E. Darque-Ceretti, A. Etcheberry, and A. Aucouturier. Appl. Surface Sci. 52: 205–213 (1991).
341. R. N. Spitz, J. E. Barton, M. A. Barteau, R. H. Staley, and A. W. Sleight. J. Phys. Chem. 90: 4067–4075 (1986).
342. T. Yamanaka, and K. Tanabe. J. Phys. Chem. 80: 1723–1727 (1976).
343. M. I. Zaki, A. K. H. Nohman, G. A. M. Hussein, and Y. E. Nashed. Colloids Surf. A 99: 247–253 (1995).
344. T. Hamieh, M. Nardin, M. Raguel-Lescouët, H. Haidara, and J. Schultz. Colloids Surf. A 125: 155–161 (1997).
345. K. Meguro, and K. Esumi. In Acid-Base Interactions, (K. L. Mittal, and H. R. Anderson, eds.), VSP 1991, pp. 117–134.
346. W. M. Mullins, and B. L. Averbach. Surface Sci. 206: 41–51 (1988).
347. A. Carre, F. Roger, and C. Varinot, J. Colloid Int. Sci. 154: 174–183 (1992).
348. G. Deo, and I. E. Wachs, J. Phys. Chem. 95: 5889–5895 (1991).
349. H. Hu, I. E. Wachs, and S. R. Bare. J. Phys. Chem. 99: 10897–10910 (1995).
350. G. H. Kelsall, Y. Zhu, and H. A. Spikes. J. Chem. Soc. Faraday Trans. 89: 267–272 (1993).
351. Y. Hirata, and H. Wakita. J. Ceram. Soc. Jpn. 107: 303–307 (1999).
352. K. F. Hayes, G. Redden, W. Ela, and J. O. Leckie. J. Colloid Int. Sci. 142: 448–469 (1991).
353. G. Furrer, C. Ludwig, and P. W. Schindler. J. Colloid Interf. Sci. 149: 56–67 (1992).
354. B. V. Zhmud and J. Sonnefeld. J. Chem. Soc. Faraday Trans. 91: 2965–2970 (1995).
355. P. Hesleitner, D. Babic, N. Kallay, and E. Matijevic. Langmuir 3: 815–820 (1987).
356. N. Sahai, and D. A. Sverjensky. Geochim. Cosmochim. Acta 61: 2801–2826 (1997).
357. L. J. G. Fokkink, A. de Keizer, J. M. Kleijn and J. Lyklema. J. Electroanal. Chem. 208: 401–403 (1986).
358. J. Lyklema. Croat. Chem. Acta 43: 249–260 (1971).
359. M. D. Petkovic, S. K. Milonjic, and V. T. Dondur. Sep. Sci. Tech. 29: 627–638 (1994).
360. B. Venkataramani, and A. R. Gupta. Indian J. Chem. 27 A: 290–296 (1988).
361. R. W. Lahman. Clays Clay Min. 24: 320–326 (1976).
362. J. Xue, and P. M. Huang. Geoderma 64: 343–356 (1995).
363. R. Paterson, and H. Rahman, J. Colloid Interf. Sci. 94: 60–69 (1983).
364. S. Ardizzone, D. Lettieri, and S. Trasatti. J. Electroanal. Chem. 146: 431–437 (1983).
365. O. A. Petrii, and A. Vitins. Elektrokhimiya 27: 461–476 (1991).
366. O. A. Petrii. Electrochem. Acta 41: 2307–2312 (1996).
367. A. Vitins, O. A. Petri, B. B. Damaskin, Y. A. Ermakov, A. Grzejdziak, E. L. Kolomnikova, S. A. Sukhishvili, and A. A. Yaroslavov. Elektrokhimiya 28: 404–413 (1991).

368. J. W. Murray. J. Colloid Int. Sci. 46: 357–371 (1974).
369. L. S. Balistrieri, and J. W. Murray. Geochim. Cosmochim. Acta 46: 1041–1052 (1982).
370. K. Ooi, Y. Miyai, and S. Katoh. Solv. Extr. Ion Exch. 5: 561–572 (1987).
371. K. Ooi, Y. Miyai, S. Katoh, H. Maeda, and M. Abe. Bull. Chem. Soc. Jpn. 61: 407–411 (1988).
372. H. Tamura, N. Katayama, and R. Furuichi. Environ. Sci. Technol. 30: 1198–1204 (1996).
373. Y. Ran, J. Fu, R. J. Gilkes, and R. W. Rate. Sci. China D 42: 172–181 (1999).
374. S. M. Ahmed. Can. J. Chem. 44: 1663–1670 (1966).
375. R. P. Abendroth. J. Colloid Interf. Sci. 34: 591–596 (1970).
376. A. Komura, K. Hatsutori, and H. Imanaga. Nippon Kagaku Kaishi 779–784 (1978).
377. H. L. Michael, and D. J. Williams. J. Electroanal. Chem. 179: 131–139 (1984).
378. D. B. Kent, and M. Kastner. Geochim. Cosmochim. Acta 49: 1123–1136 (1985).
379. S. K. Milonjic. Colloids Surf. 23: 301–312 (1987).
380. H. Sonntag, V. Itschenskij, and R. Koleznikova. Croat. Chim. Acta 60: 383–393 (1987).
381. C. P. Schulthess, and D. L. Sparks. Soil Sci. Soc. Am. J. 53: 366–373 (1989).
382. M. P. Rigney, E. F. Funkenbusch, and P. W. Carr. J. Chromatogr. 499: 291–304 (1990).
383. W. H. Casey, A. C. Lasaga, and G. V. Gibbs. Geochim. Cosmochim. Acta 54: 3369–3378 (1990).
384. S. K. Milonjic. Colloids Surf. 63: 113–119 (1992).
385. W. H. Casey. J. Colloid Int. Sci. 163: 407–419 (1994).
386. J. Sonnefeld, A. Gobel, and W. Vogelsberger. Colloid Polym. Sci. 273: 926–931 (1995).
387. K. M. Spark, B. B. Johnson, and J. D. Wells. Eur. J. Soil Sci. 46: 621–631 (1995).
388. T. P. Goloub, and L. K. Koopal. Langmuir 13: 673–681 (1997).
389. M. Löbbus, W. Vogelsberger, J. Sonnefeld, and A. Seidel. Langmuir 14: 4386–4396 (1998).
390. D. B. Ward, and P. V. Brady. Clays Clay Min. 46: 453–465 (1998).
391. K. Csoban, and P. Joo. Colloids Surf. A 151: 97–112 (1999).
392. F. A. Rodrigues, P. J. M. Monteiro, and G. Sposito. J. Colloid Interf. Sci. 211: 408–409 (1999).
393. M. P. Sidorova, H. Zastrow, L. E. Ermakova, N. F. Bogdanova, and V. M. Smirnov. Kolloid. Zh. 61: 113–117 (1999).
394. E. Yu. Chardymskaya, M. P. Sidorova. O. S. Semenova, and M. D. Khanin. Kolloidn. Zh. 51: 607–610 (1989).
395. J. T. Webb, P. D. Bhatnagar, and D. G. Williams. J. Colloid Interf. Sci. 49: 346–361 (1974).
396. S. K. Milonjic, Z. E. Ilic, and M. M. Kopecni. Colloids Surf. 6: 167–174 (1983).
397. C. Contescu, J. Jagiello, and J. A. Schwarz. Langmuir 9: 1754–1765 (1993).
398. Q. Du, Z. Sun, W. Forsling, and H. Tang. J. Colloid Interf. Sci. 187: 221–231 (1997).
399. W. Liu, Z. Sun, W. Forsling, Q. Du, and H. Tang. J. Colloid Interf. Sci. 219: 48–61 (1999).
400. L. Wang, A. Maes, P. de Canniere, and J. van der Lee. Radiochim. Acta 82: 233–237 (1999).
401. P. V. Brady, R. T. Cygan, and K. L. Nagy. J. Colloid Interf. Sci. 183: 356–364 (1996).
402. P. V. Brady, R. T. Cygan, and K. L. Nagy. In Adsorption of Metals by Geomedia (E. A. Jenne, ed.), Academic Press, San Diego 1998. pp. 371–382.
403. B. Baeyens, and M. H. Bradbury. J. Contam. Hydrol. 27: 199–222 (1997).
404. T. R. Holm, and X. F. Zhu. J. Contam. Hydrol. 16: 271–287 (1994).
405. J. Szczypa, I. Kajdewicz, and M. Kosmulski. J. Colloid Interf. Sci. 137: 157–162 (1990).
406. L. E. Ermakova, M. Sidorova, N. Jura, and I. Savina. J. Membr. Sci. 131: 125–141 (1997).

407. J. Lyklema. Fundamentals of Interface and Colloid Science, Vol. I. Academic Press, London 1991, pp. 5.75–5.76.
408. M. J. Avena, O. R. Camara, and C. P. dePauli. Colloids Surf. 69: 217–228 (1993).
409. R. W. O'Brien, and L. R. White. J. Chem. Soc. Faraday Trans. II 74: 1607 (1978).
410. C. Grosse and V. N. Shilov. J. Colloid Interf. Sci. 211: 160–170 (1999).
411. A. S. Dukhin, V. N. Shilov, H. Ohshima, and P. J. Goetz. Langmuir 15: 6692–6706 (1999).
412. E. Matijevic. Acc. Chem. Res. 14: 22–29 (1981).
413. E. Matijevic. Pure Appl. Chem. 50: 1193–1210 (1978).
414. N. Spanos, and A. Lycourghiotis. J. Colloid Interf. Sci. 171: 306–318 (1995).
415. N. H. Sagert, C. H. Ho, and N. H. Miller. J Colloid Interf. Sci. 130: 283–287 (1989).
416. F. Dumont, D. van Tan, and A. Watillon. J. Colloid Int. Sci. 55: 678–687 (1976).
417. C. S. Luo, and S. D. Huang. Separ. Sci. Technol. 28: 1253–1271 (1993).
418. Z. Zhang, C. Boxal, and G. H. Kelsall. Colloids Surf. A 73: 145–163 (1993).
419. K. Subramaniam, S. Yiacoumi, and C. Tsouris. Separ. Sci. Technol. 34: 1301–1318 (1999).
420. S. V. Krishnan, and I. Iwasaki. Environ. Sci. Technol. 20: 1224–1229 (1986).
421. T. Morimoto, H. Nakahata, and S. Kittaka. Bull. Chem. Soc. Jpn. 51: 3387–3388 (1978).
422. R. J. Crawford, I. H. Harding, and D. E. Mainwaring. Langmuir 9: 3050–3056 (1993).
423. R. J. Crawford, I. H. Harding, and D. E. Mainwaring. J. Colloid Interf. Sci. 181: 561–570 (1996).
424. D. W. Fuerstenau, and K. Osseo-Asare. J. Colloid Interf. Sci. 118: 524–542 (1987).
425. P. Jayawera, S. Hettiarachchi, and H. Ocken. Colloids Surf. A 85: 19–27 (1994).
426. S. Ardizzone, A. Daghetti, L. Franceschi, and S. Trasatti. Colloids Surf. 35: 85–96 (1989).
427. R. O. James, and T. W. Healy. J. Colloid Int. Sci. 40: 53–64 (1972).
428. M. G. MacNaughton, and R. O. James. J. Colloid Interf. Sci. 47: 431–440 (1974).
429. J. Jednacak, V. Pravdic, and W. Haller. J. Colloid Int. Sci. 49: 16–23 (1974).
430. W. Smit, and H. S. Stein. J. Colloid Interf Sci. 60: 299–307 (1977).
431. W. Smit, C. L. M. Holten, H. N. Stein, J. J. M. de Goeij, and H. M. J. Theelen. J. Coll. Interf. Sci. 63: 120–128 (1978).
432. P. J. Scales, F. Grieser, and T. W. Healy. Langmuir 6: 582–589 (1990).
433. S. Nishimura, H. Tateyama, K. Tsunematsu, and K. Jinnai. J. Colloid Interf. Sci. 152: 359–367 (1992).
434. R. T. Klingbiel, H. Coll, R. O. James, and J. Texter. Colloids Surf. 68: 103–109 (1992).
435. H. M. Jang, and S. H. Lee. Langmuir 8: 1698–1708 (1992).
436. P. J. Scales, and E. Jones. Langmuir 8: 385–389 (1992).
437. P. J. Scales, F. Grieser, T. W. Healy, L.R. White and D.Y.C. Chan Langmuir 8: 965–974 (1992).
438. G. M. Litton, and T. M. Olson. Colloids Surf. A 87: 39–48 (1994).
439. D. E. Dustan. J. Chem. Soc. Faraday Trans. 90: 1261–1263 (1994).
440. I. Larson, C. J. Drummond, D. Y. C. Chan, and F. Grieser. J. Phys. Chem. 99: 2114–2118 (1995).
441. W. Janusz. Ads. Sci. Technol. 14: 151–161 (1996).
442. M. Yasrebi, M. Ziomek-Moroz, W. Kemp, and D. H. Sturgis. J. Am. Ceram. Soc. 79: 1223–1227 (1996).
443. P. G. Hartley, I. Larson, and P. J. Scales. Langmuir 13: 2207–2214 (1997).
444. I. Sondi, O. Milat, and V. Pravdic. J. Colloid Interf. Sci. 189: 66–73 (1997).
445. N. F. Bogdanova, M. P. Sidorova, L. E. Ermakova, and I. A. Savina. Kolloid. Zh. 59: 452–459 (1997).

446. L. E. Ermakova, M. P. Sidorova, and V. M. Smirnov. Kolloid. Zh. 59: 563–565 (1997).
447. M. P. Sidorova, N. F. Bogdanova, I. E. Ermakova, P. V. Bobrov. Kolloid. Zh. 59: 568–571 (1997).
448. P. B. Wright, A. S. Lister, J. G. Dorsey. Anal. Chem. 69: 3251–3259 (1997).
449. I. Larson, and R. J. Pugh. J. Colloid Interf. Sci. 208: 399–404 (1998).
450. G. Tari, I. Bobos, C. S. F. Gomes, and J. M. F. Ferreira. J. Colloid Interf. Sci. 210: 360–366 (1999).
451. L. Yezek, R. L. Rowell, M. Larwa, and E. Chibowski. Colloids Surf. A 141: 67–72 (1998).
452. A. Fernandez-Nieves, and F. J. de las Nieves. Colloids Surf. A 148: 231–243 (1999).
453. N. Spanos, and P. G. Koutsoukos. J. Colloid Interf. Sci. 214: 85–90 (1999).
454. Ye. A. Nechayev, and V. N. Sheyin. Kolloidn. Zh. 41: 361 363 (1979).
455. S. Mathur, and B. M. Moudgil. J. Colloid Interf. Sci. 196: 92–98 (1997).
456. H. F. A. Trimbos, and H. N. Stein. J. Colloid Int. Sci. 77: 397–406 (1980).
457. M. Kagawa, Y. Syono, Y. Imamura, and S. Usui. J. Am. Ceram. Soc. 69: C50–C51 (1986).
458. S. N. Omenyi, B. J. Herren, R. S. Snyder, and G. V. F. Seaman. J. Colloid Int. Sci. 110: 130–136 (1986).
459. T. Mehrian, A. de Keizer, and J. Lyklema. Langmuir 7: 3094–3098 (1991).
460. J. Wang, B. Han, M. Dai, H. Yan, Z. Li, and R. K. Thomas. J. Colloid Interf. Sci. 213: 596–601 (1999).
461. J. S. Lyons, D. N. Furlong, and T. W. Healy. Aust. J. Chem. 34: 1177–1187 (1981).
462. J. A. Yopps, and D. W. Fuerstenau. J. Colloid Sci. 19: 61–71 (1964).
463. A. W. Gibb, and L. K. Koopal. J. Colloid Interf. Sci. 134: 122–138 (1990).
464. J. Depasse. J. Colloid Interf. Sci. 194: 260–262 (1997).
465. K. Ooi, Y. Miyai, and S. Katoh. Separ. Sci. Technol. 22: 1779–1789 (1987).
466. K. Ooi, Y. Miyai, and S. Katoh. Separ. Sci. Technol. 21: 755–766 (1986).
467. S. Y. Shiao, and R. E. Meyer, J. Inorg. Nucl. Chem. 43: 3301–3307 (1981).
468. R. Sprycha. J. Colloid Interf. Sci. 96: 551–554 (1983).
469. A. R. Bowers, and C. P. Huang. J. Colloid Int. Sci. 105: 197–214 (1985).
470. D. Ballion, and N. Jaffrezic-Renault. J. Radioanal. Nucl. Chem. 92: 133–150 (1985).
471. C. P. Schulthess, and D. L. Sparks. Soil Sci. Soc. Am. J. 51: 1136–1144 (1987).
472. W. Janusz, and J. Szczypa. Mater. Sci. Forum 25–26: 427–430 (1988).
473. R. Sprycha. J. Colloid Int. Sci. 127: 12–25 (1989).
474. F. Thomas, J. Y. Bottero, and J. M. Cases. Colloids Surf. 37: 281–294 (1989).
475. M. Kosmulski. J. Colloid Interf. Sci. 135: 590–593 (1990).
476. C. P. Schulthess, and J. F. McCarthy. Soil Sci. Soc. Am. J. 54: 688–694 (1990).
477. M. Kosmulski. J. Colloid Interf. Sci. 156: 305–310 (1993).
478. M. Kosmulski. Bull. Pol. Acad. Sci. Chem. 41: 325–331 (1993).
479. M. Kosmulski. Pol. J. Chem. 67: 1831–1839 (1993).
480. M. Kosmulski. Colloids Surf. A 83: 237–243 (1994).
481. S. Mustafa, B. Dilara, Z. Neelofer, A. Naeem, and S. Tasleem. J. Colloid Interf. Sci. 204: 284–293 (1998).
482. S. B. Johnson, P. J. Scales, and T. W. Healy. Langmuir 15: 2836–2843 (1999).
483. K. Nagashima, and F. Blum. J. Colloid Interf. Sci. 217: 28–36 (1999).
484. M. D. Petkovic, S. K. Milonjic, and V. T. Dondur. Bull. Chem. Soc. Jpn. 68: 2133–2136 (1995).
485. M. Nanzyo. J. Soil Sci. 35: 63–69 (1984).
486. W. N. Rowlands, R. W. O'Brien, R. J. Hunter, and V. Patrick. J. Colloid Interf. Sci. 188: 325–335 (1997).

487. R. Wendelbow, thesis, cited after T. Hiemstra, H. Yong, and W. H. van Riemsdijk. Langmuir 15: 5942–5955 (1999).
488. B. Venkataramani, K. S. Venkateswarlu, J. Shankar and L.H. Baestle. J. Colloid Int. Sci. 76: 1–6 (1980).
489. W. Janusz. J. Colloid Interf. Sci. 145: 119–126 (1991).
490. N. Z. Misak, I. M. El-Naggar, H. B. Maghrawy, and N. S. Petro. J. Coll. Int. Sci. 135: 6–15 (1990).
491. N. Z. Misak, and I. M. El-Naggar. React. Polym. 10: 67–72 (1989).
492. G. A. Kokarev, A. F. Gubin, V. A. Kolesnikov, and S. A. Skobelev. Zh. Fiz. Khim. 59: 1660–1663 (1985).
493. N. I. Ampelogova. Radiokhimiya 25: 579–584 (1983).
494. B. Venkataramani, A. R. Gupta, and R. M. Iyer. J. Radioanal. Nucl. Chem. 96: 129–136 (1985).
495. H. Catalette, J. Dumonceau, and P. Ollar. J. Contamin. Hydrol. 35: 151–159 (1998).
496. A. Breeuwsma, and J. Lyklema. Disc. Faraday Soc. 52: 324–333 (1972).
497. S. Chibowski. Pol. J. Chem. 59: 1193–1199 (1985).
498. M. Colic, D. W. Feurstenau, N. Kallay, and E. Matijevic. Colloids Surf. 59: 169–185 (1991).
499. H. Amhamdi, F. Dumont, and C. Buess-Herman. Colloids Surf. A 125: 1–3 (1997).
500. W. Janusz, A. Sworska, and J. Szczypa. Colloids Surf. A 149: 421–426 (1999).
501. F. J. Hingston, A. M. Posner, and J. P. Quirk. J. Soil Sci. 23: 177–192 (1972).
502. M. I. Tejedor-Tejedor, and M. A. Anderson. Langmuir 2: 203–210 (1986).
503. R. S. Rundberg, Y. Albinsson, and K. Vannerberg. Radiochim. Acta 66/67: 333–339 (1994).
504. W. Janusz. Ads. Sci. Technol. 18: 117–134 (2000).
505. Y. Inoue, O. Tochiyama, H. Yamazaki, and A. Sakurada. J. Radioanal. Nucl. Chem. 124: 361–382 (1988).
506. J. M. Kleijn, and J. Lyklema. J. Colloid Interf. Sci. 120: 511–522 (1987).
507. D. N. Strazhesko, V. B. Strelko, V. N. Belyakov, and S. C. Rubanik. J. Chromat. 102: 191–195 (1974).
508. E. A. Nechaev, and V. P. Romanov. Kolloid. Zh. 36: 1095–1100 (1974).
509. Yu. G. Frolov, S. K. Milonich, and V. L. Razin. Kolloid. Zh. 41: 516–521 (1979).
510. G. Peschel, and P. Ludwig. Ber. Bunsenges. Phys. Chem. 91: 536–541 (1987).
511. M. Kosmulski, and E. Matijevic. Colloid Polym. Sci. 270: 1046–1048 (1992).
512. J. C. J. van der Donck, G. E. J. Vaessen, and H. N. Stein. Langmuir 9: 3553–3557 (1993).
513. A. Kitamura, T. Yamamoto, H. Moriyama, and S. Nishikawa. J. Nucl. Sci. Techn. 33: 840–845 (1996).
514. M. Kosmulski. J. Colloid Interf. Sci. 208: 543–545 (1998).
515. J. Depasse. J. Colloid Interf. Sci. 220: 174–176 (1999).
516. N. Jaffrezic-Renault, N. Karisa, H. Andrade-Martins, and N. Deschamps. Radiochem. Radioanal. Letters 37: 257–266 (1979).
517. N. Z. Misak, N. S. Petro, and E. S. I. Shabana. Colloids Surf. 55: 289–296 (1991).
518. R. Rautiu, and D. A. White. Solv. Extr. Ion Exch. 14: 721–738 (1996).
519. B. Venkataramani, K. S. Venkateswarlu, and J. Shankar. Proc. Indian Acad. Sci. (Chem). 87 A: 409–414 (1978).
520. H. S. Mahal, and B. Venkataramani. Indian J. Chem. 37A: 993–1001 (1998).
521. N. Jaffrezic-Renault, and H. Andrade-Martins. J. Radioanal. Chem. 55: 307–316 (1980).
522. A. Foissy, A. M. Pandou, J. M. Lamarche, and N. Jaffrezic-Renault. Colloids Surf. 5: 363–368 (1982).
523. R. Sprycha. J. Colloid Interf. Sci. 102: 173–185 (1984).

524. W. Janusz. Mater. Chem. Phys. 24: 39–50 (1989).

525. F. Dumont, J. Warlus, and A. Watillon. J. Colloid Int. Sci. 138: 543–554 (1990).

526. W. Janusz. Pol. J. Chem. 68: 1871–1880 (1994).

527. M. Kosmulski, and J. B. Rosenholm. J. Phys. Chem. 100: 11681–11687 (1996).

528. W. Janusz, I. Kobal, A. Sworska, and J. Szczypa. J. Colloid Interf. Sci. 187: 381–387 (1997).

529. W. Rudzinski, R. Charmas, W. Piasecki, A. J. Groszek, F. Thomas, F. Villieras, B. Prelot, and J. M. Cases. Langmuir 15: 5921–5931 (1999).

530. W. Janusz, A. Sworska, and J. Szczypa. Colloids Surf. A 152: 223–233 (1999).

531. M. Kosmulski, S. Durand-Vidal, J. Gustafsson, and J. B. Rosenholm. Colloids Surf. A 157: 245–259 (1999).

532. F. Dumont, P. Verbeiren, and C. Buess-Herman. Colloids Surf. A 154: 149–156 (1999).

533. A. J. Shor, K. A. Kraus, W. T. Smith, and J. S. Johnson. J. Phys. Chem 72: 2200–2206 (1968).

534. Y. Inoue, and H. Yamazaki. Bull. Chem. Soc. Jpn. 60: 891–897 (1987).

535. W. Janusz. J. Radioanal. Nucl. Chem. 125: 393–401 (1989).

536. M. A. Osman, C. Moor, W. R. Caseri, and U. W. Suter. J. Colloid Interf. Sci. 209: 232–239 (1999).

537. E. M. Mikhail, H. F. Ghoneimy, and N. Z. Misak. Colloids Surf. 66: 231–239 (1992).

538. H. Yamazaki, Y. Inoue, N. Kikuchi, and H. Kurihara. Bull. Chem. Soc. Jpn. 64: 566–575 (1991).

539. I. Altug, and M. L. Hair. J. Phys. Chem. 71: 4260–4263 (1967).

540. M. A. A. Schoonen, Geochim. Cosmochim. Acta 58: 2845–2851 (1994).

541. S. Zalac, and N. Kallay. Croat. Chem. Acta. 69: 119–124 (1996).

542. W. Vogelsberger, T. Mittelbach, and A. Seidel. Ber. Bunsenges. Phys. Chem. 100: 1118–1127 (1996).

543. M. A. Blesa, A. J. G. Maroto, A. E. Regazzoni. J. Colloid Interf. Sci. 140: 287–290 (1990).

544. M. L. Machesky. In Chemical Modeling in Aqueous Systems. II (D. C. Melchior, and R. L. Basset, eds.), ACS Symp. Series 416, 1990, pp. 282–292.

545. J. Lyklema. Fundamentals of Interface and Colloid Science, Vol. II. Academic Press, London 1995, Fig. 3.61.

546. W. Rudzinski, R. Charmas, and W. Piasecki. Langmuir, 15: 8553–8557 (1999).

547. D. A. Sverjensky, and N. Sahai. Geochim. Cosmochim. Acta 62: 3703–3716 (1998).

548. A. Revil, P. A. Pezard, and P. W. J. Glover. J. Geoph. Res. B 104: 20021–20031 (1999).

549. H. van der Vlekkert, L. Bousse, and N. de Rooij. J. Coll. Int. Sci. 122: 336–345 (1988).

550. P. H. Tewari, and A. W. McLean. J. Colloid Interf. Sci. 40: 267–272 (1972).

551. P. Somasundaran, and R. D. Kulkarni. J. Colloid Interf. Sci. 45: 591–600 (1973).

552. K. Ch. Akratopulu, L. Vordonis, and A. Lycourghiotis. J. Chem. Soc. Faraday Trans. I. 82: 3697–3708 (1986).

553. S. Subramanian, J. A. Schwarz, and Z. Hejase. J. Catalysis. 117: 512–518 (1989).

554. M. L. Machesky, and P. F. Jacobs. Colloids Surf. 53: 297–314 (1991).

555. M. Kosmulski, J. Matysiak, and J. Szczypa. J. Colloid Interf. Sci. 164: 280–284 (1994).

556. P. V. Brady. Geochim. Cosmochim. Acta 58: 1213–1217 (1994).

557. C. Contescu, A. Contescu, and J. A. Schwarz. J. Phys. Chem. 98: 4327–4335 (1994).

558. P. H. Tewari, and A. B. Campbell. J. Colloid Int. Sci. 55: 531–539 (1976).

559. G. Y. Onoda, and P. L. de Bruyn. Surf. Sci. 4: 48–63 (1966).

560. M. A. Blesa, N. M. Figliolia, A. J. G. Maroto, and A. E. Regazzoni. J. Colloid Int. Sci. 101: 410–418 (1984).

561. L. G. J. Fokkink, A. de Keizer, and J. Lyklema. J. Colloid Int. Sci. 127: 116–131 (1989).

562. A de Keizer, L. G. J. Fokkink, and J. Lyklema. Colloids Surf. 49: 149–163 (1990).

563. M. Kosmulski, J. Matysiak, and J. Szczypa. Bull. Pol. Acad. Sci. Chem. 41: 333–337 (1993).
564. N. Kallay, S. Zalac, J. Culin, U. Bieger, A. Pohlmeier, and H. D. Narres. Progr. Coll. Polym. Sci. 95: 108–112 (1994).
565. S. Pivovarov. J. Colloid Interf. Sci. 206: 122–130 (1998).
566. N. Kallay, T. Preocanin, S. Zalac, H. Lewandowski, and H. D. Narres. J. Colloid Interf. Sci. 211: 401–407 (1999).
567. M. L. Machesky, and M. A. Anderson. Langmuir 2: 582–587 (1986).
568. W. A. Zeltner, E. C. Yost, M. L. Machesky, M. I. Tejedor-Tejedor, and M. A. Anderson. Geochemical Processes at Mineral Surfaces, (J. A. Davis, K. F. Hayes, eds.), ACS Symp. Series 323: 142–161 (1986).
569. B. B. Johnson. Environ. Sci. Technol. 24: 112–118 (1990).
570. D. P. Rodda, B. B. Johnson, and J. D. Wells. J. Colloid Interf. Sci. 161: 57–62 (1993).
571. D. P. Rodda, J. D. Wells, and B. B. Johnson. J. Colloid Interf. Sci. 184: 564–569 (1996).
572. M. J. Gray, M. A. Malati, and M. W. Rophael. J. Electroanal. Chem. 89: 135–140 (1978).
573. N. Kallay, and S. Zalac. Croat. Chem. Acta. 67: 467–479 (1994).
574. Y. G. Berube, and P. L. de Bruyn. J. Colloid Int. Sci. 27: 305–318 (1968).
575. S. R. Mehr, D. J. Eatough, L. D. Hansen, E. A. Lewis, and J. A. Davis. Thermochim. Acta. 154: 129–143 (1989).
576. K. C. Akratopulu, C. Kordulis, and A. Lycourghiotis. J. Chem. Soc. Faraday Trans. 86: 3437–3440 (1990).
577. N. Kallay, S. Zalac, and G. Stefanic. Langmuir 9: 3457–3460 (1993).
578. M. L. Machesky, D. J. Wesolowski, D. A. Palmer, and K. Ichiro-Hayashi. J. Colloid Interf. Sci. 200: 298–309 (1998).
579. A. Husain. J. Colloid Int. Sci. 102: 389–399 (1984).
580. S. Mustafa, S. Parveen, A. Ahmad, and D. Begum. Ads. Sci. Technol. 15: 789–802 (1997).
581. N. M. Figliolia, thesis, quoted after, M. A. Blesa, A. J. G. Maroto, and A. E. Regazzoni. J. Colloid Interf. Sci. 140: 287–290 (1990).
582. L. Ricq, A. Pierre, J. C. Reggiani, J. Pagetti, and A. Foissy. Colloids Surf. A 138: 301–308 (1998).
583. S. N. Lvov, X. Y. Zhou, and D. D. Macdonald. J. Electroanal. Chem. 463: 146–156 (1999).
584. M. Kosmulski. Colloids Surf. A 95: 81–100 (1995).
585. M. Kosmulski. In Interfacial Dynamics, (N. Kallay, ed), Elsevier New York 1999, pp. 273–312.
586. M. Kosmulski. Colloids Surf. A. 159: 277–281 (1999).
587. M. Kosmulski. In Adsorption on Silica Surfaces, (E. Papirer, ed), Marcel Dekker, NY 2000, pp. 343–368.
588. G. Wang, P. Sarkar, and P. S. Nicholson. J. Am. Cer. Soc. 80: 965–972 (1997).
589. M. Kosmulski, and A. Plak. Colloids Surf. A 149: 409–412 (1999).
590. G. F. Hudson, and S. Raghavan. Colloid Polym. Sci. 266: 77–81 (1988).
591. P. Hesleitner, N. Kallay, and E. Matijevic. Langmuir 7: 178–184; 1554 (1991).
592. E. M. Andrade, F. V. Molina, and D. Posadas. J. Colloid Interf. Sci. 215: 370–380 (1999).
593. H. F. Ghoneimy. Arab. J. Nucl. Sci. 26: 219–234 (1993).
594. Y. Xue, and S. J. Traina. Environ. Sci. Technol. 30: 3161–3166 (1996).
595. S. Spange, F. Simon, G. Heublein, H. J. Jacobasch, and M. Borner. Colloid Polym. Sci. 269: 173–178 (1991).
596. C. Schwer, and E. Kenndler. Anal. Chem. 63: 1801–1807 (1991).
597. M. Kosmulski, and E. Matijevic. Langmuir 7: 2066–2071 (1991).

598. M. Kosmulski, and E. Matijevic. Langmuir 8: 1060–1064 (1992).

599. M. Kosmulski. J. Colloid Interf. Sci. 179: 128–135 (1996).

600. H. A. Ketelson, R. Pelton, and M. A. Brook. J. Colloid Interf. Sci. 179: 600–607 (1996).

601. M. Kosmulski. J. Colloid Interf. Sci. 190: 212–223 (1997).

602. Y. Kanda, T. Nakamura, and K. Higashitani. Colloids Surf. A 139: 55–62 (1998).

603. Y. Kanda, T. Murata, and K. Higashitani. Colloids Surf. A 154: 157–166 (1999).

604. Y. Kanda, S. Iwasaki, and K. Higashitani. J. Colloid Interf. Sci. 216: 394–400 (1999).

605. M. Kosmulski. P. Erikson, C. Brancewicz and J. B. Rosenholm. Colloids Surf. A 162: 37–48 (2000).

606. O. Griot. Trans. Faraday Soc. 62: 2904–2908 (1966).

607. P. Jackson, and G. D. Parfitt. Kolloid Z. Z. Polymer. 244: 240–245 (1971).

608. J. Szczypa, L. Wasowska, and M. Kosmulski. J. Colloid Interf. Sci. 126: 592–595 (1988).

609. M. Kosmulski, E. Matijevic. Colloids Surf. 64: 57–65 (1992).

610. M. Kosmulski. In Fine Particles Science and Technology, (E. Pelizzetti, ed.), Kluwer, Dordrecht 1996. pp. 185–196.

611. D. M. Nikipanchuk, Z. M. Yaremko, and L. B. Fedushinskaya. Kolloid. Zh. 59: 350–354 (1997).

612. M. Kosmulski, P. Eriksson, and J. B. Rosenholm. Anal. Chem. 71: 2518–2522 (1999).

613. E. H. P. Logtenberg, and H. N. Stein. Colloids Surf. 17: 305–312 (1986).

614. A. Delagado, F. Gonzalez-Caballero, and J. M. Bruque. Colloid Polym. Sci. 264: 435–438 (1986).

4

Strongly Adsorbing Species

The list of compounds whose presence does not induce a shift in the CIP and in the IEP of materials showing pH dependent surface charging is confined to a few 1–1 salts, lower alcohols, and a few other organic solvents. The pH dependent surface charging in absence of strongly adsorbing species was discussed in Chapter 3. Even the salts known as inert electrolytes can induce a substantial shift in the IEP (typical accuracy in IEP determination is about 0.1 pH unit, thus a shift by 0.5 pH unit can be considered as substantial) when their concentration in solution exceeds 0.1 mol dm^{-3} (Section 3.III.C). Other compounds can induce a substantial shift in the IEP and charging curves when their concentrations are below 10^{-3} mol dm^{-3} (for some systems even below 10^{-6} mol dm^{-3}). These compounds or products of their dissociation in aqueous solutions are referred to as strongly adsorbing species, and their sorption is discussed in the present chapter. The affinities of these "strongly adsorbing species" to specific surface are diverse, and they can differ by many orders of magnitude. This chapter is restricted to discussion of empirical data, e.g. some common patterns observed in sorption behavior are presented. Interpretation of these results in terms of various models is discussed in Chapter 5.

The lack of coincidence between CIP and IEP of (hydr)oxides and related materials indicates the presence of strongly adsorbing species, added intentionally or present as impurities. Specific adsorption induces shifts in the IEP on the one hand and in the point of zero proton charge on the other in two opposite directions. When a shift in the CIP and IEP is addressed, the pristine PZC of given material (cf. Tables

310

3.1 and 3.3) is meant as a natural reference point. In the seventies these shifts were the sole criterion to distinguish between specific and nonspecific adsorption. Recently the results obtained by means of spectroscopic methods play more and more significant role.

The solution chemistry of inert electrolytes is rather simple: they tend to be fully dissociated, e.g. in NaCl solutions only four ionic species are present, namely, Na^+, Cl^-, H^+ and OH^- ions. In contrast, solution chemistry of ions that tend to be specifically adsorbed is more complex and pH dependent. Cations form hydroxocomplexes, e.g.

$$Zn^{2+} \rightarrow Zn(OH)^+ \rightarrow Zn(OH)_2 \rightarrow [Zn(OH)_3]^- \rightarrow [Zn(OH)_4]^{2-}$$

and carbonate complexes (in systems that are not CO_2 free), e.g.

$$UO_2^{2+} \rightarrow UO_2CO_3 \rightarrow [UO_2(CO_3)_2)]^{2-} \rightarrow [UO_2(CO_3)_3]^{4-}$$

and acids occur at different degrees of deprotonation, e.g.

$$H_3PO_4 \rightarrow H_2PO_4^- \rightarrow HPO_4^{2-} \rightarrow PO_4^{3-}$$

moreover, both cations and anions tend to form mono- and polynuclear water soluble complexes, e.g.

$$[Cu(NH_3)_4]^{2+}, \log K\ 11.8$$

and/or sparingly soluble salts, e.g.

$$BaSO_4, \log K_{sp} - 10$$

with other ions. Certain anions form polyacids, e.g.

$$CrO_4^{2-} \rightarrow Cr_2O_7^{2-} \rightarrow Cr_3O_{10}^{2-} \rightarrow Cr_4O_{13}^{2-}$$

and the degree of condensation is pH dependent. Both complexes and sparingly soluble salts can stabilize atypical oxidation states of some metals, e.g. Co(III) is a very strong oxidizer in absence of complexing agents,

$$Co^{3+} + e = Co^{2+}\ E^0 = 1.92\ V$$

but it is a moderately strong oxidizer in the presence of ammonia

$$[Co(NH_3)_6]^{3+} + e = [Co(NH_3)_6]^{2+} E^0 = 0.11\ V$$

Solution chemistry of electrolytes is discussed in detail in handbooks of inorganic and analytical chemistry.

Chemical formulas of crystalline salts whose ions tend to be specifically adsorbed do not reflect complex solution chemistry of these salts. Apparently the adsorption systems are simple, but in fact the solutions are multicomponent systems, and the coexisting species can substantially differ in their affinities to the surface. The speciation in the interfacial region can be completely different from that in bulk solution. Many analytical methods do not distinguish between particular species, and the results representing overall sorption behavior of all species involving the element of interest are obtained. Fortunately, some results obtained by means of spectroscopic methods can be resolved into pieces of information regarding

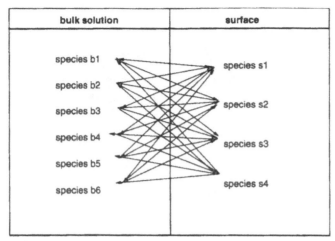

FIG. 4.1 Equilibria between bulk and surface species in specific adsorption.

particular surface species. The onset of uptake often coincides with the pH, at which certain ionic species in solution (e.g. the $[MeOH]^+$ ion for divalent metal cations) becomes dominant. However, according to contemporary theories of adsorption from solution, discussed in more detail in Chapter 5, the surface species do not univocally correspond to species in bulk solution, i.e. each surface species is in equilibrium with all bulk species, as illustrated in Fig. 4.1, even the number of solution and surface species can be different. In other words, the surface species must not be identified with "adsorbed $[MeOH]^+$ ions", "adsorbed Me^{2+} ions" etc. Certainly, surface and bulk species show some analogies, e.g. elements than form polymeric species in solution are more likely to adsorb as polymeric surface species.

An important and often overlooked aspect of complicated solution chemistry of salts other than inert electrolytes is formation of water soluble complex species involving products of dissolution of sparingly soluble materials (adsorbents). For instance, indirect spectroscopic evidence has been presented [1] that a silicate complex is the dominating soluble species of Eu (total Eu concentration of 10^{-9} mol dm^{-3}) in the presence of silicates (2.7×10^{-4} mol Si dm^{-3}) over the pH range 6–8. The solubility of silica is about 10^{-3} mol dm^{-3}, thus, water soluble silicate complexes can play an important role in the partition of Eu (and other trivalent lanthanides) between aqueous solution and silica.

Most solutions contain colloidal particles, e.g. dust, that do not cause visible turbidity and that sustain multiple filtrations and centrifugations. With low initial concentrations of the adsorbate of interest ($< 10^{-9}$ mol dm^{-3}) a substantial fraction of the adsorbate is associated with these colloids rather than homogeneously distributed in solution. Uptake of adsorbate from solution in such systems occurs partially by attachment of colloidal particles originally present in solution (with the adsorbate of interest sorbed by these particles) to the surface of the adsorbent. Then the adsorbate is slowly redistributed. This aspect of sorption has been widely discussed (e.g. [2]) with respect to radionuclides (radiocolloids), but this phenomenon is not limited to radioactive elements. Filtration of water used in the experiments can partially reduce but not completely eliminate this problem. With lower concentration of dust particles, their role in the sorption process is negligible down to lower concentration of the adsorbate. Namely, when the concentration of adsorbate exceeds the sorption capacity of the colloids present in water by many orders of magnitude, the fraction of adsorbate associated with these colloids cannot be too high. On the other hand with higher concentrations of the adsorbate of interest ($> 10^{-6}$ mol dm^{-3}), the solubility product of hydroxides or certain salts is often exceeded (especially at high pH) and these sparingly soluble compounds can produce colloids, whose concentration is not related to the original concentration of dust particles. Such colloids can be very fine and stable, and their attachment to the surfaces of adsorbents complicates the sorption process.

The uptake of strongly adsorbing species from solution involves different processes, i.e. adsorption, diffusion in solid phase, formation of new phases by coprecipitation and/or surface precipitation, etc. The difference between adsorption and surface precipitation was discussed by Sposito [3]. The term "sorption" will be used in this book when no evidence is available to assess the contribution of particular processes to the overall uptake. In some publications "adsorption" has been used as a generic term for different sorption processes. These phraseological subtleties are less significant in surface charging studies under pristine conditions (Chapter 3), and even less significant in studies of adsorption of gases or binary mixtures of liquids, and definitions of adsorption in textbooks of colloid chemistry usually do not focus on problems characteristic for specific adsorption from aqueous solution.

The uptake of certain species by coprecipitation (the adsorbent is formed in the presence of adsorbate of interest) is usually higher than adsorption on pre-formed precipitate at otherwise identical conditions (pH, concentration of sorbate, amount of adsorbent). In many older publications, e.g. [4], the terms "sorption" and "adsorption" are used in the title and abstract, while in fact coprecipitation was studied. The term "adsorption" is used in the present book only when the solid phase was formed in absence of the adsorbate, and the adsorbate was added afterwards. Concentrated solutions used to precipitate sparingly soluble materials are only stable in the presence of complexing agents or at extreme pH values. Thus, the history of the system in the coprecipitation experiment on the one hand and in sorption experiment with a pre-formed adsorbent on the other is very different, and with identical final conditions, the solid phases obtained in these two experiments are not necessarily the same, even for very long equilibration times. The results of thermochemical calculations (cf. Table 2.2 and examples in Section 2.II) are of limited practical importance due to kinetic limitations. Thermodynamic data relevant to fixation of metal cations on marine Mn-Fe nodules and sediments, especially formation of complex oxides have been collected by Ruppert [5].

A few examples of surface charging of semiconductors polarized to certain potential by an external potentiostat under pristine conditions are presented in Chapter 3. Uptake of strongly adsorbing species by such electrodes has been also studied [6], but these results will not be discussed in this book. Adsorption of ions on sufficiently conductive materials can be enhanced or depressed by external electric potentials. This property would be beneficial in wastewater treatment, namely, the adsorbents can be regenerated without use of chemicals. Model calculations demonstrated that the potentials sufficient to enhance or depress the uptake by over an order of magnitude are low enough to avoid electroplating or electrolysis of water [7].

Specific adsorption on well defined materials has been the subject of many reviews [8–13]. Specific adsorption plays a key role in transport of nutrients and contaminants in the natural environment, and many studies with natural, complex, and ill defined materials have been carried out. Specific adsorption of ions by soils and other materials was reviewed by Barrow [14,15]. The components of complex mineral assemblies can differ in specific surface area and in affinity to certain solutes by many orders of magnitude. For example, in soils and rocks, (hydr)oxides of Fe(III) and Mn(IV) are the main scavengers of metal cations and certain anions, even when their concentration expressed as mass fraction is very low. Traces of TiO_2 present as impurities are responsible for the enhanced uptake of U by some natural kaolinites. In general, complex materials whose chemical composition seems very similar can substantially differ in their sorption properties due to different nature and concentration of "impurities", which are dispersed in a relatively inert matrix, and which play a crucial role in the sorption process. In this respect the significance of parameters characterizing overall sorption properties of complex materials is limited. On the other hand the assessment of the contributions of particular components of a complex material to the overall sorption properties would be very tedious.

Examples of studies of specific adsorption are presented in Tables 4.1 (small cations), 4.2 (small anions), 4.4 (organic compounds), 4.5 (surfactants), and 4.6 (polymers). Studies presenting new experimental data are compiled together with discussions, interpretations and model calculations based on experimental results taken from other publications. The tables are sorted primarily by the chemical formula of the adsorbent, according to the rules outlined in the Introduction, and then by the adsorbate. The entries with a single adsorbate are followed by entries with multiple adsorbates (for the same adsorbent) sorted by the name or chemical formula of the first adsorbate, then of the second adsorbate, etc. Each Table is accompanied by an alphabetical index of adsorbates.

The experimental systems, goals, and methods in particular publications are very diverse, but there is a general agreement as to what variables have to be controlled to obtain meaningful results. Accordingly, Tables 4.1–4.6 have the following columns.

- Adsorbent
- Adsorbate
- Total concentration
- Solid to liquid ratio
- Equilibration time
- Temperature
- pH
- Electrolyte
- Method
- Result
- Reference

Some cells in Tables 4.1–4.6 are empty because of incomplete information about experimental conditions. Ranges are often reported rather than specific values of particular parameters, because the variables change in the series of experiments and even in the course of one experiment. With three or more values of given variable a range is reported, e.g. "pH 2–8", while with two different values a comma is used as a separator, e.g. "pH 2, 8".

Adsorbent: the chemical formula and characterization (cf. explanations for Table 3.1) are reported in this column. Certainly the chemical formula is only available for well defined materials. Naturally occurring adsorbents are sometimes referred to as geomedia. Complex adsorbents often contain substantial (and variable) amounts of adsorbate of interest, similar adsorbates that can compete with the adsorbate of interest for the surface sites, and other species that strongly affect adsorption of the species of interest, e.g. humic substances present in soil enhance the uptake of heavy metal cations. These strongly interacting substances are firmly bound with the adsorbent, and their concentration in solution is often beyond the range of common analytical methods. The problem of strongly interacting species originally present in the adsorbent concerns also apparently pure and well defined adsorbents. Studies of surface charging in inert electrolytes (Chapter 3) have been used as a test for presence of specifically adsorbed species in the original adsorbent (*vide infra*).

Adsorbate: adsorption of ions of inert electrolytes has been discussed in Chapter 3. Specific adsorption of anions on the one hand and cations on the other show many common features and it is expedient to discuss these two groups separately. It is also well known that the same compound (e.g. amino acids) can be present in solution as anions or cations at different pH. The small ions and molecules whose sorption is discussed in this chapter have been sorted into three groups corresponding roughly to the sign of the electric charge of the specifically adsorbing species. Results referring to sorption of metallic elements (including metaloorganic compounds, e.g. CH_3Hg^+) are listed in Table 4.1 (cations) disregarding the actual speciation, i.e. the amphoteric character of some metals and formation of anionic species with certain ligands. For example sorption of Ni in the presence of equal amount of EDTA is reported in Table 4.1 (cations) although Ni is present predominantly in form of an anionic Ni-EDTA complex. A few metallic elements show only acidic behavior at high oxidation state(s), and basic or amphoteric behavior at low oxidation state(s). For example sorption of Cr(VI) on different materials is reported in Table 4.2 (anions) and sorption behavior of Cr(III) is reported in Table 4.1 (cations). Many metallic elements have one stable oxidation state, and this oxidation state is not specified in the column "adsorbate". Some other metals have more than one relatively stable oxidation state, and these oxidation states are specified. For example Pu can be present in natural waters in four different oxidation states (III, IV, V, VI). Atypical oxidation states can be stabilized by certain ligands. With trace amount of the element of interest the oxidation state is often unknown.

In some publications sorption of several adsorbates was studied under the same conditions. The following abbreviation: adsorbate 1 > adsorbate 2 > adsorbate 3, etc. is used when the affinity of the adsorbate 1 to the surface is greater than the affinity of adsorbate 2, etc. This may be manifested by higher uptake, more substantial shift in the IEP, etc.

Sorption behavior of inorganic and carboxylic acids is reported in Table 4.2 (anions), but sorption behavior of amino acids and phenols is reported in Table 4.4 (organic compounds). This classification is arbitrary, e.g. some derivatives of phenol whose sorption is reported in Table 4.4 have higher acidic dissociation constants than carbonic acid (whose sorption is reported in Table 4.2). Also sorption behavior of humic and fulvic acids is reported in Table 4.4. Humic substances play a special role in sorption of metal ions, and their interactions with metal cations in solution were discussed in Refs. [16–19]. This term is used for a broad class of substances, whose molecular weight and nature of acidic groups can differ substantially, e.g. aquatic humic substances have lower molecular weight than soil humic substances, and the latter contain more aromatic carbon. These differences certainly affect the binding of metal cations. Typical molecular weight for humic acid is several thousand and it is precipitated at pH about 2. The term fulvic acid is used for low molecular weight (< 1000) substances, which are soluble also in acidic media.

Concentration in solution: usually initial (total) concentration or concentration range is reported and the same units (molar or mass concentration of certain compound or element) are used as in the original publications. The concentration can be easily converted from one unit to another when the chemical formula of the adsorbate is known. In a few original publications the molal concentration of the adsorbate was reported (mol per kg of solution). These concentrations were converted into molar concentration assuming that the specific density of the solution was 1 kg/dm^3.

In some publications only "equilibrium" concentrations (concentration in solution equilibrated with the solid for certain time) are reported, and these concentrations are shown in Tables 4.1–4.6. They are somewhat lower than the initial (total) concentrations, but the difference between initial and equilibrium concentration usually does not exceed one order of magnitude. In some publications the amount of adsorbate per gram of per m^2 of adsorbent is reported. Such concentrations cannot be converted into mol dm^{-3} unless the solid to liquid ratio is known. The mass or molar ratio between an element – constituent of the adsorbate and another element – constituent of the adsorbent has been reported in some publications. There are a few other examples of concentration reported in original paper that cannot be easily converted into mol dm^{-3} without additional information, e.g. solutions were prepared by dilution of saturated solution of a salt by known factor, but the concentration of saturated solution is not reported, and the temperature is unknown.

Solid to liquid ratio: the available range of solid to liquid ratios is limited to a few orders of magnitude. When this ratio is too high, a paste like material is obtained, in which diffusion is very slow, and separation of clear liquid is difficult. When the solid to liquid ratio is too low, sorption by glassware, filter paper etc., becomes significant with respect to the sorption by the material whose adsorption properties are studied. Thus, for low solid to liquid ratios control experiments without sorbent are recommended to assess the significance of these artifacts. Most often the solid to liquid ratio is given as a combination of two numbers, namely, the ratio of the mass of solid to volume of solution and the specific surface area of solid. The latter is usually obtained by means of BET (Brunauer–Emmet–Teller) method. An interesting but rather rarely employed idea is to compare different adsorbents (having similar or dissimilar chemical formula) by performing experiments at the same solid to liquid ratio (in m^2/dm^3) and other experimental conditions (concentration of the adsorbate, pH, ionic strength, time of equilibration). Surface saturation degree (ratio of total number of specifically adsorbed ions to the number of surface sites) was used instead of solid to liquid ratio in some papers. This approach assumes that the site density is known (cf. Chapter 5).

Microelectrophoresis requires a relatively low solid to liquid ratio. When the results obtained by means of microelectrophoresis and another method are presented as one entry, the data in the column "solid to liquid ratio" refers to this other method, and the solid to liquid ratio in microelectrophoresis (not reported) is usually lower by many orders of magnitude.

Equilibration time chosen by different authors ranges from a few minutes to many weeks. Different arguments were offered to support these choices, and as a matter of fact the choice is rather arbitrary. The results obtained for different equilibration times at otherwise the same conditions can be very different, and all these results are in some sense valuable and interesting. Even for very long equilibration times the adsorption systems do not reach thermodynamic equilibrium. Some systems reach pseudoequilibrium, i.e. the distribution of the element of interest between the solid and liquid phase does not change on further equilibration. The equilibration time necessary to reach pseudoequilibrium depends on the experimental conditions, e.g. pH, especially when the sorption process involves formation of new phases. Thus, in a series of data points corresponding to the same equilibration time some results represent pseudoequilibrium while some other results represent a state far from pseudoequilibrium. Results of kinetic studies of sorption are compiled in Table 4.8. The sorption process is a combination of processes whose rate is very diverse. With a short equilibration time pseudoequilibrium can be reached with respect to the fastest process(es) while the progress of the slower processes is negligible. With a longer equilibration time pseudoequilibrium with respect to slower process(es) can also be reached. Terminology like "sufficient equilibration time" or "incorrect procedure, because the equilibration time was too short" is often encountered in scientific publications, although in view of the above discussion these terms are rather vague. In acid-base titrations hysteresis is commonplace, i.e. the ions presorbed at favorable electrostatic conditions (high pH for cations, and low pH for anions) do not desorb when the electrostatic conditions become less favorable. In order to quantify the sorption hysteresis, a hysteresis coefficient was introduced [20]. This coefficient equals zero for a reversible process and one for completely irreversible sorption.

Apparently the problems with slow equilibration can be avoided by increasing the equilibration time, but it is not always the case. For example, the presence of complexing anions often enhances dissolution of metal oxides and related adsorbents. This results in the following kinetic pattern: the uptake as a function of time rises, peaks and then declines. In order to account for this effect, the declining segment was extrapolated to $t = 0$ [21], and the intersection with the axis of ordinates was interpreted as uptake corrected for dissolution of the adsorbent. Anyway, negligible dissolution of the adsorbent is a substantial advantage of experiments with relatively short equilibration times.

Tables 4.1–4.6 report the final equilibration time, i.e. the time elapsed after all reagents were added. In some studies the reagents were added in two or more phases with certain equilibration time after each phase. The scatter of results obtained at apparently identical experimental conditions in some studies can be partially due to uncontrolled prequilibration. The time and conditions of preequilibration are not reported in Tables 4.1–4.6 and the readers interested in details are referred to original publications. Probably the time and conditions of preequilibration can affect the results in certain systems, but systematic studies of such effects are scarce. A few examples of substantial effects of preequilibration on the adsorption are reported in the column "results".

Temperature: the results compiled in Tables 4.1–4.6 were obtained at different temperatures, and in some studies the temperature was not controlled. The results reported in Table 3.11 and Fig. 3.104 indicate that the PZC of oxides and related materials shifts to low pH when the temperature increases (with a few exceptions). Most surfaces carry more negative charge at elevated temperature (at given pH), and this creates favorable conditions for adsorption of cations and unfavorable conditions for adsorption of anions. Therefore elevated temperature would enhance uptake of cations, and low temperature would enhance uptake of anions at constant pH, if the electrostatic interaction was the only factor. On the other hand, the rate of chemical reactions and diffusion is enhanced at elevated temperatures. Thus, the kinetic and electrostatic effect on cation adsorption add up and the uptake increases with temperature. With anions these effects act in opposite directions: the uptake increases with temperature when the kinetic factor prevails; the uptake decreases with temperature when the electrostatic factor prevails, finally the both effects can completely cancel out.

The quantification of temperature effects on surface charging has been discussed in Section 3.IV, and many problems and limitations considered in that section concern also specific adsorption. Interpretation of temperature effects on specific adsorption is even more complex, e.g. due to the discussed above complicated solution chemistry, which is also temperature dependent. While literature data relevant to speciation of solutions involving hydrolyzable cations and/or weak acids at one temperature (usually 20°C or 25°C) are readily available [22], the information on temperature effects on stability constants of water soluble complexes is rather incomplete.

Unfortunately there is no model-free definition of ΔH_{ads} analogous to Eq. (3.43) for specifically adsorbing ions. Namely, it is not possible to define the surface reaction responsible for specific adsorption in such way that the equilibrium constant of this reaction would be equal to a directly measured quantity (cf. Eq. (3.42)). Many publications report ΔH_{ads} for specifically adsorbing ions, based on combination of Eqs. (2.20) and (2.24). Comparison of ΔH_{ads} values reported in different publications in similar systems and critical evaluation of these results is rather difficult. First the value and even sign of the enthalpy of adsorption depends how the surface reaction responsible for the adsorption is defined. Sometimes the uptake increases with the temperature, but negative enthalpy of adsorption is reported, e.g. when metal hydroxocomplexes are considered as reagents. Many adsorption models involve more than one surface species, and ΔH of reactions representing their formation can assume different values and even have opposite signs. These models are discussed in detail in Chapter 5. The numerical instability of the model parameters calculated from experimental data is a substantial disadvantage of complicated models in characterization of temperature effect on specific adsorption. Therefore simplified approach was preferred by many authors. The adsorption "reaction" can be written as

$$\text{ion in solution} \Leftrightarrow \text{ion adsorbed} \tag{4.1}$$

and the "equilibrium constant" of reaction (4.1) equals

$$K_{ads} = c_{ads}/a \tag{4.2}$$

where a is the activity of the ion of interest in solution (the theory of activity of ions in dilute aqueous solutions of salts is well established) and c_{ads} is the concentration of adsorbed ion (definition of activity of surface species requires some model assumptions). Combination of Eq. (4.2), (2.20) and (2.24) with the derivative replaced by a differential quotient leads to

$$\Delta H_{ads} = R \ln(a_2/a_1)_{C_{ads}}/(1/T_2 - 1/T_1) \tag{4.3}$$

where a_1 and a_2 are the activities of the ion of interest at temperatures T_1 and T_2 at constant c_{ads}. When the activities in solution are low and the adsorption isotherm is linear (Eq. (4.2)) the ratio a_2/a_1 at constant c_{ads} can be estimated as:

$$a_2/a_1 = U_1(100 - U_2)/[U_2(100 - U_1)] \tag{4.4}$$

where U is the uptake (in %) at constant total concentration (which can be directly adjusted in contrast with c_{ads}). The K_{ads} defined by Eq. (4.2) is a "conditional constant", and first of all it is pH dependent. Therefore in Eq. (4.3) not only c_{ads} but also factors affecting K_{ads} must be kept constant. These factors are interrelated, and only some of them can be freely adjusted. The most common definition of ΔH_{ads} based on Eq. (4.3) is

$$\Delta H_{ads} = R \ln(a_2/a_1)_{C_{ads}, \, pH}/(1/T_2 - 1/T_1) \tag{4.5}$$

at constant ionic strength and in absence of competing species. This definition leads to positive (endothermic) ΔH_{ads} for cations (including cations of inert electrolytes) due to discussed above combination of favorable electrostatic conditions and higher rate of physical and chemical processes at elevated temperatures. Positive and negative values of ΔH_{ads} defined by Eq. (4.5) have been reported for anions. The ΔH_{ads} value calculated from Eq. (4.5) depends on the experimental conditions, i.e. pH and c_{ads}. Systematic studies of these effects are rare, and most often values of ΔH_{ads} referring to some arbitrary pH and c_{ads} are reported. This is chiefly because combination of Eq. (4.4) and (4.5) gives accurate results only when U_1 and U_2 are between 10% and 90%. When U_1 or U_2 is too close to 0% or 100% the relative error in Eq. (4.4) becomes unacceptable, and considering that the uptake of ions often increases (decreases) from 10% to 90% over a pH range about one pH unit wide (cf. Fig. 4.7), this approach is not suitable for systematic studies of ΔH_{ads} as a function of pH.

In order to account for the solutes other than protons, Eq. (4.5) must be rewritten as

$$\Delta H_{ads} = R \ln(a_2/a_1)_{C_{ads}, \, a_k}/(1/T_2 - 1/T_1) \tag{4.6}$$

where the subscript a_k denotes that the activities of all ionic species in solution but two (the ion of interest and another ion) are constant. Activities of all ions cannot be freely adjusted because of the condition of electroneutrality. Certainly the ΔH_{ads} defined by Eq. (4.6) is also a configurational enthalpy, i.e. it depends on c_{ads} and a_k. By analogy with solution chemistry it would be expedient to define some standard conditions and split the ΔH_{ads} into standard and configurational terms (cf. Eq. (2.6)), but this task is anything but trivial.

Another, less common approach to definition of ΔH_{ads} is by equations analogous to Eq. (4.5) and (4.6) in which the surface concentrations of ionic species are constant, i.e.

$$\Delta H_{ads} = R \ln (a_2/a_1)_{C_{ads},\sigma_0}/(1/T_2 - 1/T_1) \tag{4.7}$$

and

$$\Delta H_{ads} = R \ln (a_2/a_1)_{c_{ads,k}}/(1/T_2 - 1/T_1) \tag{4.8}$$

where the subscript $c_{ads,k}$ indicates that surface concentrations of the species of interest and all other species but one are kept constant. A similar definition with the surface concentration of all ionic species kept constant [23] is erroneous, namely, it violates the condition of electroneutrality. In order to adjust concentrations of all ions at the surface, their concentrations in solution must be adjusted, and this cannot be done freely for all ions. Some experimentally determined ΔH_{ads} based on Eq. (4.5) are reported in the column "results" of Tables 4.1 and 4.2. The ΔH_{ads} based on Eq. (4.7) are less positive for cations and less negative for anions with respect to the values calculated from Eq. (4.5) at the same experimental conditions, and they are referred to as "ΔH_{ads} at constant σ_0". The interpretation of temperature effects on specific adsorption in terms of more sophisticated models is addressed in Chapter 5.

The temperature effects on the magnitude of specific adsorption can be also calculated from the ΔH_{ads} obtained calorimetrically. It should be emphasized that interpretation of calorimetric measurements in complex, multicomponent systems is very difficult.

The **pH** effects on sorption of ions have been interpreted in terms of:

- pH dependent σ_0, creating more favorable electrostatic conditions for adsorption of cations and less favorable conditions for adsorption of anions at high pH
- pH dependent speciation in solution: hydrolyzed metal cations and neutral (or partially deprotonated) molecules of acids have higher affinity to the surface than unhydrolyzed metal cations and fully deprotonated anions, respectively.
- competition of protons with metal cations and competition of hydroxyl anions with other anions for the same surface sites.

Each of these approaches qualitatively explains the observed trends, namely the uptake of metal cations in enhanced and the uptake of anions is most often depressed at high pH. For kinetic reasons the observed pH effects on specific adsorption depend on the equilibration time.

The difference between initial and final pH is often substantial and these two quantities have to be clearly distinguished. It is certainly easier to carry out experiments at constant initial pH than at constant final pH, but significance of adsorption isotherms at constant initial pH is questionable. In some publications initial and final pH values are reported, but only the final value is given in Tables 4.1–4.6. In a few publications only the initial pH is reported.

The pH effect on the surface charging of materials has been discussed in Chapter 3. Detailed discussion of the pH effect on speciation in solution is beyond the scope of the present book. In order to illustrate some common trends Fig. 4.2 shows the calculated speciation of Pb at a low ionic strength and Fig. 4.3 shows the Pb speciation in the presence of 0.1 mol dm^{-3} NaCl as a function of the pH. Figure 4.4 shows the calculated speciation of phosphoric acid at a low ionic strength. The calculations were carried out using the following values of stability constants [22]: $\log K(PbOH^+) = 6.3$; $\log K[Pb(OH)_2] = 10.9$; $\log K[Pb(OH)_3^-] = 13.9$; $\log K(PbCl^+) = 1.59$; $\log K(PbCl_2) = 1.8$, and the following values of dissociation constants of phosphoric acid: $pK_{a1} = 2.16$; $pK_{a2} = 7.21$; $pK_{a3} = 12.32$. The total Pb concentration was assumed to be so low, that the solubility product of hydroxide or basic chloride is not exceeded. The results presented in Fig. 4.2–4.4 can be summarized as follows.

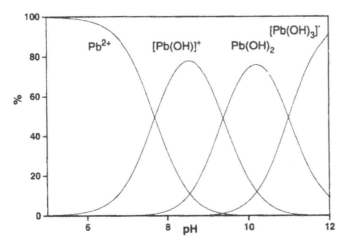

FIG. 4.2   Speciation of Pb(II) at a low ionic strength.

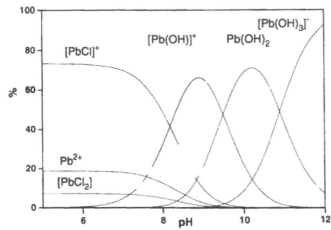

FIG. 4.3   Speciation of Pb(II) in the presence of 0.1 mol dm^{-3} NaCl.

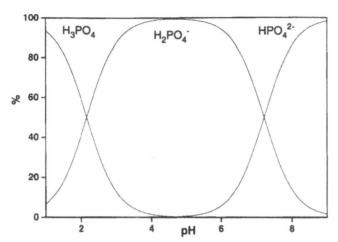

FIG. 4.4 Speciation of phosphoric acid at a low ionic strength.

- except at extreme pH values there are at least two coexisting species at comparable concentrations
- hydrolysis of fully deprotonated anions in acidic and neutral pH range leads to mixture of anions and neutral molecules
- hydrolysis and complexation of cations leads to mixture of cations, neutral molecules and often also of anionic species
- the effect of the same concentration of anionic ligands on the speciation of metals can be more or less significant at different pH.

Similar trends are observed for other metal ions and weak acids, respectively.

Electrolyte. The surface charge density and thus the surface potential and sorption properties of materials discussed in this book are very sensitive to the pH, namely, a shift of the pH by a fraction of one pH unit can induce increase or decrease of uptake of certain solutes by an order of magnitude (cf. Fig. 4.5) and/or entirely change the structure of the surface complex detected by spectroscopic methods (Sections 4.I.E–H). The effect of inert electrolytes on the sorption properties of materials is less significant. Dzombak and Morel [12] reviewed sorption properties of hydrous ferric oxide and found that the ionic strength effect on cation adsorption was insignificant. A few other authors have erroneously generalized this conclusion, i.e. the absence of inert electrolyte effect on the specific adsorption in other systems has been claimed without experimental evidence. The inert electrolytes affect the absolute value of the surface charge but not its sign (Chapter 3), thus their effect on the electrostatic interaction ion-surface is not very dramatic. On the other hand, the composition of ionic medium can affect specific adsorption via speciation in solution, via adsorption kinetics, and via sorption competition (the latter is emphasized in the ion-exchange approach), thus the overall effect of an inert electrolyte on specific adsorption can be substantial. Many multivalent cations form nitrate and halide complexes. The stability of these complexes is usually low, but at ionic strengths > 0.1 mol dm^{-3} the speciation can be significantly influenced by complexation (cf. Fig. 3.2 and 3.3). Even chlorates (VII) whose ability to form complexes with metal cations is limited,

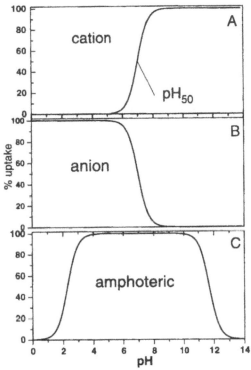

FIG. 4.5 Typical percentage of uptake (pH) curves for cations (A); anions (B); and (C) amphoteric substances.

significantly depress the thermodynamic activity of salts involving specifically adsorbing ions. Thus, the Gibbs energy of the ions of interest (at constant concentration of their salt) is depressed at high ionic strength (Eq. (2.6)), and this leads to lower adsorption. Moreover, the ions of inert electrolytes can reduce the uptake of specifically adsorbing ions by certain materials as a result of competition for ion exchange sites. In this respect the absence of ionic strength effect on specific adsorption in a few specific systems seems to be rather a result of fortuitous cancellation of different effects than a general rule.

The presence of ions other than the ions of interest in adsorption experiments cannot be avoided. They are present as counterions (electroneutrality condition) and they are introduced by the pH adjustment. All ions interact with other ions and with the surface, and a complex system of equations must be solved to account for all these interactions. Experimental studies of specific adsorption are often conducted at a high ionic strength, i.e. in the presence of excess of inert electrolyte with respect to the ions whose specific adsorption is studied. The formula and molar concentration of inert electrolyte is given in the column "electrolyte" in Tables 4.1–4.6. With ionic strengths on the order of 0.1 mol dm^{-3} the variations in the ionic strength caused by pH adjustment and adsorption of inert electrolyte are negligible. Some publications report the salinity expressed in less common units, e.g. as mass concentration. It should be emphasized that the adsorption behavior observed at high ionic strengths can be very different from adsorption behavior at a low ionic strength and otherwise the same conditions. The difficulty in experiments at a low ionic strength is that in such systems the ionic strength varies from one data point to another due to the pH adjustment and adsorption of inert electrolyte. For studies carried out without addition of inert electrolyte the name or formula of the counterion is reported in the column "electrolyte". The counterions often belong to one of the following groups:

- anions: nitrate V, chlorate VII, halides
- alkali metal cations.

The interaction of these ions with the ions of interest on the one hand, and with the surface on the other is chiefly electrostatic, and the effect of the nature of these counterions on specific adsorption is rather insignificant.

In addition to the studies conducted in the presence of inert electrolytes, some sorption experiments were carried out in the presence of other ions that can also be specifically adsorbed. The studies of specific adsorption of cations in the presence of strongly adsorbing anions, with or without addition of an inert electrolyte (e.g. adsorption of Ni in the presence of EDTA) are reported in Table 4.1, and the studies of specific adsorption of anions in the presence of strongly adsorbing cations (e.g. adsorption of EDTA in the presence of Ni) are reported in Table 4.2. On the other hand, sorption of cations in the presence of other strongly adsorbing cations (e.g. Ni in the presence of Cu), or anions in the presence of other strongly adsorbing anions (e.g. arsenate in the presence of phosphate) is reported in Table 4.7 and discussed in Section 4. VI devoted to sorption competition. Multivalent cations and weak acids often form stable complexes in solution, thus the effect of strongly adsorbing anions on specific adsorption of metal cations and *vice versa* is much more significant than the corresponding effect of inert electrolytes. This type of studies is often undertaken to simulate the effects of adsorption on migration of pollutants in the nature, namely the natural systems contain strongly adsorbing anions and cations simultaneously. Many sorption studies have been carried out in natural water or in synthetic solutions whose composition simulates natural water, e.g., in artificial sea water. The composition of these solutions varies from one publication to another, but it is not reported in Tables 4.1 and 4.2.

Some studies of sorption of heavy metal ions were conducted in the presence of standard pH buffer solutions. Acetates, phthalates and particularly phosphates present in these buffers significantly affect the adsorption of most anions and some cations. It should be emphasized that the adsorbents act themselves as pH buffers when the solid to liquid ratio is sufficiently high. Therefore additional buffering is usually not necessary.

The ionic strength effects on specific adsorption in particular systems can be very diverse. Some examples are reported in the column "results". In many systems the uptake of anions and cations is depressed at a high ionic strength. For example sorption of metal cations on negatively charged surfaces is often depressed in the presence of anions that form water soluble complexes with these cations. This can be interpreted as competition for the cations between the surface and ligands in solution. In other, apparently similar systems the uptake is rather insensitive to the ionic strength. A few examples of enhanced uptake at a high ionic strength have been also reported. The latter type of behavior is characteristic for adsorption occurring against electrostatic forces, e.g. when cations adsorb on a positively charged surface.

The ionic strength effect on the uptake curves can be used as a test of different adsorption models (Chapter 5).

Turner et al. [24] described two types of sorption behavior of clay minerals toward U(VI). At low ionic strengths the uptake is dominated by ion exchange and it is rather insensitive to pH, while at high ionic strengths typical adsorption edges (cf. Fig. 4.5) are observed. The latter behavior of the adsorbent was called oxide like.

Barrow reports common intersection points of uptake (pH) curves obtained for sorption of phosphates (constant total concentration) by soils at different NaCl concentrations [25]. The slope of these curves decreases when the ionic strength increases, and the uptake of phosphates from 1 mol dm^{-3} NaCl is almost independent of pH. The position of these intersection points on the pH scale is a function of the initial phosphate concentration. Similar effect was reported for selenates (IV). Interestingly for borates [26] whose uptake increased with pH over the pH range of interest, a common intersection point (pH of zero salt effect) was also observed but in this case the slope was higher for higher ionic strengths. This type of behavior has not been reported for well defined adsorbents.

A few **other variables**, not addressed in Tables 4.1–4.6 have a significant effect on specific adsorption. Many experimental studies were carried out in an atmosphere of inert gas, e.g. nitrogen or argon, and outgased solutions were used. Other investigators did not pay attention to the problem of dissolved CO_2 and they used solutions saturated with atmospheric air. The "CO_2 error" related to primary surface charging has been discussed in Chapter 3. The CO_2 effect on proton adsorption can induce an indirect CO_2 effect on adsorption of all ionic species. Moreover, carbonates compete with other anions for the surface sites. Finally certain cations, e.g. actinides form stable water soluble carbonate complexes, thus the carbonate ligands in solution compete for these cations with the surface sites. Some publications report comparison of results obtained under conditions designed to reduce the concentration of CO_2 on the one hand, and at a non-zero, controlled CO_2 partial pressure on the other. These studies have been undertaken chiefly in systems with stable water soluble carbonate complexes, and indeed, the results obtained in the presence and absence of CO_2 are significantly different. The partial pressures of CO_2 in such studies are reported in the column "electrolyte" (partial pressure of CO_2 is related to concentration of carbonates in solution). The exclusion of CO_2 is nowadays a standard element of experimental procedure in studies of specific adsorption, but the CO_2 effect can very well be insignificant in specific systems, e.g. when the affinity of the adsorbate to the surface is very high.

The method of addition of particular reagents can affect the results obtained at otherwise the same experimental conditions. For example the uptake of metal cations by surfaces with preadsorbed humic acid is different from that observed when the solution containing metal cations and humic acid is aged before being contacted with the adsorbent. This aspect of specific adsorption is related to the discussed above issue of preequilibration. Many studies report a fraction of pre-sorbed ions released into the solution on addition of excess of competing/complexing ions or when unfavorable electrostatic conditions are generated by pH adjustment. Even a long treatment with concentrated acids does not induce release of pre-sorbed metal cations, specially when the time elapsed after the first contact between adsorbent and adsorbate was long. A few specific examples of such behavior are reported in the column "results".

Many adsorbates and/or adsorbents are redox sensitive. The specific adsorption in such systems depends on the redox potential, which is very difficult to measure or control, thus, systematic studies in this direction are rare. On the other hand some practical implications are well known, e.g. the uptake of chromates by soils and sediments in enhanced on addition of Fe(II) salts [27] as an effect of a redox reaction, in which Cr(VI) is reduced to Cr(III). A few examples of redox reactions accompanying sorption processes are reported in the column "results". The changes of oxidation state in the sorption process are probably more common than it is apparent from literature reports, but they are often overlooked, namely, analytical methods must be specially tailored to observe these changes.

I. EXPERIMENTAL METHODS

A. Uptake (Static)

Most studies of adsorption from solution involve measurements of the overall distribution of the element of interest between the liquid and solid phase. Such results are presented alone or together with results obtained by means of other methods. The distribution between the liquid and solid phase is often determined by addition of the solid to the solution containing certain amount of the adsorbate, and the magnitude of adsorption is calculated from depletion of the solution. Alternatively the adsorbent can be preequilibrated with solution not containing the adsorbate, and known amount of adsorbate is added to the dispersion afterwards. Different analytical methods (atomic absorption spectroscopy, radio-tracer method, ion selective electrodes, etc.) can be used to determine the initial and final concentration of the solute. Specific experimental setups will not be presented here; many detailed descriptions can be found in the cited literature. Adaptation of a setup designed for certain experimental conditions to another system requires caution, and probably there is no universal experimental procedure that can be recommended for studies of uptake regardless of the system of interest. The distribution of the adsorbate depends on experimental conditions, and the main variables have been discussed above. The control and/or measurement of these variables can be a source of problems and errors. The poor reproducibility in some studies of specific adsorption can be also explained in terms of other factors that remain beyond control. For example many solids have air-filled micropores that become partially water-filled during the sorption experiment, thus the surface area available for specific adsorption is in fact variable for the same solid to liquid ratio. Studies of sorption of common ions by natural materials (soils, rocks) are complicated by the fact that the sorbent already contains substantial and variable amount of sorbate. Thus, contact of such sorbents with sufficiently diluted solution of sorbate results in increase of the concentration of sorbate in solution. Separation of solid and liquid phase is usually achieved by centrifugation or filtration, and very small particies cause problems in both methods. Colloidal particles are often held at the top of centrifuged solution due to surface tension, and they are collected with the supernatant. Such errors are avoided when pipette tip is inserted sufficient distance below the surface to collect the clear supernatant. Fine colloids pass membranes and the elements associated with these colloids are included in the filtrate. Filters often adsorb solutes, and this results in overestimated uptake from very dilute solutions.

The effects of sorption by filters can be minimized when the first aliquot of filtrate is discarded.

The distribution of the solute between the solid and liquid phase can be determined not only from depletion of solution, but also by analysis of the adsorbent separated from the solution. The latter approach is not too popular because of serious experimental difficulties. Namely, such measurements are severely affected by the liquid trapped in pores and cracks of the solid and intergrain space. Washing procedures aimed at removal of residual liquid usually induce partial release of pre-sorbed substances.

Some publications report a single data point, representing the distribution of the adsorbate of interest between the solid and the liquid at certain conditions (initial concentration of adsorbate, solid to liquid ratio, equilibration time, temperature, pH, ionic strength, etc.), which are more or less precisely described. The disadvantage of such an approach is that the result is only valid for these particular experimental conditions, and a change in any of the above variables can lead to a completely different result. Therefore systematic studies are preferred in which one or a few parameters vary while the other parameters are kept constant. Such results are often presented in graphical form. The kinetic studies in which the equilibration time is the independent variable and the other parameters are kept constant are discussed separately in Section VII.

Adsorption isotherm, i.e. a plot of the sorption density as a function of the concentration of the solute in solution is one of the most common representations. Although the term "adsorption isotherm" is widely used, these plots can in fact represent uptake by surface precipitation, etc. Different data points are obtained by changing the initial concentration of the solute and/or the solid to liquid ratio while the other parameters, particularly the final pH are kept constant. Sometimes sets of adsorption isotherms (each one corresponding to certain pH) are reported. There are also publications, even quite recent reporting adsorption isotherms at variable and/ or uncontrolled pH. Interpretation of such results is rather difficult. The adsorption isotherms at constant pH (and other parameters) are often interpreted in terms of Langmuir or Freundlich equations. The popularity of the latter is probably due to the log–log scale which is used to plot adsorption isotherms covering many orders of magnitude of concentrations, and Freundlich equation gives a straight line in such coordinates. These and other types of adsorption isotherms are discussed in Chapter 5, and Tables 4.1 and 4.2 report the parameters of Langmuir or Freundlich equations in the column "results". It should be emphasized that some of these isotherms consisted of a few data points, and the match between the actual data points and the theoretical equations was often far from perfect. Fokkink et al. [28] plotted adsorption isotherms at constant surface charge density (rather than at constant pH).

The other common representation of the uptake studies is in form of percentage of uptake (pH) or distribution coefficient (pH) curves at constant initial concentration of the adsorbate and solid to liquid ratio. The typical course of the percentage of uptake (pH) curves for cations and anions is presented in Fig. 4.5–4.7. In a few publications the dissolved fraction rather than uptake is plotted as a function of pH. In such a representation "anion type" curves look like the "cation type" curves shown in Fig. 4.5 (A) and *vice versa*. The typical uptake curves for cations (A) and anions (B) (for low concentration of the solute) are shown in Fig. 4.5. The uptake increases from practically 0% to practically 100% over a pH range about 2 pH units wide. These curves are referred to as "adsorption edges" or "adsorption envelopes". Figure 4.5 shows also the adsorption behavior typical for amphoteric elements (C), namely, a cation type adsorption edge is observed at low pH and an anion type adsorption edge at high pH, and the uptake between these two adsorption edges is practically 100%. The experimental studies often cover a narrow pH range, and only one of these two adsorption edges is actually observed. The term "adsorption edge" does not imply that adsorption actually occurs (cf. "adsorption isotherm"), e.g. the uptake due to precipitation can also result in similar curves as these shown in Fig. 4.5. Some speciation curves shown in Figs. 4.2–4.4 have similar shapes as the adsorption edges. Therefore, sorption has been associated with predominance of certain species in solution. Tables 4.1 and 4.2 report the pH_{50} values (50% uptake, Fig. 4.5 (A)) in the column "results". It should be emphasized that the data points representing about 50% uptake are much more reliable than the data points representing nearly 0% or nearly 100% uptake. In vast majority of publications, the uptake was determined from concentration in solution. With nearly 100% uptake the concentration in solution often approaches the detection limit of the method, and it cannot be precisely determined. With nearly 0% uptake, the uptake is calculated as a difference of two large and almost equal numbers, and this results in a huge relative error in the amount sorbed. With the uptake of about 50% both solution concentration and amount sorbed can be determined with reasonable precision. Certainly the pH_{50} depends on the experimental conditions, especially the solid to liquid ratio and initial concentration of the solute, thus, the difference in pH_{50} values obtained in similar systems (absorbent, adsorbate), but at different conditions must not be treated as a discrepancy. For typical adsorption edges (cation type in Table 4.1 and anion type in Table 4.2) the pH_{50} value is reported without further comments. Most often this value is a result of interpolation, i.e. a data point representing an exactly 50–50 distribution can be only obtained by lucky coincidence.

Figure 4.6 shows some common deviations of actually observed uptake curves from "ideal" curves presented in Fig. 4.5. The following two types of behavior are commonplace:

- "normal" behavior except 100% uptake is not reached, namely almost constant uptake (about 90%) is observed even very far from pH_{50} (Fig 4.6(A))
- "normal" behavior except 0% uptake is not reached, namely almost constant uptake (about 10%) is observed even very far from pH_{50} (Fig 4.6 (B)).

Figure 4.6 shows uptake curves for cations, and analogous curves for anions are their mirror images. There is no generally accepted explanation why the uptake does not reach 0% at sufficiently unfavorable, or the uptake does not reach 100% at sufficiently favorable electrostatic conditions, even at low concentrations of the solute. Formation of very stable ternary surface complexes involving impurities on the one hand and formation of complexes with products of dissolution of the adsorbent on the other have been discussed as possible rationale, but some examples of unusual results can be also due to experimental errors (inadequate phase separation). Figure 4.6(C) shows two types of uptake curves with a maximum. Such uptake curves are observed for cations in the presence of carbonates (or other weak acids), which form stable complexes with a metal cation of interest. At low pH these ligands are fully protonated, and they do not compete with the surface for metal

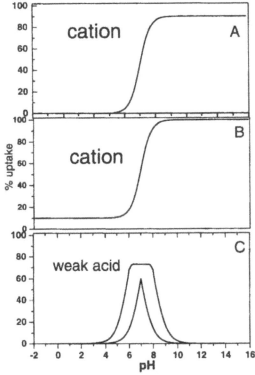

FIG. 4.6 Less common types of percentage of uptake (pH) curves: (A) the uptake does not reach 100% (cations); (B) the uptake does not reach 0% (cations); (C) two uptake curves with a maximum (anions or cations).

cations. Thus, the presence or absence of a maximum in uptake curves depends on the completeness of removal of CO_2 from the system. For anions uptake curves with a maximum are more common than for cations, and they are observed also in absence of any specifically adsorbing species except for the anion of interest. This behavior reflects the difference in solubility between products of hydrolysis of strongly adsorbing cations on the one hand and of anions on the other: weak acids in fully protonated form are relatively soluble, while some products of hydrolysis of heavy metal cations (hydroxides, basic salts) are practically insoluble. Hingston et al. [29] noticed that the pH of maximum of uptake of weak acids on goethite often matches their pK_1. The shapes of the uptake curves with a maximum are very diverse, e.g. they often resemble the uptake curve of an amphoteric substance shown in Fig. 4.5(C) except the range of 100% uptake in not so wide. Not always is the uptake curve symmetrical with respect to its maximum.

A few publications report staircase uptake curves. The curve shown in Fig. 4.7(A) consists of three stairs (each stair is a miniature of a "normal" adsorption edge) and two plateaus, but the number of stairs can be different. In view of the scatter of the data points there is no sharp difference between "normal" and staircase uptake curves, also the number of stairs in the uptake curve can be a matter of subjective judgment.

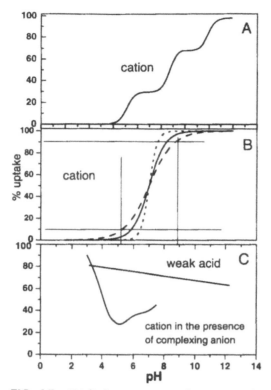

FIG. 4.7 (A) Staircase type uptake curve (cations); (B) different steepness of uptake curves (cations) and $\Delta pH(10-90\%)$; (C) a type of percentage of uptake (pH) curve observed for some anions.

Figure 4.7(B) shows three cation type uptake curves corresponding to the same pH_{50} but of different slopes. The steepness of the uptake curves can be quantified as ΔpH 10–90%, i.e. the width of the pH range from 10% to 90% uptake. Unfortunately, precise determination of this parameter is difficult because of the scatter of experimental results and to deviations from "ideal" behavior (Fig. 4.6 (A) and (B)). Experimental uptake curves of cations are rather steep (ΔpH 10–90% of about 1 pH unit) while for anions ΔpH 10–90% >2 pH units is commonplace. Uptake of some anions is very little sensitive to the pH, and such uptake curves are not s-shaped, but the uptake nearly linearly decreases with the pH (Fig. 4.7 (C)).

Finally, a few uptake curves with a minimum have been reported (cf. Fig. 4.7 (C)), chiefly for sorption of cations in the presence of anions that form very stable water soluble complexes with these cations.

The s-shaped curves shown in Fig. 4.5 (A) and (B) can be linearized when the distribution coefficient defined as

$$K_D = C_{ads}/C_w \tag{4.9}$$

is plotted as a function of pH. The percentage of uptake can be easily recalculated into K_D when the total concentration of the adsorbate and the solid to liquid ratio are known. The C_w in Eq. (4.9) is the concentration of the adsorbate in solution and

C_{ads} is the concentration of the adsorbate bound by the adsorbent, and both concentrations represent sums of concentrations of various species (cf. Figs 4.1–4.4). The C_{ads} can be expressed in mol/kg or in mol/m^2, and consequently K_D can be expressed in dm^3/kg or in dm^3/m^2 (1 dm^3/m^2 = 10^{-3} m).

   An example of actual experimental data (Ni sorption on alumina, low initial Ni concentration, different solid to liquid ratios) plotted as percentage of uptake on the one hand and as K_D on the other is presented in Fig. 4.8. The results plotted as

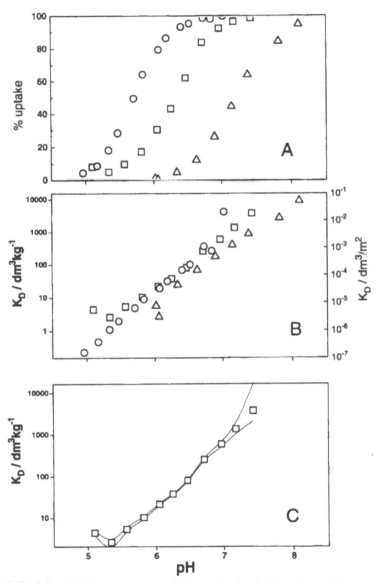

FIG. 4.8   (A) The percentage of Ni uptake for 2 (triangles), 20 (squares), and 200 g (circles) of alumina/dm^3; (B) the same results plotted as K_D (pH); (C) the same results as in Fig. 4 (B) (20 g alumina/dm^3), the lines indicate the error in K_D induced by a 1% error in the percentage of uptake.

percentage of uptake (A) give three separate curves for different solid to liquid ratios. On the other hand, the K_D results obtained at different solid to liquid ratios (B) are relatively consistent. The most serious discrepancies (high relative error in the calculated K_D) are observed when the uptake is close to 0% or 100%. The significance of the calculated K_D values is illustrated in Fig. 4.8 (C), namely, the lines represent the K_D range corresponding to a $\pm 1\%$ error in the percentage of uptake.

The value of K_D depends on the experimental conditions, but it is widely believed that a K_D determined at certain experimental conditions can be used to predict sorption for given adsorbent–adsorbate combination. The variability of K_D was emphasized by Krupka et al. [30], who recently collected K_D values related to migration of radionuclides in geosphere. A dozen of less recent compilations of K_D was quoted by Jenne [31].

Although in specific systems the K_D can be nearly constant over a wide range of experimental conditions, generally it is not the case, and even constant pH does not assure a constant K_D. Systematic studies with different materials of the same chemical composition and different specific surface area showed, that K_D expressed in dm^3/kg was higher for materials having higher specific surface area, but for K_D expressed in dm^3/m^2 the order was reversed [32]. Perhaps this is partially due to application of the BET surface area which is not necessarily appropriate in such calculations. An increase in the initial concentration of adsorbate by 2 orders of magnitude can result in decrease in K_D by one order of magnitude [32]. Even more serious discrepancies in K_D (dm^3/m^2) are observed when results from different sources are compared. Rabung et al. [33] demonstrated that their K_D (dm^3/m^2) for Eu (III)–hematite system at pH 4 was lower by 2 orders of magnitude than a value taken from literature for the same system, but their K_D at pH 5.5 was higher by 3 orders of magnitude than a value taken from another literature source. In other words the K_D values at given pH (extrapolated, because different pH ranges were covered) obtained in two publications differ by a factor of $> 10^5$. Also the difference between the lowest and the highest slope of log K_D(pH) lines reported in literature for Eu(III)–hematite system is substantial: the results range from < 1 to ≈ 2.

The K_D is often reported in form of empirical equation K_D (pH, concentration of competing or complexing ions). Allegations can be found in literature that in absence of complexing anions or competing cations log K_D for metal cations is a linear function of pH. These empirical observations have no theoretical grounds, and they are valid for certain systems over a limited pH range (cf. Fig. 4.8 (B)). Radovanovic and Koelmans [34] tried to rationalize experimental K_D of Cu and Zn in natural water systems. Their model requires the following input data: for solid phase: organic carbon, Fe and Mn content; for liquid phase: pH, electric conductivity, [Cl$^-$], [SO$_4^{2-}$], [HCO$_3^-$], dissolved organic carbon, and [Ca^{2+}].

Advantages of presentation of the uptake curves in terms of K_D over sigmoidal percentage of uptake (pH) curves have been emphasized by Jenne [31]. The main problem with K_D is an illusion of general character of this parameter (the symbol K_D suggests a sort of equilibrium constant) while in fact it is only valid for specific experimental conditions.

The discussed above methods of presentation of uptake data (adsorption isotherms at constant pH and percentage of uptake or K_D (pH) curves) use the

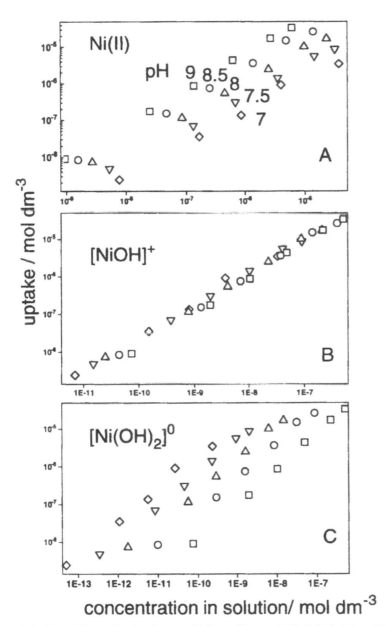

FIG. 4.9 Adsorption isotherms of Ni on silica at pH 7; 7.5; 8; 8.5 and 9, the uptake is plotted as a function of: (A) total Ni(II) concentration in solution; (B) concentration of $NiOH^+$; (C) concentration of $Ni(OH)_2^0$.

overall concentration of the adsorbate in solution as the independent variable and ignore the speciation in solution (cf. Figs 4.2–4.4).

Alternatively the concentration or activity of particular species in solution can be used as the independent variable, although this method of presentation of data is rare. Figure 4.9 (A) shows the uptake of Ni on silica as a function of the overall Ni(II) concentration in solution. A shift in the pH by 0.5 pH unit causes a substantial

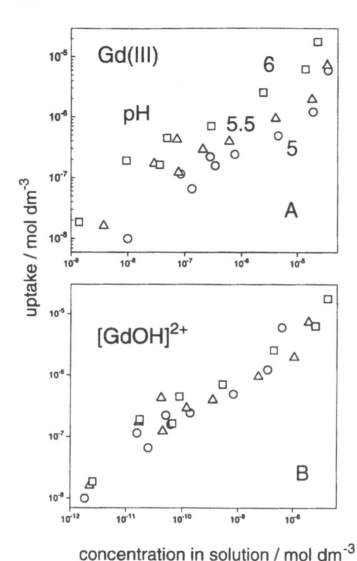

FIG. 4.10 Adsorption isotherms of Gd on silica at pH 5; 5.5 and 6, the uptake is plotted as a function of: (A) total Gd(III) concentration in solution; (B) concentration of GdOH^2 .

shift of the adsorption isotherms. Very similar results (not shown in the figure) are observed when the concentration of the Ni^{2+} ions is plotted as the independent variable. On the other hand with the concentration of the NiOH$^+$ ions as the independent variable the adsorption isotherms corresponding to different pH values overlap (Fig. 4.9(B)). When the uptake is plotted as a function of the concentration of the Ni(OH)$_2^0$ (Fig. 4.9 (C)) or Ni(OH)$^-_3$ species in solution, again a series of different adsorption isotherms corresponding to different pH values is obtained. Figure 4.10 shows uptake of Gd (III) on silica, and again the adsorption isotherms plotted as a function the overall Gd (III) concentration are different (A) and the adsorption isotherms overlap when the concentration of GdOH^{2+} ions is the independent variable (B). In the latter example the concentration of GdOH^{2+} ions in solution is a fraction of 1% of the concentration of Gd^{3+} ions over the entire pH range of interest, thus the overlap of the adsorption isotherms does not necessarily indicate any particular role of the $GdOH^{2+}$ ions in the sorption process.

B. Uptake (Dynamic)

The control over the uptake experiment in a chromatographic column and interpretation of results is even more difficult than with static experiments. For example the pH in dynamic experiments varies not only as a function of time but also spatially (as a function of the position in a column). On the other hand, chromatography is an efficient method to entirely remove certain solutes from solution, it is also used to simulate migration of pollutants in natural systems.

In dynamic studies the chemical retardation is defined as the ratio of velocities of water (w) and of the species of interest (M) through a control volume,

$$R_f = V_w / V_M \tag{4.10}$$

and it is related to K_D

$$R_f = 1 + K_D(\rho/\varepsilon_b) \tag{4.11}$$

where ρ and ε_b are the bulk density and bulk porosity of the material. The latter is interrelated with the surface area

$$A = \varepsilon_b/[r_h\rho/(1 - \varepsilon_b)] \tag{4.12}$$

where r_h is the mean hydraulic radius of pores. The parameter ε_b is not applicable for fractured systems, for which the relationship between R_f and K_D is more complicated.

The surface based distribution coefficient used in chromatography is defined as

$$K_a = V(t_r - t_o)/(Am) \tag{4.13}$$

where V is the volume flow rate, subscripts denote retention and dead time, A is the specific surface area and m is the mass of the adsorbent, so the K_a has a dimension of length.

C. Proton Stoichiometry

The course of acid-base titrations conducted in the presence of specifically adsorbing species is different from that observed at pristine conditions. The experimental limitations and difficulties encountered by potentiometric titration have been discussed in Section 3.1.B.1 and 3.1.B.2. Cancellation of some errors simplifies the assessment of the effect of specific adsorption on surface charging (comparison of titrations in the presence and absence of specific adsorption), but the compensation of errors is not complete, e.g. the solubility of materials (adsorbents) is strongly affected by specific adsorption.

To avoid semantic problems the term σ_0 is not used in this section. Some authors identify σ_0 with the proton charge while the others argue that the σ_0 is a sum of proton charge and the charge of specifically adsorbed ions. More recently a model has been proposed in which only a fraction of the charge of specifically adsorbed ions contributes to the σ_0. The location of the charge of specifically adsorbing ions is discussed in Chapter 5 (adsorption models).

In view of the pH effects induced by hydrolysis in solution, the blank curve method (titration of the initial solution without the absorbent, Fig. 3.3) is only applicable, when the titration curves of the initial solution without the adsorbent and of the supernatant are identical, i.e. when the hydrolysis is negligible over the pH range of interest. Otherwise the proton adsorption can be obtained by back titration of the supernatant (Section 3.I.B. 2). In both methods (blank curve, back titration) the results need a correction for the acid or base associated with the original adsorbent, which is obtained from titrations at different ionic strengths under pristine conditions (Fig. 3.3). The description of the experimental procedure in the papers on the proton stoichiometry of specific adsorption is often not complete enough to assess if all necessary precautions have been taken into account, and the discrepancies in the results reported by different authors for similar systems are probably due in part to differences in the experimental procedure and interpretation of results.

Specific adsorption of anions is accompanied by adsorption of protons and specific adsorption of cations is accompanied by proton release. Consequently the point of zero proton charge is shifted to high pH in the presence of specifically adsorbed anions, and to low pH in the presence of specifically adsorbed cations. This is illustrated in Figs 4.11 (anions) and 4.12 (cations) for typical experimental conditions at which titrations are carried out (alumina, 2600 m^2/dm^3).

The charging curves for the initial concentration of the strongly adsorbing species of 10^{-5} mol dm^{-3} are practically identical with the charging curves at pristine conditions, which are not shown in Figs 4.11 (A) and 4.12 (A). A substantial effect of specific adsorption on the surface charge (the charging curve obtained at pristine conditions serves as the reference) and a substantial shift in the PZC is only observed at a sufficiently high ($> 10^{-4}$ mol dm^{-3}) initial concentration of the strongly adsorbing species. The concentration range in the present examples (10^{-5} to 10^{-3} mol dm^{-3}) covers the transformation from absence of any effect to a shift in the PZC by about one pH unit for specific solid to liquid ratio. In other systems (absorbent, adsorbate, solid to liquid ratio) the "critical" concentration at which specific adsorption induces a substantial shift in the PZC can be very well higher or lower by an order of magnitude. The results shown in Figs. 4.11 and 4.12 indicate that the initial concentration of specifically adsorbing species must be sufficiently high to observe a considerable effect of specific adsorption on the surface charging. However, the possibility to obtain meaningful surface charging curves by acid-base titration at high initial concentrations of adsorbates is confined, namely surface precipitation of cations and enhanced dissolution of absorbents in the presence of anions are potential sources of errors.

Figs. 4.11 (B) and 4.12(B) show that the titration curves obtained at different ionic strengths and constant initial concentration of specifically adsorbing species, can, but not necessarily do have a CIP. Such a CIP (if there is one) does not correspond to zero proton charge; the proton charge at CIP is positive (specific adsorption of anions) or negative (specific adsorption of cations, Fig. 4.12 (B)). Strongly adsorbing species associated with the surface of (allegedly pure) adsorbent are probably the reason for the discrepancies in the PZC (CIP) values reported in the literature (cf. Table 3.1 and 3.3).

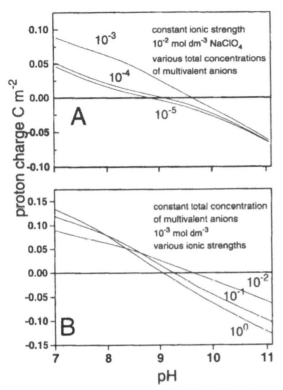

FIG. 4.11 The effect of specific adsorption of anions on the proton charge of alumina: (A) constant ionic strength, various total phosphate concentrations; (B) constant total phosphate concentration, different ionic strengths (schematically).

Not necessarily is the number of coadsorbed/released protons equivalent to the number of specifically adsorbed ions multiplied by their valence, for example specific adsorption of divalent cations is typically accompanied by release of less than two protons per one sorbed cation. The number of protons released per one adsorbed cation or adsorbed with one anion is called proton stoichiometry coefficient r. This coefficient can be directly determined only when the magnitude of specific adsorption is sufficiently high compared with the pristine surface charge, otherwise the effect of specific adsorption on the potentiometric titrations is insignificant (Figs. 4.11 (A) and 4.12(A)). Figures 4.13 (A) and 4.14(A) show the difference between the proton charge in the presence and absence of specifically adsorbed species calculated from the charging curves presented in Figs. 4.11 and 4.12, respectively. This difference becomes insignificant when the uptake is negligible, i.e. at high pH for anions and at low pH for cations. Figures 4.13 (B) and 4.14 (B) show the r calculated from the difference between the proton charge in the presence and absence of specific adsorption taken from Figs. 4.13 (A) and 4.14(A), and converted from charge density (C m^{-2}) into proton adsorption density (mol dm^{-3}) This difference is divided by the magnitude of specific adsorption. The proton stoichiometry coefficient is a function of the pH, although for specific systems the pH effect can be more or less significant. The proton stoichiometry coefficient depends also on the ionic strength

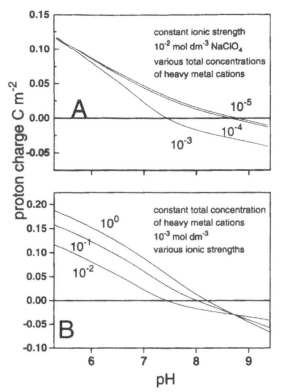

FIG. 4.12 The effect of specific adsorption of cations on the proton charge of alumina: (A) constant ionic strength, various total concentrations of heavy metal ions; (B) constant total concentration of heavy metal ions, different ionic strengths (schematically).

and on the concentration of specifically adsorbing ions, but these effects have not been systematically studied.

Indirect methods have been proposed to calculate r at low concentrations of specifically adsorbing ions, i.e. r is related to the slope of K_D (pH) curves (cf. Fig. 4.8(B)). Some publications report r values calculated as the stoichiometric coefficients of the reaction:

$$Me^{n+}(solution) + surface = Me(surface\ complex) + r\ H^+\ (solution) \qquad (4.14)$$

using different adsorption models (Chapter 5), but the physical sense of such r values is different from that of the experimentally determined ones.

Fokkink et al. [35] argue that for divalent metal cations the proton stoichiometry coefficient as a function of (PZC-pH) should give one bell shaped master curve for all oxides and all metal cations, with a maximum r value slightly below 2 for PZC-pH = 0 and $r = 1$ at PZC-pH = ±4. They support their calculations by experimental data from six sources covering seven adsorbents and six metal cations.

Some papers report r values substantially exceeding the valence of the adsorbing ions, but such results seem to be due to experimental error or inadequate interpretation. Also negative r values (protons coadsorbed with cations or protons released when anions are adsorbed) are rather unlikely.

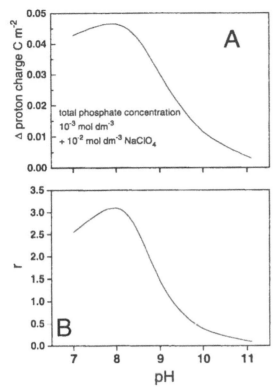

FIG. 4.13 (A) The difference in the proton charge of alumina induced by the presence of 10^{-3} mol dm^{-3} of phosphate (total concentration), calculated from the charging curves in Fig. 4.11 (A). (B) The proton stoichiometry coefficient calculated from the curves shown in Fig. 4.13 (A).

D. Electrokinetic (ζ+ "acousto", cf. Chapter 3)

Figure 3.2 shows the electrokinetic potential of a material as a function of the pH for different concentrations of inert electrolyte. Such plots are typical for simple and mixed (hydr)oxides and for many salts whose surface charging is pH dependent (Section 3.I.G). The increase in the ionic strength at constant pH depresses the absolute value of ζ potential, but the IEP is not affected. The electrokinetic curves in the presence of inert electrolytes are nearly symmetrical with respect to the pristine IEP (Figs. 3.84–3.100), when the electrolyte concentration is not too high. Very high concentrations of 1–1 electrolytes (>0.1 mol dm^{-3}) induce effects characteristic for specific adsorption of cations, i.e.

- The absolute value of negative ζ potentials (above the IEP) is considerably lower than the positive ζ potentials at the same |pH-IEP|, thus the shape of electrokinetic curves is not symmetrical.
- The IEP is shifted to higher pH values when a critical concentration of the electrolyte is exceeded (cf. Figs. 3.101–3.103).

With strongly interacting cations this critical concentration is typically about or even below 10^{-3} mol dm^{-3}, thus, it is lower than the critical concentration of alkali metal cations by several orders of magnitude. With strongly interacting anions

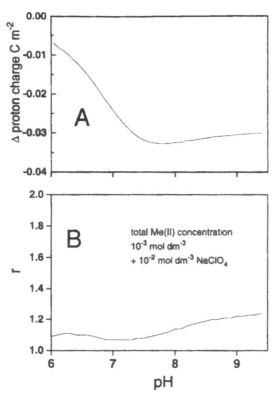

FIG. 4.14 (A) The difference in the proton charge of alumina induced by the presence of 10^{-3} mol dm^{-3} of heavy metal cations (total concentration), calculated from the charging curves in Fig. 4.12 (A). (B) The proton stoichiometry coefficient calculated from the curves shown in Fig. 4.14 (A).

opposite effects are observed, i.e. the IEP is shifted to lower pH values and the positive branch of the electrokinetic curves (below the IEP) is depressed. Some salts contain strongly interacting anions and cations, also complex systems often involve specific adsorption of anions and cations simultaneously. In such systems the shifts in the IEP due to specific adsorption of anions and cations partially cancel out. The direction of the resultant shift depends on the concentrations of specifically adsorbing anions and cations and their affinities to the surface and by fortuitous compensation of the anion and cation effect a pristine IEP is observed.

Allegedly pure materials often contain specifically adsorbing ions as impurities. These impurities induce a shift in the IEP, and special cleaning methods are necessary to remove them. Most likely some unusual pH values reported as pristine IEP for certain materials (Table 3.1, and 3.3) are caused by specific adsorption of impurities, namely, anionic impurities induce a low IEP and cationic—a high IEP. Lack of coincidence between the IEP and CIP (cf. Fig. 4.12) and unsymmetrical shape of the electrokinetic curves corroborates this assertion. The errors caused by specific adsorption of anionic impurities are more common than the cation effects. The CO_2 and SiO_2 errors in electrokinetic measurements and difficulties in removal of multivalent anions, which are present in solutions used to prepare monodispersed colloids, can serve as a few examples (cf. Section 3.I.B.1).

The effect of specific adsorption on electrokinetic behavior of materials is usually presented in form of ζ(pH) curves at constant initial (total) concentration of a specifically adsorbing salt. The electrophoretic mobility rather than the ζ potential is often plotted as a function of the pH. The mobility (directly measured quantity) is a complicated function involving the ζ potential on the one hand and particle size and shape, and concentrations of ionic species in the solution on the other (cf. Figs. 3.80 and 3.81), and exact calculation of the ζ potential in real systems (polydispersed and irregularly shaped particles) is practically impossible. This is a serious difficulty in quantitative interpretation of electrokinetic data obtained in the presence of specific adsorption. On the other hand, the zero electrophoretic mobility corresponds to zero ζ potential, and the shifts in the IEP along the pH axis can be determined with accuracy on the order of 0.1 pH unit.

The effect of specific adsorption of anions (phosphate) on the electrokinetic behavior of alumina is shown in Figs. 4.15–4.18 (experimental data from Ref. [36]). All data points correspond to the same solid to liquid ratio. The electrokinetic curve obtained at initial phosphate concentration of 2×10^{-7} mol dm^{-3} (Fig. 4.15) does not differ from the electrokinetic curve at pristine conditions (not shown). The presence of 10^{-5} mol dm^{-3} phosphate induces a substantial shift in the IEP, and this shift is more pronounced at higher phosphate concentrations. This behavior is typical for specific adsorption of anions. The results from Fig. 4.15 and a few analogous sets of data points obtained at different initial phosphate concentrations (10^{-7} to 10^{-4} mol dm^{-3}) are re-plotted in Fig. 4.16 in the coordinates total phosphate concentration in solution – electrophoretic mobility. This representation gives a random cloud of points. Also the electrophoretic mobility plotted as the function of phosphate surface concentration (not shown) does not reveal any regularity. On the other hand the electrophoretic mobility plotted as the function of $[HPO_4^{2-}]$ (Fig. 4.17) or as the function of $[PO_4^{3-}]$ (Fig. 4.18) produces one master curve containing all data points

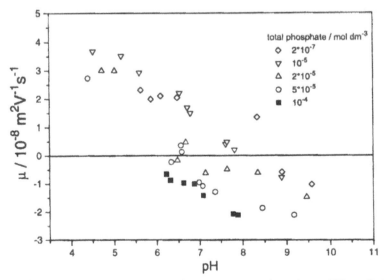

FIG. 4.15 The effect of phosphate on electrophoretic mobility of alumina, plotted as a function of pH (data from Ref. 36).

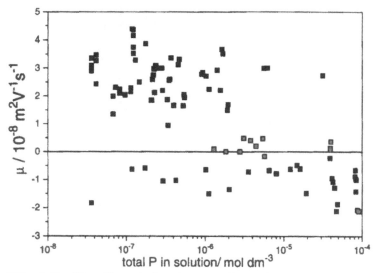

FIG. 4.16 The effect of phosphate on electrophoretic mobility of alumina, plotted as a function of total P concentration in solution (data from Ref. 36).

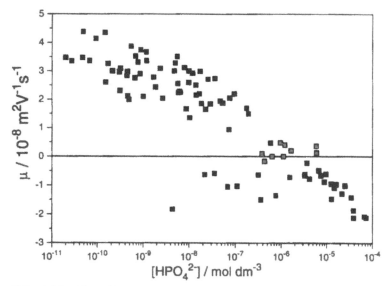

FIG. 4.17 The effect of phosphate on electrophoretic mobility of alumina, plotted as a function of [HPO$_4^{2-}$] (data from Ref. 36).

representing different pH values and different total phosphate concentrations. Systematic studies in this direction have not been carried out, and it is difficult to assess if similar regularities are observed for other systems with specific adsorption of anions. It should be emphasized that the concentration of the PO$_4^{3-}$ species in solution is only a small fraction of the total phosphate concentration over the pH range of interest, while the electrophoretic mobility plotted as the function of the concentration of the H$_2$PO$_4^-$ species (dominating in solution at pH 2–7, cf. Fig. 4.4)

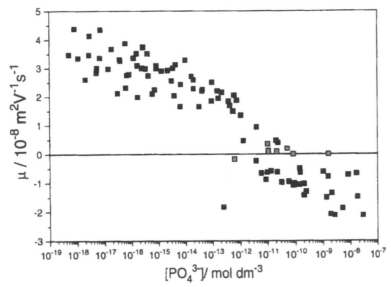

FIG. 4.18 The effect of phosphate on electrophoretic mobility of alumina, plotted as a function of $[PO_4^{3-}]$ (data from Ref. 36).

produces a random cloud of points (the graph is not shown). The scatter of data points in Figs. 4.17 and 4.18 is partially due to simplified calculation of the speciation in solution, e.g. the solubility of alumina was neglected.

Most electrokinetic studies of specific adsorption of anions were conducted with materials whose pristine IEP falls at relatively high pH. These materials carry positive proton charge over a wide pH range, thus the electrostatic conditions are favorable for adsorption of anions. Electrokinetic studies of specific adsorption of anions by materials with the pristine IEP at low pH are rare. The uptake of anions by silica (in absence of strongly interacting cations) is low, and the ζ potential is not affected by the presence of anions. It is difficult to assess if the absence of anion effect on electrokinetic potential is peculiar for silica, or common for materials having their pristine IEP at low pH.

Specific adsorption of certain cations induces shifts in the IEP to higher pH values and at sufficiently high concentration of strongly interacting cations there is no IEP and the electrokinetic potential is positive up to very high pH values. However, shifts to lower pH were also reported, namely, when the specifically adsorbing cation forms an oxide whose IEP falls below the IEP of the adsorbent.

The electrokinetic behavior of silica in the presence of specific adsorption of cations has been a subject of many studies, and it is different from the behavior of most other materials. A typical set of electrokinetic curves of silica obtained at various initial concentrations of cations is shown in Fig. 4.19. At low concentrations the strongly interacting cations do not affect the electrokinetic potential. At higher concentrations the negative ζ is depressed or even reversed to positive, but only over a limited pH range (in Fig. 4.19 the cation effect is most pronounced at pH $\approx$6) while at very low and very high pH the effect of specific adsorption of cations on the electrokinetic curves is insignificant. This may result in multiple IEP. Multiple IEP have been also reported for titania, while for most other materials the IEP is shifted

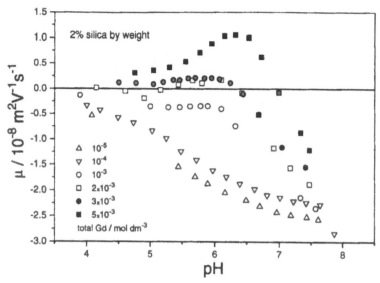

FIG. 4.19 The effect of Gd on electrophoretic mobility of silica, plotted as a function of pH.

to higher pH at sufficiently high concentration of specifically adsorbing cations. Probably multiple IEP are characteristic for systems with onset of adsorption (cf. Fig. 4.5(A) at pH far above the pristine IEP.

The minimum concentration of metal cations necessary to induce a sign reversal of the ζ potential of silica to positive is a function of the solid to liquid ratio. A few studies of the effect of specific adsorption of cations on the electrokinetic curves at different solid to liquid ratios have been reported. For example, with electrophoresis (low solid to liquid ratio) clear shifts of IEP of titania are observed at Ca concentrations on the order of 10^{-4} mol dm^{-3}, but with streaming potential (high solid to liquid ratio) there was no significant shift in IEP of alumina and titania at Ca concentration as high as 10^{-3} mol dm^{-3} [37]. Therefore, the relationship between the concentration of strongly adsorbing ions and the course of electrokinetic curves shown in Figs. 4.15 and 4.19 is only valid for specific solid to liquid ratio (which is different in these two figures). For example, with 0.05% silica by weight, the sign reversal of the ζ potential of silica to positive was observed in the presence of 10^{-4} mol dm^{-3} Gd (results not shown) and with 2% silica by weight the effect of 10^{-4} mol dm^{-3} Gd is rather insignificant, cf. Fig. 4.19. Figure 4.20 presents the data from Fig. 4.19 and analogous results obtained at different solid to liquid ratios, but the mobility is plotted as a function of Gd surface concentration. All data points form an irregular cloud, but a clear regularity is observed when the results obtained at certain pH value (interpolated when necessary) are considered separately: the minimum surface concentration of Gd that induces reversal of the sign of ζ to positive increases when the pH increases. This result is in accordance with the expectations, namely, the pristine negative surface charge (that has to be overbalanced by the positive charge of adsorbed Gd) is higher at high pH (cf. Figs. 3.61–3.66). Figure 4.21 presents the mobility plotted as a function of total Gd(III) concentration in solution. The points corresponding to different pH values and different solid to liquid ratios form one master curve, although with some

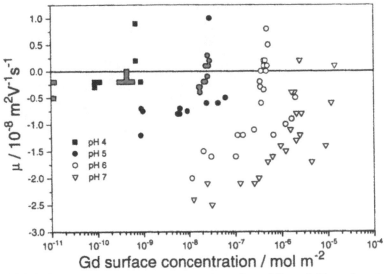

FIG. 4.20 The effect of Gd on electrophoretic mobility of silica, plotted as a function of Gd surface concentration.

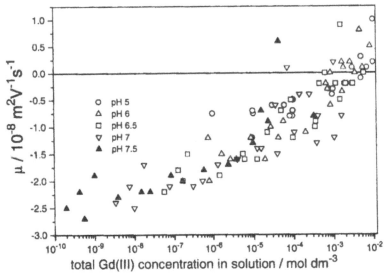

FIG. 4.21 The effect of Gd on electrophoretic mobility of silica, plotted as a function of total Gd concentration in solution.

scatter. The Gd^{3+} species prevails over the entire pH range shown in Fig. 4.21, and the concentration of other Gd species is negligible. Thus, the graph showing the mobility as a function of Gd^{3+} is practically identical with Fig. 4.21. Even when a broader pH range is taken into account (and the concentration of $GdOH^{2+}$ becomes significant) the mobilities plotted as the function of total Gd(III) concentration in solution or as a function of $[Gd^{3+}]$ are consistent with the master curve formed by the points shown in Fig. 4.21. The scatter of data points in Figs. 4.20 and 4.21 is

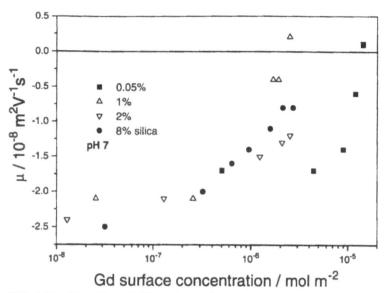

FIG. 4.22   The effect of Gd on electrophoretic mobility of silica at pH 7, plotted as a function of Gd surface concentration.

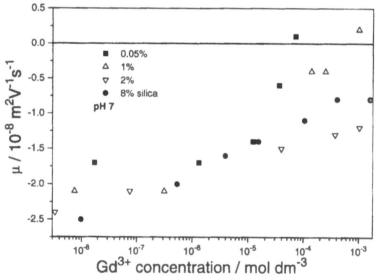

FIG. 4.23   The effect of Gd on electrophoretic mobility of silica at pH 7, plotted as a function of $[Gd^{3+}]$.

much less significant when the results obtained with different solid to liquid ratios are plotted separately. This is illustrated in Figs. 4.22 and 4.23. At pH 7 the Gd surface concentration necessary to reverse the sign of the ζ potential of silica to positive is higher for a low solid to liquid ratio, but $[Gd^{3+}]$ necessary to reverse the sign of the ζ potential is higher for a high solid to liquid ratio. Similar trends are observed at different pH values. Figure 4.24 shows the same results as Fig. 4.21 (and a few data points obtained at pH 8 and 9), but different symbols were used to plot

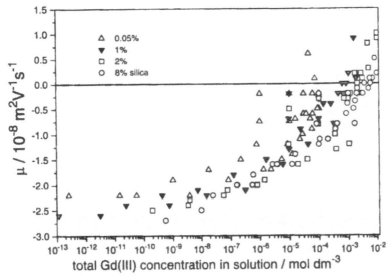

FIG. 4.24 The same results as in Fig. 4.21, different symbols were used to indicate data points obtained at different solid to liquid ratios.

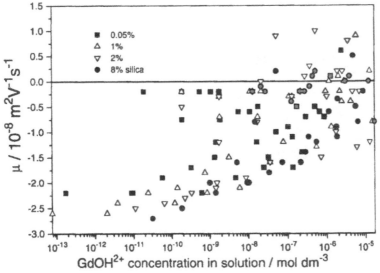

FIG. 4.25 The effect of Gd on electrophoretic mobility of silica, plotted as a function of [GdOH^{2+}].

the results corresponding to different solid to liquid ratios. The mobility plotted as the function of total Gd(III) concentration in solution (Fig. 4.24) or as a function of [Gd^{3+}] (not shown, but the graph is practically identical as Fig. 4.24) is quite consistent (independent of pH and Gd initial concentration) when only the points corresponding to one solid to liquid ratio are considered. On the other hand, a random cloud of points is obtained when the mobility is plotted as the function of the concentration of GdOH^{2+} (Fig. 4.25), even at constant solid to liquid ratio.

The effect of the solid to liquid ratio on the electrokinetic curves (Figs. 4.22–4.24) is probably due to surface precipitation. Adsorption prevails at high solid concentrations and with low solid to liquid ratios the contribution of surface precipitation is substantial.

The discussed above correlations between different quantities characterizing specific adsorption on the one hand and concentration of certain species in solution on the other are limited to specific quantity or method. For example the uptake of Gd on silica plotted as the function of $[GdOH^{2+}]$ gives one curve for different pH (Fig. 4.10(B)), but no regularity is observed when the electrophoretic mobility is plotted as the function of $[GdOH^{2+}]$ (Fig. 4.25). In contrast, adsorption isotherms plotted as the function of $[Gd^{3+}]$ are pH dependent (Fig. 4.10 (A)), but the electrophoretic mobility of silica plotted as the function of $[Gd^{3+}]$ (cf. Fig. 4.24) gives one, pH independent curve.

Surface precipitation has been also deemed responsible for multiple IEP of silica in the presence of heavy metal cations. The reversal of sign from negative to positive (cf. Fig. 4.19, 2×10^{-3} mol dm^{-3} Gd, pH 5) was interpreted as the result of increasing coverage of the surface of silica (IEP at pH about 2, if any) by the (hydr)oxide of the heavy metal (IEP at pH > 6, cf. Table 3.1). The IEP of the mixed surface falls between the IEP of the components, and it depends on the contribution of particular components to the surface, cf. Table 3.3. At sufficiently high pH the surface of silica is entirely covered by the (hydr)oxide of the heavy metal and the reversal of sign from positive to negative (cf. Fig. 4.19, 2×10^{-3} mol dm^{-3} Gd, pH 6.5) was interpreted as the IEP of the (hydr)oxide of the heavy metal. This interpretation is corroborated by the shift of the IEP to lower pH values observed when a metal whose oxide has a IEP at relatively low pH is adsorbed on an oxide whose IEP falls at higher pH.

E. IR Spectroscopy

The energies of vibrational transitions in organic molecules and in anions of oxy-acids are related to the force constants of particular chemical bonds, thus, they are sensitive to the small changes in molecular structure caused by adsorption. Therefore it is possible to identify particular surface species (cf. Fig. 4.1), and even to determine their concentrations by means of IR spectroscopy. This is a serious advantage of spectroscopic methods over the uptake measurements, which give only the overall distribution of the adsorbate between the solid and liquid phase. Unfortunately serious experimental problems are encountered in application of spectroscopic methods *in situ*, i.e. when the solid remains in contact with the surrounding solution. Many results obtained *ex situ*, i.e. after evaporation of the liquid have been published, but their significance to the mechanism of sorption from aqueous solutions is limited by possible changes in the structure of the surface species caused by the removal of the liquid phase.

The main problem in IR measurements *in situ* is a strong IR absorption by the solvent water. Therefore the measurements in the classical transmission mode are practically excluded (the path length must not exceed 15 μm). Short path lengths are provided by the attenuated total reflection (ATR) mode. The probing light beam undergoes multiple total internal reflection in the ATR element (e.g. a cylindrical rod made of ZnSe, with two cone-sharpened ends) and each reflection results in penetration of a thin layer of the medium surrounding this element. The penetration depth of the evanescent wave from the ATR element into the solution equals [38]

$$d_p = \lambda_1 (2\pi)^{-1} [\sin^2\theta - (n_2/n_1)^2]^{-1/2} \tag{4.15}$$

where the subscript 1 refers to the ATR element material, subscript 2 refers to the solution, λ is the wave length, n is the refractive index, and θ is the internal reflection angle. The number of internal reflections is defined by the shape and geometrical dimensions of the ATR element. The IR absorption is a sum of absorption by the solution species and by the surface species on the one hand and by water on the other, and the latter component dominates, thus the signal to noise ratio is rather poor after elimination of the water "background". The ATR technique would be ideal to study adsorption on the ATR material. Unfortunately, the number of materials suitable to make ATR elements (e.g. sufficiently transparent) is limited, and these materials are not particularly important as adsorbents. However, the ATR element can be immersed in a dispersion containing the adsorbent and adsorbate of interest or coated with colloidal particles of the adsorbent and then contacted with solution containing the adsorbate. The latter approach ensures constant solid to liquid ratio in the volume from which the IR spectrum is collected, thus the effective refractive index of the medium surrounding the ATR element (Eq. (4.15)) is constant. Figure 4.26 shows the absorption spectra of sulfate [39] in solution (A) and over a layer of TiO_2 (B). The sulfate concentrations in solution are 0, 2, 4, 6 and 10×10^{-2} mol dm^{-3} (in Fig. 4.26 (B) additionally 8×10^{-2} mol dm^{-3}). With these high solution concentrations the adsorbed sulfate adds a constant spectral contribution to the spectra. Figure 4.26 establishes the validity of Lambert–Beer law, proves the additivity of the spectra of the surface and the solution species, and can be used to determine the effective path length. Figure 4.27 A shows spectra of sulfate (0, 1, 2, and 5×10^{-5}, 1, 2, and 4×10^{-4}, and 1.5, 5, and 8.9×10^{-3} mol dm^{-3} sulfate; pH 3) obtained in the presence of TiO_2. The spectra presented in Fig. 4.27(A) can be resolved as linear combination of three components. Individual spectra of the components are shown in Fig. 4.27 (B). The contribution of the component 1 is a linear function of sulfate concentration, and contributions of the components 2 and 3 can be expressed in terms of Langmuir adsorption isotherm equation, thus they are interpreted as IR spectra of two surface species.

Analysis of IR spectra of weak acids is more complicated because the solution spectrum alone is a linear combination of spectra of different solution species, namely of the acid at different degrees of deprotonation, whose contributions are pH dependent.

The peak deconvolution presented in Fig. 4.27(B) is not unique, i.e. another set of components or other equations of adsorption isotherm than Langmuir can produce very similar spectra. The significance of the peak deconvolution depends on the signal to noise ratio, which is a serious problem in the IR method.

F. TRLFS

Time resolved laser induced fluorescence spectroscopy has been successfully applied to study sorption of Cm(III) on silica *in situ* [40]. Peak deconvolution of the

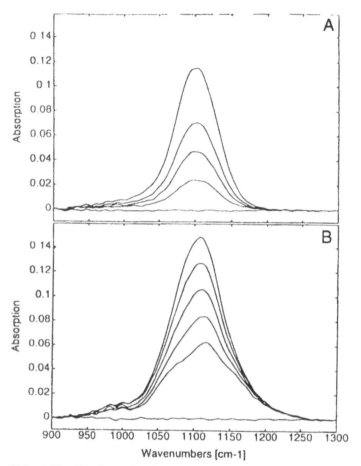

FIG. 4.26 IR absorption by sulfate in solution (A) and sulfate in contact with TiO$_2$ (B). (Reprinted from Langmuir, Ref. 39, copyright ACS.)

fluorescence spectra (similar to that shown in Fig. 4.27) suggests existence of two surface species, whose fluorescence peaks are somewhat broader than the peak of the solution Cm^{3+} species and they fall at longer wavelengths (fluorescence maximum at 594 nm for Cm^{3+}, and 602 and 605 nm for the surface species). The concentrations of these two species are pH dependent. Time dependency of emission decay can be used to estimate the number of water molecules in the first coordination shell of a metal cation. Short lifetime corresponds to high hydration. An empirical formula has been derived from the lifetimes of excited states of aqueous complexes whose hydration numbers can be determined by independent methods. According to this formula the surface complexes Cm-silica coordinate 2 and 0 water molecules, respectively, while 9 water molecules are coordinated by Cm^{3+}.

Trivalent actinides and perhaps lanthanides are potential adsorbates that can be studied using this method. Application of TRLFS to adsorbents other than silica is limited in view of the requirement that the dispersion should be transparent to visible light.

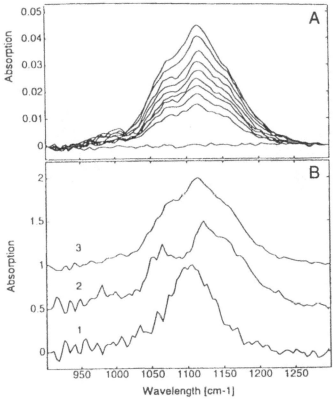

FIG. 4.27 IR absorption by sulfate in contact with TiO_2 (A), and spectra of particular surface species (B). (Reprinted from Langmuir, Ref. 39, copyright ACS.)

G. X-Ray Absorption and Fluorescence

XAS (X-ray Absorption Spectroscopy) is suitable to study sorption of heavy elements, thus its applications for cations are more common than applications in anion adsorption. X-ray absorption results in ionization of core electrons, namely, the energy required to eject a core electron falls in the X-ray region for most elements. The X-ray absorption spectrum of an element has a characteristic narrow (a few tens eV) energy range called absorption edge. The absorption increases abruptly over this energy range, which is referred to as K-edge, L-edge, etc., according to the of the principal quantum number of the ejected electron (K corresponds to $n = 1$ etc.). The fine structure of the X-ray absorption spectrum (local maxima) provides information about the chemical environment of the absorbing element. It is expedient to discuss two regions separately: the XANES (X-ray absorption Near Edge Structure) analysis refers to the absorption edge itself (a few tens eV wide energy range) while the EXAFS (Extended X-ray Absorption Fine Structure) analysis refers to energies reaching about one hundred or a few hundred eV above the absorption edge.

The absorption experiments *in situ* have been originally carried out in transmission mode. The dispersion in form of paste (high solid to liquid ratio) and suitable pH adjustment assured high percentage of uptake. Under such experimental conditions the ratio of the solution species to the surface species was low enough to neglect the former and to treat the overall spectrum as the spectrum of the surface species.

The XANES is interpreted as electronic transitions to empty valence states of the absorbing atom. XANES analysis is conducted by comparison of the absorption spectra of the paste of the adsorbent with ions of interest adsorbed on it on the one hand, and of a series of compounds involving the ions of interest on the other. This series can include:

- Aqueous solutions having different pH and concentration, thus different solution speciation.
- Insoluble metal (hydr)oxides and basic salts involving the ions of interest.
- Salts and mixed oxides involving the ions of interest and the component(s) of the adsorbent.

The match between the spectrum of the paste and one of the solid compounds may verify, e.g. surface precipitation of sparingly soluble (hydr)oxide as the mechanism of sorption. On the other hand, the match between the spectrum of the paste and the model solution spectrum (at a high percentage of uptake) suggests nonspecific (electrostatic) adsorption.

The structure of the surface species in specific adsorption (different from the tested model compounds) can be determined from the EXAFS analysis. The EXAFS spectrum results from modulation (relative change in the adsorption coefficient) due to back scatter from neighboring atoms. It is convenient to introduce the term "shell" as a hypothetical sphere with the absorbing atom (most often the metal cation whose adsorption mechanism is being studied) in the center and N identical atoms (scatterers) at a distance R from it. After conversion of the EXAFS spectra from energy into k-space (momentum space), the modulation by particular shells can be conveniently expressed as a function of N and R. This representation requires a number of empirical parameters that can be calculated from analysis of EXAFS spectra of model compounds of known structure. Once these parameters have been determined, the EXAFS spectra can be calculated for the assumed structure (several shells) and then compared with experimental results. Only the shells with $R < < 1$ nm significantly contribute to the spectra. In models with too many shells the fitted parameters (all R and N values) cannot be unequivocally determined, thus only the R and N values obtained for a few closest shells are significant. Some R and N values estimated from EXAFS analysis are reported in the column "results" of Table 4.1. These parameters depend on the experimental conditions, e.g. the pH, adsorbate to adsorbent ratio and equilibration time, and in most cases they probably represent a mixture of two or more surface species. The XAS measurements in transmission mode require relatively high adsorbate concentrations ($> 10^{-4}$ mol dm^{-3}), thus, precipitation or clustering have been often reported.

Systems with lower concentrations of heavy metal ions can be studied by means of grazing-incidence XAFS spectroscopy. In grazing-incidence geometry the X-ray beam penetrates only about 10 nm into the sample. This method can be applied to study sorption on specified crystallographic planes when single crystals are used rather than powders. Incident beam angles of a fraction of 1 assure total (external) reflection geometry.

Micro SXRF (Synchrotron-based micro-X-ray fluorescence) spectroscopy is an excellent tool to study elemental distributions on microscopic level. This method helps to understand the role of particular minerals in sorption properties of complex mineral assemblies. Iron and managanese oxide phases were detected in polished thin sections of natural zeolitic tuff by SXRF. This section was contacted with synthetic ground water containing Pu(V), and Pu was found to be predominantly associated with manganese oxide, but not with hematite [41]. Micro SXRF can be combined with micro-XANES, and micro-EXAFS to determine the oxidation state and coordination environment of Pu adsorbed at different regions of the tuff [42].

H. Nuclear Magnetic Resonance

The chemical environment of adsorbed paramagnetic ions [Cu(II), Fe(III), Mn(II)] has been studied by comparison of ESR (electron spin resonance) spectra of wet adsorbents treated with solutions of salts of these metals on the one hand and of some model compounds (*vide* Section G) on the other. For instance, clustering or surface precipitation can be verified as the sorption mechanism.

The ESR spectroscopy has been used to determine the distribution of paramagnetic ions between the solution and surface without phase separation. The ESR spectra of the adsorbate depend on the orientation of crystals of known morphology with respect to the magnetic field, thus, the role of particular crystallographic planes in the sorption process can be investigated.

I. TDPAC

Time differential perturbed angular correlation (TDPAC) spectroscopy provides information about chemical environment of nuclei whose decay results in a γ–γ cascade. This method allows to study sorbed ions at very low concentrations, but the number of suitable nuclei, i.e. isomeric nuclei having life times sufficient to prepare the sample (^{111m}Cd, ^{199m}Hg, ^{204m}Pb) is limited. Decay of β-active nuclides often results in excited nuclei, and some of them produce γ-γ cascades. The TDPAC nuclei from this group are more abundant, but the difference in chemical properties between the mother and daughter nuclide is a serious disadvantage. TDPAC has been used to study ^{181}Hf sorption by silica (the β-decay gives ^{181}Ta). The TDPAC spectra of adsorbed Hf were interpreted in terms of three different surface species whose relative fractions depend on pH [43].

J. Mössbauer

Decay of an excited nucleus to the ground state results in emission of a γ-photon. In general the energy of this photon is not equal to the transition energy, because of the recoil of the emitting nucleus. However, when the recoil is taken up by the crystal as a whole rather than by an individual atom, the recoil energy is negligible, and the γ-photons emitted by excited nuclei can be absorbed by like nuclei in ground state and bring about their excitation. The resonance is very narrow, and small shifts and/or splittings of nuclear levels due to the electrons surrounding the nucleus of interest put the emission and absorption out of step when the γ-source and the absorber are in different chemical forms. The chemical shift can be balanced by a Doppler shift when the relative velocity of the source and the absorber is on the order of a few mm/s. The experimental setup consists of a mobile γ-source, stationary γ-absorber (containing the same nuclide as the source in different chemical form) and a γ-counter. The counting rate is plotted as a function of the velocity of the emitter, and one or more minima in such a graph mark the velocities corresponding to resonance absorption. These Mössbauer spectra can be compared with spectra of model compounds, cf. Section G. Unfortunately, most nuclei which are suitable for studies of Mössbauer effect are not particularly important as components of adsorbents or adsorbates.

K. Ex Situ Studies

All surface spectroscopic techniques can be applied to adsorbents that have been contacted with solution containing certain adsorbate and then washed and dried. The general idea of these methods can be summarized as follows: a probe beam impinges on the surface and this results in another beam that exits the surface and carries the surface information to the analyzer. These beams can be photons, electrons or ions of various energies and different combinations of the nature and energy of the incident and resultant beam gives different types of information. Therefore, combination of two or more surface spectroscopies gives a more complete picture than a single method.

The results obtained *ex situ* provide a valuable supplementary information that is not available in *in situ* studies. For instance, the surface precipitate can be directly seen on images of the adsorbent particles obtained by means of high resolution transmission electron microcopy (TEM).

Only selected **results** are presented in Tables 4.1–4.6. Some trends are described qualitatively, e.g. the uptake is depressed, enhanced or unaffected as the result of changes in

- ionic strength
- concentration of strongly interacting counterions
- temperature.

Some directly measured quantities are reported, i.e. the pH_{50} values (cf. Fig. 4.5), or the pH (value or range) of maximum uptake (cf. Fig. 4.6 (C)), the shifts in the IEP (cf. Fig. 4.15), etc. Other results, i.e. the parameters of adsorption isotherms at constant pH and the R and N values derived from EXAFS spectra reported in Tables 4.1 and 4.2 are the best-fit values. The results of fitting procedures are usually not unequivocal, i.e. they depend on arbitrarily chosen criteria, and the same experimental data can be reproduced using many different sets of parameters or different types of adsorption isotherm.

Finally, some formulas and stability constants of (hypothetical) surface species (or Gibbs energies of adsorption) are reported in Tables 4.1–4.4. These quantities belong to adsorption models proposed by the authors of cited publications, but they are not sufficient to calculate the uptake curves or adsorption isotherms when the model involves an electrostatic factor. Adsorption models themselves are not discussed in the present chapter, their terminology is explained in detail in Chapter 5. In contrast with the directly measured quantities that represent the sorption properties at specific experimental conditions, the model parameters characterize the sorption process over a wide range of experimental conditions, although the match between experimental and theoretically calculated quantities was not always satisfactory. The directly measured quantities upon which the models were based can be found in the original publications. Presentation of model parameters was preferred in Tables 4.1–4.4 over directly measured quantities when relatively large sets of experimental data were modeled using relatively few model parameters or when some spectroscopic evidence of existence of surface species was offered. Otherwise, directly measured quantities are presented even when the authors of the original publications proposed some model.

II. SMALL IONS

A. Cations

Sorption of metal cations on different materials has many actual or potential technical applications, e.g. preparation of catalysts and other composite materials. Sorption on hydrous metal oxides was also considered as attractive alternative to other methods of extraction of some rare metals, e.g. U [44] from sea water.

Sorption of heavy metal cations, chiefly the topics related to environmental protection, was reviewed by Charlet [45]. An entire recent book [46] is devoted to uptake of metals by metal oxides, clay minerals and more complex systems.

Hayes and Katz [47] reviewed application of X-ray absorption spectroscopy in studies of metal ion sorption.

Selected studies of specific adsorption of cations are summarized in Table 4.1.

A few sets of macroscopic parameters (K_D, pH_{50}) characterizing sorption of metal cations on Al_2O_3, FeOOH, and SiO_2, obtained in different laboratories at various experimental conditions are presented in Figs. 4.28–4.59. Only the systems for which many different results have been published were taken into account. The results obtained at a possibly low ionic strength, in absence of salts other than inert electrolytes at $\approx 20°C$ were preferred, and results obtained in the presence of nitrates or chlorates VII were preferred over the results obtained in the presence of halides in order to avoid variations in sorption of cations induced by the temperature and ionic strength. The other variables (pH, initial concentration of metal cations, equilibration time, solid to liquid ratio) vary significantly from one data point to another. Multiple data points were taken from some publications, reflecting the effect of these variables on the magnitude of uptake.

The K_D is chiefly a function of pH while the role of the other variables is usually less significant. Systematic studies suggest that the K_D of metal cations increases by an order of magnitude per one pH unit when other variables are kept constant. Therefore K_D is plotted as the function of pH.

The pH_{50} is chiefly a function of the solid to liquid ratio while the role of the other variables is usually less significant. Systematic studies suggest that the pH_{50} of metal cations decreases by one pH unit when the solid to liquid ratio increases by an order of magnitude and the other variables are kept constant. Therefore pH_{50} is plotted as the function of the solid to liquid ratio.

The effects of other variables on K_D and pH_{50} are very diverse. Typically the K_D decreases and pH_{50} increases when the initial concentration of metal cations increases, but a few examples of constant K_D up to very high initial concentrations have been also reported. Usually the K_D increases when the equilibration time increases.

(*Text continues on pg. 412*)

Index of adsorbates for Table 4.1

Adsorbate	Typical valence (less typical valence)	Entries
Ag	I	181, 197–200, 306, 322, 329, 343, 415, 488, 587, 608, 639
Al	III	111, 120, 126, 170, 265, 266, 300, 302, 305, 307, 337, 339, 340, 344, 345, 448, 460–462, 507, 510, 524, 531
Am	III	45, 46, 335, 416, 519, 639
As	III	171 (see also Table 4.2)
Au	III (I)	218, 308, 329
Ba	II	47–49, 80, 105, 114, 116–118, 121, 125, 135, 172, 197, 201, 298, 309, 322–325, 333, 335, 336, 338, 341, 342, 346, 347, 417–421, 453, 458, 463, 464, 489–492, 520, 525, 556, 557, 604
Be	II	348, 609
Bi	III	171, 265
Ca	II	1, 47, 48, 50–52, 105–107, 116, 118, 121, 125, 139, 140, 172, 197, 202, 203, 265, 267, 322–324, 326, 333, 338, 341, 342, 349, 350, 417–420, 422–425, 453, 458, 465, 466, 489–492, 511, 515, 520, 524, 525, 528–530, 539, 556, 577, 604, 606, 610, 612
Cd	II	2–5, 50, 53–59, 63, 105, 108, 141, 173, 174, 182, 197, 204–210, 219–228, 268–276, 298, 299, 310, 322, 329, 334, 351, 352, 415, 426–431, 467–469, 488, 489, 493, 504, 505, 521, 532, 541, 542, 558–560, 578, 579, 594, 605, 606, 620, 621, 623, 627, 637, 642–644, 646
Ce	III (IV)	177, 327, 336, 447, 453, 519, 557, 613, 645
Cm	III	353
Co	II (III)	6–16, 49, 53, 54, 60–66, 86, 87, 105, 127–130, 136, 142–144, 197, 204, 211, 268–270, 277–281, 299, 311, 312, 322, 328–330, 333, 335, 336, 354–358, 426, 427, 432–439, 449, 450, 452, 453, 455, 459, 470–474, 494, 495, 506, 508, 509, 512, 515, 516, 519, 533, 540, 543–546, 558, 559, 561, 580, 583, 589–591, 595, 596, 611, 622, 625, 639
Cr	III	17–19, 60, 67, 120, 136, 177, 178, 205, 212, 213, 265, 359–361, 432, 453, 455, 506, 516, 526 (Cr VI see Table 4.2)
Cu	II	20–26, 50, 55–57, 60–63, 73–75, 78, 79, 81–83, 88–97, 105, 109, 110, 126, 137, 173–176, 179, 183–185, 197, 200, 206, 207, 214–216, 229–231, 265, 267, 269–273, 277, 279, 282–286, 291, 298, 322, 325, 328, 329, 362, 371, 426, 428, 432–437, 440, 449, 453, 459, 475–477, 493, 494, 496, 506, 522, 523, 532, 535, 540, 561, 562, 564, 578, 591, 605, 623, 628, 638, 642–644, 646
Er	III	177, 447

Index of adsorbates for Table 4.1 (Continued)

Adsorbate	Typical valence (less typical valence)	Entries
Sc	III	160, 295, 329, 335, 336, 447, 453
Sn	II	448, 639
Sr	II	48, 53, 77, 105–107, 114, 116–118, 121, 125, 132, 161, 172, 177, 190–192, 201, 202, 210, 254, 255, 298, 332, 333, 335, 336, 338, 341, 342, 398, 418–420, 443, 450, 453, 459, 489, 492, 498, 520, 534, 537, 538, 549, 550, 560, 563, 570–572, 595, 597, 604, 618, 624, 637, 639, 645, 647
Ta	V	329
Tb	III	447
Th	IV	162, 163, 180, 256, 321, 335, 399–401, 416, 424, 438, 444, 482, 490, 519, 551, 556, 582, 619, 639
Tl	III (I)	170, 197, 209, 322, 334
Tm	III	447
U	VI (UO$_2^2$)	40, 41, 100, 112, 113, 133, 134, 164–168, 180, 193, 194, 257, 260, 290, 296, 402–408, 444–446, 453, 454, 456, 457, 483–486, 499–503, 505, 513, 518, 519, 552, 555, 573–576, 582, 585, 586, 602, 626, 633, 639, 640
V	III (IV)	67, 101, 110, 335 (V(V) see Table 4.2)
Y	III	71, 335, 336, 409, 441, 447, 453
Yb	III	138, 177, 410, 447
Zn	II	42–44, 49, 50, 54, 57, 59, 61, 63, 66, 72, 78, 102–105, 108, 109, 115, 122–124, 137, 169, 176–179, 195–197, 203, 205–207, 209, 211–214, 261–265, 269–270, 272–275, 277 286, 289, 297–299, 322, 328, 329, 333, 334, 411–413, 425–427, 430, 431, 434, 435, 437, 449, 452, 453, 487, 489, 504, 505, 510, 519, 521, 532, 559, 562, 579, 581, 588, 596, 603, 605, 606, 634–636, 641, 643, 644
Zr	IV (ZrO^2)	414, 445, 446, 514

TABLE 4.1 Specific Adsorption of Cations

Adsorbent	Adsorbate	c total	S:L ratio	Eq. time	T	pH	Electrolyte	Method	Result	Ref.
1. Al_2O_3, Degussa.	Ca	10^{-1} mol dm	0.5 g (100 m²/g), 30 cm³			7 10	chloride	ζ, proton release	No shift in the IEP (streaming potential), CIP of charging curves at different $CaCl_2$ concentrations at pH 7.4	37
2. Al_2O_3,	Cd	$5 \cdot 10^{-7}$ mol dm	1 g (125 m²), dm	2-4 h	25	6-7.5	0.7 mol dm⁻³ nitrate 0.5 mol dm chloride, 0.2 mol dm sulfate, 0-0.01 mol dm thiosulfate	uptake	The uptake at low pH is enhanced in the presence of thiosulfate	48
3. Al_2O_3, α	Cd	10^{-4} mol dm	2.50 g dm³			5 10	0.1 mol dm $NaNO_3$	uptake	Linear log-log adsorption isotherms at pH 7 7.5, slope 0.7	49
4. Al_2O_3, chromatographic, Merck, washed	Cd	10^{-4} mol dm	20 g (130 m²/g), dm		15, 35	4 10	0.01-1 mol dm NaCl, NaClO₄	uptake	AlO Na Cd^2 AlOCd Na log K 7.4 (15 C), 7.1 (35 C) AlO·Na Cd^2 H Cl AlOHCdCl + Na log K 15.7 (15 C), 14.9 (35 C)	50
5. Al_2O_3, α	Cd							uptake, proton release		51
6. Al_2O_3 99.5%, reagent grade	Co	10^{-4} mol dm⁻¹			30	9 5		ζ	ζ 0 at 3 10 mol Co m	52
7. Al_2O_3, Degussa C washed	Co (II and III)	2 10 10 mol dm	1 g (100 m²), dm	1 d	25	5 10	0.01 mol dm NaClO₄, [EDTA] 0 or [Co]	uptake	Co(II) in absence of EDTA pH₅₀ 7; Co(II) in the presence of EDTA: anion like adsorption edge pH₅₀ 7.5; Co(III) in the presence of EDTA uptake lower by a factor of 3 as compared with Co(II)	53
8. Al_2O_3,	Co	6.3 10^{-4} 4.2 10^{-6} mol dm	5 g (500 m²), 200 cm		15, 35	6 14	chloride	acousto	The IEP shifts to higher pH values, positive ζ, at pH 14 for 4.2 10^{-6} mol dm⁻³ Co.	54
9. Al_2O_3	Co	carrier free·10^{-4} mol dm	2, 50			5 9	0.1 mol dm NaCl	uptake	The uptake at constant σ₀ increases with T	55
10. Al_2O_3, Houdry 415	Co	10^{-3} 10 mol dm	5 m² 14 cm	1 d	25	4 10	0.01, 0.1 mol dm NH₄NO₃	uptake, ζ	Positive, almost pH independent ζ, at 10^{-4} mol dm⁻³ Co	56
11. Al_2O_3, α, Buehler	Co	2×10^{-6} 10^{-3} mol dm	0.2 200 g (12.6 m²/g), dm	2 d	25	5 9	0.01-0.1 mol dm $NaNO_3$	uptake	AlOCo log K -1.7 to -0.6; the value depends on the assumed TLM parameters with almost the same SOS DF. The uptake is ionic strength independent. Even with surface coverages below 10% formation of substantial amount of surface precipitate.	57
12. Al_2O_3, α, Linde A	Co	10^{-4} 10^{-2} mol dm⁻³			20	8.1		EXAFS	2.7×10^{-7} mol Co m⁻²: 0.7 Al atoms at 0.363 nm, 0.6 Co atoms at 0.316 nm; 2×10^{-5} mol Co m⁻² 4 Co atoms at 0.31 nm.	58
13. Al_2O_3, JRC, ALO-4	Co	10^{-6} 10^{-3} mol dm⁻³		3 12 h	25	6-7	0.1 mol dm $Na.NO_3$	uptake	Curvilinear log-log adsorption isotherms, their slope decreases as pH increases.	59

TABLE 4.1 Continued

Adsorbent	Adsorbate	c total	S L ratio	Eq time	T	pH	Electrolyte	Method	Result	Ref.
14 Al_2O_3, Linde A, $\alpha + \gamma$, washed	Co	10^{-4} 10^{-2} mol dm^{-3}	20-50 g (15 m^2/g.) dm^3	6-9 d	21	8-8 2	nitrate	EXAFS	At low coverages multinuclear complexes are formed with 0.6-0.9 nearest Co neighbors at 0.311 nm. At high coverages double Al-Co hydroxide is formed.	60
15. Al_2O_3, Linde A, $\alpha + \gamma$	Co	10^{-4} 10^{-2} mol dm^{-3}	20-50g dm^3	6 d	23	8 8.2	nitrate	EXAFS	For Co initial concentrations $> 5 \times 10^{-4}$ mol dm^{-3}: 2.5-4 nearest Co neighbors at 0.307-0.312 nm and 2.5-4.1 second next Co neighbors at 0.61-0.618 nm.	61
16. Al_2O_3, α, Union Carbide	Co	9×10^{-5} mol dm^{-3}	single crystal	4 d		8 1	nitrate	grazing-incidence XAFS	The uptake on the (0001) plane is due to formation of $(Al_2OH)_3Co^2$ $(H_2O)_3$ species. The uptake on the (1 102) plane is due to formation of $[(Al_3O)_2(Al_2OH)(AlOH)]$ Co(H$_2$O)$_2$ species.	62
17. Al_2O_3, δ, Degussa C	Cr(III)	5×10^{-4} 3×10^{-2} mol dm^{-3}	10g dm^3	10 min 14 d		2.5-4	0.1 mol dm^{-3} KNO$_3$, [oxalate] = 3[Cr]	uptake, ESR	The pH$_{50}$ = 3.3 in absence of oxalate. The uptake curves in the presence of oxalate are anion like after a 10 min equilibration, pH independent 25% uptake was observed after a 1 d equilibration, and cation like uptake curves are reported for a 14 d equilibration.	63
18. Al_2O_3, γ, from nitrate	Cr(III)	10^{-2}-0.4 mol dm^{-3}	2 g (214 m^2/g), 60 cm^3	1 d	25	1.9 3.6	nitrate	uptake, ζ		64
19. Al_2O_3, Woelm Pharma	Cr(III)	10^{-5} 1.2×10^{-4} mol dm^{-3}	10 g (143 m^2 g), dm^3	2.5 h	25	3.5- 7	10^{-4}- 10^{-2} mol dm^{-3} KNO$_3$	uptake	The uptake in enhanced at high ionic strengths, $\equiv$AlOCrOH log K -3.96, slope 0.7. Linear log-log adsorption isotherm at pH 4.6, slope 0.7.	65
20. Al_2O_3, γ, Degussa C	Cu	10^{-4} mol dm^{-3}	1 g (100 m^2) dm^3	3 h	25	2 10	0.025 mol dm^{-3} NaClO$_4$. 0-10^{-3} mol dm^{-3} aspartate, NTA, glycine	uptake	Uptake peaks at pH 5.5 (60%) in the presence of aspartate, and at pH 7.5 in the presence of NTA (25% uptake) and glycine (15%).	66
21. Al_2O_3, δ	Cu							EPR, H and N ENDOR (ex situ)	Structure of ternary surface complexes with bipyridine and dimethylglioxyme is proposed.	67
22. Al_2O_3, α, A 16 SG, Alcoa	Cu	10^{-5} 10^{-2} mol dm^{-3}	100 g (8.4 m^2 g) dm^{-3}		25	3 10	0.01 mol dm^{-3} NaCl, 0-0.003 mol dm^{-3} glutamic and benzoic acid	uptake	In absence of organic ligands pH$_{50}$ 5.3 for 10^{-4} mol dm^{-3}; linear log-log adsorption (initial concentration) plot at pH 4.5 slope 0.8; the uptake at pH 4.5 is enhanced in the presence of glutamic and benzoic acid and it is not affected by phenol (0.05 mol dm^{-3}).	68
23. Al_2O_3, δ, Degussa C	Cu	10^{-3} mol dm^{-3}	1 g (100 m^2), 25 cm^3	1 d		8.6	0.1 mol dm^{-3} KNO$_3$ + glucose phosphate, pyrophosphate, tripolyphosphate, orthophosphate, nitrilotris(methylene)triphosphonate [ligand] = [Cu] or [ligand] = 2[Cu]	ESR		69

No.	Adsorbent	Species	Concentration	Solid	Time	T	pH	Electrolyte	Method	Comments	Ref.
24.	Al$_2$O$_3$, Degussa C	Cu	10^{-4} mol dm	2000 m^2 dm^3	1 d		4-9	0.1 mol dm^3 NaNO$_3$, + 0.2×10^{-4} mol cm^3 2,2'-dipyridine	uptake, EXAFS	pH$_{50}$ ~ 5 in absence of bipy; 7.5 in the presence of 2×10^{-4} mol dm^3 bipy. In absence of bipy adsorbed Cu has on average 1.7 closest Al neighbor at distance of 0.281 nm and there is no clustering of Cu. In the presence of bipy the surface species has a ratio Cu: bipy equal 1:1 even in the presence of excess of bipy.	70
25.	Al$_2$O$_3$, Degussa C	Cu	10^{-4} mol dm	1000-2000 m^2 dm^3	1.7 d		4-6.5	0.1 mol dm^3 NaNO$_3$	uptake, EXAFS	pH$_{50}$ 5 (2000 m^2 dm^3). Cu has 0.9-1.6 closest Al neighbors at distance of 0.283 nm and there is no clustering of Cu.	71
26.	Al$_2$O$_3$, Degussa C	Cu	5-10^{-4} mol dm	625 m^2 dm^3	1 d		4-8	0.001-1 mol dm^3 NaNO$_3$, 1-5×10^{-3} mol dm glutamate	uptake, EXAFS	In absence of glutamate, pH$_{50}$ 5.8 for 1-0.1; the uptake is depressed at high ionic strength; uptake is depressed in the presence of glutamate at pH 6 and enhanced at pH 6. Inner sphere 1:1 ternary complex is formed in alkaline pH range, outer sphere 2:1 glutamate:Cu ternary complex is formed in acidic range.	72
27.	Al$_2$O$_3$, δ, Degussa C	Ni	5×10^{-4} mol dm^3	111 m^2, 53 cm^3	1 d		4-8	10^{-2} mol dm^3 NaClO$_4$	uptake	pH$_a$ 6.6	73
28.	Al$_2$O$_3$, Houdry	Ni	10^{-4}-0.03 mol dm	3.57 g (123 m^2 g) dm		25	4-10	0.1 mol dm^3 NH$_4$NO$_3$	uptake, ζ	Only positive ζ potentials up to pH 10.5 at 10^{-4} mol dm^3 Ni and 10 mol dm NH$_4$NO$_3$.	74
29.	Al$_2$O$_3$	Ni	10 mol dm^3	1 g dm^3			3-9	0.025 mol dm^3 NaClO$_4$, 10 mol dm^3 EDTA	uptake	Anion like adsorption edge; pH$_{50}$ 7.3.	75
30.	Al$_2$O$_3$, α, NIST	Np(V)	10^{-6} mol dm	4 g (0.23 m^2/g) 25 cm^3	10 d	20 ±2	3-11	0.01 mol dm^3 NaNO$_3$	uptake	K_E 8 dm^3 kg and is rather pH independent at pH 4-10, also the difference between experiments conducted in capped vials (low CO$_2$) and in a CO$_2$ free glove box is insignificant	76
31.	Al$_2$O$_3$, Cabot, washed	Pb	1-2.9×10^{-4} mol dm^3	3.12 g (117 m^2 g) dm	2 d	25	4-6	0.1 mol dm^3 NaClO$_4$	uptake, proton release	Model without Boltzmann factor: ≡AlOPb log K-2.2 (≡AlO)$_2$Pb log K-8.1 1.5 protons released per one adsorbed Pb.	77
32.	Al$_2$O$_3$, Japan Aerosil, washed	Pb	10^{-4}-2×10 mol dm^{-3}	25 g (100 m^2 g) dm^3		20	4.5-5.4	nitrate	uptake	Langmuir type adsorption isotherms at constant pH, binding constants 0.75, 1.6 and 2×10^3 mol^{-1} dm^3 (typo in Table 1), adsorption capacity 6.2, 6.6 and 8.3×10^{-5} mol g^{-1} at pH 4.5, 5, and 5.5, respectively	78
33.	Al$_2$O$_3$, Cabot	Pb	0.005-0.015 mol dm	100 g (117 m^2 g) dm			6	0.1 mol dm^3 NaNO$_3$	XAS	0.9:1 O at 0.223 nm, 1.8-2.2 O at 0.245 nm; for the 0.015 mol dm^{-3} Pb sample also 0.5 Pb at 0.345 nm and 0.5 Al at 0.372 nm.	79
34.	Al$_2$O$_3$, Degussa C	Pb	10^{-3} mol dm^3	2.1-6.3 g (100 m^2 g) dm	2 d		5.5-5.9	0.1 mol dm^3 NaCl	XAFS	EXAFS spectra suggest precipitation of PbOHCl, and absence of ternary Pb-Cl surface complexes. Pb adsorbs as inner sphere complex.	80
35.	Al$_2$O$_3$, Alfa	Pb	2×10^{-7}-2×10^{-6} mol dm	65000 cm^{-1}	1 d	25	6	0.05 mol dm^3 NaNO$_3$	uptake	Langmuir equation does not reproduce the results.	81
36.	Al$_2$O$_3$, from isopropoxide	Pd (II)	2×10^{-4}-2×10^{-2} mol dm^3	1 g (119 m^2, 75 cm)	1 h		10.6	0.07 mol dm^3 NH$_4$Cl	uptake		82
37.	Al$_2$O$_3$, Condea	Pd (II)	5, 6×10 mol dm^3		30 min	21	7-12	NaNO$_3$, NH$_4$NO$_3$	uptake	The uptake in the presence and absence of NH$_3$ peaks at pH 10.5.	83

TABLE 4.1 Continued

Adsorbent	Adsorbate	c. total	S L ratio	Eq. time	T	pH	Electrolyte	Method	Result	Ref.
38. Al_2O_3, γ, Cyanamid. calcined	Pd	0.002–0.03 mol dm⁻³	65000 cm⁻¹	1 h	25	8 10	0.1 mol dm⁻³ NH₄	uptake	Langmuir adsorption isotherms at constant pH.	84
39. Al_2O_3, BDH	Pu (IV)	10 mg dm³	0.1 g 10 cm³	5 h		natural	0.1 1 mol dm⁻³ Na₂CO₃	uptake, chromatography	The uptake of Pu is not influenced by dibutyl phosphate (< 0.1 mol dm⁻³).	85
40. Al_2O_3, α, NIST	U	95 ppb	0.1 g (0.07 2 m²/g) 40 cm³	8 d		2 9	0.1 mol dm⁻³ NaNO₃, open to atmosphere, HCO₃	uptake	The uptake peaks at pH 6.5; ≡AlO(UO₂)₂CO₃(OH); log K 12.6-12.9.	86
41. Al_2O_3, α, NIST	U	5×10⁻⁷ mol dm⁻³	2.8 g (0.07, 0.23 m² g) dm³	10 d	20 ±2	2 9	0.1 mol dm⁻³ NaNO₃, 10⁻³·⁵ atm CO₂	uptake	K_D peaks at pH 6.3 (100 dm³ kg).	32
42. Al_2O_3, α	Zn					6-7		uptake		87
43. Al_2O_3, α, Fisher	Zn	80 ppm	0.2 g, 50 cm³	1 d			NaCl	uptake, proton release	1.6-1.8 protons released per one adsorbed Zn. Number of protons released + number of Cl⁻ adsorbed − 2× number of Zn adsorbed.	88
44. Al_2O_3, γ, Degussa C	Zn	5×10⁻⁵ 10⁻³ mol dm⁻³	0.5-5 g (100 m² g) dm³	2 d		4-8	0.01 mol dm⁻³ NaClO₄	uptake, ζ	pH₅₀ = 6.2 (10⁻⁴ mol dm⁻³ Zn, 5 g Al₂O₃ dm³), the IEP is shifted to higher pH.	89
45. Al_2O_3, γ, Degussa	Am(III), Eu	10⁻¹¹ 10⁻⁴ mol dm⁻³	3.6 g dm³			4-8	0.1-0.001 mol dm⁻³ NaClO₄	uptake, TRLFS	The uptake is rather insensitive to the ionic strength. Practically identical results for Am and Eu. The log-log adsorption isotherms at constant pH consist of two rectilinear segments, the slope of low concentration segment − 1.	90
46. Al_2O_3, γ, Degussa C	Am(III) Np(V) Pu	5×10⁻¹⁰ mol dm⁻³ 10⁻¹⁴ mol dm⁻³ 2×10⁻¹⁰ mol dm⁻³	200 ppm	4-7 d	25	3 11	0.1 mol dm⁻³ NaClO₄ + 0.50 ppm humic acid	uptake	Am(III): pH₅₀ ~ 5.5, the uptake is enhanced in the presence of humic acid; Np(V): pH₅₀ = 7.2, the uptake is enhanced in the presence of humic acid at pH < 8 and depressed at pH > 8; Pu: pH₅₀ = 7.3, the uptake is enhanced in the presence of humic acid at pH < 8.	91
47. Al_2O_3, ?	Ba, Ca	10⁻³ mol dm⁻³			22.5	6-11	chloride	ζ	Positive ζ over the entire pH range. Ba Ca.	92
48. Al_2O_3, ?	Ba, Ca, Mg, Sr	10⁻⁵ 10⁻³ mol dm⁻³	1.39 g (117 m²/g) 0.5 dm³	15 mn	25	4-11	0.1 mol dm⁻³ NaCl	uptake, ζ, proton release	Langmuir adsorption isotherms. Mg > Ca > Sr > Ba; for 10⁻³ mol dm⁻³ Ca no IEP, only positive ζ potentials. 1 1.27 proton released per 1 Ca adsorbed	93
49. Al_2O_3, α, Linde, sintered at 1000 C	Ba, Co, Mn(II), Ra, Zn	10⁹ 10³ mol dm³	1 g (1.5 m²) dm³	>2 h	25	5-10	0.1 mol dm⁻³ chloride, nitrate, sulfate	uptake	pH₅₀ 6.8 for Zn, 7.6 for Co, 8.6 for Mn and 10.1 for Ba and Ra is independent of concentration in solution up to 10⁻⁷ mol dm⁻³, and then linearly increases with log c. The effect of anions on the uptake curves is insignificant.	94

No.	Adsorbent	Species	Concentration	Solid	Time	T (°C)	Electrolyte	pH	Method	Remarks	Ref.
50.	Al_2O_3, γ, Degussa C	Ca, Cd, Cu, Ni, Pb, Zn	1×10^{-6} 2×10^{-4} mol dm^{-3}	0.3-1 g (100 m^2) dm^{-3}	30 min	25 ±2	0.025 mol dm^{-3} NaClO$_4$, 0-10^{-4} mol dm^{-3} EDTA	3 10	uptake ζ	Anion like uptake(pH) curves for [metal] [EDTA] 10^{-4} mol dm^{-2}; 1 g dm^{-3}, almost identical for Ni, Cd, Cu, Pb, Zn. Neither 0 nor 100% uptake is reached, cf. Fig. 4.6 A and B. At the same conditions the Ca uptake peaks at pH 6.5. The uptake of Ni at [Ni] [EDTA] 10^{-4} mol dm^{-3} is rather T insensitive at pH 3 and 3.5 and decreases when T increases at pH 6. The shift in the IEP is negligible when [EDTA] [Ni].	95
51.	Al_2O_3, chromatographic, Merck	Ca, Eu	10^{-5} mol dm^{-3}	40 g dm^{-3}	20 min	15, 35	0.001-0.1 mol dm^{-3} NaCl	4-10	uptake	Eu uptake is enhanced at high ionic strength; Ca uptake at constant σ$_0$ is T independent; Eu uptake at constant σ$_0$ increases with T.	96
52.	Al_2O_3	Ca,Mg	0.1 mol dm^{-3}	10 g (115 m^2/g) 50 cm^3	1 h	22 ±2	chloride	3-10	proton release, ζ	Shift of PZC to pH 7.5 (pristine value at pH 8.3) for CaCl$_2$ and MgCl$_2$, no shift for MgSO$_4$, positive, almost pH independent ζ potentials for CaCl$_2$ and MgCl$_2$. A pH independent ζ = 0 in the presence of MgSO$_4$.	97
53.	Al_2O_3, 2 samples: γ, Fisher and synthetic from nitrate	Cd,Co,Eu,Sr carrier free		5-500 mg/2 10 cm^3	6 d	..	0.01 5 mol dm^{-3} NaCl, NaNO$_3$	5-10	uptake	The uptake of Sr is depressed at high ionic strengths and the slope of log (distribution coefficient) vs. log [Na] curves increases when pH increases. The uptake of Cd is depressed in the presence of NaCl, but NaNO$_3$ does not exert any significant effect. Uptake of Co is insensitive to NaCl up to 4 mol dm^{-3}.	98
54.	Al_2O_3, Prolabo	Cd, Co, Mn(II), Zn	$< 10^{-7}$ mol dm^{-3}	1, 10 g (53 m^2 g) dm^{-3}	1 d	20 ±2	0.05 mol dm^{-3} NaNO$_3$, 2×10^{-4} mol dm^{-3} salicylate	4-9	uptake	Zn > Cd > Co > Mn(II) in absence of salicylate. Uptake of Mn and Co is rather insensitive to the presence of salicylate. Uptake of Cd and Zn is enhanced at pH < 6 and depressed at pH > 7.	99
55.	Al_2O_3, γ, Degussa	Cd, Cu	1 3 × 10^{-6} mol dm^{-3}		1 d		0.55 mol dm^{-3} NaCl, 25 mg dm^{-3} humic acid	6.4	uptake	Humic acid enhances uptake of Cu, its effect on Cd uptake is negligible.	100
56.	Al_2O_3, γ, Cabot, washed	Cd, Cu	5×10^{-7}-6×10^{-6} mol dm^{-3}	50 mg 1 g (120 m^2 g) dm^{-3}	9 h		0.01-0.1 mol dm^{-3} NaCl	4.5 8.5	uptake	The pH$_{50}$ (5×10^{-7} mol dm^{-3}, 50 mg/dm^3) 6.3 for Cu, 7.9 for Cd. Uptake in the presence of dissolved organic matter is enhanced at low pH and depressed at high pH. Cu uptake depends on the method of addition of reagents, namely, is lower when organic ligands are added first, and higher when Cu is added first, for the same total concentration of reagents.	101
57.	Al_2O_3, γ	Cd,Cu, Pb,Zn							uptake		102
58.	Al_2O_3, Degussa C, γ	Cd, Pb	10^{-4} mol dm^{-3}	1 g (110 m^2) 50 cm^3	1 min	25	1 mol dm^{-3} NaClO$_4$	1.5-7	uptake	The pH$_{50}$ 5.8 for Cd and 3.8 for Pb. ΔH_{ads} 21 kJ mol^{-1} for Cd and 82 kJ mol^{-1} for Pb.	103
59.	Al_2O_3, Mager	Cd,Zn	10^{-5} 10^{-2} mol dm^{-3}			25	10^{-3} mol dm^{-3}	6-6.5	uptake	Linear log-log adsorption isotherms, slope 0.6. Once certain concentration is exceeded, further increase of the concentration in solution does not induce higher adsorption (plateau).	104

TABLE 4.1 Continued

Adsorbent	Adsorbate	c. total	S:L ratio	Eq. time	T	pH	Electrolyte	Method	Result	Ref.
60. Al_2O_3, γ, hydrolysis of isopropoxide	Co, Cr(III), Cu, Ni	7.8×10^{-5} mol dm⁻³	0.05-0.3 g (170 m²/g) 50 cm³	2 d		4-10	0.002 mol dm⁻³ KNO₃	uptake	The residual concentrations of Co in solution with and without alumina are practically the same. The pH₅₀ (0.1 g alumina 50 cm³) 6.7 for Ni, 5.2 for Cu and 4.7 for Cr.	105
61. Al_2O_3, γ, Japan Aerosil	Co,Cu,Mn Pb, Zn	10^{-4}-3×10^{-3} mol dm⁻³	30 g dm³	1 d	25	4-7	0.0075 mol dm⁻³ NaNO₃	uptake, ζ	The pH₅₀ for 3×10^{-3} mol dm⁻³ 5.3 for Cu, 5.6 for Pb, 6.3 for Zn, and 7 for Co.	106 107
62. Al_2O_3, Linde A	Co, Cu,Ni	10^{-3} mol dm⁻³	160 m²/dm³	>2 h	22	4-11	0.5 mol dm⁻³ NH₄NO₃	uptake, ζ	The uptake has a maximum at pH 8, and a minimum at pH 9.5. The ζ potential in the presence of 0.01 mol dm⁻³ NH₃ and 2×10^{-5} mol dm⁻³ Co is positive up to pH 11	108
63. Al_2O_3, α, Queensland Alumina, washed	Cd, Co, Cu, Zn	10^{-4} mol dm⁻³	45 m²·dm⁻³	20 min	25	3.5-10	0.001-0.1 mol dm⁻³ KNO₃, 0.1 mol dm⁻³ NaCl	uptake, proton release	For 0.001 mol dm⁻³ NaNO₃, pH₅₀ ~ 6.4 for Cu, 7.7 for Zn, 8.4 for Co and 8.7 for Cd. Co uptake is slightly enhanced in the presence of 0.1 mol dm⁻³ NaCl, and Cd uptake is slightly depressed. Cd uptake is also slightly depressed in the presence of 0.1 mol dm⁻³ NaNO₃.	109
64. Al_2O_3, γ, Houdry 415, calcined at 600 C	Co,Ni	5×10^{-4}-2.5×10^{-2} mol dm⁻³	5 m²:15 cm³	20 h	25	4-7	0.1 mol dm⁻³ NH₄NO₃	uptake, ζ, proton release	Only positive and rather pH insensitive ζ in the presence of 10^{-4} mol dm⁻³ Ni and Co and 0.01 mol dm⁻³ NH₄NO₃. Uptake on alumina impregnated with NH₄F solution prior to calcination is enhanced in comparison with the original alumina.	110
65. Al_2O_3, γ	Co,Ni	0.01, 0.1 mol dm⁻³	5 g (195 m²/g). 200 cm³	0.3-3.5 h	25	7-8.2	1 mol dm⁻³ NH₄NO₃	EXAFS, IR	Formation of coprecipitates having hydrotalcite-like structure.	111
66. Al_2O_3, γ	Co,Ni,Zn	0.01 mol dm⁻³	5 g (200 m²/g). 200 cm³	3 h	25	4-11	1 mol dm⁻³ NH₄NO₃	uptake, XAS	Uptake peaks at pH 7-8, and has a minimum at pH 10.	112
67. Al_2O_3, δ, Degussa C	Cr(III) V(IV)	5×10^{-5} mol dm⁻³ 2.5×10^{-4} mol dm⁻³	10 g (100 m²/g) dm⁻³	1-4 d	25	2.4-4	0.1 mol dm⁻³ NaClO₄	uptake	pH₅₀ 3.3 (V and Cr)	113
68. Al_2O_3, δ, Degussa	Fe(III), Ni	4.6×10^{-7}-3×10^{-5} mol dm⁻³	0.46 g (100 m²/g) dm⁻³	30 min-3 h	22, 24	3-9	0.01 mol dm⁻³ NaNO₃, [EDTA], [metal], MES and HEPES buffers	uptake	AlOH + FeEDTA + H ⇌ AlEDTAFe + H₂O log K13.08 ≡AlOH + FeEDTA + OH ≡AlOFeEDTA²⁻ + H₂O log K-0.95	114
69. Al_2O_3, γ, Aerosil, washed	Ga,In	3×10^{-3} mol dm⁻³	40 g (100 m²/g)/dm³	12 d	25	1.5-4	0-0.1 mol dm⁻³ NaNO₃	uptake	The pH₅₀ 2.7 for Ga and 3.2 for In. The uptake curves are independent of the ionic strength.	115
70. Al_2O_3, chromatographic, Merck	Gd, Ni	10^{-6} mol dm⁻³	0.1-10 g (155 m²/g) 50 cm³	20 min	25	4-7	0.01, 0.1 mol dm⁻³ NaClO₄	uptake	AlONi log K -2.99, ≡AlOGd³⁺ log K 9.51; the charge of Ni and Gd distributed 1:1 between the 'and '0⁰ plane.	116
71. Al_2O_3, chromatographic, Merck	Gd, Ni, Y	2×10^{-4}-4×10^{-4} mol dm⁻³	0.1-10 g (155 m²/g) 50 cm³	20 min	25	4-8	0.01-1 mol dm⁻³ NaClO₄	uptake	ΔH_{ads} 65 kJ mol (Ni), and 90 kJ mol (Y, Gd) The shift in pH₅₀ induced by change in T (15-35 C) is independent of the solid to liquid ratio, concentration of multivalent cations and ionic strength.	117

No.	Adsorbent	Adsorbate	Concentration	Solid loading	Time	Temp.	pH	Electrolyte	Method	Comments	Ref.
72.	Al_2O_3, γ, Degussa	Ni, Pb, Zn	10^{-6} 10^{-4} mol dm⁻³	1 g (100 m²)/dm³	0.5-10 h		3 10	0.025 mol dm⁻³ NaClO₄, 0-10⁻⁴ mol dm⁻³ EDTA	uptake	The pH_{50} = 6.5 (10^{-6} mol dm⁻³ Ni, no EDTA); 6.2 (10^{-6} mol dm⁻³ Ni, 10^{-4} mol dm⁻³ EDTA, anion like adsorption edge); 6 (10^{-4} mol dm⁻³ Pb, 2.5 g Al_2O_3/dm³).	118
73.	Al_2O_3·n H_2O, freshly precipitated from sulfate	Cu	10^{-7} 10^{-4} mol dm⁻³	$10^{-3.5}$ mol Al dm⁻³	1 h	18	5 9	I 0.0135	uptake	The slope of log-log adsorption isotherm at pH 6.9 = 0.6 for [Cu] < $10^{-5.5}$ mol dm⁻³. The log [soluble Cu] is a linear function of pH over the range 6-9 with a slope of ≈-1.	119
74.	Al_2O_3·nH_2O	Cu	$10^{-7.5}$ 10 mol dm⁻³	10^{-3} mol Al/dm³	45 min	18	6.9	I = 0.0122	uptake XAS	The slope of log-log adsorption isotherm 0.5 for concentrations in solution 10^{-7} $10^{-5.5}$ mol dm⁻³. Precipitation of CuAl₂O₄.	120
75.	Al_2O_3·n H_2O	Cu	$10^{-7.5}$ $10^{-4.17}$ mol dm⁻³	$10^{-3.5}$ 10^{-3} mol Al/dm³	45 min	18	5 9	I = 0.0122	uptake	The pH_{50} = 6.3 for $10^{-3.3}$ mol Al dm⁻³ and $10^{-4.17}$ mol Cu dm⁻³; the uptake curve interpreted in terms of surface precipitation model.	121
76.	Al_2O_3·n H_2O	Pb	10^{-4} mol dm⁻³	6 25×10⁻³ mol Al/dm³	4 h 8 d		4-8		uptake	The pH_{50} 6.6 (4 h equilibration), 6 (8 d)	122
77.	Al_2O_3·n H_2O	Sr	10^{-7} 10^{-2} mol dm⁻³	1 g (411 m²)/200 cm³	4 h	25	5-9	0.06-0.6 mol dm⁻³ NaNO₃	uptake	Staircase type uptake curve. 10% uptake at pH 5 and 40% uptake at pH 6-9. Linear log-log adsorption isotherms at constant pH, slope 1. ΔH_ads 17 kJ mol.	123
78.	Al_2O_3·n H_2O	Cu, Mg, Zn	1 11×10⁻⁴ mol dm⁻³	10^{-3} mol Al/dm³	1 d	21	4-11	0.1, 1 mol dm⁻³ NaNO₃	uptake	pH_{50} 5.2 (Cu), 5.8 (Zn), 8.5 (Mg)	124
79.	Al_2O_3·n H_2O	Cu, Pb	$5×10^{-4}$ $2.5×10^{-2}$ mol dm⁻³	0.5 g (259 m²) 15 cm³	1 d		4-6.5	16% sulfate w w in the adsorbent	uptake		125
80.	AlOOH, boehmite	Ba	10^{-3} 10^{-1} mol dm⁻³	0.1 g (143 m²/g) 20 cm³	1 d	20	3 11	Cl	ζ	Only positive ζ.	126
81.	AlOOH, boehmite	Cu	10^{-8} 10^{-3} mol dm⁻³			room	4.5 7.5	0.05 mol dm⁻³ NaCl	uptake, ESR	pH_{50} 5.6 ($5×10^{-4}$ mol dm⁻³ Cu)	127
82.	AlOOH, boehmite, from sec-butoxide	Cu	$5×10^{-4}$ mol dm⁻³	1.15 g (217 m²/g)/dm³	>1 d		6.5	0.01 mol dm⁻³ KNO₃ + 0-5×10⁻⁴ mol dm⁻³ sulfate or phosphate	XAS, EPR	Cu does not form surface precipitate. Inner sphere complexes at low surface loading, outer sphere complexes at high surface loading. Sulfates and phosphates have little effect on the local environment of adsorbed Cu	128
83.	AlOOH, boehmite	Cu		1 15 g (217 m²/g)/dm³			6.5	0.01 mol dm⁻³ KNO₃	EPR		129
84.	AlOOH, boehmite	Eu	10^{-9} mol dm⁻³	500 mg (175 m²/g)/dm³	2 d	room	2 11	0.05 mol dm⁻³ NaClO₄ + 0-20 mg dm⁻³ humic acid	uptake	The pH_{50} 4.2 in absence of humic acid. In the presence of humic acid the uptake is depressed at pH > 6 and enhanced at pH < 5.	130
85.	AlOOH, boehmite	Mg, Mn	$9×10^{-5}$ $3.6×10^{-4}$ mol dm⁻³	1.5 m² 75 cm³	1 d		5-11	0.009 mol dm⁻³ NaClO₄	uptake, ζ ESR	Mn does not significantly influence the ζ potential, Mg induces a slight shift of IEP to higher pH. The pH_{50} 9.5 for Mg and 8 3 for Mn.	131
86.	Al(OH)₃, amorphous	Co	$3×10^{-6}$ $3×10^{-3}$ mol dm⁻³	100 mg/50 cm³	7 h	25	6.8 8.2	KNO₃, 0.01 mol dm⁻³ NH₃	uptake		132

TABLE 4.1 Continued

	Adsorbent	Adsorbate	c. total	S L ratio	Eq. time	T	pH	Electrolyte	Method	Result	Ref.
87.	Al(OH)$_3$, gibbsite	Co	10^{-6} mol dm^{-3}	7.5 g (4 m^2/g) dm^{-3}	30 min	25	4.5–8 5	10^{-2}, 10^{-1} mol dm^{-3} NaClO$_4$	uptake	The pH$_{50}$ 6.7, the uptake is insensitive to ionic strength.	133
88.	Al(OH)$_3$, gel, from chloride	Cu	10^{-3} 10^{-3} mol dm^{-3}	100 mg/25 cm^3	1 d		low	Cl	uptake, ESR	The uptake is enhanced in the presence of humic acid.	134
89.	Al(OH)$_3$, amorphous	Cu	10 10^{-3} mol dm^{-3}	0.1 g (111 m^2/g) 20 cm^3	1 d	room	4.5–7.5	0.05 mol dm^{-3} NaCl	uptake, ESR	pH$_{50}$ 5	127
90.	Al(OH)$_3$, gibbsite	Cu	10^{-8} 10^{-3} mol dm^{-3}	0.1 g (5.9 m^2/g) 20 cm^3	1 d	room	4.5–7.5	0.05 mol dm^{-3} NaCl	uptake, ESR	pH$_{50}$ 5.4	127
91.	Al(OH)$_3$, gibbsite	Cu	5×10^{-5}, 5×10^{-4} mol dm^{-3}	13 g (96 m^2 g) dm^3	1 d		4.5–7	nitrate	uptake, ESR, IR	The pH$_{50}$ 5 (5×10^{-5} mol dm^3 Cu)	135
92.	Al(OH)$_3$, gel	Cu							uptake, ESR		136
93.	Al(OH)$_3$, amorphous	Cu	5×10^{-4} mol dm^{-3}	200 mg/40 cm^3	1 d		4–8	0.1 mol dm^{-3} NaClO$_4$, 0–10^{-2} mol dm^{-3} aspartic and glutamic acid	uptake, ESR	The pH$_{50}$ 5.2 in absence of organic ligands. The uptake at pH 6 is enhanced in the presence of $> 10^{-3}$ mol dm^{-3} of aspartic and glutamic acid and depressed at pH > 6 in the presence of 10^{-2} mol dm^{-3} of aspartic and glutamic acid.	137
94.	Al(OH)$_3$, amorphous	Cu	5×10^{-4} mol dm^{-3}	200 mg (182 m^2/g) 40 cm^3	1 d		4–8	0.1 mol dm^{-3} NaClO$_4$, 10^{-3} 10^{-2} mol dm^{-3} O-phospho-L-serine, O-phospho-L-tyrosine	uptake	100% uptake at pH 6 in the presence of 10^{-3} mol dm^3 of organic ligands.	138
95.	Al(OH)$_3$, amorphous	Cu	5×10^{-4} mol dm^{-3}	200 mg/40 cm^3	1 d		4–8	0.1 mol dm^{-3} NaClO$_4$, 0–10^{-2} mol dm^{-3} pyridinedicarboxylic acids	uptake, ESR	The pH$_{50}$ 5.2 in absence of organic ligands. The uptake at high pH is depressed in the presence of 10^{-2} mol dm^{-3} of 2.3; 2.4; 2.5 and 2.6 pyridine dicarboxylic acids. For two former isomers the uptake is rather insensitive to the pH, for the latter two isomers the uptake decreases when pH increases.	139
96.	Al(OH)$_3$, amorphous	Cu	5×10^{-4} mol dm^{-3}	200 mg/40 cm^3	1 d		4–8	0.1 mol dm^{-3} NaClO$_4$, 0–10^{-2} mol dm^{-3} organic ligands	uptake, ESR	The uptake is depressed by a factor of 2 in the presence of 10^{-3} mol dm^{-3} and by factor > 4 in the presence of 10^{-2} mol dm^{-3} of L-; D,L,o- and D,L,m-tyrosine, and phenylalanine. With 10^{-2} mol dm^{-3} of organic ligands, the uptake is rather insensitive to pH. Uptake in the presence of 3,4 dihydroxyphenylalanine peaks at pH 6.	140
97.	Al(OH)$_3$, amorphous	Cu	5×10^{-4} mol dm^{-3}	200 mg/40 cm^3	1 d		4–8	0.1 mol dm^{-3} NaClO$_4$, 0–10 mol dm^{-3} organic ligands	uptake, ESR	The uptake is depressed by a factor of 2 in the presence of 10^{-3} mol dm^{-3} and by factor > 4 in the presence of 10^{-2} mol dm^{-3} of D,L-2,3 diaminopropionic and D,L-2,4 diaminobutyric acid. The presence of 10^{-3} mol dm^3 of	141

(continuation from preceding row) ethylenediamine almost entirely prevents uptake of Cu. The uptake is also depressed in the presence of glutamic acid.

No. & adsorbent	Ion	Metal concentration	Electrolyte	Solid	t	T (°C), pH	Method	Comments	Ref.
98. Al(OH)$_3$, amorphous	Mn	2 · 10^{-3} mol dm^{-3}	0.1 mol dm^{-3} NaCl, 0 8 × 10^3 mol Al dm^{-3}	5 × 10^3 mol Al dm^{-3}	1 d	room 5 8	ESR	Two Mn species: "free" and "bound", the bound species is more stable in the presence of sulfate.	142
99. Al(OH)$_3$, gibbsite	Ni	3 × 10^{-3} mol dm^{-3}	0.1 mol dm^{-3} NaNO$_3$	10 g dm^{-3}	2 d	25 7.5	uptake, EXAFS	5 N neighbors at distance of 0.302 nm which is shorter than in Ni(OH)$_2$, and 1 8 Al neighbor at 0.305 nm. Formation of mixed Al-Ni hydroxide similar to takovite Ni$_6$Al$_2$(OH)$_{16}$CO$_3 \cdot$H$_2$O.	143
100. Al(OH)$_3$, gibbsite, Alcoa	U	10^{-6} mol dm^{-3}	0.1 mol dm^{-3} NaCl + 0 2.5 × 10^{-3} mol dm^{-3} citrate	5 g (11.2 m^2/g) dm^{-3}	1 d	22 3 11	uptake	The pH$_{50}$ 4 8 in absence of citrate; the effect of citrate on the uptake of U is rather insignificant.	144
101. Al(OH)$_3$, amorphous	V(IV)	5 · 10^{-4} mol dm^{-3}	0.25 mol dm^{-3} NaClO$_4$ succinic, L-malic, and D,L-2-mercaptosuccinic acid; ligand to V ratios 2:1 to 500 1	200 mg/40 cm^3	1 d	4 8	uptake, ESR	The pH$_{50}$ 5 in absence of organic ligands. Uptake at pH 5 is enhanced in the presence of small (up to 1:20) amounts of organic ligands. The uptake is depressed at pH 5 at high concentrations of organic ligands Sorption of V(V) at similar conditions. cf. Table 4.2.	145
102. Al(OH)$_3$, 2 samples gibbsite and amorphous	Zn	2 64 mg Zn dm^{-3}	0.01 mol dm^{-3} Na$_2$SO$_4$	0 05 g (59, 441 m^2/g) 20 cm^3	1 d	25 2 8	uptake	Onset of uptake at pH 4	146
103. Al(OH)$_3$, amorphous, from AlCl$_3$	Zn	10^{-4}-3 · 10^{-3} mol dm^{-3}	1 mol dm^{-3} NaClO$_4$, NaCl; 0.0005 mol dm^{-3} Na$_2$SO$_4$			6-6.5	uptake	The uptake increases in the presence of chlorides and more significantly in presence of sulfates	147
104. Al(OH)$_3$, hydrargillite	Zn	10^{-6}-10^{-4} mol dm^{-3}	0.01 mol dm^{-3} NaClO$_4$, 0-20 mg fulvic acid dm^{-3}	300 mg (200 m^2/g) dm^{-3}	3 d	20 3 10	uptake	The pH$_{50}$ 7 6, the uptake is enhanced in the presence of fulvic acid at pH 8.	148
105. Al(OH)$_3$, gel	Ba, Ca, Cd, Co, Cu, Mg, Ni, Pb, Sr, Zn	1 25 · 10^{-4} mol dm^{-3}	1 mol dm^{-3} NaNO$_3$	0.093 mol Al dm^{-3}	3 h	4-10	uptake	The pH$_{50}$ ~ 4.8 (Cu), 5 2 (Pb), 5.6(Zn), 6.3(Ni), 6 5(Co), 6.6(Cd), 8 1(Mg), 9 2(Sr). Aging (205 days) enhances uptake of Mg. but depresses uptake of Ca, Sr, and Ba	149
106. Al(OH)$_3$, gibbsite	Ca, Sr	2 · 10^{-6} mol dm^{-3}	1 mol dm^{-3} NaNO$_3$	4.55 × 10^3 mol Al 25 cm^3	1 d	room 5 10	uptake	pH$_{50}$ 6.7 (Ca) and 6.4 (Sr).	150
107. Al(OH)$_3$, gel	Ca, Sr	2 × 10^{-6} mol dm^{-3}	1 mol dm^{-3} NaNO$_3$	2 33 · 10^3 mol Al 25 cm^3	3 h	room 6 10	uptake	pH$_{50}$ 8 3 (Ca) and 9 (Sr). The log K_D(pH) curves consist of two linear segments: slope 1.7 for Ca and 1 6 for Sr at pH 8 and 1 for Ca and 0.9 for Sr at pH 8	150
108. Al(OH)$_3$, fresh	Cd, Zn	10^{-6} mol dm^{-3}	0.1 mol dm^{-3} NaNO$_3$	10^{-3} mol Al dm^{-3}	2 h	6 9	uptake	pH$_{50}$ 7.8 (Cd) and 6.5 (Zn).	151
109. Al(OH)$_3$, gibbsite	Cu, Mg, Zn	1 11 × 10^4 mol dm^{-3}	0.1, 1 mol dm^{-3} NaNO$_3$	10^{-3} mol Al dm^{-3}	1 d	21 4 11	uptake	pH$_{50}$ 5.6 (Cu). 6.5 (Zn). 8.8 (Mg)	124
110. Al(OH)$_3$, gel	Cu, V(IV)	10^{-3}-0.02 mol dm^{-3}	0.1 mol dm^{-3} NaCl, sulfate	2.6 × 10^{-3} mol Al 75 cm^3	1 d	20 3-4.3	uptake, ESR	The uptake of Cu is enhanced by a factor 2 in the presence of (sulfate) [Cu], uptake of V is also enhanced in the presence of sulfate. Preadsorbed V is removed in the presence of 0.01 mol dm^{-3} EDTA at pH 6-7.	152

TABLE 4.1 Continued

Adsorbent	Adsorbate	c. total	S:L ratio	Eq. time	T	pH	Electrolyte	Method	Result	Ref.
111 BeO	Al	0.001 mol dm³			25	4-12	chloride	ζ	IEP is shifted to pH 8.5 (7.1 in KCl).	153
112. CeO₂, hydrolysis of Ce(SO₄)₂, crystalline	U	0.01-0.1 mol dm³	0.3 g (96 m²/g), 25 cm³	1 d	27	3.5 (initial)	nitrate	uptake	Sorption capacity 0.76 mmol g, uptake decreases when T increases ΔH_ads −34 kJ mol.	154
113 CeO₂, synthetic hydrous, 2 samples	U	0.01 mol dm³	0.3 g 25 cm³	1 d	27	3.5-10.7	nitrate, chloride, sulfate, carbonate (up to 0.2 mol dm³)	uptake		155
114. CeO₂	Ba,Sr	10⁻⁷ 10⁻² mol dm	0.1 g 10 cm³	3 h	30	7.24 11.42	Nitrate	uptake	A linear log-log adsorption isotherm, slope 0.8 (pH not specified). ΔG_{ads} −6.4 kJ mol⁻¹ for Ba and -5.4 kJ mol⁻¹ for Sr and ΔH_{ads} 12.5 kJ mol⁻¹ for Ba and 9.6 kJ mol⁻¹ for Sr (from temperature dependence of ΔG_{ads}). Very broad adsorption edges (cf. Fig. 4.7). The isotope exchange of preadsorbed Sr and Ba with ions in solution is very slow.	156
115. CeO₂·n H₂O	Hg, Zn	10⁻⁸ 10⁻³ mol dm³	0.1 g, 10 cm³	>1 h	30	6-11	nitrate, chloride; EDTA, glycine, oxalate, phosphate, acetate, sulfate, concentration of ligands 3×[Zn or Hg]	uptake	A linear log-log adsorption isotherm slope 0.83 for Zn and 0.75 for Hg (pH not specified). ΔH_{ads} 12 kJ mol⁻¹ for Hg and Zn. The uptake is depressed in the presence of EDTA, glycine, oxalate, phosphate, acetate, sulfate (EDTA is more efficient as depressor of Cd uptake than glycine, etc.). γ-irradiation does not affect the sorption.	157
116. Co₃O₄	Ba, Ca, Mg, Sr	0.001-0 1 mol dm³			25		Nitrate	proton release	The shift in PZC induced by the salt concentration of 0.1 mol dm³ (in mV, 1 pH unit = −59 mV): Mg 40, Ca 50, Sr 60, Ba 80.	158
117. Co₃O₄	Ba, Sr	0.001-0.01 mol dm³	0.5 g (10.7 m² g), dm³			4-10	nitrate, chloride	uptake, ζ	For 0.01 mol dm³ Ba positive ζ (>40 mV (no IEP).	159
118. HCoO₂	Ba, Ca, Mg, Sr	10⁻³ mol dm³				4-11	nitrate		Mg>Ca>Sr−Ba, with Mg positive ζ over the entire pH range.	160
119. Cr₂O₃	Sb(V)	carrier free	30 mg 40 cm³	1 h	23	3	HCl	Mossbauer	Uptake >90%	161
120. Cr₂O₃	Al, Cr	0.001 mol dm³			25	4-12	chloride	ζ	IEP is shifted to pH 8.4 in AlCl₃ and 7 1 in Cr₂(SO₄)₃ (2.25 in KCl).	153
121 Cr₂O₃, 2 samples	Ba, Ca, Mg, Sr	10⁻³ mol dm³				4-10	nitrate	ζ	Mg Ca−Sr Ba, positive ζ.	160
122. Cr₂O₃, amorphous	Ni, Zn	50 ppm	53 m²·dm³	40 mn	25	4-11	10 mol dm³ KNO₃	uptake	pH₅₀ 5.8(Zn); 6 7 (Ni)	162
123. Cr₂O₃·nH₂O	Ni, Zn	50 ppm	250 ppm Cr			4-11	0.5 mol dm³ NH₄NO₃	uptake	The Zn uptake increases up to pH 6, then decreases (with some scatter) up to pH 10, and increases again. Similar pattern for Ni with a maximum uptake at pH 7 5, minimum at pH 9, and less significant scatter	163
124 Cr₂O₃·nH₂O	Ni, Zn	50 ppm	250 ppm Cr			4.5 10.5		ζ	The shift in IEP to higher pH by 0.2 pH unit for Zn and 0.5 pH unit for Ni.	164

No. adsorbent	Adsorbate	Concentration	Solid	Time	T	pH	Electrolyte	Method	Remarks	Ref.
125. HCrO₂, α, 2 samples	Ba, Ca, Mg, Sr	10^{-6} 10^{-3} mol dm⁻³				4-11	nitrate	ζ proton release	Mg>Ca>Sr Ba positive ζ at cation concentration of 10^{-3} mol dm⁻³	160
126. Cu₂O	Al, Cu(II)	0.001 mol dm⁻³			25	4-12	chloride	ζ	IEP is shifted to pH 8.4 in AlCl₃, and 6.2 in CuCl₂ (11.5 in KCl).	153
127. Fe₃O₄	Co	3.7 6.3×10⁻⁶ mol m⁻²			25	6.5			ΔH_{ads} 30 kJ mol	52
128. Fe₃O₄, synthetic	Co	10^{-4} mol dm⁻³	0.2 g (5 m²/g). 50 cm³	1 h	30	4-10	0.1 mol dm⁻³ KNO₃	uptake, ζ	The pH_{50} 7.5, positive ζ over the entire pH range at [Co]>10 mol dm⁻³	165
129. Fe₃O₄, synthetic, spherical	Co	1.7×10⁻⁷ 1.7×10⁻⁴ mol dm⁻³	0.2 g (1.7 m²/g). 20 cm³	3 h	25	5.5-8	10^{-4} 10^{-1} mol dm⁻¹ NaNO₃	uptake	The uptake is slightly depressed at high ionic strengths, with 1.7×10⁻⁷ mol dm⁻³ Co pH_{50} 6.3 with 10^{-4} mol dm⁻³ NaNO₃, and 6.5 with 10^{-1} mol dm⁻³ NaNO₃. The uptake is enhanced when T increases. Curvilinear log-log adsorption isotherms at constant pH with a slope of 1 in the low concentration range.	166
130. Fe₃O₄, synthetic, spherical Co	Mn(II)	10^{-6} 10^{-3} mol dm⁻³	4×10⁻³ mol dm⁻³ (11.4 m²/g)	3 12 h	25	6-7.5	0.1 mol dm⁻³ NaNO₃	uptake	Curvilinear log-log adsorption isotherms at constant pH.	59
131. Fe₂.₉O₄, magnetite		carrier free		1 h		5.5 12		uptake	Uptake peaks at pH 10 (90%)	167
132. Fe₃O₄, synthetic	Sr	0.01 mol dm⁻³	0.4 g 25 cm³	2 d	25		Cl	uptake	Low uptake, not affected by γ-irradiation.	168
133. Fe₃O₄, synthetic, hydrous	U	10^{-4} mol dm⁻³	0.1 g 25 cm³	4 h	25	4-5.5	0.1 mol dm⁻³ NaClO₄	uptake	pH_{50}~4.7	169
134. Fe₃O₄, magnetite, synthetic	U	$5×10^{-6}$ 10^{-4} mol dm⁻³	8 mg (3 m²/g) 20 cm³	1 d	25	3-6 7-9	0.0005 mol dm⁻³ NaCl, 0.001 mol dm⁻³ NaHCO₃	uptake, ζ	The ζ at pH 3-6 is not affected by presence of U up to $5×10^{-5}$ mol dm⁻³.	170
135. Fe₃O₄, magnetite, natural, 2.4% SiO₂	Ba Eu	$5×10^{-5}$ mol dm⁻³ $2×10^{-4}$ mol dm⁻³	2 g (18.3 m²/g) dm⁻³		25	3 11	0.1 mol dm⁻³ NaNO₃	uptake	pH_{50} 8.8 (Ba) and 6.2(Eu), staircase uptake curves.	171
136. Fe₃O₄, synthetic	Co, Cr(III), Fe(III), Mn(II)	1 10 mg dm⁻³	0.25 g 25 cm³	1 d		natural	nitrate, chloride	uptake, chromatography	Uptake increases with T except (for Fe(III) whose uptake decreases with T (25-95 C)	172
137. Fe₃O₄	Cu, Fe (II and III), Mn, Ni, Zn	10^{-5} $5×10^{-3}$ mol dm⁻³	10-20 g dm⁻³	1 d	natural			uptake	Fe>Cu Zn Mn	173
138. Fe₃O₄, Puratronic	Ni, Yb	$2×10^{-5}$ mol dm⁻³	400-600 mg (1.8 m²/g). 50 cm³	1 7 d		4-7	0.05-0.1 mol dm⁻³ NaNO₃	uptake	≡F=OYb² log K 0.8; ≡F=OYbOH log K -6.6; ≡F=ONi(OH)₂⁻ log K -14.7	174
139. Fe₂O₃, synthetic hematite, Ca 3 samples		$5×10^{-4}$ mol dm⁻³	1-5 g (18 31 m²/g). 50 cm³	3 h	20	4-11		proton release, coagulation	The PZC is shifted to pH 6.5 in the presence of 5 × 10⁻⁴ mol dm⁻³ Ca. No further shift up to 0.5 mol dm⁻³ Ca.	175
140. Fe₂O₃, hematite Fluka	Ca	10^{-3} 10^{-1} mol dm⁻³	1 g (4.2 m²), 20 cm³	45 min	20	4-10	chloride	uptake, proton release	Linear log-log adsorption isotherms at constant pH. slope 0.6	176

TABLE 4.1 Continued

Adsorbent	Adsorbate	c. total	S:L ratio	Eq. time	T	pH	Electrolyte	Method	Result	Ref.
141. Fe_2O_3, hematite	Cd	10^{-4} 10^{-2} mol dm^{-3}		1 h	20		0.02 mol dm^{-3} KNO_3	uptake, proton release	1.3–1.8 proton released per one adsorbed Cd. Temperature effects (5–60) were also studied. ΔG^0_{ads}-42 kJ/mol. ΔH_{ads} 7 kJ/mol.	28
142 Fe_2O_3, α	Co	6×10^{-6} mol dm^{-3}	0.04 mol dm^{-3}	30 min		0–13	0.15 mol dm^{-3} NaCl, 10^{-4} 10^{-2} mol dm^{-3} EDTA glycine, L + arginine, L + cysteine	uptake	The pH$_{50}$ ~ 7 in absence of organic ligands. The uptake slightly declines at pH > 11. In the presence of 10^{-2} mol dm^{-3} EDTA the uptake is reduced by 30% (no effect at lower EDTA concentrations). The uptake peaks at pH 7.5 in the presence of glycine, and with >10^{-3} mol dm^{-3} glycine the uptake of Co at pH > 10 is completely depressed. The uptake is reduced by 40% in the presence of 10^{-2} mol dm^{-3} arginine. The uptake peaks at pH 8.5 in the presence of cysteine, and with 10^{-2} mol dm^{-3} cysteine the uptake of Co is completely depressed. (Captions for Fig. 1 and 4 have been swapped).	177
143. Fe_2O_3, hematite, monodispersed	Co	1.7×10^{-8} 1.7×10^{-3} mol dm^{-3}	2 g (13 m^2/g) dm^3	2 h	25	5.5 9.5	10^{-4} 1 mol dm^{-3} NaNO$_3$, 0.1 mol dm^{-3} NaCl, NaClO$_4$, NaHCOO, NaCH$_3$COO	uptake	The pH$_{50}$ 7.4 (1.7×10^{-7} mol dm^{-3}); the effect of ionic strength on uptake curves is negligible; the uptake increases with T (50, 75 C); the uptake is unaffected when the ionic strength is adjusted to 0.1 mol dm^{-3} by chloride or chlorate VII, and it is slightly enhanced in the presence of formates or even more enhanced in the presence of acetates (pH$_{50}$ 7). The percentage of uptake gradually decreases as the total Co concentration increases.	178, 179
144. Fe_2O_3, Kanto	Co	10^{-5} 10^{-3} mol dm^{-3}		3 12 h	25	5–6.5	0.1 mol dm^{-3} NaNO$_3$	uptake	Slope of log-log adsorption isotherms 0.7 at pH 5, and 0.3 at pH 6.5.	59
145. Fe_2O_3, α	Eu	5×10^{-7} mol dm^{-3}	0.04 mol Fe$_2$O$_3$ dm^{-3}	30 min		2 12	0.15 mol dm^{-3} NaCl + 10^{-4} 10^{-2} mol dm^{-3} EDTA, oxalate, tartrate, sulfosalicylate	uptake	In absence of complexing agents pH$_{50}$ 5.5, uptake slightly < 100% at pH > 12. Presence of > 0.001 mol dm^{-3} of oxalate induces a shift of pH$_{50}$ to higher pH. At > 0.001 mol dm^{-3} of EDTA uptake is reduced over the entire pH range (the uptake does not reach 100%, cf. Fig. 4.6(A)). Tartrate and sulfosalicylate do not exert any significant effect.	180
146. Fe_2O_3, hematite, natural	Eu	10 mol dm^{-3}	0.5 g (43 m^2 g) dm^3	1 d		3 10	0.01 mol dm^{-3} NaClO$_4$, 2 mg dm^{-3} fulvic acid	uptake	The pH$_{50}$ ~ 5.3 in absence of fulvic acid and 8.8 in the presence of fulvic acid. The uptake is enhanced by fulvic acid at pH 4.	181
147. Fe_2O_3, hematite, natural	Eu	10^{-10} 10^{-4} mol dm	0.1 g (4 m^2/g) 15 cm^3	14 d		2 9	0.1 mol dm^{-3} NaClO$_4$ buffers	uptake	Linear log-log adsorption isotherms at constant pH, slope 0.9 up to surface coverage of 10^{-4} mol kg. The CO$_2$ effect on Eu sorption at pH 6 is negligible.	33

Adsorbent	Sorbate	Sorbate concentration	Solid/amount	Time	T	pH	Electrolyte	Method	Results / Comments	Ref
148. Fe_2O_3, α, Merck	Ga	carrier free	6.4 g dm⁻³	30 min		1 12	0.15 mol dm⁻³ NaCl + 10⁻⁴ 10⁻² mol dm⁻³ EDTA, oxalate, citrate, sulfosalicylate	uptake	In absence of complexing agents 100% uptake between two adsorption edges pH₅₀ 2.5 and 10. Presence of >0.001 mol dm⁻³ of EDTA and citrate depresses the uptake over the entire pH range. Oxalate and sulfosalicylate do not exert any significant effect at pH>7 but uptake at pH < 7 is somewhat depressed. The depression of uptake in presence of ligands is correlated with stability of Ga complexes.	182
149. Fe_2O_3, α	Ga	carrier free	0.04 mol dm⁻³	30 min		1-12	0.15 mol dm⁻³ NaCl 3.09×10⁻² mol dm⁻³ citrate	uptake	100% uptake between two adsorption edges: pH₅₀ 3 and pH₅₀ 9. In the presence of citrate the uptake peaks at pH 7.5 (50%).	183
150. Fe_2O_3, hematite, 99.999, Johnson Matthey	La	2×10 mol dm⁻³	0.3 3 g (8.5 m²/g), 50 cm³	1 21 h		5 7	0.1 mol dm⁻³ NaNO₃	uptake	pH₅₀ – 6.4 (0.3 g hematite 50 cm³), 6.1 (1 g), and 5.7 (3 g).	184
151. Fe_2O_3, hematite, Fluka	Mg	10⁻³ 10⁻¹ mol dm⁻³				4, 10	chloride	uptake, proton release	Linear log uptake (log initial concentration) plots, slope 0.8	185
152. Fe_2O_3, α	Mn(II)	3.3×10⁻¹¹–4×10 mol dm⁻³	0.1 g 25 cm³	30 min	20 -22	2 14	1 mol dm⁻³ NaNO₃, 5 mol dm⁻³ NaCl	uptake	The pH₅₀ 7 (3.3×10⁻¹¹ mol dm⁻³ Mn, Mn III and IV are present in carrier free Mn); pH₅₀ 9 (4×10⁻³ mol dm⁻³ Mr). The uptake decreases with T (26 50 C). ΔH_ads -126 kJ mol.	186
153. Fe_2O_3, hematite	Np (V)					4-10	0.005-0.1 mol dm⁻³ NaClO₄, 0-2% CO₂ in gas mixture	uptake	Divalent anion was selected as a component. ≡FeONpO₂ log K -2.09 ≡FeONpO₂(HCO₃)₂²⁻ log K 24.62 The uptake is insensitive to the ionic strength.	187
154. Fe_2O_3, hematite	Np (V)	1.2×10⁻⁷ 1.2×10⁻⁶ mol dm⁻³	0.1 1 g (14.4 m²/g)/dm³	1 d	25					188
155. Fe_2O_3, α, ultrapure	Pb	4×10⁻⁶ mol dm⁻³	0.1 g 25 cm³	30 min	22 ±2	1 14	1 mol dm⁻³ NaNO₃ 1 mol dm⁻³ NaCl	uptake	100% uptake between two adsorption edges: pH₅₀ 1.5 and pH₅₀ 13. 100% uptake between two adsorption edges: pH₅₀ 4.5 and pH₅₀ ≈ 13 (0.1 mol dm⁻³ NaCl causes a less significant depression of adsorption at low pH).	189
156. Fe_2O_3, hematite, synthetic	Pb	2×10⁻³ 10⁻² mol dm⁻¹	18 g (49 m²/g) dm³	1 2 d		6-8	0.1 mol dm⁻³ NaNO₃	XAFS	0.2-0.5 Fe atoms at 0.33 nm, no Pb clustering, bidentate surface complexation.	190
157. Fe_2O_3, α	Ru	carrier free	0.04 mol dm⁻³	30 min		0-13	0.15 mol dm⁻³ NaCl 10⁻² mol dm⁻³ EDTA 10⁻² mol dm⁻³ citrate	uptake	Broad maximum (50% uptake) at pH 2 11 in absence of complexing agents. The uptake at pH 3 10 is reduced by a factor of 5 in the presence of EDTA, and by a factor of 2 in the presence of citrate.	191
158. Fe_2O_3, α	Ru	carrier free	0.04 mol dm⁻³	30 min		0-13	0.1 mol dm⁻³ Na₂SO₃	uptake	90% uptake independent of pH over the range 7 13.	191
159. Fe_2O_3, α	Sb(V)	carrier free	30 mg 40 cm³	>1h	23	4	HCl	Mossbauer	Minimum uptake (15%) at pH 5.	161
160. Fe_2O_3, α	Sc	1.2×10⁻¹⁰, 4×10⁻⁵ mol dm⁻³	0.1 g 25 cm³			1 14	1 mol dm⁻³ NaNO₃	uptake	No adsorption at pH < 3. At pH 6 14 complicated uptake curves, uptake ≈ 80%.	192

TABLE 4.1 Continued

Adsorbent	Adsorbate	c. total	S:L ratio	Eq. time	T	pH	Electrolyte	Method	Result	Ref.
161. Fe_2O_3, hematite, Reachim Sr	Sr	2×10^{-4}, 2×10^{-3} mol dm^{-3}	70 g (6 m^2 g) dm^{-3}		25	8.5-10.2	0.1 mol dm^{-3} NaCl	uptake, proton release	The pH$_{50}$ = 9.5 for 2×10^{-4} mol Sr dm^{-3}; the uptake increases with T (25-75°C).	193
162. Fe_2O_3, hematite, Baker	Th	1 ppm	500 mg (7 m^2/g) 21 cm^3	30 d	20	2.35	0.005 mol dm^{-3} H$_2$SO$_4$	uptake	5% uptake	194
163. Fe_2O_3, hematite, monodispersed, from chloride	Th	2.5×10^{-10} mol dm^{-3}	50 mg (19 m^2 g) dm^{-3}	1 d		2 10	0.01, 0.1 mol dm^{-3} NaClO$_4$	uptake	≡FeOHTh^4 log K 18.7; ≡FeOTh(OH)$_2$ log K -2; ≡FeOTh(OH)$_4$ log K 16.7	195
164. Fe_2O_3, hematite, natural and synthetic	U	10^{-5} mol dm^{-3}	1 g (15-19 m^2) dm^{-3}	7 d	25	3 9	0.1 mol dm^{-3} NaNO$_3$, 0-0.01 mol dm^{-3} carbonate	uptake	pH$_{50}$ ~ 5.2 for synthetic hematite, with the natural sample the uptake shows a maximum at pH 7; the uptake at pH > 5 is depressed in the presence of carbonate, with 10^{-3} mol dm^{-3} carbonate the uptake peaks at pH 6.5.	196
165. Fe_2O_3, hematite, from chloride	U	5×10^{-6}, 5×10^{-5} mol dm^{-3}	0.2 g.dm^{-3}	1 d	25	4.4-6.4	0.0005 mol dm^{-3} NaCl, humic acid	uptake, ζ, IR	U uptake at pH 4.4 is enhanced in the presence of preadsorbed humic acid by factor up to 10 with 20 mg humic acid dm^{-3}.	197
166. Fe_2O_3, hematite, synthetic	U	0.5-16 mg dm^{-3}	0.2 g (34 m^2/g) dm^{-3}	16 h	25	3-6	0.0005 mol dm^{-3} NaCl	uptake, ζ	The pH$_{50}$ = 5.5. With 5×10^{-5} mol U dm^{-3} the IEP is shifted to pH 6.	198
167. Fe_2O_3, hematite, from chloride	U	5×10^{-6}, 5×10^{-5} mol dm^{-3}	0.2 g (34 m^2/g) dm^{-3}	16 h	25	6-10	0.001 mol dm^{-3} NaHCO$_3$	uptake, ζ	The uptake decreases when pH increases. The uptake slightly increases when T increases (25-60°C).	199
168. Fe_2O_3, hematite, Aldrich	U	10^{-6} mol dm^{-3}	0.8g (45 m^2/g) dm^{-3}	2 d		4-8	0.001-0.1 mol dm^{-3} KNO$_3$	uptake	The pH$_{50}$ = 5.2, irrespective of the ionic strength, for the system insulated from CO$_2$, and in contact with atmospheric CO$_2$.	200
169. Fe_2O_3, α, Fisher	Zn	80 ppm	0.2 g, 50 cm^3	1 d		6-7	NaCl	uptake, proton release	1.5-1.9 protons released per one adsorbed Zn. The number of released protons + the number of coadsorbed Cl 2×the number of adsorbed Zn.	88
170. Fe_2O_3	Al, Ga, In, Tl (III)	4×10^{-4}-4×10^{-4} mol dm^{-3}	0.1 g, 25 cm^3	30 min	18 20	2 13	1 mol dm^{-3} NaClO$_4$	uptake	100% uptake between two adsorption edges: 4×10^{-4} mol dm^{-3} Al pH$_{50}$ = 4.5 and pH$_{50}$ = 9.6; 4×10^{-5} mol dm^{-3} Al pH$_{50}$ 4 and pH$_{50}$ ~ 11; 4×10^{-4} mol dm^{-3} Ga pH$_{50}$ -3.3 and pH$_{50}$ ~ 9; 4×10^{-5} mol dm^{-3} Ga pH$_{50}$ = 2.5 and pH$_{50}$ ~ 10; 4×10^{-4} mol dm^{-3} In pH$_{50}$ - 3.7 and pH$_{50}$ > 13; 4×10^{-6} mol dm^{-3} In pH$_{50}$ 3 and pH$_{50}$ > 13.	201
171. Fe_2O_3, special purity	As(III), Bi(III), Sb(III)	4×10^{-6}, 4×10^{-5} mol dm^{-3}	0.1 g 25 cm^3	30 min		-1 13	1 mol dm^{-3} NaClO$_4$	uptake	100% uptake between two adsorption edges: Sb: pH$_{50}$ ~ -0.5 and pH$_{50}$ 14.5; As: pH$_{50}$ 1.3 and pH$_{50}$ ~14.5; Bi: pH$_{50}$ 2 and pH$_{50}$ 13.5	202
172. Fe_2O_3, hematite, from nitrate, 3 samples	Ba, Ca, Mg, Sr	0.005-0.5 mol dm^{-3}	10 g (18 35 m^2/g) 500 cm^3	5 min	20	4-10	nitrate	proton release	Mg > Ca > Sr; Ba: CIP at pH 6.5 for different concentrations of Ca(NO$_3$)$_2$.	203

No. / Adsorbent	Adsorbate	Adsorbate concentration	Method	Electrolyte	pH	T (°C)	Time	Solid	Comments	Ref
173. Fe_2O_3	Cd	10^{10} mol dm^{-3}	uptake	1 mol dm^{-3} NaNO$_3$	4-9					204
174. Fe_2O_3, Polysciences	Cd, Cu	5×10^{-5} mol dm^{-3} 10^{-5} mol dm^{-3}	uptake, ζ	0.001 mol dm^{-3} NaNO$_3$	2 10	25	3 d	250–300 ppm, 3.6 m^2/g	At pH 3 5.5 pH independent uptake of Cd (65%) and Cu (40%), at pH > 8 100% uptake (cf. Fig 4.6 (B)).	205
175. Fe_2O_3, hematite, Baker	Co, Cu, Ni	10 mol dm^{-3}	uptake, ζ	0.1 1 mol dm^{-3} NH$_4$NO$_3$	4-11	22	>2h	160 m^2/dm^2	The uptake is depressed at high NH$_3$ concentrations, minumum uptake at pH 9.5. Maximum uptake at pH 6.5 for Cu and 8 for Ni	108
176. Fe_2O_3, hematite, Kanto	Co, Cu, Pb, Zn	10^{-6} 10^{-3} mol dm^{-3}	uptake	0.1 mol dm^{-3} NaNO$_3$	3.5-5.5	25	12 h	80 m^2/dm^3	pH$_{50}$ = 4 (Pb), 4.5 (Cu), 5.2 (Zn), 5.3 (Co) for 10^{-6} mol dm^{-3}.	206
177. Fe_2O_3, hematite	Cd(II), Cr(III), Er, Gd, Sr, Yb, Zn	carrier free–10^{-4} mol dm^{-3}	uptake	nitrate or chloride	5.5 5.5 12		1 h	4×10^{-3} mol Fe$_2$O$_3$ dm^{-3}	Uptake of Cr peaks at pH 6. Uptake of Er peaks at pH 8. Uptake of Zn peaks at pH 10. Fcr Ce, Gd, and Yb typical adsorption edges pH$_{50}$ = 6 (carrier free).	167
178. Fe_2O_3, amorphous	Cr(III), Ni, Zn	50 ppm	uptake	10^{-3} mol dm^{-3} KNO$_3$	4-11	25	40 min	60m^2 dm^3	pH$_{50}$ 4.3 (Cr); 6.5(Zn);7.2(Ni)	162
179. Fe_2O_3, hematite 3 samples	Cu, Mg, Zn	11 10^{-4} mol dm^{-3}	uptake	NaNO$_3$	4-11	21	1 d	10^{-3} mol Fe (9 248 m^2 g) dm^{-3}	pH$_{50}$ 4 2 5.2 for Cu, 5.5 5.7 for Zn, and 8 2 10 for Mg	124
180. Fe_2O_3, hematite, monodispersed	Th, U	10^{-13},10^{-6} mol dm^{-3}	uptake	10^{-1} mol dm^{-3} NaClO$_4$, humic acid, CO$_2$	1 11	25	12 h 2d	3 37 9 g (17 4 m^2 TEM) dm^3	Neutral molecules of acids are the components, HL humic acid. (≡Fe$_x$O)$_2$UO$_2$ log K -0.087 (≡Fe$_x$O)$_2$UO$_2$ log K -3.43 (≡Fe$_x$O)$_2$UO$_2$CO$_3^{2-}$ log K -12.14 ≡Fe$_x$OH··UO$_2$L$_2^{2-}$ log K 13 ≡Fe$_x$OH··UO$_2$L$_4^4$ log K 5 4 ≡Fe$_x$OHTh^4 log K 28.9 FeOThL^2 log K 23.73 In L, 1 is subscript indicating different organic ligands (not the number of ligands L in one molecule).	207, 208
181. $Fe_2O_3.H_2O$	Ag		uptake	thiosulfate					≡FeOAg pK -3.5 ≡FeO$^-$ Ag pK -3 9 ≡FeS$_2$OAg pK 16.8	209
182. $Fe_2O_3.H_2O$, amorphous	Cd	0 11 9.4% w w Cd Fe$_2$O$_3$;H$_2$O	EXAFS	chlorate VII	6 7 9.5	25	1 d		Reinterpretation of data taken from literature. Cc-Fe distances 0.332 and 0.35 nm. The number of nearest and 2nd nearest Fe neighbors, 0.7 and 0.8, respectively is independent of the surface coverage.	210
183. $Fe_2O_3.H_2O$, freshly precipitated, from chloride	Cu	10^{-7} 10^{-4} mol dm^{-3}	uptake	1 0.0135	5 9	18	1 h	$10^{-3.41}$ mol Fe dm^3	The slope of log-log adsorption isotherm at pH 6.9 0.6 for [Cu] $10^{-4\,5}$ $10^{-4\,5}$ mol dm^{-3}. The concentration of soluble Cu over the pH range 7.5 9 corresponds to the limit defined by bulk precipitation of Cu(OH)$_2$.	119
184. $Fe_2O_3.H_2O$, from chloride	Cu	3.4×10^{-3} mol dm^{-3}	uptake, XRD	1 0.0135	6-10.2	18		0.019 mol Fe dm^3	Precipitation of mixture of iron and copper hydrous oxides.	120

TABLE 4.1 Continued

Adsorbent	Adsorbate	c. total	S:L ratio	Eq. time	T	pH	Electrolyte	Method	Result	Ref.
185. Fe$_2$O$_3$·nH$_2$O, from chloride	Cu	10$^{-7.5}$ 10$^{-4.17}$ mol dm^{-3}	10$^{3.43}$ mol Fe dm^{-3}	45 min	18	5-9	I = 0.0135	uptake	The pH$_{50}$ ~6.2 for 10$^{3.43}$ mol Fe/dm^3 and 10$^{-4.17}$ mol Cu dm^{-3}.	121
186. Fe$_2$O$_3$·H$_2$O, amorphous	Hg	1.8×10^{-7}, 3.4×10^{-5} mol dm^{-3}	50 mg dm^{-3}			4-11	0.001-0.1 mol dm^{-3} NaCl 0.1 mol dm^{-3} NaClO$_4$	uptake	≡FeOHg log K 6.9 ≡FeOHgOH log K -0.9 ≡FeOHgCl log K 9.8 Reinterpretation of data taken from literature.	211
187. Fe$_2$O$_3$·H$_2$O, from nitrate	Ni	10^{-6} 10^{-4} mol dm^{-3}	0.011 mol Fe dm^{-3}	14h	22 ±3	8	0.1 mol dm^{-3} NaNO$_3$, NaClO$_4$	uptake	Linear log-log adsorption isotherm, slope 1. Cf. # 240.	212
188. Fe$_2$O$_3$·H$_2$O, amorphous	Np (V)	4.5×10^{-13}-4.5×10^{-11} mol dm^{-3}	0.001-0.01 mol Fe dm^{-3}	4 h	25	4-9	0.1 mol dm^{-3} NaNO$_3$, 10$^{-3.5}$ atm CO$_2$	uptake	Linear log-log adsorption isotherms at constant pH, slope 1.	213
189. Fe$_2$O$_3$·H$_2$O, from nitrate	Pb	2×10^{-7} 2×10^{-6} mol dm^{-3}	1 g (36.6 m^2) dm^{-3}	1 d	25	6	0.05 mol dm^{-3} NaNO$_3$	uptake	Rather poor match between data points and best-fit Langmuir isotherm.	81
190. Fe$_2$O$_3$·H$_2$O	Sr	10^{-8} 10^{-3} mol dm^{-3}		3h 3d	25	3-9	0.03 mol dm^{-3} NaNO$_3$	uptake	Linear log-log adsorption isotherm at pH 7, slope 1. Scattered uptake (pH) results suggesting a staircase uptake curve (Fig. 4.7(A)).	214
191. Fe$_2$O$_3$·H$_2$O	Sr	10^{-3} 10^{-2} mol g Fe$_2$O$_3$·H$_2$O (36.6m^2 g)		4 h	22	7		EXAFS	No dehydration upon adsorption, 10 oxygen atoms at 0.265 nm. With Sr loading of 10^{-2} mol g fits obtainable with Fe as the second neighbor.	215
192. Fe$_2$O$_3$·H$_2$O, from nitrate	Sr	10$^{-5.5}$ 10^{-2} mol dm^{-3}	4.18 g dm^{-3}	2 h		2.5 11	none	uptake	With 10$^{-5.5}$ mol Sr dm^{-3} pH$_{50}$ ~7.6.	216
193. Fe$_2$O$_3$·H$_2$O	U	10^{-8} 10^{-4} mol dm^{-3}	10^{-3} mol Fe dm^{-3}	2 d	25	3 10	0.004-0.5 mol dm^{-3} NaNO$_3$; 10$^{-3.5}$ 10^{-2} atm CO$_2$	uptake, EXAFS	1 Fe atom at 0.34 nm. no polynuclear surface species. 100% uptake at pH 5.5 7 for [U]<10^{-5} mol dm^{-3}. The uptake is rather insensitive to ionic strength.	217
194. Fe$_2$O$_3$·H$_2$O, from nitrate	U	10^{-6} 10^{-4} mol dm^{-3}	89 mg (600 m^2 g) dm^{-3}	2 d	25	3 10	0.1 mol dm^{-3} NaNO$_3$ + NaHCO$_3$ + 0-10^{-4} mol dm^{-3} phosphate	uptake	For samples in contact with air the uptake peaks at pH 6.5 (95% for 10^{-4} mol dm^{-3} U). At lower U concentrations 100% uptake between two adsorption edges. The center of the 100% uptake range is at pH 6.5. By increasing the partial pressure of CO$_2$ to 0.01 atm the uptake at high pH is depressed and the low pH adsorption edge is not affected. Presence of phosphates and humic acid (9 mg dm^{-3}) enhances the uptake at low pH and the high pH adsorption edge is not affected	218
195. Fe$_2$O$_3$·H$_2$O, amorphous	Zn	2-64 mg Zn dm^{-3}	0.1 g (29, 303 m^2 g) 20 cm	1 d	25	2 8	0.01 mol dm^{-3} Na$_2$SO$_4$	uptake		146
196. Fe$_2$O$_3$·H$_2$O, amorphous, hydrolysis of nitrate, in absence and presence of citrate	Zn	1 18 μg dm^{-3}	20 mg 20 cm^3	4 d	25	6, 7	0.1 mol dm^{-3} NaNO$_3$	uptake	Langmuir type adsorption isotherms	219

No. Adsorbent	Species	Method	Concentration	Electrolyte	Solid/solution	Time	T	pH	Remarks	Ref
197. Fe₂O₃·H₂O, amorphous	Ag, Ba, Ca, Cd, Co, Cu, Fe(II), Hg,Mg,Mn, Pb,Tl(I),Zn								Minus log of equilibrium constants of the reactions $\equiv$FeOH + Me^z + n H₂O $\equiv$FeO Me(OH)$_n^{(z-n)}$ + (n-1) H , n=0, 1, calculated from literature data data (Fe₂O₃·H₂O and goethite): Ag 5.1,12.3; Ba 7.8,16.3; Ca 6.6, 15.5; Cd 5, 11.3; Co 5, 11.8; Cu 4.3, 8.8; Fe(II) 5.1, 11.1; Hg 1, 3.1; Mg 6.8, 15.6; Mn(II) 5.4, 12.1; Pb 4, 7.5; Tl(I) 8.1, 16.4; Zn 5, 10.6.	220
198. Fe₂O₃·H₂O, amorphous, 4 h aged	Ag, Cu	uptake	4×10^{-7} mol dm^{-3}, 10^{-6} mol dm^{-3}	0.1 mol dm^{-3} NaNO₃	10^{-3} mol Fe/dm^3	2-4 h	25	4-7	100% uptake is not reached for Ag. 0% uptake is not reached for Cu (Fig. 4.6 A and B)	221
199. Fe₂O₃·H₂O, amorphous	Ag, Cu	uptake	4×10^{-7} mol dm^{-3}, 10^{-6} mol dm^{-3}	0.1 mol dm^{-3} NaNO₃	10^{-3} mol Fe/dm^3	2-4 h	25	4-9	pH₅₀ 5.4 (Cu); 7.1 (Ag)	222
200. Fe₂O₃·H₂O	Ag, Cu	uptake	4×10^{-7} mol dm^{-3}, 10^{-6} mol dm^{-3}	0.1 mol dm^{-3} NaNO₃ + 0-0.1 mol dm^{-3} NaCl, Na₂S₂O₃, sulfate, organic ligands	10^{-3} mol Fe/dm^3	2 h	25	4-12	The Ag uptake is depressed in presence of chlorides: pH₅₀ 7.8 and 12 with 0.005 and 0.1 mol dm^{-3} NaCl, respectively. Uptake of Ag in the presence of 4×10^{-7} or 4×10^{-6} mol dm^{-3} Na₂S₂O₃ has a minimum at pH 7-8. Uptake of Cu is rather insensitive to addition of 10^{-4} mol dm^{-3} salicylate, 5×10^{-4} mol dm^{-3} protocatechuate or 10^{-3} mol dm^{-3} sulfate, it is depressed in the presence of 4×10^{-5} mol dm^{-3} picolinate and enhanced in the presence of 10^{-4} mol dm^{-3} gluataminate.	222
201. Fe₂O₃·H₂O, from chloride	Ba, Sr	uptake	10^{-8} 10^{-2} mol dm^{-3}		0.1 g 10 cm^3	3 h	30	6-12	Linear log-log adsorption isotherms, slope 0.9, pH not specified. ΔH_{ads}(30-60 C)~18 kJ mol (Ba, pH 9.2) and 21 kJ mol (Sr, pH 9.8); neutron and γ-irradiation (11 GBq Ra-Be source) results in partial desorption.	223
202. Fe₂O₃·H₂O, from nitrate	Ca,Sr	uptake	2×10^{-6} mol dm^{-3}	1 mol dm^{-3} NaNO₃	2.33×10^3 mol Fe 25 cm^3	3 h	room	5-10	pH₅₀ ~7.1 (Ca and Sr).	150
203. Fe₂O₃·H₂O, amorphous	Ca, Zn	proton release	2.5×10^{-4} mol dm^{-3}	1 mol dm^{-3} NaNO₃	10^{-2} mol Fe/dm^3	1 d	25	7.8, 5.4	0.9% proton released per one Ca adsorbed. 1.7 proton released per one Zn adsorbed.	224
204. Fe₂O₃·H₂O, amorphous	Cd, Co, Pb	uptake	10^{-5} mol dm^{-3}	1 mol dm^{-3} NaNO₃	10^{-3} mol Fe/dm^3	1 d	25	4-10	The pH₅₀ 4.6 for Pb, and 6.7 for Co and Cd. Aging (up to 21 weeks) shifts the pH₅₀ by about +0.2 pH unit.	225
205. Fe₂O₃·H₂O, amorphous	Cd, Cr(III), Ni, Pb, Zn	uptake	10^{-8} 5×10^{-5} mol dm^{-3}	0.1 mol dm^{-3} NaNO₃, transport water	10^{-3} mol Fe/dm^3		25	3-8.5	The pH₅₀ (5×10^{-5} mol dm^{-3} of heavy metal, 0.1 mol dm^{-3} NaNO₃)=4.2 for Cr, 4.8 for Pb, 5.5 for Cu, 6.7 for Zn, 7.2 for Zn and 7.3 for Ni. Replacement of 0.1 mol dm^{-3} NaNO₃ by power plant fly ash transport water enhances the uptake of heavy metals in the acidic region.	226
206. Fe₂O₃·H₂O, amorphous	Cd, Cu, Pb, Zn	proton release	10^{-7} $10^{-4.5}$ mol dm^{-3}	0.1 mol dm^{-3}	10^{-3} mol Fe/dm^3		25	4.5 7.2	The proton stoichiometry coefficients taken from literature are compared with reciprocal slope of log-log adsorption isotherms at constant pH. The agreement was good with Cd and Cu (r=1.8 and 1.9, respectively), but rather poor with Zn.	227

TABLE 4.1 Continued

Adsorbent	Adsorbate	c. total	S L ratio	Eq. time	T	pH	Electrolyte	Method	Result	Ref.
207. $Fe_2O_3 \cdot H_2O$, from nitrate, fresh	Cd, Cu, Pb, Zn	2×10^{-4} mol dm⁻³	10 mg/25 cm³	1 d	25	4–8	nitrate	uptake	The pH₅₀ = 4.4 for Pb, 5.6 for Cu, 6 for Zn and 6.6 for Cd. A >90% release of preadsorbed metal cations: in 1 mol dm⁻³ HNO₃; Zn and Cu; in 0.5 mol dm⁻³ acetic acid: Cd and Zn. Other complexing agents: EDTA, etc. were less efficient. Results of similar experiments for goethite and two Fe ores are also reported.	228
208. $Fe_2O_3 \cdot H_2O$	Cd, Pb	$5 \times 10^{-7} - 5 \times 10^{-5}$ mol dm⁻³	10^{-3} mol Fe/dm³	4 h	25	6–7.5	0.1–0.7 mol dm⁻³ nitrate, 0.5 mol dm⁻³ chloride, 0.2 mol dm⁻³ sulfate, 0–0.01 mol dm⁻³ thiosulfate	uptake	In the presence of 0.1 mol dm⁻³ nitrate, the pH₅₀ 6.7 (5×10^{-7} mol Cd dm⁻³), and 4, 4.6, and 5 for 5×10^{-7}, 5×10^{-6}, and 5×10^{-5} mol Pb dm⁻³, respectively. The uptake of Cd is enhanced in the presence of thiosulfate. but with 10^{-2} mol dm⁻³ thiosulfate at pH >6.5 the uptake is depressed.	48
209. $Fe_2O_3 \cdot H_2O$	Cd, Pb, Tl (I), Zn	10^{-3} mol dm⁻³	6.25×10^{-3} mol Fe dm⁻³	3 h		4–8		uptake	The pH₅₀ 7.3 for Zn, 7.6 for Cd, 5.3 for Pb, no significant uptake of Tl(I). Discrepancy between Pb adsorption data in Fig.2 and 5, probably a typo.	122
210. $Fe_2O_3 \cdot H_2O$	Cd, Sr	2×10^{-7} mol dm⁻³	0.01 mol Fe/dm³	4 h	25	4–8	0.03 mol dm⁻³	uptake	Cd: pH₅₀ = 5.6. Linear adsorption isotherm at pH 7 for concentrations in solution up to 10^{-3} mol dm⁻³ Sr. Uptake increases with T. ΔH_{ads} (4–25°C) = 41 kJ/mol for Sr and 90 kJ/mol for Cd (pH 7).	229
211. $Fe_2O_3 \cdot H_2O$, from chloride, four samples	Co, Zn	carrier free, 10^{-3} mol dm⁻³	1 g/100 cm³	2 d	25	4–7	none, 0.1 mol dm⁻³ NaCl	uptake	In absence of NaCl, pH₅₀ = 6.2 for Co and 5.7–6 for Zn (different samples of the adsorbent). The uptake is enhanced in the presence of NaCl, at least for some samples of the adsorbent.	230
212. $Fe_2O_3 \cdot H_2O$	Cr (III), Ni, Zn	50 ppm	250 ppm Fe		25	3.5–10.5	0.0175 mol dm⁻³ KNO₃	ζ, uptake	The pH₅₀ = 7.5 for Ni. The shift in IEP to higher pH by 1 pH unit for Zn, 0.5 pH unit for Cr and 2 pH units for Ni.	164
213. $Fe_2O_3 \cdot H_2O$	Cr (III), Ni, Zn	50 ppm	250 ppm Fe (316 m²/g)			4–11	0.5 mol dm⁻³ NH₄NO₃	uptake	The uptake of Ni and Zn increases up to pH 7.5, then decreases up to pH 10, and increases again. pH₅₀ = 4.2 for Cr.	163
214. $Fe_2O_3 \cdot H_2O$	Cu,Mg,Zn	1.1×10^{-4} mol dm⁻³	10^{-3} mol Fe/dm³	1 d	21	4–11	0.1, 1 mol dm⁻¹ NaNO₃	uptake	The pH₅₀ = 4.3 (Cu), 5.8 (Zn), 8.5 (Mg).	124
215. $Fe_2O_3 \cdot H_2O$	Cu, Pb	10^{-5} mol dm⁻³	10^{-4} mol Fe dm⁻³			4–8	0.005–0.5 mol dm⁻³ NaCl, NaClO₄, synthetic ocean water	uptake	The pH₅₀ = 6.4 for Cu and 5.8 for Pb. The uptake of Cu is insensitive to ionic strength, the uptake of Pb is depressed at high chloride concentration.	231
216. $Fe_2O_3 \cdot n\,H_2O$	Cu, Pb	5×10^{-4}, 2.5×10^{-2} mol dm⁻³	0.5 g (341 m²/g) 15 cm³	1 d		4–6.5	5% sulfate w w in the adsorbent	uptake		125
217. $Fe_2O_3 \cdot H_2O$, from nitrate	Fe (III), Ni, Pd (II)	$4.6 \times 10^{-7} - 3 \times 10^{-5}$ mol dm⁻³	0.08 g dm⁻³	30 min –3 h	22 24	5–9	0.01 mol dm⁻³ NaNO₃, [EDTA]=[metal]; MES and HEPES buffers	uptake	Anion-like adsorption edges. ≡FeOH + NiEDTA²⁻ + H ≡FeEDTANi⁻ + H₂O log K = 9.36	114

No. Sorbent	Sorbate	Concentration	Solid	Time	T/°C	pH	Electrolyte	Method	Results	Ref.
218. FeOOH, goethite, from nitrate	Au (I and III)	10^{-5} 2×10^{-4} mol dm^{-3}	1 g (81 m²)/dm³	25, 40 min	23 ±3	4-8	0.01-0.1 mol dm^{-3} NaNO$_3$, NaCl, [S$_2$O$_3$] = 3[Au (I)]	uptake	≡FeOH + PdEDTA^{2-} + H$^+$ ≡FeEDTAPd$^-$ + H$_2$O log K 11.32; ≡FeOH + FeEDTA$^-$ + H$^+$ ≡FeEDTAFe + H$_2$O log K 10.97; ≡FeOH + FeEDTA$^-$ + OH$^-$ ≡FeoFeEDTA^{2-} + H$_2$O log K -1.44. Sorption of Au III in the presence of chloride has a broad maximum at pH 5.5, uptake at low pH increases as chloride concentration increases. Sorption of Au 1 in the presence of [S$_2$O$_3$] 3[Au I] slightly decreases as pH increases (Fig. 4.7. (C)) the uptake is depressed at higher ionic strength.	232
219. FeOOH. ?	Cd	5×10^{-7} mol dm^{-3}	0.5 g (22 m²/g) dm³	4 h	25	7, 9	0.7 mol dm^{-3} mol dm^{-3} nitrate, 0.5 mol dm^{-3} chloride, 0.2 mol dm^{-3} sulfate, 0-0.01 mol dm^{-3} thiosulfate	uptake	The pH$_{50}$=7.3 in 0.1 mol dm^{-3} NaNO$_3$. The uptake is depressed at high ionic strengths and in the presence of thiosulfate.	48
220. FeOOH, goethite, from nitrate	Cd	10^{-6} 2×10^{-4} mol dm^{-3}	55 m²/dm³	30 min	25	4-10	0.01 mol dm^{-3} KNO$_3$	uptake	The pH$_{50}$=7 for 10^{-6} mol dm^{-3} Cd, the uptake gradually decreases as the Cd concentration increases. The uptake is enhanced as T increases, ΔH$_{ads}$ ~ 13 kJ/mol (10–70 C, pH 6.5-7, from binding constants in Langmuir equation). The shift in the pH$_{50}$ induced by the change in T is more significant at high initial Cd concentrations.	233
221. FeOOH, goethite, from nitrate	Cd	10^{-5} mol dm^{-3}	1 g (34 m²), 400 cm³	3 h	20, 25	3, 8	0.001 mol dm^{-3} KNO$_3$, 10^{-5} mol dm^{-3} oxalate	uptake, ζ, proton release	Only positive ζ in the presence of Cd, with and without oxalic acid. The pH$_{50}$ 7.3 in absence and 6.8 in the presence of oxalic acid.	234
222. FeOOH, goethite, Research Organic Inorganic Chemical Corp., washed	Cd	3×10^{-6} 3×10^{-5} mol dm^{-3}	12.5 g (21.3 m²/g) dm³	1 d	21	3, 9	0.01-0.1 mol dm^{-3} NaNO$_3$; 3×10^{-4} 2×10^{-3} mol dm^{-3} Na$_2$SO$_4$	uptake	The pH$_{50}$ 5.4 for 3×10^{-6} mol dm^{-3} Cd in absence of sulfate. The uptake is slightly enhanced in the presence of sulfate. The difference between uptake curves obtained with 3×10^{-4} and 2×10^{-3} mol dm^{-3} Na$_2$SO$_4$ is insignificant.	235
223. FeOOH, goethite	Cd	2×10^{-4} 10^{-3} mol dm^{-3}	8×10^{-4} mol Fe dm³, 40.7 m²/g		25	2.7, 9	0.1 mol dm^{-3} NaCl or NaNO$_3$	uptake, proton release	≡FeOHCd^{2+} log K 6.43; ≡FeOCd$^+$ log K -2.22; ≡FeOCdOH log K -12.01; ≡FeOHCdCl log K 6.85; ≡FeOCdCl log K -2.38	236
224. FeOOH, goethite, Bayer, washed	Cd		0.16-1.5% w/w Cd FeOOH	1 d	25	7.5	chlorate VII	EXAFS	Cd-Fe distances 0.326 and 0.348 nm. The ratio of the number of 2nd nearest to nearest Fe neighbors increases when the surface coverage increases.	210
225. FeOOH, goethite	Cd	10^{-8} 3×10^{-4} mol dm^{-3}	4.4-39 g (95 m²/g) dm³	4 h	25	5-9	0.1 mol dm^{-3} NaNO$_3$	uptake, proton release	The uptake at pH 7 is enhanced at high ionic strength (0.5 mol dm^{-3} NaNO$_3$). Linear log-log adsorption isotherms at constant pH, slope 0.9 at pH 5, gradually decreasing to 0.5 at pH 8. 1.54 proton released per one Cd adsorbed.	237
226. FeOOH, goethite, from nitrate	Cd	5×10^{-5} 10^{-3} mol dm^{-3}	6 g (95 m²/g)/dm³	1 d	25	4-9	0.1 mol dm^{-3} NaNO$_3$, 0-2×10^{-3} mol dm^{-3} phosphate	uptake	In the presence of phosphate [P]>[Cd] the Cd uptake is enhanced: with 5×10^{-5} mol dm^{-3} Cd the pH$_{50}$ is shifted by -1 pH unit in the presence of 5×10^{-4} mol P dm^{-3}.	238

TABLE 4.1 Continued

Adsorbent	Adsorbate	$c.$ total	S L ratio	Eq. time	T	pH	Electrolyte	Method	Result	Ref.
227. FeOOH, goethite	Cd	5×10^{-5} mol dm^{-3}			25	3-9	0.005 mol dm^{-3} NaNO$_3$ + 0-10^{-3} mol dm^{-3} benzene carboxylic acids	uptake	The pH$_{90}$ 7 in absence of organic acids. The uptake is enhanced by mellitic = trimesic > pyromellitic > trimellitic > phtalic ≈ hemimellitic acid at concentrations > [Cd].	239
228. FeOOH, goethite, from nitrate	Cd	2.5×10^{-3} mol dm^{-3}	10 g (71 m^2/g)/dm^3 30-280 m^2/dm^3	1 d	20	4-8	0.01 mol dm^{-3} NaNO$_3$	uptake, ζ	The pH$_{90}$ = 6.	240
229. FeOOH, goethite	Cu	7×10^{-6}-10^{-4} mol dm^{-3}				4-7	0.01-0.1 mol dm^{-3} NaNO$_3$	uptake		241
230. FeOOH, goethite	Cu	10^{-5}-10^{-3} mol dm^{-3}	37 m^2/dm^3	1 d	25	5, 5.5	0.01 mol dm^{-3} KNO$_3$	uptake	Adsorption isotherms at constant pH: $\Gamma = \Gamma_{max}(1/2\ b_1c_1 + b_2c_2)/(1 + b_1c_1 + b_2c_2)$ where Γ_{max} is the maximum adsorption density, c_1 and c_2 are concentrations of CuOH and Cu$_2$(OH)$_2^{2+}$ in solution, and b_1 and b_2 are empirical parameters.	242
231. FeOOH, goethite	Cu	1×10^{-5} mol dm^{-3}	0.42 g (47.6 m^2/g)/dm^3	1 min	25	3-12	0.01 mol dm^{-3} NaNO$_3$; 10^{-5} mol dm^{-3} AMP, DTPMP, EDTMP, HEDP, IDMP, NTMP	uptake	The pH$_{90}$ = 5.2 in absence of phosphonates; the uptake is rather insensitive to the presence of AMP, IDMP, and HEDP. The uptake is enhanced at pH < 6 and depressed at pH > 7 in the presence of NTMP, EDTMP, and DTPMP.	243
232. FeOOH, goethite, synthetic	Eu	10^{-8} mol dm^{-3}	0.25 g (62 m^2/g)/dm^3	1 d		3-10	0.01 mol dm^{-3} NaClO$_4$, 2 mg/dm^3 fulvic acid	uptake	In absence of fulvic acid pH$_{90}$ = 5.5. The uptake is depressed in the presence of fulvic acid at pH > 6.	181
233. FeOOH, goethite, from nitrate	Eu	10^{-9} mol dm^{-3}	50 mg (36.4 m^2/g)/dm^3	2 d	room	2-11	0.05 mol dm^{-3} NaClO$_4$ + 1-10 mg dm^{-3} humic acid	uptake	The pH$_{90}$ = 4.5 in absence of humic acid. In the presence of humic acid the uptake is depressed at pH > 5 and enhanced at pH < 4.	130
234. FeOOH, goethite, from nitrate	Hg	5×10^{-6}-5×10^{-5} mol dm^{-3}				3-10	0-0.005 mol dm^{-3} Cl	uptake	The uptake at pH 3-6 is depressed in the presence of 5×10^{-4} mol dm^{-3} chloride, the uptake at pH 8 is insensitive to the presence of chlorides. Linear log-log adsorption isotherms at pH 4 and 9.6, slope 0.74.	244
235. FeOOH, goethite, hydrolysis of nitrate	Hg	2×10^{-4}-8×10^{-4} mol dm^{-3}	11 g (43 m^2/g)/dm^3		25	3-8	0.1 mol dm^{-3} NaNO$_3$, NaCl	proton release, uptake	≡FeOHg log K 4.45 ≡FeOHgOH log K -2.77 ≡FeOHgCl log K 7.43	245
236. FeOOH, goethite	Hg, CH$_3$Hg$^+$	10^{-4} mol dm^{-3}	9 g (40.7 m^2/g)/dm^3	2 d	25	2.7-9	0.1 mol dm^{-3} NaNO$_3$	uptake	≡FeOHgCH$_3$ log K 5.7 ≡FeOHgCH$_3$ log K -2.2	246
237. FeOOH, goethite, Bayferrox 910, Bayer	Hg	1.24×10^{-5} mol dm^{-3}	10 g (15 m^2/g)/dm^3	1 d	25	2.5-11	0.1 mol dm^{-3} NaNO$_3$ + 2×10^{-5} mol dm^{-3} Cl; 0.1 mol dm^{-3} NaCl	uptake	≡FeOHg log K 4.9 ≡FeOHgOH log K -2.3 ≡FeOHgCl log K 8	247
238. FeOOH, goethite, from nitrate	Hg	36.4 ppm	0.27 g (34.4 m^2/g)/0.4 dm^3	10 h	25	4.6	0.1 mol dm^{-3} NaClO$_4$	EXAFS	Uptake 13.7%; 1.9 oxygen atoms at 0.204 nm. 0.6 Fe atoms at 0.328 nm.	248

No., adsorbent	Element	Concentration	Solid	Time	T (°C)	pH	Electrolyte	Method	Notes	Ref
239. FeOOH, goethite	La	10^{-3} mol dm^{-3}	167 m^2 dm^{-3}	3 d	25	3–8	nitrate	uptake, high resolution TEM	pH$_{50}$ = 4.6. Precipitation of La(OH)$_3$ · n H$_2$O at pH > 8. The precipitate forms separate particles.	249
240. Fe$_2$O$_3$·H$_2$O, from nitrate	Ni	10^{-5} mol dm^{-3}	0.009 mol Fe dm^{-3}	12, 24 h		4–10	10^{-1} mol dm^{-3} NaNO$_3$, 10^{-5} mol dm^{-3} EDTA	uptake	Ni sorption in absence of EDTA is reversible. ≡FeONi log K 0.87. In the presence of EDTA (premixed Ni-EDTA solution) anion-like adsorption edge pH$_{50}$ = 6.3. ≡FeEDTANi^{-1} log K 28.76, EDTA^{4-} was selected as a component. When Ni is added first and EDTA after some delay (2h–2 d) or vice versa the uptake peaks at pH 6-7 (80–95% dependent on the delay, which was up to 50 h).	250
241. FeOOH, goethite	Np (V)	4.5×10^{-13}–4.5×10^{-11} mol dm^{-3}	0.01 mol Fe dm^{-3}		25	6–9.5	10^{-1} mol dm^{-3} NaNO$_3$	uptake	≡FeONpO$_2$ log K -0.98	251
242. FeOOH, goethite	Np (V)	1.2×10^{-7} mol dm^{-3}	1 g (45 m^2/g) dm^{-3}	1 d	25	4–9	0.1 mol dm^{-3} NaClO$_4$, 0-2% CO$_2$ in gas mixture	uptake	Reinterpretation of data taken from literature. ≡FeONpO$_2$ log K ≈ -3.2	187
243. FeOOH, goethite	Np (V)							uptake	Divalent anion was selected as a component. ≡FeONpO$_2$ log K -1.56 ≡FeONpO$_2$(HCO$_3$)$_2$$^{2-}$ log K 22.81	188
244. FeOOH, goethite, synthetic	Pb	0.002 mol dm^{-3}	30 g (52 m^2/g) dm^{-3}	>2 h	25	3–7	0.01 1 mol dm^{-3} NaNO$_3$	uptake	pH$_{50}$= 4.6.	252
245. FeOOH, goethite	Pb	2×10^{-3} 3×10^{-2} mol dm^{-3}	30 g (52 m^2/g) dm^{-3}	16 h	22	3–8	0.1 mol dm^{-3} NaNO$_3$	uptake, XAS	The ionic strength effect on uptake curves is insignificant. At low surface coverage inner sphere complex is formed; at high surface coverage polymeric Pb surface species, but no surface precipitation.	253
246. FeOOH, goethite, Bayer	Pb	2.41×10^{-7}–6.8×10^{-4} mol dm^{-3}	16–50 mg (14.7 m^2/g) dm^{-3}	30 min		6.6–8.2	10^{-3} mol dm^{-3} KHCO$_3$, 2.5×10^{-4} mol dm^{-3} tris buffer	uptake, proton release	1.25 proton released per one Pb adsorbed. ≡FeOPb log K -0.52 (≡FeO)$_2$Pb log K -6.27	254
247. FeOOH, goethite, hydrolysis of nitrate	Pb	10^{-7} 10^{-3} mol dm^{-3}	11 14 g (40 m^2/g) dm^{-3}		25	3 8	0.1 mol dm^{-3} NaNO$_3$, NaCl	proton release, uptake	≡FeOHPb2 log K 8.2 ≡FeOPb log K 0.17 ≡FeOPbOH log K -8.85 ≡FeOPbCl log K -0.35 ≡FeOHPbCl log K 7.5 ≡FeOHPbOHCl log K -8	255
248. FeOOH, goethite	Pb	2.4×10^{-3} mol dm^{-3}	8, 19 g (45 m^2/g) dm^{-3}	1 2 d	room	6.7	0.1 mol dm^{-3} NaNO$_3$	XAFS	Polynuclear surface Pb species at [Pb]>[≡FeOH]. 0.2-0.3 Fe atoms at 0.33 nm	190
249. FeOOH, goethite	Pb	0.001 mol dm^{-3}	1.6 g (22 m^2/g) dm^{-3}	24 h	25	3–9	0.1 mol dm^{-3} NaNO$_3$	uptake, XPS	The pH$_{50}$ 6; 90% uptake at pH >7.5 (100% is not reached). Linear adsorption isotherm at pH 6 up to 5×13^{-5} mol dm^{-3} Pb in solution. ≡FeOHPb^{2+} dominates at pH <5.5; ≡FeOPb dominates at 5.5<pH <6.5; ≡FeOPbOH and polynuclear complexes dominate at pH >6.5.	256
250. FeOOH goethite	Pb	2×10^{-4}–4×10^{-3} mol dm^{-3}	1–20 g (45 m^2/g) dm^{-3}	2 d		5–7	0.1 mol dm^{-3} NaCl	XAFS	Ternary Pb-Cl surface complexes are formed at pH <6 with Pb bound to a surface oxygen and Cl bound to surface Fe.	80
251. FeOOH, goethite	Pm	10^{-9} mol dm^{-3}	0.2 g (117 m^2/g) 10 cm^3	>1 d		1–11	0.01, 0.1 mol dm^{-3} NaClO$_4$	uptake	The uptake is insensitive to the ionic strength. For 1 site nm^{-2} the best fit was obtained assuming formation of ≡FeOHPm^{2+}.	257

TABLE 4.1 Continued

Adsorbent	Adsorbate	c. total	S:L ratio	Eq. time	T	pH	Electrolyte	Method	Result	Ref.
252. FeOOH, goethite	Pu (IV and V)	10^{-11}, 10^{-10} mol dm^{-3}	28.5 m^2/dm^3	4-25 d		2-9	0.1 mol dm^{-3} NaNO$_3$	uptake	The pH$_{50}$ 3.8 for Pu IV and 4.8 for Pu V (10^{-10} mol dm^{-3}). The uptake at pH 8.6 is depressed from 100% to 0 in the presence of 0.5 mol dm^{-3} carbonate.	258
253. FeOOH, goethite, from chloride	Ra	trace	5 g 10 cm^3	1 d	room	1,10	HCl, NaOH	uptake		259
254. FeOOH, goethite, from nitrate	Sr	373 ppm	12 g dm^3	1 d	room	9.2, 10.2	0.1 mol dm^{-3} NaClO$_4$	uptake EXAFS	Inner sphere complex, 1.8 Fe atoms at 0.43 nm at pH 10.2.	260
255. FeOOH, goethite, from chloride	Sr	10^{-3} mol dm^{-3}	1500 m^2/dm^3	2 d	25	6-10	0.1 mol dm^{-3} NaCl	uptake, EXAFS	Uptake is depressed at pH 7-8.5 in the presence of atmospheric CO$_2$, but it is enhanced at pH > 8.5. The SrCO$_3$ surface precipitate is formed at pH 8.5, but it was not detected at higher pH.	261
256. FeOOH, goethite	Th	$4.5,9\times10^{-6}$ mol dm^{-3}	0.54, 8.6 g dm^3	3 h	20	2.5-6	0.422 mol dm^{-3} NaCl + 0-0.028 mol dm^{-3} Na$_2$SO$_4$; sea water	uptake	The pH$_{50}$ = 3 (9×10^{-6} mol dm^{-3}, 8.6 g dm^3) in absence of sulfate, uptake is depressed in the presence of sulfates.	262
257. FeOOH, goethite	U	10^{-5} mol dm^{-3}	1 g (45 m^2) dm^3	4 h	25	3-9	0.1 mol dm^{-3} NaNO$_3$, 0-0.01 mol dm^3 carbonate	uptake	The pH$_{50}$ = 4.2; in the presence of carbonates the uptake at pH > 5 is depressed, and it shows a maximum at pH 6-7.	196
258. FeOOH, goethite	U	2 mg/dm^3	0.2 g (58.5 m^2/g) 25 cm^3	1 d		4-9	synthetic drainage water, 0.1 mol dm^{-3} Na(Cl) + 0.22 93 kPa CO$_2$	uptake	In 0.1 mol dm^{-3} Na(Cl) anion like adsorption edges with pH$_{50}$ = 8.3 at 0.22 kPa CO$_2$, 7.5 at 5 kPa and 6.4 at 93 kPa.	263
259. FeOOH, goethite, from nitrate	U	10^{-6} mol dm^{-3}	1 g (68 m^2/g) dm^3	1 d	22	3-11	0.1 mol dm^{-3} NaCl + 0-1×10^{-4} mol dm^3 citrate	uptake	The pH$_{50}$ shifts from 4.3 in absence of citrate to 3.2 in the presence of 1. $\times10^{-5}$ mol dm^3 citrate. For 10^{-4} mol dm^{-3} citrate > 90% uptake over the entire studied pH range.	144
260. FeOOH, goethite, High Purity Fine Chemical Inc.	U	10^{-4} mol dm^{-3}	1.2, 12 g/(52 m^2/g) dm^3	2 d		3-9	0.1 mol dm^{-3} KNO$_3$	uptake	With 12 g/dm^3 and in contact with atmospheric air 100% uptake between two adsorption edges pH$_{50}$ = 4 and 9.	200
261. FeOOH, goethite	Zn	2-64 mg Zn dm^3	0.1 g dm 20 cm^3	1 d	25	2-8	0.01 mol dm^{-3} Na$_2$SO$_4$	uptake		146
262. FeOOH, goethite, from nitrate	Zn	3.8×10^{-7}, 10^{-4} mol dm^{-3}	50 mg (95 m^2/g) 200 cm^3	2 d	25	5.4-8.5	0.001 mol dm^{-3} NaNO$_3$	uptake	pH$_{50}$ 5.5 (0.2 mg Zn dm^3); 6.2 (2 mg Zn dm^3)	264
263. FeOOH, goethite	Zn	10^{-6} mol dm^{-3}	70 g (50 m^2 g)/dm^3	3 d	20	3-10	0.01 mol dm^{-3} NaClO$_4$, 0-20 mg fulvic acid /dm^3	uptake	In absence of fulvic acid pH$_{50}$ = 7.6, the uptake is enhanced in the presence of fulvic acid at pH < 8.	148
264. FeOOH, goethite, 2 samples	Zn	10 mg/dm^3	1 g (130 m^2/g)/dm^3	1 d	25	2-10	none	uptake		265
265. FeOOH, goethite, from nitrate	Al, Bi(III), Ca, Cr, Co (II and III), Cu, Fe(III), In, La, Ni, Pb, Pd, Zn	2×10^{-7}, 2×10^{-5} mol dm^{-3}	0.16-1.72 g (21 m^2 g)/dm^3	30 min, 2 d		3-9	0.001-0.1 mol dm^{-3} NaNO$_3$, [EDTA] = [metal] MES and HEPES buffers	uptake	Anion-like adsorption edges for heavy metals, almost identical pH$_{50}$ = 6.7 for Pb,Ni,Cu,Zn,Co (0.46 g dm^3, 4.6×10^{-7} mol dm^{-3}, 0.01 NaNO$_3$). pH$_{50}$ = 9.5 for Pd at the same conditions. The uptake of Ca peaks at pH 6.8. High and rather pH insensitive uptake of La and Bi (cf. Fig. 4.7(C), no adsorption of Co(III)) at pH > 6.	266

No.	Adsorbent	Ion	Concentration	t (°C)	pH	Electrolyte	Solid	Time	Method	Comments	Ref.
266.	FeOOH	Al, Fe(III)	0.001 mol dm⁻³	25	4–12	chloride 0.01, 0.1 mol dm⁻³ NaCl	1.6 g (79.4 m² g) dm⁻³	1 d	ζ uptake	IEP is shifted to pH 8.4 in AlCl₃, (6.6 in KCl and FeCl₃).	153
267.	FeOOH, goethite	Ca, Cu	2.3×10^{-6}–2×10^{-4} mol dm⁻³	20	4–11	0–0.001 mol dm⁻³ phthalic and chelidamic acid				Uptake of Cu is insensitive to ionic strength, uptake of Ca is depressed at high ionic strength. The percentage of uptake of Ca and Cu gradually decreases as concentration of metal cations increases. Uptake of Cu is enhanced in the presence of phthalic and chelidamic acid at pH < 5, the uptake of Ca is not affected by phthalic and depressed in the presence of chelidamic acid at pH > 9. (≡FeOH)₂Ca² log K 6.78; ≡FeOCa log K -6.97; ≡FeOCaOH log K -16.96; ≡FeOCu log K 2.21. Divalent anions are chosen as components. H₂L = phthalic acid; H₃L = chelidamic acid; ≡FeOHCuL log K 10.91; ≡FeOHCuHL log K 13.06; ≡FeOCuL² log K -0.74; ≡FeOHCuHL₂³⁻ log K 10.24	267
268.	FeOOH, goethite	Cd, Co	5×10^{-5} mol dm⁻³	25	3–10	0.005 mol dm⁻³ KNO₃	94 m² dm⁻³	20 min	uptake, proton release	The pH_{50} = 6.6 for Cd and 7.2 for Co (staircase, plateau at pH 4–5, 10% uptake). ΔH_{ads} (10–70°C, pH 7) = 23 kJ/mol (Cd) and 26 kJ/mol (Co). Langmuir adsorption isotherm at pH 7 for Cd, binding constant 28 m³/mol, capacity 5.4×10^{-7} mol m². Two site Langmuir type adsorption isotherm at pH 7 for Co. 1 proton released per one adsorbed Cd, 1.2 protons released per one adsorbed Co.	268
269.	FeOOH, goethite	Cd, Co II and III, Cu, Hg, Pb, Zn	3×10^{-5} mol dm⁻³	20	4–8	0.075 mol dm⁻³ NaNO₃, NaCl	20 mg (89 m²/g), 40 cm³	25 min, 1 d	uptake	Co(III) in form of ammonium and ethylenediamine complexes is not adsorbed at pH 3–11. Uptake of Co (II), Cu and Zn does not change when NaNO₃ is replaced by NaCl. The uptake of Pb and Cd is slightly enhanced and uptake of Hg is severely depressed in the presence of chloride.	269
270.	FeOOH, goethite	Cd, Co, Cu, Pb, Zn	3.2×10^{-5} mol dm⁻³	20	4.7–8	0.075 mol dm⁻³ NaNO₃, KNO₃	20 mg (89 m²/g), 40 cm³	25 min	uptake, proton release	1.8–2.4 protons released per one metal cation adsorbed.	270
271.	FeOOH, 2 samples, goethite and lepidocrocite	Cd, Cu	2×10^{-5}–3×10^{-3} mol dm⁻³		4.7 7.9	sulfate nitrate	0.1 g (20, 80 m²/g) 10 cm³	1 d	uptake, XAS	Cd on goethite: 3 Fe at 0.375 nm; Cd on lepidocrocite: 2 Fe at 0.331 nm and 1 Fe at 0.377 nm; Cu on goethite: 2 Cu or Fe at 0.292 nm; Cu on lepidocrocite: 2 Cu or Fe at 0.304 nm and 2 Cu or Fe at 0.367 nm.	271

TABLE 4.1 Continued

Adsorbent	Adsorbate	c. total	S:L ratio	Eq. time	T	pH	Electrolyte	Method	Result	Ref.
272. FeOOH, goethite	Cd, Cu, Pb, Zn	3×10^{-4} 3×10^{-1} mol dm⁻³	28.5 m²/dm³	2.5 h	25	3-8	0.1 mol dm⁻³ NaNO₃, 0.53 mol dm⁻³ NaCl, sea water	uptake	The uptake curves with 0.1 mol dm⁻³ NaNO₃ and 0.53 mol dm⁻³ NaCl are practically identical. The uptake (pH) curves are independent of the initial concentration of heavy metal cations for [Cu] up to 3×10^{-5} mol dm⁻³, [Pb] and [Cd] up to 3×10^{-6} mol dm⁻³. Cu uptake is enhanced in the presence of sulfate or sea water, the uptake of Zn and Cd is depressed, and the uptake of Pb in enhanced at pH < 5 and depressed at pH > 5.	272
273. FeOOH, goethite	Cd, Cu, Pb, Zn	10^{-6} 9×10^{-4} mol dm⁻³	0.1-10 g (40 m²/g) dm³	15 min	22	3-9	0.1 mol dm⁻³ KNO₃, or NaNO₃	uptake	pH₅₀ 6.2 (2×10^{-6} mol dm⁻³ Pb, 0.107 g dm³ goethite); 4.7 (2×10^{-6} mol dm⁻³ Pb, 10.7 g dm³ goethite); 7.5 (1 9×10^{-6} mol dm⁻³ Cd, 0.9 g dm³ goethite); 6.5(2×10^{-6} mol dm⁻³ Zn, 0.9 g dm³ goethite); 4.3 (10^{-6} mol dm⁻³ Cu, 9 g dm³ goethite).	273
274. FeOOH, goethite, from chloride	Cd, Ni, Zn	10^{-5} mol dm⁻³	20 mg (73 m²/g) 10 cm³	42 d	20	4-8	0.01 mol dm⁻³ Ca(NO₃)₂	uptake	pH₅₀ 5 for Zn, 5.5 for Ni and 6 for Cd. Uptake increases when T increases.	274
275. FeOOH, goethite, synthetic, 0.86% Si	Cd, Ni, Zn	10^{-6} 10^{-4} mol dm⁻³	2 g (75 m²/g) dm³	2h–42 d	20	4-8	0.01 mol dm⁻³ Ca(NO₃)₂	uptake	The pH₅₀ (10^{-6} mol dm⁻³)= 4.9 for Zn and 5.8 for Cd and Ni, and increases by 1 pH unit when the total concentration increases to 10^{-4} mol dm⁻³. Increase of T from 5 to 35 C induces a shift in pH₅₀ to lower pH values by 1.1 pH unit for Ni, 0.9 for Zn and 0.4 for Cd. This shift is rather insensitive to total concentration and aging time.	275
276. FeOOH, goethite, synthetic	Cd, Pb	0.0001-0.002 mol dm⁻³	30 g (52 m²/g) dm³	16 h	25	3-7	0.001 1 mol dm⁻³ NaNO₃,	uptake	The pH₅₀ 5.5 for Cd, and 4.6 for Pb.	276
277. FeOOH, goethite	Co, Cu, Hg, Pb, Zn	7×10^{-6} 10^{-4} mol dm⁻³	12 280 m²/dm³			4-7	0.01-0.5 mol dm⁻³ NaNO₃	uptake	The ionic strength effect on uptake curves is insignificant. The percentage of Pb uptake gradually decreases as the total Pb concentration increases. ≡FeOCu log K 1.92 ≡FeOPb log K 1.16 ≡FeOZn log K 0.34 ≡FeOCo log K 0.54 ≡FeOHg log K 4.97	277
278. FeOOH, goethite	Co, Cu, Pb, Zn	10^{-5} mol dm⁻³				3 6.5	0.1 mol dm⁻³ NaNO₃, NaCl	uptake, desorption	Reinterpretation of data taken from literature. Cu≥Pb>Zn>Co; uptake of Zn, Co, and Pb in the presence of NaCl is higher than in the presence of NaNO₃.	278
279. FeOOH, goethite	Co, Cu, Pb, Zn	10^{-4} mol dm⁻³	45-180 m²/dm³	20 min	25	3.5 10	0.001–0.1 mol dm⁻³ KNO₃, 0.1 mol dm⁻³ NaCl	uptake	For 45 m² dm³ and 0.001 mol dm⁻³ NaNO₃, pH₅₀ 5.3 for Cu, 7.4 for Zn, 8 for Co and 8.1 for Cd. Cu and Zn uptake is slightly enhanced in the presence of 0.1 mol dm⁻³ NaCl, and Co and Cd uptake is slightly depressed. Cd uptake is also depressed at 0.1 mol dm⁻³ NaNO₃.	109

Adsorbent	Species	Concentration	Solid (surface area), volume	Time	T	pH	Electrolyte	Method	Result/remarks	Ref
280. FeOOH, akaganeite, synthetic	Co, Mn(II), Ni, Zn	2×10^{-3} mol dm^{-3}	50 mg (52 m^2/g) 50 cm^3	3 d	27	5 9	0.1–0.5 mol dm^{-3} NaCl, 0.1 mol dm^{-3} NaNO$_3$, major ion seawater	uptake	The log-log adsorption isotherms for 0.5 mol dm^{-3} NaCl at constant pH are linear with slope 1 for Mn (pH 7.2), 0.55 for Zn (pH 6.9) and 0.85 for Ni (pH 6.9) up to about $10^{-4.4}$ mol of metal ions adsorbed per gram. Then the slope decreases. The log-log adsorption isotherm for Mn(II) has a complicated shape.	279
281. FeOOH, amorphous, synthetic	Co, Mn(II), Ni, Zn	2 10 10 mol dm^{-3}	50 mg (71 m^2/g) 50 cm^3	3 d	27	5 9	0.1–0.5 mol dm^{-3} NaCl, 0.1 mol dm^{-3} NaNO$_3$, major ion seawater	uptake	The log-log adsorption isotherms for sea water at constant pH are linear with slope 1.1 for Zn (pH 7.3) and 0.95 for Ni (pH 7.1) up to about $10^{-3.7}$ mol of metal ions adsorbed per gram. Then the slope decreases.	280
282. FeOOH, goethite, 3 samples	Cu, Mg, Zn	1 1 10^{-4} mol dm^{-3}	10^{-3} mol Fe/dm^3 (58 80 m^2/g)	1 d	21	4 11	0–1 mol dm^{-3} NaNO$_3$	uptake	The ionic strength has no effect on Cu and Zn uptake, pH$_{50}$ 5 and 6.5, respectively. The uptake of Mg is slightly depressed (shift in pH$_{50}$ by 0.5 pH unit) when the ionic strength increases from 0 to 0.1 mol dm^{-3} NaNO$_3$.	124
283. FeOOH, lepidocrocite, 3 samples	Cu, Mg, Zn	1.1 10^{-4} 1 1 10 mol dm^{-3}	10^{-3} mol Fe/dm^3 (94 m^2/g)	1 d	21	4 11	0.1, 1 mol dm^{-3} NaNO$_3$	uptake	The pH$_{50}$ 4.8 for Cu, 6 for Zn, and 8.6 for Mg at initial concentrations of 1.1×10^{-4} mol dm^{-3}. The pH$_{50}$ is gradually shifted to higher pH when the total concentration of metal cations increases.	124
284. FeOOH, goethite	Cu, Pb, Zn	3×10^{-5} 1 2×10^{-3} mol dm^{-3}	0.5 g (89 m^2/g) or 2 g (74 m^2/g) dm^{-3}	1 2 d	20	4–6	0.0075–0.1 mol dm^{-3} NaCl or 0.075 mol dm^{-3} KNO$_3$	uptake, proton release	Reinterpretation of literature data. Cu and Pb uptake is enhanced in the presence of chlorides. In absence of chlorides 1.3 protons released per Pb and 1.8 per Cu adsorbed, almost pH independent. In the presence of chlorides the proton stoichiometry factor is lower and increases when pH increases.	281
285. FeOOH, goethite	Cu, Pb, Zn	10 ,10^{-4} mol dm^{-3}	0.2 g (55 m^2/g)/0.3 dm^3	20 min	25	3 11	0.01 mol dm^{-3} KNO$_3$	uptake, proton release	The pH$_{50}$ 5.3 for Cu, 6.1 for Pb, and 7.4 for Zn.	282
286. FeOOH, goethite	Cu, Pb, Zn	7 10^{-6} 10^{-4} mol dm^{-3}	12 28 g/dm^3			4 7	0.01–0.5 mol dm^{-3} NaNO$_3$	uptake		283
287. FeOOH, lepidocrocite	Fe(III), Ni	4.6×10^{-5} 3×10^{-5} mol dm^{-3}	0.46 g (78 m^2/g) dm^{-3}	30 min 3 h	22 24	5 8	0.01 mol dm^{-3} NaNO$_3$, [EDTA] [metal], MES and HEPES buffers	uptake	Anion-like adsorption edge. FeOH + NiEDTA + H ⇌ FeEDTANi + H$_2$O log K 10.42 FeOH + FeEDTA + H ⇌ FeEDTAFe + H$_2$O log K 12.62 FeOH + FeEDTA + OH ⇌ FeOFeEDTA^{2-} + H$_2$O log K -1.16	114
288. FeOOH, goethite	Mg, Mn	9×10^{-5} 3.6×10^{-4} mol dm^{-3}	1.5 m^2 75 cm^3	1 d		5 11	0.009 mol dm^{-3} NaClO$_4$	uptake, ζ, ESR	Mn does not significantly influence the ζ potential, in the presence of Mg ζ is positive and almost pH independent	131
289. FeOOH, goethite	Pb, Zn	10^{-6} 10^{-4} mol dm^{-3}	37 m^2/dm^3	1 d	25	5.5 7.5	0.01 mol dm^{-3} KNO$_3$	uptake, proton release	Proton stoichiometry: 1.2 (Pb, pH 5.5) 1.6–1.8 (Zn, pH 6.5–7.5). Two site Langmuir type adsorption isotherms for Zn and Pb. ΔH_{ads} from parameters of two site Langmuir adsorption isotherms, 10–70 C Zn, site 1: 4–10 kJ mol, site 2: 16–28 kJ mol, Pb, site 1: 6 kJ mol, site 2: 8 kJ mol.	284

TABLE 4.1 Continued

Adsorbent	Adsorbate	c. total	S:L ratio	Eq. time	T	pH	Electrolyte	Method	Result	Ref.
290. FeOOH, amorphous, from chloride	Ra	$5\times10^{-10}, 5\times10^{-7}$ mol dm^{-3}; $5\times10^{-7}-10^{-4}$ mol dm^{-3}	0.279 g Fe dm^{-3}	7 d	25	6.8-8.6	0.01 mol dm^{-3} NaCl or NaHCO$_3$	uptake	Uptake of Ra at pH 7 is depressed as T increases from 25 to 60 C.	285
291. Fe(OH)$_3$	Cu	$1.9\times10^{-4}, 7.9\times10^{-4}$ mol dm^{-3}	10^{-4} mol dm^{-3} Fe	30 min	2-14		1 mol dm^{-3} KNO$_3$, NaClO$_4$, 0.1-5 mol dm^{-3} NH$_4$NO$_3$	uptake	At 0.25-1 mol dm^{-3} NH$_4$NO$_3$ two maxima are observed at pH 6 and at pH 12 in the uptake (pH) curves. Preadsorbed Cu (pH about 7, contact times 10 min-2 d) is entirely or almost entirely desorbed within 4 min at pH about 11 in presence of ammonia	286
292. Fe(OH)$_3$, from nitrate	Eu	10^{-8} mol dm^{-3}	0.05 g (320 m^2/g) dm^{-3}	1 d		3-10	0.01 mol dm^{-3} NaClO$_4$, 2 mg dm^{-3} fulvic acid	uptake	pH$_{50}$ 5. The uptake is depressed in the presence of fulvic acid at pH > 5 and enhanced at pH < 4.	181
293. Fe(OH)$_3$	Ni	4×10^{-5} mol dm^{-3}	10^{-4} mol 20 cm^3		25	6-11	1 mol dm^{-3} KNO$_3$, 1.5 mol dm^{-3} NH$_4$NO$_3$	uptake	Uptake data obtained at different NH$_4$NO$_3$ concentrations plotted as log ((uptake/(1-uptake)) vs. log fraction of neutral hydroxo complex gave a straight line with slope 1 (reinterpretation of earlier published data).	287
294. Fe(OH)$_3$, fresh precipitated	Pb	10^{-5} mol dm^{-3}	4×10^{-3} mol Fe dm^{-3}	30 min	22±2	1-14	1 mol dm^{-3} NaNO$_3$, 1 mol dm^{-3} NaCl	uptake	100% uptake between two adsorption edges: pH$_{50}$ 4.5 and pH$_{50}$ 13. 100% uptake between two adsorption edges: pH$_{50}$ 6 and pH$_{50}$ 13.	189
295. Fe(OH)$_3$, fresh precipitated	Sc	$1.2\times10^{-10}, 1.1\times10^{-3}$ mol dm^{-3}	$4\times10^{-4}, 4\times10^{-2}$ mol Fe dm^{-3}			1-14	1 mol dm^{-3} NaNO$_3$, 0.05-0.5 mol dm^{-3} Na$_2$CO$_3$	uptake	The uptake curves are insensitive to the initial Sc(III) concentration. In absence of carbonates pH$_{50}$ 3 for 4×10^{-2} mol Fe dm^{-3} and pH$_{50}$=4.5 for 1×10^{-3} mol Fe dm^{-3} The uptake is 100% at pH 6-13 when [Fe] > > [Sc] The uptake is reduced at [Na$_2$CO$_3$] > 0.1 mol dm^{-3} at pH 7-11 (minimum uptake at pH 9.5), but it is rather insensitive to the presence of Na$_2$CO$_3$ outside this pH range.	192
296. Fe(OH)$_3$, amorphous, synthetic	U	10^{-5} mol dm^{-3}	1 g (306 m^2) dm^{-3}	4 h	25	3-9	0.1 mol dm^{-3} NaNO$_3$, 0-0.01 mol dm^{-3} carbonate	uptake	The pH$_{50}$ ~3.5; in the presence of carbonates the uptake at pH > 5 is depressed, and it shows a maximum at pH 6-7	196
297. Fe(OH)$_3$, ?	Zn	10 mg dm^{-3}	1 g (468 m^2/g) dm^{-3}	1 d	25	2-10	none	uptake		265
298. Fe(OH)$_3$, gel	Ba,Ca,Cd, Co,Cu,Mg Ni, Pb, Sr, Zn	1.25×10^{-4} mol dm^{-3}	0.093 mol dm^{-3} of Fe dm^{-3}	3 h		3-9	1 mol dm^{-3} NaNO$_3$	uptake	pH$_{50}$ 3.1 (Pb), 4.4 (Cu), 5.4(Zn), 5.6(Ni), 5.8(Co), 6 (Cd), 7.4 (Sr), 7.8(Mg)	149
299. Fe(OH)$_3$	Cd,Zn	10^{-6} mol dm^{-3}	10^{-3} mol Fe dm^{-3} (193 m^2/g)	2 h		5-9	0.1 mol dm^{-3} NaNO$_3$	uptake	pH$_{50}$ ~6.6 (Cd) and 6.2 (Zn)	151
300. HgO	Al	0.001 mol dm^{-3}			25	4-12	chloride	ζ	IEP at pH 8.4 (no IEP in KCl).	153

Adsorbent	Adsorbate	Concentration	Solid/liquid	Time	T	pH	Electrolyte	Method	Results	Ref
301. IrO₂, from IrCl₃	La	$10{-}10^{-3}$ mol dm⁻³		1 d		2-6	0.002 mol dm⁻³ KNO₃	ζ	No IEP for 5×10^{-5} mol dm⁻³ La; for 10^{-4} mol dm⁻³ La mixture of positive and negative ζ potentials, probably multiple IEP	288
302. La₂O₃	Al, La	0.001 mol dm⁻³						ζ	IEP is shifted to pH 8.2 in AlCl₃ and 9.9 in LaCl₃ (6.7 in KCl)	153
303. MgO	Mn(II)	4×10^{-4} mol dm⁻³	0.05 g 25 cm³	30 min	25	4-12	chloride	uptake		186
304. MgO	Pb	10^{-12} mol dm⁻³	0.09 g 25 cm³	30 min	20-22	2-14	1.4 mol dm⁻³ NaNO₃	uptake	100% uptake between two adsorption edges pH₅₀ 9.7 and pH₅₀ 13	189
305. MgO	Al, Mg	0.001 mol dm⁻³			25	4-12	chloride	ζ	IEP is shifted to pH 8 in AlCl₃ and 9.2 in MgCl₂ (9.8 in KCl)	153
306. MnO₂, cryptomelane	Ag	$10^{-4}{-}2\times10^{-4}$ mol dm⁻³	1 g (23 m²) dm⁻³	1 h	20	3-7	0.01,0.1 mol dm⁻³ KNO₃, LiNO₃, NaNO₃	uptake, release of K		289
307. MnO₂, δ	Al	0.001 mol dm⁻³			25	4-12	chloride	ζ	IEP is shifted to pH 8 (1.9 in KCl)	153
308. MnO₂, δ	Au (I and III)	0.1-6 mg dm⁻³	1 g (93 m²) dm⁻³	6 h	20	2-10	0.01, 0.1 mol dm⁻³ NaNO₃, [S₂O₃] 2[Au I] or [Cl] 4[Au III]	uptake	Uptake of Au decreases when pH increases Uptake of AuCl₄ is depressed at high ionic strength	290
309. MnO₂ (MnSO₄, NaOH in air)	Ba	$10{-}10$ mol cm⁻³	0.1 g 10 cm³	0.5 h	30	6.42		uptake	Linear log-log adsorption isotherm slope 0.8; the uptake increases with T	291
310. MnO₂, BDH	Cd	$4\cdot10^{-4}$ mol dm⁻³	50 mg 1.5 cm³	15 min	23	0.7-8.7 (initial)	acetate buffer	uptake	Minimum uptake at pH 3 and maximum at pH 7 Effects of 9 anions on the Cd uptake from 0.01 mol dm⁻³ HNO₃ and 2 mol dm⁻³ NH₃ were also tested, of which only EDTA substantially depressed the uptake	292
311. MnO₂	Co	trace	1 g/300 cm³	15 min	25	5 (initial)	0-5 mol dm⁻³ NaCl	uptake	The NaCl, NaHCOO, and NaBr effect on the uptake is insignificant at 0.01 mol dm⁻³ of these salts, but with 0.1 mol dm⁻³ NaCl the uptake is depressed, and it is even more depressed at higher NaCl concentrations More or less significant depression in uptake was also observed in the presence of 0.01 mol dm⁻³ of sodium iodide. molybdate, tungstate, thiosulfate, borate, tartrate, oxalate and EDTA	293
312. MnO₂, IC 12	Co	$10^{-6}{-}10^{-4}$ mol dm⁻³		3-12 h	25	4-5.5	0.1 mol dm⁻³ NaNO₃	uptake	Linear log adsorption isotherms, slope 0.5 at pH 4.5 and 0.3 at pH 5.5	59
313. MnO₂, electrodeposited,	Hg	$10^{-5}{-}2\cdot10^{-4}$ mol dm⁻³	10 mg 25 cm³	1 d	25	5	nitrate	uptake	Langmuir isotherm, capacity 0.015 mol kg⁻¹	294
314. MnO₂, α, synthetic	Hg	$10{-}2\cdot10^{-4}$ mol dm⁻³	10 mg 25 cm³	1 d	25	5	nitrate	uptake	Langmuir isotherm, capacity 0.25 mol kg⁻¹, binding constant 4440 dm³ mol⁻¹	294
315. MnO₂... δᵐ	La	10^{-3} mol dm⁻³	167 m² dm⁻³	3 d	25	3-8	nitrate	uptake, high resolution TEM	The pH₅₀ 4 Precipitation of La(OH)₃·n H₂O at pH 5	249
316. MnO₂, thermal decomposition of Mn(NO₃)₂	Mn(II)	0.0001-0.01 mol dm⁻³	5 g/100 cm³	3 h	20	5.4-7	10^{-2} mol dm⁻³ NaCl	uptake, ζ	Also 10 and 35 C Linear adsorption isotherm at pH 7 Uptake increases when T increases.	295

TABLE 4.1 Continued

Adsorbent	Adsorbate	c. total	S:L ratio	Eq. time	T	pH	Electrolyte	Method	Result	Ref.
317. MnO_2, electrolytic, IC3	Ni	10^{-6}–1 mol dm^{-3}	0.1 g (47 m^2/g), 50 cm^3	1 d	25	1–7	sulfate, 0.0024–0.04 mol dm^{-3} acetate	uptake, proton release	Linear log-log adsorption isotherms at pH 3.5, 4.5, slope 0.23; uptake at pH 5.35 increases with T (25–100 C). 1.4–1.6 protons released per adsorbed Ni at pH 4.5–6.5.	296
318. MnO_2, 3 samples: Fisher; ICN, synthetic	Pb	2×10^{-7}–2×10^{-6} mol dm^{-3}		1 d	25	6	0.05 mol dm^{-3} NaNO$_3$	uptake	Uptake by freshly precipitated sample exceeds that by commercially available reagents by 2 orders of magnitude.	81
319. MnO_2, Fisher	Ra	trace	5 g 10 cm^3	1 d	room	1, 10	HCl NaOH	uptake	100% uptake at pH 10.	259
320. MnO_2, special purity	Sb(III)	4×10^{-6}, 4×10^{-5} mol dm^{-3}	0.5g, 25 cm^3	30 min		1–7	1 mol dm^{-3} NaClO$_4$	uptake	pH$_{50}$ ~0.5	202
321. MnO_2, δ	Th	9×10^{-8}–4.5×10^{-5} mol dm^3	8.3 g (130 m^2/g), dm^3	6 h	20	2.5–6	0.422 mol dm^{-3} NaCl + 0–0.028 mol dm^{-3} Na$_2$SO$_4$; sea water	uptake	The pH$_{50}$ 4.3 in absence of sulfate, uptake is depressed in the presence of sulfate, and even more severely depressed in the presence of 8×10^{-6} mol dm^{-3} EDTA and CDTA.	262
322. MnO_2, δ	Ag, Ba, Ca, Cd, Co, Cu, Fe(II), Hg, Mg, Mn(II), Pb, Tl(I), Zn								Minus log of equilibrium constants of the reactions $MnOH + Me^z + n\ H_2O \rightleftharpoons MnO\ Me(OH)_n^{(z-n)}$ $(n+1)\ H^+$, n = 0,1,2, calculated from literature data: Ag 3.3, 13.6, -; Ba 4.4, -, -; Ca 5.3, -, -; Cd 1.9, 10.6, 19.4; Co 1.5, 9.8, 17.6; Cu 0.1, 7.5, 13.4; Fe(II) 1.7,9.9,19.4; Hg -2.4,0.5,-; Mg 5.9,-,-; Mn(II) 2, 11.1,21.1; Pb -1.8,6.5,-; Tl(I) 4.7,16,-; Zn 1.5, 8.8, 15.	220
323. MnO_2, δ, synthetic	Ba, Ca, Mg, Sr	2×10^{-5}–6×10^{-4} mol dm^{-3}	2×10^{-3} mol dm^{-3}	1.5 h	25	7	KCl	uptake, K release	Langmuir type adsorption isotherms, uptake increases when T increases.	297
324. MnO_2	Ba, Ca, Mg, Sr	0.001–0.1 mol dm^{-3}			25		nitrate	proton release	The shift in PZC induced by the salt concentration of 0.1 mol dm^{-3} (in mV, 1 pH unit = -59 mV): Mg 60, Ca 60. Sr 50, Ba 70.	158
325. MnO_2, different forms, 15 samples	Ba, Cu	8×10^{-4} mol dm^{-3}, 4×10^{-4} mol dm^{-3}	0.2 g (2-94 m^2/g), 50 cm^3	30 h	27	2.5–6	unbuffered, phthalate buffer	uptake, proton release	0.6-5.2 proton released per adsorbed Cu, 0.4-4.4 proton released per adsorbed Ba (the high values of proton stoichiometry coefficients were obtained at very low pH).	298
326. MnO_2, δ, from MnCl$_2$ + KMnO$_4$	Ca, Mg		1.27 g (74 m^2/g), dm^3	1 d	25	2-8	NaCl	uptake	Mg, Ca: uptake onset at pH 2.	299
327. MnO_2, δ, synthetic	Ce(III), Eu	10^{-4} mol dm^{-3}	20 g (30 m^2/g), dm^3	5 d	4	7.3	0.1 mol dm^{-3} NaClO$_4$	XANES	100% uptake at pH 4: oxidation of Ce (III) to Ce(IV).	300
328. MnO_2, IC 12, ?	Co, Cu, Mn, Ni, Zn	10^{-6}–10^{-3} mol dm^{-3}	80 m^2 dm^{-3}	12 h	25	3.5–5.5	0.1 mol dm^{-3} NaNO$_3$	uptake	The pH$_{50}$: 3.7 (Cu), 4 (Mn), 4.9 (Co), 5.1 (Zn), 5.3 (Ni) for 10^{-4} mol dm^{-3}	206
329. MnO_2, BDH	Ag, Au(III), Cd, Co, Cu, Eu, Hg, Mn(IV), Rh(III), Ru(III), Sc, Ta, Zn	4×10^{-4} mol dm^{-3}	50 mg 15 cm^3	15 min	23		0.01 mol dm^{-3} HNO$_3$; 2 mol dm^{-3} NH$_3$	uptake	Distribution coefficients are reported. MnO$_2$ is a particularly good scavenger of Ag (K_D 3200 cm^3/g) from 0.01 mol dm^{-3} HNO$_3$, and of Eu (K_D 10000 cm^3/g) from 2 mol dm^{-3} NH$_3$.	292

Adsorbent	Adsorbate	Concentration	Solid / amount	Time	Temp (°C)	pH	Electrolyte	Method	Description	Ref
330. $MnO_{1.93}$ n H_2O	Co	5 10 10^{-1} mol dm	0 2 130 m dm³	3 h		3 8	10^{-5} 1 mol dm⁻³ NaCl, artificial sea water	uptake, ζ	The uptake is depressed at high ionic strengths 10^{-1} mol dm⁻³ Co has no effect on electrokinetic curves (iep at pH 3 if any) At 10^{-4} mol dm⁻³ Co two isoelectric points, the 2nd at pH 11.	301
331. MnO_2 n H_2O	Hg	10 3 10 mol dm⁻³	10 mg 25 cm⁻¹	1 d	25	3 12	NaCl, 0–0.01 mol dm⁻³ Na_2SO_4	uptake	For 5 10 mol dm⁻¹ Hg pH_{50} 3 in absence of chlorides or sulfates. Chlorides (10^{-4} mol dm⁻³) depress the uptake; and for [Cl] 10^{-3} mol dm⁻³ a minimum of uptake is observed at pH 4.5. In the presence of sulfates (10^{-4} mol dm⁻³) the uptake decreases when pH increases. Preadsorbed Hg can be entirely released in 5×10^{-3} mol dm⁻³ EDTA. Other complexing agents (DTPA, phosphates, citrates) and concentrated acids release only a fraction of preadsorbed Hg.	294
332. MnO_2 n H_2O	Sr	10 10^{-4} mol dm⁻³	1 g (359 m², dm³	4 h	25	2 7	0.015 1.5 mol dm⁻³ $NaNO_3$	uptake	0 and 100% uptake is not reached, cf. Fig. 4.6. With 0.01² mol dm⁻³ $NaNO_3$, 35% uptake at pH 2 and 90% uptake at pH 7. Uptake is depressed by 20% by 1.5 mol dm⁻³ $NaNO_3$. Linear log-log adsorption isotherms at constant pH, slope 1. ΔH_{ads} 53 kJ/mol (pH 3.5, 7).	123
333. $MnO_{1.93}$ n H_2O	Ba, Ca, Co, Mg, Mn, Ni, Sr, Zn	2.5×10^{-4} 10^{-3} mol dm⁻³	0 1 g (260 m²/g) 200 cm³	12 h	25	3 8	10^{-1} mol dm⁻³ NaCl	uptake, proton release	Selectivity Co≥Mn Zn>Ni>Ba>Sr Ca Mg. Protons released per metal cation adsorbed: Zn 0.8, Ni 0.9 (Fig. 10), 1.2 (Table 2), Mn 0.9, Co 1 1.1, Cu 1.2.	302
334. MnO_2 n H_2O	Cd, Pb, Tl(I), Zn	10^{-4} mol dm⁻³	4 36 × 10⁻⁴ mol Mn dm⁻³	3 h	25	2 8		uptake	The pH_{50} ~3 for Zn and Cd (very slow increase of uptake with pH). ~2 for Pb. A V-shaped uptake curve for Tl(I), with a minimum (70% uptake) at pH 4.	122
335. Nb_2O_5 n H_2O, hydrolysis of chloride	Am, Ba, Co, Eu, Fe(III), Mn(II), Np, Sc, Sr, Th, U, Y	10^{-10} 10^{-4} mol dm⁻³	0.3 g 20 cm³	2 9 d	room	1 7	0.1 mol dm⁻³ NaCl	uptake	Ba>Sr>Ca>Mg. U≥Fe(III)>Sc>Eu≥Ce≥Y>Co>Mn(II). The d log K_D d pH ~ 1.7 for Co, 1 for Np (V).	303
336. Nb_2O_5 n H_2O, from chloride	Ba, Ca, Ce(III), Co, Eu, Fe(III), Mn(II), Sc, Sr, Y	10^{-4} mol dm⁻³	0.3 g 20 cm³		25		0.1 mol dm⁻³ Na	uptake	The d log K_D d pH 0.9 for Ca, 1 for Sr, 1.1 for Ba, Sc and Fe(III), 1 2 for Y, 1 3 for Ce(III) and Eu, and 1.4 for Mn and Co at pH 4.6	304
337. NiO	Al, Ni	0.001 mol dm⁻³			25	4 12	chloride	ζ	IEP is shifted to pH 8 in $AlCl_3$ and 7.5 in $NiCl_2$ (6.4 m KCl).	153
338. NiO	Ba, Ca, Mg, Sr	0.001–0.1 mol dm⁻³			25		nitrate	proton release	The shift in PZC induced by the salt concentration of 0.1 mol dm⁻³ (in mV, 1 pH unit = −59 mV): Mg 80, Ca 40, Sr 20, Ba 0 (no specific adsorption).	158
339. PbO	Al, Pb	0.001 mol dm⁻³			25	4 12	chloride	ζ	IEP is shifted to pH 8.2 in $AlCl_3$, and 5.2 in $Pb(NO_3)_2$ (10.7 in KCl).	153
340. PbO_2	Al	0.001 mol dm⁻³			25	4 12	chloride	ζ	IEP is shifted to pH 8 (1.75 in KCl).	158

TABLE 4.1 Continued

Adsorbent	Adsorbate	c. total	S:L ratio	Eq. time	T	pH	Electrolyte	Method	Result	Ref.
341 PbO$_2$	Ba, Ca, Mg, Sr	0.001–0.1 mol dm^{-3}			25		nitrate	proton release	The shift in PZC induced by the salt concentration of 0.1 mol dm^{-3} (in mV, 1 pH unit -59 mV): Mg 10, Ca 20, Sr 30, Ba 50.	158
342. RuO$_2$	Ba, Ca, Mg, Sr	0.001–0.1 mol dm^{-3}			25		nitrate	proton release	The shift in PZC induced by the salt concentration of 0.1 mol dm^{-3} (in mV, 1 pH unit -59 mV): Mg 10, Ca 20, Sr 20, Ba 30.	158
343. SiO$_2$ quartz	Ag	4×10^{-7} mol dm^{-3}	25 g (3.3 m^2 g) dm^3	2 h	25	7.5–10.5	0.1 mol dm^{-3} NaNO$_3$, 10^{-4} mol dm^{-3} ethylenediamine	uptake	The uptake increases from 20% at pH 7.5 to 60% at pH 10.5. The uptake is enhanced in the presence of ethylenediamine.	222
344. SiO$_2$	Al	0.001 mol dm^{-3}	16 g (571 m^2 g) dm^3	60 d	25	4–12	chloride	ζ	IEP is shifted to pH 8.5 (2.2 in KCl).	153
345. SiO$_2$ W.R. Grace 254	Al	0.1 mol dm^{-3}					nitrate	solid state NMR		305
346. SiO$_2$ quartz sand, washed	Ba	0.1 mol dm^3	1.25% moisture	>20 min	35	3–11		conductance	Specifically adsorbed Ba does not contribute to the conductance.	306
347. SiO$_2$ fumed, Aldrich	Ba	10^{-4} 10^{-2} mol dm^{-3}	6% w/w (390 m^2 g)	20 min	25	3 9	chloride	acousto	IEP shifts to pH 5 with 10^{-2} mol dm^{-3} Ba.	307
348. SiO$_2$ fibers	Be	10^{-8} g dm^3	0.035 g 440 g solution	0.5 h	20	6–9.5	0.1 mol kg	uptake		308
349. SiO$_2$ different samples	Ca	0.01 mol dm^{-3}	450 m^2 dm^3	1 h	24 25	8.1 9.5	diethanol amine buffer	uptake, coagulation		309
350. SiO$_2$ natural, washed	Ca	10^{-4} 2×10^{-2} mol dm^{-3}	0.01%	5 min		6–12	10^{-3} mol dm^{-3} KNO$_3$, 10^{-2} mol dm^3 sulfate, 2×10^{-2} mol dm^3 carbonate	ζ, SEM	In absence of sulfate or carbonate: the sign of ζ is reversed to positive at pH 10 for 10^{-2} mol dm^{-3} Ca. The ζ(pH) curves obtained in the presence of Ca and sulfate resemble the electrokinetic curves of CaSO$_4$. The ζ(pH) curves obtained in the presence of Ca and carbonate resemble the electrokinetic curves of CaCO$_3$.	310
351. SiO$_2$	Cd	10^{-7} ×10^{-4} mol dm^{-3}	30 g (3.3 m^2 g) dm^3	4 h	20	7.5–10	0.1–0.7 mol dm^{-3} nitrate, 0.5 mol dm^3 chloride, 0.2 mol dm^3 sulfate, 0–0.03 mol dm^3 thiosulfate	uptake	The uptake curves in 0.1 mol dm^{-3} nitrate are independent of initial Cd concentration over 3 decades, pH$_{50}$ 8.4. The uptake is depressed in the presence of thiosulfate.	48
352. SiO$_2$ chromatographic, Merck, washed	Cd	10^{-5} mol dm^{-3}	10 g (388 m^2 g) dm^3		15, 35	4–10	0.01 1 mol dm^{-3} NaCl, NaClO$_4$	uptake	SiO-Na + Cd^2 SiOCd + Na log K -0.9 (15 C), -1 (35 C) The uptake is depressed in the presence of 0.1 mol dm^{-3} NaCl, and even more depressed in the presence of 0.1 mol dm^{-3} CsCl.	50
353. SiO$_2$ Aerosil OX 50, Degussa	Cm(III)	7×10^{-7} mol dm^{-3}	0.01 0.5 g (44 m^2 g) dm^3	1 d		4–10	10^{-1} mol dm^{-3} NaClO$_4$	TRLFS	Two surface species: the species formed at pH 7 has lifetime 220 μs and emission maximum at 602 3 nm. At pH 7 in addition to the former another species is formed which has lifetime 740 μs and emission maximum at 604.9 nm.	40

No. Adsorbent	Adsorbate	Concentration	Solid/solution	Time	T	pH	Electrolyte	Method	Result	Ref.
354. SiO₂, quartz, various heat treatment	Co	10 mol dm⁻³	10 g (45 m²/g) 50 cm³	1 h	25	7	chloride	uptake, proton release		311
355. SiO₂	Co	1 9 10 2 5×10 mol dm⁻³	10 g (50 m²/g) 200 cm³			0 9	chloride, NH₃	acousto	The IEP shifts to higher pH values, positive ζ at pH 9 for 2 5×10⁻² mol dm⁻³ Co	54
356. SiO₂	Co	carrier free-10⁻⁴ mol dm⁻³	1 g 50 cm³		15, 35	5 9	0.1 mol dm⁻³ NaCl	uptake	The percentage of uptake curves (up to 70% uptake) are independent of Co total concentration The uptake at constant σ₀ increases with T, ΔH_ads 20 23 kJ mol (at constant σ₀).	55
357. SiO₂, Min-U-Sil 30, quartz	Co	1 5 10⁻⁴ 3×10⁻³ mol dm⁻³	190 m² dm⁻³	1 21 d	22 ±2	6 7 10.7 0 1	0.1 mol dm⁻³ NaNO₃	EXAFS uptake	6 next Co neighbors at 0 312 nm, 1 1 3 1 next Si neighbors at 0.34 nm	312
358. SiO₂, quartz acid washed	Co	3 3 10⁻¹ 1 3×10⁻³ mol dm⁻³	25 77 m² dm⁻³	2-4 d	22 4	5.9 7.6	0.1 mol dm⁻³ NaNO₃	uptake, Si release	Formation of trioctohedral clays, analogs of kerolite, chrysotile, and stevensite Co₃Si₄O₁₀(OH)₂: 6 H₂O 3 Co²⁺ 4 Si(OH)₄ log K 15.1: Co₃Si₂O₅(OH)₄ + 6 H 3 Co²⁺ 2 Si(OH)₄ H₂O log K 22: NaCo₂.₅Si₄O₁₀(OH)₂ 6 H 4 H₂O Na 2.5 Co 4 Si(OH)₄ log K 19.2 Solubility of these salts is much lower than that of Co(OH)₂	313
359. SiO₂, Aerosil 200	Cr(III)	10 0.4 mol dm	2 g 60 cm³	1 d	25	1.9 3.6	nitrate	uptake, ζ		64
360. SiO₂, NH-R, Machinery Nagel	Cr(III)	10 mol dm⁻³	0.2 g (500 m²/g), 20 cm³	3 5 h	20	4 7	0.01 mol dm⁻³ KNO₃ ; 0 8 ×10⁻³ mol dm⁻³ KF	uptake	The pH₅₀ 5 1 Linear log-log adsorption isotherms at pH 5 (log [Cr(III)aq] -5 5 to -4), slope 0 7 Somewhat different results are observed with fresh and aged Cr(III) solutions. In the presence of acetate buffer the slope of log-log adsorption isotherms equals 1 Uptake is depressed at higher ionic strengths	314
361. SiO₂	Cr(III)	10 1 2×10⁻⁴ mol dm⁻³	10 g (359 m²/g) dm⁻³	3 h	25	3.5 7	10⁻⁴ 10⁻² mol dm⁻³ KNO₃	uptake	SiCrCrOH log K -12 2	65
362. SiO₂, Cabosil	Cu	10⁻⁴ mol dm⁻³	1 g (200 m²/g) dm⁻³	3 h	25	2 10	0.025 mol dm⁻³ NaClO₄, 10⁻³ mol dm⁻³ aspartate	uptake	no uptake	66
363. SiO₂, five samples	Cu	10 2×10⁻⁴ mol dm⁻³	1 g (112 850 m²/g) 50 cm³	5 h	20	6	0.01 mol dm⁻³ KNO₃	uptake	Langmuir adsorption isotherms (also at 8 60 C) ΔH_ad 23 kJ mol	315
364. SiO₂, Aerosil 300, Degussa	Cu	10⁻³ mol dm⁻³	40 g dm⁻³	1 d	25	0.6 5	0.1 mol dm⁻³ NaNO₃, + 1 6×10⁻³ mol dm⁻³ organic ligands	uptake, ESR	The ∂H₅₀ 6.6 in absence of organic ligands. The uptake is enhanced by 2,2' bipyridine, 1,10-phenanthroline (up to 2×10⁻³ mol dm⁻³), 2,2', 6', 2'' terpyridine (up to 1×10⁻³ mol dm⁻³), cyclam 1,4,8,11 tetraazatetradecane (1×10⁻³ mol dm⁻³ pH 5), DOHDO-pn 3,9,-dimethyl-4,8, diaza-undeca-3,8-diene-2,10 dione-dioxime (1 × 10⁻³ mol dm⁻³), 2-picolinate (pH 4). The uptake is depressed by 1,2-ethylenediamine, EDTA, iminodiacetate, nitrilotriacetate, and glycinate	316

TABLE 4.1 Continued

Adsorbent	Adsorbate	c total	S:L ratio	Eq time	T	pH	Electrolyte	Method	Result	Ref.
365. SiO$_2$, Fisher	Cu	10^{-3} mol dm^{-3}	1 g (730 m^2/g) 25 cm^3	1 d	25	2 9	0.1 mol dm^{-3} NaNO$_3$, 1 3×10^{-3} mol dm^{-3} organic ligands	uptake	The pH$_{50}$ ~ 5.7 in absence of organic ligands. The uptake is enhanced in the presence of organic ligands: 2,2',6',2"-terpyridine, 2-pyridine methanol, 2-(aminomethyl) pyridine, specially in the acidic region. The effect of pyridine and 3,4-lutidine on uptake curves is insignificant. Uptake is depressed in the presence of picolinic, salicylic, and 5-sulfosalicylic acid, specially in the basic region.	317
366. SiO$_2$	Cu	1.2×10^{-4} mol dm^{-3}	10 g (300 m^2/g) dm^{-3}	1 h	20	5 2 7.2	0 1 mol dm^{-3} NaCl, 10^{-3} mol dm^{-3} 2,2' dipyridine	uptake	2,2' dipyridine was added to SiO$_2$ with preadsorbed Cu.	318
367. SiO$_2$, fumed S5130, Sigma	Cu	3.9×10^{-4}, 3.9×10^{-3} mol dm^{-3}	0.2 g (390 m^2/g) 20 cm^3	1 d	room	5 5	10^{-2} mol dm^{-3} NaNO$_3$	uptake, EXAFS, EPR	At 0.8% monolayer Cu(OH)$_2$ clusters are formed on the surface.	319
368. SiO$_2$, Degussa, Aerosil 200	Cu	10^{-4} mol dm^{-3}	2000 m^2 dm^{-3}	1 d		4 8	0 1 mol dm^{-3} NaNO$_3$, 0 2× 10^{-4} mol dm^{-3} 2,2'-dipyridine	uptake, EXAFS	The pH$_{50}$ 6 2 in absence of bipy and 4.5 in the presence of 1×10^{-4} mol dm^{-3} bipy. In absence of bipy adsorbed Cu has on average 0 8 closest Cu neighbor at distance of 0.258 nm (dimerization).	70
369. SiO$_2$, Degussa, Aerosil 200	Cu	10^{-4} mol dm^{-3}	1000, 2000 m^2 dm^{-3}	1 4 d		5 5 8	0.1 mol dm^{-3} NaNO$_3$	uptake, EXAFS	Cu has on average 1 0 closest Si neighbor at distance of 0 3 nm. The number of Cu neighbors at 0.26 nm increases as a function of equilibration time.	71
370. SiO$_2$, quartz	Cu	5×10 5×10^{-4} mol dm^{-3}				4 10		ζ, coagulation	The sign of ζ is reversed to positive at pH 7 10	320
371. SiO$_2$	Cu	6.65×10^{-7} mol m^{-2}				5 5	0.01 mol dm^{-3} KNO$_3$	EPR		129
372. SiO$_2$, quartz	Eu	10^{-8} mol dm^{-3}				3 10	0.01 mol dm^{-3} NaClO$_4$, 2 mg dm^{-3} fulvic acid	uptake	The pH$_{50}$ 7. The uptake is depressed in the presence of fulvic acid	181
373. SiO$_2$, Porasil C, Alltech	Fe(III)	10^{-4} 5×10^{-3} mol dm^{-3}	0.07 g (16 m^2/g) dm^3, 4 60 g (94 m^2/g) dm^3	20 min 8 h	25	3	fulvic acid nitrate	uptake	Polymeric Fe species on the surface	321
374. SiO$_2$, chromatographic, Merck	Gd	2×10^{-8} 2×10 mol dm^{-3}	10 100 g (144 m^2/g) dm^3	20 min	25	3 5 7	0.01 mol dm^{-3} NaClO$_4$	uptake, XPS	ΔH_{ads} 36 kJ mol^{-1} for 2×10^{-8} and 67 kJ mol^{-1} for 2×10 mol Gd dm^{-3}	322
375. SiO$_2$, fumed, Sigma	Gd	1×10^{-7} 2×10 mol dm^{-3}	0.05 8% by weight, 390 m^2/g	1 h	25	4 9	0.001 mol dm^{-3} NaNO$_3$	ζ	Figs 4 19 4 24	323
376. SiO$_2$, gel, Fuji	Hf	trace	60 mg 5 cm^3	10 d		1 12	30 mg dm^{-3} humic acid	uptake	80% uptake in the presence and in absence of humic acid	324
377. SiO$_2$, Aerosil 380, Degussa	Hf	3×10^{-6} mol dm^{-3}	250 mg 5 cm^3	1 d		1 7	0 1 mol dm^{-3}	TDPAC, EXAFS	No surface precipitation Three adsorbed species species 1 and 2 prevailing at pH 5, species 3 prevailing at pH 5.	325

No.	Adsorbent	Ion	Concentration	Solid	Time	T/°C	pH	Electrolyte	Method	Remarks	Ref.
378.	SiO₂: quartz	Hg	1.84×10^{-7}, 10^{-3} mol dm⁻³	40 g (5 m² g⁻¹) dm⁻³	1 d		1-10	10^{-1} mol dm⁻³ NaClO₄, 10^{-3}, 10^{-1} mol dm⁻³ NaCl	uptake, ζ	The sign of ζ is reversed to positive at pH 3-4 in the presence of 10^{-4} mol dm⁻³ Hg. The uptake from 10^{-1} mol dm⁻³ NaClO₄ peaks at pH 5-6. The uptake is depressed in the presence of chlorides	326
379.	SiO₂: n	Hg	1×10^{-8} mol dm⁻³	40 g dm⁻³			1-12	0.1 mol NaCl, 0.1 mol dm⁻³ NaClO₄	uptake	SiOHgOH log K -3.2; SiOHgCl log K 7, reinterpretation of data taken from [326]	211
380.	SiO₂: Aerosil 200, Degussa	Hg	1.2×10^{-10} mol dm⁻³	30 g (200 m² g⁻¹) dm⁻³, 323 m² dm⁻³	1 d	25	2.5-9	0.1 mol dm⁻³ NaNO₃ + $0 \cdot 10^{-4}$ mol dm⁻³ Cl	uptake	SiOHgOH log K -3.7; FeOHgCl log K 5.8	247
381.	SiO₂: Aerosil	Mg	$4 \cdot 10^{-6}$, $6.5 \cdot 10^{-4}$ mol dm⁻³		20 min	25	7-11	0-0.7 mol dm⁻³ NaCl, 0.1 mol dm⁻³ LiCl, CsCl	uptake	The uptake is depressed at [NaCl] 0.01 mol dm⁻³. Hydroxysilicate precipitation. Uptake at 5 C was also studied	327
382.	SiO₂: natural, quartz	Mg	10^{-4}, 10^{-1} mol dm⁻³	0.7 g (140 m² g⁻¹) 35 cm³	1 d		8-12	chloride	ζ	With 10^{-4} mol dm⁻³ Mg positive ζ at pH 9-10	328
383.	SiO₂: Zeothix 265, washed	Mn(II)	$5.3 \cdot 10^{-4}$, $5.3 \cdot 10^{-10}$ mol dm⁻³	100 m² dm⁻³			2-10	NaCl	uptake, proton release	The pH₅₀ 3.6 (no NaCl, $5.3 \cdot 10^{-4}$ mol dm⁻³ Mn)	329
384.	SiO₂: Cabot	Ni	10^{-4} mol dm⁻³		30 min	25 ±2	3-9	0.025 mol dm⁻³ EDTA, mol dm⁻³, 10^{-2} mol dm⁻³ NaClO₄	uptake	Zero uptake	95
385.	SiO₂: Zeosyl 100, Zeo 49	Ni	$5 \cdot 10^{-4}$ mol dm⁻³	111 m² 53 cm³	1 d		2-8	10^{-4} mol dm⁻³ NaClO₄	uptake	The pH₅₀ 7.6, staircase uptake curve, cf. Fig 4.7 (A)	73
386.	SiO₂: Spherosil XOA 400	Ni	0.01-0.1 mol dm⁻³	5 g (400 m², 70 cm³)	1 d	25	5-11	1 mol dm⁻³ NH₄NO₃	uptake, EXAFS, XANES, IR	The uptake peaks at pH 8.5	330
387.	SiO₂: chromatographic, Merck	Ni	$1 \cdot 10^{-4} \times 10^{-4}$ mol dm⁻³	10 g (144 m² g⁻¹) dm⁻³	20 min	25	6-10	0.01-0.3 mol dm⁻³ NaCl, NaClO₄	uptake	Linear log-log adsorption isotherms slope 0.95 at pH 8-9. At low initial concentrations the Ni uptake is rather insensitive to the ionic strength. At high initial concentrations the Ni uptake is depressed at high ionic strength	331
388.	SiO₂: fumed, Sigma	Ni	$1 \cdot 10^{-8} \cdot 2 \times 10^{-4}$ mol dm⁻¹	0.05-8% by weight, 390 m² g⁻¹	1 h	25	4-9	0.001 mol dm⁻³ NaNO₃	ζ	The sign of ζ is reversed over limited pH range depending on Ni concentration and solid to liquid ratio. The Ni effect on ζ is most pronounced at pH 7-8	332
389.	SiO₂: quartz, Wedron 510, sieved, washed	Np(V)	$10^{-7} \cdot 10^{-6}$ mol dm⁻³	4.80 g (0.03 m² g⁻¹)	10 d	20 ±2	5-11	0.1 mol dm⁻³ NaNO₃	uptake	In experiments conducted in capped vials (low CO₂) Kᴅ constantly increases with pH to reach the highest value of 20 dm³ kg at pH 10.5. In experiments conducted in contact with atmosphere the uptake peaks at pH 8	76
390.	SiO₂: quartz, Min-U-Sil 5	Np (V)	1.2×10^{-7} mol dm⁻³	30 g (6 m² g⁻¹) dm⁻³	1 d		4-12	0.1 mol dm⁻³ NaClO₄	uptake	SiONpO₂ log K -6.92	188
391.	SiO₂: amorphous	Pb	10^{-4} mol dm⁻³	2.5 g (200 m² g⁻¹) dm⁻³	10 h		3-8	0.025 mol dm⁻³ NaClO₄, 10^{-4} mol dm⁻³ EDTA	uptake	The pH₅₀ 6.1 (no EDTA); presence of EDTA depresses the uptake (10% uptake at pH 9)	333
392.	SiO₂: Cabot	Pb	10^{-4} mol dm⁻³				3-10				118
393.	SiO₂: quartz	Pb	$10 \cdot 10^{-4}$ mol dm⁻³	0.1 g (0.68 m² g⁻¹) dm⁻³	1 d	22	3-11	0.001 mol dm⁻³ KNO₃, $0-10^{-3}$ mol dm⁻³ carbonate	ζ	The sign of ζ is reversed from negative to positive at pH 7-11 in the presence of 10^{-4} mol dm⁻³ Pb, when [carbonate] [Pb].	334

TABLE 4.1 Continued

Adsorbent	Adsorbate	c total	S L ratio	Eq time	T	pH	Electrolyte	Method	Result	Ref.
394. SiO₂, natural, washed	Pb	10^{-4} mol dm⁻³	0.01%	5 min		2 12	10^{-3} mol dm⁻³ KNO_3, 10^{-4} mol dm⁻³ sulfate, 10^{-4} 1 mol dm⁻³ carbonate	ζ, SEM	In absence of sulfate or carbonate: no effect at pH < 4, positive ζ at pH 7 11. In the presence of sulfate the ζ(pH) curve is similar to that of $PbSO_4$; In the presence of 10^{-4} mol dm⁻³ carbonate (with or without sulfate) the electrokinetic curve is similar to that of $PbCO_3$.	310
395. SiO₂, Aerosil	Pd	3×10^{-3} 1 6×10^{-2} mol dm⁻³	1 g (327 m²) 75 cm³	1 h	21	5 12	$NaNO_3$, NH_4NO_3	uptake		83
396. SiO₂, Sarabhai	Pu (IV)	7 10 mg dm⁻³	0.1 g 10 cm³	5 h		natural	0.1 1 mol dm⁻³ Na_2CO_3	uptake, chromatography		85
397. SiO₂, Spherosil, BDH	Ru(II)	10^{-5} 10^{-3} mol dm⁻³	150 m²/dm³	45 min	25	4 10	chloride, chlorate VII, [dipyridine]-3[Ru]	uptake	0% uptake is not reached (cf. Fig. 4.6).	335
398. SiO₂, amorphous, Mallinckrodt	Sr	10^{-3} mol dm⁻³	10000 m²/dm³	2 d	25	6-10	0.1 mol dm⁻³ NaCl	uptake, EXAFS	Uptake at pH 7 8.5 is depressed in the presence of atmospheric CO_2, but it is enhanced at pH > 8.5.	261
399. SiO₂, Ottawa sand, EM Science	Th	1 ppm	500 mg (0.4 m²/g) 21	30 d	20	2.31	0.005 mol dm⁻³ H_2SO_4	uptake	2% uptake	194
400. SiO₂, Aerosil OX 200, Degussa	Th	10^{-6} 3.5×10^{-5} mol dm⁻³	0.037 26 g (169 m² g) dm³	> 1 d	25	0-3.5	0.1, 1 mol dm⁻³ $NaClO_4$	uptake	(≡SiO₂)₂Th² log K-1.9 The sorption is reversible	336
401. SiO₂, Aerosil OX 200, Degussa	Th	10^{-4} 1.5×10^{-2} mol dm⁻³	24 g (169 m² g)/dm³	1 d		2.8-4	1 mol dm⁻³ $NaClO_4$	EXAFS	2 Si atoms at 0.38-0.39 nm bidental inner sphere complex.	337
402. SiO₂, gel, Merck	U	2 1×10^{-6} mol dm⁻³	1 g (720 m²) 100 cm³	1 d	room	2 8	nitrate; 10^{-3} mol dm⁻³ $NaHCO_3$	uptake	The uptake peaks at pH 7 (log K_D 3.5, K_D in cm³ g) without carbonate and 6 in the presence of carbonate.	338
403. SiO₂, 3 samples, spherical	U	10^{-6}-0.05 mol dm⁻³	0.5-5 g (50-410 m²/g) dm³	14 d		3-7	nitrate	EXAFS uptake	Polynuclear species at pH 4.3 and 0.01 mol dm⁻³ U, U-U distance 0.382 nm.	339
404. SiO₂, Fluka	U	20-130 mg dm⁻³	1 g (550 m²) dm³			3-7	nitrate	uptake, chromatography		340
405. SiO₂, 160 m² g	U	2.5×10^{-3} mol dm⁻³		1 d	25	natural	sulfate	luminiscence spectroscopy (ex situ)	Uranyl ions are attached to silica surface by means of two water molecules.	341

No. & adsorbent	Ion	Concentration	Solid	Time	Temp (°C)	pH	Electrolyte / additive	Method	Remarks	Ref.
406. SiO_2: quartz, Wedron 510, U sieved, washed	U	2×10^8 2×10^{-6} mol dm^{-3}	2.50 g (0.03 m^2/g) dm^{-3}	10 d	20 ±2	2 9	$10^{-3.5}$ atm CO_2	uptake	K_D peaks at pH 6.5 (50 dm^3 kg for 2×10^{-7} mol dm^3 U).	32
407. SiO_2: quartz, Merck	U	10^{-6} mol dm^{-3}	25 g dm^{-3}	2 d	25	3 7.5	0.001-0.1 mol dm^{-3} $NaNO_3$, 0.001-0.01 mol dm^{-3} carbonate	uptake	The uptake is depressed at high ionic strength. Presence of carbonates enhances the uptake at pH < 6 and depresses the uptake at pH > 6.	342
408. SiO_2: quartz, Sigma	U	10 , 5×10^{-4} mol dm^{-3}	12 g (10 m^2/g) dm^3	2 d	3-8		0.01 mol dm^{-3} KNO_3	uptake	The uptake measured in contact with atmospheric CO_2 shows a maximum at pH 7 (95% for 10^{-5} mol U dm^3).	200
409. SiO_2: Ludox SM	Y	10^{-4}, 5×10^{-4} mol dm^{-3}			25	4-10	0.001 mol dm^{-3} $NaNO_3$	ζ	The sign of ζ is reversed from negative to positive at pH 6.5 7 for 5×10^{-4} mol dm^3 Y	343
410. SiO_2: Merck 60 H	Yb	2×10 mol dm^{-3}	10,100 mg (384 m^2/g) 50 cm^3	7 d	6-7	0.05, 0.1 mol dm^{-3} $NaNO_3$		uptake	$SiOYb(OH)_2$; log K-16 2	344
411. SiO_2: Cabosil	Zn	5×10 10^{-3} mol dm^{-3}	0.5 5 g (200 m^2/g) dm^3	2 d	4-8	0.01 mol dm^{-3} $NaClO_4$		uptake	pH$_{50}$ 5 8 (10^{-4} mol dm^{-3} Zn, 5 g SiO_2 dm^3)	89
412. SiO_2	Zn	10 mol dm^{-3}	1 g (300 m^2)/100 cm^3		3-8	0.1 mol dm^{-3} NaCl 10^{-3} mol dm^3 2,2' dipyridyl		uptake	The pH$_{50}$ 7.5 in absence, and 5.8 in the presence of 2.2' dipyridyl	345
413. SiO_2: quartz	Zn	10^{-6} mol dm^{-3}	50 mg (10 m^2/g) dm^3	3 d	20	3 10	0.01 mol dm^{-3} $NaClO_4$, 0-20 mg fulvic acid dm^3	uptake	The pH$_{50}$ 7 6 in absence of fulvic acid, 7 9, and 9 3 in the presence of 2 and 20 mg of fulvic acid dm^3	148
414. SiO_2: spray-ICP technique	Zr	0.001 mol dm^{-3}				3 5	chloride	ζ	Positive ζ.	346
415. SiO_2: BDH, and Degussa	Ag, Cd	2×10 8×10^{-3} mol dm^{-3}		5 h	25	3 7	0.1 mol dm^{-3} KNO_3	uptake	ΔH_{ads} Cd 12.5 kJ mol (25-65 C) at pH 7	347
416. SiO_2: amorphous, from Na_2SiO_3	Am(III) Np(V) Th	5×10^{10} mol dm^{-3} 10^{14} mol dm^{-3} 10^{11} mol dm^{-3}	1200 ppm	4-7 d	25	0-11	0.1 mol dm^{-3} $NaClO_4$ + 0-50 ppm humic acid	uptake	Am(III): pH$_{50}$ 6.2, the uptake is enhanced in the presence of humic acid at pH < 8 and depressed at pH 8; Np(V) the uptake peaks at pH 10 (70%); the uptake is enhanced in the presence of humic acid at pH 9 and depressed at pH 9; Th (with 60 ppm SiO_2): pH$_{50}$ 0.8 At pH 1.5 the uptake levels out at 75%.	91
417. SiO_2: Aerosil X 50	Ba, Ca	5×10^{-4}, 5×10^{-3} mol dm^{-3}	2.5% w w			3 9		proton release, ζ	The negative ζ (in 5×10^{-4} mol dm^{-3} $BaCl_2$ and $CaCl_2$ is lower than in 10^{-3} mol dm^{-3} KCl at pH 4.5 The negative σ$_0$ in 5×10^{-3} mol dm^{-3} $BaCl_2$ and $CaCl_2$ is higher than in 10^{-2} mol dm^{-3} KCl at pH > 5. Both effects are slightly more significant for Ba than for Ca.	348
418. SiO_2: BDH	Ba, Ca, Mg, Sr	0.001-0.1 mol dm^{-3}				2 9	conc. 0.1 mol dm^{-3} LiCl	uptake, proton release	Ba > Ca Sr > Mg.	349
419. SiO_2: from Na_2SiO_3	Ba, Ca, Mg, Sr	0.025 1 mol dm^{-3}	1 g (310 m^2) 250 cm^3	5 min	25	7 9.5	chloride	proton release	The uptake of Ca is depressed in the presence of LiCl.	350
420. SiO_2	Ba, Ca, Mg, Sr	5% wt (metal oxide in silica)	0.5 g 150 cm^3				KNO_3	proton release	Silica was impregnated with metal acetate solution, dried and calcined at 500 C, then redispersed. The PZC (pH) is 9.8 for Ca, 9 3 for Sr and 9.2 for Ba.	351

TABLE 4.1 Continued

Adsorbent	Adsorbate	c. total	S:L ratio	Eq. time	T	pH	Electrolyte	Method	Result	Ref.
421. SiO_2, quartz capillary	Ba, La	0.0003–0.05 mol dm^{-3}			20	2 5.5	chloride	ζ	No sign reversal to positive in 0.05 mol dm^{-3} $BaCl_2$ or 0.0003 mol dm^{-3} $LaCl_3$, sign reversal to positive in 0.03 mol dm^{-3} $LaCl_3$, at pH>3.5	352
422. SiO_2, chromatographic, Merck	Ca, Eu	10^{-5} mol dm^{-3}	10 g dm^{-3}	20 min	15, 35	4-10	0.001-0.1 mol dm^{-3} NaCl, LiCl, CsCl	uptake	Ca uptake at constant σ_0 is T independent. Eu uptake at constant σ_0 increases with T; Ca uptake is depressed at high ionic strength.	96
423. SiO_2, vitreous, quartz	Ca, La	10^{-4} 10^{-2} mol dm^{-3}		1 d		5-6	chloride	ζ		353
424. SiO_2, quartz natural, washed	Ca, La, Th	10^{-6} 10^{-2} mol dm^{-3}		1 d	25	5-6	chloride, nitrate	ζ	ζ reverses its sign from negative to positive at 3 × 10^{-5} mol dm^{-3} La and 3 × 10^{-6} mol dm^{-3} Th.	354
425. SiO_2	Ca, Zn	2 × 10^{-4} mol dm^{-3}, 3 × 10^{-5} mol dm^{-3}	1 g dm^{3}			5-10	0.04 mol dm^{-3} KNO_3	uptake	≡SiOHCa^2 log K -1.08, ≡SiOCa log K 10.75, ≡SiOCaOH log K 16.2 Reinterpretation of literature data.	355
426. SiO_2, BDH	Cd, Co, Cu, Ni, Zn	10^{-5} 10^{-4} mol dm^{-3}	1 g/50 cm^3	5 h	20	7	0.01 mol dm^{-3} KNO_3	uptake	Langmuir type adsorption isotherms. ΔH_{ads} − 23 kJ mol for Cu, 10 kJ mol for Ni and 2 kJ mol for Co from temperature dependence of the binding constant.	356
427. SiO_2, Aldrich	Cd, Co, Mn(II), Zn	< 10^{-7} mol dm^{-3}	1 10 g (325 m^2 g) dm^3	1 d	20 ±2	4-9	0.05 mol dm^{-3} $NaNO_3$, 2 × 10^{-4} mol dm^{-3} salicylate	uptake	Zn > Cd > Co ≈ Mn in absence of salicylate. Uptake of cations is depressed at pH > 7 in the presence of salicylate.	99
428. SiO_2, Aerosil 200 Degussa, and Silikagel H Merck	Cd, Cu, Fe(III), Pb		1.4 (160 m^2 g); 3.6 (372 m^2 g) mol kg solution	20 min	25	1 8	1, 3 mol dm^{-3} $NaClO_4$	uptake	The pH$_{50}$ 1.5 for Fe, 5.7 for Pb, 6 for Cu, and 7.2 for Cd.	357
429. SiO_2, quartz	Cd, Pb	5×10^{-4} mol dm^{-3}	0.1 g 100 cm^3	1 d		2 11	0.002 mol dm^{-3} KNO_3, or KCl	uptake, ζ coagulation	The pH$_{50}$ 7 for Pb and 8.2 for Cd. The sign of ζ is reversed at pH 6-9 (Pb) and 7 9 (Cd).	358
430. SiO_2, Ludox	Cd, Zn	10^{-6} mol dm^{-3}	8.3×10^{-3} mol Si dm^3	2 h		6-9.5	0.1 mol dm^{-3} $NaNO_3$	uptake	The pH$_{50}$ ~8.8 (Cd) and 7 (Zn).	151
431. SiO_2, Custer 432. SiO_2	Cd, Zn Co,Cr(III), Cu	10^{-5} 10^{-3} mol dm^{-3}	10^{-3} mol dm^{-3}		25	6-6.5	10^{-3} mol dm^{-3}	uptake IR		104 359
433. SiO_2, Merck	Co, Cu, Fe(II), Ni	10^{-3} mol dm^{-3}	1 g 25 cm^3	1 d	room	2 8	0.1 mol dm^{-3} $NaNO_3$, 3×10^{-3} mol dm^{-3} organic ligands	uptake	The pH$_{50}$ 5 for Fe(II), 6.2 for Cu, and >7 for Co and Ni in absence of organic ligands. The uptake is enhanced in the presence of organic ligands 1, 10-phenanthroline and 2,2' dipyridyl in the acidic region.	360
434. SiO_2, Sigma	Co, Cu, Fe(III), Ni, Zn	10 mg dm^3	0.5 g (420 m^2/g), 20 cm^3	1 h	room	1,2 8	acetate and phosphate buffer	uptake	Ni Zn Co ≈ Fe(III) Cu	361

Adsorbent	Adsorbate	Concentration	Solid	Time	T (°C)	pH	Electrolyte	Method	Comments	Ref.
435. SiO_2, gel, synthetic	Co, Cu, Mn(II), Ni,Zn	5×10^{-4} mol dm⁻³	2 g (310 m²/g) 25 cm³	3 h	25	4.5-7.5	acetate and borate buffers	uptake, chromatography		362
436. SiO_2, quartz	Co, Cu, Ni	10^{-3} mol dm⁻³	160 m²dm⁻³	>2h	22	4-11	0 5 mol dm⁻³ NH_4NO_3	uptake, ζ	The uptake of Cu has a minimum at pH 9.5. Onset of uptake of Co at pH 7, and Ni at pH 8. The ζ potential in 0.0⁻ mol dm⁻³ NH_4 + 2×10^{-5} mol dm⁻³ Co is reversed to positive at pH 9.5-10.5.	108
437. SiO_2	Co, Cu, Pb,Zn	10^{-4} mol dm⁻³	45 180 m²/dm³	20 min	25	3.5-10	0.001-0.1 mol dm⁻³ KNO_3, 0.1 mol dm⁻³ NaCl	uptake	For 45 m²/dm³ and 0.001 mol dm⁻³ KNO_3, $pH_{50} \sim 6.8$ for Cu. 7.5 for Zn, 8 9 for Co and 8.6 for Cd. Cu uptake is slightly enhanced in the presence of 0.1 mol dm⁻³ NaCl (relative to 0.001 mol dm⁻³ KNO_3), and Zn, Co and Cd uptake is slightly depressed.	109
438. SiO_2, quartz	Co, La, Th	10^{-6} 10^{-3} mol dm⁻³	0.02-0.5 g (5 m²g) dm³		25	2 11	10^{-3} mol dm⁻³ KNO_3	ζ, uptake	The sign of ζ is reversed to positive over a limited pH range, whose center falls at pH 9 (Co), 8 (La) and 5 (Th) when critical concentration of cations (dependent on the solid to liquid ratio) is exceeded.	363
439. SiO_2	Co, Ni	0.0014-0.046 mol dm⁻³						EXAFS	Neoformation of clay like phase; reinterpretation of literature data	364
440. SiO_2, quartz	Cu, Fe(III)	10^{-4} mol dm⁻³		1 h	25	3-10	10^{-3} mol dm⁻³ $NaNO_3$	ζ	The sign of ζ is reversed from negative to positive at pH 6 8 for Cu and for pH < 7 for Fe.	365
441. SiO_2, chromatographic, Merck	Gd, Y	2×10^{-8} 1×10^{-4} mol dm⁻³	10 g ((144 m²/g) dm³		25	3.5-7	0.003-0.1 mol dm⁻³ $NaClO_4$	uptake	≡SiOHGd³⁺ log K -3.49; ≡SiOHY³⁺ log K -3.49. Enhanced uptake at low initial Gd and Y concentration was interpreted in terms of formation of very strong ternary complexes with unidentified anionic impurities.	366
442. SiO_2, Stober	Mn, Ni	7×10^{-5} 3×10^{-4} mol dm⁻³	8.9×10^{-7} 1.78×10^{-6} mol of silica particles dm³		4	9.2-9.9		nuclear magnetic relaxation dispersion	Maximum coverage 0.43 Mn(II) ions nm². For Ni the attempt to determine maximum coverage was not successful.	367
443. SiO_2	Pm, Sr	$2 5 \times 10^{-8}$ mol dm⁻³	0.1 10 g dm³	9 d		7 11	0.01 mol dm⁻³ $NaClO_4$	uptake	K_D for Pm at pH 8 drops by an order of magnitude when the solid to liquid ratio increases by two orders of magnitude.	368
444. SiO_2	Th, U	5×10^{-6} mol dm⁻³	0.005-0.1 cm⁻¹	7d	room	4-8	0.001-0.1 mol dm⁻³ $NaClO_4$	uptake		369
445. SiO_2, Kemika	U, Zr	0.11 g dm⁻³	1 g (292 m²) 200 cm³		25	1.3-4	HNO_3	chromatography	Flow rates 0.15 21.5 cm³ min; also 35 C and 45 C	370
446. SiO_2, chromatographic, Kemika	U Zr	2×10^{-4} mol dm⁻³ 1.3×10^{-3} mol dm⁻³	0.25 g (292 m²/g) 25 cm³	2, 24 h	22	2 5	nitrate chloride	uptake	The pH_{50} 3.6 (U). With Zr, uptake decreases when the concentration of nitric acid increases, then reaches a constant value (65% uptake for >1 mol dm⁻³).	371
447. SiO_2	trivalent lanthanides							uptake	La < Ce < Nd < Gd < Tb = Y < Er < Tm < Yb < Lu < Sc	372

TABLE 4.1 Continued

Adsorbent	Adsorbate	c. total	S L ratio	Eq. time	T	pH	Electrolyte	Method	Result	Ref.
448. SnO_2	Al, Sn	0.001 mol dm^{-3}			25	4–12	chloride	ζ	IEP is shifted to pH 8.4 in $AlCl_3$ and 7.2 in $SnCl_2$ (2 in KCl, "Sm" in the original paper is probably a typo).	153
449. SnO_2	Co, Cu, Mn Ni, Zn	10^{-4} mol dm^{-3}	2 g/50 cm^3	2 h	20		0.1 mol dm^{-3} $NaNO_3$	uptake	Cu≥Zn>Co>Ni≥Mn.	373
450. SnO_2 from $SnCl_4$	Co, Eu, RuNO^3, Sr	0.01 1 mol dm^{-3}	0.2, 0.5 g 10, 40 cm^3	3 d	25	2 10	nitrate, chloride	uptake, ζ	Positive ζ in 0.01 mol dm^{-3} $Sr(NO_3)_2$	374
451. SnO_2, Aldrich, 99.999%	Fe (II, III), Mg	10^{-5}, 10^{-3} mol dm^{-3}		30 min		3–11	nitrate, chloride	ζ	10^{-3} mol dm^{-3} Mg: the negative ζ is depressed, sign reversal at pH 11. 10^{-3} mol dm^{-3} Fe(II): ζ 40 mV (almost pH independent). Fe(III): IEP at pH 7.5 (10^{-5} mol Fe dm^{-3} –8.5(10^{-3} mol Fe dm^{-3}).	375
452. SnO_2·n H_2O,	Co, Zn	carrier free, 10^{-3} mol dm^{-3}	1 g/100 cm^3	2 d	25	4–7	none, 0.1 mol dm^{-3} NaCl	uptake		230
453. SnO_2·n H_2O, from chloride	Ba, Ca, Ce (III), Co, Cr (III), Cu, Eu, Fe(III), La, Mg, Mn(II), Ni, Sc, Sr, U, Y, Zn	10^{-4} mol dm^{-3}	0.5 g, 20 cm^3	2 d	25		0.1 mol dm^{-3} Na	uptake	The d log K_D,d pH ~ 1 for Cu, 1.1 for Mg, Zn, Mn(II) and U, 1.2 for Ca and Co, 1.3 for Sr, Ni, and Sc, 1.4 for Ba, Ce and La, 1.5 for Y, Cr(III), and Fe(III), and 1.6 for Eu.	304
454. ThO_2, amorphous, hydrolysis of $Th(NO_3)_4$	U	0.01–0.1 mol dm^{-3}	0.3 g (54 m^2/g), 25 cm^3	1 d	27	3.5 (initial)	nitrate	uptake	Sorption capacity 0.26 mmol g, uptake decreases when T increases.	154
455. ThO_2, synthetic	Co, Cr(III), Fe (III), Mn(II)	1 10 mg dm^{-3}	0.25 g 25 cm^3	1 d	25	natural	nitrate, chloride	uptake, chromatography	Uptake of Fe (III) and Cr(III) increases with T, uptake of Mn(II) and Co decreases with T (25 95 C).	172
456. ThO_2·n H_2O synthetic, 2 samples	U	0.01 mol dm^{-3}	0.3 g 25 cm^3	1 d	27	3.5–10.7	nitrate, chloride, sulfate, carbonate (up to 0.2 mol dm^3)	uptake	0.2 mol dm^{-3} total carbonate prevents U uptake at pH 7.3 10.7	155
457. ThO_2·n H_2O from nitrate	U	10^{-4} mol dm^{-3}	0.1 g 25 cm^3	4 h	25	4–5	0.1 mol dm^{-3} $NaClO_4$	uptake	pH$_{50}$ 4.3	169
458. ThO_2·n H_2O hydrolysis of nitrate	Ba, Ca, Mg, Sr	10^{-3} 2×10^{-2} mol dm^{-3}	0.3 g (77 m^2/g), 25 cm^3	4 h	25	5–9	chloride	uptake	Onset of uptake at pH 5.2 5.5. The uptake decreases when T increases (25–70 C) except for Ba (maximum uptake at 50 C	376
459. ThO_2·n H_2O synthetic, 4 samples	Co, Cu, Ni	0.02–0.2 mol dm^{-3}	0.5 g 25 cm^3	1 d	25	neutral	sulfate, chloride	uptake	Cu > > Co > Zn	377
460. TiO_2	Al	4×10^{-7} 2×10^{-5} mol dm^{-3}	0.05 g dm^3	15 s 10 min	25	5–10	10^{-4} mol dm^{-3} KNO_3	ζ, coagulation	IEP is shifted to pH 9.2 at 10^{-3} mol dm^{-3} Al. Dispersions are stable when |ζ| > 14 mV for any Al concentration.	378 379
461. TiO_2	Al	0.001 mol dm^{-3}			25	4–12	chloride	ζ	IEP is shifted to pH 8.2 (2 in KCl).	153

No.	Adsorbent	Adsorbate	Concentration	Solid/liquid	Time	T (°C)	pH	Electrolyte	Method	Comments	Ref.
462.	TiO_2, rutile	Al	5×10^{-6} – 4×10^{-4} mol dm^{-3}	0.1 g (44 m^2 g)$^{-1}$ dm^{-3}	1 d	20	5	0.1 mol dm^{-3} NaNO$_3$	high resolution TEM	No surface precipitation.	380
463.	TiO_2, anatase, Aldrich	Ba	10^{-7} – 10 mol dm^{-3}			25	11	0.001 mol dm^{-3} KOH	ζ	The sign of ζ is reversed to positive in the presence of 2×10^{-5} mol dm^{-3} Ba.	381
464.	TiO_2, Aldrich, anatase	Ba	10^{-3} – 10^{-2} mol dm^{-3}				2–11	nitrate	acousto	With 0.003 mol Ba dm^{-3} the IEP shifts to pH 11, with 0.01 mol Ba dm^{-3} only positive ζ	382
465.	TiO_2, Degussa, P 25	Ca	10^{-3} – 1 mol dm^{-3}				5 8	chloride	ζ, proton release	No shift in the IEP (streaming potential). CIP of charging curves at different CaCl$_2$ concentrations at pH 4.6.	37
466.	TiO_2, rutile, Merck	Ca	10^{-5} – 10^{-2} mol dm^{-3}	20–50 g (15 m^2 g) dm^{-3}	15 min	20	4 10	chloride	uptake, proton release		383
467.	TiO_2	Cd	2×10^{-4} mol dm^{-3}	200 m^2/dm^3		25	4 7	0.01 mol dm^{-3} KNO$_3$	uptake	pH$_{50}$ 5 7	384
468.	TiO_2, rutile	Cd	10^{-4} – 10^{-2} mol dm^{-3}		1 h	20		0.02 mol dm^{-3} KNO$_3$	uptake, ζ, proton release	0 8 1.2 proton released per one Cd adsorbed for positively charged surface 1.8 proton released per one Cd adsorbed for negatively charged surface. Temperature effects (5–60) were also studied. At pH 7 8 the sign of ζ is reversed to positive in the presence of 3×10^{-5} mol dm^{-3} Cd	28
469.	TiO_2, rutile	Cd	5×10^{-3} – 8×10^{-2} mol dm^{-3}	2 g (51 m^2/g) 50 cm^3	1 d	20	3 7	0.2 mol dm^{-3} KNO$_3$	uptake calorimetry	ΔH$_{ads}$ 15 kJ mol	385
470.	TiO_2, rutile, from chloride	Co	10^{-5} – 10^{-3} mol dm^{-3}	0 1 5 g (23 m^2/g) dm^{-3}		25	2 11	10^{-3} mol dm^{-3} KNO$_3$	ζ, uptake	The pH$_{50}$ 5 5 with $1 1 \times 10^{-4}$ mol Co dm^{-3} and 5 g TiO$_2$/dm^3. Slope of log adsorption isotherms decreases as pH increases. The IEP is shifted to pH 10 with 10^{-4} mol Co dm^{-3} and 0.1 g TiO$_2$/dm^3.	363
471.	TiO_2, analytical grade	Co	5×10^{-6} – 4×10^{-4} mol m^{-2}	303, 1150 m^2 g		30	3 8		uptake EXAFS	ΔH$_{ads}$ −22–44 kJ mol (pH 6.5). 1.3 5 Co neighbors at 0.312 nm; 1.4–2.9 next Ti neighbors at 0.30 nm; 1.9 5 next Ti neighbors at 0.36–0.38 nm.	52
472.	TiO_2, rutile, from chloride	Co	5×10^{-4} – 3×10^{-3} mol dm^{-3}		1 11 d	22 ±2	5.2 7.9	0.1 mol dm^{-3} NaNO$_3$	uptake		312
473.	TiO_2, Kanto	Co	10^{-6} – 10^{-3} mol dm^{-3}		3 12 h	25	5 5 7	0.1 mol dm^{-3} NaNO$_3$	uptake	The slope of log adsorption isotherms decreases when the pH increases	59
474.	TiO_2, rutile, Kristallhandel Co Kelpin	Co	$1 5 \times 10^{-5}$ mol dm^{-3}	single crystal	13 d		8	nitrate	grazing-incidence XAFS, XPS	Surface concentration 2 1 10^{-6} mol m^{-2} on (110) and 7×0^{-6} mol m^{-2} on (001) plane No evidence of structures similar to solid Co(OH)$_2$	386
475.	TiO_2, anatase	Cu	10^{-4} mol dm^{-3}	1 g dm^{-3}	3 h	25	2 10	0.025 mol dm^{-2} NaClO$_4$, 10^{-3} mol dm^{-3} aspartate	uptake	Zero uptake	66
476.	TiO_2, anatase	Cu	8.6×10^{-4} mol dm^{-3}	9 5 g (90 m^2/g) dm^{-3}		25	3–6	0.1 mol dm^{-3} KNO$_3$	uptake	pH$_{50}$ 4 7	387
477.	TiO_2, anatase	Cu	8×10^{-4} mol dm^{-3}	9 g (90 m^2/g) dm^{-3}		25	3 9	0 1 mol dm^{-3} KNO$_3$, or NaClO$_4$	uptake	Uptake is depressed in the presence of 2.2' dipyridine, 8-aminoquinoline, and (to less extent) 1,2-phenylenediamine (ligand) [Cu] or 2 [Cu] Formation of three different ternary surface complexes for each ligand	388

TABLE 4.1 Continued

Adsorbent	Adsorbate	c. total	S:L ratio	Eq. time	T	pH	Electrolyte	Method	Result	Ref.
478. TiO$_2$ rutile	La	10^{-3} mol dm^{-3}	167 m^2 dm^{-3}	3 d	25	3–8	nitrate	uptake, high resolution TEM	The pH$_{50}$ 5. Precipitation of La(OH)$_3$·nH$_2$O at pH > 7. The new phase is not uniformly distributed over the rutile surface.	249
479. TiO$_2$ rutile	Mg	10^{-3} 10^{-1} mol dm^{-3}	(300 m^2 dm^{-3})		25	4–9	nitrate	proton release	Negative charge is enhanced by a factor of 2 in comparison with inert electrolytes, the positive branch of charging curve is not affected.	389
480. TiO$_2$, Degussa, P25	Pm	10^{-9} mol dm^{-3}		15 min		1 11	0.01–0.1 mol dm^{-3} NaClO$_4$	uptake	The uptake is insensitive to the ionic strength. TiOHPm$^{2\,5}$	257
481. TiO$_2$, rutile, from chloride	Ra	trace	5 g 10 cm^3	1 d	room	1,10	HCl NaOH	uptake	100% uptake at pH 10.	259
482. TiO$_2$, Degussa, P25	Th	10^{-9} 3×10^{-2} mol dm^{-3}	10 g (50 m^2 g) dm^{-3}	2 d	room	1 11	0.005–0.15 mol dm^{-3} NaClO$_4$, 0.04 mol dm^{-3} NaCl	uptake	100% uptake at pH > 3. Linear log-log adsorption isotherm at pH 1.3 and 2 slope 1. The uptake is insensitive to the ionic strength.	390
483. TiO$_2$, anatase	U	3 60 ppb	40 mg dm^{-3}	5 h	25	8	0.72 mol dm^{-3} NaCl + 2.3×10^{-3} mol dm^{-3} NaHCO$_3$, sea water	uptake	Uptake from 0.72 mol dm^{-3} NaCl + 2.3×10^{-3} mol dm^{-3} NaHCO$_3$, is greater than from sea water by a factor of 10. Effect of sulfates (3×10^{-2} mol dm^{-3}) on U uptake is negligible	391
484. TiO$_2$, brookite	U	0.01, 0.1 mol dm^{-3}	0.3 g (210 m^2/g) 25 cm^3	1 d	27	3 5 (initial)	nitrate	uptake	Sorption capacity 1 2 mmol g. uptake decreases when T increases ΔH_{ads} -9 kJ mol	154
485. TiO$_2$	U	10^{-5} mol dm^{-3}	0.2 g dm	16 h	25	6 9	0.001 mol dm^{-3} NaHCO$_3$	uptake	The uptake decreases when pH increases pH$_{50}$ 2.1	199
486. TiO$_2$, hydrous, from sulfate	U	10^{-4} mol dm^{-3}	0.1 g 25 cm^3	4 h	25	1.5 5	0.1 mol dm^{-3} NaClO$_4$	uptake		169
487. TiO$_2$, Degussa, P25	Zn	10 mg dm^{-3}	1 g (154 m^2/g) dm^{-3}	1 d	25	2 10	none	uptake	pH$_{50}$ 6	265
488. TiO$_2$, Tioxide, two samples: rutile and anatase	Ag, Cd	10^{-3} 5×10^{-2} mol dm^{-3}	0.4 g (91; 121 m^2/g) 10 cm^3	5 h	25	6	0.1 mol dm^{-3} KNO$_3$	uptake	Langmuir adsorption isotherms	347
489. TiO$_2$, two samples: rutile and anatase	Ba, Ca, Cd Sr, Zn	10^{-6} 10^{-4} mol dm^{-3}			25	3 11	0.001–0.1 mol dm^{-3} NaCl, NaClO$_4$	uptake, proton release	Uptake of Cd is insensitive to ionic strength, uptake of Ba is depressed when ionic strength increases.	392
490. TiO$_2$, rutile, natural, washed	Ba, Ca, La, Mg, Th	10 10^0 mol dm^{-3}		1 d	25	5.6	chloride, nitrate	ζ	ζ reverses its sign from negative to positive at 4×10^{-3} mol dm^{-3} Ba, 6×10^{-3} mol dm^{-3} Ca, 5×10^{-2} mol dm^{-3} Mg.	354
491. TiO$_2$, Aldrich, anatase	Ba, Ca, Mg	10^{-4} 10^{-2} mol dm	10% w w	20 min	25	3 10	nitrate	acousto	For 1 2×10^{-3} mol dm Mg and 10^{-3} mol dm^{-3} Ca two IEP negative ζ is only observed over limited pH range above the pristine PZC For 10 mol dm^{-3} Ba only positive ζ	393
492. TiO$_2$, rutile, Tioxide	Ba, Ca, Mg, Sr	10^{-3} 10^{-2} mol dm^{-3}	2 5% w w, 20.6 m^2 g		21	4–9	nitrate	proton release,	Proton stoichiometry coefficients increase when pH increases from 0.73 for Mg. 1.12 for Ca and 1.56 for Ba	394

No. Adsorbent	Metal	Concentration	Amount	t, °C	pH	Time	Electrolyte	Method	Remarks	Ref.
(cont.)								uptake, ζ	at pH 6.5 to 1.9 for Mg, Ca and Ba at pH 8.3. Only positive ζ potentials for 10^{-3} mol dm^{-3} Ca, Sr and Ba.	
493. TiO$_2$, Degussa, P-25	Cd, Cu	10^{-6} 10^{-4} mol dm^{-3}	2 g (55 m^2/g) dm^{-3}	22	3 8.5	10 h	0.003 mol dm^{-3} NaClO$_4$, 0-10^{-4} mol dm^{-3} EDTA	uptake	EDTA^4 was taken as the component –TiOCu log K 0.63; TiOCd log K –1.2; TiOHEDTACu^2 log K 23.3; TiOHEDTACd^2 log K 21.4	395
494. TiO$_2$, rutile, Tioxide	Co,Cu,Ni	10^{-1} mol dm^{-3}	160 m^2 dm^{-3}	22 / 25	4-11	2h	0.5 mol dm^{-3} NH$_4$NO$_3$	uptake, ζ	The uptake of metal cations has a minimum at pH 9.5. A maximum at pH 6 for Cu, and 7.5 for Ni uptake.	108
495. TiO$_2$, P-25, Degussa	Co, Np(V)	3×10 mol dm^{-3}; 2.5×10 mol dm	0.1 g (50 m^2/g) 10 cm^3	room	2 12	2 d	0.01, 0.1 mol dm^{-3} NaCl	uptake	TiOCo log K –3.4; TiOCo Cl log K –2.6; TiOHNpO$_2$OH log K –1.5. Desorption of NpO$_2^+$ gives same results as adsorption, in contrast 5-10% of the originally adsorbed Co is not released at low pH	396
496. TiO$_2$, from chloride	Cu, Mn(II)	4×10 3.6 10^{-4} mol dm	20 m^2 dm^{-3}		2 11.5		0.009 mol dm^{-3} NaClO$_4$	uptake, ζ, ESR	The IEP is shifted to pH 10 with 3.6×10^{-4} mol dm^{-3} Cu or Mn. Partial oxidation of adsorbed Mn(II) at pH 5.	397
497. TiO$_2$ n H$_2$O, synthetic	Pu (IV)	7 10 mg dm^{-1}	0.1 g 10 cm^3		natural	5 h	0.1 1 mol dm^{-3} Na$_2$CO$_3$	uptake, chromatography	The uptake is not affected by γ-irradiation	85
498. TiO$_2$ n H$_2$O, from sulfate	Sr	0.01 mol dm^{-3}	0.4 g 25 cm^3	4.5 6.5		2 d	Cl	uptake		168
499. TiO$_2$, gel, 27% water, hydrolysis of TiCl$_4$	U	10^{-4} mol dm	0.5 g (400 m^2/g) 50 cm^3	20	8 9	3 h	0.8 mol dm^{-3} NaCl, 3×10^{-4} mol dm^{-3} carbonate	uptake, chromatography	The log K_D $-7.43-0.34\log[H^+]|^2(CO_3^{2-})$? The surface complex involves two surface groups and two carbonate ligands per one UO$_2^{2+}$	398
500. TiO$_2$ n H$_2$O, hydrolysis of Ti(SO$_4$)$_2$, anatase	U	0.01-0.1 mol dm^{-3}	0.3 g (247 m^2/g) 25 cm^3	27	3.5 (initial)	1 d	nitrate	uptake	Sorption capacity 1.53 mmol g. uptake decreases when T increases ΔH$_{ads}$ –22 kJ mol.	154
501. TiO$_2$ n H$_2$O, synthetic, 3 samples	U	0.01 mol dm^{-3}	0.3 g 25 cm^3	27	3.5 10.7	1 d	nitrate, chloride, sulfate, carbonate (up to 0.2 mol dm^{-3})	uptake		155
502. TiO$_2$ n H$_2$O, synthetic, 3 samples	U	10^{-4} mol dm^{-3}	0.1 g (0.1 3.9 m^2 g) 25 cm^3	25	2 8	4 h	0.1 mol dm^{-3} NaClO$_4$; 10^{-4} 0.02 mol dm^{-3} nitrate, sulfate, chloride	uptake	Concentration of nitrate (pH 2.8) and sulfate (pH 4) has no effect on the uptake. uptake is depressed in presence of chloride and carbonate	399
503. TiO$_2$ n H$_2$O synthetic	U			30	3 10	30 min		uptake		400
504. TiO$_2$ n H$_2$O	Cd, Hg, Zn	10^{-7} 10^{-2} mol dm^{-3}	0.1 g 10 cm^3	30	3 10	30 min	nitrate, chloride; EDTA, glycine, oxalate, phosphate, acetate, sulfate; concentration of ligands $3\times$[Zn or Hg]	uptake	Linear log-log adsorption isotherms at pH 6.8, slope 0.86 for Zn, 0.81 for Cd and 0.9 for Hg. ΔH$_{ads}$ 12 kJ mol for Zn, 11 kJ mol for Cd, 10 kJ mol for Hg. Zn, Cd, Hg in presence and absence of complexing anions. The uptake is depressed in the presence of EDTA, glycine, oxalate, phosphate, acetate, sulfate (EDTA is more efficient as depressor of cation uptake than glycine, etc.).	401
505. TiO$_2$ n H$_2$O	Cd, Pb, U, Zn	1 2×10^{-4} mol dm^{-3}	0.05 g 300 cm^3	25	2.7 3.1	2 d		uptake		402
506. TiO$_2$, gel, (from Ti(III))	Co, Cr, Cu, Ni	0.01 mol dm^{-3}	0.1 g 10 cm^3	30	natural	2 d		uptake		403

TABLE 4.1 Continued

Adsorbent	Adsorbate	c total	S:L ratio	Eq. time	T	pH	Electrolyte	Method	Result	Ref.
507. WO$_3$	Al	0.001 mol dm^{-3}				4-12	chloride	ζ	IEP is shifted to pH 8.4 (1.95 m KCl).	153
508. ZnO, Aldrich	Co	3.6×10^{-4} mol dm^{-3}	20 g (6.9 m^2 g) dm^{-3}	1 d	25	7.9.5	0.05 mol dm^{-3} NaClO$_4$	uptake, EXAFS	The pH$_{90}$ =8.4 (4.15×10^{-4} mol dm^{-3} Co). At high surface coverages formation of structurally disordered Co(OH)$_2$ phase.	404
509. ZnO, Aldrich	Co	3-6×10^{-4} mol dm^{-3}	20 g dm^{-3}	1 d	25	7 9.5	0.05 mol dm^{-3} NaClO$_4$	EXAFS	2.7×10^{-6} mol Co m^2 1.5 oxygen atoms at 0.194 nm; 3.8 oxygen atoms at 0.21 nm; 1.1 Zn atoms at 0.28 nm; 3 Co atoms at 0.316 nm; 5.6×10^{-6} mol Co m^2: 1.2 oxygen atoms at 0.196 nm; 4.2 oxygen atoms at 0.211 nm; 0.5 Zn atoms at 0.281 nm; 5.4 Co atoms at 0.315 nm	405
510. ZnO	Al, Zn	0.001 mol dm^{-3}			25	4-12	chloride	ζ	IEP is shifted to pH 8.3 in AlCl$_3$ and 8.4 in ZnSO$_4$ (6.9 m KCl).	153
511. ZrO$_2$, sol-gel process	Ca	10^{-4} 10^{-2} mol dm^{-3}	0.1 g. 50 cm^3	1 d	25	3-11	chloride, sulfate	ζ	In the presence of 10^{-2} mol dm^{-3} CaCl$_2$ positive ζ over the entire pH range. In the presence of 10^{-3} mol dm^{-3} CaSO$_4$ ζ < 0 over the entire pH range.	406
512. ZrO$_2$	Co	carrier free	10^{-4} mol 20 cm^3	20 min	25	6-11	1 mol dm^{-3} KNO$_3$, 1 2.5 mol dm^{-3} NH$_4$NO$_3$	uptake	Uptake data obtained at different NH$_4$NO$_3$ concentrations plotted as log (uptake/(1-uptake))vs. log fraction of neutral hydroxo complex give a straight line with slope 1. Results published earlier for a similar system are deemed incorrect.	287
513. ZrO$_2$, amorphous, hydrolysis of ZrOCl$_2$	U	0.01-0.1 mol dm^{-3}	0.3 g (223 m^2/g). 25 cm^3	1 d	27	3.5 (initial)	nitrate	uptake	Sorption capacity 0.41 mmol/g. uptake decreases when T increases	154
514. ZrO$_2$, spray-ICP technique, heated at 1200 C	Zr	0.001 mol dm^{-3}				3.9	chloride	ζ	IEP at pH 6.7.	346
515. ZrO$_2$, 99.9%, reactor grade	Ca, Co	10 10^{-3} mol m^{-2}			25	2 12		uptake, ζ	ΔH$_{ads}$ 30 kJ mol for Co at pH 6.5 IEP is shifted to pH 12 for 10^{-3} mol dm^{-3} Co, for 10^{-3} mol dm^{-3} Ca the sign of ζ is reversed to positive at pH 10 12 (30 °C).	52
516. ZrO$_2$, synthetic	Co, Cr(III), Fe(III), Mn(II)	1 10 mg dm^{-3}	0.25 g. 25 cm^3	1 d		natural	nitrate, chloride	uptake, chromatography	Uptake of Fe(III) increases with T, uptake of Cr(III) decreases with T (25 95 C). Uptake of Co and Mn(II) is rather insensitive to T	172
517. ZrO$_2$·n H$_2$O, from ZrOCl$_2$	Ra	trace	5 g. 10 cm^3	1 d	room	1,10	HCl NaOH	uptake	100% uptake at pH 10	259
518. ZrO$_2$·n H$_2$O, synthetic, 2 samples	U	0.01 mol dm^{-3}	0.3 g. 25 cm^3	1 d	27	3.5 10.7	nitrate, chloride, sulfate, carbonate (up to 0.2 mol dm^{-3})	uptake	0.2 mol dm^{-3} total carbonate prevents U uptake at pH 7.3 9.4	155
519. ZrO$_2$·n H$_2$O, hydrolysis of ZrOCl$_2$	Am, Ce(III), Co, Eu, Fe(III), Mn, Th,U,Zn	10^{-6} 10^{-4} mol dm	0.12 g. 8 cm^3	6 20 d	25, room	2 7	0.1 mol dm^{-1} NaCl	uptake	Fe(III) U Th Am Ce(III) Zn Co Mn; Eu uptake increases with T(15-45 C), the T effect on Sr uptake is less significant	407

No.	Adsorbent	Adsorbate	Concentration	Solid / solution	Time	$t/°C$	pH	Electrolyte	Method	Results	Ref
520	ZrO_2 n H_2O, hydrolysis of $ZrOCl_2$	Ba, Ca, Mg, Sr	10^{-3} 1 mol dm^{-3}	1 g (252 m²), 200 cm³	5 min	21	7 9	chloride	proton release	Ca Sr Mg Ba	408
materials containing Al											
521	Al(OH)$_3$ + Fe(OH)$_3$, coprecipitate	Cd, Zn	10^{-6} mol dm^{-3}	10^{-3} mol, Al and Fe dm^{-3}	2 h		6-9	0.1 mol dm^{-3} NaNO$_3$	uptake		151
522	Fe$_2$O$_3$ + Al$_2$O$_3$ n H$_2$O. Fe:Al 1:3, 1 1, and 3:1	Cu, Pb	5×10^{-4} 2.5×10^{-2} mol dm^{-3}	0.5 g (270 400 m²/g) 15 cm³	1 d		4-6.5	8 13% sulfate w w in the adsorbent	uptake		125
523	Al(OH)$_3$ + Mg(OH)$_2$, gel from chloride, up to 0 1 Mg Al mole ratio	Cu	10 10^{-3} mol dm^{-3}	100 mg 25 cm³	1 d		low	Cl	uptake, ESR		134
524	Al(OH)$_3$: Mg (OH)$_2$. Al (Al+Mg) 0 466 mol	Al, Ca, Mg	2×10^{-3} 10^{-2} mol dm^{-3}				4, 8.9	Cl	ζ	Positive ζ.	409
525	Al$_2$O$_3$. 7 13 8% MoO$_3$	Ba, Ca	10^{-3} mol dm^{-3}			22 5	6-11	chloride	uptake, ζ	Positive ζ over the entire pH range	92
526	Al$_2$O$_3$ 3% SiO$_2$	Cr(III)	10^{-2} -0.4 mol dm^{-3}	2 g (460 m²/g) 50 cm³	1 d	25	1.9-3.6	nitrate	ζ		64
527	Al$_2$O$_3$, from isopropoxide + SiO$_2$, from ethoxide composites	Pd	2×10^{-4} 2×10^{-2} mol dm^{-5}	65000 cm^{-1}	1 h		10.6	0 07 mol dm^{-3} NH$_4$Cl	uptake		82
528	64% Al$_2$O$_3$ + 27% TiO$_2$ + 9% SiO$_2$	Ca	0 001 mol dm^{-3}			20	3.5 10.5	Cl	ζ	IEP shifts from 4.7 (NaCl) to 5.7.	410
529	70% Al$_2$O$_3$ + 30% TiO$_2$	Ca	0 001 mol dm^{-3}			20	3 7	Cl	ζ	IEP shifts from 4 5 (NaCl) to 5.2.	411
clay minerals											
530	chlorite	Ca, Mg	10^{-5} 1 mol dm^{-3}				6.5	chloride	ζ	The sign of ζ is reversed to positive in 10^{-3} mol dm^{-3} CaCl$_2$, no sign reversal in MgCl$_2$, up to 1 mol dm^{-3}.	412
531	cordierite, Baikowski	Al, Mg	10^{-9} 10^{-4} mol dm^{-3}				2 11	none	ζ FTIR	When no Mg or Al is added, the ζ is positive over the pH range 4.5-6 and negative outside this range. On addition of $>10^{-8}$ mol dm^{-3} Mg the lower limit of the pH range in which ζ is positive shifts to lower pH values. On addition of $>10^{-9}$ mol dm^{-3} Al the lower limit of the pH range in which ζ is positive shifts to lower pH values, and the upper limit-to higher pH values.	413
532	glauconite (Lithuanian)	Cd, Cu, Pb, Zn	0.5 50 ppm	0.6 g dm^{-3}	2 d		2 9	0.001-0.1 mol dm^{-3} NaClO$_4$	uptake	Zn sorption is ionic strength insensitive. For 6 ppm Zn pH$_{50}$ 6.	414
533	hectorite	Co	10^{-4} mol dm^{-3}	1.95 g (114 m²/g) dm^{-3}	5 d	25	6-6.5	0.01 0.3 mol dm^{-3} NaNO$_3$	uptake EXAFS	The uptake is depressed at high ionic strength. Inner sphere complex: 1.6 Mg atoms at 0.303 nm and 2.2 Si atoms at 0.327 nm. No clustering.	415 416

TABLE 4.1 Continued

Adsorbent	Adsorbate	c. total	S:L ratio	Eq. time	T	pH	Electrolyte	Method	Result	Ref.
534. hectorite SHCa-1	Sr	10^{-4} mol dm^{-3}		1 d	25	9.9	0.001, 0.01 mol dm^{-3} Na	EXAFS	40% uptake (0.01 mol dm^{-3} NaCl), 80% (0.001 mol dm^{-3} NaCl); Seven oxygen atoms at 0.256 nm.	417
535. illite, from China	Cu	10^{-5}-10^{-3} mol dm^{-3}	0.15 g 30 cm^3	1 d	room	3-8	0.1 mol dm^{-3} NaNO$_3$	uptake, FTTR, proton release	The pH$_{50}$ 5.1 (10^{-4} mol dm^{-3} Cu). In presence of carbonate Cu$_2$(OH)$_2$CO$_3$ is formed at high pH.	418
536. illite, Silver Hill Montana	Eu	10^{-8} mol dm^{-3}	1 g(16 m^2/g)/dm^3	2 d		3-7	0.02 1 mol dm^{-3} NaClO$_4$	uptake	>80% uptake with 0.02 mol dm^{-3} NaClO$_4$. The uptake at pH < 6 is depressed when the ionic strength increases, but 0% uptake is not reached even with 1 mol dm^{-3} NaClO$_4$ at pH 2.7 (cf. Fig. 4.6B).	419
537. illite	Sr	1.4×10^{-8} mol dm^{-3}	50 mg 20 cm^3	20		2-4		uptake		420
538. illite, IMt-1	Sr	10^{-4} mol dm^{-3}		1 d	25	4.5, 9	0.01 mol dm^{-3} Na	EXAFS	Uptake 42% at pH 4.5 and 9; Six oxygen atoms at R - 0.258 nm.	417
539. illite	Ca, Mg	10^{-5} 1 mol dm^{-3}				6.5	chloride	ζ	The sign of ζ is reversed to positive in 1 mol dm^{-3} CaCl$_2$, no sign reversal in MgCl$_2$ up to 1 mol dm^{-3}.	412
540. illite, American Petroleum Institute, standard 15	Co, Cu	50,100 ppb	1000 ppm	12 h		4-12	artificial river water, filtered sea water, 0.7 mol dm^{-3} NaCl	uptake	Uptake from river water is greater than from sea water.	421
541. kaolinite, 2 samples	Cd	10^{-6}-10^{-4} mol dm^{-3}	94 m^2 dm^3		25	3 10	0.005 mol dm^{-3} KNO$_3$	uptake, proton release	Staircase uptake (pH) curves, Cd has little effect on titration curves at pH < 6 (proton stoichiometry coefficient 0.2). (SO)$_2$Cd log K 7.75 X$_2$Cd log K 3.01 (XK exchange sites). One site Langmuir adsorption isotherm at pH 5.5.	422
542. kaolinite	Cd	5×10^{-5} mol dm^{-3}			25	3 9	0.005 mol dm^{-3} NaNO$_3$, + 0-10^{-3} mol dm^{-3} benzene carboxylic acids	uptake	Two site Langmuir adsorption isotherm at pH 7.5. The pH$_{50}$ - 7 in absence of organic acids (staircase uptake curve). The uptake is depressed by mellitic, pyromellitic, hemimellitic, trimellitic acid at acid concentrations > [Cd] (mellitic acid is more efficient than pyromellitic acid as depressor of Cd uptake, etc.); the effect of trimesic and phthalic acid on Cd uptake is insignificant.	239
543. kaolinite, 3 different samples	Co	2.8×10^{-4} 2.9×10^{-3} mol dm^{-3}	100 g (10-22 m^2 g). dm^3	3 d	22±3	5 8 3	0.01 1 mol dm^{-3} NaNO$_3$	EXAFS uptake	Evidence of inner sphere surface complexes, probably bidentate. Two Co-Al Si distances: 0.27 and 0.34 nm. Multinuclear surface complexes similar but not identical to Co(OH)$_2$ increasing in size with increasing uptake.	423

No.	adsorbent	adsorbate	concentration	solid	pH	T	time	electrolyte	method	result/comments	ref
544	kaolinite, 3 different samples	Co	14 10^-11 10^-10 mol dm^-3	100 g (5 20 3 m² g) dm^-3	7 8 8 6	22±3	3 d-45 d	0.1 mol dm^-3 NaNO3	XAS uptake	The number of nearest Co neighbors at 0.312 nm increases from 2.1 to 5.7 when the surface coverage increases 0.7 1.4 Si Al neighbors at 0.315 nm found only at low surface coverages	424
545	kaolinite, KGa1	Co	10^-6 mol dm	1.5 g (11.4 m² g) dm	4.5 8.5	25	30 min	10 , 10^-3 mol dm NaClO4	uptake	The pH50 7.2 The uptake is depressed in the presence of 10^-4 mol dm^-3 NaClO4 over the pH range 5.5 7 The uptake at pH 7 is enhanced in the presence of pre-adsorbed humic acid	133
546	kaolinite, KGa-1b	Co	3 10^-4 1.3 10 mol dm	100 g (11.4 m² g) dm^-3	7.5 7.8		2 h 1 y	0.1 mol dm^-3 NaNO3	uptake, EXAFS	Co is sorbed as a mixture of adsorbed species (monomer or polymer) and hydrotalcite-like precipitate Proportion of precipitate increases with equilibration time	425
547	kaolinite, KGa-1	Ni	5 10^-4 mol dm	37 m² 53 cm	4 8		1 d	10 mol dm NaClO4	uptake	pH50 6.5; staircase uptake curve (cf Fig. 47)	73
548	kaolinite	Ni	3 10 mol dm	10 g (15 m² g) dm	7 5	25	2 d	0.1 mol dm^-3 NaNO3	uptake, EXAFS	3.8 Ni neighbors at distance of 0.301 nm which is shorter than in Ni(OH)2, and 1.8 Al Si neighbor at 0.302 nm Formation of mixed Al-Ni hydroxide similar to takovite	143
549	kaolinite, KGa-1	Sr	10^-4 mol dm^-3	450 m dm	9.96	25	1 d	0.01 mol dm Na	EXAFS	Uptake 20%; Seven oxygen atoms at R 0.257 nm	417
550	kaolinite, KGa 1b	Sr	10 mol dm^-3		6 10	25	2 d	0.1 mol dm NaCl	uptake, EXAFS	The uptake is enhanced in the presence of atmospheric CO2 The uptake in the presence of air is rather insensitive to the pH	261
551	kaolinite, Fisher	Th	1 ppm	500 mg (8.54 m² g) dm	2.44	20	30 d	0.005 mol dm^-3 H2SO4	uptake	63% uptake	194
552	kaolinite, KGa-1, KGa-1B t	t	10^-6 10^-3 mol dm^-3	4 g dm	3 10	25	2 d	0.1 mol dm^-3 NaNO3, NaHCO3, 0 10^-4 mol dm phosphate	uptake	For samples in contact with air the uptake peaks at pH 6.8 (90% for 10 mol dm U) At higher solid to liquid ratio or lower total U concentration, 100% uptake between two adsorption edges The center of the 100% uptake range is at pH 6.5 Presence of phosphates enhances the uptake at low pH and the high pH adsorption edge is not affected	218
553	kaolinite, KGa-2	t	2 10 mol dm	0.5 g dm	6 7.5		1 d	0.0015 mol dm^-3 NaNO3	EXAFS	At low pH inner sphere mononuclear t species is responsible for adsorption; at higher pH multinuclear U species dominate	426
554	kaolinite, KGa-1	t	10^-6 mol dm	1.2 g (15 m² g) dm	3 11	22	1 d	0.1 mol dm^-3 NaCl 0 1.6 10 mol dm^-2 citrate	uptake	The pH50 shifts from 4.8 in absence of citrate to 8 in the presence of 1.6 10 mol dm^-3 citrate. With 2.7 10^-6 mol dm^-3 citrate the shift in the pH50 is insignificant	144
555	kaolinite K-Galb	t	10 , 10^-4 mol dm^-3	12 g (11.7 m² g) dm	3 8		2 d	0.01 mol dm^-3 KNO3	uptake	For 10 mol dm pH50 5 in the system insulated from atmosphere and in contact with atmospheric CO2	200
556	kaolinite, natural. washed	Ba, Ca, La, Mg, Th	10 10^0 mol dm^-3		5.6	25	1 d	chloride, nitrate	ζ	ζ reverses its sign from negative to positive in 4×10^-4 mol dm BaCl2 and CaCl2, no sign reversal with MgCl2 up to 1 mol dm^-3	354
557	kaolinite, Wako	Ba, Ce, Hf, Pt, Re	trace	10 mg 5 cm³	1 12		10 d	0 30 mg dm^-3 humic acid	uptake		324

TABLE 4.1   Continued

Adsorbent	Adsorbate	c. total	S:L ratio	Eq. time	T	pH	Electrolyte	Method	Result	Ref.
558. kaolinite, Ajax	Cd, Co	10^{-6} 10^{-4} mol dm^{-3}	94 m^2 dm^{-1}		25	3-10	0.005 mol dm^{-3} KNO$_3$	uptake, proton release	Two site Langmuir type adsorption isotherms at constant pH. (≡X)$_2$Co log K 3.06, XK permanently charged site, (≡SO)$_2$Co log K -6.87, SOH variably charged site. ΔH$_{ads}$ permanent 8 kJ mol for Cd and 11 kJ mol for Co, variable 71 kJ mol for Cd and 73 kJ mol for Co (10-70 C)	427
559. kaolinite, China Clay Supreme	Cd, Co, Mn(II), Zn	10^{-7} mol dm^{-3}	1, 10 g (15 m^2/g) dm^{-3}	1 d	20±1	4-9	0.05 mol dm^{-3} NaNO$_3$, 0.2×10^{-4} mol dm^{-3} salicylate	uptake	Uptake of Zn and Cd is depressed at pH > 7 in the presence of salicylate, the uptake of Mn and Co is rather insensitive to addition of salicylate.	99
560. kaolinite, KGa-1	Cd, Sr	8×10$^{-?}$ mol dm^{-3}, 1.4×10$^{-?}$ mol dm^{-3}	4 g (10 m^2/g), 50 cm^3	15 min	25	4-10	10^{-1} mol dm^{-3} NaClO$_4$, 10^{-1} mol dm^{-3} NaCl	uptake	The Sr uptake is rather pH insensitive	428
561. kaolinite, Ajax	Co, Cu, Mn(II), Pb, Zn	10^{-6} 10^{-4} mol dm^{-3}	100 m^2 dm^{-3}	30 min	25	3-10	5×10$^{-?}$ mol dm^{-3} KNO$_3$	uptake, proton release	Staircase uptake curves (pH) for Mn, Co, and Zn. Langmuir adsorption isotherms at pH 5.5. Two site Langmuir adsorption isotherms for Mn, Co, and Zn at pH 7.5. XK exchange sites form a X$_2$M complex, log K 3.55;3.31;3.15.3.02;3.05 for Pb, Cu, Zn, Co, and Mn; (SO)$_2$M, log K -4.9;-4.58;-6.7;-7.75;-8.82, respectively.	429
562. kaolinite	Cu, Pb	5×10^{-4} 2.5×10^{-2} mol dm^{-3}	0.5 g (24 m^2/g) 15 cm^3	1 d		4-6.5		uptake	Proton stoichiometry factor 0.05-0.2 at pH 5.5	125
563. kaolinite, unwashed, Na and Ca form	Eu, Sr	10^{-6} 10^{-2} mol dm^{-3}	0.5 g 20 cm^3	1 d	20	2-10 (initial)	nitrate	uptake		430
564. montmorillonite, SWy-2	Cu	10^{-5} mol dm^{-3}		5 d	25	3-11	none, 0.1 mol dm^{-3} KNO$_3$	uptake, EPR	Ion exchange at low pH. At neutral pH Cu is sorbed as dimeric surface species with Cu-Cu distance of 0.33 nm	431
565. montmorillonite	Ni	3×10^{-3} mol dm^{-3}	10 g (520 m^2/g) dm^{-3}	2 d	25	7.5	0.1 mol dm^{-3} NaNO$_3$	uptake, EXAFS	2.8 Ni neighbors at distance of 0.303 nm which is shorter than in Ni(OH)$_2$, and 2 Al Si neighbors at 0.307 nm	143
566. montmorillonite	Ni	2.3 × 10^{-3} mol dm^{-3}	10 g (15 m^2/g) dm^{-3}	40 min		7.5	0.1 mol dm^{-3} NaNO$_3$	uptake, EXAFS	Formation of mixed Al-Ni hydroxide similar to takovite	432
567. montmorillonite	Np(V)	10^{-6} mol dm^{-3}	4 g (97 m^2/g) dm^{-3}	14 d	20±2	3-10	0.1 mol dm^{-3} NaNO$_3$, various CO$_2$ pressures	uptake	At 10^{-3} atm CO$_2$ the uptake peaks at pH 8. AlONpO$_2$(OH) log K 13.97, ≡SiOHNpO$_2^+$ log K 4.05	433
568. Na-montmorillonite, converted from SAz-1 (Ca form)	Np(V)	10$^{-?}$ mol dm^{-3}	4 g (97 m^2/g) dm^{-3}	10 d	20±2	5-10	0.1 mol dm^{-3} NaNO$_3$	uptake	In experiments conducted in CO$_2$ free glove box and in capped vials (low CO$_2$) K$_D$ constantly increases with pH to reach the highest value of 300 dm^3/kg at pH 10. In experiments conducted in contact with atmosphere the uptake peaks at pH 8.2	76

No.	Mineral	Metal	Concentration	Solid/liquid	Electrolyte	pH	T (°C)	Time	Method	Remarks	Ref.
569	montmorillonite, SWy-2	Pb	0.002 mol dm⁻¹	10 g (667 m² g⁻¹ EGME) dm	0.006, 0.1 mol dm NaClO₄	4 8	25	1 d	uptake, XAFS	Uptake at pH 7 is depressed at high ionic strength Inner sphere complex at pH 6 77 in 0 1 mol dm NaClO₄ Outer sphere complex at pH 4 48 6 4 in 0 006 mol dm⁻³ NaClO₄	434
570	Na-montmorillonite, thermally altered	Sr	0 005 mol dm	0 1 g 100 cm³	natural 0 01 mol dm NaCl (total Cl 0 1 mol dm⁻³)	25	7 d	uptake			435
571	montmorillonite	Sr	10 10 mol dm⁻¹	0.1 g 20 cm³	natural ground water	25			uptake		436
572	montmorillonite, SAz-1	Sr	10⁻⁴ mol dm		0 01 mol dm Na, 0 1 mol dm³ Ca	7 6 9 8	25	1 d	EXAFS	Uptake 40 80%: 5 65 5 83 oxygen atoms at R 0 257 nm	417
573	montmorillonite	t	10 mol dm	0 5,5 g (5g dm³) dm	nitrate	4	14 d	EXAFS	99% uptake	339
574	montmorillonite	t									437
575	montmorillonite	t									438
576	Na-montmorillonite, converted from SAz-1 (Ca form)	t	2 10 mol dm	0 028 3 2 g (97 m² g) dm³	10 atm CO₂	2 9	20	10 d	uptake	K_D peaks at pH 6 1 (20000 dm³ kg) Consistent K_D are obtained at different solid to liquid ratios	32
577	Na-, Ca-, and Mg-montmorillonite	Ca, Mg	10 1 mol dm		chloride	6 5			ζ	Negative ζ is depressed at high Ca and Mg concentrations, but the sign is not reversed	412
578	montmorillonite	Cd, Cu, Pb	5 10 mol dm	5 g dm	10 0 17 mol dm NaNO₃, humic acid 104 mg C dm	4 6 5	25	2 d	uptake	Uptake of metal cations from 0 1 mol dm NaNO₃ at pH 6 is rather insensitive to addition of humic acid Uptake of Pb at pH 6 5 in the presence of humic acid increases when the ionic strength increases Uptake of Pb at pH 4 in the presence of humic acid decreases when the ionic strength increases	439
579	Na-montmorillonite, SWy-1	Cd, Ni, Pb, Zn	5 10⁻⁴ mol dm	75 m² 53 cm³	10 mol dm⁻¹ NaClO₄	2 10		1 d	uptake	Pb Zn Ni Cd Staircase uptake curves, substantial uptake (about 20%) of all metals even at pH as low as 2	73
580	Ca-montmorillonite	Co, Hg, Mn, Pb	carrier free	10 mg 20 cm	none; 5×10⁻⁴ mol dm⁻¹ EDTA	2 8	20	30 min	uptake, release of Ca	The uptake of Pb is practically zero in the presence of EDTA	420
581	Na-montmorillonite, washed	Ni, Zn	5 10 10⁻⁴ mol dm	1 1 g (35 m² g) dm	0 01 0 1 mol dm³ NaClO₄, 0 001 mol dm different buffers in different pH ranges	3 11		2 d	uptake	Sorption of Ni at pH 3 7 is severely depressed at high NaClO₄ concentrations (up to 2 orders of magnitude in distribution coefficient) Ni log-log sorption isotherms at pH 5 9 8 2 are linear with slopes 1, except at Ni concentrations 10 mol dm³ NaClO₄ Zn Sorption of Zn peaks at pH 8 (0 1 mol dm³ NaClO₄) Zn log-log sorption isotherms at pH 5.6 and 7 are not linear	440
582	muscovite mica	Th, U	5 10⁻⁶ mol dm³	0 005 0.1 cm³	0 001 0 1 mol dm³ NaClO₄	room 4 8		7d	uptake		369
583	smectite, Ca-montmorillonite SAz-1, converted into Na-form	Co	10⁻⁶ 10⁻⁴ mol dm³	0 5 g (95 m² g) dm³	0 001 0 1 mol dm³ NaNO₃	3 10		1 d	uptake, XAS	At low pH and ionic strength Co forms outer sphere complexes with interlayer sites At high pH and ionic strength polynuclear complexes or precipitate are formed	441
584	smectite, purified bentonite	Ra	1 2 × 10⁻⁸ mol dm⁻¹	2, 20 g dm³	none, 0.1 mol dm³ NaCl	room 7 11		40 d	uptake	The uptake is depressed at high ionic strength and it is rather insensitive to pH (twofold increase in K_D between pH 7 and 11).	442

TABLE 4.1 Continued

Adsorbent	Adsorbate	c total	S L ratio	Eq time	T	pH	Electrolyte	Method	Result	Ref.
585. smectites	U									443
586. smectite, treated with dithionite-citrate-bicarbonate and then H_2O_2	U	8×10^{-6} mol dm^{-3}	1 5 g dm^{-3}	1 d	25	4 9	0.001 -0.1 mol dm^{-3} NaClO$_4$	uptake	The uptake is depressed at high ionic strength at pH 5; 100% uptake at pH 6.	24
material containing Bi										
587. Bi-Th mixed oxide 4 1	Ag	0.1 mol dm^{-3}	5 g/500 cm^3	1 d		natural nitrate		uptake	0 08 mmol g	444
basic carbonate										
588. Mg$_6$Al$_2$(OH)$_{16}$CO$_3$·4 H$_2$O, Zn hydrotalkite, Crosfield		10 mg dm^{-3}	1 g (34 8 m^2/g) dm^{-3}	1 d	25	2 10	none	uptake	100% uptake at pH 4	265
materials containing Fe										
589. Al substituted goethite	Co	10^{-6} mol dm^{-3}	0 16 g (186 m^2/g) dm^{-3}	30 min	25	4 5 8 5	10^{-2}, 10^{-1} mol dm^{-3} NaClO$_4$	uptake	The pH$_{50}$ 6.8; the uptake is insensitive to ionic strength The uptake is enhanced in the presence of pre-adsorbed humic acid.	133
590. NiFe$_2$O$_4$, two samples	Co	1 6×10^{-6} 10^{-3} mol dm^{-3}			25	3 10		uptake, ζ	ΔH$_{ad}$ 23 54 kJ mol at pH 6 5. Linear log-log adsorption isotherms at pH 6.5, slope 0 5 and 0 25. The sign of ζ is reversed to positive in the presence of 10^{-3} mol dm^{-3} Co at pH 6.5 11.	52
silicates and materials containing Si										
591. sand	Co, Cu	50 ppb	2 g/50 cm^3	12 h		4 12	artificial river water, filtered sea water	uptake	Uptake from river water greater than from sea water	421
592. pyrophyllite	Ni	2 3×10 mol dm^{-3}	10 g (96 m^2/g) dm^3	40 min		7 5	0 1 mol dm^{-3} NaNO$_3$	uptake, EXAFS	Surface precipitation.	432
593. pyrophyllite	Ni	3×10 mol dm	10 g (96 m^2/g) dm	25		7 5	0.1 mol dm NaNO$_3$	uptake, EXAFS	4 8 Ni neighbors at distance of 0 300 nm which is shorter than in Ni(OH)$_2$, and 2.7 Al Si neighbor at 0 302 nm Formation of mixed Al-Ni hydroxide similar to takovite	143
594. talc	Cd, Pb	5×10^{-4} mol dm^{-3}	0 1 g/100 cm^3	1 d		2 11	0.002 mol dm KNO$_3$	uptake, ζ, coagulation	The pH$_{50}$ 7 for Pb and 9 for Cd The sign of ζ is reversed to positive over the pH range 6 11 for Pb and 7 10 for Cd	358
595. SnO$_2$, SiO$_2$	Co, Eu, RuNO$_3$, Sr	0 01 1 mol dm	0 2, 0.5 g 10 cm^3	3 d	25	2 10		uptake		374

material containing Sn

No.	material	species	concentration	solid/liquid	time	T	pH	electrolyte	method	comments	ref
596	SnO_2; $Fe_2O_3 \cdot n\,H_2O$	Co, Zn	carrier free, 10^{-1} mol dm^{-1}	1 g (100 cm^3)	2 d	25	4 7	none, 0 1 mol dm^{-1} NaCl	uptake		230

material containing Ti

No.	material	species	concentration	solid/liquid	time	T	pH	electrolyte	method	comments	ref
597	Ti-Si gel from $TiCl_4$ and Na_2SiO_3, 1 1 molar ratio	Sr	10 ppm $Sr(NO_3)_2$	50 mg 10 cm^3	6 h	25	3 12	0 1 4 mol dm^{-3} $NaNO_3$	uptake	The pH$_{50}$ 9 2 with 1 mol dm^{-3} $NaNO_3$. The uptake (NaOH added, pH not reported) is depressed when Na concentration increases	445

zeolites

No.	material	species	concentration	solid/liquid	time	T	pH	electrolyte	method	comments	ref
598	Zeo 49, blazer, mordenite, Huber	Ni	10^{-4} mol dm^{-1}	100 m^2 dm^{-3}	30 min	25	3 9	0.025 mol dm^{-3} $NaClO_4$, 10^{-4} mol dm^{-3} EDTA	uptake	Zero uptake except by blazer at pH 4 5	95
599	mordenite, Huber	Ni	5 10^{-4} mol dm^{-3}	111 m^2 53 cm^3	1 d	25	4 8	10^{-3} mol dm^{-3} $NaClO_4$	uptake	Staircase uptake curve pH$_{50}$ 3 2, at pH 4 7 the uptake levels out at 80%, but increases again at pH 7	73
600	Na-clinoptilolite, converted from natural material	Np(V)	10^{-1} mol dm^{-1}	4, 8 g (10 m^2 g^{-1}) dm^{-3}	10 d	20	3 11	0.01, 0.1 mol dm^{-3} $NaNO_3$	uptake	In experiments conducted in CO_2 free glove box and in capped vials (low CO_2) K_D constantly increases with pH to reach the highest value of 50 dm^3 kg^{-1} at pH 10 In experiments conducted in contact with atmosphere the uptake peaks at pH 8	76
601	mordenite, Huber	Pb	10^{-4} mol dm^{-1}	2 5 g (149 m^2 g^{-1}) dm^{-3}	10 h		3 10	0.025 mol dm^{-3} $NaClO_4$, 10^{-4} mol dm^{-3} EDTA	uptake	100% uptake in absence of EDTA; presence of EDTA almost entirely depresses the uptake	118
602	Na-clinopilolite, converted from natural material	L	2 10^{-2} 2 10^{-6} mol dm^{-3}	2 20 g (10 m^2 g^{-1}) dm^{-3}	10 d	20±2 9	0.1, 1 mol dm^{-3} $NaNO_3$, 10^{-3} atm CO_2	uptake	K_D peaks at pH 6 1 (1000 dm^3 kg^{-1} for 2 10^{-7} mol dm^{-3} L). The ionic strength effect on the uptake is insignificant	32
603	zeolite A, blazer, mordenite, Zeolex 23, Zeo 49, Zeosyl 100	Zn	5×10 10^{-1} mol dm^{-1}	0.5 5 g dm^{-3}	2 d	4 8	0.01 mol dm^{-3} $NaClO_4$	uptake, ζ	For zeolite A, blazer, Zeolex 23, Zeo 49, Zeosyl 100 typical adsorption edges are obtained For mordenite the uptake (20–70% for various Zn concentrations and solid to liquid ratios) is almost pH independent at pH 4 7 (cf Fig 4 6(B)) The Zn uptake by mordenite is also sensitive to the nature of supporting electrolyte, e g in the presence of Na salts it is significantly greater than in the presence of corresponding K salts	89
604	H-ZSM-5 zeolite	Ba, Ca, Mg, Sr	5% wt (metal oxide in zeolite)	0.5 g 150 cm^3				KNO_3	proton release	Zeolite was impregnated with metal acetate solution, dried and calcined at 500 C, then redispersed The uncorrected PZC is 9.4 for Mg, 9 6 for Ca, 9 for Sr and Ba	351
605	Ti-MCM-41, titanosilicate molecular sieve	Cd, Cu, Hg, Mg, Pb, Zn	0.05 10 mg dm^{-3}	1 g (500 cm^3)	3 d	20	0 11	0 1 mol dm^{-3} $NaNO_3$	uptake	For Pb linear log log adsorption isotherms at pH 5 5 9 1, slope 0 64 ΔH_{ads} -34 kJ mol. The uptake (0.05 mg Pb dm^{-3}) peaks at pH 8.6 (90%)	446

coatings

TABLE 4.1 Continued

Adsorbent	Adsorbate	c. total	S:L ratio	Eq. time	T	pH	Electrolyte	Method	Result	Ref.
606. Al(OH)₃ coating on SiO₂ 0.25 5×10⁻³ mol Al:g SiO₂	Ca, Cd, Zn	1.8×10⁻⁵ 2×10⁻⁴ mol dm⁻³	1 g dm³			5 10	0.04 mol dm³ KNO₃	uptake	≡SiOHCa²⁺ log K -1.08; ≡SiOCa⁺ log K -10.75; ≡SiOCaOH log K -16.2; ≡AlOHCa²⁺ log K 4; ≡AlOCa⁺ log K -5.65; ≡SiOCd⁺ log K -6.15; ≡AlOHCd²⁺ log K 5.52; ≡AlOCd⁺ log K -1.98.	355
607. iron oxide coating on sand	Pb	2.5×10⁻⁵ 5×10⁻⁴ mol dm⁻³	1 g/50 cm³	1 d	20		0.01-0.5 mol dm⁻³ NaNO₃, 10⁻⁵ 10⁻³ mol dm⁻³ EDTA, NTA	uptake	Reinterpretation of literature data. The pH₅₀ 4.3 and 5.5 for 2.5×10⁻⁵ and 5×10⁻⁴ mol dm⁻³ Pb. The ionic strength effect on the uptake curves is insignificant. With increasing NTA concentration, a series of cation type uptake curves without 100% uptake (cf. Fig. 4.6(A)) is obtained, and the plateau level gradually decreases, and with 10⁻⁴ mol dm³ Pb + 10⁻³ mol dm³ NTA the Pb uptake practically equals zero over the entire pH range. With Pb+EDTA solutions a similar behavior is observed when EDTA is added to the adsorbent with pre-adsorbed Pb. With aged Pb+EDTA solution, the uptake curves are anion type, without 100% uptake	447

glasses

Adsorbent	Adsorbate	c. total	S:L ratio	Eq. time	T	pH	Electrolyte	Method	Result	Ref.
608. soda glass	Ag	0.5 10 mg AgNO₃/dm³	8.9 cm²:3 cm³	4 d	25	5 3		uptake	The uptake increases with T (35 C and 45 C).	448
609. glass	Be	5×10⁻⁹ mol dm⁻³	115 cm²:100 cm³	1.5 h 7d	room	3 14	0.001-0.1 mol dm³ NaNO₃	uptake	The uptake is ionic strength insensitive and peaks at pH 7 11. Uptake after a 7 d equilibration is slightly higher than after 1.5 h.	449
610. glass fibers	Ca	10 5×10⁻⁴ mol dm⁻³				2 10		ζ	ζ at pH 5.5 is less negative in presence of Ca. The ζ (pH) dependence shows a hysteresis.	450
611. glass	Co	2×10⁻⁶ mol dm³		5 h		4 8	0.5 mol dm³ NaClO₄	uptake		451
612. Vycor and Pyrex glass	Ca, La	10⁻⁴ 10⁻² mol dm⁻³				4 5	chloride	ζ	The sign of ζ of some glasses is reversed from negative to positive on addition of 10⁻⁴ mol dm³ La.	353
613. controlled pore glass	Ce(III), Eu	10 mol dm⁻³	250 mg/50 cm³	20 min	20	6 8	10⁻³ 10⁻¹ mol dm³ NaCl	uptake	The uptake is depressed at high ionic strength; pH₅₀ (10⁻¹ mol dm³ NaCl) 6.8 for Eu, 7.3 for Ce. Hysteresis (a 0.1 pH-unit-shift in the pH₅₀ between base and acid titration).	452

clays

Adsorbent	Adsorbate	c. total	S:L ratio	Eq. time	T	pH	Electrolyte	Method	Result	Ref.
614. bentonite	Np(V)					2 12				453
615. bentonite	Np(V)				26	2 12				454

No.	Sorbent	Ion	Concentration	Solid	Time	T	pH	Electrolyte	Method	Remarks	Ref.
616	Na-bentonite, Kunimine	Np(IV)	10 mol dm⁻³	10 g, 20 cm³	14 d	25	6 10	10⁻² mol dm⁻³ NaClO₄, 10 mol dm⁻³ carbonate	uptake	The uptake peaks at pH 8.5	455
617	bentonite	Ra	1 2 10 mol dm⁻³	2, 20 g dm⁻³	40 d	room	7 11	0 1 mol dm⁻³ NaCl	uptake	The uptake is depressed at high ionic strength and rather insensitive to pH. The K_D obtained at two solid to liquid ratios are not consistent	442
618	bentonite	Sr	10⁻¹⁰ mol dm⁻³	5–42 g dm⁻³	7 d	22±2	9	0 1 mol dm⁻³ NaCl	uptake	The log K_D linearly decreases as the function of log (solid to liquid ratio), slope -0 19	456
619	bentonite Fisher	Th	1 ppm	500 mg [24 m²/g], 2? cm³	30 d	20	3	0 005 mol dm⁻³ H₂SO₄	uptake	99% uptake	194
620	bentonite, Laporte	Cd, Pb	2 4×10 mol dm⁻³	25 50 g dm⁻³	1 d	25		nitrate	uptake, proton stoichiometry	The uptake decreases as a function of amount of acid added (5×10⁻⁷ 10⁻⁵ mol HNO₃ g clay)	457
621	kaolin, acid washed	Cd	2 10 mol dm⁻³	1 10 g dm⁻³	20 h		3 13	EDTA,NTA,EGTA,DCyTA	uptake, ζ	The pH₅₀ 5 at 3 33 g dm⁻³ in absence of chelators 95% uptake at pH 10 for 1 10 g dm⁻³ (no increase in uptake when more kaolin is added. Addition of chelators induces release of stoichiometric (1 mol 1 mol) amounts of Cd over the certain pH range. With DCyTA (diaminocyclohexane, N,N,N',N' tetraacetic acid) over the entire pH range. With EDTA and EGTA [ethylene glycol-(aminoethyl) ether] N,N,N',N' tetraacetic acid] only at pH 6. For NTA only at pH about 9	458
622	kaolin	Co	10⁻⁴ mol dm⁻³	1 g (10 m²), dm⁻³			4 10	0 01 1 mol dm⁻³ NaNO₃	uptake	The pH₅₀ 8 for 0 1 and 1 mol dm⁻³ NaNO₃	459
623	kaolin, acid washed American standard	Cd, Cu, Pb	4 5 10⁻⁷ 7 9×10⁻⁶ mol dm⁻³		1 d		4 10	0 025 mol dm⁻³ NaNO₃, ground water, artificial ground water	uptake	36 10⁻⁶ mol dm⁻³ Cd pH₅₀ 7 3 (staircase) for 0 025 mol dm⁻³ NaNO₃, pH₅₀ 8 5 for ground water, 3 9×10⁻⁷ mol dm⁻³ Pb pH₅₀ 5 5 (staircase) for 0 025 mol dm⁻³ NaNO₃, pH₅₀ 6 2 for ground water, 6 3×10⁻⁶ mol dm⁻³ Cu pH₅₀ 5 5 for 0 025 mol dm⁻³ NaNO₃ and artificial ground water, pH₅₀ 7 (staircase) for ground water	460

natural mineral assemblies

No.	Sorbent	Ion	Concentration	Solid	Time	T	pH	Electrolyte	Method	Remarks	Ref.
624	granite, Finnsjon	Pm, Sr	2 5 10 mol dm⁻³	0 2 g dm⁻³	1 108 d		4 9	0 01 mol dm⁻³ NaClO₄, synthetic granitic water	uptake	The uptake depressed at high ionic strength. The uptake is enhanced in the presence of humic acid at pH 7	368
625	ultisol saprolite, original and treated to remove Fe	Co	10⁻⁶ mol dm⁻³	1.3 g (46, 68 m² g) dm⁻³	30 min	25	4 5 8.5	10⁻², 10⁻⁴ mol dm⁻³ NaClO₄	uptake		133
626	2 rock samples (containing U)	U		200 mg/100 cm³	7 d		3 10	synthetic ground water	isotope exchange	The uptake peaks at pH 7 8	461

TABLE 4.1 Continued

soils

Adsorbent	Adsorbate	c. total	S L ratio	Eq. time	T	pH	Electrolyte	Method	Result	Ref.
627. 4 soils	Cd	7×10^{-5} 2.1×10^{-4} mol dm^{-3}	50-200 mg/25 cm^3	1 d	25	3 8	0.01 mol dm^{-3} LiClO$_4$	uptake	Langmuir type adsorption isotherms at constant pH.	462
628. 2 soils	Cu	5×10^{-5} 1.5×10^{-4} mol dm^{-3}	1:100	18 d		5.5	0.0045 mol dm^{-3} Ca(NO$_3$)$_2$	uptake	Phosphate treatment of soils enhances their ability to bind Cu.	463
629. acidic sandy soil	Eu	10^{-8} 10^{-4} mol dm^{-3}	10 g 30 cm^3	1 d		3 6	0.005, 0.1 mol dm^{-3} NaNO$_3$	uptake	The pH$_{50}$ 4.2 (0.005 mol dm^{-3} NaNO$_3$, 10^{-5} mol dm^{-3} Eu). The uptake is 0% uptake is not reached. cf. Fig. 4.6(B). The uptake is depressed in the presence of 0.1 mol dm^{-3} NaNO$_3$. Linear log-log adsorption isotherms (0.1 mol dm^{-3} NaNO$_3$), slope 0.85 at pH 4 and 0.5 at pH 5.6.	464
630. loamy sand, different acid and lime treatment	Hg	5×10^{-7} mol dm^{-3}	1:20	1 d	25	4-7	0 10^{-4} mol dm^{-3} Cl	uptake	The uptake at pH 4.5-6 is depressed in the presence of 10^{-4} mol dm^{-3} of chlorides, with lower chlorides concentrations their effect on the Hg uptake is rather insignificant. Uptake curves obtained at different pH form one master curve when uptake is plotted against [HgOH], cf. Figs. 4.9, 4.10	244
631. loamy sand	Pb	4.83×10^{-5}-4.83×10^{-3} mol dm^{-3}	1:100-1:6	1 d	25	1 5		uptake		465
632. volcanic soil, washed	Pb	8.8×10^{-5} 7.3×10^{-4} mol dm^{-3}	0.5 g (57 m^2/g) 50 cm^3	1 d	25	3.5-6.5	0.1 mol dm^{-3} NaCl	uptake, chromatography		466
633. soil	U	2.5 μg U dm^3; 2.5 mg U dm^3	10 g (30 m^2 g) dm^3			3 8		uptake	pH$_{50}$ 4.3	467
634. loamy sand	Zn	0.02 100 mg dm^3	1:10	1 30 d	25	natural	0.03 mol dm^{-3} NaNO$_3$; 0.01 mol dm^{-3} Ca(NO$_3$)$_2$	uptake	The uptake increases with T (4-60 C).	468
635. two soils, different lime and acid treatment	Zn	10^{-6} 10^{-3} mol dm^{-3}	1:10	1 d	25	4-6.8	0.01 mol dm^{-3} Ca(NO$_3$)$_2$	uptake	The results obtained at different pH (which is due to different lime acid treatments) form one master curve when the uptake is plotted against calculated [ZnOH]. Similar results were obtained by analysis of results taken from literature.	469
636. loamy sand, different acid and lime treatment	Zn	1 mg dm^3	1:10	1 d		4 5-6.5	none, 1 mol dm^{-3} NaCl, NaNO$_3$	uptake	The uptake is depressed at 1 mol dm^{-3} NaNO$_3$ and even more for NaCl Linear log (uptake) vs pH curves	470
637. 2 soils	Cd,Ni,Sr	1.5×10^{-4} 2×10^{-3} mol dm^{-3}	2 5 g, 20 cm^3	2 h	25	6 1	0.01 mol dm^{-3} KNO$_3$	uptake, chromatography	Freundlich equation gives better results than Langmuir.	471

No.	System	Species	Concentration	Solid/volume	Time	T / pH	Medium	Method	Remarks	Ref.
638.	illitic soil, untreated and treated for carbonate and then oxide extraction	Cu, Pb	5×10^{-3} 1.5×10^{-1} mol dm^{-3} 1 10		1 d	3 7	none	uptake	The uptake of Cu and Pb by soil treated for carbonate and oxide extraction is considerably higher than that by untreated soil.	472
639.	4 natural samples (quartz clay minerals)	Ag, Am, Co, Eu, Fe, Nb, Ni, Np, Pu, Ra, Sb, Sn, Sr, Th, U	10^{-12} 2×10^{-7} mol dm	1 g (6.6 14.7 m^2/g) 10 cm^3	2 d	4-8	artificial ground water	uptake	For Ag, Np, Ra, and U, K_D(pH) curves are reported. For Am, Co, Eu, Fe, Nb, Ni, Pu, Sb, Sn, Sr, and Th, only K_D at natural pH are reported.	473

sediments

No.	System	Species	Concentration	Solid/volume	Time	T / pH	Medium	Method	Remarks	Ref.
640.	25 natural sediments	U	250 µg dm^{-3}	4 g/24 cm^3	3 d	25 3 6 initial	10^{-2} mol dm^{-3} Ca(ClO$_4$)$_2$	uptake	K_D at pH 5.5 is correlated with sum of Al and Fe(III) in sediments. Empirical formulas for K_D (including pH dependence) are proposed.	474
641.	natural sediment	Zn	1.7 10 10^{-4} mol dm^{-3}	50 400 g (0.44 m^2/g) dm^3	2 d	4.9 7.4	artificial ground water +buffer	uptake		475
642.	natural river sediment	Cd, Cu	4×10^{-7} 3 10 mol dm^{-1}	1 mg dm^3 (text) or 1 g dm^3 (figures)	1 d	25 2 8	0.01 mol dm^{-3} NaNO$_3$	uptake	$3\times10^{-}$ mol dm Cu pH$_{50}$ 5; 1.65 10^{-6} mol dm^{-3} Cd pH$_{50}$ 5.2	476
643.	streambed sediments	Cd, Cu, Ni, Pb, Zn	13 µg 16 mg dm^3	2.9 g dm	6 h	room 3 7	natural stream water	uptake, ζ	Pb Cu Zn Cd Ni	477
644.	11 natural sediments	Cd, Cu, Pb, Zn	0.01 1 mg dm^3	1 g dm^3	4 h	20 3 9	0.01 mol dm^{-3} KNO$_3$	uptake	Apparent surface complexation constants were estimated; Pb Cu Zn Cd for all sediments	478
645.	natural sediment, mixtures of fine and coarse fraction	Ce(III), Sr	10^{-6} mol dm	3 7.5 cm	30 d	2 6	ground water	uptake	The uptake of Sr is depressed in the presence of EDTA.	479
646.	suspended particulate matter from a river, 3 size fractions	Cd, Cu	4.4 10^{-7} mol dm^3 1.6×10^{-6} mol dm^3	5; 20 mg dm^3	3 h	4 7.5	0.01 mol dm^{-3} NaNO$_3$	uptake	The pH$_{50}$ 5.6 for Cu and 6.3 for Cd (20 mg SPM dm^3)	480

other

No.	System	Species	Concentration	Solid/volume	Time	T / pH	Medium	Method	Remarks	Ref.
647.	bacteria and bacteria-Fe$_2$O$_3$ composite	Sr	10 10^{-2} mol dm	0.18 0.27 g dm^3	2 h	2.5 11	none	uptake	With 10 mol Sr dm^3 pH$_{50}$ 5.5	216

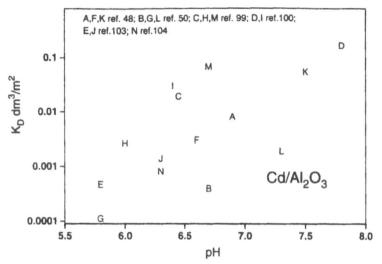

FIG. 4.28 K_D values in the Cd/Al$_2$O$_3$ system reported in the literature.

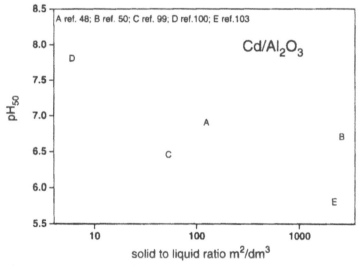

FIG. 4.29 The pH$_{50}$ values in the Cd/Al$_2$O$_3$ system reported in the literature.

The K_D and pH$_{50}$ in the system Cd-Al$_2$O$_3$ are plotted in Figs. 4.28 and 4.29. The results from Refs. 48, 103, and 104 are very consistent, but the K_D values from Ref. 50 are significantly lower. This can be explained by low T (15 C) and relatively short equilibration time. On the other hand the high K_D values from Ref. 99 are probably due to very low initial concentration of Cd ($< 10^{-7}$ mol dm^{-3}). The difference between the highest and the lowest K_D at given pH is about two orders of magnitude.

The K_D and pH$_{50}$ in the system Co-Al$_2$O$_3$ are plotted in Figs. 4.30 and 4.31. Also in this system the difference between the highest and the lowest K_D at given pH is about two orders of magnitude. Relatively low K_D are reported in Ref. 109 (short

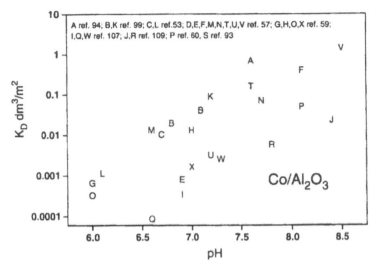

FIG. 4.30 K_D values in the Co/Al_2O_3 system reported in the literature.

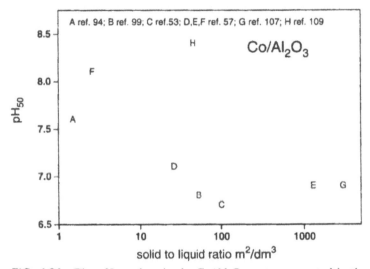

FIG. 4.31 The pH_{50} values in the Co Al_2O_3 system reported in the literature.

equilibration time) and in Ref. 60 and 107 (high initial concentration of Co, $> 10^{-4}$ mol dm^{-3}). The high K_D values from Ref. 94 are probably due to very low initial concentration of Co ($< 10^{-7}$ mol dm^{-3}).

The K_D and pH_{50} in the system Cu-Al$_2$O$_3$ are plotted in Figs. 4.32 and 4.33. The difference between the highest and the lowest K_D at given pH is by one order of magnitude. Relatively low K_D are reported in Ref. 106 (high initial concentration of Cu, $> 10^{-4}$ mol dm^{-3}).

The K_D and pH_{50} in the system Ni-Al$_2$O$_3$ are plotted in Figs. 4.34 and 4.35. The difference between the highest and the lowest K_D at given pH is by one order of

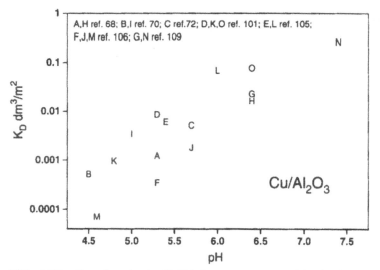

FIG. 4.32 K_D values in the Cu/Al_2O_3 system reported in the literature.

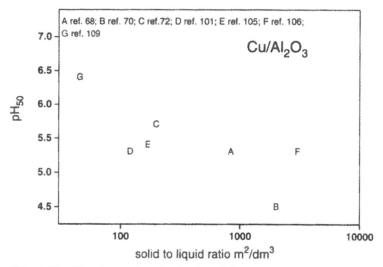

FIG. 4.33 The pH_{50} values in the Cu/Al_2O_3 system reported in the literature.

magnitude except for Ref. 73 that reports considerably lower K_D. The later are probably due to high Ni concentration of 5×10^{-4} mol dm^{-3}.

The K_D and pH_{50} in the system Pb-Al_2O_3 are plotted in Figs. 4.36 and 4.37. The results from Ref. 81 that suggest a K_D higher by 3 orders of magnitude than the values reported in other publications (at given pH) were not used in Fig. 4.36. The other results are rather consistent and the difference between the highest and the lowest K_D at given pH is by one order of magnitude. Relatively low K_D reported in Ref. 107 is due to high initial concentration of Pb ($> 10^{-4}$ mol dm^{-3}). The K_D reported in Ref. 103 are higher than in

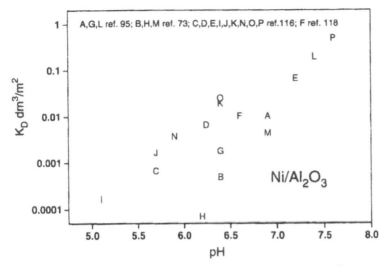

FIG. 4.34 K_D values in the Ni Al$_2$O$_3$ system reported in the literature.

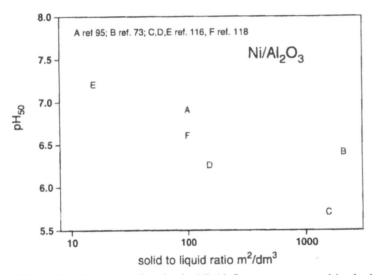

FIG. 4.35 The pH$_{50}$ values in the Ni Al$_2$O$_3$ system reported in the literature.

other publications, but this effect can be hardly explained by the experimental conditions.

The K_D and pH$_{50}$ in the system Zn-Al$_2$O$_3$ are plotted in Figs. 4.38 and 4.39. The results from Refs. 89, 104, 107, and 109 are very consistent and the difference between the highest and the lowest K_D at given pH is by less than one order of magnitude. The high K_D values from Refs. 94 and 99 are probably due to very low initial concentration of Zn ($< 10^{-7}$ mol dm^{-3}).

The K_D and pH$_{50}$ in the system Cd-FeOOH are plotted in Figs. 4.40 and 4.41. The K_D values at pH < 7 are rather consistent (the difference between the highest and

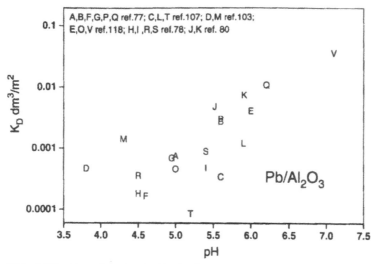

FIG. 4.36 K_D values in the Pb/Al$_2$O$_3$ system reported in the literature.

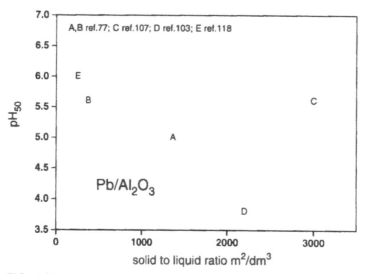

FIG. 4.37 The pH$_{50}$ values in the Pb/Al$_2$O$_3$ system reported in the literature.

the lowest K_D at given pH by less one order of magnitude), but the K_D values at pH > 7 are very scattered. Most data points taken from other publications fall within the trapezium VijW representing the range of data points in Ref. 237. Therefore, the relatively consistent K_D at low pH and the broad range of K_D at high pH (different K_D at different initial Cd concentrations) seems to be a systematic trend, but this trend was not confirmed for other systems.

The K_D and pH$_{50}$ in the system Co-FeOOH plotted in Figs. 4.42 and 4.43 are very consistent, and the difference between the highest and the lowest K_D at given pH is less than one order of magnitude.

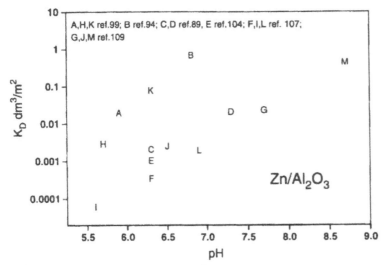

FIG. 4.38 K_D values in the Zn Al$_2$O$_3$ system reported in the literature.

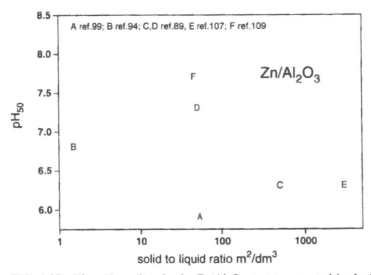

FIG. 4.39 The pH$_{50}$ values in the Zn Al$_2$O$_3$ system reported in the literature.

The K_D and pH$_{50}$ in the system Cu-FeOOH plotted in Figs. 4.44 and 4.45 are rather consistent, and the difference between the highest and the lowest K_D at given pH is by one order of magnitude except for the value from Ref. 124. The latter high value can be hardly explained by the experimental conditions.

The K_D and pH$_{50}$ in the system Ni-FeOOH are plotted in Figs. 4.46 and 4.47.

The K_D and pH$_{50}$ in the system Pb-FeOOH are plotted in Figs. 4.48 and 4.49. The difference between the highest and the lowest K_D at given pH is about two orders of magnitude. The relatively low K_D in Ref. 190 and high K_D in Ref. 272 can

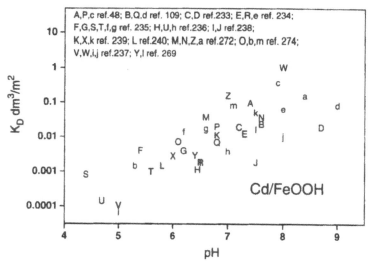

FIG. 4.40 K_D values in the Cd/FeOOH system reported in the literature.

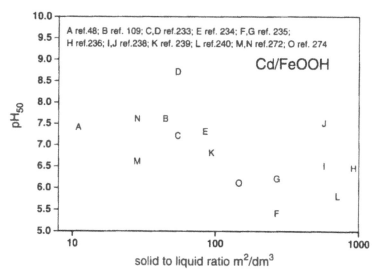

FIG. 4.41 The pH$_{50}$ values in the Cd/FeOOH system reported in the literature.

be explained by high Pb concentration in the former and low Pb concentration in the latter paper.

The K_D and pH$_{50}$ in the system Zn-FeOOH are plotted in Figs. 4.50 and 4.51. The difference between the highest and the lowest K_D at given pH is in excess of two orders of magnitude. Relatively low K_D in Refs. 282 and 109 is probably due to relatively short equilibration time. The high K_D in Ref. 265 and low K_D in Ref. 280 can be hardly explained in terms of the experimental conditions.

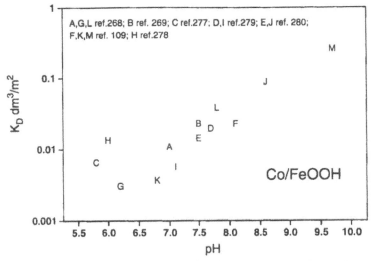

FIG. 4.42 K_D values in the Co/FeOOH system reported in the literature.

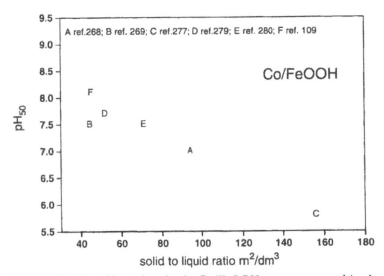

FIG. 4.43 The pH$_{50}$ values in the Co/FeOOH system reported in the literature.

The K_D in the system Ca-SiO$_2$ are plotted in Fig. 4.52.

The K_D and pH$_{50}$ in the system Co-SiO$_2$ are plotted in Figs 4.53 and 4.54. The difference between the highest and the lowest K_D at given pH is in excess of three orders of magnitude. The high K_D reported in Ref. 363 can be hardly explained in terms of the experimental conditions.

The K_D and pH$_{50}$ in the system Cu-SiO$_2$ are plotted in Figs. 4.55 and 4.56. The low K_D in Ref. 361 is probably due to the presence of pH buffer. The other results

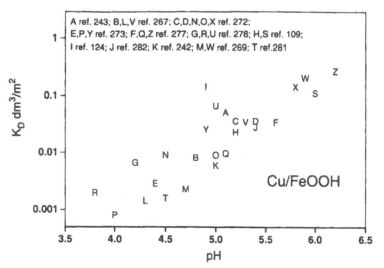

FIG. 4.44 K_D values in the Cu/FeOOH system reported in the literature.

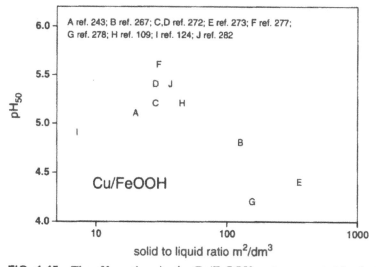

FIG. 4.45 The pH$_{50}$ values in the Cu/FeOOH system reported in the literature.

are rather consistent, and the difference between the highest and the lowest K_D at given pH is about one order of magnitude.

The K_D in the system Ni-SiO$_2$ plotted in Fig. 4.57 are remarkably consistent. Interestingly the slope of the log K_D versus pH curve is only 0.6.

The K_D and pH$_{50}$ in the system Zn-SiO$_2$ plotted in Figs. 4.58 and 4.59 are rather consistent (the difference between the highest and the lowest K_D at

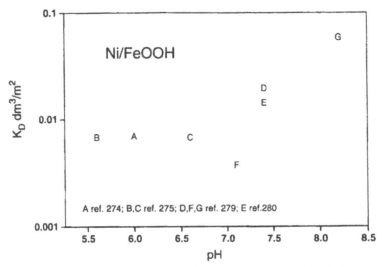

FIG. 4.46 K_D values in the Ni FeOOH system reported in the literature.

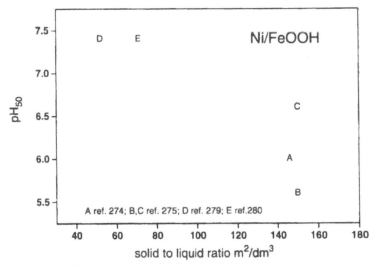

FIG. 4.47 The pH$_{50}$ values in the Ni FeOOH system reported in the literature.

given pH about one order of magnitude) except for the high K_D in Ref. 148 (3 d equilibration time) and low K_D in Ref. 361 that is probably due to the presence of pH buffer.

Comparison of the K_D and pH$_{50}$ values from different sources for a few most frequently studied systems presented in Figs. 4.28–4.59 gave very different results for apparently similar systems. Most often the results from different

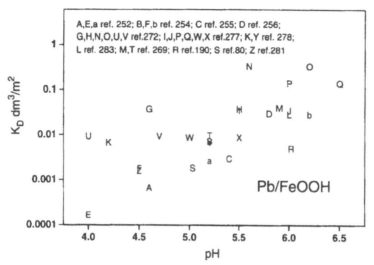

FIG. 4.48 K_D values in the Pb/FeOOH system reported in the literature.

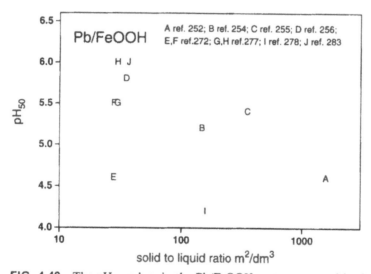

FIG. 4.49 The pH$_{50}$ values in the Pb/FeOOH system reported in the literature.

sources are rather scattered, moreover, the discrepancies can be only partially explained by the difference in the experimental conditions (temperature, time of equilibration, initial concentration of metal cations). The discrepancies can be caused by different conditions of preequilibration, and other experimental factors not shown in Table 4.1. Some of these factors might very well remain beyond the control. Therefore, the usefulness of the K_D(pH) curves taken

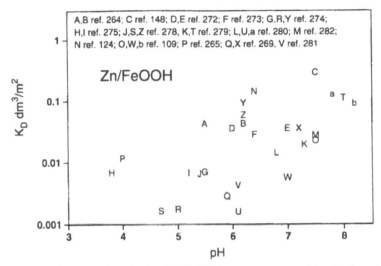

FIG. 4.50 K_D values in the Zn FeOOH system reported in the literature.

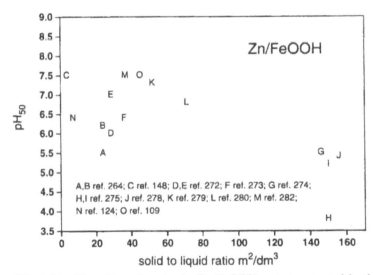

FIG. 4.51 The pH$_{50}$ values in the Zn FeOOH system reported in the literature.

from literature to predict sorption of metal cations is questionable. Some sets of results were very consistent, but more systematic studies are necessary to find out if such accordance was due to some special properties of these systems or it was a fortuitous coincidence, e.g. the experimental conditions in all relevant studies (including the factors that are not addressed in the Table 4.1) could be similar.

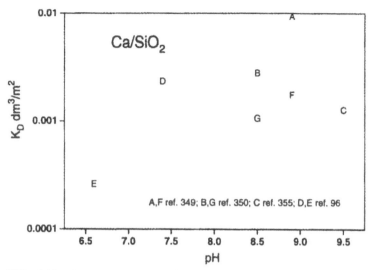

FIG. 4.52 K_D values in the Ca/SiO$_2$ system reported in the literature.

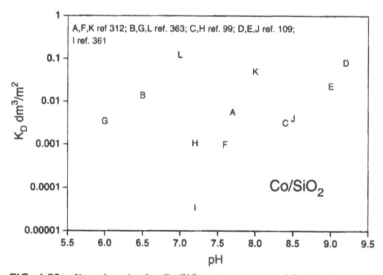

FIG. 4.53 K_D values in the Co/SiO$_2$ system reported in the literature.

B. Anions

In contrast with cations, the typical adsorption edges (cf. Fig. 4.5) for anions are less common, and more complicated shapes of uptake curves are often encountered. A few examples are given in Figs 4.6 and 4.7, moreover, some uptake curves of multivalent anions consist of a few nearly linear segments and the change in slope coincides with the pK values of consecutive dissociation steps. According to

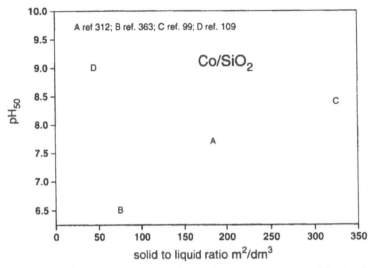

FIG. 4.54 The pH_{50} values in the Co/SiO_2 system reported in the literature.

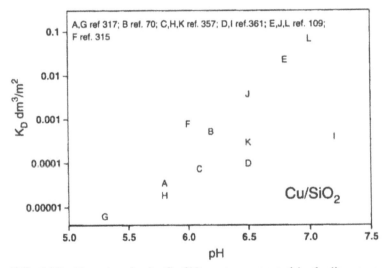

FIG. 4.55 K_D values in the $Cu\ SiO_2$ system reported in the literature.

Hingston et al. [481] the uptake of anions is proportional to $[\alpha(1-\alpha)]^{1\ 2}$ where α is the degree of dissociation of certain species (e.g. H_2SiO_4 and HPO_4^{2-} for silicate and phosphate uptake by goethite, respectively), and the pH of maximum uptake coincides with the corresponding pK value.

Examples of studies of specific adsorption of anions are summarized in Table 4.2.

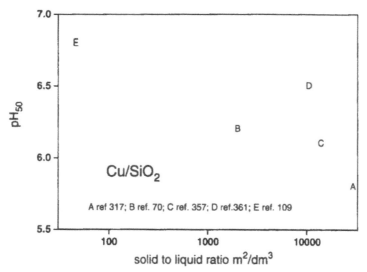

FIG. 4.56 The pH_{50} values in the Cu/SiO_2 system reported in the literature.

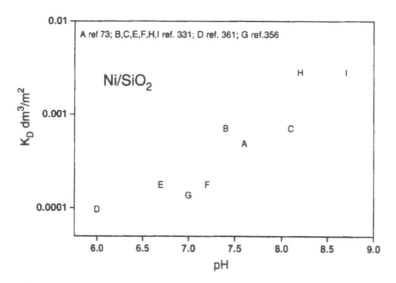

FIG. 4.57 K_D values in the Ni/SiO_2 system reported in the literature.

Most uptake curves reported in Table 4.1 for cations (in absence of complexing anions) have a typical sigmoidal shape (Fig. 4.5(A)), and there are rather few exceptions. In contrast, the shapes of uptake(pH) curves reported for anions are very diverse, thus, it is rather difficult to compare the results from multiple sources in a quantitative manner as it was done for cations in Figs. 4.28–4.59. For example, the pH_{50} as the criterion for such comparison is

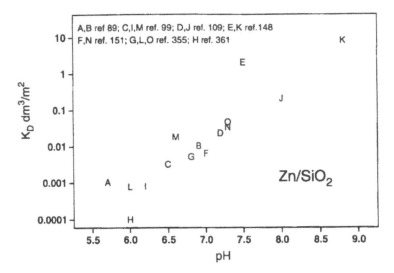

FIG. 4.58 K_D values in the Zn/SiO$_2$ system reported in the literature.

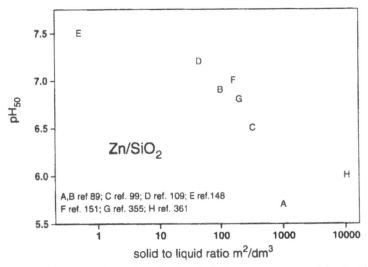

FIG. 4.59 The pH$_{50}$ values in the Zn/SiO$_2$ system reported in the literature.

not applicable for the types of uptake curves shown in Fig. 4.6(C), which are often observed for anions.

The shape of the uptake (pH) curves reported in Table 4.2 depends on the dissociation constant of the acids while the role of the nature of the adsorbent is less significant. Very weak acids (pK$_{a1}$ > 4) tend to show a maximum of uptake at pH≈pK$_{a1}$. For stronger acids some sigmoidal uptake curves (Fig. 4.5(B)) were

(*Text continues on pg. 469*)

Index of adsorbates for Table 4.2

Index of adsorbates for Table 4.2 (Continued)

Adsorbate	Entries
lactate CH₃CH(OH)COOH	235, 254
maleate HOOCCH=CHCOOH	60
malonate HOOCCH₂COOH	201
mellitic acid C₆(COOH)₆	156, 229, 243, 244, 336
methylarsenate	50, 55, 181
methylphosphonic acid	202, 245, 313
molybdate	20–27, 64, 88–91, 109, 110, 114, 140-142, 165, 175, 177, 203–206, 225, 226, 237, 238, 249, 265, 276, 304, 315 317, 323 326, 343–345, 362, 363, 369
1-naphthoic acid C₁₀H₇COOH	243, 244
nitrate III	132, 143, 207
nitrobenzoate, m-nitrobenzoate	51, 303
NTA nitrilotriacetic acid	28
NTMP nitrilotris-(methylenephosphonic acid)	224, 245, 277, 298
oxalate	112, 115, 126, 127, 154, 208, 230, 231, 234, 294, 335
PAA phosphonoacetic acid	156
p-benzoic acids	61, 302
phenylacetic acid	243, 244
6-phenylhexanoic acid	243, 244
phenylphosphate (phenylphosphonic acid)	29, 30, 58, 66, 67
3-phenylpropionic acid	243, 244
phosphate	31–40, 52, 58, 59, 62–64, 68, 92 102, 109 112, 115 118, 125, 144 147, 157, 166-168, 175, 177, 209 215, 223, 225 228, 231, 236 242, 250 253, 255, 256, 260, 261, 265, 270, 289 291, 295, 296, 299, 327 333, 337 339, 346 348, 351, 353, 356, 364 366, 369, 371 373

Index of adsorbates for Table 4.2 (Continued)

TABLE 4.2 Specific Adsorption of Anions

Adsorbent	Adsorbate	Total concentration	Solid to liquid ratio	Equilibration time	T	pH	Electrolyte	Method	Result	Ref.
1. Al_2O_3, α, Aldrich	arsenate (V)	10^{-8}–10^{-4} mol dm^{-3}	0.05, 0.5 g (0.7 m^2 g) 20 cm^3	3 d		2–10	0.1 mol dm^{-3} NaCl	uptake	Uptake peaks at pH 5.	482
2. Al_2O_3, α, A 16 SG, Alcoa	benzoate	10^{-3}–0.015 mol dm^{-3}	1.5 g (9.5 m^2 g) 20 cm^3	1 d	25	4–6	0–0.1 mol dm^{-3} NaCl	uptake		483
3. Al_2O_3,	borate	10^{-2}–0.3 mol dm^{-3}		1 d	25	5–9	0.01 mol dm^{-3} NaNO$_3$	ζ	IEP shifts to pH 6 in 0.3 mol dm^{-3} H$_3$BO$_3$.	484
4. Al_2O_3, pseudoboehmite	borate	0.2–20 mg dm^{-3}	5 g (210 m^2/g) 200 cm^3	2 d	25±2	2–10	none, diluted sea water, diluted geothermal water	uptake	Uptake peaks at pH 8, linear log-log adsorption isotherms at pH 8, slope 0.65	485
5. Al_2O_3, δ, Cabot and Degussa C	borate	5 mg dm^{-3}	10 g (70 m^2 g) dm^{-3}; 40 g (97 m^2 g) dm^{-3}	20 h		5–11	0.1 mol dm^{-3} NaCl	uptake	Uptake peaks at pH 8. Undissociated acid is a component AlH$_2$BO$_3$ log K 5.1 and 5.6.	486
6. Al_2O_3, pseudoboehmite, synthetic	borate	5 mg dm^{-3}	6.4 g (227 m^2 g) dm^{-3}	20 h		5–11	0.1 mol dm^{-3} NaCl	uptake	Uptake peaks at pH 8. Undissociated acid is a component. AlH$_2$BO$_3$, log K 5.1	486
7. Al_2O_3, δ, Degussa C, washed	carbonate	10^{-3} mol dm^{-3}	0.5 g (100 m^2 g) 35 cm^3	1 d	20	2–12	0–0.1 mol dm^{-3} NaCl	uptake, proton stoichiometry	The CIP at pH 8.3 in absence and in the presence of carbonates. The uptake peaks at pH 6 (5 · 10^{-7} mol m^{-2}).	487
8. Al_2O_3, 9 samples α,γ,τ,η	chloroplatinate	10^{-3} mol dm^{-3}	500 m^2 dm^{-3}	1 d		2–11		uptake, proton stoichiometry. Cl release	The uptake peaks at pH 4–6.	488
9. Al_2O_3,	chromate (VI)	3·10^{-3} mol dm^{-3}	30 g (100 m^2 g) dm^{-3}	1 d	25	4–9	0.008 mol dm^{-3} NaNO$_3$	uptake, ζ	pH$_{50}$ 7.2; IEP at pH 7.6.	489
10. Al_2O_3, γ, Harshaw	CrO$_3$	0.5–12% CrO$_3$, Al$_2$O$_3$						mass titration	PZC (mass titration) 7.3–3.7.	490
11. Al_2O_3, α	chromate							uptake, proton stoichiometry		51
12. Al_2O_3, pseudoboehmite, synthetic	citrate	1.3·10^{-4}–2.08·10^{-3} mol dm^{-3}	4.9 g (287 m^2 g) dm^{-3}	2–4 h	25	3–11	0.02 mol dm^{-3} NaClO$_4$	uptake	Trivalent anion (L^{3-}) is a component. ≡AlLH$_2$ log K 23.78 ≡AlLH$^-$ log K 19.71 ≡AlL^{2-} log K 12.8	491

TABLE 4.2 Continued

Adsorbent	Adsorbate	Total concentration	Solid to liquid ratio	Equilibration time	T	pH	Electrolyte	Method	Result	Ref.
13. Al$_2$O$_3$, Sumitomo, AKP-50	citrate	10^{-6} 10^{-1} mol dm^{-3}	0.005 vol %	>1 h		2 12	NH$_4$Cl	ζ	IEP shifts to pH 5.6 for 10^{-5} mol dm^{-3} citrate, 4 for 10^{-4} mol dm^{-3} citrate, 3.2 for 10^{-2} mol dm^{-3} citrate.	492
14. Al$_2$O$_3$, γ, Degussa C	EDTA	10^{-4} mol dm^{-3}	1 g (100 m^2) dm^{-3}	30 min	25±2	3 9	0.025 mol dm^{-3} NaClO$_4$, 10^{-4} mol dm^{-3} Ni	uptake, ζ	The uptake is enhanced in the presence of Ni at pH < 5. The electrokinetic curve in the presence of EDTA (no Ni) is shifted by about 20 mV over the entire pH range. The uptake of EDTA in absence of Ni is depressed when T increases.	95
15. Al$_2$O$_3$, δ, Degussa C	EDTA	2×10^{-7} 10^{-5} mol dm^{-3}	1 g (100 m^2) dm^{-3}	20 min	25	5 10	0.01 mol dm^{-3} NaClO$_4$, 0–10^{-5} mol dm^{-3} Co	uptake	With 5×10^{-7} mol dm^{-3} EDTA, in absence of Co, the pH$_{50}$~7.1 (10^{-2} mol dm^{-3} NaClO$_4$) and 6.3 (10^{-1} mol dm^{-3} NaClO$_4$). The uptake is enhanced in the presence of Co.	53
16. Al$_2$O$_3$, activated	fluoride	10–200 mg dm^{-3}	5 g 200 cm^3	2 d	25±2	2 10	none, diluted sea water, diluted geothermal water	uptake	Uptake peaks at pH 6 7 and it is independent of added electrolytes (sea water, geothermal water).	493
17. Al$_2$O$_3$, γ, Fl, Alcoa	fluoride	10^{-4} 10^{-3} mol dm^{-3}	1 10 g (287 m^2) dm^{-3}	12 h	23	3 10	0.05 mol dm^{-3} NaClO$_4$	uptake	Uptake peaks at pH 4.5 5.	494
18. Al$_2$O$_3$, α, May and Baker	fluoride	10^{-4} 10^{-3} mol dm^{-3}	0.4 g dm^{-3}	1 14 d	25	3 8	Na	uptake	Uptake peaks at pH 6.5. Substantial dissolution of the adsorbent, especially in the acidic range.	495
19. Al$_2$O$_3$, γ, Harshaw	fluoride	1.5 14 at% F Al	0.2 g (170 m^2) 25 cm^3				NH$_4$NO$_3$	proton stoichiometry	PZC (uncorrected) 7, 6.6, 6.4, 5.8, 5.7, 5.6 for F Al at % 1.5,3.5,8,11,14.	496
20. Al$_2$O$_3$, α, BDH	molybdate	8 23 mg Mo dm^{-3}	50–400 mg (4.3 m^2) 25 cm^3	1 d		2 10		uptake	Uptake peaks at pH 3.5.	497
21. Al$_2$O$_3$, γ, Degussa C	molybdate	3 10^{-3} 3×10^{-2} mol dm^{-3}	5 g (100 m^2 g) 200 cm^3	1 d		4 11	NH$_4$	acousto	With 2×10^{-3} mol dm^{-1} Mo$_7$O$_{24}^{6-}$ the IEP is shifted to pH 8.3, and ζ is reversed to negative at pH 4.8 5.5. Only negative ζ for 4×10^{-3} mol dm^{-3} Mo$_7$O$_{24}^{6-}$.	54
22. Al$_2$O$_3$, Houdry 415	molybdate	10^{-3} 3×10^{-2} mol dm^{-3}	5 m^2 14 cm^3	1 d	25	4 10	0.1 mol dm^{-3} NH$_4$NO$_3$	uptake, ζ	Negative ζ in the presence of 10^{-3} mol dm^{-3} molybdate in absence and in the presence of 10^{-4} mol dm^{-3} Co (0.01 mol dm^{-3} NH$_4$NO$_3$).	56
23. Al$_2$O$_3$, Houdry	molybdate	10^{-3} 0.03 mol dm^{-3}	3.57 g(123 m^2 g) dm^{-3}		25	4 10	0.1 mol dm^{-3} NH$_4$NO$_3$	uptake, ζ	Only negative ζ potentials at 10 mol dm^{-3} molybdate, also in the presence of 10^{-4} mol dm^{-3} Ni.	74

No.	Adsorbent	Adsorbate	Concentration	Solid loading	Time	T/°C	pH	Electrolyte	Method	Remarks	Ref.
24.	Al_2O_3, amorphous	molybdate	5×10^{-5}, 3×10^{-4} mol dm^{-3}	0.35 g (210 m^2/g) dm^3	1 d	23	3 11	0.1 mol dm^{-3} NaCl	uptake ζ	IEP is shifted to pH 8 and positive ζ is depressed by a factor >3 in the presence of 3×10^{-4} mol dm^{-3} molybdate.	498
25.	Al_2O_3, δ, Degussa C	molybdate	5×10^{-5}, 3×10^{-4} mol dm^{-3}	4 g (103 m^2/g) dm^3	1 d	23	3 11	0.1 mol dm^{-3} NaCl	uptake ζ	The sign of ζ is reversed to negative at pH 2.5-5 in the presence of 3×10^{-4} mol dm^{-3} molybdate.	498
26.	Al_2O_3, δ, Degussa C	molybdate	3×10^{-4} mol dm^{-3}	4 g (103 m^2/g) dm^3	1 d	23	3-10.5	0.01 1 mol dm^{-3} NaCl	uptake	The uptake is depressed at a high ionic strength	499
27.	Al_2O_3, γ	molybdate	5×10^{-3} mol dm^{-3}	30 g (100 m^2/g) dm^3	1 d	25	4.5-9	0-0.1 mol dm^{-3} NaNO$_3$	uptake	The effect of the ionic strength on uptake curves is insignificant. The pH$_{50}$ – 6.3.	500
28.	Al_2O_3, γ	NTA	1 5×10^{-4} mol dm^{-3}	2.5 g (100 m^2/g) dm^3	3 h	25	3-10	0.025 mol dm^{-3} NaClO$_4$. Cu:NTA 0-2	uptake	Uptake peaks at pH 6.5 (10^{-4} mol dm^{-3}) in absence of Cu. Cu enhances the uptake except at pH near 6.5.	66
29.	Al_2O_3, γ, Sumitomo	phenylphosphonic acid	8×10^{-4}, 3.6×10^{-3} mol dm^{-3}	120-900 m^2 dm^3	>8 h	25	5-9.5	0.1 mol dm^{-3} NaCl	uptake, proton stoichiometry	Uptake peaks at 8 h (dissolution of alumina). Undissociated phenylphosphonic acid H$_2$L is a component. ≡AlLH log K 9.58 ≡AlL · log K 2.94 ≡AlLOH^{2-} log K -5.30	21
30.	Al_2O_3, γ	phenylphosphonic acid	1 molecule nm^2		1 d	25	5-9.5	0.1 mol dm^{-3} NaCl	FTIR and FT Raman spectroscopy (ex situ)	Monodentate coordination, no transformation into aluminum phenylphosphate.	501
31.	Al_2O_3, pseudoboehmite	phosphate	10^{-4} 10^{-2} mol dm^{-3}	1:10000	2 d	20	3-10	K	uptake, coadsorption of K	Uptake at pH 5 increases with T. However, phosphate sorbed at 40 C is not released when the temperature decreases to 20 C.	502
32.	Al_2O_3, α, Linde	phosphate	2×10^{-7} 2×10^{-4} mol dm^{-3}	0.4 g (10 m^2/g) 120 cm^3	1 d	25	2-12	0.01 mol dm^{-3} NaCl, 1.5×10^{-5} mol dm^{-3} La, 1.8×10^{-4} mol dm^{-3} Ca	uptake	The uptake peaks at pH 4 in absence of multivalent metal cations. Langmuir type adsorption isotherms at constant pH. The uptake is enhanced in the presence of Ca and La.	503
33.	Al_2O_3, α, Linde	phosphate	3×10^{-4} mol dm^{-3}	2.5 g (10 m^2/g) dm^3	20 d	50	3 5 8.5		uptake, X-ray diffraction	Formation of AlPO$_4$.	504
34.	Al_2O_3, γ, Cabot	phosphate	10^{-4} 10^{-3} mol dm^{-3}	0.69 g (100 m^2/g) dm^3 or 0.69 g (100 m^2/g) 200 cm^3	1 d	25	2-12	0.1 mol dm^{-3} NaCl	uptake	With 10^{-3} mol P dm^{-3} the uptake peaks at pH 4.5. For lower P concentrations a plateau (100% uptake) ranges from pH 4.5 to. c.g. 6.5 (4×10^{-4} mol P dm^{-3}), or 9 (10^{-4} mol P dm^{-3}), at higher pH the uptake decreases.	505

TABLE 4.2 Continued

Adsorbent	Adsorbate	Total concentration	Solid to liquid ratio	Equilibration time	T	pH	Electrolyte	Method	Result	Ref.
35. Al_2O_3, Carlo Erba, calcined	phosphate	1 mg P 30 cm³	0.1 g (16 m² g) 30 cm³	2 d		4.2 7.5	acetate buffer 0.2 mol dm³	uptake	The uptake peaks at pH 5.2.	506
36. Al_2O_3, α, Merck ground	phosphate	$3.5\ 10^{-4}$ mol dm⁻³	5 g (3.8 m²/g) dm	1 50 d	20	5, 6	none; synthetic sewage water	uptake XRD	The amount "adsorbed" was determined from kinetic data. Long equilibration results in formation of new phase, sterretiite.	507
37. Al_2O_3, γ, Cabot	phosphate	$5\ 10^{-6}$ mol dm³	50 mg (120 m² g) dm³	10 h	22	4 11	10^{-2} mol dm⁻³ NaCl	uptake	Uptake peaks at pH 5.5.	508
38. Al_2O_3, Japan Aerosil	phosphate	$8\ 10^{-3}$ mol dm³	30 g (100 m²/g) dm³		25	4 10	$1.5\ 10^{-2}$ mol dm³ NaCl	uptake, ζ	pH_{50} 8.5; IEP at pH 5.2.	509
39. Al_2O_3, γ, Cabot	phosphate	$10^{-6}\ 10^{-3}$ mol dm³	0.13 g (98 m² g) dm³	2 d	25	4.4 10.1	0.01 mol dm³ NaClO₄	uptake, ζ	Fig. 4.15-4.18	36
40. Al_2O_3, γ, Sumitomo	phosphate	$3\ 10^{-4}$ $2\ 10^{-3}$ mol dm⁻³	1200 m² dm³	5 h	25	4.8 9.6	0.1 mol dm⁻³ NaCl	uptake, proton stoichiometry, DR FTIR	H_2PO_4 is the component ≡$AlPO_4H_2$ log K 11.49 ≡$AlPO_4H$ log K 5.14 ≡$AlPO_4^{2-}$ log K -1.82 Slow transformation into aluminum phosphate phase observed only when concentration of phosphates exceeds the concentration of surface sites.	510
41. Al_2O_3, , Sumitomo	phthalate	0.001 mol dm³	2 g (152 m² g) dm³	8 d	room	4.5 7.2	0 1 mol dm⁻³ KCl	uptake, IR, Raman	The uptake is depressed at high ionic strength. A peak at 1402 cm⁻¹ in the IR spectra was attributed to outer sphere complex; shoulder peak at 1427 was attributed to inner sphere complex.	511
42. Al_2O_3,	rhenate (VII)	0.055 mol dm³	1 g (170 m²) 58 cm³	1 d		4 10	NH₄NO₃	uptake	At pH 6 10 uptake increases when pH decreases, at pH 4 5 scattered values uptake 3 7% Rc by weight.	496
43. Al_2O_3, Degussa C and Merck	salicylate	$10^3\ 10^2$ mol dm⁻³		1 d		5 10	K	uptake, ζ, proton stoichiometry	No shift in the IEP for Degussa C Al_2O_3, IEP of the Merck Al_2O_3 shifts to pH 7.5 in the presence of 10^{-2} mol dm⁻³ salicylate.	512
44. Al_2O_3, Prolabo	salicylate	$2\ 10^{-4}$ mol dm	10 g (53 m² g) dm	1 d	20±2	5 9	0.05 mol dm³ NaNO₃	uptake	The uptake decreases linearly with pH from 80% at pH 5 to 36% at pH 9	99
45. Al_2O_3, α, Alcoa, A-14	silicate	10^{-4} mol dm⁻³	1 g dm³		20	3 10	Na	ζ	Alumina was treated with HF, washed with water, then stored in Na₂SiO₃	513

No.	Adsorbent	Adsorbate	Concentration	Solid/surface	Time	Temp.	pH	Electrolyte	Method	Result	Ref.
46	Al_2O_3	sulfate	0 05 mol dm⁻³	10 g (115 m² g) 50 cm	1 h	22±2	3 10	K	proton stoichiometry, ζ	solution at pH 3 3 for 3 6 d IEP was shifted to pH 6. Shift of PZC to pH 8 86 (pristine value at pH 8 3) Negative ζ potentials over the entire pH range	97
47	Al_2O_3	sulfate	10⁻³ mol dm⁻³			22.5	7 10	K	ζ	The positive ζ are depressed, the IEP is shifted to pH 8.45 (8 8 in KCl).	92
48.	Al_2O_3	thiosulfate	2 10⁻⁴ 10 mol dm⁻³	1 g (125 m²) dm⁻³	2 4 h	25	3 9	K	uptake	The uptake decreases from 68% at pH 3 to 20% at pH 7.5 (2 10⁻⁴ mol dm⁻³ thiosulfate).	48
49	Al_2O_3, Degussa C, washed	acetate, carbonate	10⁻³ mol dm	90 g (86 m² g) dm⁻³	1 d	room	1 9	none, 0.01 mol dm⁻³ NaCl	uptake	The uptake of acetate peaks at pH 5 and it is depressed by factor 3 in the presence of 0.01 mol dm⁻³ NaCl. The uptake of carbonate peaks at pH 6.5 and it is depressed by 30% in the presence of 0 01 mol dm⁻³ NaCl.	514
50	Al_2O_3, Aldrich	arsenate (III), arsenate (V), dimethyl- and methylarsenate V	10 10⁻³ mol dm	25 g dm⁻³	12 h 3d	23	2 10		uptake	Uptake of As(III) and methyl-derivatives is rather insensitive to pH. Uptake of As(V) peaks at pH 5	515
51.	Al_2O_3	benzoate, 3,5-dinitrobenzoate, m-nitrobenzoate, salicylate, sulfosalicylate	5 10⁻³ mol dm	1 4 g 7 cm	1 d	room	4 8	0.1 mol dm⁻³ KCl	uptake		516
52.	Al_2O_3, Degussa C	benzoate, phthalate, salicylate	10⁻⁴ 2 10⁻³ mol dm	1 8 20 g (130 m² g) dm	2 d	22	4 9	0.1 mol dm⁻³ NaClO₄	uptake, proton stoichiometry	Uptake of benzoic acid peaks at pH 4	517
53	Al_2O_3, α, A 16 SG, Alcoa	benzoate, selenate IV	10 10 mol dm	100 g (8.4 m g) dm		25	2 11	0.01 mol dm⁻³ NaCl	uptake	Uptake of benzoic acid peaks at pH 4	68
54	Al_2O_3, δ, Cabot	borate, silicate	1.85 10⁻⁴ 1 85 10 mol dm	0.25 g (70 m² g) 25 cm	20 h		4 11	0.1 mol dm NaCl	uptake	The Si uptake is rather insensitive to pH	518
55	Al_2O_3, Alcoa Fl, washed	cacodylate, methyl arsenate	10 7 10⁻⁴ mol dm	1 g (250 m²) dm	2 d	20	4 11	0.001 0 1 mol dm⁻³ NaNO₃	uptake	The effect of ionic strength on uptake curves is insignificant. For 6.7 10 mol dm methyl arsenate pH₅₀ 8.5. for higher concentrations of methyl arsenate and all studied concentrations of cacodylate 100% uptake is not reached over the studied pH range. Divalent anion MA (methyl arsenate) is the component AlMA log K 11.49 AlOHMA² log K 5.91	519

TABLE 4.2 Continued

Adsorbent	Adsorbate	Total concentration	Solid to liquid ratio	Equilibration time	T	pH	Electrolyte	Method	Result	Ref.
56. Al_2O_3, , Degussa C	CDTA, DTPA, EDTA, EGTA, HEDTA, TTHA	10^{-5} 10^{-3} mol dm^{-3}	0.3 1 g (100 m^2 g) dm^3	30 min	25±2	3 10	0.025 mol dm^{-3} $NaClO_4$	uptake, ζ	Uptake of HEDTA is rather insensitive to pH. The CDTA uptake curve has maxima at pH 5 6 and 9, the uptake of other acids decreases when pH increases. 10^{-3} mol dm^{-3} EDTA induced a shift in the IEP to pH 5.7.	520
57 Al_2O_3, α, commercial, Martinswerk, washed	citrate, tricarballylic acid	0.1 0.6% w w citric acid	1.5 g (15 m^2), 20 cm^3	2 d	25	2 11	K	ζ, uptake, ATR-FTIR	The pH_{50} at pH 10 for 0.1% citric acid and 9.2 for 0.2% citric acid. For higher concentration 100% uptake is not reached in the low pH region. 2.56×10^{-3} mol dm^{-3} citric acid shifts the IEP to pH 3.2, 2.8 10^{-3} mol dm^{-3} tricarballylic acid shifts the IEP to pH 4.	521
58. Al_2O_3, γ, Sumitomo	clodronate, phenylphosphonate, phosphate					5		uptake DR FTIR	γ-Al_2O_3 undergoes transformation into bayerite. Phenylphosphonate forms a surface complex. Clodronate dissolves alumina and than forms aluminum clodronate phase.	522
59. Al_2O_3, ~	$Fe(CN)_6^{3-}$, $Fe(CN)_6^{2-}$, phosphate, sulfate	10^{-7} 10^{-3} mol dm^{-1}	1.5 g dm^1			2 6.6		coagulation		523
60. Al_2O_3, Degussa C, δ	fumarate, maleate, succinate	5 10^{-5} 10^{-3} mol dm^{-3}	0.5 g (100 m^2 g) dm^3	1 2 d	25	3 10.5	0.05 0.5 mol dm^{-3} $NaClO_4$	uptake, ζ	The uptake peaks at pH 4 5. The uptake is depressed at high ionic strengths. The positive ζ is depressed at pH 7 in the presence of organic acids, but the IEP is not shifted.	524
61. Al_2O_3, Spherisorb column, Phenomenex	p-benzoic acids				35	4.8 9.3	8 different buffers	chromatography		525
62. Al_2O_3·n H_2O, from isoproxide	phosphate	1 500 mg dm^3	0.3 g (197 m^2 g) 15 cm^3	3 h	30	5.1, 6.2	0.01 mol dm^{-3} KCl	uptake, proton stoichiometry	Proton consumption per one phosphate anion at pH 6.2 is higher than at pH 5.1.	526
63. Al_2O_3·n H_2O Carlo Erba	phosphate	15 92 mg dm^3	4.7 9.4 g (52 m^2 g) dm^3	2 d	28	2 8		uptake	At pH 7.5, a Langmuir type adsorption isotherm, consistent results for different solid to liquid ratios; at pH 5.5 the course of adsorption isotherms depends on the solid to liquid ratio.	527

No.	Adsorbent	Species	Concentration	Solid	Time	T	pH	Electrolyte	Method	Results	Ref.
64.	$Al_2O_3 \cdot n\,H_2O$	arsenate (V), chromate (VI), molybdate, phosphate, selenate (IV)	$3.3\ 10^{-4}$ mol dm^{-3}	10^{-4} mol Al dm^{-3}	2 d	21	4-10	0.1 mol dm^{-3} NaClO$_4$	uptake	Uptake of chromate peaks at pH 5.5. pH$_{50}$ - 7.3 (molybdate), 9 (selenate, phosphate, arsenate).	528
65.	AlOOH, boehmite	benzoate	10^{-3} 0.015 mol dm^{-3}	0.5 g (147 m^2/g), 20 cm^3	1 d		4-6	none; 0.1 mol dm^{-3} NaCl	uptake		483
66.	AlOOH, boehmite	phenylphosphonic acid	10^{-3} 5×10^{-3} mol dm^{-3}	7 12 g (180 m^2/g) dm^{-3}	3 h	25	4-10	0.1 mol dm^{-3} NaCl	uptake, proton stoichiometry	H_2L, undissociated acid molecule is a component. ($\equiv$Al)$_2$OHLH log K 9.53 ($\equiv$Al)$_2$OHL log K 2.56	529
67.	AlOOH, boehmite, Condea	phenylphosphonic acid	1.7 molecule nm^{-2}		1 d	25	5 9.5	0.1 mol dm^{-3} NaCl	FTIR and F$^-$ Raman spectroscopy (ex situ)	Monodentate coordination, no transformation into aluminum phenylphosphate.	501
68.	AlOOH, boehmite	phosphate	10^{-4} mol dm^{-3}	250 1000 m^2 dm^{-3} (187 m^2/g)	1 d		3 11	0.001 mol dm^{-3} KCl	uptake, ζ, NMR	The IEP is shifted to pH 5.5. The uptake peaks at pH 4 Adsorbed phosphate species becomes deprotonated at pH 9 11 (NMR). Undissociated acid molecule is a component. $\equiv$AlH$_2$PO$_4$ log K 7.28 $\equiv$AlPO$_4^{2-}$ log K -1.05	530
69.	AlOOH, boehmite, Condea	o-phthalate	0.7 3.4 molecules nm^{-2}		2 d	25	4-10	0.1 mol dm^{-3} NaCl	uptake, proton stoichiometry, ATR IR	HL is the component. $\equiv$AlL log K 6.54 $\equiv$AlOH$_2$ $\cdots$ L^{2-} log K 4.34 The uptake at pH 4.5 is depressed by 1 3 when ionic strength increases from 10^{-2} to 1 mol dm^{-3}	531
70.	AlOOH, boehmite, Reanal	salicylate	10^{-6} 10^{-2} mol dm^{-3}		1 d		3 8	0.001-0.1 mol dm^{-3} KNO$_3$	uptake, ζ	The uptake at pH 3 and 6 is depressed at high ionic strength at equilibrium concentrations in solution < 3×10^{-4} mol dm^{-3}, but it is enhanced above this concentration.	532
71.	AlOOH, boehmite layer on Al	sulfate	2×10^{-5} 5×10^{-4} mol dm^{-3}				2 9	0.0001 mol dm^{-3} KCl	ζ	Shift in IEP to low pH or only negative ζ, also after contact with sulfate solution and rinsing.	533
72.	AlOOH, boehmite	sulfate	10^{-3} 10^{-1} mol dm^{-3}			20	3-11	Na	ζ	No shift in IEP at 10^{-3} mol dm^{-3} sulfate, at higher surface concentrations ζ 0 at pH 4 9	126
73.	Al(OH)$_3$, amorphous, from sulfate	arsenate (V)	6.7×10^{-5} 1.6×10^{-3} mol dm^{-3}	0.13–0.16 g dm^{-3}	2 d		3 10	none, 0.01 mol dm^{-3} NaClO$_4$	uptake, ζ	IEP is shifted to pH 4.5 in the presence of 3.2×10^{-3} mol dm^{-3} arsenate.	534

TABLE 4.2 Continued

	Adsorbent	Adsorbate	Total concentration	Solid to liquid ratio	Equilibration time	T	pH	Electrolyte	Method	Result	Ref.
74.	$Al(OH)_3$, amorphous, from sulfate	arsenate (V)	10^{-6}–10^{-3} mol dm^{-3}	0.13–0.16 g (690 m^2 g) dm^{-3}	2 d	25	4.5–8.5	0.01 mol dm^{-3} NaClO$_4$	uptake, ζ		36
75.	$Al(OH)_3$, amorphous, precipitated from AlCl$_3$	arsenate III and V	4×10^{-7} mol dm^{-3}	2.5 g (5.9 m^2 g) dm^{-3}	16 h		4–10	0.02–0.1 mol dm^{-3} NaCl	uptake	The ionic strength does not affect the uptake. As(III): 90% uptake between two adsorption edges: pH$_{50}$ 4.5 and 10.5. As(V) >95% uptake over the entire studied pH range.	535
76.	$Al(OH)_3$, amorphous, synthetic	borate	5 mg dm^{-3}	5.2 g (163 m^2 g) dm^{-3}	20 h		5–11	0.1 mol dm^{-3} NaCl	uptake	uptake peaks at pH 6.5. Undissociated acid is a component, ≡AlH$_2$BO$_3$ log K 5.9.	486
77.	$Al(OH)_3$, gibbsite	borate	5, 250 mg B dm^{-3}	0.375 g (66 m^2 g) 15 cm^3	1 d		3–11	0.001–1 mol dm^{-3} NaCl	uptake, ζ	The IEP is shifted to pH 7.6 at 250 mg B dm^{-3}. Uptake peaks at pH 7–8 and it is depressed at high ionic strength.	536
78.	$Al(OH)_3$, gibbsite	borate	1–100 mg B dm^{-3}	0.375 g (33 m^2 g) 15 cm^3	2 h	25	6–11	0.1 mol dm^{-3} NaCl	uptake	Uptake peaks at pH 9. The uptake is depressed when T increases.	537
79.	$Al(OH)_3$, gel	carbonate		1% w w				KCl	proton stoichiometry	11 different samples of Al(OH)$_3$ gel containing carbonates, commercial and laboratory made have CIP ranging from 6 3–7 32.	538
80.	$Al(OH)_3$, gel	carbonate		1% w w					proton stoichiometry	36 different samples of Al(OH)$_3$ gel containing carbonates, prepared by different methods give the following dependence PZC (uncorrected) - 6.86(CO$_3$/Al ratio) 9.4	539
81.	$Al(OH)_3$, gibbsite	carbonate		0.2 g dm^{-3}			3.5–10.5	10^{-2} mol dm^{-3} NaCl	ζ	IEP at pH 8.3 in dispersion titrated with Na$_2$CO$_3$ (pristine IEP at pH 9).	540
82.	$Al(OH)_3$, amorphous	carbonate		0.2 g dm^{-3}			3.5–10.5	10^{-2} mol dm^{-3} NaCl	ζ	IEP at pH 8.7 in dispersion titrated with Na$_2$CO$_3$ (pristine IEP at pH 9.2).	540
83.	$Al(OH)_3$, amorphous	carbonate	1 mol dm^{-3}	25, 50 g dm	1 d		11.2		XRD	Formation of dawsonite NaAlCO$_3$(OH)$_2$ with traces of bayerite.	540
84.	$Al(OH)_3$, amorphous	carbonate	0.01–1 mol dm^{-3}	400 g dm^{-3}	1 d		4.1–7.8	Na	ATR-FTIR	~AlOCO$_2^-$ is formed at low carbonate concentrations.	540
85.	$Al(OH)_3$, gibbsite, Alcoa	citrate	10^{-6}–10^{-3} mol dm^{-3}	5 g (11 m^2 g) dm^{-3}	1 d	22	3–11	0.1 mol dm^{-3} NaCl, 10^{-6} mol dm^{-3} U	uptake	pH$_{50}$ 10.3 in absence of U (10^{-6} mol dm citrate), the uptake is enhanced in the presence of U.	144

No.	Adsorbent	Adsorbate	Concentration	Solid	Time/Temp	pH	Electrolyte	Method	Remarks	Ref.
86.	$Al(OH)_3$, gibbsite, May and Baker	fluoride	10^{-4} 10^{-3} mol dm^{-3}	1 16 mg, 25 cm^3	1 14 d 25	3 8	Na	uptake, Al release		495
87.	$Al(OH)_3$, fresh, from nitrate	fluoride	10^{-4} 10^{-3} mol dm^{-3}	1 16 mg 25 cm^3	1 14 d 25	3 8	Na	uptake, Al release	Maximum uptake at pH 5.5 6.5, substantial dissolution of the adsorbent at pH < 6.	495
88.	$Al(OH)_3$, bayerite, from chloride	molybdate	8 28 mg Mo dm^{-3}	50–400 mg (18.5 m^2 g), 25 cm^3	1 d	2 10		uptake	Uptake peaks at pH 3.5.	497
89.	$Al(OH)_3$, gibbsite	molybdate	5 10^{-5}, 3 10^{-4} mol dm^{-3}	1.25 g (56 m^2 g) dm^{-3}	1 d	3 11	NaCl	uptake ζ	The sign of ζ is reversed to negative at pH 3–4.5 in the presence of 3 10^{-4} mol dm^{-3} molybdate.	498
90.	$Al(OH)_3$, amorphous	molybdate	3 10^{-4} mol dm^{-3}	0.35 g dm^{-3}	1 d	3 10.5	0.01 1 mol dm^{-3} NaCl	uptake	pH$_{so}$ 6.4	499
91.	$Al(OH)_3$, gibbsite	molybdate	3×10^{-4} mol dm^{-3}	1.25 g (56 m^2 g) dm^{-3}	1 d	3 10.5	0.01 1 mol dm^{-3} NaCl	uptake	The uptake peaks at pH 4 (90%).	499
92.	$Al(OH)_3$, gibbsite	phosphate	10^{-4} 10^{-2} mol dm^{-3}	1::10000	2 d	3 10	NaCl	uptake	Uptake at pH 5 increases with T.	502
93.	$Al(OH)_3$, gibbsite	phosphate	1 300 ppm P	1 g 10 cm^3	up to 2 d	4 45 7	K	uptake	Linear log log adsorption isotherms for different reaction times, slope 0.25.	541
94.	$Al(OH)_3$, gibbsite, from chloride	phosphate	3×10^{-5} 8 10^{-4} mol dm^{-3}	2.4 6 g (47 m^2 g) dm^{-3}	140- 190 d	24 4 11	0.01 mol dm^{-3} KCl	uptake		542
95.	$Al(OH)_3$, gibbsite	phosphate	10^{-7} 10^{-3} mol dm^{-3}	1%	1 d	26 5.5	0–0.02 mol dm^{-3} KCl, NaCl, MgCl$_2$, CaCl$_2$	uptake	Uptake is not affected by NaCl, KCl or MgCl$_2$. Uptake in enhanced in the presence of CaCl$_2$.	543
96.	$Al(OH)_3$, amorphous	phosphate	3.5 10^{-4} mol dm^{-3}	0.2 g (226 m^2 g) dm^{-3}	1 50 d 20	5,6	none, synthetic sewage water	uptake XRD	The amount "adsorbed" was determined from kinetic data. Long equilibration results in formation of new phase, sterrettite.	507
97.	$Al(OH)_3$, gibbsite	phosphate	10^{-5} 10^{-2} mol dm^{-3}		40 d 22	5, 9	0.1 mol dm^{-3} KCl	uptake	Uptake at pH 5 increases with T (2–46 C)	544
98.	$Al(OH)_3$, gibbsite	phosphate	10^{-7} 2 10^{-2} mol dm^{-3}		1 d 22	5	0.1 mol dm^{-3} NaCl	uptake	At high initial concentrations of phosphate and long reaction times precipitate is formed. The nature of the precipitate depends on nature and concentration of cations (Na,K,Rb,NH$_4$).	545
99.	$Al(OH)_3$, freshly precipitated	phosphate	10^{-5} 3 10^{-4} mol dm^{-3}		1 d	7 8		uptake		546
100.	$Al(OH)_3$, gel	phosphate	5×10^{-5} 2 10^{-4} mol 0.3 g Al(OH)$_3$			4.5 6.7		diffuse reflectance IR, Na and Cl uptake	Adsorption of phosphate enhances uptake of Na and reduces uptake of Cl.	547
101.	$Al(OH)_3$	phosphate	10^{-6}–0.002 mol dm^{-3}	0.2 g 25 cm^3	3 d 25	5.8	0.01 mol dm^{-3} NaCl	uptake	Linear log-log adsorption isotherm.	548

TABLE 4.2 Continued

Adsorbent	Adsorbate	Total concentration	Solid to liquid ratio	Equilibration time	T	pH	Electrolyte	Method	Result	Ref.
102. Al(OH)$_3$, bayerite, from chloride	phosphate	15-92 mg dm^{-3}	4.8-9.5 g (28 m^2 g)/dm^3	2 d	28	2 8		uptake	At pH 6.5, Langmuir type adsorption isotherms, consistent results for different solid to liquid ratios; at pH 4 the course of adsorption isotherm depends on the solid to liquid ratio.	527
103. Al(OH)$_3$, bayerite	salicylate	10^{-6}, 10^{-2} mol dm^{-3}		1 d		3 8	0.001, 0.1 mol dm^{-3} KNO$_3$	uptake, ζ	The uptake at pH 3 is depressed at high ionic strength at salicylate concentrations in solution < 10^{-4} mol dm^{-3}, but it is enhanced above this concentration. The uptake at pH 6 and 8 is enhanced at high ionic strength.	532
104. Al(OH)$_3$, reagent grade	silicate	110 ppm SiO$_2$	1:100	4 d	25	2 11	toluene added	uptake		549
105. Al(OH)$_3$, Merck	silicate	20-90 mg SiO$_2$/dm^3	1 g (4 m^2) 100 cm^3	7 d	20	4-11	0.05 mol dm^{-3} K$_2$SO$_4$	uptake, proton stoichiometry	The ionic strength (NaCl, CaCl$_2$) has no effect on uptake measured after 1 d, but the uptake measured after 50 d is enhanced at high ionic strengths, especially at high pH. 3-4 protons released per adsorbed Si(OH)$_4$ at pH 9.2.	550
106. Al(OH)$_3$, gibbsite, Martifin, washed	silicate	220 ppm SiO$_2$	1% w/w, 7.4 m^2/g	15 d	25	3-11	0.02, 1 mol dm^{-3} NaCl	uptake, ζ	IEP at pH 3 for > 10^{-5} mol Si adsorbed m^2 (4×10^{-3} mol dm^{-3} NaCl)	551
107. Al(OH)$_3$, amorphous	sulfate	10^{-4}-3×10^{-3} mol dm^{-3}	2.6 mmol Al 75 cm^3	1 d	20	6, 6.5	1 mol dm^{-3} NaCl, [Zn]=[SO$_4$]	uptake		147
108. Al(OH)$_3$, amorphous	vanadate	1.6×10^{-4} mol dm^{-3}	200 mg 40 cm^3	1 d		4-8	0.25 mol dm^{-3} NaClO$_4$; 0-0.05 mol dm^{-3} D,L-2-mercaptosuccinic acid	uptake, ESR	Cation like adsorption edge: pH$_{50}$ 4. The uptake is enhanced in the presence of 2-mercaptosuccinic acid due to reduction of V(V) to V(IV).	145
109. Al(OH)$_3$, synthetic, 2 samples	arsenate (V), chromate (VI), molybdate, phosphate, selenate (IV)	3.3×10^{-4} mol dm^{-3}	10^{-4} mol Al/dm^3	2 d	21	4-10	0.1 mol dm^{-3} NaClO$_4$	uptake	Uptake of chromate peaks at pH 5.5, of selenate at pH 5, and of arsenate at pH 4.5. The pH$_{50}$ = 7 for phosphate and 5.5 for molybdate for one sample of the adsorbent. Completely different results for another sample.	528

No.	Adsorbent	Adsorbate	Concentration	Solid	Time	Temp.	pH	Electrolyte	Method	Comments	Ref.
110.	$Al(OH)_3$, gibbsite	arsenate (V), molybdate, phosphate	1.3, 2.6×10^{-4} mol dm^{-3}	2.5 g dm^3	1–4 h	22	3–10	0.001–0.1 mol dm^{-3} NaCl	ζ, uptake	IEP shifts to pH 6 in the presence of 1.3×10^{-4} mol dm^{-3} As. The pH$_{50}$ (for 1.3×10^{-4} mol dm^{-3}) = 10 for phosphate and 6.8 for molybdate. The uptake of arsenate peaks at pH 5.5 (90%).	552
111.	$Al(OH)_3$, gibbsite	arsenate (V), phosphate	1 7×10^{-4} mol dm^{-3}	53 mg (31 m^2 g) 25 cm^3	1 d	20	3–10	0.1 mol dm^{-3} NaCl	uptake	Uptake of phosphate (1.3×10^{-4} mol dm^{-3}) peaks at pH 4.5.	553
112.	$Al(OH)_3$, gibbsite	benzoate, oxalate, phosphate	0.005, 0.05 mol dm^{-3}	0.4 g (45 m^2/g) 50 cm^3	16 h, 7 d	20	3 7	0.1 mol dm^{-3} NaCl	ζ, IR, proton stoichiometry		555
113.	$Al(OH)_3$, amorphous, fresh, from nitrate	phosphate, borate, silicate	6.5×10^{-4} 1.2×10^{-2} mol dm^{-3}	10^{-3} mol Al 30 cm^3	12 h		6.5 9.5		uptake	Borate uptake peaks at pH 7.5. Langmuir type adsorption isotherm of silicate (pH not specified).	556
114.	$Al(OH)_3$, gibbsite	fluoride, molybdate, selenate (IV), silicate, sulfate	10^{-6}–4×10^{-6} mol dm^{-3}	0.1 g (47, 58 m^2 g) 25 cm^3	1 d	20–23	2 12	0.1 mol dm^{-3} NaCl	uptake, proton stoichiometry	Uptake of fluoride peaks at pH 5, uptake of silicate peaks at pH 9. For selenate and molybdate the slope of uptake (pH) curves is more negative at pH > pK$_{a2}$ than at pH < pK$_{a2}$.	29
115.	$Al(OH)_3$, amorphous	oxalate, phosphate	10^{-4} 3.5×10^{-3} mol dm^{-3}	50 mg, 50 cm^3 (120 m^2 g)	2 d	20	4–9	0.1 mol dm^{-3} KCl	uptake	Uptake of oxalate peaks at pH 6.	557
116.	$Al(OH)_3$	phosphate, selenate (IV)	10^{-6} mol dm^{-3}	10^{-3} mol Al	2 h		5.5–10	0.1 mol dm^{-3} NaNO$_3$	uptake	Uptake of phosphates peaks at pH 7 (100%).	151
117.	$Al(OH)_3$, gibbsite	phosphate, silicate	< 100 ppm	10^{-3} mol Al dm^{-3} (41 m^2 g)				0–100 ppm Ca	laser Raman spectroscopy (ex situ)	Change in relative intensities of four lines which are attributed to surface OH groups.	558
118.	$Al(OH)_3$, gibbsite	phosphate, silicate	<200 ppm					0–111 ppm Ca	XPS (ex situ)		559
119.	Bi_2O_3, Alfa, α	chromate	carrier free–10^{-2} mol dm^{-3}				2 10	10^{-4}–0.1 mol dm^{-3} Cl, Br, I, or NO$_3$	uptake, IR	A >90% uptake at pH 5 9. The uptake at pH 7 is insensitive to the ionic strength up to 0.1 mol dm^{-3} Cl, Br, I, and NO$_3$.	560
120.	CdO	salicylate	0.001–0.1 mol dm^{-3}	1–4 g 7 cm^3	1 d	room	10.1 10.8 11.6	0.1 mol dm^{-3} KCl Na	uptake		516
121.	CdO	sulfate							uptake, pH		561
122.	Co_3O_4	sulfate	0.01 mol dm^{-3}				4–10	Na	ζ	ζ ~ -15 mV, pH independent.	159
123.	Co_3O_4	sulfate, thiocyanate	0.001–0.1 mol dm^{-3}		25			Na	proton stoichiometry	The shift in uncorrected PZC induced by the salt concentration of 0.1 mol dm^{-3} (in mV, 1 pH unit -59 mV): sulfate 70, thiocyanate 100.	158
124.	Co_2O_3	benzoate, salicylate	10^{-3} mol dm^{-3}	1–4 g 7 cm^3	1 d	room	5–10	0.1 mol dm^{-3} KCl	uptake		516
125.	Cr_2O_3	phosphate, sulfate	10^{-5} 10^{-1} mol dm^{-3}	20 m^2 100 cm^3	10 min 25		5–10	K	proton stoichiometry		562

TABLE 4.2 Continued

Adsorbent	Adsorbate	Total concentration	Solid to liquid ratio	Equilibration time	T	pH	Electrolyte	Method	Result	Ref.
126. $Cr_2O_3 \cdot nH_2O$, monodispersed, spherical, from nitrate, 3 samples	oxalate	5×10^{-5}, 5×10^{-3} mol dm^{-3}		1 d	25	1 10	10^{-2} mol dm^{-3} KCl	ζ	With 5×10^{-3} mol dm^{-3} oxalate the IEP is shifted to pH 1.5.	563
127. CrOOH, monodispersed	oxalate	10^{-6} 10^{-4} mol dm^{-3}		2 h		3	10^{-2} mol dm^{-3} KCl	ATR IR	Langmuir type adsorption isotherm	564
128. CrOOH, monodispersed	sulfate, thiosulfate	10^{-5} 10^{-3} mol dm^{-3}				3-10.8	0-10^{-2} mol dm^{-3} KCl	ATR IR	Langmuir type adsorption isotherms at constant pH.	565
129. $Cr(OH)_3$	benzoate, salicylate	10^{-3} mol dm^{-3}	1-4 g 7 cm^3	1 d	room	5 9	0.1 mol dm^{-3} KCl	uptake		516
130. Fe_3O_4, synthetic	chromate	2×10^{-3} mol dm^3	20. 33 g dm^3	1 2 d	room	6-7	0.1 mol dm^{-3} NaCl, NaNO$_3$	uptake, XAFS	Cr(VI) is reduced to Cr(III).	566
131. Fe_3O_4, natural	chromate	5×10^{-6}, 5×10^{-5} mol dm^{-3}	single crystal	1 min 1 h	room	6	0.1 mol dm^{-3} NaNO$_3$	XAS, XPS (ex situ)	Sorption by 111 surface results in reduction to Cr(III).	567
132. Fe_3O_4	nitrate (III)	0.1 cm^3 saturated NaNO$_2$ 100 cm^3	10 mg 100 cm^3	1 h		4	10^{-3} mol dm^{-3} KCl	ζ	-20 mV	568
133. Fe_2O_3, α	antimonate (V)	carrier free	45 mg (27 m^2 g) 60 cm^3	3 h	50	2 10	0.25 mol dm^{-3} LiCl	uptake	The pH$_{50}$~8; at lower T only kinetic data are published (reaction is not complete).	569
134. Fe_2O_3, hematite, Ward's	arsenate (V)	10^{-8} 10^{-4} mol dm^{-3}	0.05, 0.5 g (0.16 m^2 g) 20 cm^3	3 d		2 10	0.1 mol dm^{-3} NaCl	uptake	uptake peaks at pH 6	482
135. Fe_2O_3, Fisher	borate	5 mg dm^3	100g (16 m^2 g) dm^3	20 h		5 11	0.1 mol dm^{-3} NaCl	uptake	Uptake peaks at pH 8. Undissociated acid is a component. ≡FeH$_2$BO$_3$, log K 4.9.	486
136. Fe_2O_3, hematite	carbonate	5.9×10^{-5} mol dm^3	5.75 g (14.4 m^2 g) dm^3	1 d		4-7.5	0.1 mol dm^{-3} NaClO$_4$	uptake, ζ	Divalent anion was selected as a component. ≡FeHCO$_3$ log K 22.32 ≡FeCO$_3$ log K 12.59 or: ≡FeOH·H$_2$CO$_3$ log K 22.33 ≡FeOH·HCO$_3$ log K 13.87 The IEP under atmospheric CO$_2$ is shifted to pH 8 (originally 9.25).	188, 208

No.	Adsorbent	Adsorbate	Concentration	Solid to liquid	Time	T	pH	Electrolyte	Method	Result	Ref.
137.	Fe_2O_3, maghemite, synthetic	chromate	$2\ 10^{-3}$ mol dm^{-3}	20 33 g dm^{-3}	1 2d	room	6	0.1 mol dm^{-3} NaCl	uptake, XAFS	Outer sphere complex.	566
138.	Fe_2O_3, synthetic	chromate	$5\ 10^{-3}$ mol dm^{-3}	single crystal	1 min 1 h	room	6	0.1 mol dm^{-3} NaNO$_3$	XAS (ex situ)	Sorption by 0001 and 1(-1)02 surfaces	567
139.	Fe_2O_3, hematite, spherical	iminodiacetic acid	$10^{-4}\ 2\ 10^{-3}$ mol dm^{-3}	10 g (14.4 m^2 g) dm^{-3}	20 min	25	2 10	0.01 mol dm^{-3} NaClO$_4$	uptake, ζ	The IEP shifts to low pH (pH 3 at $2\ 10^{-3}$ mol iminodiacetic acid dm^{-3}).	570
140.	Fe_2O_3, α, Riedel de Haen	molybdate	8 27 mg Mo dm^{-3}	50 400 mg (9.1 m^2 g) 25 cm^3	1 d	23	2 10		uptake	Uptake peaks at pH 4.	497
141.	Fe_2O_3, hematite	molybdate	$3\ 10^{-4}$ mol dm^{-3}	5 g (11 m^2 g) dm^3	1 d	23	3 10.5	0.01 1 mol dm^{-3} NaCl	uptake		498
142.	Fe_2O_3, hematite	molybdate	$3\ 10^{-4}$ mol dm^{-3}	5 g dm^3	1 d	23	3 10.5	0.01 1 mol dm^{-3} NaCl	uptake		499
143.	Fe_2O_3, synthetic, 2 samples	nitrate (III)	0.1 cm^3 saturated NaNO$_2$. 100 cm^3	10 mg 100 cm^3			4	10^{-3} mol dm^{-3} KCl	ζ	ζ -20 mV for hematite and maghemite.	568
144.	Fe_2O_3, hematite	phosphate	1 300 ppm P	1 g 10 cm^3	1 h, 2 d	4 45	7	K	uptake	Linear log-log adsorption isotherms for different reaction times. slope 0.3.	541
145.	Fe_2O_3, calcined ferrihydrite	phosphate	1 mg P 30 cm^3	0.1 g (0.7 m^2 g). 30 cm^3	2 d		4.2 7.5	acetate buffer 0.2 mol dm^{-3}	uptake	The uptake increases when pH increases.	506
146.	Fe_2O_3, hematite, Aldrich	phosphate	$2.6\ 10^{-3}$ mol dm^{-3}	8 g (14 m^2/g) dm^3	1 d		1.5 8	0.1 mol dm^{-3} NaCl	DR FTIR (ex situ)	Formation of iron phosphate.	571
147.	Fe_2O_3, hematite	phosphate	1.25 13.75 mg P dm^{-3}	0.02 g (47 m^2 g) 20 cm^3	90 d	25	3.8 9	0.1 mol dm^{-3} KCl	uptake	The uptake is depressed by 20% when the pH increases from 5 to 9.	572
148.	Fe_2O_3, hematite, Alfa	salicylate	10^{-3} 0.03 mol dm^{-3}				4 10		uptake		573
149.	Fe_2O_3, hematite	salicylate	$7.2\ 10^{-3}$ mol dm^{-3}	200 g dm^3		20	3 6	$1.6\ 10^{-2}$ mol dm^{-3} NaNO$_3$	uptake, ζ	Reinterpretation of data from Ref. 575; "umbrella effect": one salicylate anion covers 4 6 surface sites, excluding them from adsorption processes.	574
150.	Fe_2O_3, hematite, Alfa	salicylate	$5\ 7.2\ 10^{-3}$ mol dm^{-3}	200 g (8.8 m^2 g) dm^3	1.5 h	22	2 5-6	0.03 1 mol dm^{-3} NaCl	uptake, ζ	The uptake peaks at pH 3. The ionic strength effect is rather insignificant.	575
151.	Fe_2O_3, synthetic, dried silicate at 500 C	silicate	110 ppm SiO$_2$	1:25		25	2 11	toluene added	uptake		549
152.	Fe_2O_3, hematite	sulfate	$10^{-6}\ 2\times10^{-3}$ mol dm^{-3}				3 5		uptake, ATR-FTIR	Absorption bands at 1128, 1060 and 976 cm^{-1}. Monodentate. inner sphere complexes.	576
153.	Fe_2O_3, hematite	sulfate	$10^{-5}\ 10^{-2}$ mol dm^{-3}				4.08	0.01 mol dm^{-3} NaClO$_4$	FTIR, STM	Inner sphere monodentate surface complex.	577

TABLE 4.2 Continued

Adsorbent	Adsorbate	Total concentration	Solid to liquid ratio	Equilibration time	T	pH	Electrolyte	Method	Result	Ref.
154. Fe_2O_3, hematite, spherical, from chloride	citrate, oxalate	10^{-6} 10^{-2} mol dm^{-3}	100 mg (12.7 m^2 g) dm^{-3}	20 min	25, 23 or 28	3 11	0.01 mol dm^{-3} NaNO$_3$	uptake, ζ, release of iron	The uptake of oxalate increases when T increases. The IEP is shifted to pH 4.5 with 10^{-6} mol dm^{-3} citrate and to pH 2.5 with 10^{-3} mol dm^{-3} oxalate (pristine IEP at pH 7.6). Adsorption isotherm at pH 5.2 has a plateau for oxalate equilibrium concentrations of 1–4×10^{-3} mol dm^{-3}, but further increase of equilibrium concentration is accompanied by increase of the uptake.	578 579
155. Fe_2O_3, hematite, hydrolysis of nitrate	fluoride, formate	10^{-3} mol dm^{-3}		1 d	25	3–10	10^{-2} mol dm^{-3} KCl	ζ	Positive ζ is depressed by 20% in presence of fluoride and formate, no shift in the IEP.	580
156. Fe_2O_3, hematite	HEDP, mellitic acid, PAA, PPA, $P_3O_{10}^{5-}$	10^{-4} 10^{-2} mol dm^{-3}	1 g (6.5 m^2), 25 cm^3		25	2 12	none	uptake, ζ	Positive ζ potentials over the entire pH range, the values are higher than in the presence of NaCl. The adsorption isotherms of $P_3O_{10}^{5-}$, mellitic acid and HEDP at pH 7 have maxima, i.e. increase in the solution concentration above 10^{-3} mol dm^{-3} leads to lower uptake. Substantial adsorption of HEDP over the entire pH range with a maximum at pH 4 and minimum at pH 9.	581
157. Fe_2O_3, synthetic, 3 samples	phosphate, sulfate	2×10^{-5} 5×10^{-3} mol dm^{-3}	1.5 g (18 31 m^2 g) 40 cm^3	3 h	20	4 11		uptake, proton stoichiometry, coagulation	The PZC is shifted to pH 9.5 in the presence of 5×10^{-4} mol dm^{-3} sulfate. Adsorption isotherms of phosphate at constant pH are not Langmuir type. Complicated uptake (pH) curves.	175
158. $Fe_2O_3 \cdot H_2O$, from chloride	arsenate (V)	5×10^{-6} 5×10^{-3} mol dm^{-3}	0.005 mol Fe dm^{-3}	7 d	room	8	0.1 mol dm^{-3} NaNO$_3$	EXAFS	2 3 Fe atoms at 0.325 nm, no evidence of solid solution formation.	582
159. $Fe_2O_3 \cdot H_2O$, from chloride	arsenate V	10^{-4} mol dm^{-3}	0.0005 mol Fe dm^{-3}	1 d	7 9		0.1 mol dm^{-3} NaNO$_3$	uptake	40% uptake (rather insensitive to pH).	583
160. $Fe_2O_3 \cdot H_2O$ (hydrolysis of nitrate)	arsenate III and V	5×10^{-5} 2×10^{-4} mol dm^{-3}	10^{-3} mol Fe dm^{-3}	2 h	25	3 11	0.005–0.1 mol dm^{-3} NaNO$_3$	uptake ζ, FTIR	The effect of ionic strength on uptake curves is insignificant. The IEP is shifted to pH 4 for 10^{-4} mol dm^{-3} As(V).	584

No.	Adsorbent	Adsorbate	Concentration	Solid	Time	pH	T	Electrolyte	Method	Comments	Ref
161.	$Fe_2O_3 \cdot H_2O$, amorphous, from nitrate	arsenate III and V	$1.33 \ 10^{-6}$ mol dm^{-3}	$5 \ 10^{-5}$ mol Fe dm^{-3}	2 h	4 9		0.01 mol dm^{-3} NaNO$_3$ 0 0.003 mol dm^{-3} Ca	uptake	Uptake of As(III) is rather pH insensitive (50 80% uptake over the entire pH range). For As (V), pH$_{50}$ 9. The uptake of As V in slightly enhanced in the presence of Ca.	585
162.	$Fe_2O_3 \cdot H_2O$, (hydrolysis of nitrate)	arsenate III and V	$5 \ 3 \ 10^{-4}$ $2 \ 7 \ 10^{-3}$ mol dm^{-3}	2.5 g (202 m^2 g) dm^{-3}	1 d	3 11		0.1 mol dm^{-3} NaCl	uptake	Linear log log adsorption isotherms of As(III) at pH 4.6 and of As (V) at pH 9.2. Langmuir type isotherms for As(V) at pH 4.6 and of As (III) at pH 9.2. Almost 100% uptake of As(III) at pH 3 10 and of As(V) at pH 3 7 for 0.8 mol As. kg ferrihydrite.	586
163.	$Fe_2O_3 \cdot H_2O$, amorphous, from nitrate	borate	5 mg dm^{-3}	6.4 g (112 m^2 g) dm^{-3}	20 h	5 11		0.1 mol dm^{-3} NaCl	uptake	Uptake peaks at pH 8.5. Undissociated acid is a component. FeH$_2$BO$_3$ log K 5.6	486
164.	$Fe_2O_3 \cdot H_2O$	EDTA	$1 \ 1 \ 10^{-4}$ mol dm^{-3}	10^{-3} mol Fe dm^{-3}		5 9	25	0.01 mol dm^{-3} NaNO$_3$		The interaction results in dissolution of the adsorbent, which is slower in the presence of multivalent cations.	587
165.	$Fe_2O_3 \cdot H_2O$, amorphous	molybdate	$1 \ 10^{-5} \ 5 \ 10^{-3}$ mol dm^{-3}	0.64 g (223 m^2 g) dm^{-3}	1 d	3 11	23	NaCl	uptake ζ	The IEP is shifted to pH 3 in the presence of $5 \ 10^{-3}$ mol dm^{-3} molybdate.	498
166.	$Fe_2O_3 \cdot H_2O$, from nitrate	phosphate	$10^{-6} \ 4 \ 10^{-4}$ mol dm^{-3}	1.25 g dm^{-3}	7 d	6 9	24	0.01 2 mol dm^{-3} NaClO$_4$	uptake	Uptake at pH 7 is enhanced at high ionic strength.	588
167.	$Fe_2O_3 \cdot H_2O$	phosphate	$10 \ 10^{-3}$ mol dm^{-3}	720 m^2 dm^{-3}	3 h	6.8		0.005 mol dm^{-3} NaCl	uptake		589
168.	$Fe_2O_3 \cdot H_2O$, from nitrate	phosphate	10^{-4} mol dm^{-3}	89 mg (600 m^2 g) dm^{-3}	2 d	3 10		0.1 mol dm^{-3} NaNO$_3$ NaHCO$_3$ 10^{-6} mol dm^{-3} U	uptake	pH$_{50}$ 9 (in equilibrium with air).	218
169.	$Fe_2O_3 \cdot H_2O$	selenate IV and VI	10^{-4} mol dm^{-3}	$5 \ 10^{-4}$ mol Fe dm^{-3}	16 h	4 9		0.01 1 mol dm^{-3} NaNO$_3$	uptake	Divalent anions were chosen as components FeSeO$_3$ log K 14.45 FeOH$_2^+$ SeO$_4$ log K 9.6 ≡FeOH$_2^+$ ·HSeO$_4$ log K 14 5	590
170.	$Fe_2O_3 \cdot H_2O$	selenate IV and VI	$1.26 \ 10^{-5}$ mol dm^{-3}	1.5% Se w w with respect to solid	1 d	2.7,3.5		0.1 mol dm^{-3} NaNO$_3$	EXAFS	Bidentate, inner sphere complexes are formed.	591
171.	$Fe_2O_3 \cdot H_2O$	silicate	$1 \ 7 \ 10^{-5} \ 10^{-3}$ mol dm^{-3}	3.33 g (269 m^2 g) dm^{-3}	14 d	3 5		0.01 mol dm^{-3} NaNO$_3$	uptake FTIR	Si(OH)$_4$ is a component. FeOSi(OH)$_3$ log K 3.86	592
172.	$Fe_2O_3 \cdot H_2O$, from nitrate, chloride and sulfate, 9 different samples	sulfate	0.01 1 mol dm^{-3}			4 10	25	0.1 mol dm^{-3} NaCl	proton stoichiometry	Uncorrected PZC values in Na$_2$SO$_4$ and Na$_2$SO$_4$ NaCl mixtures are reported.	593

TABLE 4.2 Continued

Adsorbent	Adsorbate	Total concentration	Solid to liquid ratio	Equilibration time	T	pH	Electrolyte	Method	Result	Ref.
173. $Fe_2O_3 \cdot H_2O$	thiosulfate	$4 \, 10^{-7} \, 10^{-2}$ mol dm	10^{-3} mol Fe dm^3	2 4 h	25	3 6.5	0.1 mol dm^3 NaNO$_3$	uptake		48
174. $Fe_2O_3 \, H_2O$	thiosulfate	$4 \, 10^{-7} \, 4 \, 10^{-6}$ mol dm^{-3}						uptake	Divalent anion is a component $\equiv FeS_2O_3$ log K -7 $FeOH_2^+ \cdot S_2O_3$ log K -9.7	209
175. $Fe_2O_3 \, nH_2O$	arsenate (V), chromate (VI), molybdate, phosphate, selenate	$3.3 \, 10^{-4}$ mol dm	10^{-4} mol Fe dm^3	2 d	21	4 10	0.1 mol dm^{-3} NaClO$_4$	uptake	pH$_{50}$ 6(chromate), 7.3 (molybdate); 9 (selenate, phosphate, arsenate)	528
176. $Fe_2O_3 \cdot H_2O$, amorphous	arsenate III and V, chromate, selenate IV and VI, vanadate	$7 \, 10^{-5} \, 10$ mol dm	10 mol Fe dm^3		25	4.5 12	0.1 mol dm^3 NaNO$_3$, fly ash transport water	uptake	The pH$_{50}$(5 10^{-3} mol dm^3 of anion, 0.1 mol dm^3 NaNO$_3$) 8 6 for selenate IV, 5.6 for selenate VI, 7.2 for chromate, 10.5 for vanadate, 10.8 for arsenate V. Uptake of arsenate III peaks at pH 9 (95%)	226
177. $Fe_2O_3 \, H_2O$, from nitrate	arsenate, molybdate, phosphate, selenate VI and IV, silicate, sulfate	$5 \, 10 \, 7.5 \, 10^{-4}$ mol dm^3	1:800	2 d	24	6.5	0.1 mol dm^3 NaClO$_4$	uptake	Phosphate arsenate selenate IV silicate molybdate sulfate selenate VI	594
178. $Fe_2O_3 \, H_2O$, from chloride	arsenate, selenate, sulfate, tellurate	$10 \, 2 \, 10^{-6}$ mol dm^3	25 mg 15 cm^3	1 d	25	natural	Na	uptake, IR		595
179. $Fe_2O_3 \, H_2O$, from nitrate	arsenate, silicate	$5 \, 10 \, 1 \, 8 \, 10$ mol dm	$10^{-3} \, 3 \, 9 \, 10$ mol Fe dm^3	1 7 d	25	4 12	0.1 mol dm^3 NaNO$_3$	uptake	Uptake of silicate peaks at pH 9 10, 90%; uptake of As III (5 10 mol dm^3 As, 10^{-3} mol Fe dm^3) peaks at pH 5 10.	596
180. $Fe_2O_3 \, H_2O$, from nitrate	borate, silicate	$5 \, 10^{-4} \, 1 \, 2 \, 10$ mol dm^3	10^{-1} mol Fe 30 cm	12 h				uptake	Langmuir type adsorption isotherm of borate and silicate (pH not specified)	556
181. $Fe_2O_3 \, H_2O$	cacodylate, methyl arsenate	$5 \, 10 \, 7 \, 10^{-4}$ mol dm	$5 \, 10^{-3}$ mol Fe dm	2 d	20	4 11	0.015 0.1 mol dm NaNO$_3$	uptake	The effect of ionic strength on uptake curves is insignificant for cacodylate, the uptake of methyl arsenate at pH 6 is slightly depressed at high ionic strengths.	519

No.	Adsorbent	Adsorbate	Adsorbate conc.	Solid	Time	T	pH	Electrolyte	Method	Comments	Ref.
182.	Fe$_2$O$_3$·H$_2$O	chromate, selenate VI	2 10-10 mol dm^{-3}	10^{-3} mol Fe dm^{-3}	1 h	25	4 8	0.1 mol dm^{-3} NaNO$_3$	uptake	For 6.7 10^{-5} mol dm^{-3} methyl arsenate pH$_{50}$ 10.2. For 5 10^{-5} mol dm^{-3} cacodylate pH$_{50}$ 8.3.	597
183.	Fe$_2$O$_3$·H$_2$O	protocatechuate, salicylate, sulfate, thiosulfate	4 10-10^{-3} mol dm^{-3}	10^{-3} mol Fe dm^{-3}, 182 m^2 g	2 h	25	4 10	0.1 mol dm^{-3} NaNO$_3$	uptake	The pH$_{50}$ 7.3 (5 10^{-7} mol dm^{-3} Cr); 5.8 (2 10^{-7} mol dm^{-3} Se). Broad maximum of protocatechuate adsorption at pH 5 7.	222
184.	FeOOH, 3 samples α, β and	arsenate	5 10 mol dm	30 g (50, 140, and 54 m^2 g) dm	4 d	room	8	0.1 mol dm^{-3} NaNO$_3$	EXAFS	2.16 2.45 Fe atoms at 0.33 nm. no evidence of solid solution formation.	582
185.	FeOOH, goethite	arsenate	10 mg As dm^{-3}	0.1 2 g (132 m^2 g) dm	1 d	25	3 12	0 0.05 mol dm^{-3} KNO$_3$	uptake, ζ	The IEP shifts to pH 6.2; pH$_{50}$ 10.5 (1 g dm^3), the uptake is enhanced at high ionic strength. The uptake is depressed as T increases (25 45 C).	598
186.	FeOOH, goethite	As(III)	6 5 10^{-5} 1 3 10 mol dm	2.5 g (45 m^2 g) dm^3	16 h		3 11	0.00 mol dm^{-3} NaCl	EXAFS uptake	As-Fe distance of 0.338 nm. Inner sphere complex stable towards oxidation to As(V). As(OH)$_3$ is a component. ($\equiv$Fe)$_2$HAsO$_3$, log K 9 22; ($\equiv$Fe)$_2$AsO$_3$, log K 0.51.	599
187.	FeOOH, goethite, synthetic	borate	5 mg dm^3	25 g (31 m^2 g) dm^3	20 h		5 11	0.1 mol dm^{-3} NaCl	uptake	Uptake peaks at pH 7.5 Undissociated acid is a component, $\equiv$FeH$_2$BO$_3$ log K 5.2	486
188.	FeOOH, goethite	borate	0.01 0.08 mol dm^{-3}		36 h		5 11	1 mol dm^{-3} NaCl	uptake	Uptake peaks at pH 8; the uptake at pH 6 is severely depressed in the presence of 0.05 mol dm^{-3} mannitol.	600
189.	FeOOH, goethite	borate	5,250 mg B dm^3	0.375 g (42.5 m^2 g) 15 cm^3	1 d		3 11	0.001 1 mol dm^{-3} NaCl	uptake, ζ	The IEP is shifted to pH 7 1 at 250 mg B dm^3. Uptake peaks at pH 7.5 and it is rather insensitive to the ionic strength.	536
190.	FeOOH, goethite	borate	1 100 mg dm	0.375 g (42.5 m^2 g) 15 cm^3	2 h	25	6 11	0.1 mol dm^{-3} NaCl	uptake	L ptake peaks at pH 8.5 The uptake is depressed when T increases	537
191.	FeOOH, goethite	carbonate	0.2 g dm				3.5 10.5	10^{-2} mol dm^{-1} NaCl	ζ	IEP at pH 8.0 in dispersion titrated with Na$_2$CO$_3$ (pristine IEP at pH 9.3).	540
192.	FeOOH, goethite	carbonate	1 mol dm^{-3}	100 g dm^{-3}	1 d		5 10	Na	ATR-FTIR	$\equiv$FeOCO$_2^-$ is formed.	540
193.	FeOOH, amorphous	chromate	8.7 10^{-4}, 1.7 10^{-2} mol Fe dm		4 h	25		0.1 mol dm^{-3} NaNO$_3$, 2.5 10^{-3} mol dm^{-3} Ca, Mg	uptake	CrO$_4^{2-}$ is a component. $\equiv$FeOH$_2$·CrO_4^{2-} log K 10.1 $\equiv$FeOH$_2$·$HCrO_4$ log K 19.3 The uptake is slightly enhanced in the presence of Ca or Mg.	601

TABLE 4.2 Continued

	Adsorbent	Adsorbate	Total concentration	Solid to liquid ratio	Equilibration time	T	pH	Electrolyte	Method	Result	Ref.
194.	FeOOH, goethite, 2 samples	chromate	10 mg dm^{-3}	1 g (130 m^2/g) dm^{-3}	1 d	25	2-10	none	uptake		265
195.	FeOOH, goethite, from nitrate	citrate	5×10^{-7}-10^{-4} mol dm^{-3}	1 g(29 m^2/g)/dm^3	1 d	22	3-11	0.1 mol dm^{-3} NaCl, 10^{-6} mol dm^{-3} U	uptake	The pH$_{50}$ = 10.5 (10^{-6} mol dm^{-3} citrate), the effect of U on citrate adsorption is insignificant.	144
196.	FeOOH, goethite	EDTA	10^{-5}-10^{-3} mol dm^{-3}	0.3 g (68 m^2/g)/25 cm^3	2 d	$25,30$	3-11	10^{-3}-10^{-2} mol dm^{-3} KNO$_3$	uptake, ζ	At 10^{-4} mol dm^{-3} EDTA the IEP is shifted to pH 5.5, at 10^{-3} mol dm^{-3} EDTA only negative ζ.	602
197.	FeOOH	EDTA	10^{-5} mol dm^{-3}	0.009 mol Fe dm^{-3}	1 d		4-10	0.1 mol dm^{-3} NaNO$_3$	uptake	Scattered data points at pH 6-8, pH$_{50}$ = 6.2.	250
198.	FeOOH, goethite	EDTA	4.6×10^{-7}-10^{-5} mol dm^{-3}	0.12-0.46 g(21 m^2/g)/dm^3	30 min		5-10	0.01 mol dm^{-3} NaNO$_3$	uptake	pH$_{50}$ = 7.8 (0.46 g/dm^3, 4.6×10^{-7} mol dm^{-3})	266
199.	FeOOH, goethite	EDTA	1.2×10^{-5} mol dm^{-3}	0.12 g (21 m^2/g)/dm^3		25	3-8	0.01 mol dm^{-3} NaNO$_3$	uptake	The interaction results in dissolution of the adsorbent, which is slower in the presence of multivalent cations.	587
200.	FeOOH, goethite	fluoride	10^{-6}-10^{-3} mol dm^{-3}		30 min	22.5	4-6	0.1 mol dm^{-3} NaNO$_3$	uptake	FeOH$^{-1 2}$ + H$^+$ + F$^-$ = FeF$^{-1 2}$ + H$_2$O log K 8.2	603
201.	FeOOH, goethite	malonate	10^{-4}-10^{-3} mol dm^{-3}				3-7	0.01 mol dm^{-3} NaNO$_3$	uptake	$\equiv$FeOH$^{-1 2}$ and divalent anion are the components. ($\equiv$Fe)$_2$Mal$^-$ log K 17.8 ($\equiv$Fe)$_2$MalNa log K 19.8 ($\equiv$Fe)$_3$Mal$^{0.5-}$ log K 26.7	604
202.	FeOOH, goethite	methylphosphonic acid	6×10^{-4}-2×10^{-3} mol dm^{-3}	13.3 g (80 m^2 g)/dm^3	1 d	25	3.5-9	0.01 mol dm^{-3} NaCl	uptake, ζ, ATR-FTIR	At pH 9: Langmuir constant 0.01 dm^3/μmol, maximum adsorption 84 μmol g; pH 3.5: Langmuir constant 0.33 dm^3/μmol, maximum adsorption 182 μmol g. The IEP is shifted to pH 6 at adsorption density of 130 μmol g Monodentate protonated surface species at low pH and high Γ. Bidentiate surface species at high pH and low Γ.	605
203.	FeOOH, goethite, synthetic	molybdate	1 200 mg Mo/dm^3	50-400 mg (67 m^2 g) 25 cm^3	1 d		2-10		uptake		497

No.	Adsorbent	Adsorbate	Adsorbate concentration	Adsorbent (surface area)	t	T	pH	Electrolyte	Method	Comments	Ref.
204.	FeOOH, goethite, poorly crystalline	molybdate	1×10^{-5} 5×10^{-5} mol dm^{-3}	0.64 g (149 m^2 g)/cm^3	1 d	23	3-11	NaCl	uptake	The IEP is shifted to pH 3 in the presence of 5×10^{-5} mol dm^{-3} molybdate.	498
205.	FeOOH, goethite	molybdate	1×10^{-5}, 5×10^{-5} mol dm^{-3}	1.25 g (64 m^2 g)/cm^3	1 d	23	3-11	NaCl	ζ	The sign of ζ is reversed to negative at pH 4-6 in the presence of 1×10^{-5} mol dm^{-3} molybdate.	498
206.	FeOOH, goethite, 2 samples	molybdate	3×10^{-4} mol dm^{-3}	0.64-1.25 g dm^{-3}	1 d	23	3-10.5	0.01 1 mol dm^{-3} NaCl	uptake		499
207.	FeOOH, goethite, synthetic	nitrate (III)	0.1 cm^3 saturated NaNO$_2$, 100 cm^3	10 mg, 100 cm^3			4	10^{-3} mol dm^{-3} KCl	ζ	-20 mV	568
208.	FeOOH, goethite, from nitrate	oxalate	10^{-5} mol dm^{-3}	1 g(34 m^2), 400 cm^3	3 h	20, 25	3-8	0.001 mol dm^{-3} KNO$_2$, 0-10^{-5} mol dm^{-3} Cd	uptake, ζ proton stoichiometry	The pH$_{50}$ = 7.7 in the presence and absence of cadmium. IEP is shifted to pH 7.2 in the presence of oxalic acid (pristine IEP at pH 8.6).	234
209.	FeOOH, goethite	phosphate	5×10^{-5} 10^{-4} mol g goethite	100 g dm^3 (80 m^2/g)	1 d		4	0.01 mol dm^{-3} NaCl	ATR-cylindrical internal reflection	Four possible structures of surface species (mono- bi- and tridentate) were proposed.	606
210.	FeOOH, goethite	phosphate	3×10^{-5} mol dm^{-3}	0.25 g dm^3	1 d	20±2	3-10	0.042-0.7 mol dm^{-3} NaCl	uptake	Uptake is not affected by the ionic strength or by addition of 0.05 mol kg^{-1} Mg. uptake is enhanced in the presence of 0.01 mol kg^{-1} Ca.	607
211.	FeOOH, goethite, from nitrate	phosphate	10^{-4} mol/g goethite 2.8×10^{-6} mol m^2	56 mg (84 m^2/g)/dm^3	80 d	22	3-9	0.005 mol dm^{-3} NaClO$_4$	ζ	IEP at pH 4.4 (unaged) and 4.9 (80 d aged).	608
212.	FeOOH, goethite, from nitrate	phosphate		7.2 g/39 m^2/g, dm^3	1, 8 d		3-13	0.1 mol dm^{-3} NaCl, NaNO$_3$	DR FTIR (ex situ)	Three monodentate complexes differing in degree of protonation.	571
213.	FeOOH, goethite	phosphate	10^{-5} mol dm^{-3}	10^{-3} mol Fe dm^{-3}	1 d	23	2 8.5	sea water	uptake	The uptake peaks (85%) at pH 6. Trivalent anion and FeOH^{-2} were selected as the components. $\equiv$FeO°PO$_3^q$ p+q=-2.5 log K 20.8 ($\equiv$FeO)$_2$°PO$_2^q$ p+q -2 log K 29.2 $\equiv$FeO$_2$°POOH^q p+q=-1 log K 35.4	609
214.	FeOOH, goethite	phosphate	10^{-5}-6×10^{-4} mol dm^{-3}		1 d	20	4-11	0.01 mol dm^{-3} NaNO$_3$	uptake, proton stoichiometry		610
215.	FeOOH, goethite, from nitrate	phosphate	5×10^{-4} 2×10^{-3} mol dm^{-3}	6 g (95 m^2/g) dm^3	1 d		3-10	0.1 mol dm^{-3} NaNO$_3$	uptake		238
216.	FeOOH, goethite, synthetic	selenate IV and VI	2×10^{-7} 10^{-5} mol dm^{-3}	3 300 mg (49 m^2 g)/dm^3	1 d	27	4-11	0.1 mol dm^{-3} KCl	uptake	The pH$_{50}$ ~7.5 for 2.2-6.5×10^{-7} mol dm^{-3} Se(IV) and 30 mg/dm^3. pH$_{50}$ <4.5 for 6.7×10^{-7} mol dm^{-3} Se(VI) and 300 mg/dm^3.	611

TABLE 4.2 Continued

Adsorbent	Adsorbate	Total concentration	Solid to liquid ratio	Equilib-ration time	T	pH	Electrolyte	Method	Result	Ref.
217. FeOOH, goethite	selenate IV and VI	10^{-4} mol dm^{-3}	30 g (52 m^2 g) dm^{-3}	16 h		4–9	0.001 1 mol dm^{-3} NaNO$_3$	uptake	The uptake of Se(IV) is depressed as T increases. ΔH_{ads} -20 to -80 KJ mol at pH 6.7 and depends on the surface coverage. Divalent anions were chosen as components. $\equiv FeSeO_3^-\cdots Na$ log K 14.1 $\equiv FeOH_2^+\cdots SeO_4^{2-}$ log K 8.9 $\equiv FeOH_2^+\cdots HSeO_4^-$ log K 15.7	590
218. FeOOH, goethite	selenate VI	1.26×10^{-5} mol dm^{-3}	0.1% Se w w with respect to solid (66 m^2 g)	1 d		3.5	0.1 mol dm^{-3} NaNO$_3$	EXAFS	Bidentate, inner sphere complexes are formed.	591
219. FeOOH, goethite	sulfate	3×10^{-4} 2×10^{-3} mol dm^{-3}	12.5 g (21 m^2 g) dm^{-3}	1 d	21	3 9	0.01, 0.1 mol dm^{-3} NaNO$_3$, 3×10^{-5} mol dm^{-3} Cd	uptake	The uptake is depressed when the ionic strength increases; the uptake at pH < 5 is enhanced in the presence of Cd.	235
220. FeOOH, synthetic	sulfate	3 9×10^{-4} mol dm^{-3}	11 g (39 m^2 g) dm^{-3}	2 d	25	3 8.5	0.1 mol dm^{-3} NaNO$_3$, 0.1 2 mol dm^{-3} NaCl	uptake proton stoichiometry, diffuse reflectance IR	Divalent anion is taken as a component. $\equiv FeOH_2^+\cdots L^{2-}$ log K 8.3 $\equiv FeOH_2^+\cdots HL^-$ log K 13.5 The uptake at pH 3 drops when the ionic strength (NaCl) increases.	612
221. FeOOH, goethite	sulfate	5×10^{-6} 5 10^{-4} mol dm^{-3}				3.5 9	0.005 0.1 mol dm^{-3} NaCl	ATR FTIR	Inner and outer sphere complex at pH < 6, only outer sphere at pH > 6. High ionic strength promotes formation of inner sphere complex.	613
222. FeOOH, goethite, from nitrate	sulfate	10^{-5} 10^{-1} mol dm^{-3}	0.05 15 g (96.4 m^2 g) dm^{-3}	1 d		3 5	0.01 1 mol dm^{-3} NaNO$_3$	uptake, proton stoichiometry, ζ	Divalent anion and $\equiv FeOH_2^{+\,1/2}$ are taken as components. $\equiv FeO^{1/2+}\cdots SO_3^{-1}$ log K 1.07; z0 0.35; z1 1.65.	614
223. FeOOH, goethite, synthetic	acetate, fluoride, phosphate, silicate, sulfate	2 10^{-4} 1 10^{-3} mol dm^{-3}	3.2 6 g (29 m g) dm^{-3}			3.5 10	0.1 mol dm^{-3} NaClO$_4$	proton stoichiometry, uptake	1 OH group released per 1 adsorbed F$^-$. For 2×10^{-4} mol dm^{-3} silicate or phosphate the uptake is independent of pH. The pH$_{50}$ 5.5 (fluoride), 4.5 (sulfate).	615

No.	adsorbent	adsorbate	concentration	solid	time	T	pH	electrolyte	method	comments	ref.
224.	FeOOH, goethite	AMP, DTPMP, EDTMP, HEDP, IDMP, NTMP	$1\ 10$ mol dm	0.42 g (47.6 m g) dm³	1 min	25	4 12	0.01 mol dm NaNO₃; 0.10 mol dm Ca, Cu, Zn, Fe III	uptake	The pH₅₀: 8.2 for AMP, 9.5 for DTPMP, 9.7 for EDTMP, 10 for IDMP and NTMP, and 10.2 for HEDP. The uptake of NTMP is enhanced in the presence of Ca (90% uptake at pH 5.12 with 10^{-3} mol dm³ Ca) and to less extent in the presence of Zn	243
225.	FeOOH, goethite. synthetic. 3 samples	arsenate (V), chromate (VI), molybdate, phosphate, selenate	$3.3\ 10^{-4}$ mol dm³	10^{-4} mol Fe dm³	2 d	21	4 10	0.1 mol dm⁻³ NaClO₄	uptake	Uptake of chromate peaks at pH 5, uptake of phosphate peaks at pH 4.5; pH₅₀ 9.5 for selenate and arsenate, and 6.6 for molybdate for one sample of the adsorbent. Completely different behavior of two other samples.	528
226.	FeOOH, goethite.	arsenate (V), molybdate, phosphate	$1.3, 2.6\ 10^{-4}$ mol dm³	2.5 g dm³	4 h	22	2 11	0.1 mol dm NaCl	ζ, uptake	IEP shifts to pH 3 in the presence of $1.33\ 10^{-4}$ mol dm³ As. pH₅₀ ($1.33\ 10^{-4}$ mol dm³) 10.6 for arsenate, 7 for molybdate and 11 for phosphate.	552
227.	FeOOH, goethite. 2 samples	arsenate (V), phosphate, selenite (IV)	$3\ 10^{-4}\ 5.2\ 10$ mol dm	93, 109 mg (60. 81 m g) 25 cm	1 d	20	3 10	0.1 mol dm NaCl	uptake	Uptake decreases when pH increases but the difference between pH 3 10 is only by a factor of 2 for phosphates and arsenates (V).	553
228.	FeOOH, goethite	arsenate (V), phosphate, selenite (IV)	10 mol dm	93, 109 mg (60. 81 m² g) 25 cm	1 d	20	3 10	0.1 mol dm NaCl	uptake, EXAFS	Reinterpretation of data from Ref. 552; fully dissociated anions and FeOH¹·⁵ are selected as the components. ≡FeOAsO₃ log K 20 f 0.25 (≡FeO)₂AsO₂ log K 27.9 f 0.5 (≡FeO) AsOOH⁻¹ log K 34.5 f 0.6 Analogous complexes of P with log K 21.2, 30, and 36.2 (FeO)₂SeO⁻¹ log K 24.6 f 0.67 (FeO) SeOH⁰ log K 29.7 f 0.75	616
229.	FeOOH, goethite	benzene carboxylic acids	10^{-4} mol dm		16 h, 7 d	25	3 10		uptake	The pH₅₀ 6.2 (trimesic); 6.7 (phthalic) and 9.2 (mellitic acid); the uptake of mellitic acid does not reach 100%. cf. Fig. 4.6(A).	239
230.	FeOOH, goethite	benzoate, oxalate	$10^{-4}\ 10$ mol dm	0.11 g 20 cm³		3.3		0.1 mol dm NaCl	ζ, IR		554

TABLE 4.2 Continued

Adsorbent	Adsorbate	Total concentration	Solid to liquid ratio	Equilibration time	T	pH	Electrolyte	Method	Result	Ref.
231. FeOOH, goethite	benzoate, oxalate, phosphate	3×10^{-4} mol dm^{-3}	2.5 g (34 m^2 g) dm^3	1 d	20	3 9	0.001 mol dm^{-3} KNO$_3$	uptake, ζ	Preadsorbed oxalate and benzoate is released when the pH increases, but preadsorbed phosphate is not. The IEP is not shifted in the presence of 0.0001 mol dm^{-3} benzoate, but presence of 10^{-5} mol dm^{-3} of oxalate induced a shift by -1 pH unit. Presence of 10^{-6} mol dm^{-3} of phosphate induces sign reversal of ζ to negative at pH 4.2.	617
232. FeOOH, goethite, from nitrate	carbonate, chromate	5×10^{-6}, 5×10^{-5} mol dm^{-3}	2, 10 g (45 m^2 g) dm^3	2 h		3 10	0.1 mol dm^{-3} NaClO$_4$	uptake	Uptake of carbonate peaks at pH 6. Divalent anions are components. ≡FeCO$_3^-$ log K 12.45 ≡FeCrO$_4^-$ log K 12.82	618
233. FeOOH, goethite	chelidamic acid, phthalic acid, sulfate	2.5×10^{-5} 10^{-3} mol dm^{-3}	1.6 g (79 m^2 g) dm^3	1 d	20	3 8	0.01, 0.1 mol dm^{-3} NaCl	uptake	The uptake is depressed at higher ionic strength. Divalent anions are components. ≡FeHSO$_4$ log K 13.83 ≡FeSO$_4^-$ log K 8.41 ≡FeOSO$_4^{3-}$ log K -6.04 phthalic acid H$_2$L ≡FeHL log K 15.74 ≡FeOHL^{2-} log K 2.17 chelidamic acid H$_3$L ≡FeH$_2$L log K 14.05 ≡FeHL log K 7.92 ≡FeOHL^{3-} log K -6.24	619
234. FeOOH, goethite	chromate, oxalate	5×10^{-6} 5×10^{-3} mol dm^{-3}	0.18 1.8 g (66 m^2 g) dm^3	10 h	25	4 11	0.01 0.5 mol dm^{-3} KNO$_3$	uptake, proton stoichiometry	One proton is coadsorbed with oxalate at pH 6 and 0.7 at pH 4. 0.8 proton is coadsorbed with chromate at pH 6 and 0.4 at pH 4. Increase in ionic strength depresses uptake of chromate and even more severely the uptake of oxalate. Most experimental results (except for the ionic strength effect on oxalate uptake) were successfully explained by the following model: Divalent anions are components.	620

No.	Adsorbent	Adsorbate	Concentration	Solid (surface area)	Time	T	pH	Electrolyte	Method	Comments	Ref.
235.	FeOOH, α, synthetic	citrate, lactate, l- and meso-tartrate	10^{-5} 10^{-4} mol dm^{-3}	0.1 g (70 m^2 g) 25 cm^3	16 h	25	3 10	0.1 mol dm^{-3} NaNO$_3$	uptake, IR	≡FeOxH log K 15.7 ≡FeOx⁻ log K 9.1 ≡FeOOx³⁻ log K -5.9 ≡FeCrO$_4$H log K 18.7 ≡FeCrO$_4$⁻ log K 11.6 ≡FeOHCrO$_4$²⁻ log K 3.4 Meso-tartrate L-tartrate citrate.	621
236.	FeOOH, goethite	fluoride, iodate, phosphate, salicylate	2 10^{-7} 6×10^{-3} mol dm^{-3}	10 g (81 m^2 g) dm^3	4 min	25	4	0.05 mol dm^{-3} NaNO$_3$	uptake, proton stoichiometry	Proton stoichiometry factor is strongly dependent on the surface coverage. with phosphate it varies from negative values (proton release) to above 1. Negative enthalpies of adsorption at low surface coverage (determined by calorimetry and from T dependence of adsorption, 10-40 C) become less negative and even slightly positive when the uptake is high enough.	622
237.	FeOOH, goethite	fluoride, molybdate, phosphate, selenate (IV), silicate, sulfate	10^{-6}–4×10^{-6} mol dm^{-3}	0.1 g 25 cm^3	1 d	20-23	2 12	0.1 mol dm^{-3} NaCl	uptake, proton stoichiometry	Uptake of fluoride peaks at pH 4, uptake of silicate peaks at pH 9. For selenate and molybdate the slope of the uptake (pH) curves is more negative at pH > pK$_{a2}$ than at pH < pK$_{a2}$.	29
238.	FeOOH, amorphous	molybdate, phosphate, selenate IV and VI, silicate	7×10^{-7} 10^{-5} mol dm^{-3}	4.4 264mg (600 m^2 g) dm^3			4-10	0.1 mol dm^{-3} KCl	uptake	The uptake(pH) curve for silicate was cation like. Anion like uptake curves for the other anions, pHso 7.2 (10^{-5} mol dm^{-3} Mo, 88 mg FeOOH dm^3), 10.6 (5 · 10^{-6} mol dm^{-3} P, 88 mg FeOOH dm^3), 6.3 (7×10^{-7} mol dm^3), 10.1 (7 10^{-7} mol dm^{-3} Se IV, 264 mg FeOOH dm^3) Se VI, 264 mg FeOOH dm^3).	623
239.	FeOOH, goethite, synthetic	phosphate, o-phthalate	4×10^{-4} 1.6×10^{-3} mol dm^{-3}	10 g (43 m^2 g) dm^3	1 d	25	3 8.5	0.1 mol dm^{-3} NaNO$_3$	uptake proton stoichiometry, diffuse reflectance IR	Monovalent anion is taken as component. Phthalate: ≡FeOH$_2^+$·L²⁻ log K 4.39 ≡FeOH$_2^+$·· HL⁻ log K 10.38 The uptake at pH 3.3 drops when the ionic strength increases.	624
240.	FeOOH, goethite	phosphate, salicylate		10, 100 g (81 m^2 g) dm^3			1 6-6.5	0.05 mol dm^{-3} NaNO$_3$ K	calorimetry, CIR-FTIR	Salicylate forms a chelate structure with surface iron atom.	625
241.	FeOOH, goethite	phosphate, sulfate	10^{-5} 10^{-1} mol dm^{-3}	20 m^2 100 cm^3	10 min	25	5 10	K	proton stoichiometry		562

TABLE 4.2 Continued

Adsorbent	Adsorbate	Total concentration	Solid to liquid ratio	Equilib-ration time	T	pH	Electrolyte	Method	Result	Ref.
242. FeOOH, goethite, from nitrate	phosphate, sulfate	10^{-8}–10^{-3} mol dm^{-3}	0.2–16 g (100 m^2 g) dm^3	1 d		2.5–10	0.01–0.5 mol dm^{-3} NaNO$_3$, KNO$_3$	uptake	Fully deprotonated anions and $\equiv$FeOH[1] surface groups are the components. $(\equiv FeO)_2PO_2^{2-}$ log K 30 f 0.55 $(\ FeO)_2POOH$ log K 35.5 f 0.6 $\equiv FeOPO_3^{2-}$ log K 20.5 f 0.24 $(\equiv FeO)_2SO_2^-$ log K 19.5 f 0.62 Nonlinear log-log adsorption isotherms at constant pH.	626
243. FeOOH, goethite	24 organic acids: derivatives of benzene and naphthalene	10^{-6}–10^{-3} mol dm^{-3}	1.6 g (79 m^2 g) dm^3	1 d		3–10	0.01 mol dm^{-3} NaCl	uptake	The uptake in a series of derivatives of benzene increases with the number of carboxylic groups and with the length of hydrocarbon chain. There is no systematic effect of the ring size (benzene vs. naphthalene).	627
244. FeOOH, goethite	24 organic acids: derivatives of benzene and naphthalene	10^{-6}–10^{-3} mol dm^{-3}	1.6 g (79 m^2 g) dm^3	1 d		3–10	0.01 mol dm^{-3} NaCl	uptake	The adsorption edges were successfully modeled using (two to six surface species. In the log K values listed below a monovalent anion was always chosen as a component. Mellitic acid H$_6$L $\equiv FeH_4L$ log K 12 $\equiv FeH_3L$ log K 11.4 $-FeH_2L^3$ log K 8.11 $-FeHL^4$ log K 2.16 $-FeL$ log K -2.17 Pyromellitic acid H$_4$L $\equiv FeH_3L$ log K 11.91 $\equiv FeH_2L$ log K 9.15 $\equiv FeHL$ log K 4.7 $\equiv FeL^3$ log K -1.25 $\equiv FeOHL^4$ log K -6.61 FeOL log K -13.12 Hemimellitic acid H$_3$L $\equiv FeH_2L$ log K 11.03 $\equiv FeHL$ log K 7.55	628

≡FeL²⁻ log K 2.29
≡FeOL⁴⁻ log K -14.63
Trimellitic acid H₃L
≡FeH₂L log K 11.37
≡FeHL⁻ log K 6.91
≡FeOHL³⁻ log K -5.68
≡FeOL⁴⁻ log K -13.02
Trimesic acid H₃L
≡FeH₂L log K 11.5
≡FeOHL³⁻ log K -7.35
≡FeOL⁴⁻ log K -14
Salicylic acid H₂L
≡FeHL log K 8.55
≡FeOL³⁻ log K -12.92
Dihydroxybenzoic acids H₃L
2,3;2,6; and 3,4 isomers
≡FeH₂L log K 9.17; 8.4; and 9.85
≡FeHL⁻ log K 3.36; 1.51; and 4.3
≡FeOL⁴⁻ log K -16.96; -19.72; and -16.61.
2,4; and 2,5 isomers
≡FeH₂L log K 8.79; and 8.72
≡FeOL⁴⁻ log K -19.57; and -19.5.
Trihydroxybenzoic acids H₄L
2,3,4; and 3,4,5 isomers
≡FeH₃L log K 9.63; and 9.8
≡FeH₂L log K 3.16; and 3.27
2,4,6 isomer
≡FeH₃L log K 7.79
≡FeH₂L⁻ log K 1.32
≡FeOHL⁴⁻ log K -20.02
3, 4 dihydroxyphenylacetic acid (?
probably typographical error)
hydroxy-2-naphthoic acids H₂L. I-; and
3-isomers
≡FeHL⁻ log K 9.09; and 9.46
≡FeOL³⁻ log K - 12.99; and -12.52
With benzoic, 1-naphthoic, 3,5
dihydroxybenzoic, and 6-phenylhexanoic
acids, formation of surface complexes
involving two (or three) molecules of
organic acid per one adsorption site was
considered.

TABLE 4.2 Continued

Adsorbent	Adsorbate	Total concentration	Solid to liquid ratio	Equilibration time	T	pH	Electrolyte	Method	Result	Ref.
245. FeOOH, goethite	8 different phosphonates	10^{-5}, 4×10^{-5} mol dm^{-3}	0.42 g (48 m^2/g)/dm^3	1 h	25	3.5-12	0.001 1 mol dm^{-3} NaNO$_3$	uptake	The effect of ionic strength on the uptake of nitrilotris-(methylenephosphonic) acid is negligible. Uptake of methylphosphonic, aminomethylphosphonic, hydroxymethyl-phosphonic, 1-hydroxyethane-(1, 1-diphosphonic), iminodi-(methylphosphonic), nitrilotris-(methylenephosphonic), ethylenedimitrilotetrakis-(methylenephosphonic), and diethylenetrinitrilopentakis-(methylenephosphonic) acids was interpreted in terms of formation of 2 9 different surface species whose stability constants are interrelated, namely, log K − (11.45 + 7.31 n)−(2.53 + 0.46 n) Z where n is the surface protonation level and Z is the surface complex charge, and fully deprotonated anions are components. Adsorption isotherms at constant pH were also obtained in the presence of buffers.	629
246. Fe(OH)$_3$, amorphous	carbonate		0.2 g dm^{-3}			3.5-10.5	10^{-2} mol dm^{-3} NaCl	ζ	IEP at pH 7.2 in dispersion titrated with Na$_2$CO$_3$, pristine IEP at pH 8.5.	540
247. Fe(OH)$_3$, amorphous	carbonate	0.01 1 mol dm^{-3}	100 g dm^{-3}	1 d		3.5-10.5		ATR-FTIR	≡FeOCO$_2^-$ at pH 3.5-10.5 ≡FeOCO$_2$H at pH 3.5-6.2	540
248. Fe(OH)$_3$, β	chromate	10 mg dm^{-3}	1 g (468 m^2/g) dm^3	1 d	25	2 10	none	uptake		265
249. Fe(OH)$_3$, amorphous	molybdate	3×10^{-4} mol dm^{-3}	0.64 g dm^{-3}	1 d	23	3 10.5	0.01 1 mol dm^{-3} NaCl	uptake, ATR FTIR, DRIFT	Absorption bands of adsorbed Mo species at 928 and 880 cm^{-1}	499
250. Fe(OH)$_3$, freshly precipitated and 1 d aged	phosphate	10^{-5} 3×10^{-4} mol dm^{-3}	5×10^{-4} 5×10^{-3} mol Fe dm^3	15 min		5-8		uptake		546
251. Fe(OH)$_3$	phosphate	10^{-6}-0.002 mol dm^{-3}	0.2 g 25 cm^3	3 d	25	5.8	0.01 mol dm^{-3} NaCl	uptake	Linear log-log adsorption isotherm, slope 0.1.	548

No.	Adsorbent	Adsorbate	Concentration	Solid to liquid ratio	Time	T	pH	Electrolyte	Method	Result	Ref.
252.	Fe(OH)$_3$, freshly precipitated	phosphate	6.1×10^{-6} mol dm^{-3}	2×10^{-4} 2×10^{-2} mol Fe 25 cm^3	25 min	22	2 14	1 mol cm^{-3} NaNO$_3$	uptake	Complete uptake at pH 3-11 at 2×10^{-2} mol Fe. 25 cm^3. The low pH adsorption edge is only slightly affected by the solid to liquid ratio. The high pH adsorption is severely reduced when the solid to liquid ratio < 2×10^{-3} mol Fe. 25 cm^3.	630
253.	Fe(OH)$_3$, amorphous, from sulfate	phosphate	10^{-6} 10^{-3} mol dm^{-3}	0.13 g (72 m^2/g) dm^3	2 d	25	3.4-10.2	0.01 mol dm^{-3} NaClO$_4$	uptake, ζ		36
254.	Fe(OH)$_3$, amorphous, from nitrate	citrate, lactate, 1- and meso-tartrate	10^{-5} 10^{-4} mol dm^{-3}	0.1 g (182 m^2/g) 25 cm^3	16 h	25	3-10	0.1 mol dm^{-3} NaNO$_3$	uptake, IR		621
255.	Fe(OH)$_3$	phosphate, selenate	10^{-6} mol dm^{-3}	10^{-3} mol Fe dm^{-3}	2 h		5.5-10	0.1 mol dm^{-3} NaNO$_3$	uptake	The pH$_{50}$ ~ 9.5 for phosphates.	151
256.	Fe(OH)$_3$, amorphous	phosphate, selenate VI. sulfate	0.006-0.03 mol dm^{-3}	2 3.5 g 50 cm^3	5 h		4.9-7.4	0.1 mol dm^{-3} NaClO$_4$	uptake, volume change	ΔV_{ads} +18 cm^3 mol for sulfate and selenate and +14 cm^3 mol for phosphate.	631
257.	In$_2$O$_3$	benzoate	5×10^{-3} mol dm^{-3}	1-4 g 7 cm^3	1 d	room	3-7	0.1 mol dm^{-3} KCl	uptake		516
258.	La$_2$O$_3$	benzoate, salicylate	10^{-3} mol dm^{-3}	1-4 g 7 cm^3	1 d	room	7-9	0.1 mol dm^{-3} KCl	uptake		516
259.	MgO	As(III), Sb(III), Bi(III)	4×10^{-6}, 4×10^{-5} mol dm^{-3}	0.05 g 25 cm^3	30 min		7-15	1 mol dm^{-3} NaClO$_4$	uptake	The pH$_{50}$ 14.5 for Sb, pH$_{50}$ ~ 14.5 for As, pH$_{50}$~13.5 for Bi.	202
260.	MnO$_2$	phosphate	1 mg P, 30 cm^3	0.1 g (2.2 m^2/g), 30 cm^3	2 d		4.2 7.5	acetate buffer 0.2 mol dm^{-3}	uptake	The uptake increases when pH increases.	506
261.	MnO$_2$, δ	phosphate	10^{-5} mol dm^{-3}	10^{-3} mol Mn dm^{-3}	1 d	23	2 8.5	0.07 mol dm^{-3} NaCl+2×10^{-3} mol dm^{-3} NaHCO$_3$; sea water	uptake	The uptake from 0.7 mol dm^{-3} NaCl peaks at pH 3. The uptake at pH > 5 is enhanced in the presence of 10^{-2} mol dm^{-3} Ca or 5×10^{-2} mol dm^{-3} Mg, or when 0.7 mol dm^{-3} NaCl is replaced by sea water.	609
262.	MnO$_2$	sulfate (IV)	0.01-0.02 mol dm^{-3}	5 g 100 cm^3	3 h	room. 35	7	10^{-2} mol dm^{-3} NaCl	uptake	Reductive dissolution.	295
263.	MnO$_2$	sulfate, thiocyanate	0.001-0.1 mol dm^{-3}			25		Na	proton stoichiometry	The shift in uncorrected PZC induced by the salt concentration of 0.1 mol dm^{-3} (in mV, 1 pH unit = -59 mV): sulfate -20, thiocyanate -20.	158
264.	MnO$_2$·nH$_2$O	As(III), (V)	10^{-5} mol dm^{-3}	15 mg/25 cm^3	1 d	25	2 10	0-10^{-2} mol dm^{-3} nitrate, chloride	uptake	As (III) (initially): pH$_{50}$ = 9.5, increase in ionic strength depresses the uptake. 90% of preadsorbed As is released in the presence of 0.05 mol dm^{-3} EDTA and 0.5 mol dm^{-3} oxalic acid.	632

TABLE 4.2 Continued

Adsorbent	Adsorbate	Total concentration	Solid to liquid ratio	Equilib-ration time	T	pH	Electrolyte	Method	Result	Ref.
265. $MnO_{1.93} \cdot nH_2O$, δ	molybdate, phosphate, selenate IV and VI, silicate	7×10^{-7} 10^{-5} mol dm⁻³	30, 300 mg (290 m² g) dm³			4-10	0.1 mol dm⁻³ KCl	uptake	No uptake of Se (VI), uptake of silicate peaks at pH 7.5. Se (IV): pH_{90} 8.8 (300 mg dm³, 7×10^{-7} mol dm⁻³).	623
266. Nb_2O_5	salicylate		1-4 g 7 cm³	1 d	room	3.1	0.1 mol dm⁻³ KCl	uptake		516
267. NiO	sulfate, thiocyanate	0.001-0.1 mol dm⁻³			25		Na	proton stoichiometry	The shift in uncorrected PZC induced by the salt concentration of 0.1 mol dm⁻³ (in mV, 1 pH unit = -59 mV): sulfate - 50, thiocyanate -40.	158
268. PbO_2	sulfate	0.001-0.1 mol dm⁻³			25		Na	proton stoichiometry	The shift in uncorrected PZC induced by the salt concentration of 0.1 mol dm⁻³ - 50 mV (1 pH unit = -59 mV).	158
269. PbO_2	fluoride, sulfate	0.0001-0.1 mol dm⁻³	1.5 g (1 m² g). 150 cm³	15 min	25	7 10	K	proton stoichiometry	PZC at pH 9.9 for 0.1 mol dm⁻³ K_2SO_4. irreproducible results for KF.	633
270. PbO_2, β	phosphate, sulfate	0.001 1 mol dm⁻³			25		Na	proton stoichiometry. uptake	Adsorption of anions at polarized PbO_2 electrode.	634
271. RuO_2	sulfate, thiocyanate	0.001-0.1 mol dm⁻³			25		Na	proton stoichiometry	The shift in uncorrected PZC induced by the salt concentration of 0.1 mol dm⁻³ (in mV, 1 pH unit -59 mV): sulfate -50, thiocyanate -90.	158
272. Sb_2O_3, reagent grade, washed	chromate	carrier free-10^{-2} mol dm⁻³	0.2 g (9.2 m² g) 10 cm³	15 min	30	0.5 10	Na	uptake	The uptake (carrier free chromate) peaks at pH 4-8 (95% uptake). The uptake at pH 2 decreases with T (30-60 C) ΔH_{ads} -12 kJ/mol.	635
273. Sc_2O_3	benzoate, salicylate	10^{-3} mol dm⁻³	1-4 g 7 cm³	1 d	room	4-7	0.1 mol dm⁻³ KCl	uptake		516
274. SiO_2, quartz, Merck	arsenate (V)	10^{-8} 10^{-4} mol dm⁻³	0.05, 0.5 g (0.12 m² g) 20 cm³	3 d		2 10	0.1 mol dm⁻³ NaCl	uptake	Uptake is low and pH independent.	482
275. SiO_2, 3 samples	benzoate	10^{-3} 10^{-1} mol dm⁻³	5 50 g (20-310 m² g) dm³	1 d	25	3, 7 (initial)	Na	uptake		636
276. SiO_2	molybdate	4×10^{-4}-4×10^{-3} mol dm⁻³	10 g (50 m² g). 200 cm³			1 8	NH_4	acousto	Formation of $H_3SiMo_{12}O_{40}$ at low pH.	54

No.	Adsorbent	Adsorbate	Adsorbate concentration	Solid to liquid ratio	Time	T	pH	Electrolyte	Method	Result	Ref.
277.	SiO_2	NTMP	1 10 mol dm	1 g dm³	1 min	25	5 10	0.01 mol dm⁻³ NaNO₃; 0–0.001 mol dm³ Zn	uptake	No uptake in absence of Zn. Uptake in the presence of Zn peaks at pH 7.5, 95% with 0.001 mol dm³ Zn.	243
278.	SiO_2	selenate	10^{-6} mol dm	8.3 10^{-3} mol Si dm³	2 h	25	5.5 10	0.1 mol dm³ NaNO₃	uptake	Uptake of selenates below 2%.	151
279.	SiO_2, α	thiosulfate	10^{-3} mol dm⁻³	30 g (3.3 m² g) dm³	2–4 h	25	3 9	10^{-3} mol dm⁻³	uptake	Uptake below 2%	48
280.	SiO_2, quartz	acetate, formate	10^{-4} mol dm		1 min	25	2 9	0.01 mol dm⁻³ NaCl	uptake	Zero uptake (within experimental scatter). The electrokinetic curves are not affected by HEDP and P_3O_{10}.	637
281.	SiO_2, quartz	HEDP, P_3O_{10}	0.0025 mol dm	554 m² dm⁻³		25	3 11	0.0125 mol dm³	ζ	The uptake 4%.	581
282.	SiO_2, Fisher, washed	salicylate, 5-sulfosalicylate	10^{-3} mol dm⁻³	1 g (730 m²) 25 cm³	1 d	25	3 9	10^{-3} mol dm³ NaNO₃	uptake		638
283.	SnO_2, natural, upgraded by acid leaching 94% SnO_2, 0.6% Fe	styrene phosphonic acid	0.002 0.1 g dm³	1 mg dm³	1 d	25	6	0.001 mol dm³ KNO₃	ζ coagulation	The effect of styrene phosphonic acid on the ζ potential of SnO_2 is rather insignificant.	639
284.	ThO_2, hydrous, synthetic, 4 samples	chromate	0.02 0.2 mol dm	0.5 g 25 cm³	1 d	25	neutral	K	uptake		377
285.	TiO_2, Degussa, P25	carbonate	10 mol dm⁻³	0.55 g (50 m² g) 35 cm³	1 d	20	2 9	0.001 0.1 mol dm⁻³ NaCl	uptake, proton stoichiometry	The CIP is shifted to 6.2 in the presence of carbonate (pristine CIP at pH 5.7). The uptake peaks at pH 5	640
286.	TiO_2, Degussa, P25	chromate	10 mg dm⁻³	1 g (154 m² g) dm³	1 d	25	2 10	none	uptake		265
287.	TiO_2, Degussa, P-25	EDTA	10 10^{-4} mol dm	2 g (55 m² g) dm³	10 h	22 25	3 8	0.003 mol dm NaClO₄	uptake	EDTA⁴ was taken as the component ≡TiOHEDTAH₂² log K 22.09	395
288.	TiO_2, anatase	fluoride	10 10 mol dm	40 g (90 m² g) dm	7 d	25	5	none	uptake		387
289.	TiO_2, Plutonio Argentina	phosphate	1 mg P 30 cm⁻¹	0.1 g (4.7 m² g) 30 cm³	2 d		4.2 6.5	acetate buffer 0.2 mol dm	uptake	The uptake increases when pH increases	506
290.	TiO_2, anatase	phosphate	10^{-4} 5 10 mol dm	1 g 25 cm	6 h	25	1 5	0.1 mol dm⁻³ NaClO₄	uptake	Langmuir adsorption isotherms at constant pH ΔH_ad 11 14 kJ mol	641
291.	TiO_2, anatase, Pfalz and Bauer	phosphate	10^{-6} 10^{-3} mol dm	0.13 g (8.3 m² g) dm³	2 d	25	4.4 10.6	0.01 mol dm⁻³ NaClO₄	uptake, ζ		36
292.	TiO_2	salicylate	10^{-7} 2 10 mol dm⁻¹	1 4 g 7 cm	1 d	room 25	3 6	0.1 mol dm³ KCl	uptake	The ζ is depressed from 74 mV in absence of sulfate to 54 mV (2 10^{-5} mol dm sulfate).	516
293.	TiO_2, anatase, Aldrich	sulfate					3	10^{-3} mol dm⁻³ HCl	ζ		381
294.	TiO_2, Degussa, P25 (mostly anatase)	acetate, oxalate, sulfate	10 10^{-2} mol dm⁻³	1.6 mg (50 m² g) 500 cm³	2 h	25 2	2.8 9	0.01 mol dm³ KCl	ATR-FTIR spectroscopy	Oxalate at pH 3: three adsorbed species. Langmuir type isotherms. Acetate at pH 4.5: one adsorbed species. Acetate at pH 4.5 Langmuir type isotherm. Sulfate at pH 3 two adsorbed species. Langmuir type isotherms.	39

TABLE 4.2 Continued

	Adsorbent	Adsorbate	Total concentration	Solid to liquid ratio	Equilibration time	T	pH	Electrolyte	Method	Result	Ref.
295.	TiO$_2$, synthetic, from TiCl$_4$	n-butyl phosphate, phosphate	10^{-6}–0.005 mol dm^{-3}		20 min	27	2.3 11	0–0.01 mol dm^{-3} KCl	ATR-FTIR spectroscopy	Bidentate binding of phosphate.	642
296.	TiO$_2$, anatase, from sulfate	phosphate, triphosphate	10^{-6} 3×10^{-4} mol dm^{-3}	0.7 2.5 g (125 m^2 g) dm^3	12 h		2 12		uptake		643
297.	V$_2$O$_5$	chromate	carrier free- 1.7×10^{-2} mol dm^{-3}	0.25 g (16.8 m^2 g) 12 cm^3	1 h	25	1–6	0.001–0.1 mol dm^{-3}, various cations	uptake	The uptake peaks at pH 2. Linear log-log adsorption isotherm at pH 3.2, slope 0.9. Presence of heavy metal cations (Pb, Ni, etc.) enhances the uptake. The uptake decreases when T increases, ΔH_{ads} -29 kJ mol.	644
298.	ZnO, hydrous	NTMP	1×10^{-5} mol dm^{-3}	0.1 g dm^3	1 min	25	7 10	0.01 mol dm^{-3} NaNO$_3$	uptake	pH$_{50}$ – 9.5	243
299.	ZrO$_2$	phosphate	2×10^{-5} 2×10^{-3} mol dm^{-3}	0.2 g (21 m^2 g) 30 cm^3	> 2 d	20	2 7	0–0.01 mol dm^{-3} KCl	uptake, ζ, contact angle	6.8×10^{-5} mol P adsorbed per 1 g causes a shift of IEP to pH 4.5 and reduces the contact angle from 59 to 44.	645
300.	ZrO$_2$, sol-gel process	sulfate	10^{-4} 10^{-2} mol dm^{-3}	0.1 g (66 m^2 g) 50 cm^3	1 d	25	3 11	Na, Ca	ζ	For 10^{-3} mol dm^{-3} Na$_2$SO$_4$ ζ~0 at pH 3-6. For 10^{-3} mol dm^{-3} CaSO$_4$, ζ~0 at pH 3 11.	406
301.	ZrO$_2$	H$_4$SiW$_{12}$O$_{40}$	10^{-4} mol dm^{-3}				1 10		ζ	The IEP is shifted to pH 2.7.	646
302.	ZrO$_2$	p-benzoic acids				35	4.8 9.3	7 different buffers	chromatography		525
303.	ZrO$_2$	benzoate, o-bromobenzoate, m-nitrobenzoate, o-phthalate, salicylate, sulfosalicylate,	10^{-4} 7×10^{-3} mol dm^{-3}	1–4 g 7 cm^3	1 d	room	2 8.5	0.1 mol dm^{-3} KCl	uptake		516
304.	ZrO$_2$, from ZrOCl$_2$ + NH$_3$, various heat treatment	chromate, molybdate, tungstate	10^{-3} 10^{-1} mol dm^{-3}		3 d	room	2, 8	NH$_4$	uptake, XPS (ex situ)	The uptake of Cr and Mo at pH 2 is greater than at pH 8. Cr is evenly distributed in the grains of the adsorbent (penetration depth up to 1 mm), while the penetration of Mo is limited to about 100 μm, and penetration of W to 3 μm.	647

Composite materials containing Al

No.	Material	Species	Concentration	Solid	Time	Temp	pH	Electrolyte	Method	Comments	Ref
305.	Al_2O_3, F modified	rhenate (VII)	0.05 mol dm^{-1}	1 g (170 m^2) 58 cm^3	1 d		6.5	NH_4NO_3	uptake	The uptake decreases when F Al atomic ratio increases: 0.3 atoms of Re nm^2 for 0% F; 0.08 atoms of Re nm^2 for 14 atom% F Al.	496
306.	$Al(OH)_3$ + $Fe(OH)_3$ coprecipitate	selenate	10^{-6} mol dm^{-3}	10^{-3} mol Al and Fe dm^{-3}	2 h		5.5 10	0.1 mol dm^{-3} NaNO$_3$	uptake		151
307.	Al_2O_3, MgO	rhenate (VII)	0.05 mol dm^{-3}	1 g (170 m^2) 58 cm^3	1 d		9.3	NH_4NO_3	uptake	The uptake increases when Mg Al atomic ratio increases: 0.15 atoms of Re nm^2 for 0% Mg. 0.3 atoms of Re nm^2 for 10 atom% Mg Al.	496
308.	$Al(OH)_3$ $Mg(OH)_2$, Al (Al Mg) 0.466 (mol)	carbonate, sulfate	$5\ 10^{-4}$ 10^{-3} mol dm^{-3}				9.1	Na	ζ	The sign of ζ is reversed to negative at 10^{-3} mol dm^{-3} carbonate and sulfate.	409
309.	Al_2O_3, ? 13.8% MoO_3	sulfate	10^{-3} mol dm^{-3}			22.5	7 10	K	ζ	The positive ζ are depressed, the IEP is not shifted.	92
310.	64% Al_2O_3, 27% TiO_2 9% SiO_2	sulfate	0.001 mol dm^{-3}			20	3.5 10.5	Na	ζ	IEP shifts from 4.7 (NaCl) to 4.2.	410
311.	70% Al_2O_3 30% TiO_2	sulfate	0.001 mol dm^{-3}			20	3 7	Na	ζ	IEP shifts from 4.5 (NaCl) to 3.7.	411
312.	64% Al_2O_3 27% TiO_2 9%SiO_2	sulfate	0.001 mol dm^{-3}			25	3 7	Na	ζ	IEP shifts from 4.7 (NaCl) to 4.	648
313.	TiO_2 + Al_2O_4, (flame oxidation of $TiCl_4$ $AlCl_3$ + PCl_3)	acetate, citrate, glycolate, methyl phosphate, succinate	10^{-2} 0 2 mol dm^{-3}	50% w w (6 m^2 g)	2 d	22 24	4 10	10^{-2} mol dm^{-3} KNO$_3$, NaCl	uptake, IR		649

Clay minerals

No.	Material	Species	Concentration	Solid	Time	Temp	pH	Electrolyte	Method	Comments	Ref
314.	illite	arsenate III and V	$4\ 10^{-7}$ mol dm^{-3}	25 g (24.2 m^2 g) dm^{-3}	16 h		4-10	0.005-0.1 mol dm^{-3} NaCl	uptake	The ionic strength does not affect the uptake. As(III): the uptake increases from 40% at pH 4 to 80% at pH 9; As(V): the uptake peaks at pH 7 (96%).	535
315.	illite	molybdate	0.5 150 mg Mo dm^{-3}	50 g (10.8 m^2 g) dm^{-3}	1 d	25	3.9	0.01 mol dm^{-3} NaCl	uptake		650
316.	illite, IMt	molybdate	$3\ 10^{-4}$ mol dm^{-3}	50 g (25 m^2 g) dm^{-3}	1 d	23	3 10.5	NH_4Cl	uptake		498
317.	illite, IMt	molybdate	3×10^{-4} mol dm^{-3}	50 g dm^{-3}	1 d	23	3 10.5	0.01 1 mol dm^{-3} NaCl	uptake		499
318.	kaolinite, KGa-1	arsenate III and V	4×10^{-7} mol dm^{-3}	25 g (9.1 m^2 g) dm^{-3}	16 h		4-10	0.005-0.1 mol dm^{-3} NaCl	uptake	The ionic strength does not affect the uptake. As(III) the uptake peaks at pH 9 (60%). As (V): pH$_{50}$ = 8.5.	535

TABLE 4.2 Continued

	Adsorbent	Adsorbate	Total concentration	Solid to liquid ratio	Equilibration time	T	pH	Electrolyte	Method	Result	Ref.
319	kaolinite, KGa-2	borate	1–100 mg dm³	2.5 g (19 m² g) 25 cm	2 h	25	6–11	0.1 mol dm³ NaCl	uptake	Uptake peaks at pH 8. The uptake is depressed when T increases.	537
320.	kaolinite, KGa-2	borate	5–250 mg B dm	2.5 g (19 m² g) 25 cm³	1 d		3–11	0.001–1 mol dm³ NaCl	uptake, ζ	Uptake peaks at pH 8. The electrokinetic curves are rather insensitive to the presence of borates.	536
321.	kaolinite, KGal	citrate	2.7 10⁷ 1.6 10⁵ mol dm	1.2 g (15.5 m² g) dm	1 d	22	3–11	0.1 mol dm³ NaCl, 10⁻⁶ mol dm³ U	uptake	The pH$_{50}$ 8.2 (2.7 10⁷ mol dm³ citrate), in absence of U.	144
322.	kaolinite	fluoride	10⁻⁴–10⁻¹ mol dm³	10, 15 g dm³	1 d	25	3.5–9	0.004–0.1 mol dm⁻³ NaNO₃	uptake	The pH$_{50}$ 5(15 g dm³, 10⁻⁴ mol dm⁻³), the effect of ionic strength on uptake curves is negligible.	651
323.	kaolinite	molybdate	0.5–150 mg Mo dm³	10 g (14.2 m² g) dm³	1 d	25	4.6–5.5	0.01 mol dm³ NaCl	uptake	The IEP is shifted to pH 2 in the presence of 3 10⁻⁴ mol dm⁻³ molybdate.	650
324.	kaolinite, KGa-1	molybdate	5 10⁻⁵, 3 10⁻⁴ mol dm³	200 g (9 m² g) dm³	1 d	23	3–11	NaCl	uptake ζ	The IEP is shifted to pH 2 in the presence of 3 10⁻⁴ mol dm⁻³ molybdate.	498
325.	kaolinite, KGa-2	molybdate	5 10 , 3 10⁻⁴ mol dm³	100 g (19 m² g) dm³	1 d	23	3–11	NaCl	uptake ζ	The IEP is shifted to pH 2 in the presence of 3 10⁻⁴ mol dm⁻³ molybdate.	498
326.	kaolinite, KGa-1, KGa-2	molybdate	3 10⁻⁴ mol dm⁻³	100, 200 g dm³	1 d	23	3–10.5	0.01–1 mol dm³ NaCl	uptake	Uptake at pH 5 increases with T	499
327	kaolinite	phosphate	10⁻⁴–10⁻² mol dm⁻³	1:5000	1 d	20	3–10	0.01–1 mol dm³ NaCl	uptake, coadsorption of K	Phosphate sorbed at 40 C is partially released when the temperature decreases to 2 C	502
328.	kaolinite	phosphate	2.5 10⁻⁴ mol dm³	7.5 g (11 m² g) dm³	1–40 d	50	3.5–7.5		uptake		504
329.	kaolinite	phosphate	2 10⁻⁷–2 10⁻⁴ mol dm	0.9 g (11 m² g) 110 cm	1 d	25	2–12	0.01 mol dm³ NaCl, 1.3 10⁻⁴ mol dm³ La, 1.3 10⁻⁴ mol dm Ca	uptake	The uptake peaks at pH 4, presence of Ca or La enhances the uptake at pH 4–6 Langmuir type adsorption isotherms at constant pH.	503
330	kaolinite, K and Ca form	phosphate	10–2 10⁻⁴ mol dm	1.5 g 95 cm³	7 d	24	5, 8	0.02 mol dm³ KCl	uptake	The uptake of P by the Ca form is higher than that by K form	652
331.	kaolinite	phosphate	3 10⁻⁶–2 10⁻⁴ mol dm³	0.8 g 200 cm³	1 d	27	3.8–9	0.01 mol dm³ NaCl	uptake	Uptake peaks at pH 5. ΔH$_{ads}$ 17 kJ mol.	653

No.	Adsorbent	Species	Concentration	Amount	Time	T	pH	Electrolyte	Method	Comments	Ref.
332.	kaolinite	phosphate	1 25 13 75 mg P dm^{-3}	0.02 g(12 m^2 g) 20 cm^3	90 d	25	3.8 9	0.1 mol dm^{-3} KCl	uptake	The uptake is rather insensitive to the pH.	572
333.	kaolinite, KGa-1B	phosphate	10^{-4} mol dm^{-3}	4 g dm^{-3}	2 d	25	3 10	0.1 mol dm^{-3} NaNO$_3$ + 10^{-6} mol dm^{-3} U	uptake	10% uptake at pH 3 8 (in equilibrium with air).	218
334.	kaolinite	silicate	110 ppm SiO$_2$	1:10	4 d	25	2 10	toluene added	uptake		549
335.	kaolinite, KGa-1	acetate, formate oxalate	10^{-4} mol dm^{-3}	680 1000 m^2 dm^3	1 min	25	2 9	0.01 mol dm^{-3} NaCl	uptake	Uptake of acetate and formate below 10%. Oxalate uptake peaks at pH 6 (80%). The calculated adsorption constants for oxalate at 25 C and 60 C are equal.	637
336.	kaolinite	benzene carboxylic acids	10^{-4} mol dm^{-3}			25	3 10		uptake	At pH 3 7 the uptake of trimesic, phtalic, and mellitic acid is about 10%.	239
337.	hematite-kaolinite composite	phosphate	1.25 13.75 mg P dm^{-3}	0.02 g (22 m^2 g) 20 cm^3	90 d	25	3.8 9	0.1 mol dm^{-3} KCl	uptake	The uptake is rather insensitive to the pH.	572
338.	mica	phosphate	3 10^{-6} 2 10^{-4} mol dm^{-3}	0.8 g 200 cm^3	1 d	27	4.5 8.5	0.01 mol dm^{-3} NaCl	uptake, ζ	The effect of P on the electrokinetic curves is rather insignificant	653
339.	mica	phosphate		0.6 g 150 cm^3	1 d		5	0.01 mol dm^{-3} NaCl	uptake	ΔH_{ads} 15 kJ mol	641
340.	montmorillonite	arsenate III and V	4 10^{-7} mol dm^{-3}	25 g (18.6 m^2 g) dm^{-3}	16 h		4 10	0.015-0.1 mol dm^{-3} NaCl	uptake	The uptake of As(III) at pH 3 9 (35%) is rather insensitive to the pH and the ionic strength. The uptake of As(III) at pH > 9 is enhanced at high ionic strength. The uptake of As(V) peaks at pH 5 (70%).	535
341.	montmorillonite SWy-1	borate	1 100 mg B dm^3	0.75 g (19 m^2 g) 25 cm^3	2 h	25	6-11	0.1 mol dm^{-3} NaCl	uptake	Uptake peaks at pH 10.5. The uptake is enhanced when T increases.	537
342.	montmorillonite SWy-1	borate	5 mg B dm^3	0.75 g (19 m^2 g) 25 cm^3	1 d		3 11	0.01 1 mol dm^{-3} NaCl	uptake	Uptake peaks at pH 10 and it is enhanced at high ionic strength.	536
343.	montmorillonite	molybdate	0.5 150 mg Mo dm^3	10 g (5.4 m^2 g) dm^3	1 d	25	4	0.01 mol dm^{-3} NaCl	uptake		650
344.	montmorillonite, SWy-1, SAz-1, STx-1	molybdate	3×10^{-4} mol dm^{-3}	30,40 g (19 70 m^2 g) dm^3	1 d	23	3 10.5	0.1 mol dm^{-3} NaCl	uptake		498
345.	montmorillonite, SWy-1, SAz-1, STx-1	molybdate	3×10^{-4} mol dm^{-3}	30,40 g dm^3	1 d	23	3 10.5	0.01 1 mol dm^{-3} NaCl	uptake	The uptake is rather insensitive to the ionic strength.	499
346.	montmorillonite, K and Ca form	phosphate	10^{-5} 2×10^{-4} mol dm^{-3}	1.5 g 95 cm^3	7 d	24	4.5 9.5	0.01, 0.03 mol dm^{-3} KCl	uptake	The uptake of P by the Ca form is higher than that by K form.	652
347.	smectite	phosphate	3×10^{-6} 2×10^{-4} mol dm^{-3}	0.8 g 200 cm^3	1 d	27	4.5 8.5	0.01 mol dm^{-3} NaCl	uptake		653
348.	vermiculite	phosphate	3 10^{-6} 2×10^{-4} mol dm^{-3}	0.8 g 200 cm^3	1 d	27	5 9	0.01 mol dm^{-3} NaCl	uptake		653

TABLE 4.2 Continued

Adsorbent	Adsorbate	Total concentration	Solid to liquid ratio	Equilibration time	T	pH	Electrolyte	Method	Result	Ref.
Basic carbonates										
349. $Mg_6Al_2(OH)_{16}CO_3 \cdot 4$ H_2O, hydrotalcite, Crosfield	chromate	10 mg/dm³	1 g (34.8 m²/g)/dm³	1 d	25	2 10	none	uptake	The uptake is rather insensitive to the pH.	265
Silicates										
350. $ZrSiO_4$, natural	monoalkyl phosphates	10^{-7} 3.5×10^{-3} mol dm⁻³	0.8 g (1.7 m²/g)/400 cm³	1 h	25	3 10	0.01 mol dm⁻³ KCl	ζ, uptake, DRIFT	The concentration necessary to reverse the sign of ζ from positive to negative at pH 3 ranges from 10^{-5} mol dm⁻³ (dodecyl phosphate) to 2×10^{-3} mol dm⁻³ (hexyl phosphate).	654
Coatings										
351. aluminated silica, positive sol 130M, duPont	sulfate, phosphate		0.2 g/100 cm³	1 h 30 d		2 5		ζ coagulation		655
352. aluminum oxide coating on sand	selenate IV and VI	8×10^{-4} 1.8×10^{-3} mol dm⁻³	5 g/50 cm³	3 h		2 12	0.1 mol dm⁻³ NaCl	uptake	Langmuir type adsorption isotherms at constant pH. For 8×10^{-4} mol dm⁻³: $pH_{50} = 8.8$ for Se (VI); $pH_{50} > 12$ for Se (IV).	656
Glasses										
353. porous glass, 97% SiO_2 3% Al_2O_3	phosphate	2.5×10^{-4} – 1.25×10^{-3} mol dm⁻³	52 m²/200 cm³	40 min	25	2	0.05-0.1 mol dm⁻³ KCl	uptake	The uptake is depressed by a factor 2 when [KCl] increases from 0.05 to 0.1 mol dm⁻³. The uptake is enhanced when T increases.	657
Clays										
354. bentonite	silicate	110 ppm SiO_2	1:10	4 d	25	2 10	toluene added	uptake		549
355. kaolin, Ward's	arsenate (V)	10^{-8}-10^{-4} mol dm⁻³	0.05, 0.5 g (5.1 m²/g)/20 cm³	3 d		2 10	0.1 mol dm⁻³ NaCl	uptake	Uptake peaks at pH 5.	482

No.	Soil	Species	Concentration	Ratio	Time	Temp	pH	Electrolyte		Remarks	Ref.
...	......	phosphate	10^{-6} 2×10^{-5} mol dm^{-3}	0.6 g/50 cm^3	1 d	28	4, 4.5	0.01 mol dm^{-3} NaCl	uptake	ΔH_{ads} = 13-26 kJ mol	641
	Soils										
357.	sandy loam	borate	10^{-3} 10^{-3} mol dm^{-3}	1:5	1 d	25	4-8	0.01 1 mol NaCl	uptake	Uptake (pH) curves obtained at different ionic strengths intersect at pH 5.7.	658
358.	2 soils	borate	1 100 mg B dm^{-3}	5 g/25 cm^3	2 h	25	6-11	0.1 mol dm^{-3} NaCl	uptake	Uptake peaks at pH 9 10. The uptake decreases when T increases.	537
359.	2 soils	borate	5 mg B dm^{-3}	5 g/25 cm^3	1 d	25	3-11	0.01-1 mol dm^{-3} NaCl	uptake	Uptake peaks at pH 9.5 and it is enhanced at high ionic strength.	536
360.	soil	chromate									659
361.	sandy loam, different acid and lime treatment	fluoride	0.1 10 mg dm^{-3}	1:10	1 d	25	4-7	0.01 mol dm^{-3} MgCl$_2$	uptake	The uptake peaks at pH 5.5.	660
362.	3 soils	molybdate	3 10^{-4}, 1×10^{-3} mol dm^{-3}		1 d	23	3 11	NaCl	uptake	pH$_{50}$ = 5-6	498
363.	3 soils	molybdate	3×10^{-4} mol dm^{-1}	200 g/dm^3	1 d	23	3 10.5	0.01 1 mol dm^{-3} NaCl	uptake		499
364.	soil	phosphate	3×10^{-2} 10 mg P dm^{-3}	1:10	1 d	25	4.5-7	0.01 mol dm^{-3} CaCl$_2$	uptake		661
365.	6 soils	phosphate	10^{-2} 100 mg.dm^{-3}	1:10	1 d	25	4-7	0.01 mol dm^{-3} CaCl$_2$, NaCl	uptake	Ca promotes P uptake at high pH, but at low pH the P uptake is higher in absence of Ca.	662
366.	loamy sand, different acid and lime treatment	phosphate	0.1 10 mg dm^{-3}	1:10	1 d		4-7	0-1 mol dm^{-3} NaCl	uptake	The intersection point of uptake (pH) curves at constant [P] obtained at different ionic strengths gradually shifts to lower pH as [P] increases.	470
367.	loamy sand, different acid and lime treatment	selenate IV and VI	10^{-7} 10^{-4} mol dm^{-3}	1:10	1 d	25	4.5-8	0-1 mol dm^{-3} NaCl, 0.01 mol dm^{-3} CaCl$_2$	uptake	The intersection point of uptake (pH) curves at constant [Se] and different ionic strengths at pH 5-6 (lower pH at higher [Se]). The effect of replacement of 0.01 mol dm^{-3} NaCl by 0.01 mol dm^{-3} CaCl$_2$ on Se VI uptake is insignificant, and Se IV uptake at pH > 5 is enhanced in the presence of Ca.	663
368.	12 soils	silicate	55,110 ppm SiO$_2$	1:5	4 d	25	3-9	0.0 . 0.05 mol dm^{-3} CaCl$_2$, toluene added.	uptake	The effect of CaCl$_2$ concentration on uptake of silicate is rather insignificant, also replacement of CaCl$_2$ with MgCl$_2$, NH$_4$Cl, or NaCl does not affect the results.	549

TABLE 4.2 Continued

Adsorbent	Adsorbate	Total concentration	Solid to liquid ratio	Equilibration time	T	pH	Electrolyte	Method	Result	Ref.
369. 2 soils	arsenate, molybdate, phosphate	2×10^{-5} 2×10^{-3} mol dm^{-3}	1:50	2 d	25	natural	Na	uptake		664
Other materials										
370. activated red mud	chromate (VI)	20-50 ppm				3 8	1 mol dm^{-3} KCl; acetate or phosphate buffer	uptake	The uptake peaks at pH 5.2. The uptake decreases as T increases.	665
371. activated red mud	phosphate	1.26 10^{-3} mol dm^{-3}	0.5-10 g (249 m^2 g) dm^3		30	3.5 6	1 mol dm^{-3} KCl, acetate buffer	uptake	The uptake linearly decreases from 80% at pH 3.3 to 55% at pH 6. The uptake increases as T increases. Langmuir type adsorption isotherm.	666
372. water treatment residual (WTR)	phosphate	1.25-6 mol kg WTR	2 g 18 cm^3	7 d		2 12	0.01 mol dm^{-3} NaCl 0.02% NaN$_3$	uptake	Sorption behavior was successfully modeled as a combination of interaction with hydrous ferric oxide (stability constants of surface complexes taken from literature), and quarternary polyamine used for water treatment: $\equiv Q^+ \ H_2PO_4^-$ $\equiv QH_2PO_4$ log K -22.31 $\equiv Q^+ \ HPO_4^{2-}$ $\equiv QHPO_4^-$ log K -15.66. The later mechanism of sorption prevails at pH > 8.	667
373. water treatment residual (WTR)	phosphate	0.1-10 g dm^3		2 d		6	0.01 mol dm^{-3} NaCl + 0.02 % NaN$_3$	uptake, ζ coagulation		668

observed, but other sources report uptake curves with a maximum for the same combination adsorbent – adsorbate. It should be emphasized that experimental results are often scattered, and then the assignment of an uptake curve to certain type is rather subjective. In some studies the maximum of uptake as the function of pH might not be apparent due to a limited pH range studied.

The shapes of uptake curves for acids whose $pK_{a1} > 6$ are very consistent. No single case of sigmoidal uptake curve (Fig. 4.5(B)) is reported in Table 4.2 for carbonate or borate. Uptake of carbonate peaks at pH 5–6.5, that roughly corresponds to the pK_{a1} of carbonic acid. Uptake of borate peaks at pH 6.5–10.5, and the most frequently reported pH of maximum uptake near 8 is somewhat lower then the pK_{a1} of boric acid.

The uptake of fluoride peaks at pH 4–7. These pH values are significantly higher than the pK_a of HF (3.2). However, a few examples of sigmoidal uptake curves have been also reported for fluoride.

The shapes of uptake curves reported in different sources for selenate IV and VI are rather consistent. Namely, sigmoidal uptake curves (Fig. 4.5(B)) were observed for these adsorbates with a few exceptions. Selenic VI acid is stronger than other acids considered in Table 4.2 ($pK_{a1} < 0$, pK_{a2} 1.7). Also uptake curves of sulfate VI ($pK_{a1} < 0$) are sigmoidal.

Finally, in the group of acids whose $0 < pK_{a1} < 4$, sigmoidal uptake curves and uptake curves with a maximum are reported in different sources, and there is no apparent correlation between the type of uptake curve, and the nature of the adsorbent (actually most available data on anion adsorption are for aluminum and iron III oxides and hydroxides as adsorbents) or the experimental conditions (e.g. the initial concentration of the adsorbate). Arsenate V, chromate VI, phosphate, and molybdate are typical examples of such behavior. For three former anions the number of publications reporting sigmoidal uptake curves on the one hand and uptake curves with a maximum on the other are approximately equal, but for molybdate the sigmoidal curves are more abundant. Comparison of molybdate with other anions in terms of pK_{a1} is difficult in view of tendency to form polyacids (condensation).

The correlation of the shape of uptake curves with pK_{a1} (sigmoidal uptake curves for relatively strong acids, uptake curves with a maximum for very weak acids) is limited, e.g. selenic IV acid (sigmoidal curves with a few exceptions) is weaker than arsenic V or phosphoric acid (for which adsorption maxima are often observed). The discrepancies in sorption behavior of anions can be due to dissolution of the adsorbents.

C. Linear Free Energy Relationship

It was demonstrated in Sections A and B that in spite of a few discrepancies the macroscopic parameters characterizing the distribution of ions between solution and the surface (K_D, pH_{50}) taken from different sources are rather consistent. These coefficients can be used to compare affinities between different ions and certain surface. For example silica has higher affinity to Zn (Fig. 4.58) than to Ni (Fig. 4.57), at least at pH > 6.5, and iron III (hydr) oxides have higher affinity to Se IV (Fig. 4.60) than to Se VI (Fig. 4.62). Many affinity series have been established using experimental results obtained at certain experimental conditions, but apparently the relative affinity of a surface to different ions is rather insensitive to experimental

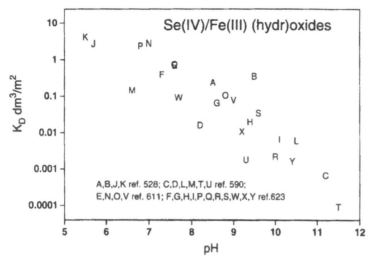

FIG. 4.60 K_D values in the Se IV/Fe III (hydr)oxide system reported in the literature.

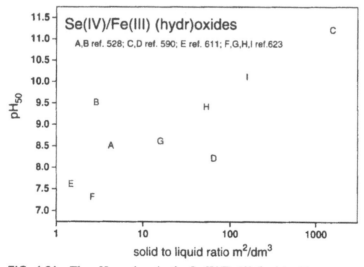

FIG. 4.61 The pH$_{50}$ values in the Se IV/Fe III (hydr)oxide system reported in the literature.

conditions. These affinity series are correlated with different physical and chemical properties of the ions. For example correlations between binding of metal cations to different adsorbents and reciprocal ionic radii have been found. However, the correlation between tendency to adsorb and tendency to react with other ions in solution is the most frequently invoked one.

The stability constants of complexes of metal ions with different ligands are to some degree correlated. Almost perfect linear correlation between stability constants of aqueous hydroxo complexes on the one hand and acetate (slope 0.27), lactate (slope 0.61) and salicylate (slope 1.29) complexes on the other has been demonstrated [669] for Fe, Pb, Cu, Zn, Co, Ni, Cd, Mn, and Mg. The adsorption process can be

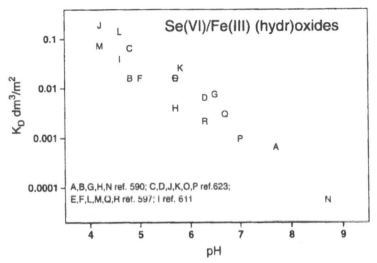

FIG. 4.62 K_D values in the Se VI/Fe III (hydr)oxide system reported in the literature.

considered as complexation of metal cations by the surface, and the stability constants of surface complexes can be established using different models (cf. Chapter 5). Since the surface oxygen atoms interact directly with the adsorbates as proved by means of spectroscopic methods (cf. Table 4.1), the correlation between stability constants of surface complexes on the one hand and solution hydroxo complexes on the other is usually demonstrated, although the latter is in turn linearly correlated with the stability constants of other solution complexes.

In order to avoid model assumptions the relationship between the stability constant of a solution complex and directly measured quantities reflecting the adsorption behavior (pH_{50} at given solid to liquid ratio or K_D at constant pH) can be also studied.

Some degree of correlation between pH_{50} obtained under the same experimental conditions for sorption of eight divalent cations on Al and Fe hydroxides on the one hand and first hydrolysis constant or log of solubility product of metal hydroxide or Pauling electronegativity on the other has been reported by Kinniburgh et al. [149]. Parks [670] collected values of pH_{50} of metal cations obtained for different experimental conditions with different materials (adsorbents) including oxides, AgI, ZnS, clays and polystyrene latex, and these results still show some degree of linear correlation with logarithm of the first hydrolysis constants. Lowson and Evans [94] report a linear correlation between pH_{50} (in the range of Henry adsorption isotherm) and the first hydrolysis constant of metal cations.

Inoue et al. [303] found that the correlation between log K_D (hydrous Nb_2O_5) at pH 3 for four trivalent and three divalent metal cations and logarithm of stability constant of 1–1 acetate complex ($r = 0.96$) was better than with the first hydrolysis constant ($r = 0.91$). An equally good correlation between log K_D (hydrous ZrO_2) at pH 4 for three trivalent and four divalent metal cations and logarithm of the stability constant of 1–1 acetate complex ($r = 0.98$) and of the first hydrolysis constant ($r = 0.99$) was found [407]. In another publication Inoue et al. [304] arranged the cations into three groups on basis of the valence and electronic structure, and the correlation between distribution coefficients (hydrous SnO_2) and the first hydrolysis constant was considered separately for each group. For alkaline earth metal cations and UO_2^{2+} the correlation was very good ($r = 0.99$). For two other groups, namely, Mn(II), Co, Zn, Cu, and Ni on the one hand and Ce(III), La,Eu,Y,Sc, and Cr(III) on the other, the correlation was less satisfactory ($r = 0.9$), chiefly due to the data point corresponding to the last-named cation in each group.

Examples of correlations between stability constants of surface complexes (calculated using different adsorption models, cf. Chapter 5) on the one hand, and the first hydrolysis constant and other constants characterizing the stability of solution complexes on the other are more numerous than the studies of correlations involving directly measured quantities. It should be emphasized that there is no generally accepted model of adsorption of ions from solution, and "stability constants of surface complexes" are defined differently in particular models, thus, the numerical values of these constants depend on the choice of the model. Moreover, some publications reporting the correlations fail to define precisely the model.

Some models invoke multiple surface species involving the specifically adsorbed ion. In these models many combinations of model parameters (stability constants of particular surface species) can fit the experimental data almost equally well. Therefore, the fitted values of stability constants of surface complexes in such models are of limited significance, and so are the correlations involving these stability constants.

Models of adsorption of ions from solutions and the meaning of their parameters is discussed in detail in Chapter 5.

Another type of correlations reported in the literature is between the properties in a series of adsorbents and their ability to bind certain adsorbate. For example the affinity of Ni to alumina (Fig. 4.34) is higher than to silica (Fig. 4.57). Generally alumina shows a higher affinity to heavy metal cations than silica in spite of a higher PZC of alumina (cf. Table 3.1) and thus less favorable electrostatic conditions for cation adsorption.

1. Anions

Stumm [671] studied the correlation between the tendency to form surface complexes with iron and aluminum oxides on the one hand and solution complexes with Fe(III) and Al(III) on the other for organic and inorganic anions. Nine data points for Fe(III) give a straight line log K (surf) = log K (soln) + 2, where K is the stability constant of surface and solution complex involving Fe and given anion. The log K (surf) and log K (soln) for Al(III) are also linearly correlated, although with some scatter.

The following correlation log K (surf) = log K (soln) + 1.5 was obtained for adsorption of five anions on goethite [615], but the K(surf) calculated from this equation for $H_2PO_4^-$ is lower than the observed one by two orders of magnitude. Another limited correlation between the stability of surface and solution complexes of anions has been reported by the same authors in Ref. [672]. It should be emphasized, that the literature data on stability constants of aqueous complexes involving Al and Fe and specifically adsorbing anions is scarce, and for a few anions whose sorption behavior has been extensively studied, the stability constants of aqueous complexes with Fe and Al are not known.

On the other hand, the dissociation constants of weak acids are readily available. A linear correlation between stability constants of surface complexes of 6

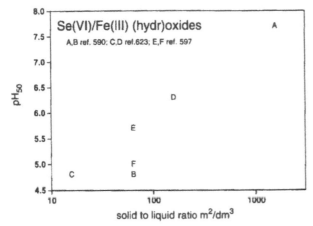

FIG. 4.63 The pH$_{50}$ values in the Se VI/Fe III (hydr)oxide system reported in the literature.

anions (four layer model) with goethite and dissociation constants of corresponding acids was reported by Barrow and Bowden [673]. A linear correlation between stability constants of surface complexes of 11 anions with goethite and protonation constants of corresponding acids was reported by Balistrieri and Chao [611]. Weaker acids have a stronger tendency to adsorb. This result is in line with absence of specific adsorption of strong acids (anions of inert electrolytes).

Correlation between Gibbs energy of adsorption of salicylic acid (at pH = PZC-0.5) and the PZC of eight oxides has been reported [516]. The same authors found a correlation between Gibbs energy of adsorption of organic acids on alumina and the C-OH distance in the COOH group (analogous dependence with zirconia showed rather limited correlation).

2. Cations

A linear correlation between the binding constant of metal cations to silica and their first hydrolysis constant was reported by Schindler et al. [357]. Similar results have been obtained for other materials. These correlations can be conveniently expressed as:

$$\log K \text{ (surf)} = a \log K \text{ (1}^{st}\text{hydrolysis)} + b \tag{4.16}$$

where a and b are empirical coefficients. A collection of straight lines representing Eq. (4.16) for sorption of multivalent cations on SiO_2, γ-Al_2O_3, goethite, amorphous Fe_2O_3, and marine particulate matter was compiled by Charlet [45]. A few examples of numerical values of the empirical coefficients a and b in Eq. (4.16) are collected in Table 4.3. In spite of different models used to calculate K (surf) all a values reported in Table 4.3 are rather consistent and they oscilate around unity. On the other hand, the b coefficients are very diverse. This is due to combination of the following factors. First the b coefficient reflects the affinity of the material (adsorbent) to metal cations, and this affinity increases in the series silica < iron III (hydr) oxides < Mn IV (hydr)oxides. Different models can produce a dramatic difference in the numerical value of K (surf), especially for surfaces carrying high

TABLE 4.3   The Coefficients a and b of Eq. (4.16) Reported in the Literature

Adsorbent	Adsorbates	a	b	Ref.
Al_2O_3, γ	Co, Cu, Mn, Pb, Zn	1	6	107
Fe_2O_3, hematite, Kanto	Co, Cu, Pb, Zn	1.16	-11.14	206
$Fe_2O_3 \cdot H_2O$	Ag, Cd, Co, Cu, Hg, Ni, Pb, Zn	1.17	-4.37	12
$Fe_2O_3 \cdot H_2O$	Ca, Cd, Co, Hg, Sr, Zn	1.3	-7.89	12
FeOOH, goethite	Cd, Co, Cu, Pb, Zn	1.2	11.1	277
FeOOH, goethite	Cd, Co, Cu, Hg, Pb, Zn	0.864	8.2	277
FeOOH, α	Ca, Co, Mg, Mn, Ni, Pb, Zn	0.75	3.97	669
iron III hydrous oxides (combined data)	Ag, Ca, Cd, Co, Cu, Mg, Pb, Pu, Zn	0.67	4	220
MnO_2, IC12	Co, Cu, (Mn), Ni, Zn	1.21	-9.26	206
MnO_2, δ	Ca, Cu, Mg, Pb, Zn	0.73	6	220
SiO_2	Cd, Cu, Fe(III), Pb	0.62	-0.09	357
ZrO_2 (log K_d at pH 4)	Ce (III), Co, Eu, Fe(III), Mn, U, Zn	0.75	6.74	407
Fe rich material isolated from natural sediment, 2 samples	Ca, Cd, Cu, Mg, Ni, Pb, Zn	1.28 .52	-0.08 0.82	674
Mn rich material isolated from natural sediment	Ca, Cd, Cu, Mg, Ni, Pb, Zn	1.25	0.34	674
11 natural sediments	Cd, Cu, Pb, Zn	0.67–1.25	2.82–7.94	478
Marine particulates as oxides	Cd, Cu, Fe, Ni, Pb, Sc, Th	0.87	6.71-log [SOH]/mol kg^{-1}	669

surface charge (far from PZC). Finally, some authors define the first hydrolysis constant of the cation Me^{n+} as

$$K(1^{st}\text{hydrolysis}) = [MeOH^{(n-1)+}][Me^{n+}]^{-1}[OH^-]^{-1} \qquad (4.17)$$

but another definition, namely

$$K(1^{st}\text{hydrolysis}) = [MeOH^{(n-1)+}][H^+][Me^{n+}]^{-1} \qquad (4.18)$$

has been also used. Division of Eq. (4.18) by Eq. (4.17) yields

$$\log K(1^{st}\text{hydrolysis}, 4.18) = \log K(1^{st}\text{hydrolysis}, 4.17) - pK_w \qquad (4.19)$$

where K_w is the ionic product of water. Consequently, the b values obtained using Eq. (4.18) are higher by 14 than those obtained with Eq. (4.17).

Smith and Jenne [220] collected cation adsorption data for goethite, δ-MnO_2, and amorphous $Fe_2O_3 \cdot n\, H_2O$. In view of substantial scatter and discrepancies between data from different sources in the plot according to Eq. (4.16), they proposed the following equation.

$$p^*K^{int} + 0.1\, g_1(zr^{-2} + g_2) = 8.6 - 0.63 \log \beta_{1n} \qquad (4.20)$$

for iron III hydrous oxides and a similar equation with the coefficients 5.2 and -0.86 on the right hand side for δ–MnO_2, where $^*K^{int}$ is the surface complexation constant, β_{1n} is the n-th hydrolysis constant of given cation, and the coefficients g_1 and g_2 are defined as follows

$$g_1 = (1 + 2S + D)(z + 2) \qquad (4.21)$$

$$g_2 = g(n)(z - 1) + 0.1\, d\, (n - 3)^2(1 - S) \qquad (4.22)$$

where S = 1 for ions with s electrons in the outermost electron shell (otherwise S = 0), D is the number of d electrons in the neutral metal atom, n is the principal quantum number of the outermost electron shell of the ion, $g(n) = 1$ when $n > 1$, otherwise $g(n) = 0$, and d is the number of d electrons in the ion.

Sverjensky [675] stated a linear correlation between log of binding constants of certain cation and reciprocal dielectric constants in a series of adsorbents. Moreover, slopes and intercepts of these straight lines obtained for different cations were found to be linearly correlated. Using this approach the binding constant of given cation to any material can be calculated from its binding constant to two materials whose dielectric constants are sufficiently different (e.g. silica and titania). Low affinity of silica to metal cations is in line with Sverjensky's hypothesis, but verification of this theory for other materials would be rather difficult. First, the experimental results vary from one source to another (cf. Fig. 3.28–3.59) thus the calculated binding constants depend on an arbitrary choice of this or another set of experimental data. The calculated binding constants are subject to uncertainties related to different model assumptions which are discussed in detail in Chapter 5. Finally, Sverjensky's hypothesis assumes that the binding constants are calculated for the same type of surface complex, and the spectroscopic results (cf. Tables 4.1 and 4.2) indicate that the mechanisms of sorption can vary not only from one system to another, but even for the same adsorbate-adsorbent pair, e.g. as the function of the solid to liquid ratio or of equilibration time.

The correlations between the surface and solution chemistry of metal cations are not limited to the equilibrium constants. The following correlation has been reported [113] for five divalent cations on alumina:

$$\log k_{ads}(int) = -4.16 + 0.92 \log k_{H_2O} \tag{4.23}$$

where the $k_{ads}(int)$ is intrinsic adsorption rate constants ($mol^{-1}s^{-1}$), and k_{H_2O} is the rate constant for water exchange for metal cations (s^{-1}). Kinetics of sorption is discussed in Section VII.

III. ORGANIC COMPOUNDS AND THEIR MIXTURES

Studies of adsorption of organic compounds other than carboxylic acids, surfactants or polymers are collected in Table 4.4. These compounds are not meant to constitute a uniform group: they differ in molecular weight and in the nature of functional groups, thus, their sorption mechanisms are diverse. Some organic compounds have distinct acidic, basic or amphoteric properties, and they are present in solution chiefly as anions or cations over certain pH range. Sorption of such ionizable compounds is affected by surface charging, and this is reflected in shapes of uptake (pH) curves characteristic for ionic species (cf. Figs. 4.5 and 4.6). Other compounds dissolve in molecular form.

Several compounds considered in Table 4.4 form very stable complexes with many metal cations, and this may result in chemical dissolution of adsorbents. Most adsorbates discussed in this section are susceptible to oxidation in course of the sorption experiment, often on expense of reduction of the adsorbent.

A few examples of adsorption experiments with complex and rather ill defined mixtures of organic compounds are presented in Table 4.4. It should be emphasized

(Text continues on pg. 491)

Index of adsorbates for Table 4.4

Adsorbate	Entries
α-alanine	21, 45
albumin	1, 72, 89, 102, 103
2-aminomethylpyridine	83
3-amino-2-naphthol	95
2-aminophenol	95
2-aminophenols	96
8-aminoquinoline	97
aspartic acid	32
catechol	2, 24–26, 58, 59, 90, 91
catechols	96
chlorobenzenes	119
chlorophenols	11, 33, 46, 60, 119, 133
cysteine	51
dicyclohexylamine	47
dicyclohexylammonium nitrate (III)	47
dihydroxynaphthalenes	58, 59
dinitronaphthalenes	119
2,4-dinitrophenol	3, 53, 95
2,6-dinitrophenol	95
2,2'-dipirydyl	27, 40, 73–75, 97, 106, 113, 121
dissolved organic carbon	4, 76, 92, 114
fulvic acid	5, 6, 12–14, 28, 29, 34, 48, 54, 58, 61, 77, 78, 85, 110, 127–129
glutamic acid	7, 15, 32, 45, 52
glycine	45
humic acid	8, 12–14, 16, 22, 23, 34, 41, 42, 49, 55, 56, 59, 61, 79, 85, 111, 112, 115–117, 122, 123, 127, 129, 132
hydroquinone	24
8-hydroxyquinoline	20, 43, 80, 107, 124, 130
3,4-lutidine	84

TABLE 4.4 Adsorption of Organic Compounds and Their Mixtures

Adsorbent	Adsorbate	Total concentration	Solid to liquid ratio	Equilibration time	T	pH	Electrolyte	Method	Result	Ref.
1 Al_2O_3, α, reagent grade	albumin (bovine serum)	0.2 5 g dm³	0.4 g (2.6 m²/g) 5 cm³	2 h	40	3.2 9.5	10^{-3} mol dm³ KNO_3	uptake	Langmuir type adsorption isotherms at constant pH. the uptake peaks at pH 5.5.	676
2 Al_2O_3, γ, Degussa C	catechol	10^{-4} 2×10^{-3} mol dm⁻³	2 20 g (130 m²/g) dm³	2 d	22	4 8	0.1 mol dm³ $NaClO_4$	uptake	Uptake increases with pH.	517
3 Al_2O_3, γ	2,4 dinitrophenol	5.6×10^{-5} mol dm³	10 g (93 m²/g) dm³	15 30 min	25	4 8	0.1 mol dm⁻³ NaCl; 0.003 mol dm⁻³ buffer (acetate, ammonia)	uptake	The uptake from acetate-ammonia buffer peaks at pH 5 (30%). The uptake from 0.1 mol dm⁻³ NaCl is lower than from buffer solutions.	677
4 Al_2O_3, γ, Cabot	dissolved organic carbon	9.4-112 mg dm³	1 g (120 m²) dm³	10 h	22	4 10	0.002-0.1 mol dm³ NaCl	uptake, ζ	Uptake peaks at pH 5; the uptake is depressed at high ionic strength. IEP is shifted to pH 6 in the presence of 0.9 mg DOC dm⁻³ addition of 4×10^{-3} mol dm⁻³ Ca depresses the uptake at pH < 6 and enhances the uptake at pH > 8.	508
5 Al_2O_3, α, Aldrich	fulvic acid	10 mg dm³	10 g (0.7 m²/g) dm³	5 d		4 9	0.1 mol dm³ $NaClO_4$	uptake	Anion like adsorption edge. pH_{50} 6.8	482
6 Al_2O_3, γ, Degussa C	fulvic acid	10^{-2} 10 mg dm³	10 mg/50 cm³			6.5	0.001 mol dm³ NaCl	ζ	The sign of ζ is reversed to negative at 4 mg dm³.	678
7 Al_2O_3, γ, Degussa C	glutamic acid	1.5×10^{-3} mol dm³	625 m² dm³	1 d		4 8	0.001 1 mol dm³ $NaNO_3$, 0-5×10^{-4} mol dm³ Cu	uptake	The uptake is depressed at high ionic strength; uptake is enhanced in the presence of Cu.	72
8 Al_2O_3, γ, Degussa C	humic acid	1 25 ppm	200 ppm		25	5 10.5	0.1 mol dm³ $NaClO_4$	ζ	The electrokinetic curves are not affected by humic acid at concentrations below 5 ppm.	91
9 Al_2O_3	methylene blue	$10^{-4.5}$ 10^{-3} mol dm³	1-4 g 7 cm³	>1 d	room	7	none, 0.1 mol dm³ KCl	uptake		679
10 Al_2O_3	nicotinate			1 d			0.1 mol dm³ KCl	uptake	The uptake is depressed by KCl.	516
11 Al_2O_3, pseudoboehmite, from chloride	chlorophenols	10^{-3} mol dm³		15 min	22	5.9	0.05 mol dm³ $NaClO_4$	IR		680
12 Al_2O_3, δ, Degussa C	fulvic and humic acid	0.2 50 mg dm³	10-800 mg (108 m²/g) dm³	4 h	20	5; 8.7	artificial sea water	uptake	Langmuir type adsorption isotherms	681

No.	Adsorbent	Adsorbate	Concentration	Solid/liquid	Time	T	pH	Electrolyte	Method	Result	Ref.
13	Al₂O₃, δ, Degussa C	fulvic and humic acid	8.4, 15 mg dm	111 m² 53 cm	1 d		2 12	0.01 mol dm NaClO₄, 10⁻⁴ mol dm Ni	uptake	The effect of Ni on uptake curves of humic and fulvic acid is rather insignificant	682
14	Al₂O₃, Degussa C	fulvic and humic acid	4 mg dm	20 400 mg (80 m² g, EGME) dm³	1 d	25	4 11	0.01 0.1 mol dm³ NaCl	uptake	HA FA, uptake decreases with pH, presence of 0.001 mol dm³ of Ca enhances the uptake of HA, especially at high pH, but the Ca effect on the uptake of FA is insignificant	683
15	Al₂O₃, α, A 16 SG, Alcoa	glutamic acid, phenol	10 10 mol dm⁻¹	100 g (8.4 m² g) dm³		25	2 11	0.01 mol cm NaCl	uptake	Uptake of glutamic acid peaks at pH 4. The uptake of phenol is rather insensitive to the pH	68
16	Al₂O₃, Degussa C	humic acid, 2 samples, phenylalanine, tyrosine	0.01 100 mg dm⁻¹	10 800 mg (108 m² g) dm	4 h	20	5; 8.7	artificial sea water	uptake	Uptake of tyrosine and phenylalanine is negligibly small	684
17	Al₂O₃, Degussa C	4-nitro-2-aminophenol, 4-nitrocatechol, 4-nitro-1,2-phenylenediamine	5 10 mol dm	0.5 15 g (90 m² g) dm³	30 min		2 11	none, 0.1 mol dm³ NaCl	uptake	The effect of ionic strength on uptake curves is rather insignificant. The uptake of 4-nitro-2-aminophenol (15 g Al₂O₃ dm³) peaks at pH 6 (10%). For 4-nitrocatechol (10 g Al₂O₃ dm³) 90% uptake between two adsorption edges pH₅₀ 4 and 11. Low uptake of 4-nitro-1,2-phenylenediamine	685
18	Al₂O₃, Degussa C	nitrobenzenes, nitrophenols	5 10 3 10⁻⁴ mol dm	5 200 g (96 m² g) dm	0.5 1 h	21	4	0.1 mol dm³ chloride	uptake		686
19	Al₂O₃, Merck, washed	Orange G, Trypan blue	10 2 10⁻³ mol dm	0.1 g (90 m² 10 cm	1 d	30	1 8.5	none, 0.1 mol dm³ NaNO₃	uptake	Sorption of Trypan blue is endothermic, uptake of Orange G is insensitive to T. The uptake of both dyes peaks at pH 2	687
20	Al₂O₃ n H₂O, Carlo Erba	8-hydroxyquinoline	4 3 10⁻⁴ 2 07 10 mol dm	0.1 0.2 g 25 cm	2 d		2 10	NaOH HCl	uptake	Uptake peaks at pH 8	688
21	AlOOH, boehmite	α-alanine	0.5 mol dm	10 g 50 cm³	1 d	25	4.4 11.1	none	uptake, IR		689
22	AlOOH, boehmite	humic acid	1 20 mg dm	500 mg (175 m² g) dm	2 d	room	2 12	0.05 mol dm NaClO₄	uptake	pH₅₀ 12 The IEP is shifted to pH 4 at 10 mg dm	130
23	AlOOH, boehmite	humic acid	10⁻⁴ 2 10 mol dm (of acidic functional groups)	2.5 g (107 m² g) dm	1 d	room	5 10	0.001 0.5 mol dm NaNO₃	uptake	The uptake is enhanced at high ionic strength	690
24	AlOOH, boehmite	catechol, hydroquinone, phenol	5 10 10⁻³ mol dm	25 mg (216 m² 25 cm	6 h		7	0.1 mol dm NaCl	uptake, IR	catechol hydroquinone phenol	691

TABLE 4.4 Continued

	Adsorbent	Adsorbate	Total concentration	Solid to liquid ratio	Equilib- ration time	T	pH	Electrolyte	Method	Result	Ref.
25	$Al(OH)_3$, gibbsite	catechol	5×10^{-4} 4 10^{-4} mol dm	52 mg (100 m g, TEM) 25 cm	6 h		7	0.1 mol dm NaCl	uptake		691
26	$Al(OH)_3$, amorphous	catechol	5×10^{-4} 4×10^{-4} mol dm	50 mg (207 m g) 25 cm	6 h		7	0.1 mol dm NaCl	uptake, IR		691
27	$Al(OH)_3$, bayerite	2 2'-dipyridyl	0.005 mol dm	0.1 1 g (28 m 25 cm	2 h		3 9		uptake	The uptake peaks at pH 5 5	692
28	$Al(OH)_3$, hydrargillite	fulvic acid	2, 20 ppm	300 mg (200 m g) dm	1 d	20	3 10	0.01 mol dm⁻¹ NaClO4	uptake	The uptake is rather insensitive to the pH 90% uptake at pH 3 7 (2 ppm)	148
29	$Al(OH)_3$, gibbsite, Baker	fulvic acid	10^{-1} 10 mg dm				6 5	0.001 mol dm NaCl	ζ	Negative ζ is enhanced	678
30	$Al(OH)_3$, bayerite	1,10-phenanthroline	0.001, 0.005 mol dm	0.1 1 g (28 m 25 cm	2 h		4 9		uptake	The uptake peaks at pH 5 5	693
31	$Al(OH)_3$, bayerite	quinoline	0.004 mol dm⁻³	0.1 0.5 g (28 m g) 25 cm³	2 h		3 7		uptake	The uptake peaks at pH 5	694
32	$Al(OH)_3$, amorphous	aspartic, glutamic acid	10, 10 mol dm	200 mg 40 cm	1 d		4 8	0.1 mol dm NaClO4, 5 10⁻⁴ mol dm Cu	uptake	The uptake of aspartic and glutamic acid in absence of Cu is rather insensitive to the pH The uptake is slightly depressed in the presence of Cu	137
33	$Al(OH)$ gibbsite, synthetic, 2 samples	chlorophenols	10 mol dm		15 min	22		0.05 mol dm NaClO4	IR		680
34	$Al(OH)_3$, gibbsite	fulvic acid, humic acid	0.1 5 g dm		1 d		3 7	0.1 mol dm NaCl	uptake, IR, proton stoichiome- try		695

No.	Adsorbent	Adsorbate	Concentration	Solid/liquid	Temp.	Time	pH	Electrolyte	Method	Remarks	Ref.
35	Al(OH)₃, gibbsite	nitrobenzenes, nitrophenols	$5\ 10^{-7}\ 3\ 10^{-4}$ mol dm	5 200 g (2 8 m²/g)	21	0.5 1 h		0 1 mol dm chloride	uptake		686
36	Al(OH)₃, amorphous	O-phospho-L-serine, O-phospho-L-tyrosine	$10^{-3}\ 10^{-}$ mol dm	200 mg (182 m²/ 40 cm)		1 d	4 8	0 1 mol dm NaClO₄, 5 10^{-4} mol dm Cu	uptake	The uptake of O-phospho-L-serine is about 95% for 10^{-3} and about 85% for 10^{-} mol dm⁻³ and it is rather insensitive to pH or Cu addition	138
37	Bi₂O₃	methylene blue	$10^{-4}\ 10^{-}$ mol dm	1 5 g 10 cm			natural	0 1 mol dm KCl	uptake		696
38	CdO	methylene blue	$10^{-4}\ 10^{-}$ mol dm	1 5 g 10 cm			natural	0 1 mol dm KCl	uptake		696
39	CuO	methylene blue	$10^{-4}\ 10^{-}$ mol dm	1 5 g 10 cm³			natural	0 1 mol dm³ KCl	uptake		696
40	Fe₂O₃, hematite, May and Baker	2,2'-dipyridyl	0 005 mol dm⁻¹	0 1 1 g (10 3 m/g) 25 cm³		2 h	2 8	0 005 mol dm⁻¹	uptake	The uptake peaks at pH 4	692
41	Fe₂O₃, hematite, from chloride	humic acid	1 20 mg dm	0 2 g dm	25	1 d	4 4 6 4	0.0005 mol dm NaCl, U	uptake, IR	The uptake is enhanced in the presence of 10⁻ mol U dm⁻³, the sign of ζ at pH 4 4 is reversed to negative with 3 mg dm of humic acid in the presence and absence of L	197
42	Fe₂O₃, hematite, May and Baker	humic acid	5 500 mg dm⁻¹		21	1 d	4 9	0 01 0 1 mol dm KNO₃ 0 10 mol dm Cd(NO₃)₂	uptake, dynamic light scattering	Thick layer of adsorbed humic acid at low pH and high ionic strength	697
43	Fe₂O₃, α, May and Baker	8-hydroxyquinoline	$4\ 2\ 10^{-4}\ 3\ 13\ 10^{-3}$ mol dm	0 1 0 2 g 25 cm		2 d	2 8	NaOH HCl	uptake	Uptake peaks at pH 5	688
44	Fe₂O₃, hematite, May and Baker	1,10-phenanthroline	0 001, 0 005 mol dm	0 1 1 g (10 3 m²/g) 25 cm³		2 h	4 9		uptake		693
45	Fe₂O₃, hematite, monodispersed	L-alanine, glutamic acid, glycine, lysine, L-threonine	$10^{-4}\ 5\ 10^{-}$ mol dm			6 h	2 12		uptake, ζ	IEP is shifted to pH 5 5 in the presence of 10^{-4} mol dm of glutamic acid, further increase of glutamic acid concentration does not affect the electrokinetic curves	698
46	Fe₂O₃, amorphous, from nitrate	chlorophenols	$2\ 10\ 10^{-}$ mol dm		22	15 min	5 4, 5 9	0 05 mol dm⁻³ NaClO₄	IR, uptake		680
47	Fe₂O₃	dicyclohexylamine, dicyclohexylammonium nitrate (III)	$10\ 10^{-}$ mol dm	10 mg 100 cm	25		2 11	10⁻³ mol dm⁻³ KCl	ζ	The IEP is shifted to pH 4 in the presence of 1 7 10^{-4} mol dm⁻³ dicyclohexylammonium nitrate (III)	568

TABLE 4.4 Continued

	Adsorbent	Adsorbate	Total concentration	Solid to liquid ratio	Equilibration time	T	pH	Electrolyte	Method	Result	Ref.
48	Fe_2O_3, hematite, Baker	fulvic acid, natural organic matter	1 50 mg C dm^{-3}	0.2% w w (10 m^2/g)	1 d	room	3 11	0.01, 0.1 mol dm^{-3} NaCl, 0.01 mol dm^3 Na$_2$SO$_4$, 0.001 mol dm^3 Na$_3$PO$_4$	uptake	The NOM and FA adsorption isotherms (at pH 4 and 6.5) are almost identical. The NOM uptake decreases as pH increases; it is insensitive to the ionic strength, but it is depressed in the presence of sulfate and phosphate.	20
49	Fe_2O_3, hematite, monodispersed	humic acid, marine colloidal matter	10, 16 mg dm^{-3}	3.3, 9 g (17.4 m^2 g TEM) dm^3	1 2 d	25	1 10	10^{-1} mol dm^{-3} NaClO$_4$, CO$_2$	uptake, ζ	Neutral molecules of acids are the components, the subscript in L indicates different organic ligands (not the number of ligands L in one molecule). humic acid $\equiv Fe_wL_1$ log K 10.5 $-Fe_wL_2$ log K 11.2 $\equiv Fe_wL_3$ log K 8.38 $\equiv Fe_wOH_2 \cdots L_a^-$ log K 11.6 $-Fe_wOH_2 \cdots L_3^-$ log K 13 marine colloidal matter $\equiv FeL_1$ log K 9.24 $\equiv FeL_2$ log K 7.26 $\equiv FeOH_2 \cdots L_3$ log K 7.51 $\equiv FeOH_2 \cdots L_a^-$ log K 9.14 In the presence of 0.05 ppm of humic acid the electrokinetic behavior in systems equilibrated with atmospheric CO$_2$ is not influenced. With 0.5 ppm humic acid the IEP is shifted to pH 6 (originally 8.5).	207, 208
50	Fe_2O_3, hematite, synthetic	4-nitro-2-aminophenol, 4-nitrocatechol, 4-nitro-1,2-phenylenediamine	5×10^{-5} mol dm^{-3}	35, 354 m^2 dm^3	> 30 min	25	2 10	0.1 mol dm^{-3} NaCl	uptake	Low uptake of 4-nitro-1,2-phenylenediamine.	685
51	$Fe_2O_3 \cdot H_2O$, from nitrate	cysteine	10^{-4} 10^{-2} mol dm^{-3}	800, 1010 m^2 dm^3	2 min	25	4-9	0.005 mol dm^{-3} NaCl	uptake	The uptake peaks at pH 7. Reductive dissolution of the adsorbent when the equilibration time is longer than 2 min.	589
52	$Fe_2O_3 \cdot H_2O$	glutamic, picolinic, 2,3-pyrazinedicarboxylic acid	10^{-5} 10^{-3} mol dm^{-3}	10^{-3} mol Fe dm^3 (182.5 m^2 g)	2 h	25	4-10	0.1 mol dm^{-3} NaNO$_3$	uptake	Uptake of picolinate peaks at pH 5. Uptake of 2,3-pyrazinedicarboxylic acid peaks at pH 4.5.	222
53	FeOOH, goethite, hydrolysis of nitrate	2,4 dinitrophenol	5 1×10^{-5} mol dm^{-3}	10 g (31 m^2/g) dm^3	15 30 min	25	3-10	0.1 mol dm^{-3} NaCl, 0.003 mol dm^{-3} buffer (acetate, ammonia)	uptake	The uptake from buffer peaks at pH 4. The uptake is depressed when the buffer is replaced by 0.1 mol dm^{-3} NaCl.	677

No.	Adsorbent	Adsorbate	Concentration	Solid/solution	Time	T	pH	Electrolyte	Method	Comments	Ref.
54	FeOOH, goethite	fulvic acid	2, 20 ppm	70 mg (50 m²/g)/dm³	1 d	20	3–10	0.01 mol dm⁻³ NaClO₄	uptake	The uptake of fulvic acid (20 ppm) is rather insensitive to the pH.	148
55	FeOOH, goethite, synthetic, 3 samples	humic acid	1.08 27 mg dm⁻³	0.1, 0.15 g (11–18 m²/g)/dm³	16 h	20	4–9	NaCl + NaHCO₃	uptake, ζ	The uptake (14 mg humic acid/dm³) decreases from 35% at pH 4 to 10% at pH 9. Uptake from MgSO₄+CaSO₄ medium was also studied.	699
56	FeOOH, goethite (or FeOOH), from nitrate	humic acid	1 20 mg dm⁻³	50 mg/dm³	2 d	room	2–12	0.05 mol dm⁻³ NaClO₄	uptake ζ	The pH₅₀ = 7 for 1 mg/dm³ and 3 for 10 mg/dm³. The IEP is shifted to pH 2 at 10 mg/dm³.	130
57	FeOOH, goethite, hydrolysis of nitrate	natural organic matter	8.3×10⁻⁵ 8.3×10⁻³ mol dm⁻³	0.06–4 g/dm³	1 h	20 ±2	3–9	0.01 mol dm⁻³ KNO₃, 10⁻⁴–10⁻² mol dm⁻³ Ca(NO₃)₂	uptake	The [Ca] does not exert significant effect on uptake of NOM.	700
58	FeOOH, goethite	catechol, dihydroxynaphthalenes, fulvic acid, phenol	10⁻⁶ 10⁻³ mol dm⁻³	1.6 g (79 m²/g)/dm³	1 d		3–10	0.01 mol dm⁻³ NaCl	uptake	In the log K values listed below a monovalent anion was always chosen as a component. catechol H₂L ≡FeH₃L log K = 14.12 fulvic acid H₄L ≡FeH₃L log K 13.4 ≡FeH₂L⁻ log K 11.2 ≡FeHL²⁻ log K 6.9 ≡FeOL³⁻ log K –14	627
59	FeOOH, goethite	catechol, dihydroxynaphthalenes, humic acid, phenol	10⁻⁶ 10⁻³ mol dm⁻³	1.6 g (79 m²/g)/dm³	1 d		3–10	0.01 mol dm⁻³ NaCl	uptake	For 5×10⁻⁵ mol dm⁻³ phenol and 2,7 dihydroxynaphthalene the uptake is < 7% over the entire studied pH range. At the same concentration the 2,3 dihydroxynaphthalene shows cation like adsorption edge with pH₅₀ < 3 and the uptake of catechol increases from 40% at pH 4 to 80% at pH 8.	628
60	FeOOH, goethite, synthetic	chlorophenols	10⁻³ mol dm⁻³		15 min	22	5.9		IR		680
61	FeOOH, goethite	fulvic acid, humic acid	0.1 5 g dm⁻³		1 d		3–7	0.1 mol dm⁻³ NaCl	uptake, IR		695
62	FeOOH, goethite, synthetic	4-nitro-2-, aminophenol 4-nitrocatechol, 4-nitro-1,2-phenylenediamine	5×10⁻⁵ mol dm⁻³	0.5–10 g (31 m²/g)/dm³	>30 min		2 10	none, 0.1 mol dm⁻³ NaCl	uptake	The uptake of 4-nitrocatechol and 4-nitro-2-aminophenol peaks at pH 6.5 and is slightly depressed at high ionic strength. Low uptake of 4-nitro-1,2-phenylenediamine.	685
63	In₂O₃	methylene blue	10⁻⁴ 10⁻³ mol dm⁻³	1–5 g 10 cm³			natural	0.1 mol dm⁻³ KCl	uptake		696
64	MnO₂	methylene blue	10⁻⁴–10⁻³ mol dm⁻³		>1 d			0.1 mol dm⁻³ KCl	uptake		679

TABLE 4.4 Continued

	Adsorbent	Adsorbate	Total concentration	Solid to liquid ratio	Equilib- ration time	T	pH	Electrolyte	Method	Result	Ref.
65	Nb_2O_3	methylene blue	10^{-4} 10^{-3} mol dm⁻³	1 5 g 10 cm³			natural	0.1 mol dm⁻³ KCl	uptake		696
66	NiO	methylene blue	10^{-4} 10^{-3} mol dm⁻³	1 5 g 10 cm³			natural	0.1 mol dm⁻³ KCl	uptake		696
67	PbO_2	methylene blue	10^{-4} 10^{-3} mol dm⁻³	1 5 g 10 cm³			natural	0.1 mol dm⁻³ KCl	uptake		696
68	PbO_2	α-naphthylamine	10^{-3} mol dm⁻³				7 10	0.01 mol dm⁻³ KCl	ζ	Only positive ζ up to pH 10 (pristine IEP at pH 8.2).	701
69	RuO_2, thermal decomposition of $RuCl_3$ at 420 C	methyl viologen	0.0005–0.05 mol dm⁻³	1 2 g (21.5 m²/g) 50 cm³		20	4-8	nitrate	proton stoichiome- try	The charging curves obtained at different concentrations of methyl viologen intersect at pH 5.3 (pristine PZC 5.75).	702
70	RuO_2, thermal decomposition of $RuCl_3$ at 420 C	methyl viologen	10^{-5} 10^{-3} mol dm⁻³	40–50 mg (21 5 m²/g) 8 cm³	1 d	20	7	0.005–0.1 mol dm⁻³ KNO_3	uptake, ζ	The uptake is depressed when the ionic strength increases	703
71	Sc_2O_3	methylene blue	10^{-4} 10^{-3} mol dm⁻³	1 5 g 10 cm³			natural	0.1 mol dm⁻³ KCl	uptake		696
72	SiO_2, quartz, reagent grade	albumin (bovine serum)	0 2 5 g dm⁻³	0.3 g (11 m²/g) 5 cm³	2 h	40	3 2 9 5	10^{-3} mol dm⁻³ KNO_3	uptake	Langmuir type adsorption isotherms at constant pH, the uptake peaks at pH 4.5.	676
73	SiO_2, Davison	dipyridyl	0 005 mol dm⁻³	0.1 1 g (600 m²/g) 25 cm³	2 h	20	3 9	0.1 mol dm⁻³ NaCl	uptake	The uptake peaks at pH 6.	692
74	SiO_2	2,2' dipyridyl	10^{-3} mol dm⁻³	1 g (300 m²) 100 cm³		22	3 8	0 1 mol dm⁻³ NaCl, 10^{-3} mol dm⁻³ Zn	uptake	In absence of Zn the uptake peaks at pH 5, the uptake is enhanced by Zn at pH 5.	345
75	SiO_2	2,2' dipyridyl	10 mol dm⁻³	10 g (300 m²/g)	1 d	20	5 2 7 2	0-0 1 mol dm NaCl, 1 2×10⁻⁴ mol dm Cu	uptake	2,2' dipyridyl was added to SiO_2 with preadsorbed Cu.	318
76	SiO_2, quartz	dissolved organic carbon	9 4 mg dm	24 g (5 m²/g) dm	10 h	22	2 6 5	0 01 mol dm NaCl	uptake	20% uptake at pH 2.2 Uptake at pH 4 is negligible	508
77	SiO_2, quartz	fulvic acid	2, 20 ppm	50 mg (10 m²/g) dm	1 d	20	3 9	0.01 mol dm⁻³ $NaClO_4$	uptake	Less than 10% uptake at pH 3 9	148

No.	adsorbent	adsorbate	concentration	solid	time	t/°C	pH	electrolyte	method	comments	ref.
78	SiO₂, amorphous, Aldrich	fulvic acid	10 10 mg dm				6 5	0 001 mol dm NaCl	ζ	The electrokinetic potential is not affect by fulvic acid	678
79	SiO₂ gel, Fuji	humic acid	30 mg dm	60 mg 5 cm³	10 d		1 12	10 mol dm NaClO₄	uptake	pH₀ 3 8	324
80	SiO₂, gel, Davison	8-hydroxyquinoline	1 7 10⁴ 1 38×10 mol dm	0 1 0 2 g 25 cm	2 d		2 6	NaOH HCl	uptake	Uptake peaks at pH 5 5	688
81	SiO₂	methylene blue	10⁴ 10 mol dm	0 1 1 g (600 m² 25 cm	1 d		4 7	0 1 mol dm KCl	uptake		679
82	SiO₂, Davison	1,10-phenanthroline	0 001,0 005 mol dm		2 h				uptake	The uptake peaks at pH 6	693
83	SiO₂, Davison	quinoline	0 004, 0 017 mol dm	0 1 0 5 g (600 m² 25 cm	2 h		2 10		uptake	The uptake peaks at pH 6 7	694
84	SiO	2 aminomethylpyridine, 3,4 lutidine, picolinic acid, pyridine, 2 pyridine methanol, 2,2', 6', 2 terpyridine	10 mol dm	1 g 25 cm	1 d	25	1 9	0 1 mol dm NaNO₃	uptake	Uptake of 2,2', 6' 2' terpyridine peaks at pH 5, of 3,4 lutidine at pH 7, of pyridine at pH 6, of 2 aminomethylpyridine at pH 8.5, of 2 pyridine methanol at pH 6, the uptake of picolinic acid is rather insensitive to pH	638
85	SiO₂, amorphous, Zeo 049, Huber	fulvic and humic acid	8 4, 15 mg dm	111 m² 53	1 d		2 12	0 001 mol dm NaClO₄, 10⁻⁴ mol dm Ni	uptake	Ni enhances the uptake of humic acid at pH 4; Ni effect on uptake of fulvic acid is rather insignificant	682
86	SiO₂, silica, Aerosil 380 and Kieselgur	nitrobenzenes, nitrophenols	5 10 3 10⁴ mol dm	5 200 g (360 m² g) dm	0 5 1 h	21	4	0 1 mol dm chloride	uptake		686
87	SnO	methylene blue	10⁴ 10 mol dm	1 5 g 10 cm	1 d		natural	0 1 mol dm KCl	uptake		679
88	ThO	methylene blue	10⁴ 10 mol dm					0 1 mol dm KCl	uptake		696
89	TiO₂, rutile, reagent grade	albumin (bovine serum)	0 2 g dm mol dm	0 3 g (5 5 m² g)	2 h	40	3 2 9 5	10 mol dm KNO	uptake	Langmuir type adsorption isotherms at constant pH, the uptake, peaks at pH 5 5	676
90	TiO₂, anatase	catechol	10⁻² 2 10 mol dm	9 6 g (90 m² g) dm	30 min	25	9 3	0 1 mol dm NaClO₄	uptake		387
91	TiO₂, P 25, Degussa	catechol	10 10 mol dm	0 8 g (51 m² g 50 cm	30 min	25	3 11	none 0 01 mol dm KCl	uptake, ζ	The IEP is shifted to pH 5 5 in the presence of 5 10 mol dm catechol Catechol H₂L and TiOH are the components TiL log K -3; (Ti)₂L log K 3 2	704

TABLE 4.4 Continued

	Adsorbent	Adsorbate	Total concentration	Solid to liquid ratio	Equilibration time	T	pH	Electrolyte	Method	Result	Ref.
92	TiO_2, rutile, Tioxide	dissolved organic carbon	9.4 mg dm^3	5 g (9 m^2/g) dm^3	10 h	22	3-8	0.01 mol dm^3 NaCl	uptake	Uptake peaks at pH 4.	508
93	TiO_2, synthetic, from $TiCl_4$	lysine	0.0001–0.003 mol dm^3		30 min		3.6, 11.5	0.005 mol dm^3 $(CH_3)_4NClO_4$ or 0.05 mol dm^3 KCl + HEPES buffer	ATR-FTIR spectroscopy	Maximum uptake at pH 9.8 (IEP of lysine). There is no substantial difference between the spectrum of adsorbed and dissolved lysine .	705
94	TiO_2	methylene blue	10^4 2×10^3 mol dm^3		>1 d		2 7	0.1 mol dm^3 KCl	uptake		679
95	TiO_2, P 25, Degussa	3-amino-2-naphthol, 2-aminophenol, 2,4-dinitrophenol, 2,6-dinitrophenol, 4-methyl-2-aminophenol, 4-nitro-2-aminophenol, 2-aminophenol, 2-pyridinemethanol	4-5×10^5 mol dm^3	7, 10 g (55 m^2/g) dm^3	15 30 min	25	3 10	0-0.11 mol dm^3 NaCl, NaNO$_3$ + buffer (acetate or ammonia)	uptake	The uptake is depressed at high ionic strengths.	677
96	TiO_2, P 25, Degussa	substituted 2-aminophenols, catechols, 1,2-phenylenediamines	10 5×10^3 mol dm^3	0.5 10 g (40 m^2/g) dm^3	30 min		2 10	0.001, 0.1 mol dm^3 NaCl	uptake	Uptake peaks at pH ≈6.	706
97	TiO_2, anatase	8-aminoquinoline, 2,2' dipyridyl, o-phenylenediamine	8×10^{-4} mol dm^3	9 g (90 m^2/g) dm^3		25	3 10	0 1 mol dm^3 KNO$_3$ or NaClO$_4$+ 0 1.6×10^3 mol dm^3 Cu	uptake	In absence of Cu uptake of 2,2' dipyridyl and 8-aminoquinoline increases from 10% at pH 3 to 40% at pH 5 10, for o-phenylenediamine about 5% uptake at pH 5 10. Cu enhances the uptake of 2,2' dipyridyl and 8-aminoquinoline at pH 7- 9. but at pH 4 6 the uptake is depressed.	388
98	TiO_2, 3 samples	4-nitro-2-aminophenol, 4-nitrocatechol, 4-nitro 1,2-phenylenediamine	5×10 mol dm^3	40 m^2 dm^3	> 30 min		2 10	none, 0 1 mol dm^3 NaCl	uptake	The uptake is rather insensitive to the high ionic strength. The uptake of 4 -nitro-2-aminophenol peaks at pH 6. For 4-nitrocatechol different results for different samples of TiO$_2$, cation like adsorption edge. pH$_{50}$ 2 or maximum of uptake at pH 6. Low uptake of 4-nitro-1, 2-phenylenediamine	685
99	Y_2O_3	methylene blue	10^4 10^3 mol dm^3	1 5 g 10 cm			natural	0 1 mol dm^3 KCl	uptake		696
100	V_2O_3	methylene blue	10^4 10 mol dm					0.1 mol dm KCl	uptake		679
101	WO_3	methylene blue	10^4 10^3 mol dm^3	1 5 g 10 cm^3	1 d		natural	0.1 mol dm^3 KCl	uptake		696

No.	Adsorbent	Adsorbate	Concentration	Solid/amount	Time	T	pH	Electrolyte	Method	Result/Comments	Ref.
102	ZrO_2, monoclinic, reagent grade	albumin (bovine serum)	0.2 5 g dm^{-3}	0.3 g (10.3 m^2/g), 5 cm^3	2 h	40	3.2 9.5	10^{-3} mol dm^{-3} KNO$_3$	uptake	Langmuir type adsorption isotherms at constant pH, the uptake peaks at pH 5	676
103	ZrO_2	albumin (bovine serum)	50 400 mg dm^{-3}	50 mg (21 m^2/g), 10 cm^3	2 h	20	4.15, 5.35	0.01 mol dm^{-3} acetate buffer	uptake	Pre-adsorbed phosphate reduces the uptake by factor 2.	645
104	ZrO_2	methylene blue	10^{-4} 10^{-3} mol dm^{-3}	1 5 g 10 cm^3			natural	0.1 mol dm^{-3} KCl	uptake		696
105	ZrO_2	nicotinate	10^{-4} 7 10^{-3} mol dm^{-3}	1 4 g 7 cm^3	1 d	room	2 8.5	0 1 mol dm^{-3} KCl	uptake		516

Composite materials containing Al

No.	Adsorbent	Adsorbate	Concentration	Solid/amount	Time	T	pH	Electrolyte	Method	Result/Comments	Ref.
106	76% SiO$_2$ 14% Al$_2$O$_3$ 10% water, synthetic, amorphous	2,2'-dipyridyl	0 005 mol dm^{-3}	0.1 1 g (95 m^2/g) 25 cm^3	2 h		4-7		uptake	The uptake peaks at pH 6.	692
107	76% SiO$_2$ 14% Al$_2$O$_3$ + 10% water, synthetic, amorphous	8-hydroxyquinoline	4 3×10^{-4} 2 07×10^{-3} mol dm^{-3}	0.1-0.2 g 25 cm^3	2 d		5 7.5	NaOH HCl	uptake	The uptake is rather insensitive to pH, broad maximum at pH 7.	688
108	76% SiO$_2$ 14% Al$_2$O$_3$ 10% water, synthetic, amorphous	1, 10-phenanthroline	0.001, 0.005 mol dm^{-3}	0.1 1 g (95 m^2/g) 25 cm^3	2 h		4-7		uptake	The uptake peaks at pH 6.	693
109	76% SiO$_2$ 14% Al$_2$O$_3$ + 10% water, synthetic, amorphous	quinoline	0.004, 0 017 mol dm^{-3}	0.1 0.5 g (95 m^2/g) 25 cm^3	2 h		3 11		uptake	The uptake peaks at pH 6.	694

Clay minerals

No.	Adsorbent	Adsorbate	Concentration	Solid/amount	Time	T	pH	Electrolyte	Method	Result/Comments	Ref.
110	beidellite, SBCa-1, natural and milled	fulvic acid	0.01 10 mg dm^{-3}				6.5	0.001 mol dm^{-3} NaCl	ζ	The electrokinetic curves are not significantly affected by fulvic acid.	707
111	chlorite	humic acid	0.01 10 mg dm^{-3}				natural	artificial sea water, different salinities	ζ		412

TABLE 4.4 Continued

Adsorbent	Adsorbate	Total concentration	Solid to liquid ratio	Equilibration time	T	pH	Electrolyte	Method	Result	Ref
112 illite	humic acid	0.01 10 mg dm				natural	artificial sea water, different salinities	ζ		412
113 Na kaolinite	2,2'-dipyridyl	0 005 mol dm	0 1 1 g (7 9 m²/g) 25 cm³	2 h		2 8		uptake	The uptake peaks at pH 4	692
114 kaolinite	dissolved organic carbon	9 4 mg dm	5 g (12 2 m²/g) d m	10 h	22	3 7 5	0 01 mol dm NaCl	uptake	The uptake decreases from 40% at pH 3 to 10% at pH 7 5	508
115 kaolinite, Wako	humic acid	30 mg dm	10 mg 5 cm³	10 d		1 12	NaOH HClO₄	uptake	pH_{50} 4 8	324
116 Na-kaolinite, KGa-2	humic acid	1 6 mg OC dm	0 2, 0 4 g (23 m² g dm³	1 d	25	3 11	0 001 0.1 mol dm NaClO₄	uptake, ζ, coagulation	The uptake is enhanced at high ionic strengths, uptake decreases as pH increases but there is substantial uptake even at pH 11 Langmuir type adsorption isotherm at constant pH The ζ is negative over the entire pH range in the presence of 0 5 mg OC dm and the kaolinite dispersions are stabilized by adsorbed humic acid at pH 6	708
117 Na-kaolinite, KGa-2	humic acid	1 6 mg OC dm	1 5, 200 mg (23 m²/g) dm³	1 d	25	3 11	0 001 1 mol dm NaClO₄	uptake, ζ, coagulation, dynamic light scattering	Average particle radius is stable (not increasing in time) at pH 4 in the presence of 1 mg dm³ TOC at 0 001 mol dm³ NaClO₄ With 0.01 mol dm³ NaClO₄ 2 mg/dm³ TOC is required to stabilize the value of particle radius, and with 0 1 NaClO₄ the corresponding concentration is 4 mg/dm³ TOC	709
118 Na-kaolinite	quinoline	0 004 mol dm	0.1-0 5 g (7.9 m²/g) 25 cm³	2 h	21	2 8		uptake	The uptake peaks at pH 6	694
119 K, NH₄ and Cs kaolinite	chlorobenzenes, chlorophenols, dinitronaphthalenes, nitrobenzenes, nitrophenols, phenols	5×10^{-7} 3×10^{-4} mol dm	5 200 g dm	0 5 1 h		3 9	0 1 mol dm chloride	uptake	K_D at pH 4 for 55 compounds are listed for Cs-kaolinite They range from 0 1 to 9000 Temperature dependence of K_D of 4 methyl 2 nitrophenol over the range 3 60 C gives ΔH_{ads} -42 kJ mol	686

No.	Mineral	Species	Concentration	Sample	Time	pH	T (°C)	Electrolyte	Quantity	Remarks	Ref.
120	kaolinite, KGa-1	fulvic and humic acid	8 4, 15 mg dm⁻³	37 m² 53 cm³	1 d	2 12		0.01 mol dm⁻³ NaClO₄, 10⁻⁴ mol dm⁻³ Ni	uptake	Ni effect on the uptake curves of fulvic and humic acid is rather insignificant.	682
121	Na montmorillonite	2,2'-dipyridyl	0.005 mol dm⁻³	0.1 1 g (808 m²/g) 25 cm³	2 h	2 9			uptake, Na release	The uptake peaks at pH 5.	692
122	montmorillonite	humic acid	0.01 10 mg dm⁻³			natural		artificial sea water, different salinities	ζ		412
123	montmorillonite	humic acid	45 245 mg dm⁻³	5 g dm⁻³	2 d	2 8	25	10⁻¹ mol dm⁻³ NaNO₃	uptake	The uptake decreases from 93% at pH 2.3 to 50% at pH 8 The uptake of HA at pH 6 is enhanced in the presence of Pb, Cu or Cd.	439
124	Na and Ca montmorillonite	8-hydroxyquinoline	3.4×10⁻⁴ 3.98×10⁻³ mol dm⁻³	0.1 0.2 g 25 cm³	2 d	2 10		NaOH HCl	uptake, release of Na and Ca	Uptake peaks at pH 5 at high initial concentrations, at low initial concentrations the uptake is rather insensitive to the pH.	688
125	NH₄ montmorillonite	1,10,-phenanthroline	0.005 mol dm⁻³	0.1 1 g (808 m²/g) 25 cm³	2 h	4 9			uptake	The uptake peaks at pH 7.	693
126	Na-montmorillonite	quinoline	0.004, 0.017 mol dm⁻³	0.1-0.5g (808 m² g, glycerol) 25 cm³	2 h	2 12			uptake	The uptake peaks at pH 6.	694
127	Na-montmorillonite, SWy-1	fulvic and humic acid	8 4, 15 mg dm⁻³	75 m² 53 cm³	1 d	2 12		0.01 mol dm⁻³ NaClO₄, 0-5×10⁻⁴ mol dm⁻³ Ni, Pb, Zn, Cd	uptake	Ni enhances the uptake of fulvic and humic acid at pH 7. The enhancement of uptake in the presence of Pb is even more significant.	682
128	ripidolite, CCa-1, natural and milled	fulvic acid	0.01 10 mg dm⁻³			6 5		0.001 mol dm⁻³ NaCl	ζ	The effect of fulvic acid on the ζ potential of natural sample is rather insignificant. The negative ζ potential of milled sample is enhanced in the presence of fulvic acid.	707

Zeolites

No.	Mineral	Species	Concentration	Sample	Time	pH	T (°C)	Electrolyte	Quantity	Remarks	Ref.
129	mordenite, Huber	humic and fulvic acid	8 4, 15 mg dm	111 m. 53 cm	1 d	2 12		0.01 mol dm⁻³ NaClO₄, 10⁻⁴ mol dm⁻³ Ni	uptake	Ni enhances the uptake of humic acid at pH 4; Ni effect on uptake of fulvic acid is rather insignificant.	682

Clays

No.	Mineral	Species	Concentration	Sample	Time	pH	T (°C)	Electrolyte	Quantity	Remarks	Ref.
130	Na-kaolin	8-hydroxyquinoline	3.4×10⁻⁴ 3.37×10⁻³ mol dm	0.1 -0.2 g 25 cm³	2 d	2 8		NaCH HCl	uptake	Uptake peaks at pH 7	688
131	kaolin	1,10,-phenanthroline	0.001, 0.005 mol dm⁻³	0.1 1 g (8 m²/g) 25 cm³	2 h	4 8			uptake	The uptake peaks at pH 6	693

TABLE 4.4 Continued

Adsorbent	Adsorbate	Total concentration	Solid to liquid ratio	Equilib- ration time	T	pH	Electrolyte	Method	Result	Ref.
Other materials										
132 ultisol saprolite, original and treated to remove Fe	humic acid	11 mg C dm^{-3}	1.3 g (46, 68 m^2/g) dm^3	30 min	25	4.5–8.5	10^{-1} mol dm^{-3} NaClO$_4$	uptake, ζ	90% uptake rather insensitive to pH.	133
133 zeolitic tuff	chlorophenols, phenol	40 ppm	0.5 g(75 m^2 g. methylene blue) 25 cm^3	1 d	20			uptake	The uptake increases with T (20–60 C; with some scatter).	710

that, e.g. particular samples referred to as humic acid can differ in their chemical structure, thus the differences in sorption properties must not be considered as a discrepancy.

The number of publications reporting adsorption of organic compounds (other than carboxylic acids) on inorganic materials from aqueous solutions is rather limited. There are single or very few entries for each compound (except for humic substances), thus, it is rather difficult to assess the consistence of the results from different sources, obtained at similar or different experimental conditions. Organic compounds can be more efficiently removed from aqueous solution using activated carbon and other organic adsorbents (Chapter 6) rather than inorganic adsorbents.

The uptake (pH) curves with a maximum (Fig. 4.6(C)) are often encountered with adsorbates having weak acidic (e.g. catechols), weak basic (amines) or amphoteric (amino acids) character. It was already discussed in section IIB, that very weak acids show a maximum of uptake at $pH \approx pK_{a1}$ (cf. Table 4.2). Apparently this type of behavior is common for weak Brønsted acids.

For humic acid many types of uptake curves were reported ranging from anion type adsorption edge (Fig. 4.5(B)) and uptake curves with a maximum (Fig. 4.6(C)) to almost pH independent uptake. These discrepancies might be partially due to difference in the chemical nature (molecular weight, functional groups) between particular samples referred to as humic acid. Humic acid is rather a mixture than single compound. Fractionation occurs by adsorption from such mixtures, i.e. the proportions of particular components in solution after equilibration with the adsorbent are different from those in the original solution.

IV. SURFACTANTS

Surfactants are customarily considered as a separate class of adsorbates, because of their unique properties. A surfactant molecule consists of a hydrophobic (organophilic) part (tail) and a hydrophilic part (head). This structure implies the amphiphilic character of surfactants (affinity to water and to hydrocarbons) and their strong tendency to adsorb irrespective of the nature of the adsorbent. The tail is a linear hydrocarbon chain C_8–C_{20}, and the head groups define the nonionic, cationic or anionic character of the surfactants. When the tail is too long, its hydrophobic character prevails and the molecule looses amphiphilic character. When the tail is too short the hydrophilic character of the head group prevails and the entire molecule becomes hydrophilic. The number of chains and head groups is not limited to one, e.g. double chained and zwitterionic surfactants are known. Adsorption properties in a series of analogs (surfactants having the same head group and different chain length) change gradually with the number of carbon atoms, and many publication were devoted to systematic studies of such effects.

In contrast with inorganic salts whose adsorption at water–air interface is negative, surfactants are strongly adsorbed at water–air interface, and this results in depression of the surface tension. In very dilute solutions of surfactants, adsorption at water–air interface can lead to substantial depletion of the solution. Orientation of surfactant molecules at air–water interface is illustrated in Fig. 4.64.

The hydrophilic part of nonionic surfactants usually consists of a few ethylene oxide (-CH_2- CH_2-O-) units, also derivatives of polyalcohols (e.g. anhydrosorbitol)

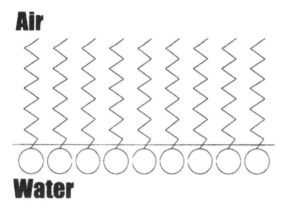

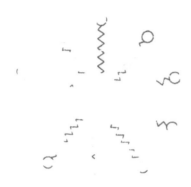

FIG. 4.64 Surfactant molecules at water/air interface and in a micelle (schematically).

are widely applied as nonionic surfactants. Sodium dodecyl sulfate $CH_3(CH_2)_{11}O$-SO_3Na is a typical example of anionic surfactant. In dilute aqueous solutions this compound is partially dissociated into amphiphilic anions $CH_3(CH_2)_{11}OSO_3^-$ and small hydrophilic cations Na^+.

Cetyltrimethylammonium bromide $CH_3(CH_2)_{15}N(CH_3)_3Br$ is a typical cationic surfactant. Dissociation of this compound results in amphiphilic cations and small hydrophilic anions. This classification of surfactants is similar to the classification of strongly interacting compounds (other than surfactants) in Sections II and III into cations (Section II A), anions (Section II B) and electroneutral and zwitterionic organic compounds (Section III). The adsorption of anionic surfactants is indeed enhanced, when the adsorbent carries high positive surface charge (at low pH for materials listed in Tables 3.1, and 3.3–3.5), and adsorption of cationic surfactants is more pronounced at high pH, and the adsorption of nonionic surfactants is often rather insensitive to the pH. However, the mechanisms of surfactant adsorption, and experimentally observed adsorption isotherms of surfactants are very different from the compounds discussed in Sections II and III.

Surfactants are widely used as modifiers of surface properties of materials, and many surfactants are known under their commercial names. Aerosols and Igepons are series of anionic surfactants, Hyamines are a series of cationic surfactants, and

Spans, Tritons, and Tweens are series of nonionic surfactants. Some of these products do not represent specific chemical compounds but rather mixtures of similar compounds having different number of ethylene oxide segments and/or different lengths of hydrocarbon chain, and the apparent adsorption isotherm is a result of interaction of particular components of the mixture in solution and on the surface.

In sufficiently dilute aqueous solutions surfactants are present as monomeric particles or ions. Above critical micellization concentration CMC, monomers are in equilibrium with micelles. In this chapter the term "micelle" is used to denote spherical aggregates, each containing a few dozens of monomeric units, whose structure is illustrated in Fig. 4.64. The CMC of common surfactants are on the order of 10^{-4}–10^{-2} mol dm^{-3}. The CMC is not sharply defined and different methods (e.g. breakpoints in the curves expressing the conductivity, surface tension, viscosity and turbidity of surfactant solutions as the function of concentration) lead to somewhat different values. Moreover, CMC depends on the experimental conditions (temperature, presence of other solutes), thus the CMC relevant for the experimental system of interest is not necessarily readily available from the literature. For example, the CMC is depressed in the presence of inert electrolytes and in the presence of apolar solutes, and it increases when the temperature increases. These shifts in the CMC reflect the effect of cosolutes on the activity of monomer species in surfactant solution, and consequently the factors affecting the CMC (e.g. salinity) affect also the surfactant adsorption.

It should be emphasized that the micelles are very dynamic, namely, the monomeric units undergo constant exchange between a micelle and solution, and between particular micelles, and the life time of a single micelle is also limited.

Dimers, trimers, etc. are practically absent in surfactant solutions. Spherical micelles are relatively monodispersed, and increase in the surfactant concentration leads to increase in the number of aggregates, but not in their size. The aggregation number N is characteristic for given surfactant (N increases with the tail length in a series of analogs, and decreases in a series of nonionic surfactants with the same tail length, when the number of ethylene oxide units increases), and it depends on the experimental conditions. The factors depressing the CMC usually induce an increase in N.

A spherical micelle should be rather considered as idealized picture, and non-spherical aggregates are often formed due to geometric packing requirements. Very high surfactant concentrations induce transition of spherical micelles into other shapes (rods, disks, vesicles, lamellar structures).

The role of speciation in solution in adsorption is discussed in Section II (cf. Also Fig. 4.1) and in Chapter 5 for small ions. Also with of ionic surfactants the presence of different species (incomplete dissociation) has been taken into account in adsorption modeling [711]. In solutions of ionic surfactants above the CMC a large fraction of the countercharge is firmly bound with the micelle, and a small fraction of the countercharge constitutes diffuse charge. It should be emphasized that most studies with ionic surfactants are carried out in absence of strongly interacting counterions, namely anionic surfactants are used in form of alkali metal salts, and cationic surfactants—in form of halides. Presence of strongly interacting counterions completely changes the solution chemistry of surfactants, e.g. interaction of anionic surfactants with multivalent metal cations usually results in precipitation of sparingly soluble salts. This effect is mutual, i.e. surfactants (added intentionally or present as impurities) affect sorption of inorganic ions.

The relationship between adsorption of surfactants on the one hand and of small organic molecules on the other has been also studied. At sufficiently high adsorption density, the adsorbed surfactant is able to "solubilize" water insoluble organic compounds, and this phenomenon is termed adsolubilization. This may result in uptake of certain adsorbates which do not adsorb on the material of interest in the absence of surfactant. Adsolubilization of weak acids and bases is strongly pH dependent and it often shows a maximum at pH close to the pK_a of the solute.

It is convenient to discuss the adsorption of ionic and nonionic surfactants separately [712]. The adsorption of nonionic surfactants is rather insensitive to the surface charging and many publications report results obtained without pH control. On the other hand the uptake of ionic surfactants on inorganic materials is pH dependent, and log–log adsorption isotherms at constant pH often assume the shape schematically presented in Fig. 4.65. Such behavior is referred to as four-region adsorption isotherm (I-IV) [713]. Region I is the Henry range (slope 1) and represents adsorption of isolated molecules. The onset of associative adsorption at surfactant concentration in solution on the order of 0.1 CMC and surface coverage on the order of 0.001 of saturation coverage (defined by the plateau level, region IV) results in a sharp increase in the slope of the log–log adsorption isotherm. In region II aggregates called hemimicelles are formed on the surface "head-on". The idea of hemimicelles dates back to the 1950s. The difference in slope between region II and III is not always very clear. The decrease in slope is interpreted as transition from "head-on" aggregates to aggregates containing "head-on" and "head-out" surfactant molecules. The transition from region II to III coincides with the common intersection point of adsorption isotherms obtained at different ionic strengths. The uptake in region II is depressed and the uptake in region III is enhanced at high ionic strength. Finally, above the CMC further increase in surfactant concentration in solution does not induce increase in adsorption density (region IV, plateau).

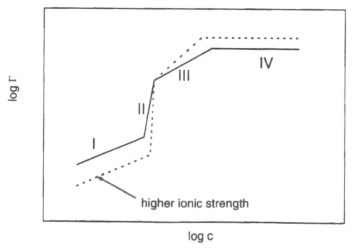

FIG. 4.65 Typical log-log adsorption isotherm of a ionic surfactant (schematically).

The four-region adsorption isotherm reflects an important difference between adsorption of ionic surfactants on the one hand and of small ions on the other. Namely, the Gibbs energy of adsorption of surfactants is a sum of contributions due to head group–solid, tail–solid and tail–tail interactions. The terms involving the tail are approximately proportional to the chain length. Adsorption of surfactants from aqueous solution on nonpolar materials (Chapter 6) is strongly affected by chemical interaction tail–solid, but in adsorption on inorganic adsorbents, the tail–tail interaction prevails, and the interaction of the surfactant with the surface is chiefly electrostatic.

The adsorption isotherms of nonionic surfactants on inorganic materials are similar to those shown in Fig. 4.65, but the difference in slope between regions II and III is usually less distinct, and they often merge into one region with approximately constant slope.

Ionic surfactants can be used to stabilize dispersions in nonpolar solvents (e.g. hydrocarbons), namely, in contrast with inorganic electrolytes that are insoluble in nonpolar solvents, ionic surfactants show some degree of solubility. Micelization occurs also in nonpolar solvents, but unlike in Fig. 4.64, the head groups are pointed inwards. This type of micelles is referred to as reversed micelles. Reversed micelles can solubilze water and simple inorganic ions. Anionic surfactants induce positive charging of inorganic materials in apolar solvents. Namely, small counterions are adsorbed on mineral surfaces, and large surfactant ions are stable in solution. Accordingly, cationic surfactants induce negative surface charge in apolar solvents. This tendency is opposite to the trend observed in aqueous system. Adsorption of surfactants was extensively discussed in recent reviews, e.g. [714–716].

Presentation of experimental data on surfactant adsorption in this chapter is limited to a few very recent studies. These results are compiled in Table 4.5.

The four-region adsorption isotherms (Fig. 4.65) can differ in the position of transition points (I/II, II/III, and III/IV) and in slopes in the regions II and III (the slopes in regions I and IV are fixed at 1 and 0, respectively), but the dependence of these parameters on the nature of the adsorbate and the adsorbent, and on the experimental conditions has not been fully recognized. Many publications report single adsorption isotherms obtained at certain pH, ionic strength, temperature, equilibration time etc., and systematic studies of the effect of one or more of these parameters on the adsorption isotherm, or involving a series of surfactants or a series of adsorbents are rare.

Adsorption isotherms of decyl pyridinium in the presence of 0.01 mol dm^{-3} of alkali chloride, at pH 8 on alumina (Ref. 737) and on silica (Ref. 722) are compared in Fig. 4.66. The solution concentrations and adsorption density in Fig. 4.66 were normalized to the transition point II/III (Ref. 737), which was identified with the intersection of adsorption isotherms obtained at various ionic strengths (Ref. 722). The log–log adsorption isotherm obtained for silica does not show clear changes in slope over the studied concentration range, and the slope of this isotherm corresponds to the slope in region III for alumina. It should be emphasized that silica carries high negative surface charge at pH 8 while for alumina this pH value is close to the PZC. Therefore a head-on orientation of surfactant particles on silica is preferred due to the electrostatics, while for alumina the preference for head-on or head-out orientation is less obvious.

(*Text continues on pg. 502*)

TABLE 4.5 Adsorption of Surfactants

Adsorbent	Adsorbate	Total concentration	Solid to liquid ratio	Equilibration time	T	pH	Electrolyte	Method	Result	Ref.
Al$_2$O$_3$, γ, Johnson-Matthey	n-decylbenzene sulfonate	6×10^{-6} 10^{-2} mol dm^{-3}	0.91 g (79 m^2/g)/dm^3	1 d	40	4	H$_2$SO$_4$	uptake NMR	A four-region adsorption isotherm.	717
Al$_2$O$_3$, γ, Aldrich	dodecylsulfate, Triton X 100	10^{-6} 10^{-2} mol dm^{-3}	0.5 g (159 m^2/g)/30 cm^3	4 d	22	5.3	0.1 mol dm^{-3} NaCl	uptake	A four-region adsorption isotherm for SDS. Slopes in regions I-III and transition points between regions are compared with results taken from literature. Solution concentrations at breakpoints reported in different publications are more consistent than corresponding adsorption densities.	718
Al$_2$O$_3$, α, AKP-50, Sumitomo	Triton TX-100, dodecylsulfonate	10^{-4} 2×10^{-3} mol/dm^3	1–5 g (10 m^2/g)/50 cm^3	>1 h	20	6-10	0.001 mol /dm^3 NaNO$_3$	acousto uptake	Uptake of preadsorbed dodecylsulfonate (10^{-4} mol dm^{-3}) at pH 3 is not influenced by addition of TX-100 (up to 10^{-3} mol dm^{-3}). For 5×10^{-4} mol dm^{-3} dodecylsulfonate the uptake at pH 3 is enhanced in the presence of TX-100. The ζ potential becomes less positive when TX-100 is added at constant dodecyl sulfonate concentration at pH <8.	719
Al$_2$O$_3$, Degussa	cetylpyridinium	2.4×10^{-3} mol/dm^3	10 g (100 m^2/g)/dm^3	3 d	25	10	0.01 mol /dm^3 NaCl, salicylate	calorimetry		720
Fe$_2$O$_3$, hematite	dodecyl sulfate and sulfonate, SDS,	10^{-6} 10^{-2} mol dm^{-3}	0.01 g (20 m^2/g)/100 cm^3		20	3 11	10^{-3} mol /dm^3 NaNO$_3$	ζ, coagulation	When pH = 4 the minimum of stability (low turbidity) is observed at 10^{-5} mol dm^{-3} dodecyl sulfonate,	721

Adsorbent	Adsorbate	Concentration	Solid/solution	Time	T	pH	Electrolyte	Method	Result	Ref.
	with small admixtures of octanol, octadecyl sulfonate	with admixture of 0-1 % of C_{18}							but this minimum disappears on addition of 0.5% dodecanol and 0.2% C_{18}. When pH = 10.5 the CCC (2×10^{-2} mol dm^{-3} NaNO$_3$) is not influenced by dodecyl sulfonate, but it increases on addition of $>10^{-4}$ mol dm^{-3} SDS. The electrokinetic mobility is not influenced by 10^{-4} mol dm^{-3} dodecyl sulfate or sulfonate at pH >8. At pH <6 the sign is reversed to negative.	722
SiO$_2$, Stober	dodecyl pyridinium	10^{-4}, 3×10^{-2} mol/dm^3	0.2 g (13 m^2/g)/10 cm^3	1 d	room	3–9	0.001–0.1 mol /dm^3 KCl	uptake ζ	The uptake at pH 8 is depressed at a high ionic strength for surfactant concentrations in solution $<10^{-3}$ mol dm^{-3} and enhanced above this concentration. At pH 8 the sign of ζ is reversed from negative to positive on addition of 10^{-3} mol dm^{-3} dodecyl pyridinium. This concentration is rather insensitive to the ionic strength.	
SiO$_2$, Aerosil 200, Degussa	cetylpyridinium	$5\ 8\times10^{-3}$ mol/dm^3	10 g (200 m^2/g)/dm^3	1 d	25	4.5–8.5	0.01 mol /dm^3 NaCl, salicylate, 4-aminosal-icylate	calorimetry		720
SiO$_2$, Aerosil 200, Degussa	4-(N,N-dimethyl, N-(n-hexadecyl) ammonium)-	10^{-4}, 7.5×10^{-4} mol/dm^3	0.1–2% w/w, 200 m^2/g	2 d		2.8–11	bromide	EPR	High pH promotes formation of surfactant aggregates on the surface while at low pH the surfactant adsorbs flat on hydrophobic areas.	723

TABLE 4.5 Continued

Adsorbent	Adsorbate	Total concentration	Solid to liquid ratio	Equilibration time	T	pH	Electrolyte	Method	Result	Ref.
SiO$_2$, Spherosil XOB015, washed	2,2,6,6-tertamethy-lpiperidiny-l-N'-oxyl alkanediyl -α-ω-bis (docecyldimethylammonium), alkanediyl = C$_2$H$_4$, C$_4$H$_8$, C$_6$H$_{12}$, C$_{10}$H$_{20}$	10^{-4}–4×10^{-3} mol/kg	0.4 g (29 m^2/g)/10 cm^3	12 h	25	natural	bromide	ζ, uptake		724
SiO$_2$, Spherosil XOB015, raw and HCl washed	ethanediyl-1,2-bis (docecyldimethylammonium), dodecyltrimethylammonium	10^{-4}–4×10^{-3} mol/kg	0.4 g (29 m^2/g)/10 cm^3		25	natural	bromide	ζ, uptake		725
SiO$_2$, Nippon Silica Kogyo	three 2-vinylpyridine-telomers with multihydrocarbon side chains	10^{-6}–2.5×10^{-5} mol/dm^3	0.1 g (17.6 m^2/g)/10 cm^3	1 d	25	6.7	none	uptake, ζ, coagulation	The sign of ζ is reversed to positive at surfactant concentration ≈ 10^{-6} mol dm^{-3}.	726

Adsorbent	Species	Concentration	Solid/area	Time	T	pH	Electrolyte	Method	Remarks	Ref.
SiO_2, Sifraco C-600, original and washed	$C_8H_{17}Ph(OCH_2CH_2)_nOH$, $C_9H_{19}Ph(OCH_2CH_2CH_2)_nSO_4Na$, $n=4$-25	10^{-5} 10^{-3} mol/dm^3	> 1 g (5 m^2)/ 20 cm^3	1 d	25	natural	1–10 g dm^{-3} NaCl, CaCl$_2$	uptake, ζ	Addition of 1 g NaCl/dm^3 depresses the uptake of TX-100, but further addition of NaCl enhances the uptake.	727
SiO_2, layer on silicon wafer	dodecyl-1,3-propylene-pentamethyl bis(ammonium chloride)	0.01–0.08 mol dm^{-3}		1 d		9		uptake	The uptake reaches plateau at cmc.	728
TiO_2, anatase	Triton TX-100, TX-102, TX-165, TX-305, TX-405, dodecylsulfonate	10^{-6}–10^{-2} mol/dm^3	0.01 g (20 m^2/g)/100 cm^3		20	3	0.001 mol /dm^3 NaNO$_3$	ζ	With $\approx 10^{-4}$ mol dm^{-3} TX-100 and TX-165 $\zeta \approx 0$. The ζ potential is clearly positive at lower and higher surfactant concentrations.	719
TiO_2, rutile + anatase, Degussa	cetylpyridinium	1.8×10^{-3} mol/dm^3	10 g (50 m^2/g)	1 d	25	9	0.01 mol /dm^3 NaCl, salicylate, 4-aminosalicylate	calorimetry		720
TiO_2, anatase, Sakei, washed	dodecyl sulfate and sulfonate, with small admixtures of octanol, octadecyl sulfonate	10^{-6}–10^{-2} mol dm^{-3} SDS, with admixture of 0–1% of C$_{18}$	0.01 g (20 m^2/g)/100 cm^3	1 d	20	4, 10.5	10^{-3} mol /dm^3 NaNO$_3$	coagulation	At pH 4 the minimum of stability (low turbidity) is observed at 10^{-5} mol dm^{-3} dodecyl sulfonate, but this minimum disappears on addition of 0.5% dodecanol or 0.2% C$_{18}$. At pH 10.5 the CCC (3×10^{-2} mol cm^{-3} NaNO$_3$) is not influenced by dodecyl sulfonate or sulfate.	721

TABLE 4.5 Continued

Adsorbent	Adsorbate	Total concentration	Solid to liquid ratio	Equilibration time	T	pH	Electrolyte	Method	Result	Ref.
TiO$_2$, Degussa, P25	dodecyl sulfate	10^{-4} – 4×10^{-3} mol dm^{-3}	0.5 g (57 m^2/g)/50 cm^3	1 d	25	4–8	Na	uptake, ζ	When the conductivity (SDS + HCl or NaOH to control the pH) = 10^{-3} μS/cm, the electrophoretic mobility of about -4×10^{-8} m^{2}V^{-1}s^{-1} is rather insensitive to the pH.	729
ZrO$_2$, Z Tech	dodecylamine, dodecyltrimethylammonium, dodecyl sulfate, dodecylbenzylsulfate	5×10^{-5} – 10^{-2} mol /dm^3	6% w/w (15 m^2/g	1 d	25	3–11	0.1 mol /dm^3 KCl	uptake, acousto	Addition of 0.1% of dodecyl sulfate or dodecylbenzylsulfate (dry weight basis) shifts the IEP to pH 6, further addition of anionic surfactants up to 0.8% have minor effect on electrokinetic curves. Dodecyltrimethylammonium up to 0.8% does not affect the electrokinetic curves.	730

Composite materials containing Al

Adsorbent	Adsorbate	Total concentration	Solid to liquid ratio	Equilibration time	T	pH	Electrolyte	Method	Result	Ref.
Mg$_{8.92}$Al(OH)$_{5.84}$(CO$_3$)$_{0.5}$·2.31 H$_2$O, synthetic	dodecylsulfate	8×10^{-4} – 2×10^{-2} mol/dm^3	200 mg (87 m^2/g)/50 cm^3	3 d	25	7, 9	none, 0.1 mol /dm^3 NaCl	uptake, ζ, SEM	The uptake (up to CMC) is rather insensitive to T (25, 40 C) and pH and it is enhanced at high ionic strength.	731

Clay minerals and clays

Adsorbent	Adsorbate	Total concentration	Solid to liquid ratio	Equilibration time	T	pH	Electrolyte	Method	Result	Ref.
kaolin	C$_8$H$_{17}$Ph(OCH$_2$CH$_2$)$_{9.5}$OH, C$_9$H$_{19}$Ph(OCH$_2$CH$_2$)$_n$SO$_4$Na, n = 4–25	10^{-5} 10^{-3} mol/dm^3	>1 g (19.3 m^2)/20 cm^3	1 d	25	natural	1–10 g dm^{-3} NaCl, CaCl$_2$	uptake, ζ	The uptake of TX-100 is enhanced at a high ionic strength.	727

Adsorbent	Adsorbate	Concentration	Solid	Time	T	pH	Electrolyte	Method	Result	No.
Na-kaolinite	$RR_1(CH_3)_2NBr$	10^{-7}–10^{-2} mol/dm^3		10 h		6	0.001 mol/dm^3 NaCl	uptake, ζ	The sign of ζ is reversed from negative to positive at surfactant concentrations on the order of 10^{-4}–10^{-3} mol/dm^3. The logarithm of this concentration is a linear function of the number of carbon atoms in the surfactant molecule in a series of single chain surfactants and in a series of double chain surfactants (different slopes).	732
Na-montmorillonite	benzyldimethyldodecylammonium	5×10^{-4}–10^2 mol/dm^3	4 g/dm^3	1 d		natural	chloride	uptake, coagulation, Cl coadsorption		733
muscovite mica	dodecyl pyridinium	5×10^{-4} mol/dm^3	10–150 mg/25 cm^3	4 d	room	natural	chloride	uptake, release of alkali metal cations	Release of alkali metal cations is nearly stoichiometric.	734
muscovite mica	dodecyl ammonium amine surfactants	10^{-6}–10^{-2} mol/dm^3	flat specimens		25	5.3–5.8	0.001 mol/dm^3 KBr	ζ	The surfactant concentration that induces sign reversal to positive increases with the number of methyl groups from dodecyl ammonium (3×10^{-5}) to dodecyl trimethyl ammonium (10^{-3} mol dm^{-3}).	735
tungstates										
MnWO$_4$	oleate	10^{-4}–10^{-3} mol/dm^3				3–11	Na	contact angle	The highest contact angle (55–65°, dependent on the crystallographic plane) at pH 8.	736

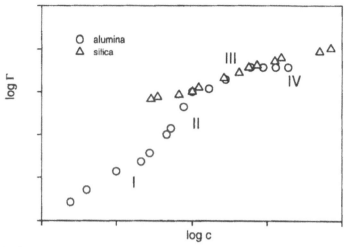

FIG. 4.66 Adsorption isotherms of decyl pyridinium in the presence of 0.01 mol dm^{-3} of alkali chloride, at pH 8 on alumina (Ref. 737) and on silica (Ref. 722).

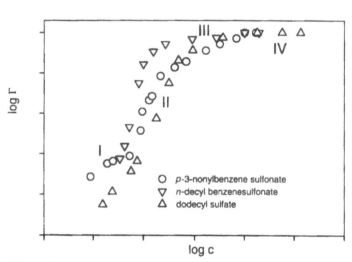

FIG. 4.67 Adsorption isotherms of p-3-nonylbenzene sulfonate in the presence of 0.01 mol dm^{-3} NaCl at pH 4.1 (Ref. 737), of n-decylbenzene sulfonate at 40 C, at pH 4 in absence of supporting electrolyte (Ref. 717), and of dodecyl sulfate at pH 5.3 in the presence of 0.1 mol dm^{-3} NaCl (Ref. 718) on alumina.

Figure 4.67 shows adsorption isotherms of three anionic surfactants on alumina at favorable electrostatic conditions, namely, of p-3-nonylbenzene sulfonate in the presence of 0.01 mol dm^{-3} NaCl at pH 4.1 (Ref. 737), of n-decylbenzene sulfonate at 40°C, at pH 4 in absence of supporting electrolyte (Ref. 717), and of dodecyl sulfate at pH 5.3 in the presence of 0.1 mol dm^{-3} NaCl (Ref. 718). The solution concentration and adsorption density in Fig. 4.67 were normalized to the transition point III/IV, which corresponds to CMC and adsorption saturation. In spite of a difference in the nature of surfactant and in experimental conditions (temperature, ionic strength), the slope and width of region II obtained in these three studies is rather consistent. The low slope in region III obtained in Ref. 717 (low ionic strength) is in line with the tendency illustrated in Fig. 4.65.

V. POLYMERS

This section is constrained to synthetic linear polymers whose molecules consist of repeating monomeric units of one or very few kinds (as opposed to natural polymeric substances). A few examples of adsorption of natural macromolecules (proteins, humic substances) are presented in Table 4.4 (organic compounds).

Synthetic polymers are widely applied to modify the surface properties of materials, and their adsorption mechanism is very different from small ions or molecules discussed in previous sections. Moreover, special methods are applied to study polymer adsorption, thus, polymer adsorption became a separate branch of colloid chemistry. Polymers that carry ionizable groups are referred to as polyelectrolytes. Their adsorption behavior is more sensitive to surface charging than adsorption of neutral polymers. Polyelectrolytes are strong or weak electrolytes, and the dissociation degree of weak polyelectrolytes is a function of the pH. The small counterions form a diffuse layer similar to that formed around a micelle of ionic surfactant.

Polymers are mixtures of macromolecules of different molecular weight, and most commercially available products have rather broad distribution of the molecular weight. Some studies were carried out with fractions of a narrow distribution of the molecular weight, separated from commercially available polymers. Adsorption leads to fractionation as discussed above for some types of surfactants. Larger polymer molecules have higher affinity to the surface than smaller molecules composed of the same type of monomeric units. The selectivity is chiefly driven by the difference in the entropy of mixing in solution. Polydispersity of polymers is also one of the factors responsible for hysteresis loops in the adsorption–desorption cycles.

The structure of polymer molecules in aqueous solution in absence of an adsorbent is an important and difficult problem by itself. The valence angles are fixed, but rotation about the C-C bonds gives the molecule some degree of flexibility. This rotation can be hindered by side groups attached to the main chain. Many conformations have nearly the same Gibbs energy, and the size and shape of a polymer molecule undergoes constant fluctuations, thus statistical methods are necessary to describe a polymer molecule in solution, and use of the time average size offers a convenient simplification. Polyelectrolytes are less flexible than neutral polymers because of electrostatic repulsion between consecutive segments.

Molecules of linear polymers in aqueous solution form coils whose size and shape depends on the ionic strength and presence of other solutes. Unlike a micelle core which is water free the polymer coils embrace molecules of water and other solutes, e.g. ions of inert electrolytes. The polymer coil in solution is nearly spherical, and the density of polymer segments is higher in the center of the coil than in peripheral regions. The size of the polymer coil depends on the Flory–Huggins parameter expressing the energy of transfer of a polymer segment from pure polymer to pure solvent, namely, in "good" solvents the coils are swollen. The application of terms "poor" and "good" solvents is not confined to nonaqueous solvents. The Flory–Huggins parameter for water depends on the ionic strength, because the interaction of polyelectrolyte segments in aqueous solution depends on screening of their mutual electrostatic repulsion by the counterions. Thus, the ionic strength affects the size of the polymer coil.

The polymer coil is flattened when the molecule is adsorbed, and the density of segments is higher near the surface than far from the surface. An adsorbed molecule looses part of its conformational entropy, but the overall entropy of the system increases in the adsorption process because of the release of water molecules (and pre-adsorbed solutes) into the solution, and the adsorption of polymers is often endothermic. Only a small fraction of monomeric units is involved in direct interaction with the surface (the polymer molecule is considered as adsorbed when at least one segment is attached to the surface). The fragments of the polymer chain in contact with the surface are referred to as trains, and the fragments surrounded by solution are called tails and loops. Although tails and loops are not directly attached to the surface their contribution to the Gibbs energy of adsorption can be substantial as a result of the change in the shape and size of the polymer coil. The fraction of trains can be estimated using spectroscopic methods. When the molecular weight increases in a series of polymers composed of the same monomeric units the fraction of loops constantly increases on expense of trains (the fraction of trains is 100% for a monomer by definition).

Contact of dilute polymer solution with an adsorbent often results in a practically 100% uptake. For high polymer concentrations in solution the adsorption density is practically constant, i.e. independent of the concentration in solution. Thus, the adsorption isotherm consists of two nearly rectilinear segments: a vertical segment at polymer concentration in solution equal to zero and a horizontal segment at polymer concentration in solution greater than zero. In practice there is always some rounded portion (knee) between the linear segments of the adsorption isotherm and the uptake slowly increases in the "plateau" region. Such an adsorption isotherm corresponds to Langmuir equation (Chapter 5) with very high binding constant, and it is called "high affinity" adsorption isotherm as opposed to "low affinity" adsorption isotherms observed for low molecular weight compounds (cf. Sections II–IV). Certainly, a regular Langmuir adsorption isotherm would look like the "high affinity" one when the studied range of solution concentrations is extended far into the plateau region. Likewise, a high affinity adsorption isotherm would probably look like a low affinity isotherm when a plot is confined to low polymer concentrations in solution. The problem with studies at very low polymer concentrations is in the detection limit of analytical methods. The "plateau" level (expressed as mass of the polymer per unit of surface area) increases when the molecular weight of the polymer increases.

Not always do the adsorption isotherms reported for polymers belong to the high affinity type. Low affinity adsorption isotherms are obtained when the molecular weight of the polymer is not too high, or for polyelectrolytes at unfavorable electrostatic conditions. Negligibly low and even negative adsorption of polymers has been also reported.

The pre-adsorbed polymer is not readily released by washing with water. Therefore the surface of inorganic material covered with a polymer can be considered as a "modified adsorbent". This property underlies many applications of polymers: the polymer layer is not washed away from the surface, thus the surface properties are retained, and the polymer does not pollute the environment.

This "irreversibility" of adsorption or rather very slow desorption kinetics is a consequence of the high affinity type adsorption isotherm, namely a surface saturated with a polymer is in equilibrium with low polymer concentration in solution, thus the release of the polymer from such a surface due to contact with water is negligibly small. Even this low "theoretical" equilibrium concentration of polymer in solution is not actually reached because of very low desorption rate, that can be explained at molecular level as follows. Release of a polymer molecule into solution requires simultaneous detachment of many polymer segments that are bound with the surface, and this is very unlikely considering that detachment of a few segments increases the driving force for attachment of some other segments. Generally the diffusion coefficients of polymers are much lower than for small molecules and this results in slower kinetics of sorption processes.

Polymers form thick adsorption layers, especially when the polymer is adsorbed from good solvent. The profile of density of polymer segments as the function of the distance from the surface has been studied by means of neutron scattering. The thickness of the polymer layers can be estimated by means of ellipsometry. The latter thickness is an average representing the continuous profile of the refractive index in the interfacial region. The adsorbed polymer affects the electrokinetic mobility of particles of inorganic adsorbents not only because of changes in the electric charge distribution in the interfacial region, but also because of changes in the hydrodynamic conditions around the particle.

Adsorption of macroanions on materials whose surface charging depends on the pH is enhanced at low pH and it results in shifts in the IEP to lower pH. Uptake of macrocations is enhanced at high pH. In this respect polyelectrolytes are similar to specifically adsorbing small ions. On the other hand, enhanced uptake at high ionic strengths is a rare phenomenon for small ions, but it is commonplace for polymers. Transfer of polyelectrolyte from bulk solution into the interfacial region (low ϵ) results in increase in the repulsive force between the segments. This unfavorable contribution to the adsorption Gibbs energy is reduced at high ionic strength, when the charge of segments is screened by the counterions. This phenomenon dominates the ionic strength effect when the surface charge density is low and the polymer adsorption is enhanced at a high ionic strength. With highly charged surfaces, the adsorbing polymer segments dislodge counterions of inert electrolyte, against the Coulombic forces that keep the counterions near the surface. This unfavorable contribution to the adsorption energy is more significant at high ionic strengths, and it can result in depression of the polymer uptake at high ionic strengths. The competition between polymer segments and counterions of inert electrolytes for surface sites shows ion specificity (e.g. lithium versus cesium), and multivalent counterions can be even stronger competitors.

Different aspects of polymer adsorption have been discussed by Lyklema [738]. Adsorption of polymers on oxides was recently reviewed by Esumi [739]. Sorption of polymers on bentonite was recently reviewed by Luckham and Rossi [740]. A few recent studies of polymer adsorption on inorganic materials are compiled in Table 4.6.

TABLE 4.6 Adsorption of Polymers

Adsorbent	Adsorbate	C total	S/L ratio	Eq. time	T	pH	Electrolyte	Method	Result	Ref.
Al_2O_3, Al6, Alcoa	polyacrylic acid, MW 5000, PVA, 88% hydrolyzed, MW 25000	0.05–200 ppm	1 g (10 m²)/100 cm³	1 d	27	3–11	10^{-3} mol dm^{-3} KNO$_3$	uptake, ζ, FTIR	In the presence of 10 ppm of PVA both positive and negative electrophoretic mobility is depressed. In the presence of 0.2 ppm of polyacrylic acid the IEP is shifted to pH 5. The uptake of PVA from 200 ppm solution is negligible at pH 3 and linearly increases with pH. The uptake of polyacrylic acid from 200 ppm solution is negligible at pH 10 and linearly increases as pH decreases.	741
Al_2O_3, α, A-11, Alcoa	polyacrylic acid, MW 2000–90000	1–100 mg/dm³	1 g (9 m²)/100 cm³	1 d	27	3–12	0.001 mol dm^{-3} KNO$_3$	uptake, ζ	Uptake of the MW 90000 polymer is higher by a factor of 1.5 than that of the MW 2000 polymer. The uptake decreases as pH increases (no adsorption at pH > 10). The IEP is shifted to low pH (5.1 for MW 90000, 6.2 for MW 5000) at 0.1 ppm polymer. For 1 ppm polymer ζ is negative over the entire pH range.	742
Al_2O_3, α, reagent grade	Duramax D3021, Rohm and Haas, polycarbonate ammonium salt, MW 2000	0.2–0.4% w/w on dry powder basis	15% v/v		25	3–12	0.01 mol dm^{-3} KCl	acousto	The IEP is shifted to pH 5 (0.2% dispersant) and <4 (0.4%).	743

Adsorbent	Adsorbate	Concentration	Solid/solution	Time	T	pH	Electrolyte	Method	Result	Ref.
SiO$_2$, Stober and quartz, Sikron 6000 SF, Quarzwerke GmbH	homo- and copolymers of diallyl-dimethylam-monium chloride and N-methyl-N-vinylaceta-mide MW 5000–428000	10^{-4}–10^{-3} mol of monomeric units dm^{-3}	3 g (5 m^2/g)/50 cm^3	15 min–12 h		4–11	0–1 mol dm^{-3} NaCl	uptake, ζ, coagulation	The IEP is shifted in the presence of poly(diallyl-dimethylammonium chloride), MW 372 000: to pH 6 at 0.074 mg/m^2 and to pH 10 at 0.314 mg/m^2. At pH 5.8 the sign of ζ is reversed to positive at certain concentration of polymer or copolymer (in mg/dm^3) which is ionic strength independent, for example, 0.05 mg/dm^3 of poly(diallyl-dimethylammonium chloride), MW 428 000.	744
SiO$_2$, Aerosil 200	poly (vinylimidazole), partially quaternized poly (vinylimid-azole)	10–2000 ppm	0.5% w/w, 200 m^2/g	24 h	20	3–7	10^{-3}–10^{-1} mol dm^{-3} NaCl	uptake, ζ	The uptake plotted as the function of quaternization ratio peaks at 2–4%. The uptake is higher at high pH (according to data shown in figures which contradict the statements in the abstract and conclusions).	745
SiO$_2$, Aerosil 130, Degussa	poly(N-3-sulfopropyl-N-methacryl-ooxyethyl-N, N, dimethyl ammonium betaine) MW 700000, 1500000	0.01–0.3 g/100 cm^3	0.16 g (130 m^2/g)/20 cm^3	2 d	25	natural	0.06–1 mo·dm^{-3} NaCl	uptake, ellipsometry	The uptake of unquaternized polymer is insensitive to NaCl concentration and uptake of quaternized polymer is enhanced at high salt concentrations.	746

TABLE 4.6 Continued

Adsorbent	Adsorbate	C total	S/L ratio	Eq. time	T	pH	Electrolyte	Method	Result	Ref.
SiO₂, spheres, Catalyst & Chemical Ind.	poly(styrene sulfonate) MW 18 000, poly-4-vinyl-N-butylpyridinium	1–10 layers	volume fraction 7×10^{-5}		25			ζ, coagulation	On alternative addition of layers of poly-4-vinyl-N-butylpyridinium (first) and poly(styrene sulfonate), ζ is positive with odd and negative with even number of layers, and the particle expands with odd and shrinks with even number of layers.	747, 748
MgAl₂O₄	Darvan 821 A, derivative of polyacrylic acid		volume fraction 0.1	5–30 min	25	natural	0.01 mol dm⁻³ KCl	acousto	The sign of ζ is reversed from positive to negative on addition of dispersant. Further addition of dispersant leads to more negative ζ, up to -85 mV.	749
bentonite	copolymer: dimethyla-minoethyl-methacryl-ate + acrylamide + sodium carboxymethyl cellulose	1–5 g/ dm³		2 h	25	6	0–1.2 mol dm⁻³ NaCl		The uptake is depressed by 25% at 1.2 mol dm⁻³ NaCl. The uptake is depressed at high T (20–60 C). The uptake is rather insensitive to addition of SDS (up to 0.005 mol dm⁻³ but it is depressed in the presence of hexadecyl trimethylammonium chloride.	750

Adsorbent	Adsorbate						Method/measurement	Remarks	Ref.
Na and Ca montmorillonite, from Wyoming bentonite	poly (ethylene oxide); TMA modified poly (ethylene oxide), MW 1550–35000	0.1 22 g/dm^3	1%			None	uptake, release of Na and Ca, viscosity		751
Y stabilized ZrO$_2$	ammonium polyacrylate, MW 10000	0.01–4%	1.8% v/v (31.6 m^2/g)	25	2–12		uptake, ζ	The IEP is shifted to pH 2.3 (0.01 % v/v of zirconia and 0.01% v/v of ammonium polyacrylate).	752
ZrO$_2$ + 3% mol Y$_2$O$_3$, nanoparticles, coprecipitation method	polyethylene imine, MW 50000	0.2–2.2% dry weight basis	4 vol% (31.5 m^2/g)	2 d	25	5–8	0.001 mol dm^{-3} KCl	uptake	Uptake increases with pH. "low affinity" sorption isotherms at constant pH.	753

VI. COMPETITION

Most results presented in Sections II–V refer to relatively simple systems with only one strongly interacting adsorbate representing certain class, e.g. studies of adsorption of phosphate in absence of other strongly interacting anions. Not necessarily can the results obtained in simple systems be generalized for more complicated systems with two or more adsorbates representing certain class. Many studies in multicomponent systems were undertaken in order to simulate sorption phenomena in natural systems.

This section reports studies of the effect of strongly interacting adsorbates on the sorption of other adsorbates representing the same class (small cations, small anions, surfactants, polymers), e.g. sorption of copper is studied in absence and in the presence of other heavy metal cations at otherwise identical conditions (solid to liquid ratio, pH, equilibration time, etc.). A few examples of adsorption competition between anions or cations of inert electrolytes are also presented. This limitation does not imply that actual adsorption competition occurs only between adsorbates representing the same class.

Generally, any two adsorbates potentially affect one another's sorption. For example sorption of heavy metal cations at different ionic strengths (which is discussed in Section II.A) has been considered in terms of competition between the heavy metal cation and the cation of inert electrolyte by some authors. On the other hand the adsorption mechanism can be different within the same group of adsorbates. For example heavy metal cations and alkaline earth metal cations are grouped together in this book (as strongly interacting cations), but the alkaline earth metal cations show some similarity in their adsorption behavior with alkali metal cations, and suggestions can be found in literature that they should be considered as a separate class of adsorbates. The adsorbates referred to as organic compounds (Section III) do not form an uniform group, namely, studies of sorption of all compounds that do not fit the specified above categories are compiled in Section III, and the underlying sorption mechanisms are very different. Classification of adsorbates proposed in this book is simplified and it only gives a general idea about possible adsorption competition.

Many publications report on specific adsorption of anions and cations from sea water, i.e. in the presence of Mg, sulfate and other strongly interacting species. Such studies are reported in this section when results obtained in absence of competing species are available for comparison. Otherwise studies of sorption from sea water and other natural waters (adsorption data obtained in absence of competing species are not available) are reported in Sections II.A (cations) and II.B (anions).

Apparently the presence of another species representing the same class should lower the uptake of the species of interest. First the two species compete for surface sites (the number of surface sites available for sorption of the species of interest is lower in the presence of competitor). Moreover, with ionic species, the presence of the competitor in the interfacial region leads to less favorable electrostatic conditions for the adsorption of the species of interest. It should be emphasized that these mechanisms of adsorption competition are effective only when a significant amount of the competitor is adsorbed. For instance, adsorbed species contribute to electrostatic potential of the surface, but the unadsorbed ions in solution do not.

Therefore, the presence of potential competitors significantly depresses the uptake of small ions at favorable electrostatic conditions, but at unfavorable electrostatic conditions, the uptake in the presence and in absence of potential competitors is practically identical.

A few examples of synergism (the uptake of the adsorbate of interest in enhanced in the presence of other adsorbates representing the same class) have been also reported. In view of the above discussion, adsorption synergism appears to be a paradox, but specific cases have their rational explanation. For instance sorption of a metal cation in the presence of complexing anion can be enhanced on addition of another metal cation. The uptake in absence of "competing" cation is low when the ligand in solution successfully competes with the surface for the cation of interest. The "competing" cation binds the ligand (the concentration of the "competitor" and the stability constant of the corresponding complex must be sufficiently high), and the sorption of free cation of interest (not depressed by the complexation) is higher than the sorption in absence of the "competing" cation.

Also surface precipitation can result in sorption synergism, namely, the uptake of the cation of interest by coprecipitation with hydroxide or basic salt of the competitor can be higher than the adsorption on the original adsorbent. This example illustrates the dependence of the adsorption competition/synergism on the experimental conditions: coprecipitation can only occur at sufficiently high concentration of the competitor.

Basically the experimental parameters affecting the sorption behavior in simple systems (concentration of the adsorbate, solid to liquid ratio, equilibration time, temperature, pH, and ionic strength) are also important in sorption competition. Thus, Table 4.7 has the same columns as Tables 4.1 and 4.2 (specific adsorption in absence of competitors). In simple systems little attention is usually paid to the sequence of addition of reagents, but in studies of sorption competition this factor must not be neglected. The fact that some adsorbate is added (brought in contact with the adsorbent) first makes it a more successful competitor. Let us consider the following experiment. The adsorbent is equilibrated with solution of adsorbate 1 (adsorbate of interest), then the adsorbent is separated from the solution and contacted with solution of adsorbate 2 (competitor), and then the concentration of adsorbate 1 in solution is determined after certain equilibration time. Only a fraction of adsorbate 1 is released even if adsorbate 2 is a very strong competitor. However, adsorbate 2 added first or simultaneously with adsorbate 1 would completely prevent sorption of adsorbate 1 at otherwise identical experimental conditions.

Generally the statement whether (and to what degree) the presence of adsorbate 2 affects the sorption of adsorbate 1 is only valid for certain experimental conditions, including the discussed above history of the system.

Different experimental approaches to the sorption competition produce somewhat different results. For example some experiments were carried out in the presence of a big excess of the competitor over the adsorbate of interest, and in other studies the concentrations of adsorbate 1 and adsorbate 2 were of the same order of magnitude. Some authors performed their experiments at constant concentration of adsorbate 1 and variable concentration of adsorbate 2, while in other studies the sum of concentrations (adsorbate 1 + adsorbate 2) was constant. When the distribution

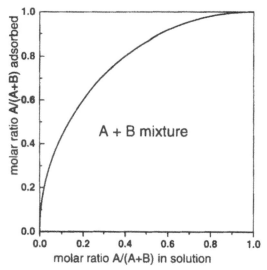

FIG. 4.68   Representation of adsorption competition (schematically).

of the both competitors between the solution and the surface is known, the selectivity coefficient can be defined as follows.

$$\frac{[\text{adsorbate 1 adsorbed}]\ [\text{adsorbate 2 in solution}]}{[\text{adsorbate 2 adsorbed}]\ [\text{adsorbate 1 in solution}]} \qquad (4.24)$$

A value of the selectivity coefficient defined by Eq. (4.24) higher than 1 means that adsorbate 1 is the more successful competitor. This selectivity coefficient is a close relative of K_D (Eq. (4.9)), and the limitations of significance of K_D are also important for the selectivity coefficient. Briefly, the values of both coefficients are only valid for certain experimental conditions. The graph in Fig. 4.68 illustrates the variability of the selectivity coefficient as the function of the concentrations of the competitors. A convex curve means that adsorbate 1 is the more successful competitor. A family of similar (but not necessarily identical) graphs is obtained when the experimental conditions (e.g. equilibration time) are varied.

Experimental studies on sorption competition in specific systems are compiled in Table 4.7. The entries are sorted by the adsorbent and then by the adsorbate (cations first, in alphabetical order, then anions, organic compounds, surfactants and polymers).

A few authors tested their models of adsorption (cf. Chapter 5) in multicomponent systems. Some results are reported in Table 4.7.

VII.   KINETICS

The "equilibrium" approach is a useful simplification in studies of specific adsorption. The sorption system is assumed to behave as if it were in state of thermodynamic equilibrium, and the adsorption models (discussed in detail in Chapter 5) usually describe the sorption process in terms of surface reactions, whose

(*Text continues on pg. 531*)

TABLE 4.7 Sorption Competition

Adsorbent	Adsorbate	Total concentration	Solid to liquid ratio	Equilibration time	T	pH	Electrolyte	Method	Result	Ref.
Al_2O_3, γ	Cs	carrier free	variable	15 d		5–10	0.01–0.25 mol dm^{-3} NaCl	uptake	Cs uptake is depressed by one order of magnitude on increase of NaCl concentration from 0.01 to 0.25 mol dm^{-3} at pH 9, but at pH 5 the effect of Na on Cs uptake is insignificant.	98
Al_2O_3, α	Zn	80 ppm	0.2 g/50 cm^3	1 d		6–7	1 mol dm^{-3} BaCl$_2$	uptake	Addition of BaCl$_2$ induces release of 17–35% of Zn pre-sorbed in absence of Ba.	88
Al_2O_3, chromatographic, Merck	Gd, Ni	5×10^{-6} mol dm^{-3}	1 g (155 m^2/g)/50 cm^3	20 min	25	4–8	0.01 mol dm^{-3} Ba(ClO$_4$)$_2$	uptake	Replacement of 0.01 mol dm^{-3} NaClO$_4$ by 0.01 mol dm^{-3} Ba(ClO$_4$)$_2$ has no significant effect on uptake curves of Gd or Ni.	117
Al_2O_3, γ	Ni, Zn, Pb	10^{-6}–10^{-4} mol dm^{-3}	1, 2.5 g (100 m^2/g)/dm^3	30 min–10 h		3–10	0.025 mol dm^{-3} NaClO$_4$; 10^{-6}, 10^{-4} mol dm^{-3} EDTA; 10^{-6}, 10^{-4} mol dm^{-3} Fe(III)	uptake	With [Fe(III)] = [Ni] = [EDTA] = 10^{-4} mol dm^{-3} the Fe(III) effect on Ni uptake is negligible. With [Fe(III)] = [Ni] = [EDTA] = 10^{-6} mol dm^{-3} the Ni sorption edge is cation like in the presence of Fe(III) and anion like in absence of Fe(III).	118
Al_2O_3, α, Aldrich	arsenate (V)	10^{-8} 10^{-4} mol dm^{-3}	0.05, 0.5 g (0.7 m^2/g)/20 cm^3	3 d		3–9	0.1 mol dm^{-3} NaCl + 0–80 mg/dm^3 sulfate; 0–25 mg/dm^3 fulvic acid	uptake	Presence of sulfate or fulvic acid (>20 mg/dm^3) depresses the uptake of As(V) by 30% at pH < 5. No sulfate effect at pH > 7.	482
Al_2O_3, Alcoa, Fl	borate	20 mg B/dm^3	25 g (210 m^2/g)/dm^3	2 d		9.4, 9.6	sulfate up to 2 g/dm^3; silica up to 80 mg Si/dm^3	uptake	The effect of sulfate and silicate on B uptake is rather insignificant.	485

TABLE 4.7 Continued

Adsorbent	Adsorbate	Total concentration	Solid to liquid ratio	Equilib-ration time	T	pH	Electrolyte	Method	Result	Ref.
Al_2O_3, α	phosphate	1.15×10^{-4} mol dm^{-3}	0.6 g (10 m^2/g)/130 cm^3	1 d	25	2–12	0.01 mol dm^{-3} NaCl	uptake	The uptake of phosphate is not significantly influenced by 3.75×10^{-4} mol dm^{-3} succinate, acetate, or glycine, 1.15×10^{-4} mol dm^{-3} silicate, 3.75×10^{-3} mol dm^{-3} bicarbonate or 31 mg/dm^3 of humic acid. 3.75×10^{-4} mol dm^{-3} fluoride depresses the uptake at pH 3–7 by 30%; even more significant depression is observed in the presence of 3.75×10^{-4} mol dm^{-3} citrate, oxalate, EDTA or tartrate at pH < 8.	503, 504
Al_2O_3, α, A 16 SG, Alcoa	selenate IV	10^{-5}–10^{-2} mol dm^{-3}	100 g (8.4 m^2/g)/dm^3		25	4	0.01 mol dm^{-3} NaCl, 0–0.003 mol dm^{-3} glutamic and benzoic acid	uptake	The uptake of selenate is not affected by glutamic or benzoic acid or phenol.	68
Al_2O_3, Degussa C, washed	acetate, carbonate	$10^{-3.48}$ mol dm^{-3}	0.36 g (86 m^2/g)/35 cm^3	1 d		1–9	0–0.01 mol dm^{-3} NaCl	uptake	Presence of $10^{-3.48}$ mol dm^{-3} acetate does not affect the uptake of carbonate. Presence of $10^{-3.48}$ mol dm^{-3} carbonate depresses the uptake of acetate at pH > 4.	514
Al_2O_3, δ, Cabot	borate, silicate	2–9×10^{-4} mol dm^{-3}	0.25 g (70 m^2/g)/25 cm^3	2–20 h		5–11	0.1 mol dm^{-3} NaCl	uptake	The Si uptake is rather insensitive to B, but B uptake is somewhat reduced in the presence of Si. The effect is more significant when Si is added before addition of B.	518

Adsorbent	Adsorbate	Concentration	Solid/solution	Time	Temp.	pH	Electrolyte	Method	Result	Ref.
Al_2O_3, γ, Aldrich	dodecylsulfate, Triton X 100	10^{-6} 10^{-2} mol dm^{-3}	0.5 g (159 m^2/g)/30 g solution	4 d	room	5.3	0.1 mol dm^{-3} NaCl	uptake	Uptake of Triton X 100 is enhanced in the presence of SDS.	718
Al_2O_3, A16, Alcoa	polyacrylic acid MW 5000, poly (vinyl alcohol) 88% hydrolysed, MW 25 000	0.05–200 ppm	1 g (10 m^2/g)/100 cm^3	1 d	27	3–11	10^{-3} mol dm^{-3} KNO$_3$	uptake, ζ	Uptake of polyacrylic acid (up to 200 ppm) is rather insensitive to addition of PVA (200 ppm). Also the ζ potential in the presence of 0.2 ppm of polyacrylic acid is not affected by PVA up to 100 ppm. The uptake of PVA is depressed in the presence of polyacrylic acid.	741
$Al(OH)_3$, gel	Cu, V(IV)	10^{-3}–0.02 mol dm^{-3}	2.6 mmol Al/75 cm^3	1 d	20	4.3	0.1 mol dm^{-3} NaCl	uptake	V is not released when Cu is added to the gel with pre-sorbed V.	152
$Al(OH)_3$, gibbsite	arsenate (V)	2.7×10^{-4} mol dm^{-3}	53 mg (31 m^2/g)/25 cm^3	1 d	20	3–11	0.1 mol dm^{-3} NaCl + 0–6.5 $\times10^{-4}$ mol dm^{-3} phosphate	uptake	The uptake is depressed by a factor of 4 in the presence of 6.5×10^{-4} mol dm^{-3} phosphate.	553
$Al(OH)_3$, amorphous, from AlCl$_3$	arsenate III, V	4×10^{-7} mol dm^{-3}	2.5 g dm^{-3}	3 h	20	4–10 10	10^{-3} mol dm^{-3} phosphate	uptake	1/3 of initially sorbed As(III) is released after addition of phosphate.	535
$Al(OH)_3$, amorphous, fresh, from nitrate	borate	1.8×10^{-3}, 1.8×10^{-2} mol dm^{-3}	10^{-3} mol Al/30 cm^3	12 h			silicate up to 0.002 mol/g	uptake	Uptake of borate is depressed in the presence of silicate.	556
$Al(OH)_3$, gibbsite	phosphate	1–300 ppm P	1 g/10 cm^3	1 d		7	0.0125–0.0025 mol dm^{-3} EDTA, oxalate, fluoride	uptake	Release of pre-sorbed phosphate in consecutive extractions by EDTA was more efficient than with other anions.	541
$Al(OH)_3$, gibbsite	arsenate (V), molybdate, phosphate	1.33×10^{-4} mol dm^{-3}	2.5 g/dm^3	4 h	22	2–11	0.1 mol dm^{-3} NaCl	uptake	The uptake in binary systems is depressed compared with single ion systems by 10–30% except uptake of arsenate is enhanced in the presence of molybdate at pH > 9.	552

TABLE 4.7 Continued

Adsorbent	Adsorbate	Total concentration	Solid to liquid ratio	Equilib-ration time	T	pH	Electrolyte	Method	Result	Ref.
Al(OH)$_3$, amorphous	oxalate, phosphate	10^{-3} mol dm^{-3}	50 mg/50 cm^3	4 d or 2 d + 2 d	20	4–9	0.1 mol dm^{-3} KCl, toluene added	uptake	Addition of phosphate does not remove pre-sorbed oxalate and addition of oxalate has rather insignificant effect on pre-sorbed phosphate. However the uptake is oxalate and phosphate from mixture is significantly lower than in absence of competing anion. The uptake of phosphate is even lower when the adsorbent is equilibrated with oxalate for 2 d first (in absence of phosphate), and then phosphate is added. Similarly the surface with pre-adsorbed phosphate adsorbs less oxalate than when the both ions are added simultaneously.	557
Bi$_2$O$_3$, Alfa, α	chromate	carrier free				7	13 anions, 10^{-4}–0.1 mol dm^{-3}	uptake	The uptake is depressed in the presence of (the most efficient competitors are listed first) arsenate > phosphate > molybdate > citrate > tungstate > EDTA > tartrate > > oxalate > thiosulphate. The effect of acetate, thiocyanate, borate, sulphate IV, fluoride, and iodate on uptake of Cr(VI) is rather insignificant.	560
CeO$_2$	Ba, Sr	10^{-5} mol dm^{-3}	0.1 g/10 cm^3	3 h	30	11.4	nitrate	uptake	No significant decrease in adsorption of Ba and Sr on	156

Adsorbent	Species	Concentration	Solid	Time	T	pH	Electrolyte	Mode	Comments	Ref.
Cr$_2$O$_3$, amorphous	Ni, Zn	50 ppm	250 ppm	40 min	25	4–11	10^{-3} mol dm^{-3} KNO$_3$	uptake	addition of 6×10^{-5} mol dm^{-3} of other alkaline earth metal cations. Presence of Zn does not affect the uptake of Ni.	754
Fe$_3$O$_4$, natural magnetite, 2.4 % SiO$_2$	Cs	2×10^{-5} mol dm^{-3}	2 g (18.3 m^2/g)/dm^3		25	3–9	0.01 mol dm^{-3} NaNO$_3$	uptake	Cs uptake is depressed by a factor of 5 in the presence of 0.01 mol dm^{-3} Na with respect to a low ionic strength system.	171
Fe$_2$O$_3$, α	Zn	80 ppm	0.2 g/50 cm^3	1 d		5–7	1 mol dm^{-3} BaCl$_2$	uptake	Addition of BaCl$_2$ induces release of 10–37% of Zn pre-sorbed in absence of Ba.	88
Fe$_2$O$_3$, amorphous	Cr(III), Ni, Zn	50 ppm	250 ppm	40 min	25	4–11	10^{-3} mol dm^{-3} KNO$_3$	uptake	Presence of Ni or Zn does not affect the uptake of Cr. Uptake of Ni is not affected by Zn, but it is enhanced in the presence of Cr: the pH$_{50}$ is shifted by 1.5 pH unit.	754
Fe$_2$O$_3$, gel, from sulfate, 9 samples	Cs, K, Li, Na				25	12–13	0.01–0.1 mol dm^{-3} NaOH	uptake	Li > Na, other selectivity coefficients can be < or > 1 for different samples.	755
Fe$_2$O$_3$, hematite	mellitic acid	3×10^{-3} mol dm^{-3}	3 g (6.5 m^2/g)/25 cm^3	3 d	25	7	$10^{-4}-5\times10^{-5}$ mcl dm^{-3} P$_3$O$_{10}^{5-}$	uptake	Uptake decreases linearly on addition of P$_3$O$_{10}^{5-}$, presence of 5 $\times10^{-3}$ mol dm^{-3} of P$_3$O$_{10}^{5-}$ reduces the uptake by factor 2.	581
Fe$_2$O$_3$, hematite	phosphate	40 ppm	1 g/10 cm^3	1 d		7	0.00125–0.0025 mol dm^{-3} EDTA, oxalate, fluoride	uptake	Release of pre-sorbed phosphate in consecutive extractions by EDTA was more efficient than with other anions.	541
Fe$_2$O$_3$·H$_2$O	Cu, Mg, Zn	1.1×10^{-4}, 1.1×10^{-3} mol dm^{-3}	10^{-3} mol Fe/dm^3	1 d	21	4–11	NaNO$_3$	uptake	No competition effects in solution containing simultaneously 1.1×10^{-4} mol dm^{-3} of Cu, Mg and Zn; the uptake of Mg in the presence of 1.1×10^{-3} mol dm^{-3} of Cu and Zn is slightly depressed (shift in pH$_{50}$ by 0.3 pH unit) as compared with a single ion system.	124

TABLE 4.7 Continued

Adsorbent	Adsorbate	Total concentration	Solid to liquid ratio	Equilibration time	T	pH	Electrolyte	Method	Result	Ref.
$Fe_2O_3 \cdot H_2O$	arsenate III and V	$\leq 1.33 \times 10^{-6}$ mol dm^{-3}	5×10^{-5} mol Fe/ dm^3	2 h		4–9	0.01 mol dm^{-3} NaNO$_3$, carbonate, sulfate	uptake	Presence of 0.01 mol dm^{-3} carbonate does not affect the uptake curves. The uptake of As V at pH 4 is slightly (by 10%) depressed in the presence of 0.01 mol dm^{-3} sulfate. The uptake of As(III) at pH < 5 is depressed by a factor 3 in the presence of 2.6×10^{-6} mol dm^{-3} sulfate.	585
$Fe_2O_3 \cdot H_2O$, amorphous, fresh, from nitrate	borate	1.8×10^{-3}, 1.8×10^{-2} mol dm^{-3}	10^{-3} mol Fe/30 cm^3	12 h		silicate up to 0.002 mol/g		uptake	Uptake of borate is depressed in the presence of silicate.	556
$Fe_2O_3 \cdot H_2O$, from chlorate VII	phosphate	0.5–1.7 mol kg^{-1} (of the adsorbent)	1:800	2 d	24	6.5	0.1 mol dm^{-3} NaClO$_4$; arsenate molybdate selenate VI and IV silicate sulfate 0.5–1.7 mol kg^{-1} (of the adsorbent)	uptake	The uptake of phosphate is depressed by (the most efficient competitors are listed first) arsenate > selenate IV > silicate > molybdate. The presence of sulfate and selenate VI does not affect the uptake of phosphate.	594
$Fe_2O_3 \cdot H_2O$, from nitrate	arsenate, silicate	1.7×10^{-7} 5×10^{-5} mol dm^{-3}	$1.7 \times 10^{-7} - 10^{-3}$ mol Fe/dm^3	1–7 d	25	4–12	0.1 mol dm^{-3} NaNO$_3$	uptake	Uptake of arsenate V is depressed in the presence of silicate, the silicate effect on uptake of arsenate III is insignificant.	596
FeOOH, goethite	Th	9×10^{-6} mol dm^{-3}	0.54–8.6 g/dm^3	3 h	20	2.5–6	0.422 mol dm^{-3} NaCl + 0.054 mol	uptake	No significant effect of Mg or Ca on Th uptake.	262

Solid	Species	Concentration	Surface area/solid	Time	Temp (°C)	pH	Medium	Process	Remarks	Ref.
FeOOH, goethite, from nitrate	Zn	3.82×10^{-7} 10^{-4} mol dm^{-3}	50 mg (95 m^2/g)/200 cm^3	2 d	25	5.4–8.5	0.001 mol dm^{-3} MgCl$_2$, 0.01 mol dm^{-3} CaCl$_2$	uptake	The uptake is depressed in the presence of Mg by 20% except at pH > 7.8 (100% uptake).	264
FeOOH, goethite	Cd, Cu, Pb, Zn	3×10^{-6} mol dm^{-3}	28.5 m^2/dm^3	2.5 h	25	3–8	NaNO$_3$ + 25 mg Mg/dm^3 0.53 mol dm^{-3} NaCl + 0–0.054 mol dm^{-3} MgCl$_2$, sea water	uptake	Uptake of Zn and Cd is depressed in the presence of Mg. The uptake curves of Cd and Zn obtained with sea water and with 0.53 mol dm^{-3} NaCl + 0.054 mol dm^{-3} MgCl$_2$ are identical. Uptake of Cu, Pb, and Zn from major ion sea water in single ion system and in the presence of equimolar amounts of Cd, Cu, Pb and Zn are identical; the uptake of Cd in the presence of equimolar amounts of Cu, Pb and Zn is depressed by about 30% with respect to a single ion system.	272
FeOOH, β, akageneite, synthetic	Co, Mn(II), Ni, Zn	$3–8 \times 10^{-5}$ mol dm^{-3}	50 mg/50 cm^3	3 d	27	6–9	sea water	uptake	Uptake of Co is depressed by a factor of 2 in the presence of Ni, Zn and Mn at equimolar concentrations. Uptake of Zn is unaffected when Ni, Co, and Mn at equimolar concentrations are present.	279
FeOOH, amorphous, synthetic	Co, Mn(II), Ni, Zn	$1.6–6 \times 10^{-5}$ mol dm^{-3}	0.2, 1 g/dm^3	3 d	27	6–9	sea water	uptake	Uptake of Co and Ni is depressed in the presence of Mn and Zn at equimolar concentrations. Uptake of Zn and Mn is unaffected by Ni and Co at 1 g FeOOH/dm^3 but is depressed at 0.2 g/dm^3.	280

TABLE 4.7 Continued

Adsorbent	Adsorbate	Total concentration	Solid to liquid ratio	Equilibration time	T	pH	Electrolyte	Method	Result	Ref.
FeOOH, goethite	Cu, Pb, Zn	$10^{-6}-10^{-3}$ mol dm^{-3}	0.09–9 g (40 m^2/g)/dm^3	2 d	25	3.5–8.5	0.1 mol dm^{-3} NaNO$_3$	uptake, proton release	≡FeOHCu^{2+} log K 8.8 ≡FeOCu log K 0.9 ≡FeOCuOH log K -6.6 ≡FeOHPb^{2+} log K 8.6 ≡FeOPb^{+} log K 0.17 ≡FeOPbOH log K -8.85 ≡FeOZn log K -2 (≡FeO)$_2$Zn^{2+} log K 10.67 ≡FeOZn(OH)$_2^-$ log K -18.2	756
FeOOH, goethite	Cu, Zn, Mg	1.1×10^{-4}, 1.1×10^{-3} mol dm^{-3}	10^{-3} mol Fe/dm^3	1 d	21	4–11	NaNO$_3$	uptake	No competition effects of Mg and Zn on Cu sorption. The uptake of Mg from 1.1×10^{-4} mol dm^{-3} solution is depressed over the entire pH range (shift in pH$_{50}$ by 0.8 pH unit) in the presence of equimolar concentrations of Cu and Zn. The uptake of Mg from 1.1×10^{-3} mol dm^{-3} solution is rather insensitive to the presence of equimolar concentrations of Cu and Zn.	124
FeOOH, goethite	borate	0.001 mol dm^{-3}		36 h		5–11	1 mol dm^{-3} NaCl + 0.001 mol dm^{-3} sulfate or phosphate	uptake	In the presence of phosphate the uptake is depressed by 1/3, the effect of sulfate is insignificant.	600
FeOOH	chromate	5×10^{-6} mol dm^{-3}	$8.7\times10^{-4}-1.74 \times10^{-2}$ mol dm^{-3}	4 h	25	5–10	0.1 mol dm^{-3} NaNO$_3$, $0-2.5\times10^{-3}$	uptake	The uptake is depressed in the presence of CaSO$_4$ and H$_4$SiO$_4$.	601

Sorbent	Species	Concentration	total iron	Time	Temp.	Electrolyte	Method	Comments	Ref.
FeOOH, goethite, from nitrate	chromate	5×10^{-6} mol dm^{-3}	10 g (45 m^2/g)/dm^3	2 h	6–10	0.1 mol dm^{-3} NaClO$_4$	uptake	The pH$_{50}$ is shifted by -0.5 pH unit under 4.5×10^{-4} atm CO$_2$ and by -1.5 pH unit under 4×10^{-2} atm CO$_2$.	618
FeOOH, goethite	EDTA	9×10^{-6} mol dm^{-3}	0.12 g (21 m^2/g)/dm^3	30 min	22–24	0.01 mol dm^{-3} NaNO$_3$ + 10^{-6}–4×10^{-5} mol dm^{-3} phosphate	uptake	The uptake of EDTA in presence and absence of Pb is depressed by a factor >3 by 4×10^{-5} mol dm^{-3} phosphate.	266
FeOOH, goethite	phosphate	3×10^{-5} mol dm^{-3}	0.25 g/dm^3	1 d	20±2	0.42 mol dm^{-3} NaCl, sulfate, fluoride, sea water	uptake	Uptake is depressed in the presence of 0.0014 mol dm^{-3} sulfate or fluoride at pH<5 and in the presence of 50 mg/kg humic acid at pH<8. The uptake from major ion sea water is lower than from NaCl solution at pH<7.	607
FeOOH, goethite, synthetic	selenate IV	10^{-6} mol dm^{-3}	200 mg (49 m^2/g)/dm^3	1 d	22	0.1 mol dm^{-3} KCl + different anions up to 0.1 mol dm^{-3}	uptake	The uptake is 100% in absence of competing anions. It is reduced to 50% in the presence of 3×10^{-5} mol dm^{-3} phosphate, 3×10^{-3} mol dm^{-3} silicate, 5×10^{-2} mol dm^{-3} citrate or molybdate. The effect of carbonates, oxalates, fluorides and sulfates at concentrations below 10^{-3} mol dm^{-3} is insignificant.	611
FeOOH, goethite	arsenate (V), molybdate, phosphate	1.33×10^{-4} mol dm^{-3}	2.5 g/dm^3	4 h	22	0.1 mol dm^{-3} NaCl	uptake	The uptake in binary systems is depressed compared with single ion systems by 10–50% except uptake of arsenate is enhanced in the presence of molybdate at pH>7.	552

TABLE 4.7 Continued

Adsorbent	Adsorbate	Total concentration	Solid to liquid ratio	Equilibration time	T	pH	Electrolyte	Method	Result	Ref.
FeOOH, goethite	arsenate (V) selenate (IV)	10^{-3} mol dm^{-3}	93, 109 mg (60, 81 m^2/g)/ 25 cm^3	1 d	20	3–11	0.1 mol dm^{-3} NaCl + 0–2.6×10^{-3} mol dm^{-3} phosphate	uptake	The uptake of arsenate is depressed by a factor of 4 in the presence of 2.6×10^{-4} mol dm^{-3} phosphate.	553
FeOOH, goethite	chelidamic acid, phthalate, sulfate	2.5×10^{-5} –10^{-3} mol dm^{-3}	1.6 g (79 m^2/g)/ dm^3	1 d	20	3–8	0.01, 0.1 mol dm^{-3} NaCl	uptake	The uptake in single-anion system is higher that in double-anion system, especially at acidic pH. The model derived from single component sorption behavior works in double-anion systems, but the values of log K need some adjustment.	619
FeOOH, goethite	chromate, oxalate	10^{-5}–10^{-3} mol dm^{-3}	1.8 g (66 m^2/g)/ dm^3	12 h	25	4–11	0.05 mol dm^{-3} KNO$_3$	uptake	At chromate and oxalate concentrations up to 5×10^{-5} mol dm^{-3} presence of oxalate does not affect sorption of chromate and vice versa. At 2×10^{-4} and 8×10^{-4} mol dm^{-3} of both ions presence of oxalate does not affect sorption of chromate, but sorption of oxalate is depressed in the presence of chromate. The uptake of chromate is depressed at a 4× excess of oxalate.	620
FeOOH, goethite, synthetic	phosphate, phthalate	6.5×10^{-4} mol/dm^3	6.5×10^{-4} mol Fe/dm^3, 43 m^2/g	1 d	25	3–8.5	0.1 mol dm^{-3} NaNO$_3$, 0–3.25×10^{-3} mol dm^{-3} o-phthalate	uptake	The presence of phthalate does not affect the uptake of phosphate, but the uptake of phthalate is depressed on addition of equimolar concentration of phosphate.	624

Sorbent	Species	Concentration	Solid	Time	pH	Temp (°C)	Electrolyte	Property	Remarks	Ref.
FeOOH, goethite, from nitrate	phosphate, sulfate	2×10^{-5}–10^{-3} mol dm^{-3}	0.5 g (100 m^2/g)/dm^3	1 d	2.5–10		0.01 mol dm^{-3} KNO$_3$	uptake	Uptake of phosphate is rather insensitive to the presence of sulfate (except at pH < 4) but uptake of sulfate is depressed in the presence of phosphate.	626
MnO$_2$	Cs	trace	1 g/300 cm^3	15 min	5 (initial)	25	0–5 mol dm^{-3} NaCl	uptake	Cs uptake is depressed in the presence of 0.01 mol dm^{-3} Na salts (the effects depend on the nature of the anion). With 5 mol dm^{-3} NaCl no Cs is adsorbed.	293
MnO$_2$, δ	Th	9×10^{-6} mol dm^{-3}	8.3 g (130 m^2/g)/dm^3	6 h	2.5–6	20	0.422 mol dm^{-3} NaCl + 0.054 mol dm^{-3} MgCl$_2$, 0.01 mol dm^{-3} CaCl$_2$	uptake	No significant effect of Mg or Ca on Th uptake.	262
MnO$_2$, δ	phosphate	10^{-5} mol dm^{-3}	10^{-3} mol Mn dm^{-3}	1 d	2–8.5	23	0–0.7 mol dm^{-3} NaCl + 0–3×10^{-2} mol dm^{-3} sulfate	uptake	The uptake is depressed at pH < 3, at pH > 4 the effect of sulfate is insignificant. With 10 g of humic acid/dm^3 the uptake is depressed at pH < 5.	609
MnO$_2\cdot n$ H$_2$O	Cd, Pb, Tl, Zn	10^{-3} mol dm^{-3}	4.36×10^{-3} mol Mn dm^{-3}	3 h	6			uptake	Presence Cd, Zn and Tl (0.001 mol dm^{-3}) does not induce desorption of preadsorbed Pb. Presence of Pb (0.001 mol dm^{-3}) reduces sorption of Tl by a factor of 3 and sorption of Cd and Zn by a factor of 4. The history of the system (which of the heavy metal cations is added first) does not affect the uptake in double-cation systems.	122

TABLE 4.7 Continued

Adsorbent	Adsorbate	Total concentration	Solid to liquid ratio	Equilibration time	T	pH	Electrolyte	Method	Result	Ref.
$MnO_2 \cdot n\ H_2O$	As(III)	10^{-5} mol dm^{-3}	15 mg/ 25 cm^3	1 d	25	3–11	phosphate	uptake	The uptake is reduced by 10% in the presence of 10^{-4} mol dm^{-3} phosphate and by 20% in the presence of 10^{-2} mol dm^{-3} phosphate. The effect of sulfate on the uptake of As III is insignificant.	632
SiO_2	Cr	4×10^{-5} mol dm^{-3}	10 g (500 m^2/g)/dm^3	3.5 h	20	4.8	$CaCl_2$, $LaCl_3$, 10^{-4}–10^{-2} mol dm^{-3} ionic strength	uptake	The uptake of Cr is depressed when the ionic strength increases but effects of Ca and La are not more significant than the effect of KCl (at the same ionic strength).	314
SiO_2, Merck, 60 H	Cs	4×10^{-5} mol dm^{-3}	200, 400 mg (384 m^2/g)/50 cm^3	7 d		4–10	0.001–0.1 mol dm^{-3} $NaNO_3$	uptake	Increase of Na concentration from 0.001 to 0.1 mol dm^{-3} depresses the Cs uptake by a factor of 5.	344
SiO_2, quartz	Hg	1.84×10^{-7} mol dm^{-3}	40 g (5 m^2/g)/dm^3	1 d		6–10	2×10^{-3}–10^{-1} mol dm^{-3} $Mg(NO_3)_2$	uptake	No significant difference between uptake from $NaClO_4$ and $Mg(NO_3)_2$	326
SiO_2, Cabot	Pb	10^{-4} mol dm^{-3}	2.5 g (200 m^2/g)/dm^3	10 h		3–10	0.025 mol dm^{-3} $NaClO_4$; 10^{-4} mol dm^{-3} EDTA; 0–10^{-4} mol dm^{-3} Fe(III)	uptake	Presence of Fe (III) enhances the uptake of Pb at pH 6–10.	118
SiO_2, gel, Davison	Cs, K, Li, Na, Rb	0.1 mol dm^{-3}			25	6.7	Cl	uptake	Selectivity factor with respect to Na: Li 0.65, K 1.8, Rb 2.4, Cs 3.2.	757
SnO_2, hydrous, Sn + HNO_3	Cs, Li, Na	0.1 mol dm^{-3}		1 d	25	11		uptake	Exchange of alkali metal cations studied also at 45 and 65°C.	758

Adsorbent	Species	Concentration	Solid/liquid	Time	Temp.	pH	Electrolyte	Method	Comments	Ref.
ThO₂, hydrous, different samples	Cl NO₃ SO₄ - CNS	0-0.1 mol dm⁻³	0.5 g/25 cm³	1 d		2,3		uptake	Hysteresis.	759
TiO₂	U	20 ppb	40 mg/dm³	5 h	25	8	0.72 mol dm⁻³ NaCl-2.3 ×10⁻³ mol dm⁻³ Na₂CO₃, 0-10⁻² mol dm⁻³ Ca	uptake	U uptake is depressed by 40% in the presence of 0.01 mol dm⁻³ Ca, 5×10⁻² mol dm⁻³ Mg depresses the uptake of U by 10%.	391
TiO₂, Degussa P-25	Cd, Cu	10⁻⁴ mol dm⁻³ (Cu+Cd)	2 g (55 m²/g)/dm³	10 h	22-25	3-8.5	0.003 mol dm⁻³ NaClO₄, 10⁻⁴ mol dm⁻³ EDTA	uptake	The % uptake of Cd is enhanced (at constant total concentration Cd+Cu) when Cd is partially replaced by Cu. The % uptake of Cu is unaffected (at constant total concentration Cd+Cu) when Cu is partially replaced by Cd.	395
TiO₂, anatase, commercial	Cs, Na	10⁻³ mol dm⁻³			25	3-10	chloride	uptake	The uptake of Cs is unaffected by 10⁻³ mol dm⁻³ Na but it is significantly depressed in the presence of >10⁻² mol dm⁻³ Na. The uptake of Na is significantly depressed in the presence of >10⁻² mol dm⁻³ Cs.	760
TiO₂·n H₂O	Cd, Hg, Zn	10⁻⁵ mol dm⁻³	0.1 g/10 cm³		30		6×10⁻⁵ mol dm⁻³ Mg, Ca, Ba, Sr	uptake	The uptake of heavy metal cations is slightly depressed in the presence of alkaline earth metal cation, the effect is most significant with Hg (10%) and least significant for Zn.	401
V₂O₅	chromate	carrier free	0.25 g (16.8 m²/g)/ 12 cm³	1 h	25	1-6	10⁻³-10⁻¹ mol dm⁻³, 12 anions	uptake	The uptake is severely depressed in the presence of 10⁻³ mol dm⁻³ phosphate, thiosulfate and citrate (factor >2), and also in the presence of molybdate and arsenate.	644

TABLE 4.7 Continued

Adsorbent	Adsorbate	Total concentration	Solid to liquid ratio	Equilib-ration time	T	pH	Electrolyte	Method	Result	Ref.
ZrO_2, from $ZrOCl_2$	Cs, Li, Na	0.01–0.1 mol dm^{-3}	0.1 g/25 cm^3		25	13	hydroxide	uptake	Exchange between alkali metal cations, also studied at 45 C and 65°C, and in aqueous methanol.	761

Clay minerals

illite	arsenate III, V	4×10^{-7} mol dm^{-3}	25 g dm^{-3}	3 h		7	10^{-3} mol dm^{-3} phosphate	uptake	1/4 of the initially sorbed As (III) and (V) is released in the presence of phosphate. The released As(III) is partially oxidized to As(V).	535
kaolinite	arsenate III, V	4×10^{-7} mol dm^{-3}	25 g dm^{-3}	3 h		7	10^{-3} mol dm^{-3} phosphate	uptake	The As (III) initially sorbed at pH < 6 is released in the presence of phosphate. The As(III) initially sorbed at pH > 6 is only partially released in the presence of phosphate. The released As is partially oxidized to As(V). The As(V) initially sorbed at pH > 8 is released in the presence of phosphate. The As(V) initially sorbed at pH < 8 is only partially released in the presence of phosphate.	535
kaolinite	iodide	10^{-4} mol dm^{-3}	15 g/dm^3	1 d	25	3–10	0.004–0.1 mol dm^{-3} NaNO$_3$	uptake	The uptake of iodide is depressed at high ionic strength.	651
kaolinite	phosphate	8.75×10^{-5} mol dm^{-3}	0.9 g (11 m^2/g)/115 cm^3	1 d	25	2–12	0.01 mol dm^{-3} NaCl	uptake	The uptake is depressed in the presence of 34 mg dm^{-3} of humic acid and in the presence of	503

adsorbent	sorbate	concentration	loading	time	temperature	pH	electrolyte	method	comments	ref
kaolinite	phosphate	10^{-5} mol dm^{-3}	1 10%		24	5 9	KCl	uptake	1.33×10^{-4} mol dm^{-3} fluoride (at pH <6), oxalate, tartrate, EDTA, and citrate (at pH <8). The uptake of phosphate is enhanced in the presence of acetate and amino acids and depressed in the presence of oxalate, citrate and bicarbonate.	762
Na-montmorillonite, thermally altered	Cs, Na	0–0.01 mol dm^{-3}	0.1 g/ 100 cm^3	7 d	25	natural	0–0.01 mol dm^{-3} NaCl (total Cl 0.01 mol dm^{-3})	uptake	Cs > Na except for 300–600 C dried samples at high Cs concentration. Distribution coefficient depends on Na:Cs ratio.	435
montmorillonite	arsenate III, V	4×10^{-7} mol dm^{-3}	25 g dm^{-3}	3 h		7	10^{-3} mol dm^{-3} phosphate	uptake	Only 1/2 of the initially sorbed As(III) is released in the presence of phosphate. The released As(III) is partially oxidized to As(V). The As(V) initially sorbed at pH > 8 is released in the presence of phosphate. The As(V) initially sorbed at pH < 8 is only partially released in the presence of phosphate.	535
montmorillonite	phosphate	10^{-5} mol dm^{-3}	1–10%		24	5–9	KCl	uptake	The uptake of phosphate is enhanced in the presence of acetate and amino acids and depressed in the presence of oxalate, citrate and bicarbonate.	762
montmorillonite	diquat, paraquat, acriflavine (organic cations)	0.01 mol dm^{-3}	1.7 g (756 m^2/g)/dm^3	7 d	25	6.9–7.4	none	uptake		763

TABLE 4.7 Continued

Adsorbent	Adsorbate	Total concentration	Solid to liquid ratio	Equilibration time	T	pH	Electrolyte	Method	Result	Ref.
smectite, treated with dithionite-citrate-bicarbonate and then with H_2O_2	U	8×10^{-6} mol dm⁻³	1.5 g/dm³	1 d	25	4–9	$Ca(ClO_4)_2$ ionic strength $10^{-2.17}$, $10^{-1.13}$	uptake	The uptake from $Ca(ClO_4)_2$, ionic strength $10^{-1.13}$ is comparable with uptake from 0.1 mol dm⁻³ $NaClO_4$. The uptake from $Ca(ClO_4)_2$, ionic strength $10^{-2.17}$ is significantly lower than uptake from 0.01 mol dm⁻³ $NaClO_4$.	24
Zeolites										
mordenite, Huber	Pb	10^{-4} mol dm⁻³	2.5 g (200 m²/g)/dm³	10 h		3–10	0.025 mol dm⁻³ $NaClO_4$ + 10^{-4} mol dm⁻³ EDTA, 0–10^{-4} mol dm⁻³ Fe(III)	uptake	About 50% uptake (rather insensitive to pH) in the presence of Fe(III), no uptake in absence of Fe(III).	118
Coatings										
aluminum oxide coating on sand	selenate IV and VI	8×10^{-4}, 1.2×10^{-3} mol dm⁻³	5 g/50 cm³	3 h		6.8, 7.8	0–1.5×10^{-3} mol dm⁻³ sulfate, carbonate	uptake	The uptake of selenate is depressed in the presence of sulfate and carbonate. This effect is more pronounced for selenate (IV) than for selenate (VI) and sulfate depresses the uptake stronger than carbonate.	656

Material	Species	Concentration	Ratio	Time	Temp	pH	Medium	Method	Remarks	Ref
iron oxide coating on sand	Pb	2×10^{-4} mol dm⁻³	50 g (68 m²/g)/dm³	1 d	20	4–8	0.05–0.2 mol dm⁻³ Ca(NO₃)₂	uptake	The uptake of Pb is not affected by the presence of Ca.	447
Glasses										
controlled pore glass	Ce(III), Eu	carrier free	250 mg/50 cm³	20 min	20	6–8	10^{-3}–10^{-1} mol dm⁻³ NaCl; 10^{-5} mol dm⁻³ Ce, Eu	uptake	Uptake of Ce 155 and Eu (152+154) from 10^{-5} mol dm⁻³ Ce is higher than that from 10^{-5} mol dm⁻³ Eu.	452
Clays										
bentonite, Laporte	Cd, Pb	2–4×10^{-5} mol dm⁻³	50 g/dm³	1 d	25		nitrate	uptake	The uptake is plotted as a function of amount of acid added (5×10^{-7}–10^{-3} mol HNO₃/g clay). Cd-bentonite is partially converted into Pb-form in the presence of Pb, the conversion of Pb-bentonite into Cd-form in the presence of Cd is less pronounced.	457
Soils										
loamy sand	Zn	400 mg/g (of solid)	1:10	7 d	15	natural	0–0.01 mol dm⁻³ Ca(NO₃)₂	uptake	The uptake from Ca(NO₃)₂ solution is lower than that from 0.03 mol dm³ NaNO₃.	468
soils	silicate	110 ppm SiO₂	1:5, 1:25		25	2–8	0.01 mol dm⁻³ CaCl₂ toluene added, 0–0.1 mol dm⁻³ citrate	uptake	Presence of citrates reduces the uptake of silicate.	549

TABLE 4.7 Continued

Adsorbent	Adsorbate	Total concentration	Solid to liquid ratio	Equilib- ration time	T	pH	Electrolyte	Method	Result	Ref.
3 soils	arsenate, molybdate, phosphate	2×10^{-5}– 2×10^{-3} mol dm^{-3}	1:50, 1:2.5	2 d	25	natural	Na	uptake	Uptake of phosphate (initial concentration up to 10^{-4} mol dm^{-3}) is insensitive to the presence of arsenate. For other binary mixtures of anions the uptake was depressed in the presence of competitor.	664
loamy sand	phosphate, selenate IV	5–100 mg/dm^3	1:10	30 d	25	5.6 initial	0.01 mol dm^{-3} $CaCl_2$	uptake	Phosphate is more successful competitor.	764
Other materials										
activated red mud	chromate (VI)						phosphate, sulfate	uptake	The uptake of Cr is reduced in the presence of phosphate and sulfate.	665
6 rock forming minerals	Np	2×10^{-11} mol dm^{-3}				6–9.5	artificial ground water	uptake	Np-Ca competition is discussed on basis of literature data.	251

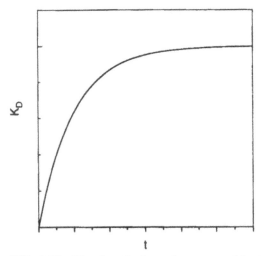

FIG. 4.69   Kinetics of adsorption governed by first order reaction.

equilibrium constants are employed to calculate the distribution of the adsorbate between the surface and the solution at various experimental conditions. Most experimental K_D values reported in Sections II–VI were measured after equilibration times (ranging from a few minutes to many days) that allegedly exclude changes in distribution of the adsorbate when further equilibration of the system is allowed. Figure 4.69 shows an idealized picture of adsorption kinetics. The curve was calculated assuming a first order reversible reaction to be the rate determining process. The net rate (resultant of the forward and backward process) is high in the beginning, it is getting slower in course of time, and finally the increase in the observed K_D as the function of time becomes negligible. Adsorption systems with stable K_D are often referred to as being in state of equilibrium, but actually they are not. Most likely the K_D would change on equilibration exceeding the studied time range by a few orders of magnitude. Some kinetic studies report a slow increase in uptake over the entire time scale studied. These results suggest that the choice of equilibration time in "equilibrium" experiments is in fact rather arbitrary. Even if a constant K_D is actually established after certain time, the knowledge of the kinetic behavior of the adsorption system is still important when an adsorption system of interest does not have enough time to equilibrate (e.g. dynamic studies at high flow rate). Therefore, theories describing not only "equilibrium", but also kinetic behavior of adsorption systems would be much-desired. Certainly the theoretical model cannot be better than the experimental data upon which the model is based, and the results of experimental studies of sorption kinetics reported in Table 4.8 are often contradictory, i.e. very different kinetic behavior has been reported in apparently similar systems. Reproducible kinetic results can be obtained provided that the parameters affecting adsorption kinetics are controlled, and in this respect kinetic studies are more demanding than "equilibrium" experiments. Actually, even the parameters that are well known to affect the "equilibrium" adsorption, e.g. the temperature were not always controlled in kinetic experiments. On the other hand, a few systematic studies of temperature effect on sorption kinetics have been published. The temperature effect on the rate of chemical reaction can be quantified as the activation energy defined as

$$E_{act} = RT^2 \, \mathrm{d}\ln k / \mathrm{d}T \tag{4.25}$$

where k is the rate constant of the reaction of interest. A few examples of activation energies calculated from experimental kinetic data are reported in this section. Also the diffusion coefficient increases with temperature, e.g. for spheres $D = kT/6\pi\eta a$, where η is the viscosity of the medium and a is the sphere radius.

Most experimental kinetic curves are rather smooth, i.e. the concentration of adsorbate in solution monotonically decreases, but some kinetic curves reported in the literature have multiple minima and maxima, which are rather unlikely to be reproducible. Such minima and maxima represent probably the scatter of results due to insufficient control over the experimental conditions. For instance use of a specific type of shaker or stirrer at constant speed and amplitude does not necessarily assure reproducible conditions of mass transfer. Some publications report only kinetic data—results of experiments aimed merely at establishing the sufficient equilibration time in "equilibrium" experiments. Other authors studied adherence of the experimentally observed kinetic behavior to theoretical kinetic equations derived from different models describing the transport of the adsorbate. Design of a kinetic experiment aimed at testing kinetic models is much more demanding, and full control over all parameters that potentially affect the sorption kinetics is hardly possible.

In most kinetic studies the adsorbate of interest is present only in solution in the beginning of the kinetic run ($t = 0$) and the concentration of the adsorbate in solution is followed as a function of time. A series of data points corresponding to different equilibration times can be obtained by the following means.

- A series of n identical dispersions is prepared, and the adsorbate is brought into contact with the adsorbent at the same $t = 0$ in all dispersions. The dispersions are allowed to equilibrate under the same conditions (stirring, shaking, etc.). A sample of solution is withdrawn (filtration, centrifugation) from dispersion #1 after equilibration time t_1, another sample of solution is withdrawn from dispersion #2 after equilibration time t_2, ..., finally a sample of solution is withdrawn from dispersion #n after equilibration time t_n.
- One dispersion is prepared and samples of solution are withdrawn from this dispersion after equilibration times $t_1, \ldots t_n$.

Apparently the choice of either method should not affect the course of kinetic curves, but each one has potential pitfalls that are probably responsible for the discussed above discrepancies. In the first method a series of dispersions with the same solid to liquid ratio can be easily prepared by mixing certain mass of the adsorbent with known mass or volume of the solution, but these dispersions can differ in particle size distribution, thus, the adsorption kinetics in particular dispersions can be different. Use of one dispersion eliminates this problem, but, e.g. stopping the shaker to collect one sample affects the further kinetics of adsorption. Moreover, unless the collected samples are very small, the solid to liquid ratio and the amount of adsorbate are not constant in course of the kinetic experiment, and this is in conflict with assumptions made in derivation of the kinetic equations being tested. Continuous monitoring of the concentration of the adsorbate as the function of time without necessity of sample collecting would be an ideal solution in this respect. Moreover, since the sample collecting takes at least a few seconds, very fast processes can only be followed by continuous monitoring. The review by Hachiya et al. [765] emphasizes advantages of conductometric monitoring in kinetic studies.

Conductometric monitoring was used in combination with the pressure jump method described in detail in another review by Hachiya et al. [766]. The pressure jump method has been extensively used to study kinetics of adsorption of heavy metal cations on oxides. The conductance in such system increases at elevated pressure and this was interpreted as desorption of metal cations. Once the pressure suddenly drops from initial value (on the order of 10 MPa) to atmospheric pressure, the metal cations re-adsorb and the kinetics of this process can be followed using conductometric sensor. Handling of the conductometric data is rather difficult since the uptake of metal cations is accompanied by release of protons. Thus adsorption of metal cations can result in increase or decrease in the conductance of the solution depending on the number of protons released per one adsorbed metal cation, and fortuitous cancellation is also possible. Moreover the pressure jump results in changes in primary surface charging (adsorption-desorption of protons) which also contribute to the changes in the conductance of the solution. Hachiya et al. observed constant (time independent) conductance a fraction of one second after the pressure jump. This suggests that the surface reactions are rather fast. The interpretation of the conductance (t) curves in terms of rate constants of one or two surface reactions offered by these authors requires some model assumptions. The adherence of experimental kinetic curves to a model does not prove that the model is correct, namely, many models can result in practically identical kinetic curves. Yiacoumi and Tien [767] in principle agree that the surface reactions responsible for adsorption are complete within a fraction of one second, but they permit that some adsorbates, namely, ions which exchange ligands in solution very slowly, may also react very slowly with the surface. Such reactions need many hours to be completed. However for most adsorbates, the kinetics of sorption on the time scale of hours or days is governed by a transport process (e.g. slow diffusion in narrow pores of the adsorbent) rather than a chemical reaction.

Practically any experimental kinetic curve can be reproduced using a model with a few parallel (competitive) or consecutive surface reactions or a more complicated network of chemical reactions (Fig. 4.70) with properly fitted forward and backward rate constants. For example, Hachiya et al. used a model with two parallel reactions when they were unable to reproduce their experimental curves using a model with one reaction. In view of the discussed above results, such models are likely to represent the actual sorption mechanism on time scale of a fraction of one second (with exception of some adsorbates, e.g. Cr that exchange their ligands very slowly). Nevertheless, models based on kinetic equations of chemical reactions were also used to model slow processes. For example, the kinetic model proposed by Amacher et al. [768] for sorption of multivalent cations and anions by soils involves several types of surface sites, which differ in rate constants of forward and backward reaction. These hypothetical reactions are consecutive or concurrent, some reactions are also irreversible. Model parameters were calculated for two and three

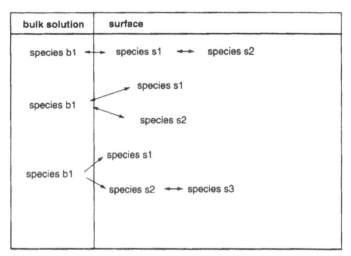

FIG. 4.70 Schemes of reactions used in modeling of adsorption kinetics.

concurrent reactions, one of which is irreversible (three or five adjustable parameters) for experimental data representing Cd and chromate uptake by eight soils. Interestingly, the actual speciation in solution (existence of metal cations at different degrees of hydrolysis or weak acids at different levels of deprotonation), and difference in rate of reactions leading from different solution species to the same surface species (Fig. 4.1) is usually neglected in kinetic models based on chemical kinetics.

Even if the sorption process is actually controlled by a chemical reaction, the adherence of the experimental data to a simple kinetic equation is rather unlikely, when the kinetic experiments are carried out at variable pH (the pH changes as the function of time), because the driving force for the adsorption–desorption process in not controlled merely by the concentrations of the adsorbate in solution and on the surface. For low concentrations of the adsorbate the problem of variable pH can be avoided by pre-equilibration of the adsorbent with solution containing all components but the adsorbate before the adsorbate is added. Even at high concentrations of adsorbate a kinetic experiment can be carried out at constant pH. To this end the adsorption system is equilibrated until the distribution of the adsorbate of interest between the solution and the adsorption layer reaches a constant value. Then a minute amount of adsorbate tagged with a radionuclide is added to the solution at $t = 0$. Addition of radionuclide does not significantly affect the pH or the total amount of the adsorbate in the system. Then small samples of dispersion are taken at $t = t_1, \ldots, t_n$, and the concentration of the radionuclide in solution is followed as the function of time. Such kinetic experiment is referred to as pure isotope exchange [769] to emphasize the pre-equilibration with respect to macrocomponents (isotope exchange experiments can be also carried out without pre-equilibration). The theory of pure isotope exchange is simpler than in more general case of sorption kinetics. The redistribution of isotopes of one element is the only net change in the system, irrespectively how complex is the actual mechanism of isotope exchange.

t = 0   t = ∞

FIG. 4.71 The profile of specific activity in the solid (gray field) in course of isotope exchange.

With exception of a few light elements the isotope effects (differences in physical and chemical properties between isotopes) can be neglected, and the rate constants of reactions and diffusion coefficients are practically identical for the radioactive and inactive isotope. Some theoretical kinetic equations of heterogeneous isotope exchange have been compiled in Ref. 769. Heterogeneous isotope exchange is a multistep process and the effective rate is governed by the slowest step when the other processes are much faster. Diffusion in the solid particle is often regarded as the slowest step, and then kinetic equations representing the spatial distribution of the radionuclide in a highly symmetrical solid particle (sphere, plate, cylinder) as the function of time can be adopted from the theory of heat transfer (identical differential equations govern the mass and heat transfer kinetics). Figure 4.71 illustrates the profile of specific activity in a solid surrounded by liquid (only the liquid is radioactive at $t=0$) in course of isotope exchange. The exchange fraction F is defined as

$$F = \frac{i(t) - i(0)}{i(\infty) - i(0)} \tag{4.26}$$

where i is the (radio) activity (proportional to the concentration of radionuclide) in particular state (solution or adsorbed), $t=0$ is the initial state, and $t=\infty$ is the equilibrium state. $F=0$ for the initial state and $F = 1$ at equilibrium. For highly symmetrical solid particles the solutions of the kinetic equations, $F(t)$ are obtained in form of slowly converging infinite series of terms involving only two independent variables, namely, C, the ratio of the amounts of the element of interest in two states (adsorbed and in solution) and dimensionless time $\tau = Dt/r^2$, where D is the effective diffusion coefficient and r is the radius of the sphere or cylinder or the half-thickness of the plate. This means that all kinetic curves obtained for given geometry (e.g. spherical) and given C can be superimposed on one another by scaling of the t-axis, irrespective of the r and D. At certain conditions these series reduce to simple equations, e.g. $F \propto t^{1/2}$ for spherical geometry and small t. Kinetic equations of isotope exchange in assemblies of spheres of different radii have been also published. It has been shown [770] that an assembly of plates (distribution of plate thickness) can be found that gives a kinetic equation identical as the kinetic equation for spheres. The particle diffusion (as rate limiting step) is suitable to describe the kinetics of ion exchange (the ions of interest uniformly distributed in a ion exchanger bead), but not necessarily suitable for adsorption on the external surface of the particle. However, high surface area of adsorbents is usually due to porous structure, and slow diffusion in narrow and tortuous pores is potentially the rate determining step of the sorption process. Putting aside the actual route of mass transfer and actual distribution of the radionuclide on atomic scale, the diffusion of the adsorbate in a porous particle on macroscopic scale can be looked at in terms of diffusion in homogeneous medium. Effective diffusion coefficient in a porous particle is a function of the pore radius, and there is an experimental evidence that the isotope exchange is slower in narrow pore adsorbents than in their wide pore counterparts at otherwise identical conditions. The diffusion in pores is a combination of diffusion in pore filling fluid and multiple adsorption–desorption cycles (surface reactions). Most likely the surface reaction involves not only the adsorbate of interest, but also ions of inert electrolyte, namely, accelerated equilibration at high ionic strength has been reported.

Diffusion in the liquid film surrounding the solid particle is another process potentially affecting the rate of heterogeneous isotope exchange. It is expedient to assume that a solid particle in liquid medium carries a thin layer of liquid that is not disrupted even by vigorous stirring, and the mass transfer in this film occurs only by diffusion (similar model is used in interpretation of microelectrophoresis, cf. Chapter 3). This type of rate control leads to linear dependence between $\ln(1-F)$ and t. The kinetic curves of isotope exchange are often presented in $\ln(1-F)$-t coordinates. Linear dependence between $\ln(1-F)$ and t is also obtained when the rate is controlled by a first order reversible surface reaction. Thus, the shape of kinetic curves does not unequivocally indicate the rate determining process. Combination of two or more surface reactions (cf. Fig. 4.70) leads to curvilinear $\ln(1-F)$-t plots. Curvilinear $\ln(1-F)$-t plots are also obtained when the rate of two or three of discussed above processes (diffusion in liquid film, surface reaction, intraparticle diffusion) are comparable. Kinetic equations for spherical geometry with two and three of the above processes simultaneously affecting the overall rate of isotope exchange have been derived.

Considering the mechanisms of sorption discussed in Section II, e.g. surface precipitation or formation of new phases involving the adsorbent and the adsorbate, the above kinetic model is not sufficient to describe isotope exchange in all relevant systems, although most experimental kinetic curves can be reproduced by proper adjustment of parameters (effective D in the solid and in the liquid film, rate constants of surface reactions) within the discussed above model. Spectroscopic studies (Section I and II) suggest that the uptake of adsorbate is often due to simultaneous formation of surface complex and surface precipitation. Single $F(t)$ curve based on the radioactivity of the solution is not sufficient to describe sorption kinetics in such systems.

The actual rate determining step cannot be infered from the course of the kinetic curves, but some assessment is possible by means of specially arranged experiment. Interruption test makes it possible to assess the role intraparticle diffusion (diffusion in pores of the adsorbent) as potential rate determining step. The isotope exchange experiment is started in usual way (*vide infra*), and k kinetic data points are collected (equilibration times t_1, ..., t_k), but at the time t_k the solid is separated from the liquid, then both phases are stored separately for a time much longer than t_k, and then contacted again, and the time when the phases were separated is not counted as the time of exchange. This kinetic run is compared with a control run without interruption. When the both runs are identical, the rate of the isotope exchange is not limited by the intraparticle diffusion. When the intraparticle diffusion plays an important role in the rate control the F after interruption is higher than in the control dispersion (without interruption). This is because the concentration of radionuclide in the solid (cf. Fig. 4.71) has leveled out during the interruption giving rise to a higher gradient of the specific activity in the outermost part of solid particle after the interruption.

The same steps as discussed above for the case of isotope exchange (diffusion in liquid film, surface reaction, intraparticle diffusion) were considered in a kinetic model [771] of metal ion adsorption from solution. This model was presented in a book with diskettes (FORTRAN program, rate controlled by reaction, by transport or mixed control).

Barrow [772] derived a kinetic model for sorption of ions on soils. This model considers two steps: adsorption on heterogeneous surface and diffusive penetration. Eight parameters were used to model sorption kinetics at constant temperature and another parameter (activation energy of diffusion) was necessary to model kinetics at variable T. Normal distribution of initial surface potential was used with its mean value and standard deviation as adjustable parameters. This surface potential was assumed to decrease linearly with the amount adsorbed and amount transferred to the interior (diffusion), and the proportionality factors were two other adjustable parameters. The other model parameters were sorption capacity, binding constant and one rate constant of reaction representing the adsorption, and diffusion coefficient of the adsorbate in the solid. The results used to test the model cover a broad range of T (3–80°C) and reaction times (1–75 days with uptake steadily increasing). The pH was not recorded or controlled.

The discussed above theoretical models predict continuous increase of K_D in time. Kinetic curves of adsorption with a maximum after certain equilibration time (rather than plateau or constant increase in uptake) observed for some systems (most often with anions as the adsorbate) can be due to

- Dissolution of the adsorbate.
- Recrystallization (the surface area is reduced in course of time).
- Variable pH (the electrostatic conditions are more favorable in the beginning of the adsorption process).

Desorption kinetics has been also studied: the adsorbate is released because of unfavorable electrostatic conditions, change in temperature, or addition of reagents forming very stable water soluble complexes with the adsorbate of interest.

Examples of kinetic studies in specific adsorption systems are compiled in Table 4.8. Only studies reporting series of data points corresponding to different equilibration times are reported. In some other publications the description of kinetic study is limited to the statement that certain equilibration time was sufficient to attain stable K_D, but these results are not reported.

Recent review of sorption kinetics in soils [796] reports a number of theoretical and empirical kinetic equations that have been used in complex and ill defined systems.

(*Text continues on pg. 555*)

TABLE 4.8 Adsorption Kinetics

Adsorbent	Adsorbate	Total concentration	Solid to liquid ratio	Equilibration time	T	pH	Electrolyte	Method	Result	Ref.
Al_2O_3, δ, Degussa C	Cr	5×10^{-4}–3×10^{-3} mol dm^{-3}	10 g/dm^3	> 30 d		2.5–7.5 [oxalate] = 3	0.1 mol dm^{-3} KNO$_3$, [Cr]	ESR	For a short equilibration time (10 min) uptake increases when pH decreases when pH increases (anion like behavior). For an equilibration time of 1 d, the uptake is pH independent, for 14 d aged sample cation like behavior is observed.	63
Al_2O_3, γ, hydrolysis of isopropoxide	Cu	7.8×10^{-5} mol dm^{-3}	0.2 g (170 m^2/g)/100 cm^3	1 d		6.3	0.002 mol dm^{-3} KNO$_3$	uptake, pH, Al release	The reaction is not complete.	105
Al_2O_3, γ, Japan Aerosil C	Pb	10^{-4}–1.2×10^{-3} mol dm^{-3}	15, 25 g (100 m^2/g)/dm^3	20 μs–20 ms	20	4.5–5.4	nitrate or chloride	uptake	Fast process: $k_1 = 1.4\times10^8$ mol^{-1} dm^3 s^{-1}, $k_{-1} = 10^4$s^{-1} and slow process: $k_2 = 13$ s^{-1}, $k_{-2} = 1.5\times10^6$ mol^{-1} dm^{3}s^{-1}.	78
Al_2O_3, α	U	95 ppm	0.1 g (0.07–2 m^2/g)/40 cm^3	20 d		4–6.5	0.1 mol dm^{-3} NaNO$_3$	uptake, proton release	Constant pH and uptake after a 8 d equilibration.	86
Al_2O_3, γ, Japan Aerosil C	Co, Cu, Mn, Pb, Zn	1–3×10^{-3} mol dm^{-3}	30 g (100 m^2/g)/dm^3	0.2 s	25	4–7	0.0075 mol dm^{-3} NaNO$_3$	pressure jump	Intrinsic adsorption rate constants (mol^{-1} dm^3 s^{-1}): Pb 6.4×10^4, Cu 7.4×10^3. Zn 5.1×10^2, Mn 32, Co 15.	773, 106
Al_2O_3, δ	Cr, V	5×10^{-5}, 1×10^{-4} mol dm^{-3}	10 g (100 m^2/g)/dm^3	6 d	25	2.9–4.4	0.1 mol dm^{-3} NaClO$_4$	uptake	The Cr uptake steadily increases after 6 d reaction time.	113

Adsorbent	Adsorbate	Concentration	Solid/solution	Equilibration time	T (°C)	pH	Electrolyte	Method	Comments	Ref.
Al$_2$O$_3$, γ, Aerosil	Ga, In	3×10^{-3} mol dm^{-3}	40 g (100 m^2/g)/dm^3		25	1.5-4	0-0.1 mol dm^{-3} NaNO$_3$	pressure jump		115
Al$_2$O$_3$, α, Aldrich	arsenate (V)	10^{-6} mol dm^{-3}	25 g (0.7 m^2/g)/dm^3	19 d		4-9	0.1 mol dm^{-3} NaCl	uptake	No difference in results obtained after 5 and 19 d.	482
Al$_2$O$_3$, Alcoa F-1; Merck; Kaiser	borate	20 mg B/dm^3	5 g (210-350 m^2/g)/200 cm^3	2 d	25±2	not controlled	none	uptake	The reaction is not complete.	485
Al$_2$O$_3$, δ, Cabot	borate	9×10^{-4} mol dm^{-3}	0.25 g (70 m^2/g)/dm^3	24 d		5-11	0.1 mol dm^{-3} NaCl	uptake	The 1 and 24 d results are practically identical. The 10 min and 2 h results are practically identical. Uptake increases for longer equilibration times at pH <7, but at pH >8 the uptake decreases at longer equilibration times.	518
Al$_2$O$_3$, γ,	chromate (VI)	1 5×10^{-3} mol dm^{-3}	30 g (100 m^2/g)/ dm^3	200 ms	25	7.5-9	0.008 mol dm^{-3} NaNO$_3$	uptake	Intrinsic values for adsorption/desorption of monovalent chromate $k_1 = 5.3×10^4$ mol^{-1} dm^3 s^{-1}, $k_{-1} = 19$ s^{-1}; for divalent chromate $k_2 = 9.9×10^4$ mol^{-1} dm^3 s^{-1}, $k_{-2} = 52$ s^{-1}.	489
Al$_2$O$_3$, 3 samples	fluoride	25 mg/dm^3	25 g/dm^3	2 d	25	not controlled	Na	uptake	No further uptake after 10 h equilibration.	493
Al$_2$O$_3$, γ	fluoride	10^{-3} mol dm^{-3}	10 g (287 m^2/g)/dm^3	1 h	25	5-5.2	0.05 mol dm^{-3} NaClO$_4$	uptake	Reaction not complete within 1 h (almost 100% uptake).	494
Al$_2$O$_3$, α, BDH	molybdate	14 mg Mo/dm^3	4 g(4.3 m^2/g)/dm^3	1 d		7		uptake	No further uptake after a 2 h equilibration.	497
Al$_2$O$_3$, γ	molybdate	5×10^{-3} mol dm^{-3}	30 g (100 m^2/g)/dm^3		25	5.8-6.8	0.01-0.1 mol dm^{-3} NaNO$_3$	pressure jump	Fast process (relaxation time about 1/100 s) followed by slow process (relaxation time about 1/10 s).	500

TABLE 4.8 Continued

Adsorbent	Adsorbate	Total concentration	Solid to liquid ratio	Equilibration time	T	pH	Electrolyte	Method	Result	Ref.
Al_2O_3, α	phosphate	3×10^{-4} mol dm^{-3}	2.5 g(10 m^2/g)/dm^3	20 d	50	3.5–8.5		uptake, XRD	The uptake is still increasing after 15 day reaction.	504
Al_2O_3, α, Merck	phosphate	3.5×10^{-4} mol dm^{-3}	5 g(3.8 m^2/g)/dm^3	62 d	20	5, 6	none; synthetic sewage water	uptake, XRD	The log-log plot uptake(t) consists of two nearly linear segments. The high t segment corresponds to formation of a new phase, sterrettite.	507
Al_2O_3, γ, Japan Aerosil C	phosphate	8×10^{-3} mol dm^{-3}	30 g (100 m^2/g)/ dm^3	200 ms	25	6.5–8.2	1.5×10^{-2} mol dm^{-3} NaCl	uptake	Intrinsic values for adsorption/desorption of monovalent phosphate $k_1 = 4.1 \times 10^5$ mol^{-1} dm^3 s^{-1}, $k_{-1} = 2.3$ s^{-1}; for divalent phosphate $k_2 = 1.1 \times 10^7$ mol^{-1} dm^3 s^{-1}, $k_{-2} = 2.7$ s^{-1}.	509
Al_2O_3, γ, Sumitomo	phosphate	3.4 molecules/nm^2		1 y	25	6 initial	0.1 mol dm^{-3} NaCl	uptake	Adsorption is complete within 5 h, then slow phase transformation which is not complete even after 100 d.	510
Al_2O_3, γ, Sumitomo	phthalate	0.001 mol dm^{-3}	2 g(152 m^2/g)/dm^3	8 d	room	4.5–7.2	0–1 mol dm^{-3} KCl	uptake, IR, Raman	At pH > 5 no further uptake after 5 d equilibration.	511
Al_2O_3, γ	benzoate, salicylate	2×10^{-4} mol dm^{-3}	2.2–3.7 g (130 m^2/g)/ dm^3	10 d	22		0.1 mol dm^{-3} NaClO$_4$	uptake	Fast process (complete within 6 h) followed by slow process, the uptake steadily increases even	517

Adsorbent	Adsorbate	Concentration	Solid/solution	Time	Temp	pH	Electrolyte	Method	Result	Ref.
Al_2O_3, γ	clodronate, phosphate	3.4 molecules/nm²		6 d 1 y		slightly acidic		uptake	for equilibration times > 2 d. The reaction with clodronate is complete (100% uptake) after 6 days.	522
Al_2O_3, Degussa C	fumarate, maleate, succinate	10^{-4} mol dm⁻³	0.5 g (100 m²/g)/dm³	5 d	25	4.3–4.6	0.05 mol dm⁻³ $NaClO_4$	uptake	No further uptake of fumarate and maleate after a 1 d equilibration.	524
Al_2O_3, Carlo Erba	8-hydroxyquinoline	4.5×10^{-4} mol dm⁻³	1 g/dm³	1 d		6.45–6.7		uptake	Uptake steadily increases.	688
Al_2O_3, A16, Alcoa	polyacrylic acid MW 5000, PVA 88% hydrolysed, MW 25 000	25, 100 ppm	1 g (10 m²)/100 cm³	32 h	27	7.3–7.8	10^{-3} mol dm⁻³ KNO_3	uptake	Uptake of PVA is complete within 1 d; uptake of polyacrylic acid is complete within 8 h.	741
$Al(OH)_3$, amorphous	Cu		0.1 g (111 m²/g)/20 cm³	60 d	room	5	0.05 mol dm⁻³ NaCl	uptake, pH	The uptake peaks after 2 weeks.	127
$Al(OH)_3$, gibbsite	Ni	3×10^{-3} mol dm⁻³	10 g/dm³	5 d	25	7.5	0.1 mol dm⁻³ $NaNO_3$	uptake	The uptake is steadily increasing.	143
$Al(OH)_3$, amorphous, from sulfate	arsenate (V)	10^{-3} mol dm⁻³	1.29 g/dm³	3 d		6.5		uptake, pH	Constant uptake after 1 d.	534
$Al(OH)_3$, amorphous, fresh, from nitrate	borate	4.6×10^{-4}, 2.3×10^{-3} mol dm⁻³	10^{-3} mol Al/30 cm³	3 d				uptake	Borate preadsorbed on fresh precipitate is released on aging.	556
$Al(OH)_3$, gibbsite	phosphate	1–300 ppm P	1 g/10 cm³	2 d	4–45	7	K	uptake	Linear log (uptake) – log t curves, uptake is steadily increasing after 2 d equilibration.	541
$Al(OH)_3$, gibbsite	phosphate	3×10^{-5}–8×10^{-4} mol dm⁻³	6 g (47 m²/g)/dm³	6 d	24	4–11	0.01 mol dm⁻³ KCl	isotope exchange	Slowly exchangeable phosphate is absent at pH > 9.	542

TABLE 4.8 Continued

Adsorbent	Adsorbate	Total concentration	Solid to liquid ratio	Equilibration time	T	pH	Electrolyte	Method	Result	Ref.
Al(OH)₃, gibbsite	phosphate	2.42×10^{-4} mol dm^{-3}	1%	4 d	26	5.5	0.001 mol dm^{-3} CaCl$_2$	uptake	Reaction is not complete after 4 d equilibration.	543
Al(OH)₃, amorphous	phosphate	3.5×10^{-4} mol dm^{-3}	0.2 g (3.8 m²/g)/dm³	62 d	20	5, 6	none; synthetic sewage water	uptake, XRD	The log-log plot uptake(t) consists of two nearly linear segments. The high t segment corresponds to formation of a new phase, sterrettite.	507
Al(OH)₃, gibbsite	phosphate	0.009 mol dm^{-3}		4 d	2–46	5	0.1 mol dm^{-3} KCl + NaCl	uptake	Uptake is steadily increasing. Replacement of Na by K accelerates the reaction.	544
Al(OH)₃, gibbsite	phosphate	0.009 mol dm^{-3}		27 d	22	5	0.1 mol dm^{-3} NaCl, KCl, RbCl, NH₄Cl	uptake	Uptake steadily increases, the kinetic curves of phosphate sorption are rather insensitive to the nature of the supporting electrolyte for the first few hours, then the uptake of phosphate depends on the cation (K > Rb > NH₄ > Na).	545
Al(OH)₃, amorphous, from chloride	phosphate	0.001, 0.002 mol dm^{-3}	0.2 g/25 cm³	3 d	25	5.8	0.01 mol dm^{-3} NaCl	uptake	The uptake steadily increases.	548
Al(OH)₃, Merck,	silicate	80 mg SiO$_2$/dm³	1 g (4 m²)/100	25 d	20	9.2	0.05 mol dm^{-3} K₂SO₄	uptake	Uptake steadily increases; also at 10 and	550

Solid	Species	Concentration	(cm^3)	Time	Temp	pH	Medium		Remarks (35 C.)	Ref.
bayerite + gibbsite Al(OH)₃, Martifin, gibbsite	silicate	160 ppm SiO_2	1% w/w, 7.4 m^2/g	15 d	25	3–11	0.02 mol dm^{-3} NaCl	uptake	Uptake is stable after 8 d.	551
Al(OH)₃, gel, various neutralization ratios	decanoate dodecanoate	10^{-3} mol dm^3 5×10^{-4} mol dm^{-3}	3.9 ppm of Al(OH)₃	2 h	25	6.5, 7.5	Na	uptake	For a few systems the uptake goes through a maximum (at $t=3$–10 min); no substantial changes after 20 min.	774
Al(OH)₃, gel, various neutralization ratios	dodecylsulfonate, tetradecylsulfonate	10^{-4} mol dm^3 4×10^{-5} mol dm^{-3}	12 ppm of Al(OH)₃	1 h	25	6.5, 7.5	0.125 g NaCl + 0.025 g NaHCO₃ /dm^3 Na	uptake	No substantial changes after 20 min.	775, 776
Bi₂O₃, Alfa, α	chromate	10^{-6} 10^{-2} mol dm^{-3}		6 h	27	7.5	Na	uptake	No further uptake after a 1 h equilibration (Fig. 2). The uptake steadily increases (Fig. 3, 35 50 C).	560
CeO₂	Ba, Sr	10^{-2} 10^{-2} mol dm^{-3}	0.1 g/10 cm^3	3 h	30-60	11.4	nitrate	uptake	At 50 C: adsorption rate constants for $c=10^{-5}$ mol dm^{-3}: 7.55×10^{-2} min^{-1} for Ba and 4.26×10^{-2} min^{-1} for Sr. Activation energies (from temperature dependence) 10.5, and 13.7 kJ mol^{-1} for Ba and Sr, respectively.	156
CeO₂·n H₂O	Hg, Zn	10^{-7} 10^{-2} mol dm^{-3}	0.1 g/10 cm^3	3 h	30	7.5	nitrate, chloride	uptake	No further uptake of Zn after a 1 h equilibration.	157
CuO	sulfate		30 mg cm^{-2} (layer)	40 min		0	1 mol dm^{-3} HClO₄	uptake	The uptake peaks after a 8 min reaction.	777
Fe₃O₄	Cs	carrier free	0.1 g 20 cm^3	7 d	20–25	6.7 8	none, 3 g H₃BO₃ dm^{-3}	uptake		778

TABLE 4.8 Continued

Adsorbent	Adsorbate	Total concentration	Solid to liquid ratio	Equilibration time	T	pH	Electrolyte	Method	Result	Ref.
Fe_3O_4	H		30 g dm^{-3}		25	2.5–5	2×10^{-3} mol dm^{-3} N(CH$_3$)$_4$ClO$_4$	pressure jump	Adsorption rate constant = 1.4×10^5 mol^{-1} dm^{3}s^{-1}; desorption rate constant – 0.34 s^{-1}.	779
Fe_3O_4, synthetic	chromate	2×10^{-3} mol dm^{-3}	20 g (23 m^2/g)/dm^3	1 d	room	7	0.1 mol dm^{-3} NaCl, NaNO$_3$	uptake	Cr(VI) is completely removed from solution within 8 h (and reduced to Cr(III).	566
Fe_2O_3, hematite, monodispersed	Co	10^{-5} mol dm^{-3}	0.55–5 g (13 m^2/g)/dm^3	22 h	25	8	10^{-2} mol dm^{-3} NaNO$_3$	uptake	With 0.55 g/dm^3 the uptake is stable after 1 h, with higher solid to liquid ratio, a fast process is complete within 2 h, then slow drift is observed.	178
Fe_2O_3, hematite, natural	Eu	10^{-8} mol dm^{-3}	0.5 g (43 m^2/g)/dm^3	3 d		3–10	0.01 mol dm^{-3} NaClO$_4$, 2 mg/dm^3 fulvic acid	uptake	In absence of fulvic acid the uptake curves obtained after 1 h and 3d equilibration are identical. In the presence of fulvic acid the uptake steadily increases.	181
Fe_2O_3	H		30 g dm^{-3}		25	2.5 5	2×10^3 mol dm^{-3} N(CH$_3$)$_4$ClO$_4$	pressure jump	Adsorption rate constant = 2.4×10^5 mol^{-1} dm^{3}s^{-1}; desorption rate constant = 0.16 s^{-1}.	779
Fe_2O_3, natural hematite	H		2 g/60 cm^3	1 min	20	8 9	0.01–1 mol dm^{-3} NaCl, KCl,	pH electrode	The rate constant of proton desorption in 1 mol dm^{-3} solutions of 1–1	780

Adsorbent	Species	Concentration	Solid/solution	Time	T/°C	pH	NaNO₃, CaCl₂	Method	Comments	Ref.
Fe_2O_3	La	2×10^{-5} mol dm^{-3}	0.3 g (8.5 m²/g)/50 cm³	1 h		6.6	0.1 mol dm^{-3} NaNO₃	uptake, proton release	salts is higher by a factor of 1.5 than at lower ionic strengths or in CaCl₂. Stable pH and La concentration after 10 min.	184
Fe_2O_3, gel, from sulfate, 8 samples	Na	trace	1 g/225 cm³	8 min	25		0.1 mol dm^{-3} LiOH	uptake	The uptake steadily increases.	755
Fe_2O_3, α	antimonate (V)	carrier free	45 mg (27 m²/g)/60 cm³	10 min	30	4	0.25 mol dm^{-3} LiCl	uptake	Reciprocal Sb concentration in solution is a linear function of t. Also at 40, 50, and 60 C.	569
Fe_2O_3, gel, from chloride, 2 samples	chloride	0.01 mol dm^{-3}	1 g/160 cm³	0.5 h	25	2	0, 0.09 mol dm^{-3} NaCl	uptake	Also kinetic study in water – lower n-alcohol mixtures.	781
Fe_2O_3, hematite	molybdate	13.5 mg Mo/dm³	8 g (9.1 m²/g)/dm³	1 d		7		uptake	No further uptake after a 1 h equilibration.	497
Fe_2O_3, hematite	phosphate	12, 20 ppm P	1 g/10cm³	2 d	4-45	7	K	uptake	Linear log (uptake) – log t plots.	541
Fe_2O_3, hematite	sulfate	3×10^{-5}, 5×10^{-4} mol dm^{-3}	0.15 m²/g, 20 mg cm⁻² (layer)	3 h		0	1 mol dm^{-3} HClO₄	uptake	The reaction is not complete.	777
Fe_2O_3, gel, from chloride	bromide, chloride, nitrate, sulfate	0.01 mol dm^{-3}		3 h	25	2		uptake		782
Fe_2O_3, hematite	HEDP, $P_3O_{10}^{-5}$	0.01 mol dm^{-3}	3 g (6.5 m²/g)/25 cm³	3 d	25	2-11	none	uptake	In acidic range, the uptake steadily increases; no further uptake of HEDP at pH 11 after 2 h equilibration.	581
Fe_2O_3	8-hydroxyquinoline	4.2×10^{-4} mol dm^{-3}	8 g/dm³	1 d		8.4		uptake	Uptake steadily increases after a 1 d equilibration.	688

TABLE 4.8 Continued

Adsorbent	Adsorbate	Total concentration	Solid to liquid ratio	Equilibration time	T	pH	Electrolyte	Method	Result	Ref.
$Fe_2O_3 \cdot H_2O$, from chloride	Ba	10^{-7}–10^{-2} mol dm⁻³	0.1 g/10 cm³	12 h	30	9.2		uptake	Constant uptake after a 3 h equilibration.	223
$Fe_2O_3 \cdot H_2O$,	Cd	10^{-7}, 10^{-8} mol dm⁻³	1 g/dm³	100 d	25	6,7	0.03 mol dm⁻³	uptake	Systematic increase in the uptake during the first 20 d.	229
$Fe_2O_3 \cdot H_2O$, from nitrate	Ni	10^{-6}–10^{-4} mol dm⁻³	0.011 mol Fe/dm³	12 d	22 ±3	8	0.1 mol dm⁻³ NaNO₃, NaClO₄	uptake	Desorption kinetics of preadsorbed Ni is studied after addition of EDTA (10^{-5} mol dm⁻³). With 10^{-4} mol dm⁻³ Ni the reaction is complete within 1 d; with 10^{-5} mol dm⁻³ Ni the reaction is not complete within 12 d.	212
$Fe_2O_3 \cdot H_2O$, hydrolysis of nitrate in presence and absence of citrate	Zn	10^{-4} mol dm⁻³	20 mg/20 cm³	1 d	15–45	7	0.2 mol dm⁻³ NaNO₃	uptake	No further uptake after a 12 h equilibration at 25°C. Kinetics of release of pre-adsorbed Zn on addition of diethylene triamine pentaacetic acid was also studied.	219
$Fe_2O_3 \cdot H_2O$, amorphous	Ca Zn	2.5×10^{-4} mol dm⁻³	10^{-2} mol Fe/dm³	2 d	25	7.8 5.4	1 mol dm⁻³ NaNO₃	uptake, proton release	At fast process (complete within 6 min or less) is followed by a slow process (continuous increase in uptake over many hours).	224
$Fe_2O_3 \cdot H_2O$	silicic acid	10^{-4}–7×10^{-4} mol dm⁻³	3.33 g/dm³	20 d		3–6	0.01 mol dm⁻³ NaNO₃	uptake	Uptake steadily increases.	592

Adsorbent	Species	Concentration	Solid	Time	Temp.	pH	Medium	Method	Remarks	Ref.
$Fe_2O_3 \cdot H_2O$ from nitrate	silicate	10^{-3}–1.8×10^{-3} mol dm^{-3}	10^{-3}, 3.7×10^{-2} mol Fe/dm^3	7 d	25	4–12	0.1 mol dm^{-3} NaNO$_3$	uptake	With 10^{-3} mol Fe/dm^3 and 1.8×10^{-3} mol dm^{-3} silicate no further uptake after a 1 d equilibration, with 3.7×10^{-2} mol Fe/dm^3 and pH < 6 the reaction is not complete within 4 d.	596
FeOOH, goethite, synthetic	Eu	10^{-8} mol dm^{-3}	0.25 g (62 m^2/g)/dm^3	3 d		3–10	0.01 mol dm^{-3} NaClO$_4$, 2 mg/dm^3 fulvic acid	uptake	In absence of fulvic acid the uptake curves obtained after 1 h and 3d equilibration are identical.	181
FeOOH, goethite, synthetic	Pb(II)	0.002 mol dm^{-3}	30 g(52 m^2/g)/dm^3	0.4 s	25	4–5.5	0.01 mol dm^{-3} NaNO$_3$	pressure jump	The $k_1^{int} = 2$–4×10^5 mol^{-1} dm^3 s^{-1} (varies with Δp).	252
FeOOH, goethite, from chloride	Cd, Ni, Zn	10^{-5} mol dm^{-3}	20 mg (73 m^2/g)/10 cm^3	42 d	20	5–6	0.01 mol dm^{-3} Ca(NO$_3$)$_2$	uptake	Linear uptake vs. $t^{1/2}$ plots for 1 h < t < 21 d.	274
FeOOH, goethite, synthetic	molybdate	16 mg Mo/dm^3	0.5 g(67 m^2/g)/dm^3	1 d		6.5		uptake	The uptake does not change after a 1.5 h equilibration.	497
FeOOH, goethite, synthetic	phosphate, o-phthalate	6.4×10^{-4} mol dm^{-3}	8×10^{-4} mol Fe/dm^3, 43 m^2/g	1 d	25	4–4.3	0.1 mol dm^{-3} NaNO$_3$	uptake	Release of preadsorbed phthalate is practically complete 3 h after addition of phosphate.	624
Fe(OH)$_3$	Eu	5×10^{-7} mol dm^{-3}	0.01 mol Fe dm^{-3}	6 d		4–12	0.15 mol dm^{-3} NaCl	uptake		180
Fe(OH)$_3$, from nitrate	Eu	10^{-8} mol dm^{-3}	0.05 g (320 m^2/g)/dm^3	3 d		3–10	0.01 mol dm^{-3} NaClO$_4$, 2 mg/dm^3 fulvic acid	uptake	In absence of fulvic acid the uptake curves obtained after 6 h and 3d equilibration are identical.	181

TABLE 4.8 Continued

Adsorbent	Adsorbate	Total concentration	Solid to liquid ratio	Equilib- ration time	T	pH	Electrolyte	Method	Result	Ref.
Fe(OH)$_3$, precipitate aged 1–15 days	Ga	carrier free	0.005 mol dm^3	2 d		5–11	0.15 mol dm^{-3} NaCl	uptake	For fresh precipitate at pH 5–7 the uptake of 100% is reached within 3 min. Otherwise the uptake steadily increases.	183
Fe(OH)$_3$, freshly precipitated and 1 d aged	phosphate	3 mg P/dm^3	5 mg Fe /dm^3	10 d				uptake	Uptake steadily increases.	546
Fe(OH)$_3$	phosphate	0.001, 0.002 mol dm^{-3}	0.2 g/25 cm^3	3 d	25	5.8	0.01 mol dm^{-3} NaCl	uptake	The uptake is constantly increasing.	548
MnO$_2$, cryptomelane	Ag		1 g/dm^3	1.5 h		6	LiNO$_3$, NaNO$_3$	uptake, release of K isotope exchange	The reaction is not complete.	289
MnO$_2$	H/T		0.42 g/5 cm^3	10 d	48–80	3.3	HCl			783
MnO$_2$ (MnSO$_4$ + NaOH in air)	Ba	10^{-7}–10^{-2} mol dm^{-3}	0.1 g/10 cm^3	160 min	30	6.42		uptake	Reaction is complete within 30 min. 1st order rate constant 10^{-1} min^{-1}. Activation energy (from temperature dependence of rate constant 6.7 kJ mol^{-1}.	291
MnO$_2$, BDH	Cd	4×10^{-4} mol dm^3	50 mg/1.5 cm^3	1 h	23	2	HNO$_3$	uptake	Maximum uptake after 15 min then the uptake decreases to 1/3 of the maximum value.	292

Adsorbent	Species	Concentration	Solid/solution	Time	Temp (°C)	pH	Electrolyte	Method	Comments	Ref.
MnO$_2$, synthetic, λ	Li	0.1 mol dm^{-3}	3 g/300 cm^3	2 d	25	12	Cl	uptake, H release	Uptake steadily increases.	784
MnO$_2$, electrolytic	Ni	0.04 mol dm^{-3}	0.1 g (47 m^2/g)/50 cm^3	4 d	25	3.5, 5.3	0.04 mol dm^{-3} sulfate, 0.0026 mol dm^{-3} acetate	uptake	After a 1 d equilibration the uptake increases very slowly.	296
MnO$_2$, δ	phosphate	10^{-5} mol dm^{-3}	10^{-3} mol Mn dm^{-3}	2 d	25	8	sea water	uptake	The fast step is complete within 1 d, then the reaction is very slow. Also at 5, 15 and 35 C.	609
Nb$_2$O$_5$, hydrous	Na	0.01 1 mol dm^{-3}	0.05–0.17 g (320 m^2/g)/200 cm^3	5 min	5.2–29.6	6.6–11	Cl	isotope exchange	The F(t) at pH 11 increases when Na concentration increases. The F(t) curve is pH independent. Interruption test suggests particle diffusion control. Even ^{22}Na distribution between the particle and solution after 5 min.	785
Sb$_2$O$_3$, reagent grade, washed	chromate	10^{-8} 10^{-4} mol dm^{-3}	0.2 g (9.2 m^2/g), 10 cm^3	1 h	30–60	2	Na	uptake	No further uptake after a 15 min equilibration.	635
SnO$_2$	Co	0.05–0.1 mol dm^{-3}	0.06 g/6 cm^3	2 d	25	2–4	chloride	isotope exchange	Linear log (1-F) versus t plots with different slopes in the range below 2 h and above 8 h, particle size has no effect on the kinetics.	786
SiO$_2$	Cr	4×10^{-5} mol dm^{-3}	10 g (500 m^2/g)/dm^3	12 h	20	3.2–3.5	0.01 mol dm^{-3}	uptake	The release of Cr pre-sorbed at pH 4.8–5.	314
SiO$_2$, quartz	Eu	10^{-8} mol dm^{-3}	0.07 g (16 m^2/g)/dm^3	3 d		3–10	NaClO$_4$, 2 mg/dm^3 fulvic acid	uptake	In absence of fulvic acid the uptake curves obtained after 1 h and 3d equilibration are identical.	181

TABLE 4.8 Continued

Adsorbent	Adsorbate	Total concentration	Solid to liquid ratio	Equilibration time	T	pH	Electrolyte	Method	Result	Ref.
SiO$_2$, five samples	H/T		130 m^2/50 cm^3	6 d	70			isotope exchange	No further exchange after a 3 d equilibration for 4 samples.	315
SiO$_2$, Fluka	U	50 mg/dm^3	1 g (550 m^2)/dm^3	1.5 h		4.5 6.2		uptake	No further uptake after a 30 min equilibration.	340
SiO$_2$, Aerosil 200	poly (vinylimidazole)		0.5% w/w, 200 m^2/g	60 h	20	7	10^{-2} mol dm^{-3} NaCl	uptake	No further uptake after a 1 d equilibration.	745
SnO$_2$, hydrous, different samples	Na	0.1 mol dm^{-3}		20 min	25		Cl	isotope exchange		787
TiO$_2$, anatase	H		5, 9.9 g (144 m^2/g)/dm^3	50 ms	25	3 5	4×10^3 mol dm^{-3} NaCl	pressure jump	Adsorption rate constant = 4×10^{-4} dm^3 cm^{-2} s^{-1}; desorption rate constant = 8×10^{-9} mol cm^{-2} s^{-1}.	788
TiO$_2$, anatase, from sulfate	triphosphate	2×10^{-5} 10^{-4} mol dm^{-3}	22–75 mg/30 cm^3	15 h				conductivity	With 10^{-4} mol dm^{-3} triphosphate the uptake steadily increases, with lower concentrations the reaction is complete within 6 h.	643
TiO$_2$, rutile, Merck	PVA, MW 72 000, 2% acetate groups	500 ppm	50 g (14 m^2/g)/dm^3	35 min	20	4–10		uptake	Reaction is complete within 15 min	789
TiO$_2$·n H$_2$O	Zn	10^{-7}–10^{-2} mol dm^{-3}	0.1 g/10 cm^3	2 h	30	6.8	chloride	uptake	No further uptake after 30 min equilibration.	401
V$_2$O$_5$	chromate	10^{-6}–10^{-2} mol dm^{-3}		2 h	25	3.2	Na	uptake	No further uptake after a 1 h equilibration for all chromate concentrations.	644

Adsorbent	Species	Concentration	Solid	Time	T	pH	Electrolyte	Method	Comment	Ref
ZnO	sulfate		30 mg cm^{-2} (layer)	40 min	0		1 mol dm^{-3} HClO$_4$	uptake	The uptake peaks after a 15 min reaction.	777
Composite materials containing Al										
MgAl$_2$O$_4$, 3 different commercial samples	H		volume fraction 0.1	5 h	25	natural	0.01 mol dm^{-3} KCl	acousto	78% Alumina spinel from Alcoa shows the same ζ after 5 and 300 min equilibration. The other two samples (90% Alcoa and 70% Unitec) show a constant shift in ζ, even after a 300 min equilibration.	749
Clay minerals										
hectorite	Co	10^{-4} mol dm^{-3}	1.95 g (114 m^2/g)/dm^3	5 d	25	6.5	0.01, 0.3 mol dm^{-3} NaNO$_3$	uptake	No further uptake after a 2 day equilibration in 0.01 mol dm^{-3} NaNO$_3$.	415, 790
kaolinite, KGa-1b	Co	3×10^{-4} 1.3×10^{-3} mol dm^{-3}	100 g/dm^3	1 y		7.5–7.8	0.1 mol dm^{-3} NaNO$_3$	uptake, EXAFS	No further increase in the number of second-neighbor cobalt atoms after 10 h equilibration.	791
kaolinite	Ni	3×10^{-3} mol dm^{-3}	10 g/dm^3	2 d	25	7.5	0.1 mol dm^{-3} NaNO$_3$	uptake, Si release	Ni uptake and Si release are constant after 1 d equilibration.	143
kaolinite	Cs, Sr	10^{-8}, 10^{-5} mol dm^{-3}	0.1 g/20 cm^3	27 d	natural		ground water	uptake	No further uptake after a 4 d equilibration.	436
kaolinite	phosphate	2.5×10^{-4} mol dm^{-3}	7.5 g (11 m^2/g)/dm^3	40 d	50	3.5–7.5		uptake	The uptake at pH <5.3 is steadily increasing.	504
kaolinite, Ca form	phosphate	10^{-5} mol dm^{-3}	1%	7 d	24	7.5	KCl	uptake		652
mica	phosphate	3–6×10^{-5} mol dm^{-3}	0.8 g/200 cm^3	1 d	25	5.6–8.1	0.01 mol dm^{-3} NaCl	uptake	The uptake is constant after a 3 h equilibration.	653

TABLE 4.8 Continued

Adsorbent	Adsorbate	Total concentration	Solid to liquid ratio	Equilibration time	T	pH	Electrolyte	Method	Result	Ref.
montmorillonite	Ni	3×10^{-3} mol dm^{-3}	10 g/dm^3	3 d	25	7.5	0.1 mol dm^{-3} NaNO$_3$	uptake, Si release	The uptake steadily increases.	143
montmorillonite	Sr		0.1 g/20 cm^3	27 d		natural	ground water	uptake		436
montmorillonite, Ca form	phosphate	2×10^{-5} mol dm^{-3}	1%	7 d	24	5–8	KCl	uptake	Maximum uptake after 2–4 days, then desorption or dissolution.	652
montmorillonite, Na and Ca form	8-hydroxyqui-noline	5.4×10^{-4} mol dm^{-3}	2 g/dm^3	1 d		6.6–6.8		uptake	Uptake by the Ca form steadily increases.	688
muscovite mica	dodecyl pyrydinium	5×10^{-4} mol dm^{-3}	50 mg/25 cm^3	5 d		natural		uptake	Fast step complete within a few hours, then slow increase in uptake.	734
smectite	phosphate	3×10^{-5} mol dm^{-3}	0.8 g/200 cm^3	1 d	25		0.01 mol dm^{-3} NaCl		Release of preadsorbed phosphate was studied.	653

Silicates and composite materials containing Si

Adsorbent	Adsorbate	Total concentration	Solid to liquid ratio	Equilibration time	T	pH	Electrolyte	Method	Result	Ref.
76% SiO$_2$+ 14% Al$_2$O$_3$ + 10% water	8-hydroxyqui-noline	3.3×10^{-4} mol dm^{-3}	1 g/dm^3	1 d		7.2		uptake	Fast step complete within 1 h, then slow increase in uptake	688
pyrophyllite	Ni	3×10^{-3} mol dm^{-3}	10 g/dm^3	8 d	25	7.5	0.1 mol dm^{-3} NaNO$_3$	uptake, Si release	Ni uptake and Si release are constant after a 2 d equilibration.	143

Zeolites

Adsorbent	Adsorbate	Total concentration	Solid to liquid ratio	Equilibration time	T	pH	Electrolyte	Method	Result	Ref.
clinoptilolite	Pb	500 mg/dm^3		3 h	16, 35	4.2–5.2		uptake	Uptake steadily increases.	792

	Species	Concentration	Solid/solution	Time	Temp (°C)	pH	Electrolyte	Method	Comments	Ref.
Coatings										
aluminum oxide coating on sand	selenate IV and VI	8×10^{-4} mol dm^{-3}	5 g/50 cm^3	8 h		4.8	0.1 mol dm^{-3} NaCl	uptake	100% uptake is reached within 1 h.	656
iron oxide coating on sand	Pb	10^{-4} mol dm^{-3}	10 g (68 m^2/g)/dm^3	60 h	20	6	0.05 mol dm^{-3} NaNO$_3$	uptake	Constant uptake after 20 h.	447
Glasses										
soda glass	Ag	0.5 mg AgNO$_3$/dm^3	8.9 cm^2/3 cm^3	110 d	25	5.3		uptake		448
glass	Co	2×10^{-6} mol dm^{-3}		2 h		7.6	0.5 mol dm^{-3} NaClO$_4$; borate buffer	uptake, isotope exchange	Release of Co presorbed at pH 6.8 is complete in 1 h at pH 2.	451
porous glass, 97% SiO$_2$ 3% Al$_2$O$_3$	phosphate	2.5×10^{-4} – 1.25×10^{-3} mol dm^{-3}	52 m^2/200 cm^3	40 min	25	2	0.05, 0.1 mol dm^{-3} KCl	uptake	No further uptake after a 10 min equilibration (but the uptake steadily increases at 40 C).	657
Clays										
kaolin, Ward's	arsenate (V)	10^{-6} mol dm^{-3}	25 g (5.1 m^2/g)/dm^3	19 d		2–10	0.1 mol dm^{-3} NaCl	uptake	Uptake steadily increases at pH 2.	482
kaolin, K form	8-hydroxyquinoline	4.2×10^{-4} mol dm^{-3}	8 g/dm^3	1 d		6.4		uptake	The fast step is complete within 1 h, then very slow increase in uptake.	688
Soils										
2 soils	Cu	5×10^{-5}, 1.5×10^{-4} mol dm^{-3}	1:100	18 d		5.5	Ca(NO$_3$)$_2$, ionic strength 0.015	uptake	Uptake steadily increases.	463
loamy sand	Zn	10^{-5}–10^2 mg/dm^3	1:10	30 d	25	natural	0.01 mol dm^{-3} Ca(NO$_3$)$_2$	uptake	Uptake steadily increases.	468

TABLE 4.8 Continued

Adsorbent	Adsorbate	Total concentration	Solid to liquid ratio	Equilib-ration time	T	pH	Electrolyte	Method	Result	Ref.
2 soils	phosphate	500–2000 mg P/kg soil	1:10–1:600	10 d	25		0.01 mol dm^{-3} Ca(NO$_3$)$_2$	uptake	Uptake steadily increases.	793
soil	phosphate	400 mg P/kg soil		4 d	25			isotope exchange	Reinterpretation of literature data.	794
loamy sand	selenate IV and VI	10^{-7}–10^{-4} mol dm^{-3}	1:5–1:100	38 d	25	natural		uptake	Uptake steadily increases.	795
3 soils	silicate	110, 135 ppm SiO$_2$	1:5, 1:25	6 d	25	4.8, 6.4	toluene added	uptake		549
loamy sand	phosphate, selenate IV	5–100 mg/dm^3	1:10	30 d	25	5.6 initial	0.01 mol dm^{-3} CaCl$_2$	uptake	The concentration in solution is continuously decreasing; Preferential sorption of phosphate (over selenate) is more pronounced for longer equilibration times.	764
Other materials										
activated red mud	chromate (VI)	20–50 ppm		1 d		5.2	1 mol dm^{-3} KCl; acetate buffer	uptake	Fast process which is complete within 2 h followed by slow process: uptake is still increasing after 15 h.	665
activated red mud	phosphate (VI)	1.26×10^{-3} mol dm^{-3}	2 g (249 m^2/g)/dm^3	1 d	30	5.2	1 mol dm^{-3} KCl	uptake	The reaction is complete within 8 h.	666

REFERENCES

1. M. P. Jensen, and G. R. Choppin. Radiochim. Acta 72: 143–150 (1996).
2. F. Kepak. Atom. Energy Rev. 5–62 Suppl. 2 1981, pp. 5–62.
3. G. Sposito. In: Geochemical Processes at Mineral Surfaces, J. A. Davis and K. F. Hayes (Eds.) ACS Symp. Series. 323 ACS 1986, pp. 217–228.
4. H. S. Posselt, F. J. Anderson and W. J. Weber. Environ. Sci. Technol. 2: 1087–1093 (1968).
5. H. Ruppert. Chem. Erde. 39: 97–132 (1980).
6. V. E. Kazarinov, V. N. Andreev and A. P. Mayorov. J. Electroanal. Chem. 130: 277–285 (1981).
7. A. D. Ebner, J. A. Ritter, and B. N. Popov. J. Colloid Interf. Sci. 203: 488–492 (1998).
8. P. W. Schindler. In: Adsorption of Inorganics at Solid–Liquid Interfaces, M. A. Anderson, and A. J. Rubin, (Eds.), Ann Arbor 1981, Chapter 1.
9. D. G. Kinniburgh and M. L. Jackson. In: Adsorption of Inorganics at Solid–Liquid Interfaces; M. A. Anderson and A. J. Rubin, (Eds.), Ann Arbor 1981, p. 91.
10. J. Lyklema, Croat. Chem. Acta, 60: 371–381 (1987).
11. P. W. Schindler and W. Stumm. In: Aquatic Surface Chemistry W. Stumm, (Ed.), Wiley 1987, pp. 83–110.
12. D. A. Dzombak and F. M. Morel. Surface Complexation Modeling: Hydrous Ferric Oxide, Wiley, N. Y. 1990.
13. M. L. Machesky. In: Chemical Modeling in Aqueous Systems II. D. C. Melchior, and R. L. Bassett, (Eds.), ACS Symposium Series; 416: 282–292 (1990).
14. N. J. Barrow. In: Adsorption on New and Modified Inorganic Sorbents. A. Dabrowski and V. A. Tetrykh (Eds.), Elsevier 1996, pp. 829–856.
15. N. J. Barrow. Aust. J. Soil Res. 37: 787–829 (1999).
16. G. Buckau (Ed.). Effects of Humic Substances on the Migration of Radionuclides: Complexation and Transport of Actinides. Report FZKA 6124. Forschungszentrum Karlsruhe 1998.
17. W. Hummel. In: Modelling in Aquatic Chemistry. I. Grenthe and I. Puigdomenech (Eds.), Nuclear Energy Agency. OECD Paris 1997, pp. 153–206.
18. D. R. Turner. In: Metal Speciation and Bioavailability in Aquatic Systems. A. Tessier and D. R. Turner (Eds.), Wiley, NY 1995, pp. 149–203.
19. J. H. Ephraim and B. Allard. In: Modelling in Aquatic Chemistry. I. Grenthe and I. Puigdomenech (Eds.), Nuclear Energy Agency. OECD Paris 1997, pp. 207–244.
20. B. Gu, J. Schmitt, Z. Chen, L. Liang and J. F. McCarthy. Environ. Sci. Technol. 28: 38–46 (1994).
21. E. Laiti, L. O. Ohman, J. Nordin and S. Sjoberg. J. Colloid Interf. Sci. 175: 230–238 (1995).
22. R. M. Smith, and A. E. Martell, Critical Stability Constants, Vol. 1–6, Plenum, New York 1976–1986; E. Hogfeldt. Stability Constants of Metal-Ion Complexes. Part A. Inorganic Ligands. IUPAC Chemical Data Series, No. 21 Pergamon. New York 1982; L. G. Sillen, and A. E. Martell. Stability Constants. Chem Soc. London. Special Publ. 17, London 1971.
23. D. G. Hall, Langmur 13: 91–99(1997).
24. G. D. Turner, J. M. Zachara, J. P. McKinley and S. C. Smith. Geochim. Cosmochim. Acta 60: 3399–3414 (1996).
25. N. J. Barrow. Reactions with Variable Charge Soils. Nijhoff, Dordrecht 1987.
26. N. J. Barrow. Aust. J. Soil Res. 27: 475–492 (1989).
27. J. P. Seaman, P. M. Bertsch and L. Schwallie. Environ. Sci. Technol. 33: 938–944 (1999).
28. L. G. J. Fokkink, A de Keizer and J. Lyklema. J. Colloid Int. Sci. 135: 118–131 (1990).
29. F. J. Hingston, A. M. Posner and J. P. Quirk. J. Soil Sci. 23: 177–192 (1972).

30. Understanding Variation in Partition Coefficient, K_D, Values, US Environmental Protection Agency, and Office of Air and Radiation, EPA 402-R-99-004A, Washington, DC, August 1999.

31. E. A. Jenne. In: Adsorption of Metals by Geomedia. E. A. Jenne, (Ed.), Academic Press, San Diego 1998, pp. 1–73.

32. R. T. Pabalan, D. R. Turner, F. P. Bertetti and J. D. Prikryl. In: Adsorption of Metals by Geomedia. E. A. Jenne, (Ed.), Academic Press, San Diego 1998, pp. 99–130.

33. T. Rabung, H. Geckeis, J. I. Kim and H. P. Beck. J. Colloid Interf. Sci. 208: 153–161 (1998).

34. H. Radovanovic and A. A. Koelmans. Environ. Sci. Technol. 32: 753–759 (1998).

35. L. G. J. Fokkink, A de Keizer and J. Lyklema. J. Colloid Int. Sci. 118: 454–462 (1987).

36. M. A. Anderson and D. T. Malotky. J. Colloid Int. Sci. 72: 413–427 (1979).

37. M. Tschapek, C. Wasowski and R. M. Torres Sanchez. J. Electroanal. Chem. 74: 167–176 (1976).

38. N. J. Harrick. Internal Reflection Spectroscopy. Wiley New York 1967.

39. S. J. Hug and B. Sulzberger. Langmuir 10: 3587–3597 (1994).

40. K. H. Chung, R. Klenze, K. K. Park, P. Paviet-Hartmann and J. I. Kim. Radiochim. Acta 82: 215–219 (1999).

41. M. C. Duff, D. B. Hunter, I. R. Triay, P. M. Bertsch, D. T. Reed, S. R. Sutton, G. Shea-McCarthy, J. Kitten, P. Eng, S. J. Chipera, and D. T. Vaniman. Environ. Sci. Technol. 33: 2163–2169 (1999).

42. M. C. Duff, M. Newville, D. B. Hunter, P. M. Bertsch, S. R. Sutton, I. R. Triay, D. T. Vaniman, P. Eng and M. L. Rivers. J. Synchrotron Rad. 6: 350–352 (1999).

43. C. Pohlmann, D. Degering, H. Geckeis and W. G. Thies. Hyperfine Inter. 120/121: 313–318 (1999).

44. H. Geckeis, R. Klenze and J. I. Kim. Radiochim. Acta 87: 13–21 (1999).

44. B. Venkataramani. Radiochemistry and Radiation Chemistry Symposium. Kanpur Dec. 9–13, 1985, pp. 50–62.

45. L. Charlet. In: Chemistry of Aquatic Systems: Local and Global Perspectives, G. Bidoglio and W. Stumm, (Eds.), Kluwer Dordrecht 1994, pp. 273–305.

46. E. A. Jenne, (Ed.), Adsorption of Metals by Geomedia. Academic Press, San Diego 1998.

47. K. F. Hayes and L. E. Katz. In: Physics and Chemistry of Mineral Surfaces, P. V. Brady, (Ed), CRC Boca Raton 1996, pp. 147–223.

48. M. M. Benjamin and J. O. Leckie. Environ. Sci. Technol. 16: 162–170 (1982).

49. B. D. Honeyman and J. O. Leckie. In: Geochemical Processes at Mineral Surfaces, J. A. Davis and K. F. Hayes, (Eds.), ACS Symp. Series 323: 162–190 (1986).

50. M. Kosmulski. Colloids Surf. A 117: 201–214 (1996).

51. B. D. Honeyman. Thesis, quoted after S. Goldberg, J. A. Davis and J. D. Hem. In: The Environmental Chemistry of Aluminum. G. Sposito, (Ed.), CRC Press, 1996, pp. 271–331.

52. P. H. Tewari and W. Lee. J. Colloid Int. Sci. 52: 77–88 (1975).

53. D. C. Girvin, P. L. Gassman and H. Bolton. Soil Sci. Soc. Am. J. 57: 47–57 (1993).

54. M. de Boer, R. G. Leliveld, A. J. van Dillen, J. W. Geus and H. G. Bruil. Appl. Catalysis A. 102: 35–51 (1993).

55. M. Kosmulski. Colloids Surf. A 83: 237–243 (1994).

56. N. Spanos and A. Lycourghiotis. Langmuir 10: 2351–2362 (1994).

57. L. E. Katz and K. F. Hayes. J. Colloid Interf. Sci. 170: 477–490; 491–501 (1995).

58. S. N. Towle, J. A. Bargar, P. Persson, G. E. Brown. and G. A. Parks Physica B 208–209: 439–440 (1995).

59. H. Tamura, N. Katayama and R. Furuichi. J. Colloid Interf. Sci. 195: 192–202 (1997).

60. S. N. Towle, J. A. Bargar, G. E. Brown and G. A. Parks. J. Colloid Interf. Sci. 187: 62–82 (1997).
61. S. N. Towle, J. R. Bargar, P. Persson, G. E. Brown and G. A. Parks. Mat. Res. Soc. Symp. Proc. 432: 237–242 (1997).
62. S. N. Towle, J. R. Bargar, G. E. Brown and G. A. Parks. J. Colloid Interf. Sci. 217: 312–321 (1999).
63. R. Karthein, H. Motschi, A. Schweiger, S. Ibric, B. Sulzberger and W. Stumm. Inorg. Chem. 30: 1606–1611 (1991).
64. M. I. Zaki, S. A. A. Mansour, F. Taha and G. A. H. Mekhemer. Langmuir 8: 727–732 (1992).
65. K. Csoban and P. Joo. Colloids Surf. A 151: 97–112 (1999).
66. H. A. Elliott and C. P. Huang. J. Colloid Int. Sci. 70: 29–45 (1979).
67. M. Rudin and H. Motschi. J. Colloid Int. Sci. 98: 385–393 (1984).
68. D. Ballion and N. Jaffrezic-Renault. J. Radioanal. Nucl. Chem. 92: 133–150 (1985).
69. W. Möhl, A. Schweiger and H. Motschi. Inorg. Chem. 29: 1536–1543 (1990).
70. S. F. Cheah, G. E. Brown and G. E. Parks. Mat. Res. Soc. Symp. Proc. 432: 231–236 (1997).
71. S. F. Cheah, G. E. Brown and G. E. Parks. J. Colloid Interf. Sci. 208: 110–128 (1998).
72. J. P. Fitts, P. Persson, G. E. Brown and G. A. Parks. J. Colloid Interf. Sci. 220: 133–147 (1999).
73. C. P. Schulthess and C. P. Huang. Soil Sci. Soc. Am. J. 54: 679–688 (1990).
74. N. Spanos and A. Lycourghiotis. J. Colloid Interf. Sci. 171: 306–318 (1995).
75. A. R. Bowers. Thesis, quoted after B. Nowack, J. Lützenkirchen, P. Behra and L. Sigg. Environ. Sci. Technol. 30: 2397–2405 (1996).
76. F. P. Bertetti, R. T. Pabalan and M. G. Almendarez. In: Adsorption of Metals by Geomedia E. Jenne, (Ed.), Academic Press, NY, 1998 pp. 131–148.
77. H. Hohl and W. Stumm. J. Colloid Int. Sci. 55: 281–288 (1976).
78. K. Hachiya, M. Ashida, M. Sasaki, H. Kan, T. Inoue and T. Yasunaga. J. Phys. Chem. 83: 1866–1871 (1979).
79. C. J. Chisholm-Brause, K. F. Hayes, A. L. Roe, G. E. Brown, G. A. Parks and J. O. Leckie. Geochim. Cosmochim. Acta 54: 1897–1909 (1990).
80. J. R. Bargar, G. E. Brown and G. A. Parks. Geochim. Cosmochim. Acta 62: 193–207 (1998).
81. Y. M. Nelson, L. W. Lion, M. L. Shuler and W. C. Ghiorse. Limnol. Oceanorg. 44: 1715–1729 (1999).
82. J. A. Schwarz, C. T. Ugbor and R. Zhang. J. Catalysis 138: 38–54 (1992).
83. D. Spielbauer, H. Zeilinger and H. Knözinger. Langmuir 9: 460–466 (1993).
84. C. Contescu, J. Hu and J. A. Schwarz. J. Chem. Soc. Faraday Trans. 89: 4091–4099 (1993).
85. I. C. Pius, M. M. Charyulu, B. Venkataramani, C. K. Sivaramakrishnan and S. K. Patil. J. Radioanal. Nucl. Chem. 199: 1–7 (1995).
86. J. D. Prikryl, R. T. Pabalan, D. R. Turner and B. W. Leslie. Radiochim. Acta 66/67: 291–296 (1994).
87. R. O. James and M. G. MacNaughton. Geochim. Cosmochim. Acta 41: 1549 (1977)
88. M. Kalbasi, G. J. Racz and L. A. Loewen-Rudgers. Soil Sci. 125: 146–150 (1978).
89. C. P. Huang and E. A. Rhoads. J. Colloid Int. Sci. 131: 289–306 (1989).
90. T. Rabung, T. Stumpf, H. Geckeis, R. Klenze and J. I. Kim. Radiochim Acta 88: 711–716 (2000).
91. L. Righetto, G. Bidoglio, G. Azimonti and I. R. Bellobono. Env. Sci. Technol. 25: 1913–1919 (1991).
92. M. Escudey and F. Gil-Llambias. J. Colloid Int. Sci. 107: 272–275 (1985).
93. C. P. Huang and W. Stumm. J. Colloid Int. Sci. 43: 409–420 (1973).
94. R. T. Lowson and J. V. Evans. Aust. J. Chem. 37: 2165–2178 (1984).

95.   A. R. Bowers and C. P. Huang. J. Colloid Int. Sci. 110: 575–590 (1986).
96.   M. Kosmulski. Ber. Bunsenges. Phys. Chem. 98: 1062–1067 (1994).
97.   E. A. Nechaev and T. B. Golovanova. Kolloid. Zh. 36: 889–894 (1974).
98.   S. Y. Shiao, Y. Egozy and R. E. Meyer. J. Inorg. Nucl. Chem. 43: 3309–3315 (1981).
99.   L. Benyahya and J. M. Garnier. Environ. Sci. Technol. 33: 1398–1407 (1999).
100. M. Plavsic and B. Cosovic. Colloid Surf. A 151: 189–200 (1999).
101. J. A. Davis. Geochim. Cosmochim. Acta 48: 679–691 (1984).
102. M. M. Benjamin. Thesis, quoted after S. Goldberg, J. A. Davis and J. D. Hem. In: The
      Environmental Chemistry of Aluminum. G. Sposito (Ed.), CRC Press, Boca Raton
      1996, pp. 271–331
103. P. V. Brady. Geochim. Cosmochim. Acta 58: 1213–1217 (1994).
104. J. Liu, S. M. Howard and K. N. Han. Langmuir 9: 3635–3639 (1993).
105. E. Baumgarten and U. Kirchhausen-Düsing. J. Colloid Int. Sci. 194: 1–9 (1997).
106. T. Yasunaga and T. Ikeda. Geochemical Processes at Mineral Surfaces. J. A. Davis, K.
      F. Hayes, (Eds.), ACS Symp. Series 323: 230–253 (1986).
107. K. Hachiya, M. Sasaki, Y. Saruta, N. Mikami and T. Yasunaga J. Phys. Chem. 88: 23–
      27 (1984).
108. D. W. Fuerstenau and K. Osseo-Asare. J. Colloid Interf. Sci. 118: 524–542 (1987).
109. K. M. Spark, B. B. Johnson and J. D. Wells. Eur. J. Soil Sci. 46: 621–631 (1995).
110. L. Vordonis, N. Spanos, P. G. Koutsoukos and A. Lycourghiotis. Langmuir 8: 1736–
      1743 (1992).
111. J. B. d'Espinose de la Caillerie, M. Kermarec and O. Clause. J. Am. Chem. Soc. 117:
      11471–11481 (1995).
112. J. L. Paulhiac, and O. Clause. J. Am. Chem. Soc. 115: 11602–11603 (1993).
113. B. Wehrli, S. Ibric and W. Stumm Colloids Surf. 51: 77–88 (1990).
114. B. Nowack, J. Lützenkirchen, P. Behra and L. Sigg. Environ. Sci. Technol. 30: 2397–
      2405 (1996).
115. C. F. Lin, K. S. Chang, C. W. Tsay, D. Y. Lee, S. L. Lo and T. Yasunaga. J. Colloid
      Interf. Sci. 188: 201–208 (1997).
116. M. Kosmulski. Colloids Surf. A 149: 397–408 (1999).
117. M. Kosmulski. J. Colloid Interf. Sci. 192: 215–227 (1997).
118. A. R. Bowers and C. P. Huang. Water Res. 21: 757–764 (1987).
119. K. G. Karthikeyan, H. A. Elliott and F. S. Cannon. Environ. Sci. Technol. 31: 2721–
      2725 (1997).
120. K. G. Karthikeyan, H. A. Elliott and J. Chorover. J. Colloid Interf. Sci. 209: 72–78
      (1999).
121. K. G. Karthikeyan and H. A. Elliott. J. Colloid Interf. Sci. 220: 88–95 (1999).
122. R. Rao Gadde and H. A. Laitinen. Anal. Chem. 46: 2022–2026 (1974).
123. P. Trivedi and L. Axe. J. Colloid Interf. Sci. 218: 554–563 (1999).
124. M. Okazaki, K. Takamidoh and I. Yamane. Soil Sci. Plant Nutr. 32: 523–533 (1986).
125. H. A. B. Potter and R. N. Yong. Appl. Clay Sci. 14: 1–26 (1999).
126. M. P. Sidorova, L. E. Ermakova, I. A. Savina and I. A. Kavokina. Kolloid Zh. 59: 533–
      537 (1997).
127. M. B. McBride. Clays Clay Min. 30: 21–28 (1982).
128. F. J. Weesner and W. F. Bleam. J. Colloid Interf. Sci. 196: 79–86 (1997).
129. K. Xia, R. W. Taylor, W. F. Bleam and P. A. Helmke. J. Colloid Interf. Sci. 199: 77–82
      (1998).
130. A. J. Fairhurst and P. Warwick. Colloids Surf. A 145: 229–234 (1998).
131. W. F. Bleam and M. B. McBride. J. Colloid Interf. Sci. 103: 124–132 (1985).
132. J. S. Redinha and P. F. P. Pires Ferreira. Rev. Port. Quim. 14: 193–200 (1972).
133. J. M. Zachara, C. T. Resch and S. C. Smith. Geochim. Cosmochim. Acta 58: 553–566
      (1994).

134. M. B. McBride. Soil Sci. Soc. Am. J. 42: 27–31 (1978).
135. M. B. McBride, A. R. Fraser and W. J. McHardy. Clays Clay Min. 32: 12–18 (1984).
136. M. B. McBride. Soil Sci. Soc. Am. J. 49: 843 (1985).
137. G. Micera, L. Strinna Erre and R. Dallocchio Colloids Surf. 28: 147–157 (1987).
138. G. Micera, L. Strinna Erre and R. Dallocchio Colloids Surf. 32: 249–256 (1988).
139. G. Micera, L. Strinna Erre and R. Dallocchio Colloids Surf. 32: 237–248 (1988).
140. G. Micera, L. Strinna Erre and R. Dallocchio Colloids Surf. 44: 237–245 (1990).
141. G. Micera and R. Dallocchio. Colloids Surf. 45: 167–175 (1990).
142. G. Micera, R. Dallocchio, S. Deiana, C. Gessa, P. Melis and A. Premoli. Colloids Surf. 17: 395–400 (1986).
143. A. M. Scheidegger, G. M. Lamble and D. L. Sparks. J. Colloid Interf. Sci. 186: 118–128 (1997).
144. G. D. Redden, J. Li and J. Leckie. In: Adsorption of Metals by Geomedia. (E. Jenne, (Ed.), Academic Press, NY, 1998 pp. 291–315.
145. G. Micera and R. Dallocchio. Colloids Surf. 34: 185–196 (1988).
146. L. M. Shuman. Soil Sci. Soc. Am. J. 41: 703–706 (1977).
147. G. Micera, C. Gessa, P. Melis, A. Premoli, R. Dallocchio and S. Deiana. Colloids Surf. 17: 389–394 (1986).
148. A. Duker, A. Ledin, S. Karlsson and B. Allard. Appl. Geochem. 10: 197–205 (1995).
149. D. G. Kinniburgh, M. L. Jackson and J. K. Syers. Soil Sci. Soc. Am. J. 40: 796–799 (1976).
150. D. G. Kinniburgh, J. K. Syers and M. L. Jackson. Soil Sci. Soc. Amer. Proc. 39: 464–470 (1975).
151. P. R. Anderson and M. M. Benjamin. Environ. Sci. Technol. 24: 692–698 1586–1592 (1990).
152. C. Gessa, M. L. deCherchi, P. Melis, G. Micera and L. Strinna Erre. Colloids Surf. 11: 109–117 (1984).
153. F. Taha, A. M. El-Roudi, A. A. Abd El Gaber and F. M. Zahran. Rev. Roum. Chim. 35: 503–509 (1990).
154. H. S. Mahal, B. Venkataramani and K. S. Venkateswarlu. J. Inorg. Nucl. Chem. 43: 3335–3342 (1981).
155. H. S. Mahal, B. Venkataramani and K. S. Venkateswarlu. Proc. Indian Acad. Sci. (chem.) 91: 321–327 (1982).
156. S. P. Mishra and V. K. Singh. Appl. Radiation Isot. 46: 75–81 (1995).
157. S. P. Mishra and V. K. Singh. J. Radioanal. Nucl. Chem. 241: 145–149 (1999).
158. G. A. Kokarev, V. A. Kolesnikov, A. F. Gubin and A. Korobanov. Elektrokhimiya 18: 466–470 (1982).
159. G. A. Kokarev, A. F. Gubin, V. A. Kolesnikov and S. A. Skobelev. Zh. Fiz. Khim. 59: 1660–1663. (1985).
160. S. Kittaka, S. Yamanaka, N. Yanagawa and T. Okabe. Bull. Chem. Soc. Jpn. 63: 1381–1388 (1990).
161. T. Okada, S. Ambe, F. Ambe and H. Sekizawa J. Phys. Chem. 86: 4726–4733 (1982).
162. R. J. Crawford, I. H. Harding and D. E. Mainwaring. Langmuir 9: 3050–3056 (1993).
163. R. J. Crawford, D. E. Mainwaring and I. H. Harding. Colloids Surf. A 126: 167–179 (1997).
164. R. J. Crawford, I. H. Harding and D. E. Mainwaring. J. Colloid Interf. Sci. 181: 561–570 (1996).
165. M. A. Blesa, R. M. Larotonda, A. J. G. Maroto and A. E. Regazzoni. Colloids Surf. 5: 197–208 (1982).
166. H. Tamura, E. Matijevic and L. Meites. J. Colloid Interf. Sci. 92: 303–314 (1983).
167. S. Music and M. Ristric. J. Radioanal. Nucl. Chem. 120: 289–304 (1988).

168. B. Venkataramani, A. R. Gupta and R. M. Iyer. J. Radioanal. Nucl. Chem. 96: 129–136 (1985).
169. A. R. Gupta and B. Venkataramani. Bull. Chem. Soc. Jpn. 61: 1357–1362 (1988).
170. N. H. Sagert, C. H. Ho and N. H. Miller. J. Colloid Interf. Sci. 130: 283–287 (1989).
171. H. Catalette, J. Dumonceau and P. Ollar. J. Contamin. Hydrol. 35: 151–159 (1998).
172. B. Venkataramani, K. S. Venkateswarlu, J. Shankar and L. H. Baestle. Proc. Indian Acad. Sci. (Chem). 87A: 415–428 (1978).
173. B. Venkataramani, K. S. Venkateswarlu and J. Shankar. J. Colloid Int. Sci. 67: 187–194 (1978).
174. N. Marmier, A. Delisee and F. Fromage. J. Colloid Interf. Sci. 211: 54–60 (1999).
175. A. Breeuwsma and J. Lyklema. J. Colloid Interf. Sci. 43: 437–448 (1973).
176. S. Chibowski. Pol. J. Chem. 59: 1193–1199 (1985).
177. S. Music, M. Gessner and R. H. H. Wolf. Microchimica Acta 1979 I: 105–112 (1979).
178. I. Kobal, P. Hesleitner and E. Matijevic. Colloids Surf. 33: 167–174 (1988).
179. N. Ogrinc, I. Kobal, E. Matijevic and N. Kallay. In: Fine Particles Science and Technology. E. Pelizzetti (Ed.). Kluwer Dordrecht 1996, pp. 85–96.
180. S. Music, M. Gessner and R. H. H. Wolf. J. Radioanal. Chem. 50: 91–100 (1979).
181. A. Ledin, S. Karlsson, A. Duker and B. Allard. Radiochim. Acta 66/67: 213–220 (1994).
182. S. Music, M. Gessner and R. H. H. Wolf. Radiochim. Acta 26: 51–53 (1979).
183. S. Music and R. H. H. Wolf. Microchimica Acta 1979 I: 87–94 (1979).
184. N. Marmier and F. Fromage. J. Colloid Interf. Sci. 212: 252–263 (1999).
185. S. Chibowski and J. Szczypa. Pol. J. Chem. 61: 171–176 (1987).
186. A. I. Novikov and A. I. Shaffert. Radiokhimiya 29: 356–360 (1986).
187. R. Rundberg and Y. Albinsson, LA-UR 92–824, Los Alamos National Laboratory, USA (1992), quoted after A. M. Jakobsson and Y. Albinsson. Radiochim. Acta 82: 257–262 (1998).
188. M. Kohler, B. D. Honeyman and J. O. Leckie. Radiochim. Acta 85: 33–48 (1999).
189. A. I. Novikov and V. F. Samoilova. Radiokhimiya 31: 116–125 (1989).
190. J. R. Bargar, G. E. Brown and G. A. Parks. Geochim. Cosmochim. Acta 61: 2639–2652 (1997).
191. S. Music, M. Gessner and R. H. H. Wolf. Microchimica Acta 1979 I: 95–104 (1979).
192. A. I. Novikov and T. M. Zakrevskaya. Radiokhimiya 31: 64–71 (1989).
193. O. N. Karasyova, L. I. Ivanova, L. Z. Lakshtanov and L. Lövgren, J. Colloid Interf. Sci. 220: 419–428 (1999).
194. E. R. Landa, A. H. Le, R. L. Luck and P. J. Yeich. Inorg. Chim. Acta 229: 247–252 (1995).
195. L. Cromieres, V. Moulin, B. Fourest, R. Guillaumont and E. Giffaut. Radiochim. Acta 82: 249–255 (1999).
196. C. K. D. Hsi and D. Langmuir. Geochim. Cosmochim. Acta 49: 1931–1941 (1985).
197. C. H. Ho and H. N. Miller. J. Colloid Interf. Sci. 106: 281–288 (1985).
198. C. H. Ho and D. C. Doern. Can. J. Chem. 63: 1100–1104 (1985).
199. C. H. Ho and H. N. Miller. J. Colloid Interf. Sci. 110: 165–171 (1986).
200. J. Jung, S. P. Hyun, J. K. Lee, Y. H. Cho and P. S. Hahn. J. Radioanal. Nucl. Chem. 242: 405–412 (1999).
201. A. I. Novikov, L. A. Egorova and G. Pak. Zh. Anal. Khim. 32: 2162–2168 (1977).
202. L. A. Egorova and A. I. Novikov. Zh. Anal. Khim. 34: 237–242 (1980).
203. A. Breeuwsma and J. Lyklema. Disc. Faraday Soc. 52: 324–333 (1972).
204. A. I. Novikov and V. Ya. Stetsenko. Radiokhimiya 28: 367–371 (1986).
205. K. Subramaniam, S. Yiacoumi and C. Tsouris. Separ. Sci. Technol. 34: 1301–1318 (1999).
206. H. Tamura and R. Furuichi. J. Colloid Interf. Sci. 195: 241–249 (1997).

207. R. J. Murphy, J. J. Lenhart and B. D. Honeyman. Colloids Surf. A 157: 47–62 (1999).
208. J. J. Lenhart and B. D. Honeyman. Geochim. Cosmochim. Acta. 63: 2891–2901 (1999).
209. C. G. Ong and J. O. Leckie. In: Adsorption of Metals by Geomedia. E. A. Jenne, (Ed.), Academic Press, San Diego 1998. pp. 317–332.
210. L. Spadini, A. Manceau, P. W. Schindler and L. Charlet. J. Colloid Interf. Sci. 168: 73–86 (1994).
211. C. Tiffreau, J. Lutzenkirchen and P. Behra. J. Colloid Int. Sci. 172: 82–93 (1995).
212. A. L. Bryce and S. B. Clark. Colloids Surf. A 107: 123–130 (1996).
213. D. Girvin, L. Ames, A. Schwab and J. McGarrah. J. Colloid Int. Sci. 141: 67–78 (1991).
214. L. Axe and P. R. Anderson. J. Colloid Interf. Sci. 175: 157–165 (1995).
215. L. Axe, G. B. Bunker, P. R. Anderson and T. A. Tyson. J. Colloid Interf. Sci. 199: 44–52 (1998).
216. T. D. Small, L. A. Warren, E. E. Roden and F. G. Ferris. Environ. Sci. Technol. 33: 4465–4470 (1999).
217. T. D. Waite, J. A. Davis, T. E. Payne, G. A. Waychunas and N. Xu. Geochim. Cosmochim. Acta. 58: 5465–5478 (1994).
218. T. E. Payne, G. R. Lumpkin and T. D. Waite. In: Adsorption of Metals by Geomedia. E. A. Jenne, (Ed.), Academic Press, San Diego 1998, pp. 75–97.
219. J. Xue and P. M. Huang. Geoderma 64: 343–356 (1995).
220. R. W. Smith and E. A. Jenne. Environ. Sci. Technol. 25: 525–531 (1991).
221. J. A. Davis and J. O. Leckie. J. Colloid Interf. Sci. 67: 90–107 (1978).
222. J. A. Davis and J. O. Leckie. Environ. Sci. Technol. 12: 1309–1315 (1978).
223. S. P. Mishra and D. Tiwary. Appl. Radiat. Isot. 51: 359–366 (1999).
224. D. G. Kinniburgh. J. Soil Sci. 34: 759–768 (1983).
225. C. C. Ainsworth, J. L. Pilon, P. L. Gassman and W. G. van der Sluys. Soil Sci. Soc. Am. J. 58: 1615–1623 (1994).
226. J. O. Leckie, D. T. Merrill and W. Chow. AIChE Symposium Series. No. 243. 81: 28–42 1985.
227. M. J. Perona and J. O. Leckie. J. Colloid Interf. Sci. 106: 64–69 (1985).
228. J. Slavek and W. F. Pickering. Water, Air Soil Poll. 28: 151–162 (1986).
229. L. Axe and P. R. Anderson. J. Colloid Interf. Sci. 185: 436–448 (1997).
230. N. Z. Misak, H. F. Ghoneimy and T. N. Morcos. J. Colloid Interf. Sci. 184: 31–43 (1996).
231. K. C. Swallow, D. N. Hume and F. M. M. Morel. Environ. Sci. Technol. 14: 1326–1331 (1980).
232. M. L. Machesky, W. O. Andrade and A. W. Rose. Geochim. Cosmochim. Acta 55: 769–776 (1991).
233. B. B. Johnson. Environ. Sci. Technol. 24: 112–118 (1990).
234. I. Lamy, M. Djafer and M. Terce. Water, Air Soil Pol. 57–58: 457–465 (1991).
235. U. Hoins, L. Charlet and H. Sticher. Water Air Soil Poll. 68: 241–255 (1993).
236. L. Gunneriusson. J. Colloid Interf. Sci. 163: 484–492 (1994).
237. P. Venema, T. Hiemstra and W. H. van Riemsdijk. J. Colloid Interf. Sci. 183: 515–527 (1996).
238. P. Venema, T. Hiemstra and W. H. van Riemsdijk. J. Colloid Interf. Sci. 192: 94–103 (1997).
239. M. J. Angove, J. D. Wells and B. B. Johnson. Colloids Surf. A 146: 243–251 (1999).
240. D. Kovacevic, A. Pohlmeier, G. Ozbas, H. D. Narres and N. Kallay. Progr. Colloid Polym. Sci. 112: 183–187 (1999).
241. Z. S. Kooner. Environ. Geol. Water Sci. 20: 205–212 (1992).
242. D. P. Rodda, J. D. Wells and B. B. Johnson. J. Colloid Interf. Sci. 184: 564–569 (1996).
243. B. Nowack and A. T. Stone. Environ. Sci. Technol. 33: 3627–3633 (1999).

244. N. J. Barrow and V. C. Cox. J. Soil Sci. 43: 295–304 (1992). N. J. Barrow and V. C. Cox. J. Soil Sci. 43: 305–312 (1992).

245. L. Gunneriusson and S. Sjöberg. J. Colloid Int. Sci. 156: 121–128 (1993).

246. L. Gunneriusson, D. Baxter and H. Emteborg. J. Colloid Int. Sci. 169: 262–266 (1995).

247. P. Bonnissel-Gissinger, M. Alnot, J. P. Lickes, J. J. Ehrhardt and P. Behra. J. Colloid Interf. Sci. 215: 313–322 (1999).

248. C. R. Collins, D. M. Sherman and K. V. Ragnarsdottir. J. Colloid Interf. Sci. 219: 345–350 (1999).

249. S. Fendorf and M. Fendorf. Clays Clay Min. 44: 220–227 (1996).

250. A. L. Bryce, W. A. Kornicker, A. W. Elzerman and S.B. Clark. Environ. Sci. Technol. 28: 2353–2359 (1994).

251. M. H. Bradbury and B. Baeyens. J. Colloid Int. Sci. 158: 364–371 (1993).

252. K. F. Hayes and J. O. Leckie. In: ACS Symposium Series No 323 Geochemical Processes at Mineral Surfaces, J. A. Davis and K. F. Hayes. (Eds.), ACS 1986, pp. 114–141.

253. A. L. Roe, K. F. Hayes, C. Chisholm-Brause, G. E. Brown, G. A. Parks, K. O. Hodgson and J. O. Leckie. Langmuir 7: 367–373 (1991).

254. B. Müller and L. Sigg. J. Colloid Int. Sci. 148: 517–532 (1992).

255. L. Gunneriusson, L. Lovgren and S. Sjoberg. Geochim. Cosmochim. Acta 58: 4973–4983 (1994).

256. H. Abdel-Samad and P. R. Watson. Appl. Surf. Sci. 136: 46–54 (1998).

257. A. M. Jakobsson, Y. Albinsson and R. S. Rundberg. submitted.

258. A. L. Sanchez, J. W. Murray and T. H. Sibley. Geochim. Cosmochim. Acta 49: 2297–2307 (1985).

259. I. Nirdosh, W. B. Trembley and C. R. Johnson. Hydrometallurgy 24: 237–248 (1990).

260. C. R. Collins, D. M. Sherman and K. V. Ragnarsdottir. Radiochim. Acta 81: 201–206 (1998).

261. N. Sahai, S. A. Carroll, S. Roberts and P. A. O'Day. J. Colloid Interf. Sci. 222: 198–212 (2000).

262. K. A. Hunter, D. J. Hawke and L. K. Choo. Geochim. Cosmochim. Acta 52: 627–636 (1988).

263. M. Duff and C. Amrhein. Soil Sci. Soc. Am. J. 60: 1393–1400 (1996).

264. A. B. Ankomah. Soil Sci. 154: 206–213 (1992).

265. M. Lehmann, A. I. Zouboulis and K. A. Matis. Chemosphere 39: 881–892 (1999).

266. B. Nowack and L. Sigg. J. Colloid Interf. Sci. 177: 106–121 (1996).

267. M. A. Ali and D. A. Dzombak. Geochim. Cosmochim. Acta 60: 291–304 (1996).

268. M. J. Angove, J. D. Wells and B. B. Johnson. J. Colloid Interf. Sci. 211: 281–290 (1999).

269. E. A. Forbes, A. M. Posner and J. P. Quirk. J. Colloid Interf. Sci. 49: 403–409 (1974).

270. E. A. Forbes, A. M. Posner and J. P. Quirk. J. Soil Sci. 27: 154–166 (1976).

271. R. H. Parkman, J. M. Charnock, N. D. Bryan, F. R. Livens and D. J. Vaughan. Am. Mineralogist 84: 407–419 (1999).

272. L. S. Balistrieri and J. W. Murray. Geochim. Cosmochim. Acta 46: 1253–1265 (1982).

273. U. Palmqvist, E. Ahlberg, L. Lövgren and S. Sjöberg. J. Colloid Interf. Sci. 196: 254–266 (1997).

274. J. Gerth and G. Brummer. Fresenius Z. Anal. Chem. 316: 616–620 (1983).

275. G. W. Bruemmer, J. Gerth and K. G. Tiller. J. Soil Sci. 39: 37–52 (1988).

276. K. F. Hayes and J. O. Leckie. J. Colloid Interf. Sci. 115: 564–572 (1987).

277. Z. S. Kooner, C. D. Cox and J. L. Smoot. Environ. Toxicol. Chem. 14: 2077–2083 (1995).

278. M. Padmanabham. Aust. J. Soil Res. 21: 515–525 (1983).

279. S. B. Kanungo. J. Colloid Interf. Sci. 162: 103–109 (1994).

280. S. B. Kanungo. J. Colloid Interf. Sci. 162: 93–102 (1994).

281. N. J. Barrow, J. W. Bowden, A. M. Posner and J. P. Quirk. Aust. J. Soil Res. 19: 309–321 (1981).
282. D. P. Rodda, B. B. Johnson and J. D. Wells. J. Colloid Interf. Sci. 161: 57–62 (1993).
283. Z. S. Kooner. Environ. Geol. 21: 242–250 (1993).
284. D. P. Rodda, B. B. Johnson and J. D. Wells. J. Colloid Interf. Sci. 184: 365–377 (1996).
285. L. L. Ames, J. E. McGarrah, B. A. Walker and P. F. Salter. Chem. Geology 40: 135–148 (1983).
286. A. I. Novikov, A. A. Shaffert and E. K. Shchekoturova. Zh. Anal. Khim. 32: 1108–1115 (1977).
287. P. I. Artyukhin and S. V. Filatov. Radiokhimiya 28: 371–375 (1986).
288. S. Ardizzone, D. Lettieri and S. Trasatti. J. Electroanal. Chem. 146: 431–437 (1983).
289. R. Ravikumar and D. W. Fuerstenau. Mat. Res. Soc. Symp. Proc. 432: 243–248 (1997).
290. Y. Ran, J. Fu, R. J. Gilkes and R. W. Rate. Sci. China D 42: 172–181 (1999).
291. S. P. Mishra and D. Tiwary. J. Radioanal. Nucl. Chem. 170: 133–141 (1993).
292. S. M. Hasany and M. H. Chaudhary J. Radioanal. Nucl. Chem. 89: 353–363 (1985).
293. M. E. Mikhail and N. Z. Misak. Appl. Radioat. Isot. 39: 1121–1124 (1988).
294. P. Thanabalasingam and W. F. Pickering. Environ. Poll. B 10: 115–128 (1985).
295. A. A. Yousef, M. A. Arafa and M. A. Malati. J. Appl. Chem. Biotech. 21: 200–207 (1971).
296. H. A. Laitinen and H. Zhou. J. Colloid Int. Sci. 125: 45–50 (1988).
297. M. J. Gray and M. A. Malati. J. Chem. Tech. Biotechnol. 29: 127–134 (1979).
298. S. B. Kanungo and K. M. Parida. J. Colloid Interf. Sci. 98: 252–260 (1984).
299. L. S. Balistrieri and J. W. Murray. Geochim. Cosmochim. Acta 46: 1041–1052 (1982).
300. G. Bidoglio, P. N. Gibson, E. Haltier, N. Omenetto and M. Lipponen. Radiochim. Acta 58/59: 191–197 (1992).
301. J. W. Murray. Geochim. Cosmochim. Acta 39: 635–647 (1975).
302. J. W. Murray. Geochim. Cosmochim. Acta 39: 505–519 (1975).
303. Y. Inoue, H. Yamazaki, K. Okada and K. Morita. Bull. Chem. Soc. Jpn. 58: 2955–2959 (1985).
304. Y. Inoue, O. Tochiyama, H. Yamazaki and A. Sakurada. J. Radioanal. Nucl. Chem. 124: 361–382 (1988).
305. W. E. E. Stone, G. M. S. El-Shafei, J. Sanz and S. A. Selim. J. Phys. Chem. 97: 10127–10132 (1993).
306. E. A. Nechaev and V. P. Romanov. Kolloid. Zh. 36: 1095–1100 (1974).
307. M. Kosmulski. J. Colloid Interf. Sci. 208: 543–545 (1998).
308. I. Nukatsuka, K. Sakai, R. Kudo and K. Ohzeki. Analyst 120: 2819–2822 (1995).
309. R. K. Iler J. Colloid Interf. Sci. 53: 476–488 (1975).
310. F. Rashchi, Z. Xu and J. A. Finch. Colloids Surf. A 132: 159–171 (1998).
311. C. R. A. Clauss and K. Weiss. J. Colloid Int. Sci. 61: 577–581 (1977).
312. P. A. O'Day, C. J. Chisholm-Brause, S. N. Towle, G. A. Parks and G. E. Brown. Geochim. Cosmochim. Acta 60: 2515–2532 (1996).
313. A. Manceau, M. Schlegel, K. L. Nagy and L. Charlet. J. Colloid Interf. Sci. 220: 181–197 (1999).
314. K. Csoban, M. Parkanyi-Berka, P. Joo and P. Behra. Colloids Surf. A 141: 347–364 (1998).
315. G. C. Bye, M. McEvoy and M. A. Malati. J. Chem. Tech. Biotechnol. 32: 781–789 (1982).
316. A. von Zelewsky and J. M. Bemtgen. Inorg. Chem. 21: 1771–1777 (1982).
317. Y. J. Park, K. H. Jung and K. K. Park. J. Colloid Interf. Sci. 172: 447–458 (1995).
318. N. N. Vlasova and N. K. Davidenko Colloids Surf. A 119: 23–28 (1996).
319. K. Xia, A. Mehadi, R. W. Taylor and W. F. Bleam. J. Colloid Interf. Sci. 185: 252–257 (1997).

320. I. Larson and R. J. Pugh. J. Colloid Interf. Sci. 208: 399–404 (1998).
321. M. A. Anderson, M. H. Palm-Gennen, P. N. Renard, C. Defosse and P. G. Rouxhet. J. Colloid Interf. Sci. 102: 328–336 (1984).
322. M. Kosmulski. J. Colloid Interf. Sci. 211: 410–412 (1999).
323. M. Kosmulski, P. Eriksson, J. Gustafsson and J. B. Rosenholm. Radiochim. Acta 88: 701–704 (2000).
324. T. Yakahashi, Y. Minai, T. Ozaki, S. Ambe, M. Iwamoto, H. Maeda, F. Ambe and T. Tominaga. J. Radioanal. Nucl. Chem. 205: 255–260 (1996).
325. M. A. Denecke, H. Geckeis, C. Pohlmann, J. Rothe and D. Degering. Radiochim. Acta 88: 639–643 (2000).
326. M. G. MacNaughton and R. O. James. J. Colloid Interf. Sci. 47: 431–440 (1974).
327. D. B. Kent and M. Kastner. Geochim. Cosmochim. Acta 49: 1123–1136 (1985).
328. S. V. Krishnan and I. Iwasaki. Environ. Sci. Technol. 20: 1224–1229 (1986).
329. C. C. Schulthess and D. L. Sparks. Soil Sci. Soc. Am. J. 53: 366–373 (1989).
330. O. Clause, M. Karmarec, L. Bonneviot, F. Villain and M. Che. J. Am. Chem. Soc. 114: 4709–4717 (1992).
331. M. Kosmulski. J. Colloid Interf. Sci. 190: 212–223 (1997).
332. M. Kosmulski, P. Eriksson, J. Gustafsson and J. B. Rosenholm. J. Colloid Interf. Sci. 220: 128–132 (1999).
333. Z. J. Wang and W. Stumm. Netherlands J. Agric. Sci. 35: 231–240 (1987).
334. R. Herrera-Urbina and D. W. Fuerstenau. Colloids Surf. A 98: 25–33 (1995).
335. D. N. Furlong. Aust. J. Chem. 35: 911–917 (1982).
336. E. Östhols. Geochim. Cosmochim. Acta 59: 1235–1249 (1995).
337. E. Östhols, A. Manceau, F. Farges and L. Charlet. J. Colloid Interf. Sci. 194: 10–21 (1997).
338. K. H. Lieser, S. Quandt-Klenk and B. Thybusch. Radiochim. Acta 57: 45–50 (1992).
339. A. J. Dent, J. D. F. Ramsay and S. W. Swanton. J. Coll. Interf. Sci. 150: 45–60 (1992).
340. E. Guibal, R. Lorenzelli, T. Vincent and P. LeCloirec. Environ. Technol. 16: 101–114 (1995).
341. Yu. D. Glinka, M. Jaroniec and V. M. Rozenbaum. J. Colloid Interf. Sci. 194: 455–469 (1997).
342. G. Mignot, M. Del Nero, R. Barillon and C. Guy. Migration '99, Lake Tahoe Sept 26-Oct 1, 1999 PB 2–23.
343. M. Yasrebi, M. Ziomek-Moroz, W. Kemp and D. H. Sturgis. J. Am. Ceram. Soc. 79: 1223–1227 (1996).
344. N. Marmier, A. Delisee and F. Fromage. J. Colloid Interf. Sci. 212: 228–233 (1999).
345. N. N. Vlasova, N. K. Davidenko, V. I. Bogomaz and A. A. Chuiko. Colloids Surf. A 104: 53–56 (1995).
346. M. Kagawa, Y. Syono, Y. Imamura and S. Usui. J. Am. Ceram. Soc. 69: C 50–C51 (1986).
347. M. A. Malati, M. Mc Evoy and C. R. Harvey. Surf. Technol. 17: 165–174 (1982).
348. H. Sonntag, V. Itschenskij and R. Koleznikova. Croat. Chim. Acta 60: 383–393 (1987).
349. T. F. Tadros and J. Lyklema. J. Electroanal. Chem. 22: 1–7 (1969).
350. Yu. G. Frolov, S. K. Milonjic and V. L. Razin. Kolloid. Zh. 41: 516–521 (1979).
351. P. K. Ahn, S. Nishiyama, S. Tsuruya and M. Masai. Appl. Catalysis A. 101: 207–219 (1993).
352. N. F. Bogdanova, M. P. Sidorova, L. E. Ermakova and I. A. Savina. Kolloid. Zh. 59: 452–459 (1997).
353. J. Jednacak, V. Pravdic and W. Haller. J. Colloid Int. Sci. 49: 16–23 (1974).
354. S. N. Omenyi, B. J. Herren, R. S. Snyder and G. V. F. Seaman. J. Colloid Int. Sci. 110: 130–136 (1986).

355. J. Lutzenkirchen and P. Behra. J. Contamin. Hydrol. 26: 257–268 (1997).

356. G. C. Bye, M McEvoy and M. A. Malati. J. Chem. Soc. Faraday Trans. I 79: 2311–2318 (1983).

357. P. W. Schindler, B. Furst, R. Dick and P. U. Wolf. J. Coll. Interf. Sci. 55: 469–475 (1976).

358. P. Huang and D. W. Fuerstenau. Mat. Res. Soc. Symp. Proc. 432: 81–86 (1997).

359. J. H. Anderson. J. Catalysis 28: 76–82 (1973).

360. Y. J. Park, K. H. Jung, K. K. Park and T. Y. Eom. J. Colloid Interf. Sci. 160: 324–331 (1993).

361. A. R. Sarkar, P. K. Datta and M. Sarkar. Talanta 43: 1857–1862 (1996).

362. F. Vydra and J. Galba. Coll. Czech. Chem. Comm. 34: 3471–3478 (1969).

363. R. O. James and T. W. Healy. J. Colloid Int. Sci. 40: 42–52; 53–64 (1972).

364. L. Charlet and A. Manceau. Geochim. Cosmochim. Acta 58: 2577–2582 (1994).

365. K. K. Das, Pradip and K. A. Natarajan. J. Colloid Interf. Sci. 196: 1–11 (1997).

366. M. Kosmulski. J. Colloid Interf. Sci. 195: 395–403 (1997).

367. P. Roose, J. van Craen, C. Pathmamanoharan and H. Eisendrath. J. Colloid Interf. Sci. 188: 115–120 (1997).

368. B. Satmark and Y. Albinsson. Radiochim. Acta 58/59: 155–161 (1992).

369. G. Della Mea, J. C. Dran, V. Moulin, J. C. Petit, J. D. F. Ramsay and M. Theyssier. Radiochim. Acta 58/59: 219–223 (1992).

370. S. K. Milonjic, D. M. Cokesa and R. V. Stevanovic. J. Radioanal. Nucl. Chem. 158: 79–90 (1992).

371. S. K. Milonjic, M. Boskovic and T. S. Ceranic. Separ. Sci. Technol. 27: 1643–1653 (1992).

372. D. N. Strazhesko, V. B. Strelko, V. N. Belyakov and S. C. Rubanik. J. Chromat. 102: 191–195 (1974).

373. N. Jaffrezic-Renault. J. Inorg. Nucl. Chem. 40: 539–544 (1978).

374. R. Rautiu and D. A. White. Solv. Extr. Ion Exch. 14: 721–738 (1996).

375. C. Biegler and M. R. Houchin. Colloids Surf. 21: 267–278 (1986).

376. H. S. Mahal and B. Venkatarmani. Indian J. Chem. 37A: 993–1001 (1998).

377. B. Venkataramani, K. S. Venkateswarlu and J. Shankar. Proc. Indian Acad. Sci. (Chem). 87 A: 409–414 (1978).

378. G. R. Wiese and T. W. Healy. J. Colloid Int. Sci. 51: 434–442 (1975).

379. G. R. Wiese. Thesis, quoted after R. O. James, G. R. Wiese and T. W. Healy. J. Colloid Int. Sci. 59: 381–385 (1977).

380. S. E. Fendorf, D. L. Sparks, M. Fendorf and R. Gronsky. J. Coll. Int. Sci. 148: 295–298 (1992).

381. M. Kosmulski, E. Matijevic. Colloids Surf. 64: 57–65 (1992).

382. M. Kosmulski, J. Gustafsson and J. B. Rosenholm. J. Colloid Int. Sci. 209: 200–206 (1999).

383. S. Chibowski and J. Szczypa. Pol. J. Chem. 58: 1155–1160 (1984).

384. P. J. Stiglich. Thesis, quoted after J. A. Davis and J. O. Leckie. J. Colloid Int. Sci. 67: 90–107 (1978).

385. L. G. J. Fokkink, A. G. Rhebergen, A. de Keizer and J. Lyklema. J. Electroanal. Chem. 329: 187–199 (1992).

386. S. N. Towle, G. E. Brown and G. A. Parks. J. Colloid Interf. Sci. 217: 299–311 (1999).

387. C. Ludwig and P. W. Schindler. J. Colloid Interf. Sci. 169: 284–290 (1995).

388. C. Ludwig and P. W. Schindler. J. Colloid Interf. Sci. 169: 291–299 (1995).

389. D. E. Yates and T. W. Healy. J. Chem. Soc. Faraday I 76: 9–18 (1980).

390. A. M. Jakobsson. J. Colloid Interf. Sci. 220: 367–373 (1999).

391. H. Yamashita, Y. Ozawa, F. Nakajima and T. Murata. Bull. Chem. Soc. Jpn. 53: 1331–1334 (1980).

392. J. Jablonski, W. Janusz, R. Sprycha and M. Reszka. Fizykochem. Probl. Min. 22: 21–31 (1990).
393. M. Kosmulski, J. Gustafsson and J. B. Rosenholm. Colloid Polym. Sci. 277: 550–556 (1999).
394. H. M. Jang and D. W. Fuerstenau. Colloids Surf. 21: 235–257 (1986).
395. J. K. Yang and A. P. Davis. J. Colloid Interf. Sci. 216: 77–85 (1999).
396. A. M. Jakobsson and Y. Albinsson. Radiochim. Acta 82: 257–262 (1998).
397. W. F. Bleam and M. B. McBride. J. Colloid Interf. Sci. 110: 335–346 (1986).
398. N. Jaffrezic-Renault and H. Andrade-Martins. J. Radioanal. Chem. 55: 307–316 (1980).
399. B. Venkataramani and A. R. Gupta. Recovery of U from Seawater, Tokyo 17–19 Oct. 1983. pp. 313–324.
400. R. M. Iyer, A. R. Gupta and B. Venkataramani. In: Inorg. Ion Exch. and Ads. for Chem. Proc. in the Nucl. Fuel Cycle. IAEA TECDOC-337, Vienna 1985. pp. 249–262.
401. S. P. Mishra and V. K. Singh. J. Radioanal. Nucl. Chem. 241: 341–346 (1999).
402. A. Suzuki, H. Seki and H. Maruyama. J. Chem. Eng. Jpn. 27: 505–511 (1994).
403. J. Ragai and S. I. Selim. J. Colloid Int. Sci. 115: 139–146 (1987).
404. P. Persson, G. A. Parks and G. E. Brown. Langmuir 11: 3782–3794 (1995).
405. P. Persson, G. A. Parks and G. E. Brown. Physica B 208–209: 453–454 (1995).
406. J. Randon, A. Larbot, C. Guizard, L. Cot, M. Lindheimer and S. Partyka. Colloids Surf. 52: 241–255 (1991).
407. Y. Inoue and H. Yamazaki. Bull. Chem. Soc. Jpn. 60: 891–897 (1987).
408. J. B. Stankovic, S. K. Milonjic, M. M. Kopecni and T. S. Ceranic. Colloids Surf. 46: 283–296 (1990).
409. S. H. Han, W. G. Hou, Q. Dong, D. J. Sun, X. R. Huang and C. G. Zhang. Chem. Res. Chinese Univ. 15: 58–62 (1998).
410. M. Mullet, P. Fievet, J. C. Reggiani and J. Pagetti. J. Membr. Sci. 123: 255–265 (1997).
411. A. Szymczyk, A. Pierre, J. C. Reggiani and J. Pagetti. J. Membrane Sci. 134: 59–66 (1997).
412. I. Sondi, J. Biscan and V. Pravdic. J. Colloid Interf. Sci. 178: 514–522 (1996).
413. H. M. Jang and S. H. Lee. Langmuir 8: 1698–1708 (1992).
414. E. H. Smith and T. Vengris. Crit. Rev. Anal. Chem. 28: SI13–SI18 (1998).
415. M. L. Schlegel, A. Manceau, D. Chateigner and L. Charlet. J. Colloid Interf. Sci. 215: 140–158 (1999).
416. M. L. Schlegel, L. Charlet and A. Manceau. J. Colloid Interf. Sci. 220: 392–405 (1999).
417. C. C. Chen, C. Papelis and K. F. Hayes. In: Adsorption of Metals by Geomedia. E. A. Jenne, (Ed.), Academic Press, San Diego 1998, pp. 333–348.
418. Q. Du, Z. Sun, W. Forsling and H. Tang. J. Colloid Interf. Sci. 187: 232–242 (1997).
419. L. Wang, A. Maes, P. de Canniere and J. van der Lee. Radiochim. Acta 82: 233–237 (1998).
420. N. M. Nagy, J. Konya and G. Wazelischen-Kun. Colloids Surf. A 152: 245–250 (1998).
421. T. P. O'Connor and D. N. Kester. Geochim. Cosmochim. Acta 39: 1531–1543 (1975).
422. M. J. Angove, B. B. Johnson and J. D. Wells. Colloids Surf. A 126: 137–147 (1997).
423. P. A. O'Day, G. A. Parks and G. E. Brown. Clays Clay Min. 42: 337–355 (1994).
424. P. A. O'Day, G. E. Brown and G. A. Parks. J. Colloid Interf. Sci. 165: 269–289 (1994).
425. H. A. Thompson, G. A. Parks and G. E. Brown. J. Colloid Interf. Sci. 222: 241–253 (2000).
426. H. A. Thompson, G. A. Parks and G. E. Brown. In: Adsorption of Metals by Geomedia. E. A. Jenne, (Ed.), Academic Press, San Diego 1998. pp. 349–370.
427. M. J. Angove, B. B. Johnson and J. D. Wells. J. Colloid Interf. Sci. 204: 93–103 (1998).
428. P. V. Brady, R. T. Cygan and K. L. Nagy. In: Adsorption of Metals by Geomedia. E. A. Jenne, (Ed.), Academic Press, San Diego 1998. pp. 371–382.

429. J. Ikhsan, B. B. Johnson and J. D. Wells. J. Colloid Interf. Sci. 217: 403–410 (1999).
430. S. A. Adeleye, P. G. Clay and M. O. A. Olapido. J. Mater. Sci. 29: 954–958 (1994).
431. S. P. Hyun, Y. H. Cho, S. J. Kim and P. S. Hahn. J. Colloid Interf. Sci. 222: 254–261 (2000).
432. E. J. Elzinga and D. L. Sparks. J. Colloid Interf. Sci. 213: 506–512 (1999).
433. D. R. Turner, R. T. Pabalan and F. P. Bertetti. Clays Clay Min. 46: 256–269 (1998).
434. D. G. Strawn and D. L. Sparks. J. Colloid Interf. Sci. 216: 257–269 (1999).
435. A. Inoue. Clay Sci. 6: 251–260 (1987).
436. H. N. Erten, S. Aksoyoglu, S. Hatipoglu and H. Gokturk. Radiochim. Acta 44/45: 147–151 (1988).
437. J. P. McKinley, J. M. Zachara, S. C. Smith and G. D. Turner. Clays Clay Miner. 43: 586–598 (1995).
438. R. T. Pabalan and G. D. Turner. Aquatic Geochem. 2: 203–226 (1997).
439. A. Liu and R. D. Gonzalez. J. Colloid Interf. Sci. 218: 225–232 (1999).
440. B. Baeyens and M. H. Bradbury. J. Contam. Hydrol. 27: 199–222; 223–248 (1997).
441. C. Papelis and K. F. Hayes. Colloids Surf. A 107: 89–96 (1996).
442. Y. Tachi, T. Shibutani, H. Sato and M. Yui. J. Cont. Hydrol. 47: 171–186 (2001).
443. J. M. Zachara and J. P. McKinley. Aquatic Sci. 55: 250–261 (1993).
444. B. Venkataramani and K. S. Venkateswarlu. Proc. Indian Acad. Sci. (Chem.) 89: 241–245 (1980).
445. K. A. Venkatesan, N. Sathi Sasidharan and P. K. Wattal. J. Radioanal. Nucl. Chem. 220: 55–58 (1997).
446. Y. M. Xu, R. S. Wang and F. Wu. J. Colloid Interf. Sci. 209: 380–385 (1999).
447. M. F. Azizian, P. O. Nelson. In: Adsorption of Metals by Geomedia. E. Jenne, (Ed.), Academic Press, NY, 1998 pp. 165–180.
448. R. W. Hayes, M. W. Wharmby, R. W. C. Broadbank and K. W. Morcom. Talanta 29: 149–153 (1982).
449. P. Benes and V. Jiranek. Radiochim. Acta 21: 49–53 (1974).
450. M. Nardin, E. Papirer and J. Schultz. J. Colloid Int. Sci. 88: 204–213 (1982).
451. A. Kolics, E. Maleczki, K. Varga and G. Horanyi. J. Radioanal. Nucl. Chem. 158: 121–137 (1992).
452. M. Kosmulski and J. Szczypa. J. Radioanal. Nucl. Chem. 144: 73–77 (1990).
453. Y. Sakamoto, M. Konishi, K. Shirahashi, M. Senoo and N. Moriyama. Radioact. Waste Manag. Nucl. Fuel Cyc. 15: 13–25 (1990).
454. T. Ohe, M. Tsukamoto, T. Fujita, R. Hesbol and H. P. Hermansson, Int. Conf. Nucl. Waste Managem. Envir. Remed. Vol. 1 pp. 197–205; Prague. quoted after D. R. Turner, R. T. Pabalan and F. P. Bertetti. Clays Clay Min. 46: 256–269 (1998).
455. S. Nagasaki, S. Tanaka and A. Suzuki. Colloids Surf. A 155: 137–143 (1999).
456. D. W. Oscarson and H. B. Hume. In: Adsorption of Metals by Geomedia. E. Jenne, (Ed.), Academic Press, NY, 1998, pp. 277–289.
457. C. Breen, C. M. Bejarano-Bravo, L. Madrid, G. Thompson and B. E. Mann. Colloids Surf. A 155: 211–219 (1999).
458. J. Hong and P. N. Pintauro. Water Air Soil Pol. 86: 35–50 (1994).
459. P. A. O'Day. Thesis, quoted after P. A. O'Day, G. A. Parks and G. E. Brown. Clays Clay Min. 42: 337–355 (1994).
460. T. R. Holm and X. F. Zhu. J. Contam. Hydrol. 16: 271–287 (1994).
461. T. E. Payne and T. D. Waite. Radiochim. Acta 52/53: 487–493 (1991).
462. K. A. Bolton and L. J. Evans. Can. J. Soil Sci. 76: 183–189 (1996).
463. L. R. G. Guilherme and S. J. Anderson. In: Adsorption of Metals by Geomedia. E. Jenne, (Ed.), Academic Press, NY, 1998, pp. 209–228.
464. S. B. Clark, A. L. Bryce, A. D. Lueking, J. Gariboldi and S. M. Serkiz. In: Adsorption of Metals by Geomedia. E. Jenne, (Ed.), Academic Press, NY, 1998, pp. 149–164.

465. B. E. Reed and S. R. Cline. Separ. Sci. Technol. 29: 1529–1551 (1994).
466. M. P. Papini, Y. D. Kahie, B. Troia and M. Majone. Environ. Sci. Technol. 33: 4457–4464 (1999).
467. P. L. Brown, M. Guerin, S. I. Hankin and R. T. Lowson. J. Contam. Hydrol. 35: 295–303 (1998).
468. N. J. Barrow. J. Soil Sci. 37: 277–286 (1986).
469. N. J. Barrow. J. Soil Sci. 37: 295–302 (1986).
470. N. J. Barrow and A. S. Ellis. J. Soil Sci. 37: 303–310 (1986).
471. W. Z. Wang, M. L. Brusseau and J. F. Artiola. In: Adsorption of Metals by Geomedia. E. Jenne, (Ed.), Academic Press, NY, 1998, pp. 427–443.
472. R. N. Yong and E. M. Macdonald. In: Adsorption of Metals by Geomedia. E. Jenne, (Ed.), Academic Press, NY, 1998, pp. 229–253.
473. Y. Legoux, G. Blain, R. Guillaumount, G. Ouzounian, L. Brillard and M. Hussonnois. Radiochim. Acta 58/59: 211–218 (1992).
474. J. J. Rosentreter, H. Swantje Quarder, R. W. Smith and T. McLing. In: Adsorption of Metals by Geomedia. E. Jenne, (Ed.), Academic Press, NY, 1998, pp. 181–192.
475. J. A. Davis, J. A. Coston, D. B. Kent and C. C. Fuller. Environ. Sci. Technol. 32: 2820–2828 (1998).
476. X. Wen, Q. Du and H. Tang. Environ. Sci. Technol. 32: 870–875 (1998).
477. K. S. Smith, J. F. Ranville, G. S. Plumlee, D. L. Macalady. In: Adsorption of Metals by Geomedia. E. Jenne, (Ed.), Academic Press, NY, 1998, pp. 521–547.
478. F. Wang, J. Chen and W. Forsling. Environ. Sci. Technol. 31: 448–453 (1997).
479. K. H. Lieser and T. Steinkopff. Radiochim. Acta 47: 55–61 (1989).
480. J. R. Lead, J. Hamilton-Taylor, W. Davison and M. Harper. Geochim. Cosmochim. Acta 63: 1661–1670 (1999).
481. F. J. Hingston, R. J. Atkinson, A. M. Posner and J. P. Quirk. Nature 215: 1459–1461 (1967).
482. H. Xu, B. Allard and A. Grimvall. Water Air Soil Pollut. 40: 293–305 (1988).
483. L. Madsen and A. M. Blokhus. J. Colloid Interf. Sci. 166: 259–262 (1994).
484. R. S. Alwitt. J. Colloid Int. Sci. 40: 195–198 (1972).
485. W. W. Choi and K. Y. Chen. Environ. Sci. Technol. 13: 189–196 (1979).
486. S. Goldberg and R. A. Glaubig. Soil Sci. Soc. Am. J. 49: 1374–1379 (1985).
487. C. P. Schulthess, K. Swanson and H. Wijnja. Soil Sci. Soc. Am. J. 62: 136–141 (1998).
488. J. R. Regalbuto, A. Navada, S. Shadid, M. L. Bricker and Q. Chen. J. Catalysis. 184: 335–348 (1999).
489. N. Mikami, M. Sasaki, T. Kikuchi and T. Yasunaga. J. Phys. Chem. 87: 5245–5248 (1983).
490. M. A. Vuurman, F. D. Hardcastle and I. E. Wachs. J. Mol. Catalysis 84: 193–205 (1993).
491. P. Cambier and G. Sposito. Clays Clay Miner. 39: 369–374 (1991).
492. E. P. Luther, J. A. Yanez, G. V. Franks, F. E. Lange and D. S. Pearson. J. Am. Ceram. Soc. 78: 1495–1500 (1995).
493. W. W. Choi and K. Y. Chen. J. Am. Water Works Assoc. 71: 562–570 (1979).
494. O. J. Hao and C. P. Huang. J. Environ. Eng. 112: 1054–1069 (1986).
495. H. Farrah, J. Slavek and W. F. Pickering. Aust. J. Soil Res. 25: 55–69 (1987).
496. F. M. Mulcahy, M. Houalla and D. M. Hercules. J. Catal. 148: 654–659 (1994).
497. E. A. Ferreiro, A. K. Helmy and S. G. Bussetti. Z. Pflanzenernaehr. Bodek. 148: 559–566 (1985).
498. S. Goldberg, H. S. Forster and C. L. Godfrey. Soil Sci. Soc. Am. J. 60: 425–432 (1996).
499. S. Goldberg, C. Su and H. S. Forster. In: Adsorption of Metals by Geomedia. E. A. Jenne, (Ed.), Academic Press, San Diego, 1998, pp. 401–426.

500. C. H. Wu, C. F. Lin, S. L. Lo and T. Yasunaga. J. Colloid Interf. Sci. 208: 430–438 (1998).
501. P. Persson, E. Laiti and L. O. Ohman. J. Colloid Interf. Sci. 190: 341–349 (1997).
502. D. Muljadi, A. M. Posner and J. P. Quirk. J. Soil Sci. 17: 212–229, 230–237, 238–247 (1966).
503. Y. S. R. Chen, J. N. Butler and W. Stumm. J. Colloid Int. Sci. 43: 421–436 (1973).
504. Y. S. R. Chen, J. N. Butler and W. Stumm. Environ. Sci. Technol. 7: 327–332 (1973).
505. C. P. Huang. J. Colloid Int. Sci. 53: 178–186 (1975).
506. N. Peinemann and A. K. Helmy. J. Electroanal. Chem 78: 325–330 (1977).
507. W. H. van Riemsdijk, F. A. Weststrate and J. Beek. J. Environ. Qual. 6: 26–29 (1977).
508. J. A. Davis. Geochim. Cosmochim. Acta 46: 2381–2393 (1982).
509. N. Mikami, M. Sasaki, K. Hachiya, R. D. Astumian, T. Ikeda and T. Yasunaga. J. Phys. Chem. 87: 1454–1458 (1983).
510. E. Laiti, P. Persson and L. O. Ohman. Langmuir 12: 2969–2975 (1996).
511. O. Klug and W. Forsling. Langmuir 15: 6961–6968 (1999).
512. F. Thomas, J. Y. Bottero and J. M. Cases. Colloids Surf. 37: 281–294 (1989).
513. J. S. Moya, J. Rubio and J. A. Pask. Ceramic Bull. 59: 1198–1200 (1980).
514. C. P. Schulthess and J. F. McCarthy. Soil Sci. Soc. Am. J. 54: 688–694 (1990).
515. H. Xu, B. Allard and A. Grimvall. Water Air Soil Pollut. 57–58: 269–278 (1991).
516. E. A. Nechaev and G. V. Zvonareva. Kolloidn. Zh. 42: 511–516 (1980).
517. R. Kummert and W. Stumm. J. Colloid Int. Sci. 75: 373–385 (1980).
518. S. Goldberg and R. A. Glaubig. Soil Sci. Soc. Am. J. 52: 87–91 (1988).
519. C. D. Cox and M. M. Ghosh. Water Res. 28: 1181–1188 (1994).
520. A. R. Bowers and C. P. Huang. J. Colloid Int. Sci. 105: 197–214 (1985).
521. P. C. Hidber, T. J. Graule and L. J. Gauckler. J. Am. Ceramic Soc. 79: 1857–1867 (1996).
522. E. Laiti, P. Persson and L. O. Ohman. Langmuir 14: 825–831 (1998).
523. W. P. Cheng, C. Huang and J. R. Pan. J. Colloid Interf. Sci. 213: 204–207 (1999).
524. H. L. Yao and H. H. Yeh. Langmuir 12: 2981–2988; 2989–2994 (1996).
525. J. A. Blackwell. Chromatographia 35: 133–138 (1993).
526. S. S. S. Rajan, K. W. Perrott and W. M. H. Saunders. J. Soil Sci. 25: 438–447 (1974).
527. A. K. Helmy, S. G. de Bussetti and E. A. Ferreiro. Colloids Surf. 58: 9–16 (1991).
528. M. Okazaki, K. Sakaidani, T. Saigusa and N. Sakaida. Soil Sci. Plant Nutr. 35: 337–346 (1989).
529. E. Laiti and L. O. Ohman. J. Colloid Interf. Sci. 183: 441–452 (1996).
530. W. F. Bleam, P. E. Pfeffer, S. Goldberg, R. W. Taylor and R. Dudley. Langmuir 7: 1702–1712 (1991).
531. J. Nordin, P. Persson, E. Laiti and S. Sjoberg. Langmuir 13: 4085–4093 (1997).
532. M. Szekeres, E. Tombacz, K Ferencz and I. Dekany. Colloids Surf. A 141: 319–325 (1998).
533. B. Lovrecek, Z. Bolanca and O. Korelic. Surf. Coat. Technol. 31: 351–364 (1987).
534. M. A. Anderson, J. F. Ferguson and J. Gavis. J. Colloid Int. Sci. 54: 391–399 (1976).
535. B. A. Manning and S. Goldberg. Environ. Sci. Technol. 31: 2005–2011 (1977).
536. S. Goldberg, H. S. Forster and E. L. Heick. Soil Sci. Soc. Am. J. 57: 704–708 (1993).
537. S. Goldberg, H. S. Forster and E. L. Heick. Soil Sci. 156: 316–321 (1993).
538. J. R. Feldkamp, D. N. Shah, S. L. Meyer, J. L. White and S. L. Hem. J. Pharm. Sci. 70: 638–640 (1981).
539. E. C. Scholtz, J. R. Feldkamp, J. L. White and S. L. Hem. J. Pharm. Sci. 74: 478–481 (1985).
540. C. Su and D. L. Suarez. Clays Clay Min. 45: 814–825 (1997).
541. S. Kuo and E. G. Lotse. Soil Sci. 116: 400–406 (1974).
542. J. H. Kyle, A. M. Posner and J. P. Quirk. J. Soil Sci. 26: 32–43 (1975).

543. K. R. Helyar, D. N. Munns and R. G. Burau. J. Soil Sci. 27: 307–314 (1976).
544. W. H. van Riemsdijk and J. Lyklema. J. Colloid Int. Sci. 76: 55–66 (1980).
545. W. H. van Riemsdijk and J. Lyklema. Colloids Surf. 1: 33–44 (1980).
546. L. Lijklema. Environ. Sci. Technol. 14: 537–541 (1980).
547. M. Nanzyo. J. Soil Sci. 35: 63–69 (1984).
548. N. S. Bolan, N. J. Barrow and A. M. Posner. J. Soil Sci. 36: 187–197 (1985).
549. R. S. Beckwith and R. Reeve. Aust. J. Soil Res. 1: 157–168 (1963).
550. F. J. Hingston and M. Raupach. Aust. J. Soil Res. 5: 295–309 (1967).
551. W. B. Jepson, D. G. Jeffs and A. P. Ferris. J. Colloid Int. Sci. 55: 454–461 (1976).
552. B. A. Manning and S. Goldberg. Soil Sci. Soc. Am. J. 60: 121–131 (1996).
553. F. J. Hingston, A. M. Posner and J. P. Quirk. Discuss Faraday Soc. 52: 334–342 (1971).
554. R. L. Parfitt, V. C. Farmer and J. D. Russell. J. Soil Sci. 28: 29–39 (1977).
555. R. L. Parfitt, A. R. Fraser, J. D. Russell and V. C. Farmer. J. Soil Sci. 28: 40–47 (1977).
556. M. McPhail, A. L. Page and F. T. Bingham. Soil Sci. Soc. Am. Proc. 36: 510–514 (1972).
557. A. Violante, C. Colombo and A. Buondonno. Soil Sci. Soc. Am. J. 55: 65–70 (1991).
558. R. Alvarez, R. E. Cramer and J. A. Silva. Soil Sci. Soc. Am. J. 40: 317–319 (1976).
559. R. Alvarez, C. S. Fadley and J. A. Silva. Soil Sci. Soc. Am. J. 44: 422–425 (1980).
560. M. Bhutani and R. Kumari. J. Radioanal. Nucl. Chem. 180: 145–153 (1994).
561. W. Janusz, W. Staszczuk, A. Sworska and J. Szczypa. J. Radioanal. Nucl. Chem. 174: 83–91 (1993).
562. D. E. Yates and T. W. Healy. J. Colloid Interf. Sci. 52: 222–228 (1975).
563. G. E. Magaz, L. G. Rodenas, P. J. Morando and M. A. Blesa. Croat. Chem. Acta 71: 917–927 (1998).
564. J. Degenhardt and A. J. McQuillan. Chem. Phys. Letters 311: 179–184 (1999).
565. J. Degenhardt and A. J. McQuillan. Langmuir 15: 4595–4602 (1999).
566. M. L. Peterson, G. E. Brown and G. A. Parks. Colloids Surf. A 107: 77–88 (1996).
567. T. Kendlewicz, P. Liu, C. S. Doyle, G. E. Brown, E. J. Nelson and S. A. Chambers. Surface Sci. 424: 219–231 (1999).
568. U. Künzelmann, H. J. Jacobasch and G. Reinhard. Werkst. Korrosion 40: 723–728 (1989).
569. S. Ambe. Langmuir 3: 489–493 (1987).
570. R. Torres, N. Kallay and E. Matijevic. Langmuir 4: 706–710 (1988).
571. P. Persson, N. Nilsson and S. Sjoberg. J. Colloid Interf. Sci. 177: 263–275 (1996).
572. A. Ioannou and A. Dimirkou. J. Colloid Interf. Sci. 192: 119–128 (1997).
573. N. Kallay, S. Zalac, J. Culin, U. Bieger, A. Pohlmeier and H. D. Narres. Progr. Coll. Polym. Sci. 95: 108–112 (1994).
574. D. Kovacevic, I. Kobal and N. Kallay. Croat. Chem. Acta 71: 1139–1153 (1998).
575. D. Kovacevic, N. Kallay, I. Antol, A. Pohlmeier, H. Lewandowski and H. D. Narres. Colloids Surf. A 140: 261–267 (1998).
576. S. J. Hug. J. Colloid Interf. Sci. 188: 415–422 (1997).
577. C. M. Eggleston, S. Hug, W. Stumm, B. Sulzberger and M. D. S. Afonso. Geochim. Cosmochim. Acta. 62: 585–593 (1998).
578. N. Kallay and E. Matijevic. Langmuir 1: 195–201 (1985).
579. Y. Zhang, N. Kallay and E. Matijevic. Langmuir 1: 201–206 (1985).
580. A. E. Regazzoni, M. A. Blesa and A. J. G. Maroto. J. Colloid Interf. Sci. 122: 315–325 (1988).
581. D. Balzer and H. Lange. Colloid Polym. Sci. 255: 140–152 (1977).
582. G. A. Waychunas, B. A. Rea, C. C. Fuller and J. A. Davis. Geochim. Cosmochim. Acta 57: 2251–2269 (1993).
583. C. C. Fuller, J. A. Davis and G. A. Waychunas. Geochim. Cosmochim. Acta 57: 2271–2282 (1993).

584. T. H. Hsia, S. L. Lo, C. F. Lin and D. Y. Lee. Colloid Surf. A 85: 1–7 (1994).

585. J. A. Wilkie and J. G. Hering. Colloids Surf. A 107: 97–110 (1996).

586. K. P. Raven, A. Jain and R. H. Loeppert. Environ. Sci. Technol. 32: 344–349 (1998).

587. B. Nowack and L. Sigg. Geochim. Cosmochim. Acta 61: 951–963 (1997).

588. J. C. Ryden, J. R. McLaughlin and J. K. Syers. J. Soil Sci. 28: 72–92 (1977).

589. A. Amirbahman, L. Sigg and U. von Gunten. J. Colloid Interf. Sci. 194: 194–206 (1997).

590. K. F. Hayes, C. Papelis and J. O. Leckie. J. Colloid Interf. Sci. 125: 717–726 (1988).

591. A. Manceau and L. Charlet. J. Colloid Int. Sci. 168: 87–93 (1994).

592. H. C. B. Hansen, T. P. Wetche, K. Raulund-Rasmussen and O. K. Borggaard. Clay Miner. 29: 341–350 (1994).

593. S. I. Pechenyuk and E. V. Kalinkina. Kolloidn. Zh. 52: 716–721 (1990).

594. J. C. Ryden, J. K. Syers and R. W. Tillman. J. Soil Sci. 38: 211–217 (1987).

595. J. B. Harrison and V. E. Berkheiser. Clays Clay Min. 30: 97–102 (1982).

596. P. J. Swedlund and J. G. Webster. Wat. Res. 33: 3413–3422 (1999).

597. J. A. Davis and J. O. Leckie. J. Colloid Interf. Sci. 74: 32–43 (1980).

598. K. A. Matis, A. I. Zouboulis, D. Zamboulis and A. V. Valatdorou. Water, Air Soil Poll. 111: 297–316 (1999).

599. B. A. Manning, S. E. Fendorf and S. Goldberg. Environ. Sci. Technol. 32: 2383–2388 (1998).

600. P. M. Bloesch, L. C. Bell and J. D. Hughes. Aust. J. Soil Res. 25: 377–390 (1987).

601. J. M. Zachara, D. C. Girvin, R. L. Schmidt and C. T. Resch. Environ. Sci. Technol. 21: 589–594 (1987).

602. E. H. Rueda, R. L. Grassi and M. A. Blesa. J. Colloid Int. Sci. 106: 243–246 (1985).

603. J. C. L. Meeussen, A. Scheidegger, T. Hiemstra, W. H. van Riemsdijk and M. Borkovec. Environ. Sci. Technol. 30: 481–488 (1996).

604. J. D. Filius, J. C. L. Meeussen and W. H. van Riemsdijk. Colloids Surf. A 151: 245–253 (1999).

605. B. C. Barja, M. I. Tejedor-Tejedor and M. A. Anderson. Langmuir 15: 2316–2321 (1999).

606. M. I. Tejedor-Tejedor and M. A. Anderson. Langmuir 2: 203–210 (1986).

607. D. Hawke, P. D. Carpenter and K. A. Hunter. Environ. Sci. Technol. 23: 187–191 (1989).

608. V. A. Hackley, R. S. Premachandran, S. G. Malghan and S. B. Schiller. Colloids Surf. A 98: 209–224 (1995).

609. W. Yao and F. J. Millero. Environ. Sci. Technol. 30: 536–541 (1996).

610. T. Hiemstra and W. H. van Riemsdijk. J. Colloid Interf. Sci. 179: 488–508 (1996).

611. L. S. Balistrieri and T. T. Chao. Soil Sci. Soc. Am. J. 51: 1145–1151 (1987).

612. P. Persson and L. Lovgren. Geochim. Cosmochim. Acta. 60: 2789–2799 (1996).

613. D. Peak, R. G. Ford and D. L. Sparks. J. Colloid Interf. Sci. 218: 289–299 (1999).

614. R. P. J. J. Rietra, T. Hiemstra and W. H. van Riemsdijk. J. Colloid Interf. Sci. 218: 511–521 (1999).

615. L. Sigg and W. Stumm. Colloids Surf. 2: 101–117 (1980).

616. T. Hiemstra and W. H. van Riemsdijk. J. Colloid Interf. Sci. 210: 182–193 (1999).

617. M. Djafer, R. K. Khandal and M. Terce Colloids Surf. 54: 209–218 (1991).

618. A. van Geen, A. P. Robertson and J. O. Leckie. Geochim. Cosmochim. Acta 58: 2073–2086 (1994).

619. M. A. Ali and D. A. Dzombak. Environ. Sci. Technol 30: 1061–1071 (1996).

620. K. Mesuere and W. Fish. Environ. Sci. Technol. 26: 2357–2364; 2365–2370 (1992).

621. R. M. Cornell and P. W. Schindler. Colloid Polym. Sci. 258: 1171–1175 (1980).

622. M. L. Machesky, B. L. Bischoff and M. A. Anderson. Env. Sci. Technol. 23: 580–587 (1989).

623. L. S. Balistrieri and T. T. Chao. Geochim. Cosmochim. Acta 54: 739–751 (1990).
624. N. Nilsson, P. Persson, L. Lovgren and S. Sjoberg. Geochim. Cosmochim. Acta. 60: 4385–4395 (1996).
625. W. A. Zeltner, E. C. Yost, M. L. Machesky, M. I. Tejedor-Tejedor and M. A. Anderson. Geochemical Processes at Mineral Surfaces, J. A. Davis, K. F. Hayes, (Eds.), ACS Symp. Series 323: 142–161 (1986).
626. J. S. Geelhoed, T. Hiemstra and W. H. van Riemsdijk. Geochim. Cosmochim. Acta 61: 2389–2396 (1997).
627. C. R. Evanko and D. A. Dzombak. Environ. Sci. Technol. 32: 2846–2855 (1998).
628. C. R. Evanko and D. A. Dzombak. J. Colloid Interf. Sci. 214: 189–206 (1999).
629. B. Nowack and A. T. Stone. J. Colloid Interf. Sci. 214: 20–30 (1999).
630. A. I. Novikov and V. F. Samoilova. Radiokhimiya 31: 72–78 (1989).
631. N. U. Yamaguchi, M. Okazaki and T. Hashitani. J. Colloid Interf. Sci. 209: 386–391 (1999).
632. P. Thanabalasingam and W. F. Pickering. Water Air Soil Poll. 29: 205–216 (1986).
633. N. Munichandraiah. J. Electroanal. Chem. 266: 179–184 (1989).
634. G. A. Kokarev, V. A. Kolesnikov, L. T. Gorokhova, M. Ya. Fioshin and V.A. Kazarinov. Elektrokhimiya 20: 1155–1158 (1984).
635. M. M. Bhutani, P. N. Reddy, A. K. Mitra and R. Kumari. Langmuir 8: 1974–1979 (1992).
636. K. Spildo, H. Hoiland and M. K. Olsen. J. Colloid Interf. Sci. 221: 124–132 (2000).
637. D. B. Ward and P. V. Brady. Clays Clay Min. 46: 453–465 (1998).
638. Y. J. Park, K. H. Jung and K. K. Park. J. Colloid Interf. Sci. 171: 205–210 (1995).
639. L. J. Warren. Colloids Surf. 5: 301–319 (1982).
640. C. P. Schulthess and J. Z. Belek. Soil Sci. Soc. Am. J. 62: 348–353 (1998).
641. M. A. Malati, R. A. Fassam and I. R. Henderson. J. Chem. Tech. Biotechnol. 58: 387–389 (1993).
642. P. A. Connor and A. J. McQuillan. Langmuir 15: 2916–2921 (1999).
643. M. Miura, H. Naono and T. Iwaki. J. Sci. Hiroshima Univ. 30: 57–63 (1966).
644. M. M. Bhutani, A. K. Mitra and R. Kumari. Radiochim. Acta 56: 153–158 (1992).
645. B. Putman, P. van der Meeren and D. Thierens. Colloids Surf. A 121: 81–88 (1997).
646. L. A. Perez-Maqueda and E. Matijevic. J. Mater. Res. 12: 3286–3292 (1997).
647. M. Valigi, D. Gazzoli, A. Cimino and E. Proverbio. J. Phys. Chem. B. 103: 11318–11326 (1999).
648. A. Szymczyk, P. Fievet, M. Mullet, J. C. Reggiani and J. Pagetti. Desalination 119: 309–314 (1998).
649. W. H. Morrison. J. Colloid Int. Sci. 100: 121–127 (1984).
650. M. M. Motta and C. F. Miranda. Soil Sci. Soc. Am. J. 53: 380–385 (1989).
651. R. Weerasooriya and H. U. S. Wickramarathna. J. Colloid Interf. Sci. 213: 395–399 (1999).
652. B. Bar-Yosef, U. Kafkafi, R. Rosenberg and G. Sposito. Soil Sci. Soc. Am. J. 52: 1580–1585 (1988).
653. I. Fox and M. A. Malati. J. Chem. Tech. Biotechnol. 57: 97–107 (1993).
654. M. Bjelopavlic, J. Ralston and G. Reynolds. J. Colloid Interf. Sci. 208: 183–190 (1998).
655. E. P. Katsanis and E. Matijevic. Colloid Polym. Sci. 261: 255–264 (1983).
656. W. H. Kuan, S. L. Lo, M. K. Wang and C. F. Lin. Wat. Res. 32: 915–923 (1998).
657. E. Dalas and P. G. Koutsoukos. J. Colloid Interf. Sci. 134: 299–304 (1990).
658. N. J. Barrow. J. Soil Sci. 40: 427–435 (1989).
659. J. M. Zachara, C. C. Ainsworth, C. E. Cowan and C. T. Resch. Soil Sci. Soc. Am. J. 53: 418–428 (1989).
660. N. J. Barrow and A. S. Ellis. J. Soil Sci. 37: 287–293 (1986).
661. N. J. Barrow. J. Soil Sci. 34: 733–750 (1983).

662. N. J. Barrow. J. Soil Sci. 35: 283–297 (1984).
663. N. J. Barrow and B. R. Whelan. J. Soil Sci. 40: 17–28 (1989).
664. N. J. Barrow. J. Soil Sci. 40: 415–425 (1989).
665. J. Pradhan, S. Das and R. S. Thakur. J. Colloid Interf. Sci. 217: 137–141 (1999).
666. J. Pradhan, J. Das, S. Das and R. S. Thakur. J. Colloid Interf. Sci. 204: 169–172 (1998).
667. M. A. Butkus, D. Grasso, C. Schulthess, and H. Wijnja. J. Environ. Qual. 27: 1055–1063 (1998).
668. M. A. Butkus and D. Grasso. Environ. Eng. Sci. 16: 117–129 (1999).
669. L. Balistrieri, P. G. Brewer and J. W. Murray. Deep Sea Res. 28A: 101–121 (1981).
670. G. A. Parks. Rev. Mineral. 23: 133–175 (1990).
671. W. Stumm. Colloids Surf. A 73: 1–18 (1993).
672. W. Stumm, R. Kummert and L. Sigg. Croat. Chem. Acta 53: 291–312 (1980).
673. N. J. Barrow and J. W. Bowden. J. Colloid Interf. Sci. 119: 236–250 (1987).
674. A. Tessier, D. Fortin, N. Belzile, R. R. De Vitre and G. G. Leppard. Geochim. Cosmochim. Acta 60: 387–404 (1996).
675. D. A. Sverjensky. Nature 364: 776–780 (1993).
676. S. Fukuzaki, H. Urano and K. Nagata. J. Fermentation Bioeng. 81: 163–167 (1996).
677. A. T. Stone, A. Torrents, J. Smolen, D. Vasudevan and J. Hadley. Environ. Sci. Technol. 27: 895–909 (1993).
678. I. Sondi, O. Milat and V. Pravdic. J. Colloid Interf. Sci. 189: 66–73 (1997).
679. E. A. Nechaev and L. M. Smirnova. Kolloid. Zh. 39: 186–190 (1977).
680. K. H. S. Kung and M. B. McBride. Environ. Sci. Technol. 25: 702–709 (1991).
681. J. Tomaic and V. Zutic. J. Colloid Int. Sci. 126: 482–492 (1988).
682. C. P. Schulthess and C. P. Huang. Soil Sci. Soc. Am. J. 55: 34–42 (1991).
683. M. A. Schlautman and J. J. Morgan. Geochim. Cosmochim. Acta 58: 4293–4303 (1994).
684. V. Zutic and J. Tomaic. Mar. Chem. 23: 51–67 (1988).
685. D. Vasudevan and A. T. Stone. J. Colloid Interf. Sci. 202: 1–19 (1998).
686. S. B. Haderlein and R. P. Schwarzenbach. Environ. Sci. Technol. 27: 316–326 (1993).
687. R. K. Mishra, G. L. Mundhara and J. S. Tiwari. J. Colloid Interf. Sci. 129: 41–52 (1989).
688. E. A. Ferreiro, S. G. deBussetti and A. K. Helmy. Clays Clay Min. 36: 61–67 (1988).
689. G. M. S. El Shafei and C. A. Philip. J. Colloid Interf. Sci. 185: 140–146 (1997).
690. E. Tombacz, A. Dobos, M. Szekeres, H. D. Narres, E. Klumpp and I. Dekany. Colloid Polym. Sci. 278: 337–345 (2000).
691. M. B. McBride and L. G. Wesselink. Environ. Sci. Technol. 22: 703–708 (1988).
692. E. A. Ferreiro, S. G. deBussetti and A. K. Helmy. Z. Pflanzenernaehr. Bodenk. 146: 369–378 (1983).
693. S. G. deBussetti, E. A. Ferreiro and A. K. Helmy. Clays Clay Min. 28: 149–154 (1980).
694. A. K. Helmy, S. G. deBussetti and E. A. Ferreiro. Clays Clay Min. 31: 29–36 (1983).
695. R. L. Parfitt, A. R. Fraser and V. C. Farmer. J. Soil Sci. 28: 289–296 (1977).
696. E. A. Nechaev. Kolloidn. Zh. 42: 371–373 (1980).
697. A. W. P. Vermeer, W. H. van Riemsdijk and L. K. Koopal. Langmuir 14: 2810–2819 (1998).
698. A. Ben-Taleb, P. Vera, A. V. Delgado and V. Gallardo. Mater. Chem. Phys. 37: 68–75 (1994).
699. E. Tipping. Geochim. Cosmochim. Acta 45: 191–199 (1981).
700. G. McD. Day, B. T. Hart, I. D. McKelvie and R. Beckett. Colloids Surf. A 89: 1–13 (1994).
701. E. A. Nechaev and V. A. Volgina. Elektrokhimiya 15: 1564–1568 (1979).
702. J. M. Kleijn and J. Lyklema. J. Colloid Interf. Sci. 120: 511–522 (1987).

703. J. M. Kleijn and J. Lyklema. Colloid Polym. Sci. 265: 1105–1113 (1987).
704. R. Rodriguez, M. A. Blesa and A. E. Regazzoni. J. Colloid Interf. Sci. 177: 122–131 (1996).
705. A. D. Roddick-Lanzilotta, P. A. Connor and A. J. McQuillan. Langmuir 14: 6479–6484 (1998).
706. D. Vasudevan and A. T. Stone. Environ. Sci. Technol. 30: 1604–1613 (1996).
707. I. Sondi and V. Pravdic. J. Colloid Int. Sci. 181: 463–469 (1996).
708. R. Kretzschmar, D. Hesterberg and H. Sticher. Soil Sci. Soc. Am. J. 61: 101–108 (1997).
709. R. Kretzchmar, H. Holthoff and H. Sticher. J. Colloid Interf. Sci. 202: 95–103 (1998).
710. R. I. Yousef, M. F. Tutunji, G. A. W. Derwish and S. M. Musleh. J. Colloid Interf. Sci. 216: 348–359 (1999).
711. L. K. Koopal and L. Keltjens. Colloids Surf. 17: 371–388 (1986).
712. J. S. Clunie and B. T. Ingram. In: Adsoption from Solution at the Solid/Liquid Interface. C. D. Parfitt and C. H. Rochester, (Eds.), Academic Press, NY, 1983, pp. 105–152; D. B. Hough and H. M. Rendall. In: Adsoption from Solution at the Solid/Liquid Interface. C. D. Parfitt and C. H. Rochester, (Eds.), Academic Press, NY, 1983, pp. 247–318.
713. M. R. Bohmer and L. K. Koopal. Langmuir 8: 2649, 2660 (1992).
714. R. Sharma (Ed.), Surfactant Adsorption and Surface Solubilization. Oxford University Press, New York, 1996.
715. L. K. Koopal. In: Structure-Performance Relationships in Surfactants. K. Esumi and M. Ueno, (Eds.), Marcel Dekker, N. Y., 1997.
716. J. Zajac and S. Partyka. In: Adsorption on New and Modified Inorganic Sorbents. A. Dabrowski and V. A. Tetrykh, (Eds.), Elsevier 1996, pp. 797–828.
717. K. Nagashima and F. D. Blum. J. Colloid Interf. Sci. 214: 8–15 (1999).
718. W. Wang and J. C. T. Kwak. Colloids Surf. A 156: 95–110 (1999).
719. M. Colic, M. L. Fisher and D. W. Fuerstenau. Colloid Polym. Sci. 276: 72–80 (1998).
720. R. Bury, P. Favoriti and C. Treiner. Colloids Surf. A 139: 99–107 (1998).
721. M. Colic and D. W. Fuerstenau. Powder Technol. 97: 129–138 (1998).
722. M. Szekeres, I. Dekany and A. de Keizer. Colloids Surf. A 141: 327–336 (1998).
723. M. G. Bakker, G. L. Turner and C. Treiner. Langmuir 15: 3078–3085 (1999).
724. C. Chorro, M. Chorro, O. Dolladille, S. Partyka and R. Zana. J. Colloid Interf. Sci. 199: 169–176 (1998).
725. M. Chorro, C. Chorro, O. Dolladille, S. Partyka and R. Zana. J. Colloid Interf. Sci. 210: 134–143 (1999).
726. K. Esumi, H. Mizutani, K. Shoji, M. Miyazaki, K. Torigoe, T. Yoshimura, Y. Koide and H. Shosenji. J. Colloid Interf. Sci. 220: 170–173 (1999).
727. D. M. Nevskaia, A. Guerrero-Ruiz and J. D. Lopez-Gonzalez. J. Colloid Int. Sci. 205: 97–105 (1998).
728. C. Strom, B. Jonsson, O. Soderman and P. Hansson. Colloids Surf. A. 159: 109–120 (1999).
729. L. Yezek, R. L. Rowell, L. Holysz and E. Chibowski. J. Colloid Interf. Sci. 225: 227–232 (2000).
730. M. J. Solomon, T. Saeki, M. Wan, P. J. Scales, D. V. Boger and H. Usui. Langmuir 15: 20–26 (1999).
731. P. C. Pavan, E. L. Crepaldi, G. A. Gomez and J. B. Valim. Colloids Surf. A. 154: 399–410 (1999).
732. J. Wang, B. Han, M. Dai, H. Yan, Z. Li and R. K. Thomas. J. Colloid Interf. Sci. 213: 596–601 (1999).
733. A. Tahani, M. Karroua, H. van Damme, P. Levitz and F. Bergaya. J. Colloid Interf. Sci. 216: 242–249 (1999).

734. M. A. Osman and U. W. Suter. J. Colloid Interf. Sci. 214: 400–406 (1999).
735. S. Nishimura, P. J. Scales, S. Biggs and T. W. Healy. Langmuir 16: 690–694 (2000).
736. Z. Xu, Y. Hu and Y. Li. J. Colloid Interf. Sci. 198: 209–215 (1998).
737. L. K. Koopal, E. M. Lee and M. R. Bohmer. J. Colloid Interf. Sci. 170: 85–97 (1995).
738. J. Lyklema. Fundamentals of Interface and Colloid Science, Vol II. Academic Press, New York 1995.
739. K. Esumi. In: Polymer Interfaces and Emulsions, K. Esumi, (Ed.), Marcel Dekker, New York 1999.
740. P. F. Luckham and S. Rossi. Adv. Colloid Interf. Sci. 82: 43–92 (1999).
741. D. Santhiya, S. Subramanian, K. A. Natarajan and S. G. Malghan. J. Colloid Interf. Sci. 216: 143–153 (1999).
742. D. Santhiya, G. Nandini, S. Subramanian, K. A. Natarajan and S. G. Malghan. Colloids Surf. A 133: 157–163 (1998).
743. A. L. Costa, C. Galassi and R. Greenwood. J. Colloid Interf. Sci. 212: 350–356 (1999).
744. D. Bauer, H. Buchhammer, A. Fuchs, W. Jaeger, E. Killmann, K. Lunkwitz, R. Rehmet and S. Schwarz. Colloids Surf. A 156: 291 305 (1999).
745. B. Cabot, A. Deratani and A. Foissy. Colloids Surf. A 139: 287–297 (1998).
746. T. Kato, M. Kawaguchi, A. Takahashi, T. Onabe and H Tanaka. Langmuir 15: 4302–4305 (1999).
747. T. Okubo and M. Suda. J. Colloid Interf. Sci. 213: 565–571 (1999).
748. T. Okubo and M. Suda. Colloid Polym. Sci. 277: 813–817 (1999).
749. R. Greenwood and K. Kendall. Brit. Ceram. T. 97: 174–179 (1998).
750. L. M. Zhang, Y. B. Tan and Z. M. Li. Colloid Polym. Sci. 277: 499–502 (1999).
751. J. Dau and G. Lagaly. Croat. Chem. Acta 71: 983–1004 (1998).
752. J. Wang, L. Gao, J. Sun and Q. Li. J. Colloid Interf. Sci. 213: 552 556 (1999).
753. J. Wang and L. Gao. J. Colloid Int. Sci. 216: 436–439 (1999).
754. R. J. Crawford, I. H. Harding and D. E. Mainwaring. Langmuir 9: 3057–3062 (1993).
755. N. Z. Misak, N. S. Petro, H. F. Ghoneimy and H. N. Salama. React. Polym. 8: 69–77 (1988).
756. U. Palmqvist, E. Ahlberg, L. Lovgren and S. Sjoberg. J. Colloid Interf. Sci. 218: 388–396 (1999).
757. H. T. Tien. J. Phys. Chem. 69: 350–352 (1965).
758. N. Z. Misak and I. M. El-Naggar. React. Polym. 8: 161 171 (1988).
759. B. Venkataramani and K. S. Venkateswarlu. J. Inorg. Nucl. Chem. 42: 909–912 (1980).
760. W. Janusz, I. Kobal, A. Sworska and J. Szczypa. J. Colloid Interf. Sci. 187: 381–387 (1997).
761. N. Z. Misak and H. F. Ghoneimy. Colloids Surf. 7: 89–104 (1983).
762. U. Kafkafi, B. Bar-Yosef, R. Rosenberg and G. Sposito. Soil Sci. Soc. Am. J. 52: 1585–1589 (1988).
763. G. Rytwo, S. Nir and L. Margulies. J. Colloid Interf. Sci. 181: 551–560 (1996).
764. N. J. Barrow. J. Soil Sci. 43: 421–428 (1992).
765. K. Hachiya, Y. Moriyama and K. Takeda. In: Interfacial Dynamics, N. Kallay, (Ed.), Dekker, NY, 1999, pp. 351–403.
766. K. Hachiya, K. Takeda and T. Yasunaga. Ads. Sci. Technol. 4: 25–44 (1987).
767. S. Yiacoumi and C. Tien. J. Colloid Interf. Sci. 175: 333–346, 347–357 (1995).
768. M. C. Amacher, H. M. Selim and I. K. Iskandar. Soil Sci. Soc. Am. J. 52: 398–408 (1988).
769. M. Kosmulski, M. Jaroniec and J. Szczypa. Ads. Sci. Technol. 2: 97–119 (1985).
770. M. Kosmulski. J. Radioanal. Nucl. Chem. 117: 311–319 (1987).
771. S. Yiacoumi and C. Tien. Kinetics of Metal Ion Adsorption from Aqueous Solution: Models, Algorithims, and Applications, Kluwer, Norwell, MA, 1995.

772. N. J. Barrow. J. Soil Sci. 37: 267–276 (1986).
773. K. Hachiya, M. Sasaki, T. Ikeda, N. Mikami and T. Yasunaga. J. Phys. Chem. 88: 27–31 (1984).
774. E. Rakotonarivo, J. Y. Bottero, J. M. Cases and A. Leprince. Colloids Surf. 16: 153–173 (1985).
775. E. Rakotonarivo, J. Y. Bottero, J. M. Cases and F. Fiessinger. Colloids Surf. 9: 273–292 (1984).
776. J. Y. Bottero and J. M. Cases. In: Adsorption on New and Modified Inorganic Sorbents, A. Dabrowski and V. A. Tetrykh, (Eds.), Elsevier, 1996, pp. 319–331.
777. P. Joo and G. Horanyi. J. Colloid Interf. Sci. 223: 308–310 (2000).
778. N. I. Ampelogova. Radiokhimiya 25: 579–584 (1983).
779. R. D. Astumian, M. Sasaki, T. Yasunaga and Z. A. Schelly. J. Phys. Chem. 85: 3832–3835 (1981).
780. R. Aringhieri. Soil Sci. 144: 242–249 (1987).
781. E. M. Mikhail, H. F. Ghoneimy and N. Z. Misak. Colloids Surf. A 92: 209–220 (1994).
782. H. F. Ghoneimy, S. S. Shafik, E. M. Mikhail and N. Z. Misak. Colloids Surf. A 71: 91–97 (1993).
783. M. J. Gray and M. A. Malati. Radiochem. Radioanal. Letters 35: 307–312 (1978).
784. K. Ooi, Y. Miyai, S. Katoh, H. Maeda and M. Abe. Chem. Lett. (Japan) 989–992 (1988).
785. Y. Inoue, H. Yamazaki and M. Ikeda. Bull. Chem. Soc. Jpn. 61: 1147–1151 (1988).
786. N. Z. Misak, E. S. I. Shabana, E. M. Mikhail and H. F. Ghoneimy. React. Polym. 16: 261–269 (1992).
787. N. Z. Misak, N. S. Petro and E. S. I. Shabana. Colloids Surf. 55: 289–296 (1991).
788. M. Ashida, M. Sasaki, H. Kan, T. Yasunaga, K. Hachiya and T. Inoue. J. Colloid Int. Sci. 67: 219–225 (1978).
789. S. Chibowski. Mater. Chem. Phys. 14: 471–479 (1986).
790. M. L. Schlegel, L. Charlet and A. Manceau. J. Colloid Interf. Sci. 220: 392–405 (1999).
791. H. A. Thompson, G. A. Parks and G. E. Brown. J. Colloid Interf. Sci. 222: 241–253 (2000).
792. V. Inglezakis, N. A. Diamandis, M. D. Loizidou and H. P. Grigoropoulou. J. Colloid Interf. Sci. 215: 54–57 (1999).
793. N. J. Barrow. J. Soil Sci. 34: 751–758 (1983).
794. N. J. Barrow. J. Soil Sci. 42: 277–288 (1991).
795. N. J. Barrow and B. R. Whelan. J. Soil Sci. 40: 29–37 (1989).
796. D. L. Sparks. In: Structure and Surface Reactions of Soil Particles, P. M. Huang, N. Senesi and J. Buffle, (Eds.), J. Wiley, New York, 1998, pp. 413–448.

5

Adsorption Modeling

Experimental studies of surface charging of materials under pristine conditions on the one hand and of specific adsorption on the other are reviewed in Chapter 3 and 4, respectively. The adsorption depends on the experimental conditions and in view of continuous character of the physical quantities potentially affecting the magnitude of adsorption it is impossible to cover all combinations of these quantities in direct experiments. A multidimensional grid can be constructed to limit the number of data points, e.g. the experiments are carried out for concentration of the adsorbate every decade, temperature every 5 C, etc., but even then the number of data points necessary to cover the entire range of potential interest would be enormous. Moreover, the usefulness of such data set to calculate the magnitude of adsorption between the grid knots is questionable, namely, there is no unequivocal method of interpolation in such multidimensional space. Therefore, it would be much desired to find the relationship between the magnitude of adsorption and the experimental conditions in form of mathematical equation. Once the general form of such an equation is known, its parameters can be found by mathematical fitting procedures that minimize

$$\Sigma(X_{measured} - X_{calculated})^2$$

where the sum is taken over all experimental data points and X is the quantity of interest (surface charge density, adsorption density, ζ potential). Certainly the reliability of the fitted (adjustable) parameters depends on the number and range of data points taken into account. When the parameters of adsorption equation are known, the adsorption can be calculated for any conditions of interest, including the conditions for which experiments have not been carried out.

Relatively simple equations for adsorption as the function of experimental conditions can be found when the number of independent variables is reduced by keeping some experimental conditions constant, but significance of such equations is also limited, namely, they are valid only for certain value of the experimental quantity that was set constant. A few examples of such approach were presented in Chapter 3 and 4 (e.g. temperature effect on specific adsorption of ions, Eq. (4.3)–(4.8). The mathematical equations used in sorption modeling often have empirical character, i.e. physical ground of certain relationship is not apparent, and they are based on experimentally observed trends. Also the physical sense of the parameters of empirical equations is not known, and their values are fully adjustable. On the other hand, adsorption equations can be derived from certain physical models, i.e. assuming specific adsorption mechanism. Most often the adsorption is assumed to be due to surface reactions. These reactions are governed by Mass Law like the reactions in solution, except in contrast with bulk solution the ions are not evenly distributed near the electrically charged surface. The cations are attracted to negatively charged surfaces and repelled from positively charged surfaces, and anions are attracted to positively charged surfaces and repelled from negatively charged surfaces. Thus, the counterions are in excess over the coions near the surface (unless the surface potential equals zero). The concentration of ions of either sign at some point whose electric potential with respect to bulk solution equals ψ is expressed by Boltzmann equation:

$$n_\psi = n_0 \exp\left(-ze\psi/kT\right) \tag{5.1}$$

where n_0 is the bulk concentration, and z is the number of elementary charges carried by the ion (positive for cations, negative for anions). Boltzmann equation is only valid for dilute solutions, when the ion–ion interactions are negligible (the bulk activity coefficient equals 1).

The distribution of the electric potentials in the interfacial region (that is not accessible to direct measurements) plays a key role in adsorption modeling. The Coulombic attraction can be the main driving force of adsorption in certain systems. Only when the surface potential equals zero, the effect of the electric charge of ions on their adsorption behavior can be neglected. On the other hand, when ψ is constant, the entire exponential term in Eq. (5.1) (which can be considered as a sort of activity coefficient) is also constant, and it can be incorporated into the (conditional) equilibrium constant of respective surface reaction. The ψ in Eq. (5.1) cannot be directly measured or controlled, but for absorbents whose surface charging is pH dependent (Chapter 3), the distribution of electric potential in vicinity of the solid surface is a function of pH and ionic strength and the effect of solutes other than inert electrolyte is rather insignificant when their concentrations are not too high. Thus, in studies of adsorption properties of these materials at constant pH and constant ionic strength, the specifically adsorbing ions can be formally treated as uncharged molecules (the activity coefficient is calculated from Eq. (5.1). In turn dilute aqueous solutions of uncharged molecules behave like ideal gases, and this justifies application of theories of adsorption of gases in studies of adsorption from solution. Naturally these theories are incapable of explaining the pH effect on specific adsorption of ions. It was demonstrated in Chapter 4 that a shift in the pH by a fraction of one pH unit can enhance or depress the specific adsorption by an order of magnitude, thus the models neglecting the pH effects are of limited significance.

Theories of gas adsorption are based on firm experimental grounds. In comparison with adsorption from solution, the experimental results obtained for the gas-solid adsorption are relatively consistent. This is not surprising in view of the number and character of the experimental variables governing the sorption in "dry" and "wet" systems. Adsorption of gases depends on a few variables (partial pressure, temperature) that can be easily controlled and quantified, and it is not complicated by such phenomena as dissolution and recrystallization of the absorbent. These phenomena play a major role in adsorption from solution, they can be controlled (to some degree) by the choice of equilibration time and/or pre-equilibration, but it is very difficult to quantify such effects, especially in terms of formalism borrowed from studies of gas adsorption. The main variables potentially affecting the sorption from solution were discussed in Chapter 4, and their values are reported in Tables 4.1–4.7 (cf. the headings of the columns in these tables). However, this list of experimental variables is apparently not exhaustive, in view of discrepancies in published results. The first difficulty in adsorption modeling is then in choice of the set of experimental data among many (often contradictory) sets available in the literature. Obviously the model derived from one of two contradictory sets of experimental data will not work for the other data set. The multitude of models of adsorption from solution presented in the literature is in part due to use of different (and contradictory) sets of experimental data in modeling exercises.

Many empirical and theoretically derived equations of adsorption isotherms of gases in mono- and multicomponent systems have been published [1,2] and a few of these equations are frequently used in studies of adsorption from solution.

For example, the concept of K_D (Eq. (4.9)) corresponds to Henry adsorption isotherm (adsorption is proportional to the equilibrium concentration/pressure of the adsorbate), which can be derived from the adsorption reaction 4.1, whose "equilibrium constant" defined by Eq. (4.2) depends only on the nature of the adsorbent and the adsorbate, but it is independent of the experimental conditions (over certain limited range). It is well known that in principle K_D is variable, e.g. the effect of the pH on K_D is demonstrated in Figs. 4.28–4.63. These figures show that the pH is an important but not unique factor affecting the distribution of the adsorbate, e.g. the K_D usually decreases when the concentration of the adsorbate increases at constant pH. However, a few cases of constant K_D over a broad range of concentrations of the adsorbate are also reported in Tables 4.1 and 4.2.

I. SURFACE SITES

The Langmuir adsorption isotherm can be derived from the following surface reaction:

$$A + S \Leftrightarrow AS \tag{5.2}$$

where A represents the adsorbate of interest in solution, S is an empty surface site, and AS is the surface species responsible for the adsorption. The activity of the surface species is defined as dimensionless coverage $\theta = [AS]/([AS]+[S])$ for the species AS and as $1-\theta$ for the species S, thus the equilibrium constant of reaction (5.2) equals

$$K_{ads} = \theta(1-\theta)^{-1}a^{-1} \tag{5.3}$$

where a represents the activity of the adsorbate in solution. Solution of Eq. (5.3) with respect to θ gives the Langmuir adsorption isotherm:

$$\theta = K_{ads}a/(1+K_{ads}a) \tag{5.4}$$

The magnitude of adsorption depends on two parameters: the binding constant K_{ads} and the sorption capacity (the sum $[AS]+[S]$, which is constant). These parameters can be calculated directly from experimental data, namely at very low a, Eq. (5.4) reduces to Henry equation $\theta = K_{ads}a$, and $\theta = 1$ at very high a. Thus the level of the plateau in adsorption isotherms represents the sorption capacity, and the initial slope of the adsorption isotherm represents the binding constant. A few examples of Langmuir adsorption isotherms for different values of K_{ads} are presented in Fig. 5.1. For very high values of K_{ads} a "high affinity" adsorption isotherm is obtained (cf. Section 4.V), and the initial slope is not accessible to direct measurement (practically 100% adsorption at low a). With very low K_{ads} the plateau level is not reached over the experimentally accessible a range. In such systems, the adsorption capacity and the binding constant can be determined from the linearized equation of Langmuir adsorption isotherm, namely $a/[AS]$ is plotted as the function of a

$$a/[AS] = a([S]+[AS])^{-1} + K_{ads}^{-1}([S]+[AS])^{-1} \tag{5.5}$$

and the binding constant and sorption capacity can be calculated from the slope and the intercept. Eq. (5.5) can be also used to test the adherence of experimental data to Langmuir equation.

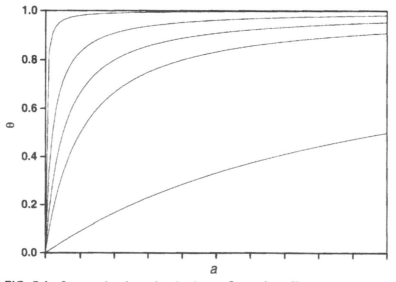

FIG. 5.1   Langmuir adsorption isotherms for various K_{ads}.

Many examples of adherence of experimental data representing adsorption of ions at constant pH to Langmuir equation are reported in Tables 4.1 and 4.2. The apparent adsorption capacity (from the plateau level or from the slope of the linearized Langmuir plot, Eq. (5.5) depends on the pH.

This result is not surprising, namely, the original Langmuir equation was derived for the case of single-gas adsorption, and adsorption from solution occurs in the presence of solvent molecules and of other solutes that are potential competitors for the adsorption sites. A reaction analogous to Eq. (5.2) occurs for each competitor, and the apparent adsorption capacity for the adsorbate of interest is a result of occupation of surface sites by the competitors. The apparent adsorption capacity, i.e. the plateau in the adsorption isotherm (measured in the presence of competitors) does not have specific physical sense. The concentration of surface sites is seldom considered as a fully adjustable parameter, because there are actual atoms or groups of atoms on the surface of the adsorbent responsible for adsorption. The number of these atoms or groups is proportional to the concentration of the adsorbent (solid to liquid ratio).

Not necessarily is the binding constant (cf. Eq. (5.3)) identical for all surface sites. The models taking into account the surface heterogeneity will be discussed later in this Chapter. Even though actual surfaces are heterogeneous, it is often expedient to assume that they behave as if they were homogeneous. The surface sites are located on the surface of the solid, and their number is nearly proportional to the surface area. In principle the number of surface sites per unit of surface area depends on the contribution of particular crystallographic faces to the total surface area, namely, different crystallographic faces have somewhat different surface site densities. However, unless crystals of known morphology are used as the adsorbent, such subtleties are usually neglected and the surface is assumed to be evenly covered with the surface sites. In such approach the number of surface sites participating in surface reaction can be calculated as the product of the exposed surface area by N_s the number of sites per unit of surface area (usually one nm^2). Most often the surface area is obtained by means of the N_2-BET (Brunauer–Emmett–Teller) method (for specific values, cf. tables in Chapter 4), but other methods are also known, and not necessarily they give the same results as the BET method. This is because the probe molecules (N_2 in case of BET) have different shapes and sizes and thus different abilities to enter narrow pores or interlayer spaces. For layer clay minerals the difference between the BET surface area and the surface area determined by glycerol method [3] by a factor of 5 or more is not unusual. For microporous materials a difference between N_2-BET surface area and the value determined from heat of immersion by a factor > 7 was reported [4]. Therefore (at least for layer and microporous materials) the value of N_s does not have absolute character, but it is associated with specific method of surface area measurement. Another fundamental question is about the relationship between the N_s and the nature of the adsorbate. It is convenient to assume that the same surface sites are responsible for adsorption of various adsorbates, and such approach has been tacitly accepted in many modeling exercises, but it is not necessarily correct.

The adsorbate-specific N_s can be found in a "saturation" experiment, i.e. from the course of the adsorption isotherm. In absence of competing species the adsorption plotted as the function of activity of the adsorbate should level out when all surface sites are occupied (adherence to Langmuir equation is not essential). Unfortunately in adsorption of ions, presence of other species (potential competitors) in solution cannot be avoided (pH adjustment, electroneutrality condition), and the adsorption at high concentrations of adsorbate is often complicated by such phenomena as dissolution of the adsorbent, formation of polynuclear species in solution leading to multilayer adsorption, surface precipitation, etc. Moreover, sorption of certain solutes may involve more than one surface site. Therefore "saturation" experiments with specific adsorbates often fail, and the concentration of proton-active sites (the sites responsible for primary surface charging) is commonly accepted in studies with other adsorbates.

Several independent methods to determine the concentration of proton-active sites (surface hydroxyl groups that can dissociate or accept proton) have been proposed. Unfortunately different methods lead to contradictory results. The "saturation" experiments (N_s is calculated from the surface charge density at very high or very low pH values) lead to relatively low N_s compared with other methods. This can be interpreted in terms of small fraction of surface sites engaged in formation of surface charge even at extreme pH values. The other possibility is that charged groups of opposite sign are present on the both sides of the PZC and their charges nearly cancel out, thus the net surface charge is small compared with concentration of surface groups of either sign. On the other hand, the hydroxyl groups detected by other methods are not necessarily all proton-active. The N_s values obtained by means of IR or tritium exchange methods give sample-specific N_s representing the total number of surface hydroxyl groups in specific sample of the adsorbent. Different types of surface hydroxyl groups can be distinguished by means of the IR method. These groups can have different binding constants (surface heterogeneity). It has been shown that the concentration of different types of silanol groups can be modified by hydrothermal treatment of silica at different pH [5]. Probably similar effects can be induced for other materials. Rutile sample outgased at 295 K has 16 exchangeable hydrogen atoms nm^{-2}. This number drops to 4.3 for the same sample outgased at 423 K and to 1.9 for 523 K. Outgasing at temperatures 723–1073 K leaves 0.2 exchangeable H atoms nm^{-2} [6]. In other words, different N_s values can be obtained for the same core material. This would explain the discrepancies in surface charge densities observed at the same experimental conditions (pH, ionic strength) between different samples having the same chemical formula (cf. Figs. 3.34–3.79). Probably the samples having relatively high σ_0 have proportionally high N_s. Unfortunately, the studies simultaneously reporting the surface charge density on the one hand and the N_s determined for the same sample of adsorbent by independent methods on the other are scarce, and this hypothesis can hardly be verified. Also the discrepancies in K_D observed in studies of specific adsorption (Figs. 4.28–4.63) can be partially due to the difference in N_s between different samples.

The opinion that the N_s is sample-specific is not generally accepted. In many publications the N_s is treated as a generic property of different samples having the same chemical composition and crystallographic structure. Attempts to derive the N_s from crystallographic data can serve as an example of such approach. For certain crystallographic structure and face, the surface sites are associated with broken bonds and/or coordinatively unsaturated atoms, and their number (per unit of surface area) can be calculated. Naturally, not necessarily all broken bonds and coordinatively unsaturated atoms produce proton active sites. Koretsky et al. [7] considered five different approaches:

- Each broken bond is one site
- Each coordinatively unsaturated atoms is a site
- Each atom less than 0.14 nm from the surface is a site
- Each oxygen atom less than 0.14 nm from the surface is a site
- Each cation with a charge of $+0.5$ to $+1$ and each anion with a charge of -0.5 to -1 is a site, each cation with a charge of $+1.5$ to $+2$ and each anion with a charge of -1.5 to -2 is counted as two sites.

Koretsky et al. calculated site densities for particular low index faces of six oxides and six silicate minerals and average site densities for cleavage and growth faces, and showed that these methods lead to very different results. Full documentation of depth, length of broken bond and Brown bond strengths of broken bonds for particular types of sites is presented. The ranges of N_s calculated by Koretsky et al. [7] and in their literature data collection (tritium exchange, acid-base titrations, NMR, adsorption and desorption of water at various conditions, and "saturation" experiments with different adsorbates in solution) for six oxides are listed in Table 5.1.

The ranges of N_s in the second and third column of Table 5.1 overlap, but one N_s value for given material can be hardly recommended. The most consistent N_s are reported for SiO_2 (difference between the highest and the lowest value by a factor of 4), and for other materials, the difference is even more significant. The N_s values reported in the last column of Table 5.1 are significantly higher than those proposed by Pivovarov [8] (Table 5.2). Pivovarov's calculations were also based on crystallographic data. Different low-index faces were taken into account, but the difference in N_s between these faces was rather insignificant. Accordance of results presented in Table 5.2 with selected results of "saturation" experiments with different adsorbates has been demonstrated.

An extensive list of N_s for quartz, silica gel, rutile, anatase, SnO_2, CeO_2, α and χ Al_2O_3, boehmite, hematite, amorphous $Fe(OH)_3$, α-$FeOOH$, MgO and ZnO determined by means of isotope exchange, IR spectroscopy, adsorption–desorption of water, acid-base titration (in this case the number of acidic and basic sites can substantially differ) and chemical reactions of surface OH groups with different species after removal of physisorbed water has been collected by James and Parks [9].

TABLE 5.1 Ranges of N_s Values Compiled and Calculated by Koretsky et al. [7]

Material	Literature N_s range nm^{-2}	Calculated N_s range for cleavage faces nm^{-2}
Al_2O_3 corundum	1.7–13	5.1–30.5
Fe_2O_3 hematite	4.3–22	4.5–27.3
FeOOH goethite	1.68–20.1	8.7–17.4
MgO periclase	8–36	0–23.6
SiO_2 quartz	3.5–11.4	7.5–13.8
TiO_2 rutile	2.7–13	11.5–21.7

TABLE 5.2 Site Densities Recommended by Pivovarov [8]

Material	Sites nm^{-2}
Al_2O_3 corundum	2.7
AlOOH boehmite	2.3
AlOOH diaspore	1.9
$Al(OH)_3$ gibbsite	2.3
Fe_2O_3 hematite	2.2
FeOOH lepidocrocite	2.1
FeOOH goethite	1.7
TiO_2 rutile	2.2

Most results (with three exceptions of much lower site densities obtained from acid-base titration) range from 2 to 22.4 sites nm^{-2}. There is no rule that one method gives systematically higher or lower N_s than another.

The oxygen (1 s) photoelectron spectra of oxidized metal surfaces can be deconvoluted into three peaks: oxygen in oxide, hydroxyl, and water, and relative intensities of these peaks make it possible to estimate the concentration of surface hydroxyl groups. For example 2.6 hydroxyl groups per nm^2 were reported for Fe_2O_3 and 4.3 groups per nm^2 for TiO_2 [10]. Ten OH groups nm^{-2} were found [11] in air formed oxide film on titanium by means of XPS.

Among less common methods, the Grignard method was adopted to determine the N_s [12]. The volume of methane produced as the result of the following reaction is measured,

$$\equiv SOH + CH_3MgI == \equiv SOMgI + CH_4 \tag{5.6}$$

where $\equiv SOH$ represents a proton-active site. Methane can be also produced as a result of reaction with water adsorbed on oxides (not necessarily as surface OH groups), so the obtained site densities can be overestimated. As a matter of fact the experimentally obtained values are rather high, cf. Tables 5.1 and 5.2.

The above few examples can be summarized as follows:

- The N_s (proton active sites) for simple oxides and related materials is on the order of a few sites nm^{-2}, but the difference in results obtained by means of different methods for the same material is often in excess of one order of magnitude.
- Different surface treatment can produce different N_s for the same core material.
- There is no dramatic difference in N_s between different materials, especially when these values are obtained using the same method.

The practical solutions of the "N_s problem" in specific publications devoted to adsorption modeling can be divided into three main groups:

- A sample-specific N_s is used, obtained by means of one of the mentioned above methods. The choice of the method is usually limited by the availability of facilities in certain laboratory.
- A literature N_s value is used, arbitrarily chosen among many values reported for the same core material.

TABLE 5.3 Site Densities Determined Using Grignard
Method, Adopted from Ref. [12]

Material	Sites nm^{-2}
Al_2O_3 JRC ALO-4	19.3
Cr_2O_3 Kanto	16.2
Fe_3O_4 Kanto	19.9
Fe_2O_3 Kanto	14.2
MnO_2 IC1	13.5
MnO_2 IC12	14.1
MnO_2 IC22	14.5
MnO_2 λ	8.8
SiO_2 JRC SIO-1	9.6
SiO_2 Kanto	5.7
TiO_2 JRC TIO-5	10.8
TiO_2 Kanto	7

- The N_s is an adjustable parameter, but only a limited range ($\approx$1–30 sites nm^{-2}) is taken into account.

The balance between pros and cons of the above approaches depends on the specific system, and one universal solution of the "N_s problem" does not exist. The accordance between the N_s obtained for different samples and by means of different methods is expected for materials in form of well defined crystals of known morphology, but not necessarily for less well defined materials.

II. SURFACE REACTIONS AND SPECIATION

The adsorption model leading to Langmuir equation is the simplest case of a general model presented in Fig. 4.1, namely, it involves only one solution species and one surface species. Most specifically adsorbing compounds produce multiple species in solution (cf. Figs. 4.2–4.4) and their speciation in solution is well documented. The identification of solution species, and their stability constants have been studied by means of potentiometry (pH electrode and other ion selective electrodes), conductance, and different spectroscopic techniques, and different methods produce rather consistent results. The surface speciation is less well documented. Some spectroscopic evidence for existence of multiple surface species in adsorption of heavy metal cations is presented in Chapter 4, but the formulae of the surface compounds can be hardly inferred directly from these results, thus the formulae of the surface compounds used in adsorption modeling have more or less speculative character. Very often the spectroscopic information is not available at all, and the number and formulae of the surface compounds are based chiefly on curve fitting exercises, but even though the choice of the formulae of surface species is not limited to trial-and-error method. For instance, it is very likely that adsorbed molecules of multivalent acids undergo stepwise deprotonation as the pH increases similarly to their counterparts in solution. Moreover, some degree of correlation between equilibrium constants characterizing deprotonation in solution and on the surface is expected. Analogically, most likely the adsorbed heavy metal cations tend to exchange one or more water ligands into an OH anion at high pH, and the equilibrium constant(s) of surface hydrolysis reaction(s) are correlated with the bulk hydrolysis constant(s). Metal cations that form stable water soluble complexes with certain anions may tend to form stable ternary surface complexes involving these anions, and again the stability constants of surface and bulk complexes are probably correlated. On the other hand formation of stable ternary surface complexes that do not have their solution counterparts is rather unlikely.

Figure 4.1 shows that with n solution species and m surface species involving the adsorbate of interest (the species are distinguished in terms of their stoichiometry; surface heterogeneity is not taken into account) there are $n \times m$ possible reversible adsorption-desorption reactions. Moreover, there are $n \times (n-1)/2$ possible reversible reactions between the solution species, and $m \times (m-1)/2$ possible reversible reactions between the surface species. This makes together $(m+n) \times (m+n-1)/2$ reversible reactions.

For common adsorbates the equilibrium constants of reactions involving only solution species are available from literature; for less common adsorbates they can be determined in separate experiments that do not involve the adsorbent. The equilibrium constants of (hypothetical) surface reactions are the adjustable parameters of the model, and they are determined from the adsorption data by means of appropriate fitting procedure. With simple models (e.g. the model leading to Langmuir equation which has two adjustable parameters) the analytical equations exist for least-square best-fit model parameters as the function of directly measured quantities, but more complicated models require numerical methods to calculate their parameters.

The equilibrium constants of $(m+n) \times (m+n-1)/2$ reversible reactions involving the adsorbate of interest are interrelated and only $m+n-1$ of them are independent. On the other hand the availability of surface sites and solution species involved in these $m+n-1$ independent reactions is affected by other surface and solution reactions (not involving the adsorbate of interest, e.g. adsorption of the competitors), thus the number of independent reactions whose equilibrium constants have to be pre-determined or fitted is usually much greater than $m+n-1$.

The following general formulation is used to describe a chemical equilibrium problem (not necessarily related to adsorption). The system contains a number of chemical species. For each species i the stability constant K_i and the concentration c_i are defined. In order to define K_i a set of components must be selected in such manner that:

- Each component represents an actually occurring species.
- Each species can be written as a product of reaction involving only the components.
- No component can be written as a product of reaction involving only the other components.
- The water molecules play a special role: their activity is equal to 1 by definition, thus, they occur in chemical reactions, but not in Mass Law expressions. Therefore, in terminology of the present section water is not a component.

There is no unique set of components for given problem. For example when the equilibrium between H_3PO_4, $H_2PO_4^-$, HPO_4^{2-}, PO_4^{3-}, and H^+ is considered (cf. Fig. 4.4), any two (exactly two) of these five species can be selected as components.

Mass Law equation is written for each species:

$$\log c_i = \log K_i + \Sigma v_{ij} \log X_j \tag{5.7}$$

where v_{ij} is the stoichiometric coefficient of component j in species i, and X_j is free concentration of component j (concentration of the species representing the component j). For species that are also components $v_{ij} = 1$ for the component corresponding to given species and 0 for all other components. The v_{ij} are not necessarily natural numbers; negative and fractional values are also possible. For example by selecting H_3PO_4, and $PO_4{}^{3-}$ as the components in the H_3PO_4 - $H_2PO_4{}^-$ $HPO_4{}^{2-}$ - $PO_4{}^{3-}$ - H^+ system one has:

$$H^+ = 1/3\, H_3PO_4 - 1/3\, PO_4{}^{3-} \tag{5.8}$$

thus, the stoichiometric coefficient of the component $PO_4{}^{3-}$ in the species H^+ equals -1/3. For each component the following mass balance equation is written

$$Y_j = \Sigma v_{ij}\, c_i - T_j \tag{5.9}$$

where Y_j is the residual in material balance for component j, and T_j is the total concentration of the component j ($\Sigma\, v_{ij}$, c_i, where c_i is the actual concentration). A non-zero value of Y_j result from the difference between actual concentrations of particular species and the concentrations calculated from Eq. (5.7) that are used in Eq. (5.9). The following notation:

N matrix of v_{ij}
C^* vector of $\log c_i$
K^* vector of $\log K_i$
X^* vector of $\log X_j$
Y vector of Y_j
T vector of T_j

allows to write Eqs (5.7) and (5.9) as:

$$C^* = K^* + NX^* \tag{5.10}$$
$$Y = \Sigma v_{ij}\, c_i - T \tag{5.11}$$

An experimental data point is represented by vectors T and X^*. In a typical set of experimental data most T_j and X_j values are common for the entire set, and only a few T_j and X_j values are variable. Not necessarily both T_j and X_j have to be known for each component.

The model of chemical equilibrium is represented by the matrix N and vector K^*. Typical approach to the adsorption modeling can be described as follows. The v_{ij} for solution species are usually known from literature, and the v_{ij} for surface species have to be pre-assumed. The K_i of solution species are usually known from literature, and the K_i for surface species have to be fitted. The goal of the fitting procedure is to minimize Y in Eq. (5.11) for certain experimentally determined T and X*. The method of solution of chemical equilibrium problem was discussed in detail by Herbelin and Westall [13], and many computer programs with user friendly interfaces are commercially available to perform this task. Once the fitting procedure is complete and the vector K^* is known, the K_D of the adsorbate, and its full speciation can be calculated for any experimental conditions (using the same computer programs). Surface species are often hypothetical, and surface speciation has a speculative character, but some models with hypothetical surface species turn out to be very successful in prediction of adsorption under different conditions.

Leckie [14] emphasized the advantage of chemical speciation over "overall" distribution coefficients in adsorption modeling. On the other hand, in many "theoretical" studies of adsorption even the speciation in solution is neglected and only the total concentration of dissolved species is taken into account. One probable reason of paying no attention to well-known experimental facts is that some authors use adsorption equations borrowed from gas adsorption, and obviously these equations are not suitable to deal with multiple solution species involving the adsorbate.

In the framework presented above [Eqs. (5.7) and (5.9)] surface and solution species are not distinguished, and their concentrations must be expressed in the same units (mol dm^{-3}), otherwise the summation in Eq. (5.9) would have no sense. Most surface reactions used in adsorption modeling involve exactly one surface species on each side of the reaction (cf. reaction 5.2) and the choice of units to express the concentration of surface species does not affect the numerical value of the equilibrium constant of the reaction.

It is well known from solution chemistry that the "equilibrium constants" defined in terms of concentrations (Eq. (5.7) have conditional character (they are constant as long as the quotient of activity coefficients of reagents remains constant), and the real equilibrium constants (Eq. (2.23)) are defined in terms of activities. Use of the same variable c_i in Mass Law and mass balance equations is essential in the algorithm solving the problem of chemical equilibrium, but the same algorithm can be applied after replacement of Eq. (5.7) by Mass Law written in terms of activities:

$$\log a_i = \log K_i^{act} + \Sigma v_{ij} \log a_j \tag{5.12}$$

where the superscript in $\log K^{act}_i$ indicates that the equilibrium constant is defined in terms of activities. The following substitution

$$a_i = \gamma_i c_i \quad a_j = \gamma_j X_j \tag{5.13}$$

where γ is the activity coefficient, leads to the equation

$$\log c_i = \log K_i^{act} + \Sigma v_{ij} \log X_j + \Sigma v_{ij} \log \gamma_j - \log \gamma_i \tag{5.14}$$

Equation (5.14) describes the relationship between real (thermodynamic) equilibrium constant and concentrations of the reagents. Equations for activity coefficients of aqueous ionic species as the function of concentrations of all ionic species in solution (at least at ionic strengths up to 0.1 mol dm^{-3} are well known and generally accepted. It should be emphasized that these equations apply only to the solution species. When $\Sigma v_{ij} \log \gamma_j - \log \gamma_i$ in Eq. (5.14) is constant for each i over the entire data set, one can simply use Eqs. (5.7) and (5.9) to calculate K_i, and then calculate K^{act}_i using the following relationship

$$\log K_i^{act} = \log K_i - \Sigma v_{ij} \log \gamma_j + \log \gamma_i \tag{5.15}$$

When $\Sigma v_{ij} \log \gamma_j - \log \gamma_i$ is variable, the following substitution is convenient in Eq. (5.14)

$$\Sigma v_{ij} \log \gamma_j - \log \gamma_i = \log X_{fi} \tag{5.16}$$

where the auxiliary variable X_{fi} can be interpreted as the free concentration of an additional (fictional) component, and it is responsible for the correction for the activity coefficient. The numerical procedure handles the X_{fi} in the same way as X_j of the actual X-type components (components for which only free concentration is known). Procedures allowing for correction for activity coefficients are built into certain existing speciation programs. Typical values of X_{fi} in the systems of interest in adsorption modeling are close to unity (they seldom fall beyond the range 0.5–2), thus, the difference between K_i and K^{act}_i is not dramatic. The difference in calculated K_i induced by change in experimental conditions (e.g. equilibration time) is often much more substantial.

In spite of the outlined above formulation of chemical equilibrium problem in terms of rigorous thermodynamics (equilibrium constants defined as quotients of activities) which is well known and does not pose any special difficulty when it is compared with formulation in terms of conditional equilibrium constants (defined as quotients of concentrations), the former approach is not very popular, and many equilibrium constants of surface reactions reported in published papers were defined in terms of concentrations. Even praise of use of concentrations rather than activities in modeling of adsorption can be found in recent literature. Many publications do not address this question explicit, and then it is difficult to figure out how the equilibrium constants of surface species were defined (K_i or K^{act}_i). Accordingly, the equilibrium constants of surface species reported in tables of Chapter 4 constitute a mixture of constants defined in different ways (K_i or K^{act}_i). The details regarding the definition of equilibrium constants can be found (but not always) in the original publications.

The approach to the chemical equilibrium problem presented in this section is not suitable for the modeling of the solution chemistry or adsorption of polymers (Section 4.V) or surfactants (Section 4.IV). Also protonation and binding of metal ions to humic and fulvic acid in solution and adsorption of these species (cf. Section 4.III) cannot be handled within the present framework. Modeling of the latter systems was discussed, e.g. by van Riemsdijk et al. [15].

III. PRIMARY SURFACE CHARGING

Most adsorption systems of practical importance contain strongly adsorbing species (multivalent cations and anions, surfactants, polymers). Systems without specific adsorption are difficult to realize even under laboratory conditions due to omnipresent strongly adsorbing impurities (cf. Chapter 3). On the other hand, the primary surface charging occurs also in more complex systems and it must be taken into account in modeling of specific adsorption.

Speciation programs fail by attempts of simultaneous fitting of too many adjustable parameters. Therefore, it is expedient to model the primary surface charging first in a possibly simple system (only H^+/OH^- ions and ions of inert electrolyte). The parameters of the model of primary surface charging are then used as fixed values in modeling of specific adsorption. Putting aside the practical aspects, the adsorption modeling may help to explain the mechanism of primary surface charging, which is an important and yet not fully understood fundamental question.

The interpretation of potentiometric titration data in absence of strongly adsorbing species in Chapter 3 already involves some model of adsorption: the protons are chemisorbed at the surface and their charge is balanced by the excess of counterions in solution in the interfacial region (diffuse layer). This model qualitatively explains the coincidence between the IEP and the CIP of potentiometric titration curves, and the ionic strength effects on the values of σ_0 and ζ. The adsorption model leading to specific mathematical relationship between the independent variables (pH, ionic strength) and the measured quantities (σ_0, ζ) requires:

- Surface reaction(s) responsible for proton adsorption.
- Method to calculate the surface potential (Eq. 5.1).

Solutions of these two problems are to some degree independent, and their combination results in different adsorption models. Detailed discussion of all possible combinations is not intended. This chapter is limited to presentation of selected examples, and to discussion, how different (often contradictory) assumptions affect the resultant model curves.

The surface reaction responsible for pH dependent surface charging can be written as

$$\equiv SO_mH_n{}^z + H^+ = \equiv SO_mH_{n+1}{}^{z+1} \tag{5.17}$$

where $\equiv S$ represents surface atoms other than O or H, and z is the number of elementary charges carried by the surface complex on the left hand side of reaction (5.17), and it can be integer or fractional, positive or negative number or zero. The physical interpretation of fractional values of z will be discussed in Sections C and E.

The equilibrium constant of reaction (5.17) can be defined as

$$K = [\equiv SO_mH_{n+1}{}^{z+1}][\equiv SO_mH_n{}^z]^{-1} a_{H+s}{}^{-1} \tag{5.18}$$

where a_{H+s} is the proton activity at the surface. Equation (5.18) is based on the assumption that the activity of the surface species i equals to the surface converge θ_i, defined as [surface species i]/[all surface species]. Combination of Eqs. (5.18) and (5.1) yields:

$$K = [\equiv SO_mH_{n+1}{}^{z+1}][\equiv SO_mH_n{}^z]^{-1} a_H{}^{-1} \exp(e\psi/kT) \tag{5.19}$$

where a_H is the bulk activity of protons. In contrast with a_{H+s}, the a_H can be experimentally determined. However, some model assumptions are necessary to calculate ψ in Eq. (5.19).

A. Is the Surface Potential Nernstian?

One of the simplest possibilities to calculate the surface potential by pH dependent surface charging is offered by the Nernst equation:

$$\psi = \ln 10 \times (pH_0 - pH) kT/e \tag{5.20}$$

(pH_0 is the pristine PZC) whose applicability to pH dependent surface charging of material surfaces has been debated in the literature [16]. Combination of Eqs. (5.19) and (5.20) gives:

$$K = [\equiv SO_mH_{n+1}{}^{z+1}][\equiv SO_mH_n^z]^{-1} 10^{pH_0} \tag{5.21}$$

and this would suggest that the concentration ratio $[\equiv SO_mH_{n+1}{}^{z+1}]/[\equiv SO_mH_n^z]$ is independent of pH, in contradiction with experimental facts, namely the positive surface charge at low pH and negative surface charge at high pH indicates that the ratio $[\equiv SO_mH_{n+1}{}^{z+1}]/[\equiv SO_mH_n{}^z]$ decreases when the pH increases. Therefore many authors argue that Eq. (5.20) is not applicable to pH dependent surface charging, and the actual slope $-d\psi/dpH$ is lower than the "theoretical" $\ln 10 \times kT/e$ (about 59 mV/pH unit at room temperature). One possible explanation of this apparent discrepancy is that the net surface charge density is a sum of two large and almost equal charges of positively charged $\equiv SO_mH_{n_1}{}^{z+1}$ surface species and of negatively charged $\equiv SO_mH_n{}^z$ surface species. Thus the ratio $[\equiv SO_mH_{n+1}{}^{z+1}]/[\equiv SO_mH_n{}^z]$ is nearly constant over the entire pH range, and the slope $-d\psi/dpH$ can be nearly Nernstian. Similar argument is used in explanation of Nernstian behavior of the AgI surface and it can also apply to pH dependent surface charging.

However simultaneous presence of a high concentration of positively and negatively charged surface species at the PZC is not the pre-requisite of applicability of the Nernst equation. The surface coverage does not necessarily reflect the activity of charged surface species. Thus, Equation (5.21) should be rather rewritten as:

$$K = a(\equiv SO_mH_{n+1}{}^{z+1}) a^{-1} (\equiv SO_mH_n{}^z) 10^{pH_0} \qquad (5.22)$$

where the activities on the right hand side refer to the species in brackets. Equation (5.22) allows the concentrations of surface species to vary with the pH when the activity coefficients of positively charged surface species on the one hand and of negatively charged surface species on the other are both pH dependent, and they change in opposite directions when the pH increases. Indeed, the activity coefficients of charged surface species depend on the concentrations of anions and cations of inert electrolyte near the surface. At the pH_0 the concentrations of anions and cations near the surface are equal, but far from the pH_0 the concentrations of counterions can be greater than the concentration of coions by several orders of magnitude (cf. Eq. (5.1)). The activity coefficients of aqueous ionic species at low ionic strengths are calculated using the Debye–Hückel equation. Modified Debye–Hückel equation (with concentration of coions calculated from Eq. (5.1) was used to calculate the activity coefficients of charged surface species [17]. Then, the concentrations of the surface species were calculated by dividing the activities of surface species by the corresponding activity coefficients (Eq. (5.13), and finally the surface charge densities were calculated as a sum of the products of concentrations of the surface species by their charges. The model calculations were in agreement with experimentally observed surface charge densities for TiO_2 (three ionic strenghts, experimental data by Yates) and hematite (four ionic strengths, experimental data by Lyklema et al.) over a broad pH range. The basic model involves only two adjustable parameters: the concentration of negatively charged surface species at the PZC, and the ion size parameter in the Debye–Hückel equation. However, this approach has not been accepted by the mainstream surface chemists, and in models based on reaction (5.17) (or combination of two or more reactions of this type with different values of m and n) the ratio $a(\equiv SO_mH_{n+1}{}^{z+1}) a(\equiv SO_mH_n{}^z)$ is a function of the pH, and the slope $-d\psi\, dpH$ is significantly lower than $\ln 10 \times kT/e$. It should be emphasized, that presence of surface species other than $\equiv SO_mH_n{}^z$ has been often invoked in modeling of primary surface charging. For instance, models with counterion binding are discussed in detail in Sections F and G. In models with counterion binding there is no discrepancy between the Nernst equation and the experimentally observed pH dependent surface charging (cf. Eq. (5.21)). Namely, the positive surface charge is chiefly due to the surface species involving the anion of inert electrolyte, cf. Eq. (5.45). Formation of this species is favored at high positive surface potential, because then the concentration of the anions of inert electrolyte in the interfacial region is high, cf. Eq. (5.1). The negative surface charge is chiefly due to the surface species involving the cation of inert electrolyte, cf. Eq. (5.44), respectively, and the role of the $\equiv SO_m H_n{}^z$ and $\equiv SO_m H_{n+1}{}^{z+1}$ surface species in the net surface charge is rather insignificant. For example Bousse et al. [18] interpreted their ISFET surface potential (pH) curves with nearly Nernstian slope as the proof for counterion binding. The surface potential as the function of the pH and ionic strength in models with counterion binding is discussed in more detail in Section G.

B. Is the Silica Surface Amphoteric?

Most material surfaces discussed in Chapter 3 are amphoteric, i.e. they carry positive net surface charge at low pH or negative net surface charge at high pH. For some other materials only surface charge of one sign was observed over the studied pH range (cf. Table 3.8). Many attempts to find PZC (or IEP, cf. Table 3.9) of silica were unsuccessful. This may be interpreted as a PZC at very low pH, but it is rather difficult to check experimentally if these materials have a PZC at all (for discussion of the experimental difficulties cf. Chapter 3). Recently Koopal [19] postulated existence of two types of pH dependent surface charging. Silica type surfaces do not have PZC, and the σ_0 asymptotically tends to zero at low pH. In contrast gibbsite type surfaces have a PZC and they can carry positive or negative σ_0. Only the latter type of surfaces displays approximately linear and nearly Nernstian dependence of surface potential as a function of pH over a pH range around the PZC (this range is broader at lower ionic strengths). With silica type surfaces the slope $d\psi_0/dpH$ is considerably lower than the Nernstian one.

In view of experimental difficulties in determination of surface charge density at extreme pH values by potentiometric titration, other arguments must be considered to resolve the problem of the possibility to charge the silica surface positively. For example a molecular dynamic study led to conclusion that silica surface is not protonated [20]. On the other hand, according to Seidel et al. [21] protonated monosilicic acid is the most abundant silicon species in solution at pH < 4.5. (Attention: Fig. 3 in Ref. 21 which shows this result has an erroneous legend. Actually the protonated species is represented by the _...._ line.) Assuming that the silica surface behaves similarly as monosilicic acid in solution, this result is in favor of protonation of silica at low pH.

Contradictory results regarding the possibility of protonation of silica are probably due to different affinities of monovalent anions to the silica surface at low pH. Recent electroacoustic study showed that silica carries positive electrokinetic charge in solutions of HCl at pH < 2, but the sign is negative even at very low pH in the presence of nitric of chloric VII acid [22]. This difference suggests some specific interaction between silica and monovalent anions at low pH, thus, the IEP and PZC do not necessarily coincide.

In the surface charge modeling presented in the next sections, silica was selected as an example of silica-type surface. This does not imply that the present author subscribes to the opinion that silica cannot carry positive surface charge. In view of different contradictory clues this remains an open question.

C. 1-pK Model

It was already mentioned in Chapter 3 that the reaction responsible for the surface charging can be defined in such manner that log K = pH$_0$ (Eq. (3.42)). Indeed, when $z = -1/2$ in reaction (5.17), e.g.

$$\equiv AlOH^{-1/2} + H^+ = \equiv AlOH_2^{+1/2} \qquad (5.23)$$

the equilibrium constant is defined as:

$$K = [\equiv AlOH_2^{+1/2}][\equiv AlOH^{-1/2}]^{-1} a_H^{-1} \exp(e\psi/kT) \qquad (5.24)$$

and at pH$_0$ [$\equiv AlOH_2^{+1\,2}$] = [AlOH$^{-1\,2}$] (the net charge equals zero) and $\psi = 0$ by definition, therefore Eq. (5.24) reduces to $K = 1/a_H$, thus log $K = pH_0$.

Equations (5.23) and (5.24) correspond to the 1-pK model originally introduced by Bolt and van Riemsdijk [23], who discussed the edge surface of gibbsite in contact with inert electrolyte solution at PZC. Two singly coordinated surface oxygen atoms share 3 protons at the PZC and this can be interpreted as the presence of equal numbers of surface $\equiv AlOH_2^{+1\,2}$ and $\equiv AlOH^{-1/2}$ groups.

Below or above the PZC the concentration of one kind of groups increases at the expense of the other, and this results in positive or negative net surface charge, respectively. The acronym 1-pK emphasizes that one surface reaction and thus one equilibrium constant describes the surface charging. Historically the 1-pK model follows the model nowadays called 2-pK model, which was introduced by Stumm et al. [24]. In the 2-pK model the surface charging is described by two surface reactions and thus two equilibrium constants are necessary. Although considered obsolete by many authors, the 2-pK model is still quite popular, and it will be discussed in some detail in Section D.

The equilibrium constant defined by Eq. (5.24) can be determined directly from experimental σ_0(pH) curves obtained at different ionic strengths as their CIP, but Eq. (5.24) is not sufficient to calculate the σ_0(pH) curves when only the pH$_0$ is known, namely, some model assumption is necessary to calculate ψ beyond the pH$_0$. Generally the surface potential makes the changes in σ_0 with pH less steep than they would be (assuming a fixed N_s value) without the exponential term in Eq. (5.24). It was already discussed above (Eq. (5.21)) that the Nernstian ψ leads to constant [$\equiv AlOH_2^{+1\,2}$]/[$\equiv AlOH^{-1\,2}$], and consequently constant σ_0 over the entire pH range, and this is in conflict with experimental facts.

Another extreme is to neglect the exponential term in Eq. (5.24) at all. This leads to overestimated effect of the pH on σ_0 (assuming a fixed N_s value). A model neglecting the surface potential is physically unrealistic, but non-electrostatic models of adsorption at solid–aqueous solution interface can be found even in very recent literature. According to the prevailing opinion the actual surface potential is between the above two extremes (Nernst potential and $\psi = 0$). The electrostatic models of oxide – inert electrolyte solution interface were discussed in detail by Westall and Hohl [25]. In this section the most common electrostatic models are combined with the 1-pK model in order to illustrate their ability to simulate the actual surface charging data.

Contradictory surface charging results were reported for common materials in the literature (cf. Figs. 3.43–3.73). In this respect one can hardly assess a model using just one arbitrarily selected set of experimental data. Therefore, different electrostatic models will be used to interpret multiple sets of experimental data.

Aluminum oxide was selected as an example of amphoteric surface. The results collected in Figs. 3.11–3.16 suggest that the crystallographic form of Al_2O_3 or possible hydration of the oxide (cf. Section 2.II, example 3) do not have a dramatic effect on the PZC. The $\sigma_0(pH)$ curves of Al_2O_3 in 0.001, 0.01 and 0.1 mol dm^{-3} inert electrolyte were reported (among others) by Sprycha [26], Cox and Ghosh [27], Mustafa et al. [28], and Csoban and Joo [29], and all these studies resulted in a sharp CIP at pH between 8.1 and 8.4, i.e. near the average and median PZC of Al_2O_3 calculated from data of over hundred publications, cf. Fig. 3.11. In spite of rather consistent PZC (a minor difference may be caused by the temperature effect, cf. Section 3.IV), the absolute values of σ_0 and the shapes of the charging curves in the above four publications are quite different. The absolute values of σ_0 are affected by the nature of inert electrolyte (K versus Na, chloride versus nitrate), but the adherence (or lack of adherence) of the charging curves to certain electrostatic model as the effect of switch from one inert electrolyte to another is rather unlikely. The differences in charging behavior reported in the four publications selected to test different electrostatic models are predominantly due to the differences in structure of the materials and in the experimental conditions (e.g. time of equilibration).

The above four sets of charging curves will be referred to as "Sprycha", "Cox", "Mustafa", and "Csoban", respectively, and they will be used to test different electrostatic models (combined with 1-pK model). In one of the original papers [28] the charging curves were not corrected to get PZC = CIP (cf. Fig. 3.1), but such correction was introduced in the present study.

The 1-pK model can be also applied to model surface charging of the "silica-type" materials (cf. Section B). For these materials the surface charging is due to the following reaction.

$$\equiv SiO^- + H^+ == \equiv SiOH \qquad\qquad (5.25)$$

Consequently, the surface can only carry negative charge. In constrast with reaction (5.23), the equilibrium constant of reaction (5.25) is not equal to the pH_0 (the pH_0 does not exist for silica-type surfaces, by definition), but it is an adjustable parameter. Thus, using analogous electrostatic models one needs one adjustable parameter more with silica-type surfaces than with amphoteric surfaces. The $\sigma_0(pH)$ curves of SiO_2 in 0.001, 0.01 and 0.1 mol dm^{-3} inert electrolyte were reported (among others) by Milonjic [30], Kosmulski [31], Löbbus et al. [32], Szekeres et al. [4], and Sidorova et al. [33]. These five sets of charging curves will be referred to as "Milonjic", "Kosmulski", "Löbbus", "Szekeres", and "Sidorova", respectively. Szekeres et al. report their σ_0 in C/g in view of contradictory specific surface areas of the studied silica obtained by different methods. The specific surface area of 318 m^2/g obtained from the heat of immersion was used here to convert the σ_0 into C/m^2. The BET surface area of 43 m^2/g would produce the σ_0 much higher than results reported by other authors.

In contrast with the Al_2O_3 results the position of the pH-axis ($\sigma_0 = 0$) for silica is less certain, because there is no CIP as the reference point. The other factors responsible for the difference between charging curves of silica obtained by different authors are the same as discussed above for alumina.

Differences in surface charge densities and ζ potentials between different samples having the same formal chemical formula were seldom addressed in modeling exercises. Sonnefeld [34] observed higher negative σ_0 for Aerosil 300 (mean radius of

4 nm) than for Aerosil OX 50 (mean radius of 20 nm), and explained the difference in terms of systematic effect of the particle size on the surface acidity constants.

The experimental and model charging curves for alumina and silica are compared in Figs. 5.2–5.104.

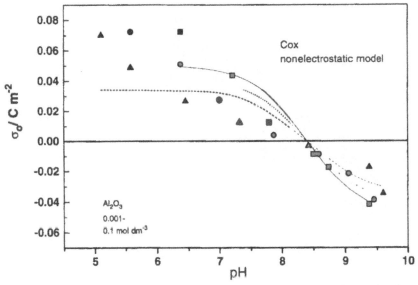

FIG. 5.2 Experimental (symbols) and calculated (lines) surface charge density of alumina as the function of the pH and ionic strength: 0.1 mol dm^{-3} squares, solid; 0.01 mol dm^{-3} circles, dash 0.001 mol dm^{-3} triangles, dot. For model parameters, cf. Table 5.4.

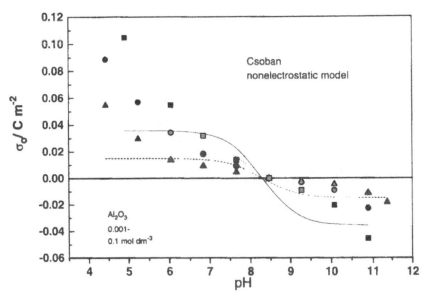

FIG. 5.3 Experimental (symbols) and calculated (lines) surface charge density of alumina as the function of the pH and ionic strength: 0.1 mol dm^{-3} squares, solid; 0.01 mol dm^{-3} circles, dash; 0.001 mol dm^{-3} triangles, dot. For model parameters, cf. Table 5.4.

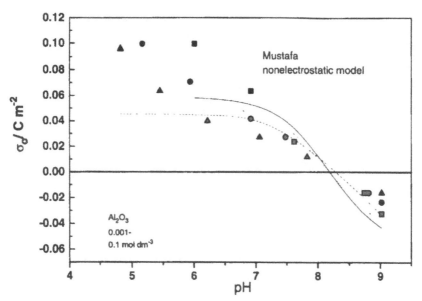

FIG. 5.4 Experimental (symbols) and calculated (lines) surface charge density of alumina as the function of the pH and ionic strength: 0.1 mol dm^{-3} squares, solid; 0.01 mol dm^{-3} circles, dash; 0.001 mol dm^{-3} triangles, dot. For model parameters, cf. Table 5.4.

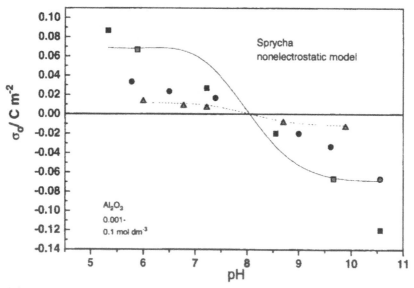

FIG. 5.5 Experimental (symbols) and calculated (lines) surface charge density of alumina as the function of the pH and ionic strength: 0.1 mol dm^{-3} squares, solid; 0.01 mol dm^{-3} circles, dash; 0.001 mol dm^{-3} triangles, dot. For model parameters, cf. Table 5.4.

1.   Non-Electrostatic Model

Let us first inspect the ability of non-electrostatic model ($\psi = 0$) to simulate the course of the surface charging curves. This model requires one adjustable parameter

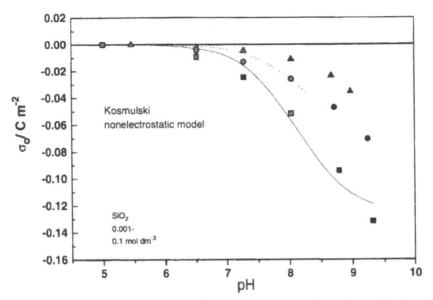

FIG. 5.6 Experimental (symbols) and calculated (lines) surface charge density of silica as the function of the pH and ionic strength: 0.1 mol dm^{-3} squares, solid; 0.01 mol dm^{-3} circles, dash; 0.001 mol dm^{-3} triangles, dot. For model parameters, cf. Table 5.5.

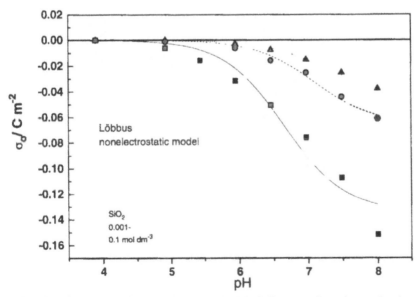

FIG. 5.7 Experimental (symbols) and calculated (lines) surface charge density of silica as the function of the pH and ionic strength: 0.1 mol dm^{-3} squares, solid; 0.01 mol dm^{-3} circles, dash; 0.001 mol dm^{-3} triangles, dot. For model parameters, cf. Table 5.5.

for amphoteric surfaces (total number of surface sites) and two adjustable parameters for silica-type surfaces. Equations (5.23) or (5.25) do not explicitly

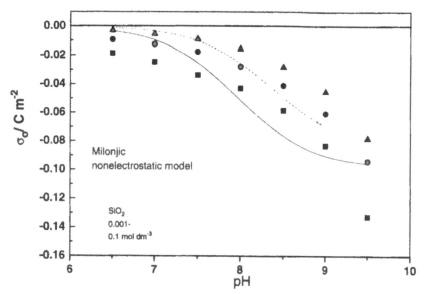

FIG. 5.8 Experimental (symbols) and calculated (lines) surface charge density of silica as the function of the pH and ionic strength: 0.1 mol dm^{-3} squares, solid; 0.01 mol dm^{-3} circles, dash; 0.001 mol dm^{-3} triangles, dot. For model parameters, cf. Table 5.5.

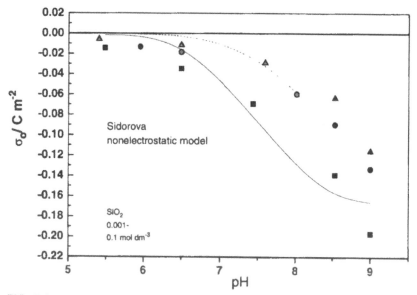

FIG. 5.9 Experimental (symbols) and calculated (lines) surface charge density of silica as the function of the pH and ionic strength: 0.1 mol dm^{-3} squares, solid; 0.01 mol dm^{-3} circles, dash; 0.001 mol dm^{-3} triangles, dot. For model parameters, cf. Table 5.5.

anticipate any ionic strength effect on the σ_0(pH) curves. Actually the surface charge at given pH (except at PZC) depends on the inert electrolyte concentration (with very

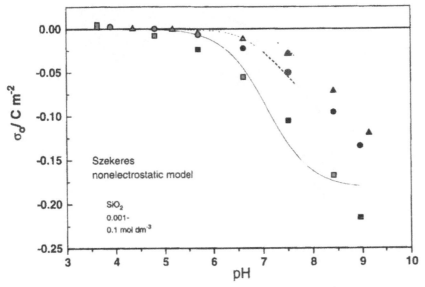

FIG. 5.10 Experimental (symbols) and calculated (lines) surface charge density of silica as the function of the pH and ionic strength: 0.1 mol dm^{-3} squares, solid; 0.01 mol dm^{-3} circles, dash; 0.001 mol dm^{-3} triangles, dot. For model parameters, cf. Table 5.5.

TABLE 5.4 The Best-fit N_s in C/m^2 Used in Figs. 5.2–5.5

Ionic strength/mol dm^{-3} Fig.	10^{-1}	10^{-2}	10^{-3}
5.2	0.1	0.083	0.068
5.3	0.071	0.053	0.03
5.4	0.12	0.1	0.091
5.5	0.14	0.068	0.023

TABLE 5.5 The Best-fit log K (reaction 5.25) and N_s in C/m^2 Used in Figs. 5.6–5.10

Ionic strength/mol dm^{-3} Fig.	10^{-1}		10^{-2}		10^{-3}	
	log K	N_s	log K	N_s	log K	N_s
5.6	8.11	0.13	8.26	0.072	8.57	0.047
5.7	6.66	0.13	7.11	0.066	7.24	0.043
5.8	7.96	0.1	8.36	0.086	8.81	0.088
5.9	7.47	0.17	8.17	0.14	8.27	0.12
5.10	7.08	0.18	7.61	0.12	8.1	0.12

few exceptions, cf. Chapter 3). Thus, the adjustable parameters assume different values at different ionic strengths. The best-fit model curves are presented in Figs. 5.2–5.5 for alumina and in Figs. 5.6–5.10 for silica, and the best-fit parameters are summarized in Tables 5.4 and 5.5. The total concentration of proton active sites in these and next tables is expressed in C/m^2 to facilitate comparison with the surface charge densities plotted in the figures. One C/m^2 corresponds to about six sites nm^{-2}.

Figures 5.2–5.5 show that the non-electrostatic model completely fails for alumina (one example of relatively good agreement between the calculated and experimental charging curve in Fig. 5.5 is probably a fortuitous coincidence). On the other hand the sigmoidal model curves roughly reflect the charging behavior of silica, especially at low ionic strengths. For silica, the number of adjustable parameters in the model can be reduced to one by fixing the K or N_s. Figures 5.11–5.15 show the model curves calculated for log K (reaction 5.25) = 8 (fixed value). The best-fit N_s values are summarized in Table 5.6.

TABLE 5.6 The Best-fit N_s in C/m^2 Used in Figs. 5.11–5.15

Ionic strength/mol dm^{-3} Fig.	10^{-1}	10^{-2}	10^{-3}
5.11	0.12	0.063	0.031
5.12	0.4	0.15	0.087
5.13	0.1	0.07	0.051
5.14	0.21	0.13	0.1
5.15	0.26	0.15	0.11

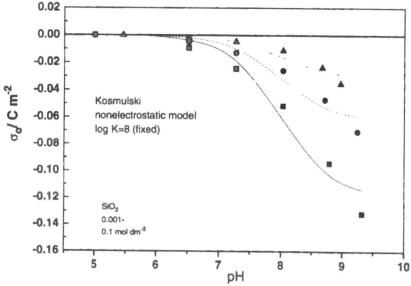

FIG. 5.11 Experimental (symbols) and calculated (lines) surface charge density of silica as the function of the pH and ionic strength: 0.1 mol dm^{-3} squares, solid; 0.01 mol dm^{-3} circles, dash; 0.001 mol dm^{-3} triangles, dot. For model parameters, cf. Table 5.6.

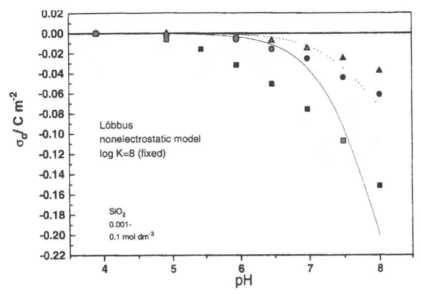

FIG. 5.12 Experimental (symbols) and calculated (lines) surface charge density of silica as the function of the pH and ionic strength: 0.1 mol dm^{-3} squares, solid; 0.01 mol dm^{-3} circles, dash; 0.001 mol dm^{-3} triangles, dot. For model parameters, cf. Table 5.6.

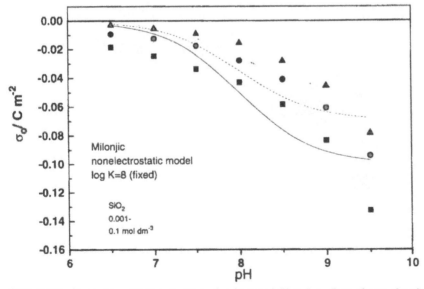

FIG. 5.13 Experimental (symbols) and calculated (lines) surface charge density of silica as the function of the pH and ionic strength: 0.1 mol dm^{-3} squares, solid; 0.01 mol dm^{-3} circles, dash; 0.001 mol dm^{-3} triangles, dot. For model parameters, cf. Table 5.6.

Figures 5.16–5.20 show the model curves calculated for $N_s = 0.2$ C/m^2 (about 1.2 sites nm^{-2}, fixed value). The best-fit K values are summarized in Table 5.7.

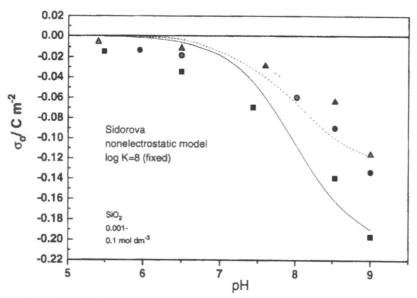

FIG. 5.14 Experimental (symbols) and calculated (lines) surface charge density of silica as the function of the pH and ionic strength: 0.1 mol dm^{-3} squares, solid; 0.01 mol dm^{-3} circles, dash; 0.001 mol dm^{-3} triangles, dot. For model parameters, cf. Table 5.6.

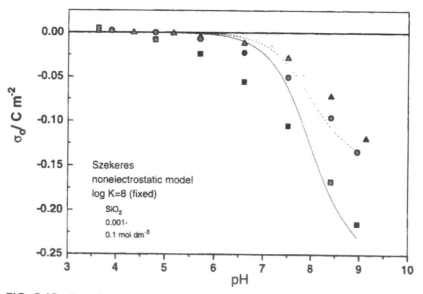

FIG. 5.15 Experimental (symbols) and calculated (lines) surface charge density of silica as the function of the pH and ionic strength: 0.1 mol dm^{-3} squares, solid; 0.01 mol dm^{-3} circles, dash; 0.001 mol dm^{-3} triangles, dot. For model parameters, cf. Table 5.6.

The fixed values of K (Figs. 5.11–5.15) and N_s (Figs. 5.16–5.20) were arbitrarily selected to illustrate the general trend, namely, with only one adjustable parameter the agreement between the calculated and experimental charging curves becomes significantly worse than with two adjustable parameters.

TABLE 5.7 The Best-fit log K (reaction 5.25) Used in Figs 5.16–5.20

Ionic strength/mol dm^{-3} Fig.	10^{-1}	10^{-2}	10^{-3}
5.11	8.68	9.31	10
5.12	7.11	8.14	8.49
5.13	8.89	9.28	9.52
5.14	7.63	8.5	8.77
5.15	7.17	8.28	8.66

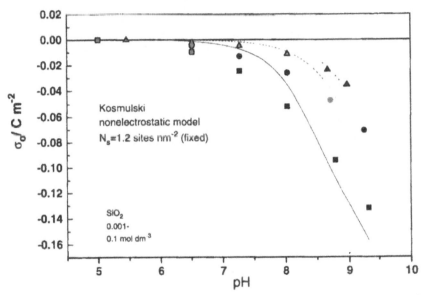

FIG. 5.16 Experimental (symbols) and calculated (lines) surface charge density of silica as the function of the pH and ionic strength: 0.1 mol dm^{-3} squares, solid; 0.01 mol dm^{-3} circles, dash; 0.001 mol dm^{-3} triangles, dot. For model parameters, cf. Table 5.7.

The failure of non-electrostatic model is not surprising in view of its physical incorrectness. The model curves in Figs. 5.2–5.20 are presented as a reference for the model curves calculated using different electrostatic models, and the non-electrostatic model itself is not recommended. A few examples of agreement between the model curves and experimental results can serve as an example that the experiment can be successfully simulated by a physically incorrect model.

2. Constant Capacitance Model

The constant capacitance model was originally introduced by Helmholtz for the surface charging of mercury. In this section the model is used in combination with the 1-pK model (reaction (5.23) or (5.25)). In the constant capacitance model the surface potential (used in expressions for the equilibrium constants, cf. Eq. (5.24)) is proportional to the surface charge density

$$\sigma_0 = C\psi \tag{5.26}$$

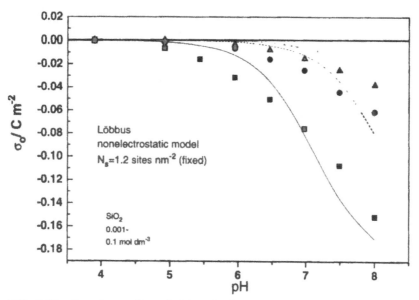

FIG. 5.17 Experimental (symbols) and calculated (lines) surface charge density of silica as the function of the pH and ionic strength: 0.1 mol dm⁻³ squares, solid; 0.01 mol dm⁻³ circles, dash; 0.001 mol dm⁻³ triangles, dot. For model parameters, cf. Table 5.7.

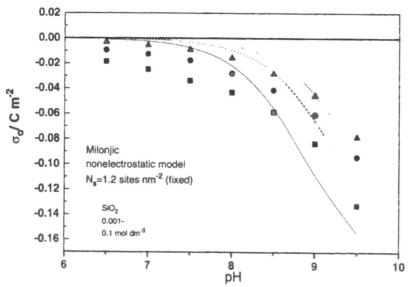

FIG. 5.18 Experimental (symbols) and calculated (lines) surface charge density of silica as the function of the pH and ionic strength: 0.1 mol dm⁻³ squares, solid; 0.01 mol dm⁻³ circles, dash; 0.001 mol dm⁻³ triangles, dot. For model parameters, cf. Table 5.7.

where C (in F/m^2) is an adjustable parameter, but it also has certain physical sense. The interfacial region can be considered as a condenser (referred to as electric double

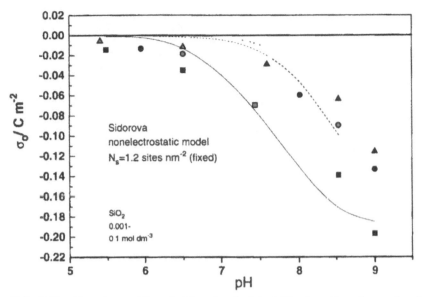

FIG. 5.19 Experimental (symbols) and calculated (lines) surface charge density of silica as the function of the pII and ionic strength: 0.1 mol dm^{-3} squares, solid; 0.01 mol dm^{-3} circles, dash; 0.001 mol dm^{-3} triangles, dot. For model parameters, cf. Table 5.7.

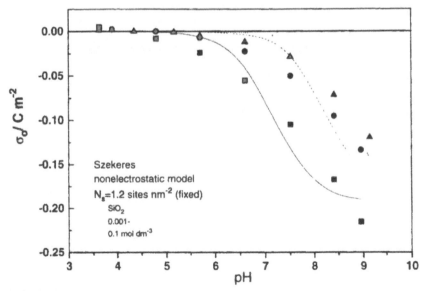

FIG. 5.20 Experimental (symbols) and calculated (lines) surface charge density of silica as the function of the pH and ionic strength: 0.1 mol dm^{-3} squares, solid; 0.01 mol dm^{-3} circles, dash; 0.001 mol dm^{-3} triangles, dot. For model parameters, cf. Table 5.7.

layer) with potential determining ions (H$^+$/OH$^-$ in case of pH dependent surface charging) on the surface side and equivalent number of counterions on the solution side. When the radius of surface curvature is much greater than the thickness of the interfacial region, C can be estimated as

$$C = \epsilon/d \qquad (5.27)$$

where ϵ is the permittivity and d is the thickness of the electric double layer. For the permittivity equal to that of bulk water and $d = 0.3$ nm (typical size of hydrated ion), Eq. (5.27) yields $C = 2.3$ F/m^2. The permittivity in the interfacial region is not necessarily equal to the bulk one, and different values of relative permittivity in the interfacial region down to 6 (the relative permittivity of bulk water is ≈ 80) can be found in the literature. This is explained in terms of orientation of water dipoles in the electrostatic field of the surface.

The 1-pK model combined with the constant capacitance model requires two adjustable parameters, namely N_s and C for amphoteric surfaces. However, the goodness of fit does not get substantially worse when a fixed value of N_s (within certain limits) is assumed, and only C is adjusted. Thus, one arbitrarily chosen N_s value of 0.3 C/m^2 (about 1.8 sites nm^{-2}) was used in the model curves shown in Figs. 5.21–5.24, and the best-fit C values are summarized in Table 5.8. One charging curve (Fig. 5.24, ionic strength of 10^{-3} mol dm^{-3}) requires a N_s lower by an order of magnitude.

Table 5.8 indicates that the best-fit C increases with the ionic strength. Assuming constant ϵ, this can be interpreted as decrease in the apparent electric double layer thickness when the ionic strength increases (Eq. (5.27)). The best-fit C values are all in relatively narrow range, and comparison with the above value of 2.3 F/m^2 suggests that ϵ in the interfacial region is lower than 80 or the double layer is thicker than 0.3 nm. The model curves in Figs. 5.21–5.24 are considerably more realistic than in Figs. 5.2–5.5 (non-electrostatic model). Constant capacitance model offers a possibility to calculate the charging curves at ionic strengths, for which experiments were not carried out, namely C, can be estimated by interpolation. The results presented in Table 5.8 do not indicate any apparent functional relationship between the best-fit C and the ionic strength.

Lutzenkirchen [35] found a linear correlation between log I and C in the constant capacitance model combined with the 2-pK model (cf. Section D), but in his model not only C but also the surface acidity constants were allowed to vary with I. The fitted capacitances were too high to have physical sense, so the constant capacitance model was categorized as data fitting model whose parameters have no physical meaning.

In the 1-pK model combined with the constant capacitance model for silica-type surfaces there are three adjustable parameters, namely K, N_s and C. However

TABLE 5.8 The Best-fit C in F/m^2 Used in Figs. 5.21–5.24

Ionic strength/mol dm^{-3} Fig.	10^{-1}	10^{-2}	10^{-3}
5.21	0.79	0.505	0.348
5.22	0.405	0.262	0.132
5.23	1.06	0.692	0.49
5.24	0.805	0.372	—

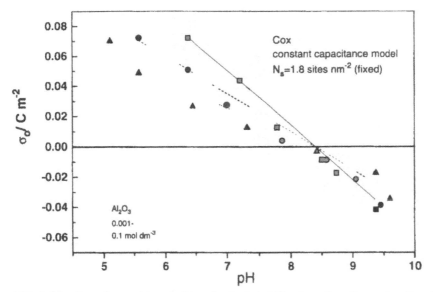

FIG. 5.21 Experimental (symbols) and calculated (lines) surface charge density of alumina as the function of the pH and ionic strength: 0.1 mol dm^{-3} squares, solid; 0.01 mol dm^{-3} circles, dash; 0.001 mol dm^{-3} triangles, dot. For model parameters, cf. Table 5.8.

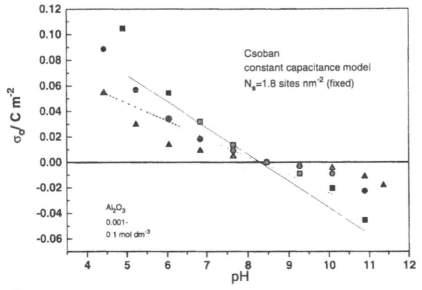

FIG. 5.22 Experimental (symbols) and calculated (lines) surface charge density of alumina as the function of the pH and ionic strength: 0.1 mol dm^{-3} squares, solid; 0.01 mol dm^{-3} circles, dash; 0.001 mol dm^{-3} triangles, dot. For model parameters, cf. Table 5.8.

either K or N_s can be fixed without substantially worsening the goodness of the fit. To avoid too many adjustable parameters models with two parameters were used to simulate the experimental charging curves.

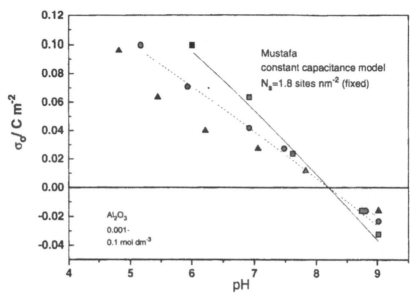

FIG. 5.23 Experimental (symbols) and calculated (lines) surface charge density of alumina as the function of the pH and ionic strength: 0.1 mol dm^{-3} squares, solid; 0.01 mol dm^{-3} circles, dash; 0.001 mol dm^{-3} triangles, dot. For model parameters, cf. Table 5.8.

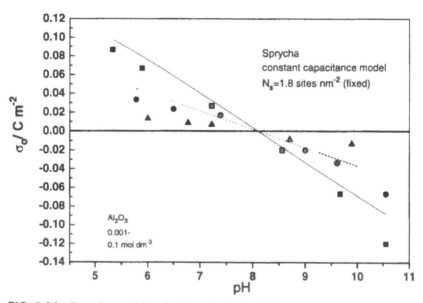

FIG. 5.24 Experimental (symbols) and calculated (lines) surface charge density of alumina as the function of the pH and ionic strength: 0.1 mol dm^{-3} squares, solid; 0.01 mol dm^{-3} circles, dash; 0.001 mol dm^{-3} triangles, dot. For model parameters, cf. Table 5.8.

Figures 5.25–5.29 show the model curves calculated for log K (reaction (5.25)) = 8 (fixed value). The best-fit C and N_s values are summarized in Table 5.9.

Figures 5.30–5.34 show the model curves calculated for $N_s = 1$ C/m^2 (about 6 sites nm^2, fixed value). The best-fit C and K values are summarized in Table 5.10.

TABLE 5.9 The Best-fit C in F/m^2 and N_s in C/m^2 Used in Figs. 5.25–5.29

Ionic strength/mol dm^{-3}	10^{-1}		10^{-2}		10^{-3}	
Fig.	C	N_s	C	N_s	C	N_s
5.25	1.6	0.26	1	0.11	0.75	0.046
5.26	1.42	7.6	0.9	0.95	0.6	0.46
5.27	0.9	0.5	0.95	0.15	0.9	0.085
5.28	1.4	2.7	1	4.2	1.7	0.2
5.29	1.4	7.3	1.2	1	1.7	0.23

TABLE 5.10 The Best-fit C in F/m^2 and log K (reaction 5.25) Used in Figs. 5.30–5.34

Ionic strength/mol dm^{-3}	10^{-1}		10^{-2}		10^{-3}	
Fig.	C	log K	C	log K	C	log K
5.30	1.404	8.59	0.834	8.99	0.814	9.71
5.31	1.48	7.11	0.885	8	0.622	8.37
5.32	0.871	8.27	0.898	8.92	1.262	9.59
5.33	1.484	7.59	—	—	1.607	8.82
5.34	1.463	7.12	1.225	8	1.43	8.63

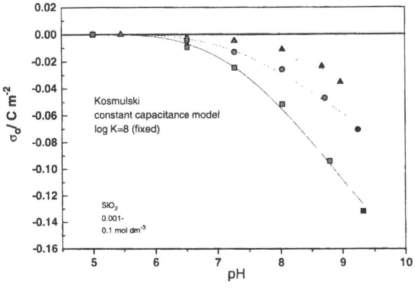

FIG. 5.25 Experimental (symbols) and calculated (lines) surface charge density of silica as the function of the pH and ionic strength: 0.1 mol dm^{-3} squares, solid; 0.01 mol dm^{-3} circles, dash; 0.001 mol dm^{-3} triangles, dot. For model parameters, cf. Table 5.9.

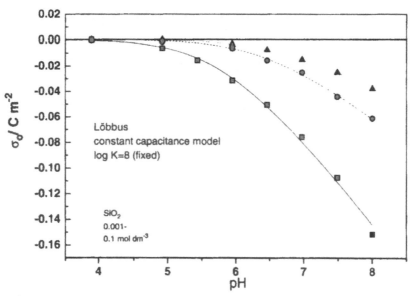

FIG. 5.26 Experimental (symbols) and calculated (lines) surface charge density of silica as the function of the pH and ionic strength: 0.1 mol dm^{-3} squares, solid; 0.01 mol dm^{-3} circles, dash; 0.001 mol dm^{-3} triangles, dot. For model parameters, cf. Table 5.9.

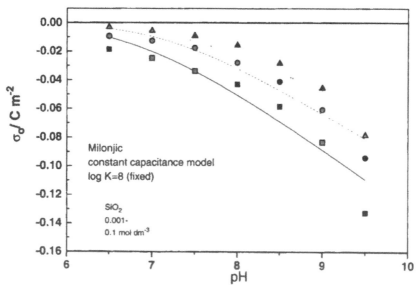

FIG. 5.27 Experimental (symbols) and calculated (lines) surface charge density of silica as the function of the pH and ionic strength: 0.1 mol dm^{-3} squares, solid; 0.01 mol dm^{-3} circles, dash; 0.001 mol dm^{-3} triangles, dot. For model parameters, cf. Table 5.9.

One charging curve (Fig. 5.33, ionic strength of 10^{-2} mol dm^{-3}) requires a lower N_s.

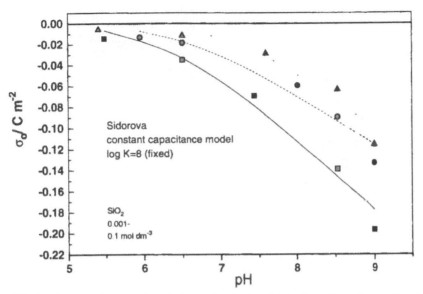

FIG. 5.28 Experimental (symbols) and calculated (lines) surface charge density of silica as the function of the pH and ionic strength: 0.1 mol dm^{-3} squares, solid; 0.01 mol dm^{-3} circles, dash; 0.001 mol dm^{-3} triangles, dot. For model parameters, cf. Table 5.9.

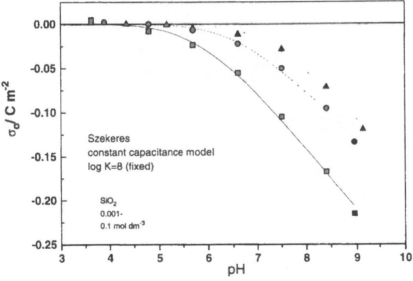

FIG. 5.29 Experimental (symbols) and calculated (lines) surface charge density of silica as the function of the pH and ionic strength: 0.1 mol dm^{-3} squares, solid; 0.01 mol dm^{-3} circles, dash; 0.001 mol dm^{-3} triangles, dot. For model parameters, cf. Table 5.9.

All calculated C values for silica fall in relatively narrow and physically acceptable range, but the relationship between C and the ionic strength is not so apparent as it was for alumina (cf. Table 5.8). In spite of spectacular accordance

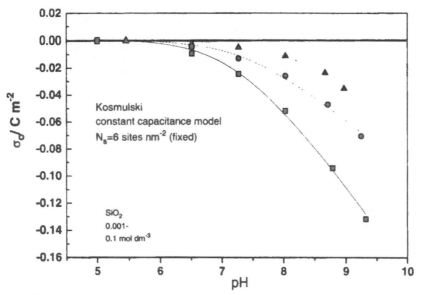

FIG. 5.30 Experimental (symbols) and calculated (lines) surface charge density of silica as the function of the pH and ionic strength: 0.1 mol dm^{-3} squares, solid; 0.01 mol dm^{-3} circles, dash; 0.001 mol dm^{-3} triangles, dot. For model parameters, cf. Table 5.10.

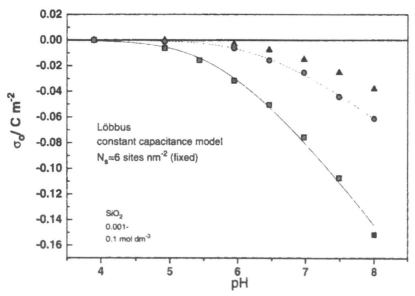

FIG. 5.31 Experimental (symbols) and calculated (lines) surface charge density of silica as the function of the pH and ionic strength: 0.1 mol dm^{-3} squares, solid; 0.01 mol dm^{-3} circles, dash; 0.001 mol dm^{-3} triangles, dot. For model parameters, cf. Table 5.10.

between the experimental and calculated curves for some sets of data (cf. Figs. 5.25, 5.26, 5.29–5.31 and 5.34) the physical meaning of the calculated best-fit parameters is

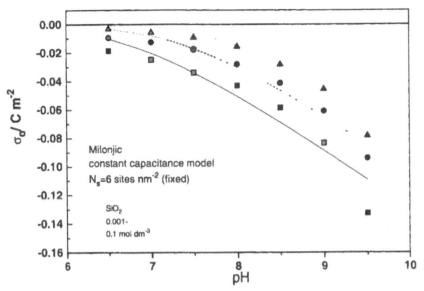

FIG. 5.32 Experimental (symbols) and calculated (lines) surface charge density of silica as the function of the pH and ionic strength: 0.1 mol dm^{-3} squares, solid; 0.01 mol dm^{-3} circles, dash; 0.001 mol dm^{-3} triangles, dot. For model parameters, cf. Table 5.10.

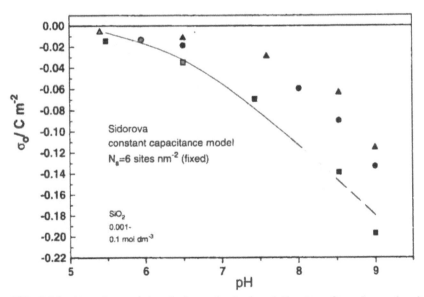

FIG. 5.33 Experimental (symbols) and calculated (lines) surface charge density of silica as the function of the pH and ionic strength: 0.1 mol dm^{-3} squares, solid; 0.01 mol dm^{-3} circles; 0.001 mol dm^{-3} triangles, dot. For model parameters, cf. Table 5.10.

problematic in view of their irregular changes as the function of the ionic strength (Table 5.9 and 5.10).

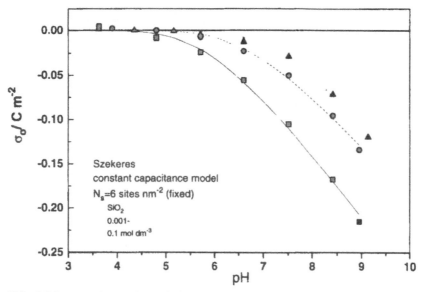

FIG. 5.34 Experimental (symbols) and calculated (lines) surface charge density of silica as the function of the pH and ionic strength: 0.1 mol dm^{-3} squares, solid; 0.01 mol dm^{-3} circles, dash; 0.001 mol dm^{-3} triangles, dot. For model parameters, cf. Table 5.10.

TABLE 5.11 The Best-fit C^* in F/m^2 used in Figs 5.35–5.38

Ionic strength/mol dm^{-3}	10^{-1}		10^{-2}		10^{-3}	
Fig.	C	C.	C	C.	C	C.
5.35	0.74	1.05	0.47	0.78	0.333	0.47
5.36	0.555	0.265	0.355	0.122	0.181	0.0795
5.37	1.16	0.78	0.705	0.56	0.495	0.377
5.38	0.655	0.802	0.28	0.5	—	—

*The subscript denotes the sign of the σ_0

Since the C value is related to the size of hydrated counterions its value may be different at the both sides of the PZC. Thus with amphoteric surfaces the goodness of fit within the constant capacitance model can be improved by allowing different C values for positive and negative branch of the charging curves. This possibility was explored in Figs. 5.35–5.38. The N_s was assumed to be equal to 0.3 C/m^2 (1.8 sites/nm^2, fixed value). The best-fit C_+ and C_- (the subscript denotes the sign of the σ_0) are summarized in Table 5.11.

The best-fit capacitances on each side of the PZC show the same trend as the capacitances calculated for the entire data sets (cf. Table 5.8), i.e. C increases with the ionic strength. Table 5.11 does not indicate any systematic relationship between C_+ and C_-, namely, in two sets C_+ is greater and in two other sets the C_- is greater. The additional adjustable parameter somewhat improves the fit (compare Figs. 5.35–5.38 on the one hand and Figs. 5.21–5.24 on the other).

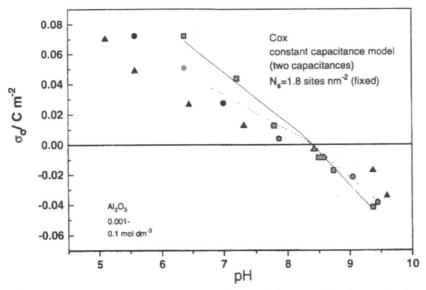

FIG. 5.35 Experimental (symbols) and calculated (lines) surface charge density of alumina as the function of the pH and ionic strength: 0.1 mol dm^{-3} squares, solid; 0.01 mol dm^{-3} circles, dash; 0.001 mol dm^{-3} triangles, dot. For model parameters, cf. Table 5.11.

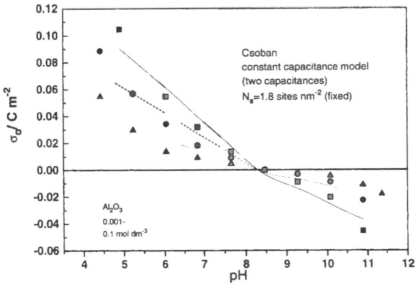

FIG. 5.36 Experimental (symbols) and calculated (lines) surface charge density of alumina as the function of the pH and ionic strength: 0.1 mol dm^{-3} squares, solid; 0.01 mol dm^{-3} circles, dash; 0.001 mol dm^{-3} triangles, dot. For model parameters, cf. Table 5.11.

3. Diffuse Layer Model

The diffuse layer model was originally introduced for surface charging of mercury by Gouy and Chapman in the early 1900s. Combination of

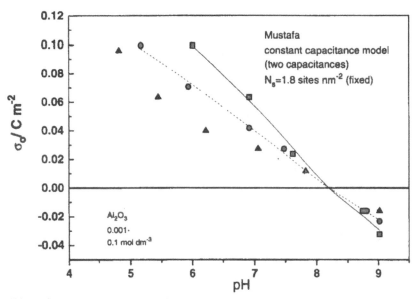

FIG. 5.37 Experimental (symbols) and calculated (lines) surface charge density of alumina as the function of the pH and ionic strength: 0.1 mol dm^{-3} squares, solid; 0.01 mol dm^{-3} circles, dash; 0.001 mol dm^{-3} triangles, dot. For model parameters, cf. Table 5.11.

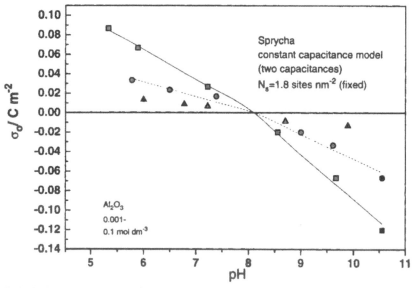

FIG. 5.38 Experimental (symbols) and calculated (lines) surface charge density of alumina as the function of the pH and ionic strength: 0.1 mol dm^{-3} squares, solid; 0.01 mol dm^{-3} circles, dash; 0.001 mol dm^{-3} triangles. For model parameters, cf. Table 5.11.

Poisson equation and Boltzmann equation (5.1) leads to the following expression:

$$\sigma_d = \varepsilon \kappa kT/e[\exp(e\psi_d/2kT) - \exp(-e\psi_d/2kT)] \qquad (5.28)$$

where

$$\kappa = (1000N_A e^2 \Sigma c_i z_i^2 / \varepsilon kT)^{1/2} \tag{5.29}$$

e is the electric charge of proton, the subscript d refers to diffuse layer and the sum in Eq. (5.29) is taken over all ions in the solution. Equation (5.28) does not introduce any adjustable parameters to the model. Therefore, N_s is the only adjustable parameter in the 1-pK model combined with the diffuse layer model for amphoteric surfaces, and with silica-type surfaces K (reaction (5.25)) is the second adjustable parameter. Moreover, according to Eq. (5.28) the surface charging depends on the ionic strength (cf. Eq. (5.29)). This means that in contrast with the constant capacitance model (which requires a separate set of model parameters for each ionic strength), in diffuse layer model the σ_0(pH) curves for various ionic strengths can be modeled using the same model parameter(s).

This is a big advantage of the diffuse layer model, namely the σ_0(pH) can be predicted for the ionic strengths, for which the experimental data are not available, without resorting to empirical interpolation or extrapolation.

Application of diffuse layer model to surface charging of alumina is illustrated in Figs. 5.39–5.42. The best-fit N_s are summarized in Table 5.12.

For the data sets presented in Figs. 5.40 and 5.41 better fit can be obtained taking K (reaction (5.23)) somewhat (by a few tenths) higher than the PZC and for the data sets presented in Figs. 5.39 and 5.42 better fit can be obtained taking K (reaction (5.23)) somewhat lower than the PZC. This is a consequence of asymmetrical shape of the charging curves (cf. Figs. 5.35–5.38). However in this section the K(reaction(5.23)) is treated as fixed (not adjustable) parameter. With the data presented in Fig. 5.42 the ionic strength effect on the surface charging curves is properly reflected by the diffuse layer model. With the data presented in Fig. 5.39 the ionic strength effect on the surface charging curves is slightly overestimated for positively charged surface and severely overestimated for negatively charged surface. The charging curves presented in Fig. 5.40 are too asymmetrical to receive good fit in the diffuse layer model. The absolute values of σ_0 are underestimated by the model below the PZC and overestimated above the PZC for all three ionic strengths. With the data presented in Fig. 5.41 the ionic strength effect on the surface charging curves is severely overestimated by the diffuse layer model on the both sides of the PZC. The significantly worse fit in the diffuse layer model with respect to the constant capacitance model for the Cox and Mustafa data sets (cf. Figs. 5.39 and 5.41 on the

TABLE 5.12 The Best-fit N_s in C m^2 Used in Figs. 5.39–5.42 (σ_0), 5.43–5.46 (Ψ_0), and 5.120, 5.133, and 5.134 (specific adsorption)

Fig.		N_s
5.39	5.43	0.26
5.40	5.44	0.11
5.41	5.45	0.42
5.42	5.46	0.21

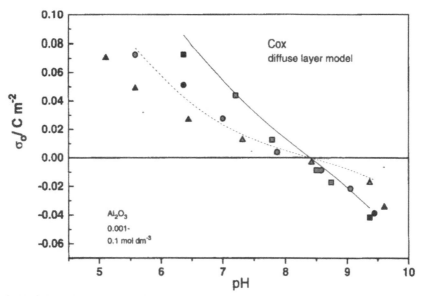

FIG. 5.39 Experimental (symbols) and calculated (lines) surface charge density of alumina as the function of the pH and ionic strength: 0.1 mol dm^{-3} squares, solid; 0.01 mol dm^{-3} circles, dash; 0.001 mol dm^{-3} triangles, dot. For model parameters, cf. Table 5.12.

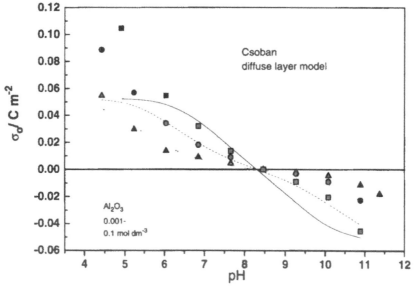

FIG. 5.40 Experimental (symbols) and calculated (lines) surface charge density of alumina as the function of the pH and ionic strength: 0.1 mol dm^{-3} squares, solid; 0.01 mol dm^{-3} circles, dash; 0.001 mol dm^{-3} triangles, dot. For model parameters, cf. Table 5.12.

one hand and 5.21 and 5.23 on the other) is somewhat misleading, namely, in the constant capacitance model the parameters (C) were fitted for each ionic strength separately. Certainly the fit in diffuse layer model could be significantly improved by allowing different N_s for different ionic strengths for these two data sets, and even

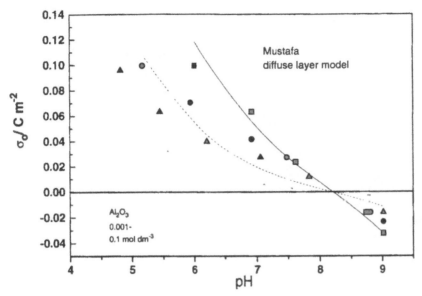

FIG. 5.41 Experimental (symbols) and calculated (lines) surface charge density of alumina as the function of the pH and ionic strength: 0.1 mol dm^{-3} squares, solid; 0.01 mol dm^{-3} circles, dash; 0.001 mol dm^{-3} triangles, dot. For model parameters, cf. Table 5.12.

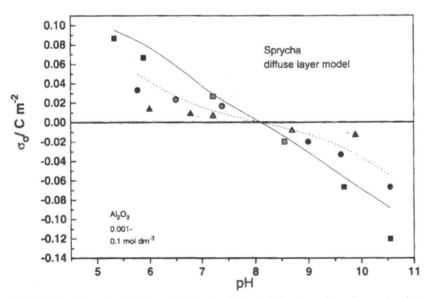

FIG. 5.42 Experimental (symbols) and calculated (lines) surface charge density of alumina as the function of the pH and ionic strength: 0.1 mol dm^{-3} squares, solid; 0.01 mol dm^{-3} circles, dash; 0.001 mol dm^{-3} triangles, dot. The model lines are identical for the 1-pK and 2-pK models. For model parameters, cf. Table 5.12 and 5.17.

more improved by allowing different N_s on the positive and negative side of the PZC, but such (problematic) refinements will not be discussed here.

The model parameters summarized in Table 5.12 were also used to calculate the surface potential as the function of the pH and ionic strength. The results are compared with Nernst potential in Figs. 5.43–5.46. The surface potential in Figs.

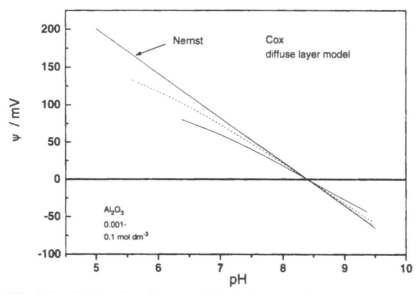

FIG. 5.43 Calculated surface potential of alumina as the function of the pH and ionic strength: 0.1 mol dm^{-3} solid; 0.01 mol dm^{-3} dash; 0.001 mol dm^{-3} dot. For model parameters, cf. Table 5.12.

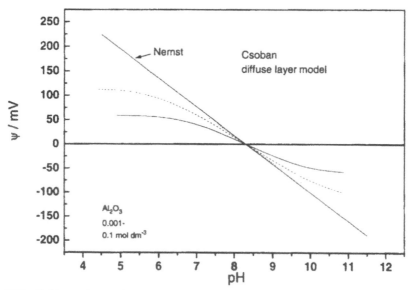

FIG. 5.44 Calculated surface potential of alumina as the function of the pH and ionic strength: 0.1 mol dm^{-3} solid; 0.01 mol dm^{-3} dash; 0.001 mol dm^{-3} dot. For model parameters, cf. Table 5.12.

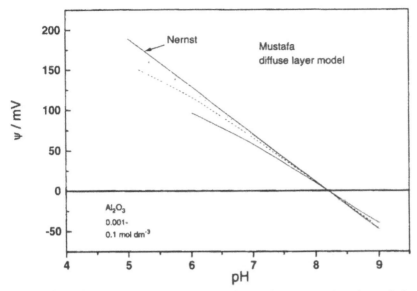

FIG. 5.45 Calculated surface potential of alumina as the function of the pH and ionic strength: 0.1 mol dm^{-3} solid; 0.01 mol dm^{-3} dash; 0.001 mol dm^{-3} dot. For model parameters, cf. Table 5.12.

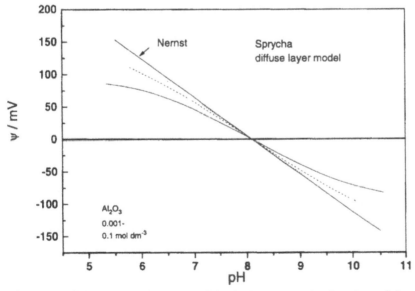

FIG. 5.46 Calculated surface potential of alumina as the function of the pH and ionic strength: 0.1 mol dm^{-3} solid; 0.01 mol dm^{-3} dash; 0.001 mol dm^{-3} dot. For model parameters, cf. Table 5.12.

5.43, 5.45, and 5.46 is nearly Nernstian at low ionic strength and near the PZC, and significantly lower than Nernstian farther from PZC and at high ionic strengths. The absolute value of the surface potential at given pH decreases when the ionic strength increases. In this respect the calculated surface potential resembles the ζ potential (cf. Figs. 3.84–3.86), but the absolute values of the calculated surface potential are higher. The deviations from Nernstian behavior in Fig. 5.44 are significant even at the lowest ionic strength and near the PZC. This is because relatively low N_s was used in the model calculations. But even in this figure the calculated surface potentials are higher than typical ζ potentials determined at the same pH and ionic strength. Such result is in accordance with expectations, namely, some potential drop between the surface and the shear plane is expected. This potential drop can be estimated using diffuse layer theory, $\ln \tanh (F\zeta/4RT) = \ln \tanh (F\psi_d/4RT) - \kappa x$, but an additional model parameter, namely x the shear plane distance is required to calculate the ζ potential.

Application of diffuse layer model to silica surface is illustrated in Figs. 5.47–5.51. The best-fit log K and N_s are summarized in Table 5.13.

TABLE 5.13 The Best-fit log K and N_s in C/m^2 Used in Figs. 5.47–5.51 (σ_0), 5.52–5.56 (Ψ_0), and 5.132 (specific adsorption)

Fig.		log K	N_s
5.47	5.52	7.46	0.29
5.48	5.53	7 (fixed)	1.1
5.49	5.54	7.08	0.13
5.50	5.55	6.23	0.21
5.51	5.56	6.09	0.25

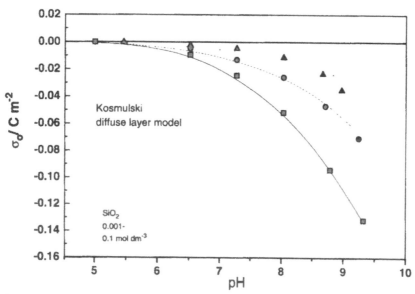

FIG. 5.47 Experimental (symbols) and calculated (lines) surface charge density of silica as the function of the pH and ionic strength: 0.1 mol dm^{-3} squares, solid; 0.01 mol dm^{-3} circles, dash; 0.001 mol dm^{-3} triangles, dot. For model parameters, cf. Table 5.13.

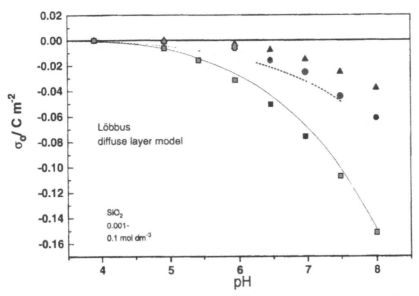

FIG. 5.48 Experimental (symbols) and calculated (lines) surface charge density of silica as the function of the pH and ionic strength: 0.1 mol dm^{-3} squares, solid; 0.01 mol dm^{-3} circles, dash; 0.001 mol dm^{-3} triangles, dot. For model parameters, cf. Table 5.13.

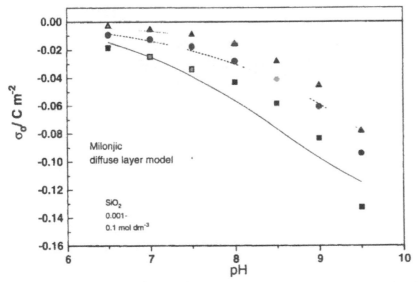

FIG. 5.49 Experimental (symbols) and calculated (lines) surface charge density of silica as the function of the pH and ionic strength: 0.1 mol dm^{-3} squares, solid; 0.01 mol dm^{-3} circles, dash; 0.001 mol dm^{-3} triangles, dot. For model parameters, cf. Table 5.13.

For the data presented in Fig. 5.48, the goodness of the fit was rather insensitive to the K value, thus log $K = 7$ was arbitrarily selected. The model

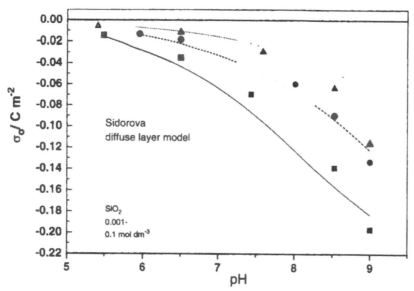

FIG. 5.50 Experimental (symbols) and calculated (lines) surface charge density of silica as the function of the pH and ionic strength: 0.1 mol dm^{-3} squares, solid; 0.01 mol dm^{-3} circles, dash; 0.001 mol dm^{-3} triangles, dot. For model parameters, cf. Table 5.13.

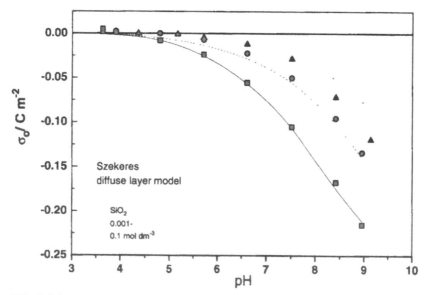

FIG. 5.51 Experimental (symbols) and calculated (lines) surface charge density of silica as the function of the pH and ionic strength: 0.1 mol dm^{-3} squares, solid; 0.01 mol dm^{-3} circles, dash; 0.001 mol dm^{-3} triangles, dot. For model parameters, cf. Table 5.13.

parameters summarized in Table 5.13 were also used to calculate the surface potential as the function of the pH and ionic strength. The surface potentials are reported in Figs. 5.52–5.56. The slope of the model curves is significantly lower than

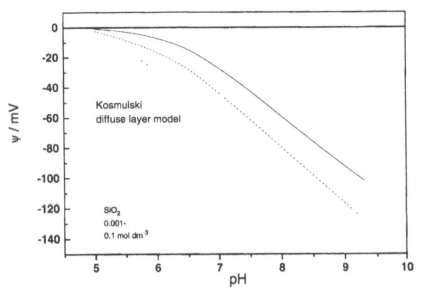

FIG. 5.52 Calculated surface potential of silica as the function of the pH and ionic strength: 0.1 mol dm⁻³ solid; 0.01 mol dm⁻³ dash; 0.001 mol dm⁻³ dot. For model parameters, cf. Table 5.13.

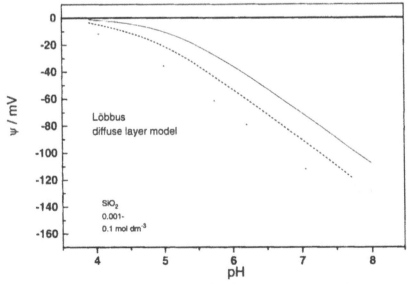

FIG. 5.53 Calculated surface potential of silica as the function of the pH and ionic strength: 0.1 mol dm⁻³ solid; 0.01 mol dm⁻³ dash; 0.001 mol dm⁻³ dot. For model parameters, cf. Table 5.13.

59 mV/pH unit, and the absolute value of the surface potential decreases when the ionic strength increases. The calculated surface potentials at pH > 8 are higher than typical ζ potentials reported for silica (Figs. 3.90–3.93). However, the calculated

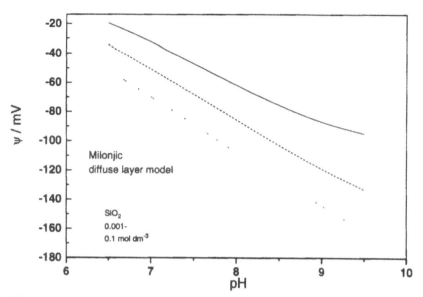

FIG. 5.54 Calculated surface potential of silica as the function of the pH and ionic strength: 0.1 mol dm^{-3} solid; 0.01 mol dm^{-3} dash; 0.001 mol dm^{-3} dot. For model parameters, cf. Table 5.13.

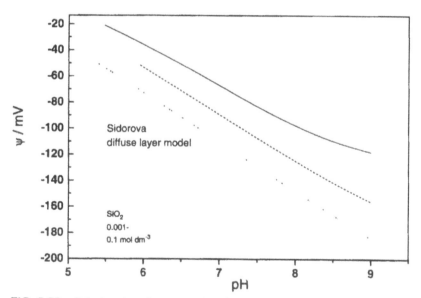

FIG. 5.55 Calculated surface potential of silica as the function of the pH and ionic strength: 0.1 mol dm^{-3} solid; 0.01 mol dm^{-3} dash; 0.001 mol dm^{-3} dot. For model parameters, cf. Table 5.13.

surface potentials at low pH are similar or even lower than ζ potentials reported for silica for corresponding pH and ionic strengths. In the presence of inert electrolyte the absolute value of the surface potential is higher than that of ζ potential (the potential is a monotonic function of the distance from the surface), thus the

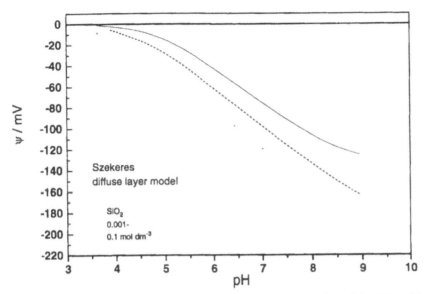

FIG. 5.56 Calculated surface potential of silica as the function of the pH and ionic strength: 0.1 mol dm^{-3} solid; 0.01 mol dm^{-3} dash; 0.001 mol dm^{-3} dot. For model parameters, cf. Table 5.13.

TABLE 5.14 The Best-fit C in F/m^2 and N_s in C/m^2 Used in Fig. 5.58 (σ_0), 5.61 (Ψ_0), 5.62 (Ψ_β), and 5.123 (specific adsorption)

C	N_s
0.7	0.65

TABLE 5.15 The Fixed C in F/m^2 and the Best-fit N_s in C/m^2 Used in Figs. 5.57, 5.59, and 5.60 (σ_0), and 5.121–5.123 (specific adsorption)

Fig.	C	N_s
5.57	4	0.59
5.59	5	1.7
5.60	4	0.27

electrokinetic results for silica at low pH do not support the presented model. Explanation of low surface charge density and simultaneously high electrokinetic potential of silica at pH about 5 is a challenging problem.

4. Stern Model

The constant capacitance model and diffuse layer model are combined in the Stern model. This model was originally introduced for the mercury electrode. The interfacial region is modeled as two capacitors in series. A part of the surface charge is balanced

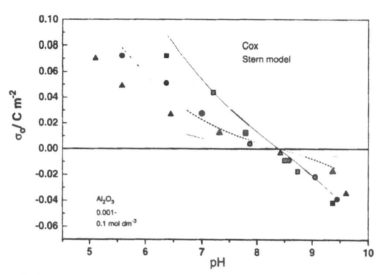

FIG. 5.57 Experimental (symbols) and calculated (lines) surface charge density of alumina as the function of the pH and ionic strength: 0.1 mol dm^{-3} squares, solid; 0.01 mol dm^{-3} circles, dash; 0.001 mol dm^{-3} triangles, dot. For model parameters, cf. Table 5.15.

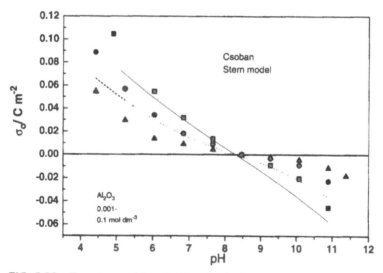

FIG. 5.58 Experimental (symbols) and calculated (lines) surface charge density of alumina as the function of the pH and ionic strength: 0.1 mol dm^{-3} squares, solid; 0.01 mol dm^{-3} circles, dash; 0.001 mol dm^{-3} triangles, dot. For model parameters, cf. Table 5.14.

by a layer of counterions near the surface. This capacitor has a constant capacitance (independent of the ionic strength). The rest of the surface charge is balanced by the diffuse layer. The relationship between the charge and potential for the diffuse layer is expressed by Eq. (5.28). The Stern model requires two adjustable parameters for amphoteric surfaces and three adjustable parameters for silica-type surfaces.

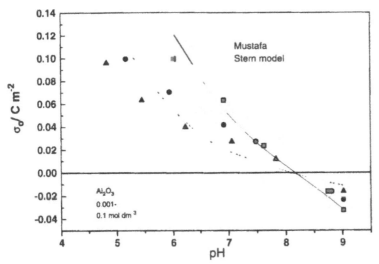

FIG. 5.59 Experimental (symbols) and calculated (lines) surface charge density of alumina as the function of the pH and ionic strength: 0.1 mol dm^{-3} squares, solid; 0.01 mol dm^{-3} circles, dash; 0.001 mol dm^{-3} triangles, dot. For model parameters, cf. Table 5.15.

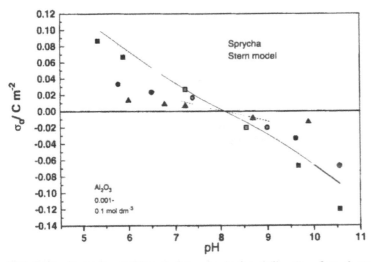

FIG. 5.60 Experimental (symbols) and calculated (lines) surface charge density of alumina as the function of the pH and ionic strength: 0.1 mol dm^{-3} squares, solid; 0.01 mol dm^{-3} circles, dash; 0.001 mol dm^{-3} triangles, dot. For model parameters, cf. Table 5.15.

Application of the Stern model to alumina surface is illustrated in Figs. 5.57–5.60. The best-fit log C and N_s are summarized in Table 5.14.

Similarly as with the diffuse layer model, for the data sets presented in Figs. 5.58 and 5.59 better fit can be obtained taking K (reaction (5.23)) somewhat higher than the PZC, and for the data sets presented in Figs. 5.57 and 5.60 better fit can be obtained taking K (reaction (5.23)) somewhat lower than the PZC. With the data sets presented in Figs. 5.57, 5.59, and 5.60, the goodness of fit is rather insensitive to the value of C, and it is slowly improving when C increases. Thus $C = 4$ or 5 F/m^2 was arbitrarily selected. These fixed C and the best-fit N_s values are listed in Table 5.15. In view of the discussion above very high C do not have physical sense. The physical meaning of different Stern layer capacitances obtained for various materials is discussed in more detail in Ref. [36]. Only for the data set presented in Fig. 5.58 the replacement of diffuse layer model by the Stern model leads to significant improvement of the fit. In this respect, the application of the diffuse layer model to data sets presented in Figs. 5.57, 5.59, and 5.60 is more attractive since this model involves only one adjustable parameter. Thus we drop the analysis of these data sets in terms of Stern layer model at this point.

Figures 5.61 and 5.62 show the surface potential and the Stern layer potential as the function of the pH and ionic strength for the model used in Fig. 5.58. The calculated surface potential is nearly Nernstian even at high ionic strengths and far from the PZC. The calculated Stern layer potentials are significantly lower than Nernstian, but higher than typical ζ potentials reported for alumina at the same pH and ionic strength (Figs. 3.84–3.86). The ζ potentials can be calculated using the Stern model, but an additional parameter, namely, distance between the Stern layer and the shear plane is required.

The Stern model was also tested for silica. With the Kosmulski, Löbbus, and Szekeres experimental data the goodness of fit with physically realistic C values was satisfactory, but worse than for diffuse layer model (cf. Figs. 5.47, 5.48 and 5.51). Since C is the additional adjustable parameter with respect to the diffuse layer model, application of Stern model to these systems is not justified. Somewhat improved with respect to the diffuse layer model, but still rather poor modeling results were obtained for the Milonjic and Sidorova experimental data. The best-fit

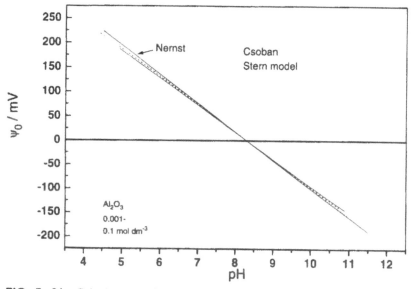

FIG. 5. 61 Calculated surface potential of alumina as the function of the pH and ionic strength: 0.1 mol dm^{-3} solid; 0.01 mol dm^{-3} dash; 0.001 mol dm^{-3} dot. For model parameters, cf. Table 5.14.

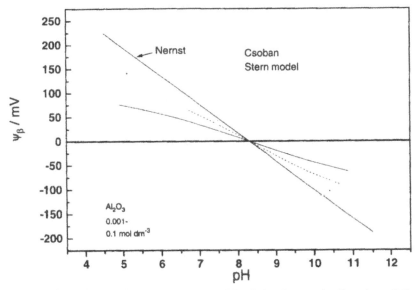

FIG. 5.62 Calculated Stern layer potential of alumina as the function of the pH and ionic strength: 0.1 mol dm^{-3} solid; 0.01 mol dm^{-3} dash; 0.001 mol dm^{-3} dot. For model parameters, cf. Table 5.14.

TABLE 5.16 The Best-fit C in F/m^2, log K, and N_s in C/m^2 Used in Figs. 5.63–5.64

Fig.	C	log K	N_s
5.63	2.1	7.98	1.6
5.64	3.6	8.53	58

model parameters are summarized in Table 5.16. The model curves are plotted in Figs. 5.63 and 5.64.

The N_s value used in Fig. 5.64 is higher by an order of magnitude than the highest values obtained by different methods (cf. Tables 5.1 and 5.3), thus, the physical meaning of the model curves presented in Fig. 5.64 is doubtful.

The above attempts to apply the Stern model to five sets of charging curves of silica were rather unsuccessful. On the other hand, Sonnefeld [37] reports successful application of the Stern model to the surface charging of silica at five concentrations (0.005–0.3 mol dm^{-3}) of five alkali chlorides. Two adjustable parameters (N_s and the acidity constant) were independent of the nature of alkali metal cation and one parameter, C was dependent on the nature of the alkali metal cation. This result was explained in terms of the correlation between the Stern layer thickness and the size of the counterions.

Is C related to the size of the counterions, than it should be different on the both sides of the PZC for amphoteric surfaces. An analogous modification of the constant capacitance model was discussed above (Figs. 5.35–5.38). The Stern model modified by allowing different C values for positive and negative branch of the

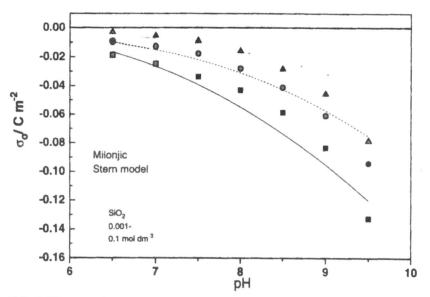

FIG. 5.63 Experimental (symbols) and calculated (lines) surface charge density of silica as the function of the pH and ionic strength: 0.1 mol dm^{-3} squares, solid; 0.01 mol dm^{-3} circles, dash; 0.001 mol dm^{-3} triangles, dot. For model parameters, cf. Table 5.16.

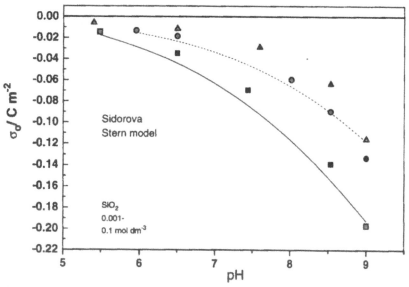

FIG. 5.64 Experimental (symbols) and calculated (lines) surface charge density of silica as the function of the pH and ionic strength: 0.1 mol dm^{-3} squares, solid; 0.01 mol dm^{-3} circles, dash; 0.001 mol dm^{-3} triangles, dot. For model parameters, cf. Table 5.16.

charging curves will be only tested for the Csoban results, since with the other three data sets the diffuse layer model was superior over the Stern model. For the assumed N_s of 0.3 C/m^2 (1.8 sites/nm^2) the best fit capacitances were 1.39 F/m^2 for the positive and 0.335 F/m^2 for the negative branch of the charging curves, respectively,

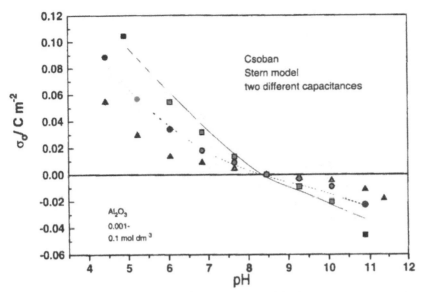

FIG. 5.65 Experimental (symbols) and calculated (lines) surface charge density of alumina as the function of the pH and ionic strength: 0.1 mol dm^{-3} squares, solid; 0.01 mol dm^{-3} circles, dash; 0.001 mol dm^{-3} triangles, dot. $N_s = 0.3$ C/m^2 (1.8 sites/nm^2), $C_+ = 1.39$ F/m^2, $C_- = 0.335$ F/m^2.

and the results of model calculations are presented in Fig. 5.65. The possibility to obtain charging curves asymmetrical with respect to the PZC is one of the advantages of the Stern model over the diffuse layer model.

The above few examples do not exhaust all modifications of the 1-pK model and their combinations with different electrostatic models. For example Eq. (5.23) can be generalized as

$$\equiv SO^{x-} + H^+ = \equiv SOH^{(1-x)+} \tag{5.30}$$

where $\equiv S$ is a surface atom, and $0 < x < 1$ is an adjustable parameter. It was shown [38] that for certain data sets the best-fit x can be different from 1/2 (Eq. (5.23)). In the generalized model

$$\log K(\text{reaction}(5.30)) = \text{PZC} - \log((1 - x)/x). \tag{5.31}$$

The physical meaning of x different from 1/2 will be discussed in Section E.

The better fit in some modifications of the 1-pK model and with more sophisticated electrostatic models (with respect to the models presented above) is usually achieved on expense of the increase in the number of adjustable parameters.

D. 2-pK Model

The model describing the surface charging of amphoteric surfaces in terms of the following two reactions

$$\equiv SOH_2^+ = \equiv SOH + H^+ \qquad K_{a1} \tag{5.32}$$

$$\equiv SOH = \equiv SO^- + H^+ \qquad K_{a2} \tag{5.33}$$

is called 2-pK model. Reaction (5.33) is identical with reaction (5.25) for silica-like surfaces, and with Eq. (5.30) with $x = 1$ (except the l.h.s. and r.h.s. are swapped). The 2-pK model precedes the 1-pK model (discussed in detail in Section C) and it is still quite popular.

The equilibrium constants of reactions (5.32) and (5.33) are interrelated, namely

$$pK_{a1} + pK_{a2} = 2\,PZC \tag{5.34}$$

With respect to the 1-pK model the 2-pK model requires one additional adjustable parameter, namely, ΔpK_a, defined as

$$\Delta pK_a = pK_{a2} - pK_{a1} \tag{5.35}$$

Combination of Eqs. (5.34) and (5.35) yields

$$pK_{a1} = PZC - 1/2\,\Delta pK_a \qquad pK_{a2} = PZC + 1/2\,pK_a \tag{5.36}$$

Use of additional adjustable parameter is justified when significant improvement of the fit is achieved. Let us explore then if introduction ΔpK_a (treated as fully adjustable parameter) improves the fit over the 1-pK model when the same electrostatic model is used. Extensive discussion on the combination of the 2-pK model with the constant capacitance model was recently published by Lützenkirchen [39]. Comparison of the combinations of the 2-pK model with different electrostatic models was recently published by Kosmulski et al. [38]. The present analysis is limited to the diffuse layer and Stern models.

Figures 5.66–5.68 (and Fig. 5.42) compare the best fit model curves for alumina (four sets of experimental data) calculated by means of the diffuse layer model combined with the 1-pK model on the one hand and the 2-pK model on the other. The best-fit N_s in the 1-pK model are presented in Table 5.12. The best-fit parameters of the 2-pK model are presented in Table 5.17.

For the data sets presented in Figs. 5.66 and 5.68 the fit steadily improves as the ΔpK_a decreases, but at $\Delta pK_a < -4$ large decrease in ΔpK_a has rather insignificant effect on the shape of the model curves. The results calculated for a fixed $\Delta pK_a = -4$ are presented as an example. The fit for these two data sets is slightly better than in the 1-pK model. With results presented in Fig. 5.42 the best-fit ΔpK_a gives the fit slightly worse than the 1-pK model. Since the difference between the 1-pK and 2-pK best-fit curves is less than the thickness of the lines in Fig. 5.42, a separate figure was not drawn. Also the difference in the shape of model curves between the 1-pK and 2-pK model for two other sets of data (Figs. 5.66 and 5.68) is

TABLE 5.17 The Best-fit N_s in C/m^2 and ΔpK_a Used in Figs. 5.66–5.68 and 5.42

Fig.	N_s	ΔpK_a
5.66	0.085	-4 (fixed)
5.67	0.2	3.62
5.68	0.13	-4 (fixed)
5.42	0.1	0.567

rather insignificant. Such results are in favor of the 1-pK model over the 2-pK model, because the additional adjustable parameter did not significantly improve the

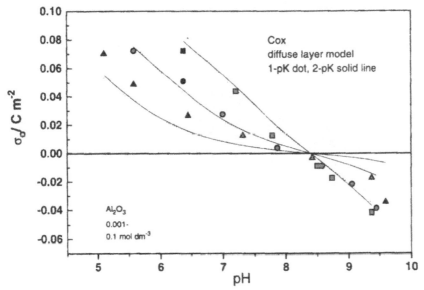

FIG. 5.66 Experimental (symbols) and calculated (1 pK dot lines; 2-pK solid lines) surface charge density of alumina as the function of the pH and ionic strength: 0.1 mol dm^{-3} squares; 0.01 mol dm^{-3} circles, 0.001 mol dm^{-3} triangles. For model parameters, cf. Tables 5.12 and 5.17.

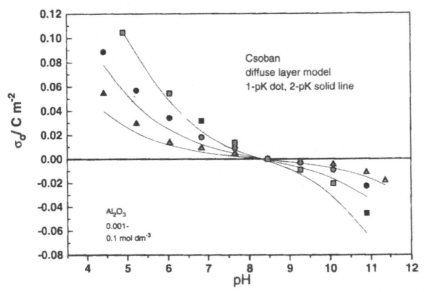

FIG. 5.67 Experimental (symbols) and calculated (1 pK dot lines; 2-pK solid lines) surface charge density of alumina as the function of the pH and ionic strength: 0.1 mol dm^{-3} squares; 0.01 mol dm^{-3} circles, 0.001 mol dm^{-3} triangles. For model parameters, cf. Tables 5.12 and 5.17.

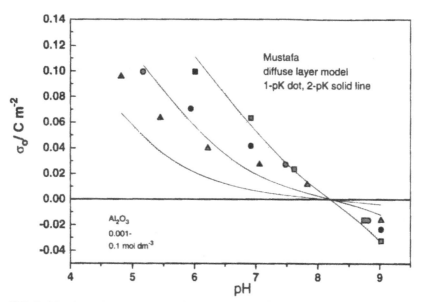

FIG.5. 68 Experimental (symbols) and calculated (1 pK dot lines; 2-pK solid lines) surface charge density of alumina as the function of the pH and ionic strength: 0.1 mol dm^{-3} squares; 0.01 mol dm^{-3} circles; 0.001 mol dm^{-3} triangles. For model parameters, cf. Tables 5.12 and 5.17.

fit. On the other hand, for the data set presented in Fig. 5.67 the 2-pK model produced completely different model curves and the fit was significantly better than in the 1-pK model combined with the diffuse layer model, and even better than in the 1-pK model combined with the Stern model (Fig. 5.58). Thus the model with two adjustable parameters N_s and ΔpK_a is more successful in simulation of these charging curves than the model with two adjustable parameters N_s and C. An attempt to get even better fit by combination of the 2-pK model with the Stern model was unsuccessful. With sufficiently high C (>6 F/m^2) practically identical model curves as these presented in Fig. 5.67 are obtained for the best-fit N_s and ΔpK_a.

Thus, for one of the analyzed above combinations of electrostatic models and sets of experimental data (diffuse layer + Csoban) the 2-pK model was superior over the 1-pK model in simulation of surface charging curves. For other combinations 1-pK model produced fits nearly equally good as (or better than) the 2-pK model.

In the above calculations the equilibrium constants of reactions (5.32) and (5.33) were treated as fully adjustable parameters. Although the fitting procedure for the data presented in Fig. 5.67 was successful in the mathematical sense, the physical sense of the best-fit ΔpK_a value (cf. Table 5.17) is problematic. The equilibrium constants characterizing consecutive steps of protonation/deprotonation of hydro-complexes in solution usually differ by over ten orders of magnitude (cf. Section E). Probably the same applies to protonation of surface metal ions, thus, only high ΔpK_a values (>10) are physically realistic.

At a high ΔpK_a only one of two consecutive protonation/deprotonation reactions (5.32) and (5.33) can occur to significant degree in normally studied pH range (3–11), and the other reaction has negligible effect on the overall surface charge. But then the model fails to explain the amphoteric behavior of the surface. Borkovec [40] proposed the following physical interpretation of low ΔpK_a in the 2-pK model. The surface site in reactions (5.32) and (5.33) is a pair of neighboring surface OH groups, which may coordinate the same or different metal centers. One of the groups is responsible for reaction (5.32) and another for (5.33). These two groups substantially differ in their proton affinities because of different coordination environment (cf. Section E). This idea can be generalized, i.e. assemblies of three, four etc. neighboring surface hydroxyl groups can be considered to build 3-pK, 4-pK, etc. models. Comparison between 1-pK and 2-pK models was also discussed by Tao and Dong [41].

E. MUSIC

Some models presented in Sections C and D were quite successful in simulation of certain sets of experimental data, but the coincidence of the calculated curves with the experimental results was only approximate. It is then tempting to improve the fit by refining these models. The most obvious extension of the existing models is to consider different types of surface groups having different protonation constants. Certainly, this requires new adjustable parameters, namely, concentrations and protonation constants of particular types of surface sites. Even with relatively simple models discussed in Sections C and D two different sets of model parameters or even two different models are likely to give practically identical model curves. This is even more likely with more complex models. Although some combinations of model parameters can be ruled out (for discussion of the physical sense of N_s, C and ΔpK_a, cf. Sections C and D) the refinement of the model on purely curve-fitting basis inevitably leads to multiple solutions (identical model curves produced by different sets of model parameters) whose assessment might be difficult.

Therefore it would be much desired to derive the concentrations and acidity constants of particular types of surface sites from some first principles, rather than to fit them as adjustable parameters. This would produce generic charging curves for certain material (rather than for specific sample). Unfortunately verification of such a model is rather difficult in view of contradictory surface charging curves reported in the literature for different samples of the same material (cf. Figs. 3.43–3.73). The surface charging depends on many variables, e.g. the PZC is temperature dependent, and the absolute value of σ_0 depends on the nature of the counterions (K versus Na, etc.). Consideration of all these variables would be very tedious, and such effects are mostly ignored in the attempts to predict the surface charging behavior of materials.

The most successful attempt to derive the surface charging behavior from the properties of the material is known as MUSIC [42] (MUltiSIte Complexation) model. Because of many refinements following the original model, MUSIC should be rather considered as a general framework.

The MUSIC approach invokes the acid-base properties of solution monomeric species, namely, logarithms of their protonation constants are linearly correlated with $\nu\,L$, where $\nu = Z\,CN$, Z is the charge of the cation, CN is its coordination number, and L is the metal–hydrogen distance.

Actually, two parallel lines are obtained for hydroxo- and oxo-complexes indicating a difference by 14 orders of magnitude between equilibrium constants characterizing two (hypothetical) protonation steps. The scatter of the data points around the best-fit straight line can be reduced by separately considering the central cations having the rare gas electron structure on the one hand and those with ten d electrons on the other. This results in four parallel $\log K$ (v/L) lines.

In the MUSIC approach the protonation of the surface oxygen sites is assumed to follow the same rule as discussed above for solution monomers. In other words, $\log K$ for particular surface sites can be calculated from the following equation:

$$\log K = A - B n v / L \qquad (5.37)$$

where n is the number of metal cations coordinated with the surface group, and the constant B equal to 5.27 nm is the slope of the discussed above straight lines obtained for solution monomers. Theoretically

$$B = (Z_H N_A e^2)/(4\pi R T \varepsilon_2 \ln 10) \qquad (5.38)$$

where $Z_H = 1$ is the valence of adsorbed proton (L in denominator of Eq. (8) in Ref. [42] is a typographical error), but the effective dielectric constant ε_2 is not directly related to any quantity that can be measured or theoretically calculated, so an empirical value of B has to be used. A is another empirical constant which depends on the degree of protonation (oxo- vs. hydroxo-complexes) on the one hand and the electron structure on the other. This approach is analogous to that presented in Section 3.II.A devoted to correlation of the pristine PZC with ionic radii and bond valence (e.g. Eq. (3.19)). The values of parameter A for surface species were obtained by adding 4 to the corresponding values for particular types of solution monomers. This results in $A = 18.4$ for hydroxocomplexes of cations with 10 d electrons, 19.7 for hydroxocomplexes of cations with rare gas configuration, 32.2 for oxocomplexes of cations with 10 d electrons, and 34.5 for oxocomplexes of cations with rare gas configuration. The above A values are based on the hypothesis that the difference in $\log K$ between surface and solution species is independent of the nature of the cation. This difference was estimated from the PZC of gibbsite whose surface charging is believed to be due to a single surface reaction (5.23). The PZC and thus $\log K$ (reaction (5.23)) is equal to 10 (cf. Section C). On the other hand, the experimental $\log K$ for protonation of the analogous solution monomer equals 5.7, and the calculated value (from the linear relationship between v/L and $\log K$ obtained for other cations) equals 6.3. Thus the difference in $\log K$ between the surface species and solution monomer is about 4 for aluminum, and this value is assumed to be valid for other elements. Additionally the A values for surface species were adjusted to give a difference of 13.8 in A between $\log K$ of analogous oxo- and hydroxocomplexes. In view of this high difference in $\log K$ between oxo- and hydroxo-complexes, a multiple protonation of the same surface site over normally encountered pH range can be excluded. Thus, only one protonation step (if any) is taken into account for particular type of surface group. In this respect the MUSIC approach can be considered as generalization of the 1-pK model.

Equation (5.37) produces different $\log K$ values for different types of surface groups in the same material, thus the overall σ_0 is very sensitive to concentrations of particular types of groups on the surface. Formal charge of the surface $(\equiv S)_N OH$ group was calculated as $N \times v - 1$. This often leads to fractional values. Thus, MUSIC rationalizes such fractional values in Eq. (5.30). Zero formal charge of a surface OH group (Eqs (5.32) and (5.33)) is obtained when $N \times Z = CN$, and this is the case, e.g. for a $\equiv$SiOH group. Densities of particular types of surface sites on specific crystal planes are derived from crystallographic data, and this is probably the most controversial step in the modeling, namely, very divergent results are obtained by different methods (cf. Section I and Ref. [7]). Finally, L in Eq. (3.37) in the original MUSIC model was calculated as one side of the triangle whose two other sides are the sum of metal cation and oxygen radius (from crystallographic data), and the OH distance in water molecule, and the angle between these two sides equals 127.75 (π-1/2 HOH angle in water molecule).

The MUSIC model calculations [43] gave PZC at pH 7.5 for the (100) face of goethite, and 10.7 for the (010) and (001) faces. Charging behavior of gibbsite, hematite, rutile and silica have been also discussed in terms of the MUSIC model. The surface charging of particular faces can be used to calculate the overall charging behavior for materials whose crystal morphology is well defined. Such well-defined crystals can be obtained for goethite using special preparation methods. Hiemstra et al. [44] discussed the effect of morphology of gibbsite on the surface charging. The crystals having higher contribution of the edge faces exhibit higher positive charge density (related to the total surface area) at pH 4–10. Theoretically the planar 001 face can be only negatively charged at pH > 10. However in less well crystallized gibbsite the defects in this face contribute to positive surface charge.

Bleam [45] argues, that the linear correlation

$$\log K = A - Bn\nu \tag{5.39}$$

which differs from Eq. (5.37) in using ν rather than ν/L (naturally the A and B values in this and subsequent equations are different than in Eq. (5.37)) gives an equally high correlation coefficient for solution monomers as Eq. (5.37). Moreover, replacement of $n\ \nu/L$ in Eq. (5.37) (or $n\ \nu$ in Eq. (5.39)) by a sum taking into account proton-proton interaction was postulated. The equation

$$\log K = A - B(n\nu/L + \nu_H/L_{HH}) \tag{5.40}$$

where $\nu_H/L_{HH} = 3.31$ nm^{-1} leads to a common A value for oxo and hydroxocomplexes. Likewise in the equation

$$\log K = A - B(n\nu + \nu_H) \tag{5.41}$$

with $\nu_H = 0.8$ the A value is common for oxo and hydroxocomplexes. Further refinement of the model was postulated by considering contribution of other hydrogen atoms (forming hydrogen bonds with the oxygen atom) to the sum in brackets in Eq. (5.40) or (5.41).

In the refined version of MUSIC [46] the protonation constant for specific site was calculated from the equation

$$\log K = -19.8\,[-2 + m\,s_H + n(1 - s_H) + \Sigma s_i] \tag{5.42}$$

where s_H the bond valence for proton bond to surface oxygen is set to 0.8, m is the number of donating and n the number of accepting hydrogen bridges with water molecules (for hydroxides and oxohydroxides internal hydrogen bridges belonging to the crystal structure are not counted), and the sum in Eq. (5.42) is taken over all structural Me-O and H-O bonds. This implies that protonation results in an increase in the charge of surface group by 0.6 rather than by an intuitive value of 1 used in the classical version of MUSIC.

The sum $m + n$ equals 2 for singly coordinated surface oxygens, and 1 for triply coordinated surface oxygens. For doubly coordinated surface oxygens $m + n$ equals 1 or 2. Silica is an exception with $m + n = 3$ in view of rather open structure. The sum (last term in brackets in Eq. (5.42)) rather that a product $n \times s_i$ allows asymmetric distribution (different bond valence for particular Me-O bonds). The expression in square brackets expresses the undersaturation of valence of surface oxygen. Eq. (5.42) is also valid for solution monomers ($m + n = 3$).

The calculated PZC of TiO_2 was not face specific (110 versus 100 which are considered the dominant ones for rutile). This explains rather consistent PZC values reported in literature for TiO_2 (cf. Table 3.1 and Figs. 3.31–3.34). The refined MUSIC model was used to predict PZC of quartz, TiO_2 (rutile and anatase), goethite and gibbsite, and this resulted in values 1.9, 5.9, 9.5, and 9.9, respectively (cf. Table 3.1 for experimental values).

Venema et al. [47] pointed at a contradiction between the prediction of the PZC of iron (hydr)oxides on the one hand and of the slope of σ_0(pH) curves on the other in the classical version of MUSIC. Namely, in this model among different possible surface hydroxyl groups only the $FeOH^{1/2-}$ and $Fe_3O^{1\,2-}$ groups were found to contribute to the pH dependent surface charging because the model predicts the other groups to be fully protonated or fully deprotonated at pH 3–11. Considering the calculated pK values for these two types of groups and experimental PZC for iron (hydr)oxides (cf. Table 3.1 and Figs. 3.19–3.26) the ratio of these groups on the surface seems to be about 1:1. However such ratios give a slope of σ_0(pH) curves about zero and this is in conflict with the experimentally observed slopes.

This problem was solved as follows. The original version of the MUSIC model assumed the charge of the central ion to be equally distributed over the ligands. In the refined model the bond valence s_{i-j} is calculated from the empirical formula

$$s_{i-j} = \exp[(R_{i-j} - R_{0,i-j})/0.037\,(\text{nm})] \tag{5.43}$$

where R_{i-j} is the bond length and $R_{0,\,i-j}$ is an empirical constant for given couple of ions i and j.

The value of $R_{0,Fe-O} = 0.1759$ nm was cited as optimized for all iron oxides and oxohydroxides, but Venema et al. used an individually optimized value for each compound. Three different surface groups were considered for the hematite (001) face and four different surface groups for the (110) face. For the (001) face a difference in two Fe-O bond lengths for a doubly coordinated surface group resulted in a dramatic difference in the calculated bond strength (0.388 versus 0.612) for different Fe atoms bound with the same oxygen. With oxyhydroxides the calculations are more difficult because of internal hydrogen bonds. Some of these bonds are broken when the surface is created, and new bonds are formed with the solvent. Depending on the three dimensional configuration it is sometimes clear that an internal hydrogen bond must disappear, because the (other) iron atom to which the OH was attached has been removed in the "cutting" of the phase.

However for some internal hydrogen bonds it is difficult to assess whether or not they persist when a surface is created. Depending on the orientation of the surface group both situations are likely and they lead to substantial difference in the calculated protonation constants of the surface groups and consequently in the predicted surface charging curves of particular faces.

Van Hal et al. [48] used the 2-pK and MUSIC models combined with diffuse layer and Stern electrostatic models (with pre-assumed site-density and surface acidity constants) to calculate the surface potential, the intrinsic buffer capacity - $(d\sigma_o/dpH_s)/e$ where pH_s is the pH at the surface, the sensitivity factor $-(d\psi_o/dpH) \times [e/(kTln\ 10)]$, which equals unity for Nernstian response, and the differential capacitance for three ionic strengths as a function of pH. The calculated surface potentials were compared with the experimentally measured ISFET response.

Rusted et al. [49] calculated charging curves of hematite using MUSIC approach, with only one adjustable parameter, namely, Stern layer capacitance of 0.25 F/m^2. The acidity constants of eight surface species at the (012) face (whose densities are derived from crystallographic data) were calculated from linear relationship between pK and gas phase acidities proposed by the same authors for goethite. The agreement with the selected set of experimental data was spectacular. It is noteworthy, however, that the calculated surface charge densities were lower by a factor > 3 than the experimental σ_0 reported by other authors for hematite (cf. Figs. 3.51–3.54).

Nagashima and Blum [50] calculated overall proton association constant for alumina as a function of pH to change by 12 orders of magnitude between pH 4 and 5. This result was interpreted in terms of different acidity constants of $\equiv AlO^{-3\ 2}$, $\equiv Al_2O^-$ and $\equiv Al_3O^{-1\ 2}$ surface groups (MUSIC). The PZC for alumina at pH 4 reported by these authors is substantially different from most results reported in the literature (cf. Table 3.1 and Fig. 3.11).

In view of the discussed above discrepancies, some spectroscopic evidence of particular surface species used in the MUSIC model would be much desired. Connor et al. [51] studied internal reflection infrared spectra of TiO_2 (amorphous + anatase) films, and obtained the following speciation. The $(\equiv Ti)_2OH^+$ species is formed at $pH < 4.3$, and $\equiv TiOH_2^+$ at $pH < 5$. The $\equiv TiOH$ species was detected at pH 4.3–10.7 (maximum abundance at pH 8) and $\equiv TiOH_2$ species at $pH < 10.7$.

For most materials the surface charging in the MUSIC approach is due to several different types of surface groups. This explains limited significance of the correlations between the PZC and other physical quantities discussed in Section 3.II.

F. Inert Electrolyte Binding

The models of primary surface charging discussed above can be expanded by considering two additional surface species resulting from binding of counterions (Y^+ and X^-) to the charged surface species. For example in the 1-pK model [52]

$$\equiv AlOH^{-1/2} + Y^+ = \equiv AlOH^{-1/2} \cdots Y^+ \tag{5.44}$$

$$\equiv AlOH_2^{+1\ 2} + X = \equiv AlOH_2^{+1\ 2} \cdots X \tag{5.45}$$

The symbol "…" emphasizes that the counterions are at some distance from the surface.

Lützenkirchen et al. [53] emphasized that in the approach with inert electrolyte binding, the activity coefficients of the ions in the solution must be taken into account (cf. Section II). Many other authors define the equilibrium constants of reactions involving ions of inert electrolyte in terms of concentrations, and this results in different values of model parameters.

Reactions (5.44) and (5.45) require special electrostatic model to calculate the electric potential experienced by the bound counterions. The $\equiv AlOH^{-1/2} \cdots Y^+$ group contributes -1/2 of one elementary charge to the surface charge and + 1 elementary charge to the charge in the "β" plane, which plays a similar role as the Stern layer. The $\equiv AlOH_2^{+1/2} \cdots X^-$ group contributes + 1/2 of one elementary charge to the surface charge and -1 elementary charge to the charge in the "β" plane.

The equilibrium constants of reactions (5.44) and (5.45) must be equal, otherwise the PZC depends on the ionic strength. Thus, one additional adjustable parameter is necessary. The electrostatic model used with reactions (5.44) and (5.45) also requires some adjustable parameters.

Hiemstra et al. [44] suggested a complicated electrostatic model, which was combined with the 1-pK concept with inert electrolyte binding (reactions (5.23), (5.44) and (5.45)). This model was used to explain the asymmetry in charging curves of aluminum (hydr)oxides. Different penetration of the surface by anions and cations of inert electrolyte was modeled by three different electrostatic planes. The counterions are adsorbed mostly in the "2" plane (which corresponds to the "β" plane), but cations of inert electrolyte are bound by one type of subsurface sites in the surface plane. In contrast, anions bound by these sites split their charge 50–50 between the surface plane and the "1" plane (placed between the surface and the "2" plane. Two capacitances (between the surface and the "1" plane on the one hand, and between the "1" and "2" planes on the other) were adjustable parameters of the electrostatic model.

G. Triple Layer Model

Triple layer model TLM is very popular and it is build into many commercial speciation programs. It combines the 2-pK model, i.e. reactions (5.32) and (5.33) with binding of inert electrolyte:

$$\equiv SOH_2^+ + X^- = \equiv SOH_2^+ \cdots X^- \tag{5.46}$$

$$\equiv SO^- + Y^+ = \equiv SO^- \cdots Y^+ \tag{5.47}$$

As discussed above for the 1-pK model with binding of inert electrolyte (cf. Section F) the equilibrium constants of reactions (5.46) and (5.47) must be equal (although a TLM version with asymmetrical inert electrolyte binding was also considered), and some electrostatic model is required to calculate the potential in the β-plane. Historically TLM precedes the 1-pK model with inert electrolyte binding, and the original version published by Davis et al. [54] is still widely used.

Reactions (5.32), (5.33), (5.46), and (5.47) require three adjustable parameters, namely, N_s, ΔpK_a, and K(reaction (5.46)). Two additional adjustable parameters, namely, the capacitance between the surface and the β-plane (inner capacitance, C_1), and the capacitance between the β-plane and the d-plane (center of the charge of diffuse layer) (outer capacitance, C_2) are required by the electrostatic model.

Davis et al. [54] designed a graphical extrapolation procedure to calculate parameters of TLM from surface charging data. Similar graphical procedure was used to estimate the ΔpK_a from electrokinetic data [55]. Bousse and Meind [56] postulate combination of potentiometric titration data with surface potential data

(ISFET) in modeling of surface charging. Modified procedure to optimize the TLM parameters was proposed (among others) by Righetto et al. [57]. Use of adsorption data of ions of inert electrolyte was also postulated. However, most often the TLM parameters are optimized by means of specially designed computer programs using only the surface charging data. Usually a set of TLM parameters complying (more or less exactly) the input surface charging data can be readily found. However, such a set is not unambiguous, i.e. many other sets comply the same surface charging data nearly equally well. Hayes et al. [58] (among others) demonstrated the possibility to obtain practically identical sets of charging curves for three ionic strengths, using different sets of TLM parameters. This aspect of the TLM model is probably not generally realized, at least reading some papers reporting the fitted TLM parameters one can hardly avoid such an impression. Therefore, it must be emphasized that the sets of TLM parameters used in the model calculations presented below are not unique, and many other sets give nearly equally good fits. Systematic studies of the effect of particular variables on the fit error are not difficult to conduct, but rather difficult to present, namely the fit error is a function of five variables. Therefore the analysis presented here is limited to a few characteristic features.

The ability of the TLM model to simulate the surface charging of alumina is illustrated in Figs. 5.69–5.72, and the sets of TLM parameters used in adsorption modeling are listed in Table 5.18.

With one exception of Fig. 5.72 addition of four adjustable parameters resulted in substantially better fit than in the 1-pK-diffuse layer model. The fit in Fig. 5.70 is equally good as in 2-pK- diffuse layer model. The fit in Figs. 5.69 and 5.71 is better than for any combination of 1-pK or 2-pK model with diffuse layer or Stern model. This improvement of fit is correlated with the values of the $\log K$ (reaction (5.46)) in Table 5.18. The $\log K$(reaction (5.46)) > 1 suggests that over the studied range of ionic strengths, the surface species involving the ions of inert electrolyte substantially contribute to the surface charging. On the other hand $\log K$(reaction (5.46)) = 0 suggests that even for the highest ionic strength (0.1 mol dm^{-3}) considered here, the contribution of the surface species involving the ions of inert electrolyte is rather insignificant.

The goodness of fit for the data set presented in Fig. 5.69 was rather insensitive to ΔpK_a and C_2 (although low C_2 about 0.2 F/m^2 gives systematically better fit than higher C_2). This means that although TLM has in principle five adjustable parameters, actually only three parameters were fitted. The TLM simulation in Fig. 5.69 properly reflects the ionic strength effect on the surface charging.

TABLE 5.18 N_s in C m^2, ΔpK_a, $\log K$(reaction 5.46), C_1, and ΨC_2 (both in F/m^2) Used to Calculate the Model Curves in Figs. 5.69–5.72 (σ_0), 5.73–5.76 (Ψ_0), 5.77–5.80 (Ψ_β), 5.81–5.84 (Ψ_d) and 5.105–5.119 (specific adsorption)

Fig.				N_s	ΔpK_a	$\log K$ (reaction 5.46)	C_1	C_2
5.69	5.73	5.77	5.81	0.1	-4 (fixed)	1.76	0.7	0.2 (fixed)
5.70	5.74	5.78	5.82	1	3.76	0	1.2	0.3
5.71	5.75	5.79	5.83	0.17	-4 (fixed)	1.72	0.8	0.2 (fixed)
5.72	5.76	5.80	5.84	0.1	1.36	0.06	1.6	4 (fixed)

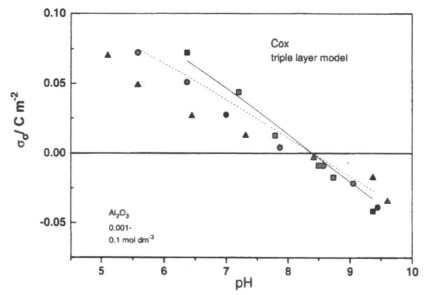

FIG. 5.69 Experimental (symbols) and calculated (lines) surface charge density of alumina as the function of the pH and ionic strength: 0.1 mol dm^{-3} squares, solid; 0.01 mol dm^{-3} circles, dash; 0.001 mol dm^{-3} triangles, dot. For model parameters, cf. Table 5.18.

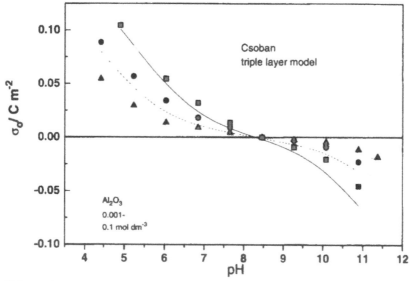

FIG. 5.70 Experimental (symbols) and calculated (lines) surface charge density of alumina as the function of the pH and ionic strength: 0.1 mol dm^{-3} squares, solid; 0.01 mol dm^{-3} circles, dash; 0.001 mol dm^{-3} triangles, dot. For model parameters, cf. Table 5.18.

The ionic strength effect on the surface charging is slightly overestimated in Fig. 5.70. In contrast with other sets of experimental data analyzed here, the range of the model parameters allowing to obtain equally good fit was rather narrow in this case.

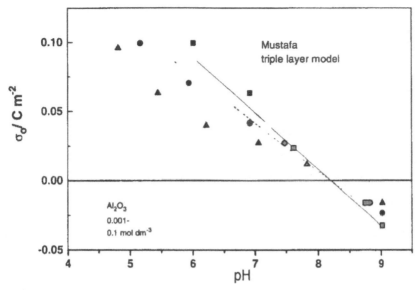

FIG. 5.71 Experimental (symbols) and calculated (lines) surface charge density of alumina as the function of the pH and ionic strength: 0.1 mol dm^{-3} squares, solid; 0.01 mol dm^{-3} circles, dash; 0.001 mol dm^{-3} triangles, dot. For model parameters, cf. Table 5.18.

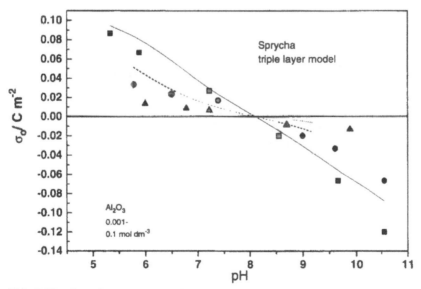

FIG. 5.72 Experimental (symbols) and calculated (lines) surface charge density of alumina as the function of the pH and ionic strength: 0.1 mol dm^{-3} squares, solid; 0.01 mol dm^{-3} circles, dash; 0.001 mol dm^{-3} triangles, dot. For model parameters, cf. Table 5.18.

Analogously as in Fig. 5.69, also the goodness of fit for the data set presented in Fig. 5.71 was rather insensitive to ΔpK_a and C_2. Also with the data set presented in Fig. 5.72, the goodness of fit was rather insensitive to C_2, but in this case high C_2 about 4 F/m^2, gave systematically better fit than low C_2. The ionic strength effect on the surface charging was slightly overestimated at low pH and slightly under-estimated at high pH.

The surface potential calculated for the model parameters from Table 5.18 is shown in Figs. 5.73–5.76. With one exception of Fig. 5.76 the surface potential is

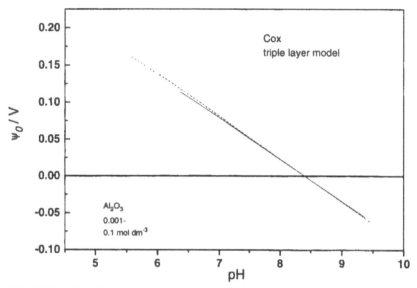

FIG. 5.73 Calculated surface potential of alumina as the function of the pH and ionic strength: 0.1 mol dm⁻³ solid; 0.01 mol dm⁻³ dash; 0.001 mol dm⁻³ dot. For model parameters, cf. Table 5.18.

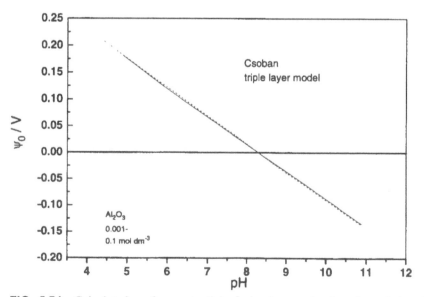

FIG. 5.74 Calculated surface potential of alumina as the function of the pH and ionic strength: 0.1 mol dm⁻³ solid; 0.01 mol dm⁻³ dash; 0.001 mol dm⁻³ dot. For model parameters, cf. Table 5.18.

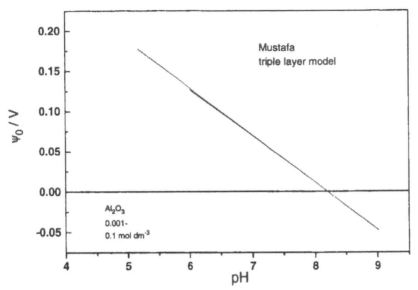

FIG. 5.75 Calculated surface potential of alumina as the function of the pH and ionic strength: 0.1 mol dm^{-3} solid; 0.01 mol dm^{-3} dash; 0.001 mol dm^{-3} dot. For model parameters, cf. Table 5.18.

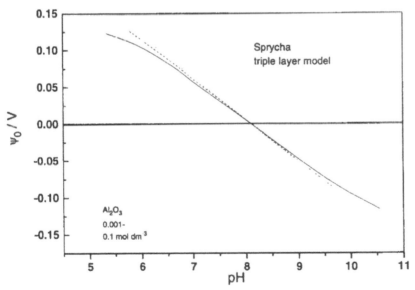

FIG. 5.76 Calculated surface potential of alumina as the function of the pH and ionic strength: 0.1 mol dm^{-3} solid; 0.01 mol dm^{-3} dash; 0.001 mol dm^{-3} dot. For model parameters, cf. Table 5.18.

nearly Nernstian even far from the PZC and at high ionic strengths. This is due to high C_2 in the latter and low C_2 in the other sets of model curves. The potential in the β-layer calculated for the model parameters from Table 5.18 is shown in Figs. 5.77–5.80. With high log K(reaction (5.46)) (Figs. 5.77 and 5.79) and high ionic strengths

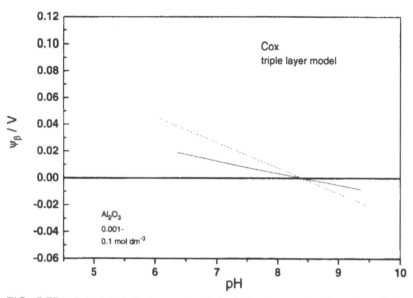

FIG. 5.77 Calculated β-plane potential of alumina as the function of the pH and ionic strength: 0.1 mol dm^{-3} solid; 0.01 mol dm^{-3} dash; 0.001 mol dm^{-3} dot. For model parameters, cf. Table 5.18.

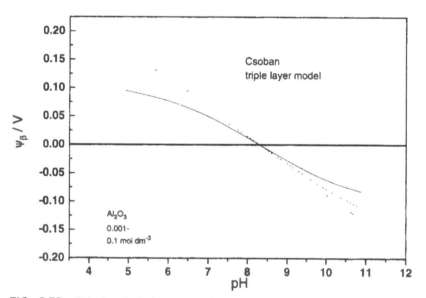

FIG. 5.78 Calculated β-plane potential of alumina as the function of the pH and ionic strength: 0.1 mol dm^{-3} solid; 0.01 mol dm^{-3} dash; 0.001 mol dm^{-3} dot. For model parameters, cf. Table 5.18.

the calculated ψ_β is lower (in absolute value) than typical ζ potentials reported for alumina at similar experimental conditions (e.g. [59]). This result does not support the presented model, namely, with inert electrolyte the potential is a monotonic

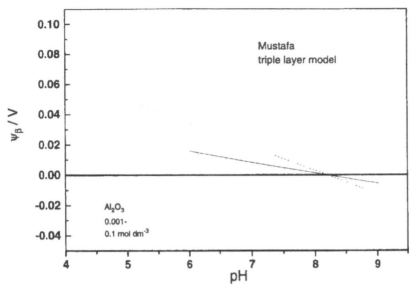

FIG. 5.79 Calculated β-plane potential of alumina as the function of the pH and ionic strength: 0.1 mol dm^{-3} solid; 0.01 mol dm^{-3} dash; 0.001 mol dm^{-3} dot. For model parameters, cf. Table 5.18.

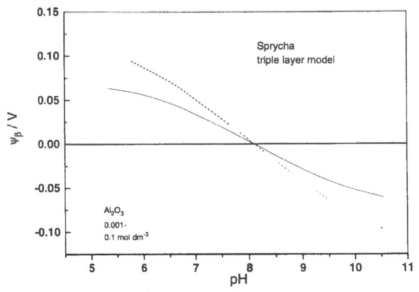

FIG. 5.80 Calculated β-plane potential of alumina as the function of the pH and ionic strength: 0.1 mol dm^{-3} solid; 0.01 mol dm^{-3} dash; 0.001 mol dm^{-3} dot. For model parameters, cf. Table 5.18.

function of the distance from the surface. In this respect, the calculated ψ_β presented in Figs. 5.78 and 5.80 are more realistic.

The potential in the d-layer calculated for the model parameters from Table 5.18 is shown in Figs. 5.81–5.84. It is often assumed that $\psi_d \approx \zeta$. Therefore the

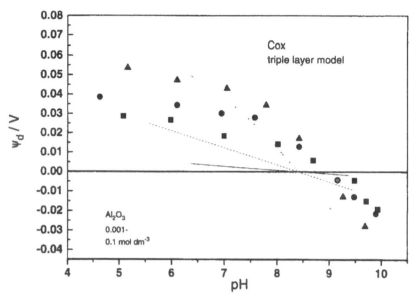

FIG. 5.81 Experimental ζ potentials (symbols) and calculated d-plane potentials (lines) of alumina as the function of the pH and ionic strength: 0.1 mol dm^{-3} squares, solid; 0.01 mol dm^{-3} circles, dash; 0.001 mol dm^{-3} triangles, dot. For model parameters, cf. Table 5.18.

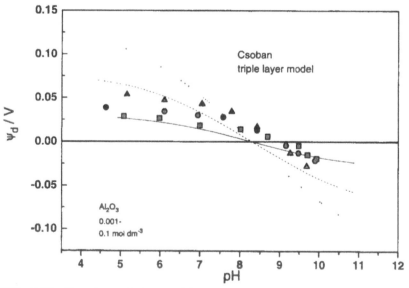

FIG. 5.82 Experimental ζ potentials (symbols) and calculated d-plane potentials (lines) of alumina as the function of the pH and ionic strength: 0.1 mol dm^{-3} squares, solid; 0.01 mol dm^{-3} circles, dash; 0.001 mol dm^{-3} triangles, dot. For model parameters, cf. Table 5.18.

calculated ψ_d is compared with ζ potentials of two aluminas taken from Refs. [59] and [60]. These ζ potentials are typical for alumina, for other representative results, cf. Figs. 3.84–3.86. The IEP of these two aluminas are somewhat higher than the

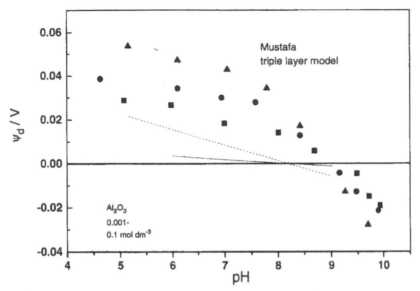

FIG. 5.83 Experimental ζ potentials (symbols) and calculated d-plane potentials (lines) of alumina as the function of the pH and ionic strength: 0.1 mol dm^{-3} squares, solid; 0.01 mol dm^{-3} circles, dash; 0.001 mol dm^{-3} triangles, dot. For model parameters, cf. Table 5.18.

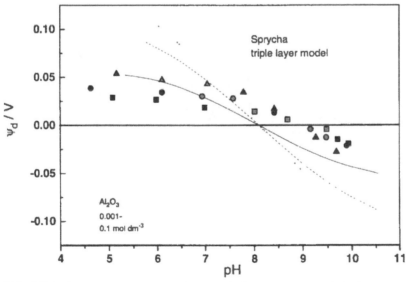

FIG. 5.84 Experimental ζ potentials (symbols) and calculated d-plane potentials (lines) of alumina as the function of the pH and ionic strength: 0.1 mol dm^{-3} squares, solid; 0.01 mol dm^{-3} circles, dash; 0.001 mol dm^{-3} triangles, dot. For model parameters, cf. Table 5.18.

PZC in the model calculations, thus, only qualitative agreement is examined. Such qualitative agreement is only obtained at one ionic strength. For all four data sets the present approach tends to overestimate the ionic strength effect on the ζ

potential. Especially, with high log K(reaction (5.46)) (Figs. 5.81 and 5.83) the ζ potential at high ionic strength is severely underestimated.

The TLM model can also be used for silica type surfaces. The number of adjustable parameters is the same as for amphoteric surfaces, namely pK(reaction (5.33)) [which is equal to log K(reaction (5.25))] is fitted instead of ΔpK_a. Limitations discussed above for the amphoteric surfaces are also valid for silica-type surfaces. The ability of the TLM model to simulate the surface charging of silica is illustrated in Figs. 5.85–5.89, and the sets of TLM parameters used in adsorption modeling are listed in Table 5.19.

The improvement of the fit with respect to models using fewer adjustable parameters was rather modest. The fit in Fig. 5.85 was slightly worse than with diffuse layer model. The fit in Fig. 5.86 was nearly as good as with diffuse layer model. The fit in Fig. 5.87 was better than with diffuse layer model, but worse than

TABLE 5.19 N_s in C/m^2, pK(reaction 5.33), log K(reaction 5.46), C_1, and C_2 (both in F/m^2) Used to Calculate the Model Curves in Figs. 5.85–5.89 (σ_0), 5.90–5.94 (Ψ_0), 5.95–5.99 (Ψ_3), 5.100–5.104 (Ψ_d), and 5.124–5.131 (Specific adsorption)

Fig.				N_s	pK(reaction 5.33)	log K(reaction 5.46)	C_1	C_2
5.85	5.90	5.95	5.100	2	8.27	0.48	1.3	0.2 (fixed)
5.86	5.91	5.96	5.101	1.5	7.05	0.52	1.7	0.2 (fixed)
5.87	5.92	5.97	5.102	0.85	7.82	0.61	0.9	4 (fixed)
5.88	5.93	5.98	5.103	4.3	7.51	0.33	1.4	2 (fixed)
5.89	5.94	5.99	5.104	1.7	7.63	1.31	1.4	1 (fixed)

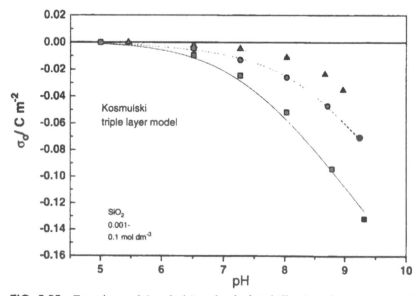

FIG. 5.85 Experimental (symbols) and calculated (lines) surface charge density of silica as the function of the pH and ionic strength: 0.1 mol dm^{-3} squares, solid; 0.01 mol dm^{-3} circles, dash; 0.001 mol dm^{-3} triangles, dot. For model parameters, cf. Table 5.19.

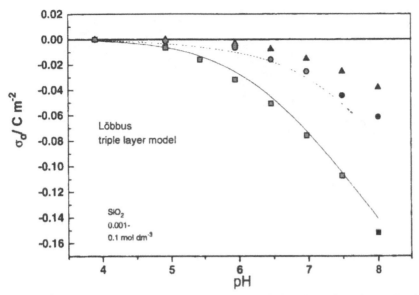

FIG. 5.86 Experimental (symbols) and calculated (lines) surface charge density of silica as the function of the pH and ionic strength: 0.1 mol dm^{-3} squares, solid; 0.01 mol dm^{-3} circles, dash; 0.001 mol dm^{-3} triangles, dot. For model parameters, cf. Table 5.19.

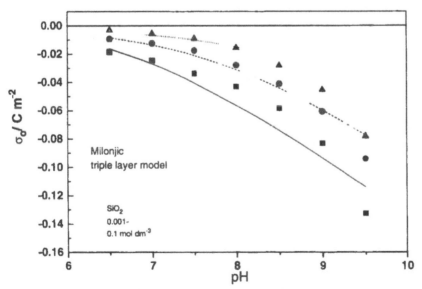

FIG. 5.87 Experimental (symbols) and calculated (lines) surface charge density of silica as the function of the pH and ionic strength: 0.1 mol dm^{-3} squares, solid; 0.01 mol dm^{-3} circles, dash; 0.001 mol dm^{-3} triangles, dot. For model parameters, cf. Table 5.19.

with Stern model. The fit in Fig. 5.88 was slightly better than with Stern model and substantially better than with diffuse layer model. The fit in Fig. 5.89 was slightly better than with diffuse layer model. With all data sets the goodness of fit was rather insensitive to C_2. The ionic strength effect on the σ_0 is overestimated at high pH in

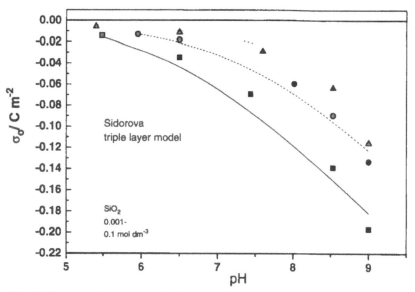

FIG. 5.88 Experimental (symbols) and calculated (lines) surface charge density of silica as the function of the pH and ionic strength: 0.1 mol dm^{-3} squares, solid; 0.01 mol dm^{-3} circles, dash; 0.001 mol dm^{-3} triangles, dot. For model parameters, cf. Table 5.19.

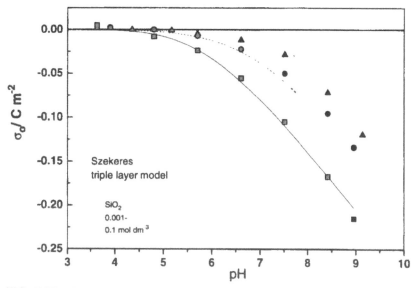

FIG. 5.89 Experimental (symbols) and calculated (lines) surface charge density of silica as the function of the pH and ionic strength: 0.1 mol dm^{-3} squares, solid; 0.01 mol dm^{-3} circles, dash; 0.001 mol dm^{-3} triangles, dot. For model parameters, cf. Table 5.19.

Figs. 5.87–5.89. The surface potential calculated using the model parameters from Table 5.19 is plotted in Figs. 5.90–5.94. This surface potential is much less sensitive to the ionic strength than the surface potential calculated using the diffuse

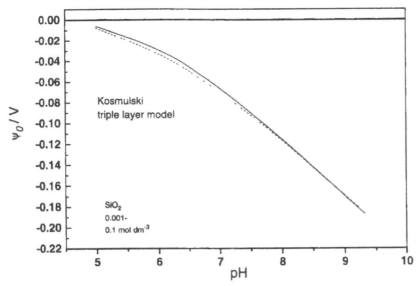

FIG. 5.90 Calculated surface potential of silica as the function of the pH and ionic strength: 0.1 mol dm^{-3} solid; 0.01 mol dm^{-3} dash; 0.001 mol dm^{-3} dot. For model parameters, cf. Table 5.19.

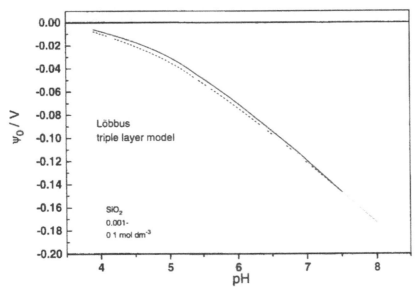

FIG. 5.91 Calculated surface potential of silica as the function of the pH and ionic strength: 0.1 mol dm^{-3} solid; 0.01 mol dm^{-3} dash; 0.001 mol dm^{-3} dot. For model parameters, cf. Table 5.19.

layer model (cf. Figs. 5.52–5.56). The potential in the β-layer calculated for the model parameters from Table 5.19 is shown in Figs. 5.95–5.99. The potential in the d-layer calculated for the model parameters from Table 5.19 is shown in Figs. 5.100–5.104.

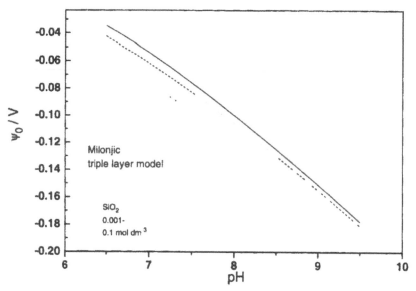

FIG. 5.92   Calculated surface potential of silica as the function of the pH and ionic strength: 0.1 mol dm^{-3} solid; 0.01 mol dm^{-3} dash; 0.001 mol dm^{-3} dot. For model parameters, cf. Table 5.19.

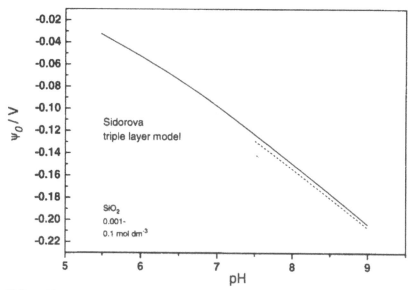

FIG. 5.93   Calculated surface potential of silica as the function of the pH and ionic strength: 0.1 mol dm^{-3} solid; 0.01 mol dm^{-3} dash; 0.001 mol dm^{-3} dot. For model parameters, cf. Table 5.19.

The calculated ψ_d is compared with ζ potentials of silica taken from Ref. [61]. These ζ potentials are typical for silica, for other representative results cf. Figs. 3.90–3.93. Reasonable qualitative agreement between ψ_d and ζ is observed in Figs. 5.101

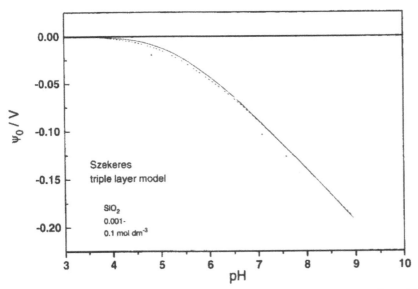

FIG. 5.94 Calculated surface potential of silica as the function of the pH and ionic strength: 0.1 mol dm^{-3} solid; 0.01 mol dm^{-3} dash; 0.001 mol dm^{-3} dot. For model parameters, cf. Table 5.19.

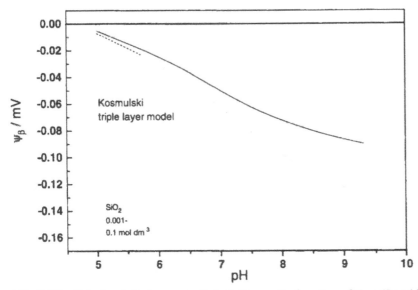

FIG. 5.95 Calculated β-plane potential of silica as the function of the pH and ionic strength: 0.1 mol dm^{-3} solid; 0.01 mol dm^{-3} dash; 0.001 mol dm^{-3} dot. For model parameters, cf. Table 5.19.

and 5.103. For the other data sets the ψ_d calculated at low pH is substantially lower than the experimental ζ. This result does not support the present model.

The calculated model parameters for silica (Table 5.19) are more consistent than for alumina (Table 5.18). Certainly, many other sets of TLM parameters can be

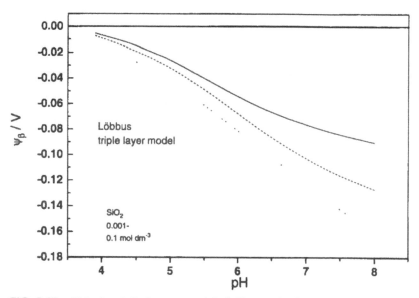

FIG. 5.96 Calculated β-plane potential of silica as the function of the pH and ionic strength: 0.1 mol dm^{-3} solid; 0.01 mol dm^{-3} dash; 0.001 mol dm^{-3} dot. For model parameters, cf. Table 5.19.

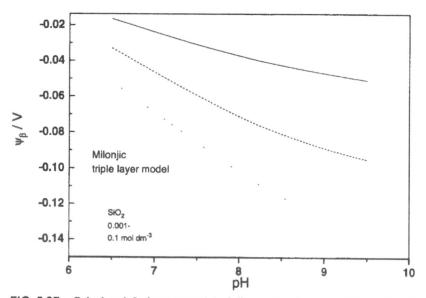

FIG. 5.97 Calculated β-plane potential of silica as the function of the pH and ionic strength: 0.1 mol dm^{-3} solid; 0.01 mol dm^{-3} dash; 0.001 mol dm^{-3} dot. For model parameters, cf. Table 5.19.

used to obtain nearly identical model curves for the both materials. The results obtained with different sets of TLM parameters can be to some degree assessed using the ζ potential data, as illustrated in Figs. 5.81–5.84 and 5.101–5.103.

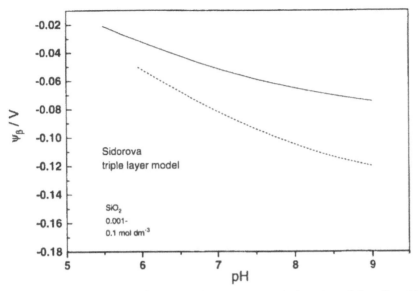

FIG. 5.98 Calculated β-plane potential of silica as the function of the pH and ionic strength: 0.1 mol dm^{-3} solid; 0.01 mol dm^{-3} dash; 0.001 mol dm^{-3} dot. For model parameters, cf. Table 5.19.

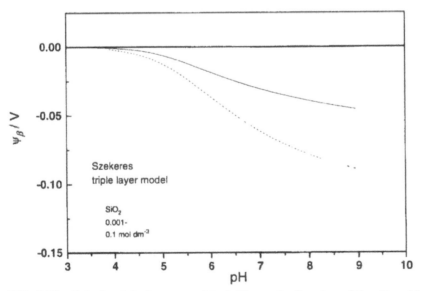

FIG. 5.99 Calculated β-plane potential of silica as the function of the pH and ionic strength: 0.1 mol dm^{-3} solid; 0.01 mol dm^{-3} dash; 0.001 mol dm^{-3} dot. For model parameters, cf. Table 5.19.

Assuming that the TLM parameters have some physical sense, attempts to find the correlation between these parameters and other physical quantities are quite natural. The significance of such correlations is limited by the discussed above difficulties in finding an unique set of TLM parameters.

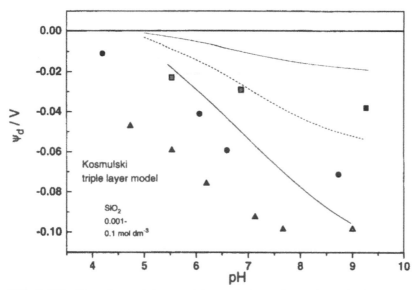

FIG. 5.100 Experimental ζ potentials (symbols) and calculated d-plane potentials (lines) of silica as the function of the pH and ionic strength: 0.1 mol dm^{-3} squares, solid; 0.01 mol dm^{-3} circles, dash; 0.001 mol dm^{-3} triangles, dot. For model parameters, cf. Table 5.19.

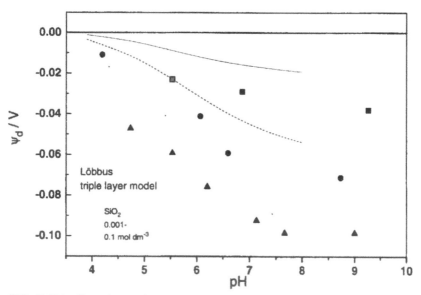

FIG. 5.101 Experimental ζ potentials (symbols) and calculated d-plane potentials (lines) of silica as the function of the pH and ionic strength: 0.1 mol dm^{-3} squares, solid; 0.01 mol dm^{-3} circles, dash; 0.001 mol dm^{-3} triangles, dot. For model parameters, cf. Table 5.19.

1.   Correlations Between TLM Parameters and Other Physical Quantities

Sverjensky and Sahai [62] found the following correlations between ΔpK_a and v Pauling bond strength:

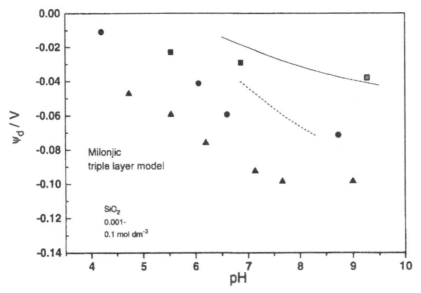

FIG. 5.102 Experimental ζ potentials (symbols) and calculated d-plane potentials (lines) of silica as the function of the pH and ionic strength: 0.1 mol dm^{-3} squares, solid; 0.01 mol dm^{-3} circles, dash; 0.001 mol dm^{-3} triangles, dot. For model parameters, cf. Table 5.19.

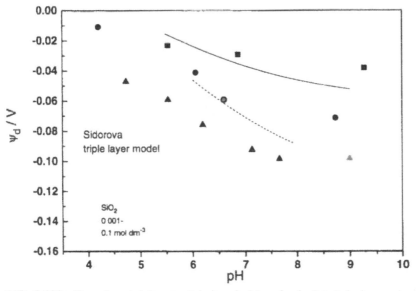

FIG. 5.103 Experimental ζ potentials (symbols) and calculated d-plane potentials (lines) of silica as the function of the pH and ionic strength: 0.1 mol dm^{-3} squares, solid; 0.01 mol dm^{-3} circles, dash; 0.001 mol dm^{-3} triangles, dot. For model parameters, cf. Table 5.19.

For constant capacitance model:

$$\Delta pK_a = (17.579 \, \nu/L) + 0.418 \tag{5.48}$$

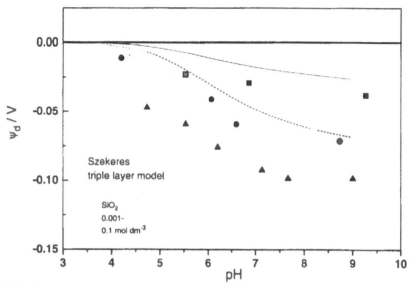

FIG. 5.104 Experimental ζ potentials (symbols) and calculated d-plane potentials (lines) of silica as the function of the pH and ionic strength: 0.1 mol dm^{-3} squares, solid; 0.01 mol dm^{-3} circles, dash; 0.001 mol dm^{-3} triangles, dot. For model parameters, cf. Table 5.19.

For constant capacitance model (10 sites nm^{-2}, 1 Fm^{-2}, another set of experimental data):

$$\Delta pK_a = (23.2 - 4.6 \log I)(\nu/L) - 1.6 \tag{5.49}$$

For diffuse layer model (10 sites nm^{-2}, 0.01–0.03 mol dm^{-3} NaNO$_3$)

$$\Delta pK_a = (30.01\,\nu/L) - 2.238 \tag{5.50}$$

For TLM:

$$\Delta pK_a = (12.692\,\nu/L) + 3.537 \tag{5.51}$$

where L is metal–hydrogen distance and I is the ionic strength. The parameters in Eq. (5.48)–(5.51) were fitted using the ΔpK_a values taken from literature.

Sahai and Sverjensky [63] calculated TLM parameters for 26 sets of charging curves taken from literature. The following procedure was used. The N_s for given material was calculated from selected tritium-exchange or infrared data or estimated from crystallographic data. The ΔpK_a was calculated from Eq. (5.51). Inert electrolyte binding constants (asymmetrical binding allowed, cf. Section 2) and the inner capacitance (outer capacitance was assumed to be 0.2 C m^{-2}) were fitted numerically. The same authors [64] claimed a correlation between the equilibrium constants of reactions (5.46) and (5.47) and the dielectric constant of the adsorbent, i.e.

$$\log K(\text{Li}) = -9.741\varepsilon^{-1} + 3.616 \tag{5.52}$$

TABLE 5.20 C_1 Values for Common 1 1 Salts According to Sahai and Sverjensky [64]

Salt	C_1/Fm^{-2}
$NaClO_4$	1.5
$LiNO_3$	1.4
$NaNO_3$	1.2
$CsCl$, KNO_3, KCl	1.1
$LiCl$, $NaCl$, NaI	1

Analogous correlations with different values of the coefficients on the r.h.s. of Eq. (5.52) were found for other monovalent metal cations. These coefficients were found to be correlated with crystallographic ionic radii, namely

$$\log K(\text{reaction (5.47)}) = -[41.5(r_i + 1.4492)^{-1} - 5.72][\ln(10)RT\varepsilon]^{-1}$$
$$+ 3.842(r_i + 1.4492)^{-1} + 1.2 \tag{5.53}$$

for monovalent cations. In Eq. (5.53), taken verbatim from the original publication, ångstroms were used as units of length and kcal were used as units of energy. Similar equation was proposed for monovalent anions except the correction term 1.4492 Å in Eq. (5.53) disappears.

Thus, according to Sahai and Sverjensky, the binding constants for monovalent cations and anions can be calculated when their crystallographic ionic radii and the dielectric constant of the adsorbent are known. Moreover, an empirical equation was proposed to calculate the inner capacitance as a function of the aqueous solvation coefficient and aqueous effective radius. These two parameters depend only on the nature of the 1–1 salt, but they are independent of the adsorbent, and the C_1 values are listed in Table 5.20.

Since Eqs. (5.51) and (5.53) can be combined with one of Eqs. (3.21)–(3.23) (for PZC) it appears that the TLM parameters for any adsorbent-1–1 electrolyte combination can be predicted when v/L, dielectric constant of the adsorbent, and crystallographic ionic radii of the anion and cation of the electrolyte are known. The inner capacity can be taken from Table 5.20, and the right choice of the N_s value (cf. Tables 5.1–5.3) seems to be the only source of uncertainty.

2. Modifications of the TLM Model

The classical version of TLM produces model charging curves symmetrical with respect to the PZC. The experimental curves are often unsymmetrical. To account for this effect two different methods were used. This certainly requires additional adjustable parameters.

One method is to allow unsymmetrical inert electrolyte binding, i.e. different equilibrium constants of reactions (5.46) and (5.47). This approach was used (among others) by Katz and Hayes [65], and many different sets of model parameters were found to simulate the surface charging of alumina equally well. With unsymmetrical inert electrolyte binding the PZC is a function of the ionic strength, and this is in conflict with the assumption that PZC = CIP (Chapter 3).

Unsymmetrical inert electrolyte binding leads to semantic problems: even at pristine conditions PZC becomes ionic strength dependent, the surface potential is not equal to zero at the PZC, etc., cf. Eqs. (3.2) and (3.3). Fortunately such problems are encountered in modeling exercises with unsymmetrical electrolyte binding, but not in reality. Anyway, this modification is not recommended. Ultimately a TLM with seven adjustable parameters (equilibrium constants of reactions (5.32), (5.33), (5.46), and (5.47) are unrelated) can be used to model raw titration data (the charging curves were not shifted to produce PZC = CIP) [66], but physical meaning of such model exercises is questionable.

Alternatively, the C_1 can be different on the both sides of the PZC. This corresponds to adsorption of anions and cations of inert electrolyte at different distances from the surface. This approach was discussed above for Stern model (Fig. 5.65), e.g. sodium cations are believed to approach closer the surface than chloride anions [67]. Asymmetry of the surface potential (pH) curves (ISFET) with respect to PZC was emphasized, but this effect was not accounted for in the calculations [18]. Charmas et al. [68] used a term "four layer model" for modification of TLM with different C_+ and C_-. This term was originally introduced by Barrow and Bowden [69] for their model of specific adsorption of anions (cf. Section IV). In the original four layer model the fourth layer is the layer of specifically adsorbed anions (between the surface and the β-layer) while the anions and cations of inert electrolyte are at the same distance from the surface. To avoid confusion the term "four layer model" is reserved here for the model by Barrow and Bowden. The TLM with different C_+ and C_- can lead to a substantial difference between calculated PZC and IEP for inert electrolytes. For example Fig. 3 in Ref. [70] suggests that the IEP is substantially shifted to lower pH with respect to PZC at ionic strength as low as 10^{-2} mol dm^{-3} when $C_{cation} = 1.2$ and $C_{anion} = 1.1$ Fm^{-2} (in contrast with the symbols used in this chapter, C_+ refers to negatively charged surface in Ref. [70] and subsequent papers of the same authors). Rudzinski et al. [71, 72] analyzed the temperature effects on the surface charging and results of calorimetric experiments in terms of TLM with different C_+ and C_-. The inner capacitance was found to be temperature dependent. TLM with different C_+ and C_- was further developed to account for surface heterogeneity [73].

A non-electrostatic model of adsorption of H$^+$/OH$^-$ ions with inert electrolyte binding has been also published [74]. Such a model is in conflict with existence of electrokinetic phenomena and can be only considered as data fitting model.

Different aspects of the TLM model were discussed in review papers by Kallay et al. [75, 76].

H. Permanent Charge

The models presented in Sections C–G are well known and generally accepted, but their application is limited to certain class of materials. Many materials carry permanent negative charge which is rather insensitive to the pH. Clay minerals are typical permanently charged minerals but some amount of permanent charge can be also present in materials in which the variable (pH dependent) charge prevails. Thus, models considering both variable and permanent charge can be useful for a broad class of materials. Acid-base chemistry of permanently charged minerals was recently discussed by Kraepiel et al. [77], who considered two models: with the permanent (pH independent) charge located on the surface on the one hand, and with permanent charge homogeneously distributed in the bulk solid on the other. In the both models amphoteric groups responsible for the variable charge are located on the surface. In the both models (2 pK, cf. Section D), the PZC can be found as

$$PZC = 0.5\,(pK_{a1} + pK_{a2}) - \log(e)\,F\psi_{PZC}/RT \tag{5.54}$$

and since ψ_{PZC} is ionic strength dependent, so is the PZC. Thus PZC cannot be found as the CIP (Fig. 3.3) for materials having substantial amount of permanent charge, and the role of the PZC in the adsorption modeling is less significant than with materials having only variable charge.

The relationship between ψ_{PZC} and ionic strength depends on the model. With fixed charge on the surface

$$F\psi_{PZC}/RT = 2\,\text{arcsinh}\,(\sigma_{fix}/0.11741^{1/2}) \tag{5.55}$$

where σ_{fix} can be fitted using titration data. With fixed charge homogeneously distributed in the bulk solid

$$F\psi_{PZC}/RT = (2I/\rho_p)[1 - \cosh\,(F\psi_p/RT)] + F\psi_p/RT \tag{5.56}$$

where

$$F\psi_p/RT = \text{arcsinh}\,(\rho_p/2I) \tag{5.57}$$

and ρ_p is a model parameter that can be fitted using titration data. The later model was used to explain adsorption of protons by montmorillonite as a function of pH at three ionic strengths, and the former was applied for kaolinite (also for three ionic strengths). The model (and experimental) charging curves of kaolinite at different ionic strengths do not have a common intersection point. The surface charge density is always more negative at high ionic strength, the charging curves merge at pH about 7 and diverge at lower and higher pH.

I. Salts

The models of surface charging presented in Sections C–G were originally designed for sparingly soluble simple metal (hydr)oxides, but nowadays these models are widely used for other materials, e.g. mixed oxides, sparingly soluble salts, and even complex and ill defined materials like soils, sediments and rocks. Indeed, some of these materials exhibit a sharp CIP of titration curves and pH independent IEP. One-site approach to these materials (e.g. 1-pK model) can be a convenient data fitting model. On the other hand more involved models are necessary to understand the surface charging mechanism. Even with simple (hydr)oxides the surface charging models with many types of surface sites (MUSIC) have been invoked.

Experimental studies of pH dependent surface charging of sparingly soluble salts were discussed in Section 3.1.G. Electrokinetic studies prevail, but some potentiometric titration data are also available. As discussed above, surface charging data are more suitable than electrokinetic data to fit the parameters of the adsorption models.

A model involving the following surface species: $\equiv CO_3H$, $\equiv CO_3^-$, $\equiv CO_3Me^+$, $\equiv MeOH$, $\equiv MeOH_2^+$, $\equiv MeO^-$, $\equiv MeHCO_3$, and $\equiv MeCO_3^-$ was proposed [78] to describe surface charging of sparingly soluble carbonates of divalent metals (Me = Mn(II), Fe(II) or Ca). In the model calculations the partial pressure of CO_2, total concentration of dissolved metal ions and pH are treated as three independent variables. This approach is different from that discussed in Section 3.I.G, where these variables were interrelated by the solubility product of sparingly soluble carbonate. Certainly, independent fitting of too many adjustable parameters (concentrations of particular types of surface sites and their stability constants, and parameters required by the electrostatic model) does not produce unequivocal results. Thus, the stability constants of the surface complexes were assigned the values of stability constants of analogous solution complexes.

Surface charging of ZnS was interpreted in terms of the following surface species: $\equiv SZn$, $\equiv SH_2$, $\equiv SZnOH^-$, $\equiv ZnS$ and $\equiv ZnSH^+$ [79]. Stability constants of the surface species were fitted to potentiometric titration data combined with the readings of the S^{2-} ion selective electrode.

It should be emphasized that the surface properties of metal sulfides are very sensitive to the redox conditions.

IV. SPECIFIC ADSORPTION

A. Adsorption as Special Case of Sorption

The experimental facts presented in Chapter 4 suggest that binding of strongly interacting ions by adsorbents is a complex multistep process. Most likely adsorption is the first step and it is followed by slower diffusion and recrystallization processes. Consequently the modeling approach should be adjusted to the time scale and other experimental conditions.

The surface complexation approach discussed in this section is suitable for adsorption, which dominates at relatively short equilibration times and relatively low concentrations of the adsorbate (up to a few hours, and up to 10^{-3} mol dm^{-3}, respectively, for typical experimental conditions). In principle this model is not suitable for long equilibration times or high adsorbate concentrations. Successful applications of surface complexation model SCM to uptake data obtained by coprecipitation (e.g., Ref. [80]) and other sorption experiments which can be hardly described as adsorption has been reported. In such instances, however, SCM should be rather regarded as a data fitting model than as a mechanistic model.

For longer equilibration times and higher adsorbate concentrations, the problems faced by sorption modeling have complete different character than the topics discussed in the present section. For instance, Manceau et al. [81] found that the saturation concentration of Co soluble species with respect to various Co clays was significantly lower than that with respect to solid $Co(OH)_2$. Thus, formation of these clays as a result of Co sorption on silica is favored over surface precipitation of $Co(OH)_2$. This result was also supported by analysis of Co K-edge EXAFS spectra of Co-rich kerolite on the one hand, and quartz with sorbed Co on the other. Calculation of Co saturation curves with respect to Co silicates is anything but trivial, e.g. the result depends on the method of assessment of the equilibrium $Si(OH)_4$ concentration. Therefore, unambiguous evidence of formation of a specific silicate phase was not found. Different clay phases can be favored over different pH ranges and at different ionic strengths.

Other similar results are reported in Tables 4.1. and 4.2, but models of sorption related to formation of new phases (e.g. surface precipitation model [82]) are beyond the scope of this chapter.

B. The Effect of Electric Potential on Specific Adsorption

As discussed in Section III, ions in the interfacial region experience a pH dependent electric potential. The concentrations of ions in the interfacial region substantially differ from their bulk concentrations (Eq. (5.1)). Agashe and Regalbuto [83] argue that sorption of heavy metal cations on oxides can be quantitatively modeled using their dehydratation energies and electrostatics, while the other effects are small or non-existing. However in general opinion adsorption modeling requires combination of Eq. (5.1) with suitable Mass Law expression. This leads to expressions similar to Eq. (5.24) for thermodynamic (pH and ionic strength independent) equilibrium constants of surface reactions responsible for the specific adsorption.

Non-electrostatic models [84–86] using only the Mass Law and neglecting Eq. (5.1) have been also proposed. These models are based on the linear relationship between log (uptake/{[free Me^{2+} in solution]×solid to liquid ratio}) and the pH (concentration of free Me^{2+} species is calculated from the total concentration of the metal in solution). The slope of such straight lines was (erroneously) interpreted as the average number of protons released per one metal cation in the following reaction:

$$\text{cation in solution} + \text{surface site} = \text{cation adsorbed} + r \text{ protons in solution}$$

(5.58)

$$K(\text{reaction}(5.58)) = \theta(1 - \theta)^{-1}a^{-1} \times 10^{-r \times pH}$$

(5.59)

where the activity on r.h.s. of Eq. (5.59) refers to free Me^{2+} cation. Eq. (5.59) should be regarded as a data fitting model. The parameter r combines the actual number of protons released and the Boltzmann factor and it depends on the experimental conditions. The significance of the parameter r in Eq. (5.59) was discussed in more detail by Kosmulski [87]. When hydrolysis in solution is negligible, the above non-electrostatic model produces a linear dependence between K_D (Eq. (4.9)) and the pH [88]. The Boltzmann factor was also neglected in the model by Tamura et al. [89], who assumed formation of two surface complexes of Co, namely, $\equiv SOCo^{+}\cdots NO_3^{-}$ and $(\equiv SO)_2Co$ with five different metal oxides. The effective stability constants of these species depend on the coverage, to account for lateral interactions (Frumkin adsorption isotherm). The electrostatic potential experienced by specifically adsorbing ions is due to a combination of the primary surface charging on the one hand, and specific adsorption itself on the other. The latter contribution may be negligible when the total concentration of the specifically adsorbing substance is sufficiently low. Properly formulated Mass Law equation(s) and electrostatic model for the primary surface charging are also valid in the presence of strongly interacting species. In this respect models reflecting the actual charging mechanism have advantage over data fitting models which are only valid at certain experimental conditions. In view of contradictory surface charging results reported in the literature for common materials (cf. Figs. 3.43–3.73) it would be much desired to use a model that reflects the primary surface charging of the specific sample of the material used in studies of specific adsorption. Indeed, many publications report an experimental study of primary surface charging and its interpretation as a preliminary work. Then specific adsorption is studied with the same lot of material. In other studies of specific adsorption, primary surface charging experiments were not performed with the sample of interest, but entire model obtained for similar material was accepted from literature. Such a model is likely to fail with another sample of material having the same chemical formula (cf. Figs. 3.43–3.73) although the surface charging data for some materials are relatively consistent. Finally in a few publications model parameters taken from different sources were combined. Such an approach can give model charging curves corresponding to the actual behavior of the specific sample of interest only by fortuitous coincidence.

Almost identical charging curves, more or less exactly reflecting the experimental σ_0, are produced by different models of primary surface charging (Figs. 5.39–5.104), and one of such models can be arbitrarily chosen for modeling of specific adsorption. This choice determines some options in modeling of specific adsorption. For example the 1-pK and 2-pK models (with or without counterion binding) involve only one type of surface sites, and the same surface sites are then assumed to bind the strongly interacting ions in the modeling of specific adsorption.

Multisite approach (MUSIC, Section III.E) to primary surface charging allows more subtle modeling of specific adsorption, e.g. the type of surface sites that do not participate in primary surface charging (because these sites are fully protonated of fully deprotonated over the pH range of interest) can very well be engaged in specific adsorption, and the sites contributing to the primary surface charge not necessarily do participate in specific adsorption. In multisite models the specific adsorption properties can differ from one sample of material having the same chemical formula to another, e.g. because different crystallographic planes are exposed in different proportions. Verification of multisite models requires well defined absorbent on the one hand, and spectroscopic evidence for surface species engaged in the adsorption on the other. Most published models of specific adsorption were based merely on the uptake data, thus, simplified approach to the surface charging, with single type of surface sites is understandable. Multisite approach can be also used for modeling of specific adsorption on less well defined materials, but then the model has rather speculative character (cf. Section V).

Also the choice of the electrostatic model for the interpretation of primary surface charging plays a key role in the modeling of specific adsorption. It is generally believed that the specific adsorption occurs at the distance from the surface shorter than the closest approach of the ions of inert electrolyte. In this respect only the electric potential in the inner part of the interfacial region is used in the modeling of specific adsorption. The surface potential can be estimated from Nernst equation, but this approach was seldom used in studies of specific adsorption. Diffuse layer model offers one well defined electrostatic position for specific adsorption, namely the surface potential calculated in this model can be used as the potential experienced by specifically adsorbed ions. The Stern model and TLM offer two different electrostatic positions each, namely, the specific adsorption of ions can be assumed to occur at the surface or in the β-plane.

Paradoxically, models of primary surface charging using greater number of adjustable parameters, are in some sense more attractive than models with fewer parameters, from the point of view of modeling of specific adsorption, because of the possibility to calculate the electric potential at two electrostatic position. The Stern or TLM models with different C_+ and C_- offer a possibility to calculate the electric potential at three different electrostatic positions, but to the best knowledge of the present author this possibility to refine electrostatic models of specific adsorption was not explored.

Many publications report models with specific adsorption on the surface or in the β-plane or mixture thereof (formation of one species with adsorbate in the surface plane and another species with adsorbate in the β-plane). Hayes and Leckie [90] successfully modeled the ionic strength effect on the uptake (pH) curves assuming adsorption of heavy metal cations on the surface and adsorption of the alkaline earth metal cations in the β-plane. Alternative interpretation of their experimental results was proposed by Lützenkirchen [91]. The surface species with heavy metal cations located in the surface plane are often referred to as inner sphere surface complexes. The surface species with heavy metal cations located in the β-plane are referred to as outer sphere surface complexes. The inner- and outer-sphere surface complexes differ in the distance between the surface atoms and the adsorbed metal ions, and this distance (for certain systems and at certain experimental conditions) can be estimated from X-ray absorption data. The outer sphere complexes in Tables in Chapter 4 are written with a separator "··" between the inner and the outer part, e.g. $\equiv FeOH_2^+ \cdot\cdot S_2O_3^{2-}$. This seperator is also used for inert electrolyte binding (cf. Section III. F). Most species used in modeling of specific adsorption (cf. Chapter 4) are inner sphere complexes, which are written without separator.

Other publications postulate specific adsorption between the surface and the β-plane. For example Barrow and Bowden [69] interpreted adsorption of anions on goethite in terms of the mentioned above four layer model. The four layers are (in order of increasing distance from the surface): surface layer (H^+ and OH^- ions), the layer of specifically adsorbed anions, the first layer of inert electrolyte counterions (analogous to the β-layer in TLM), and diffuse layer. This model requires an additional adjustable parameter, namely, the capacitance between the surface and the layer of specifically adsorbed anions. Barrow and Bowden report 2.99 F m^{-2} for phosphate and 60,000 F m^{-2} (!?) for silicate. The fit in the four layer model was substantially better than with simpler models for fluoride adsorption, but for other anions equally good fit could be obtained without introducing the additional electrostatic plane. In another paper of this series the capacitance of 3–5 F m^{-2} was used in model calculations of phosphate adsorption on aluminum and iron oxides [92]. Similar approach was used by Venema et al. [93] who applied the 1-pK model to interpret the Cd binding by goethite. The Cd^{2+} ions were assumed to adsorb in the "1"-plane, between the surface and the β-plane. The inner capacitance of 1.85 Fm^{-2} and the outer capacitance of 2.71 Fm^{-2} were the adjustable parameters of the model. This model was more successful in modeling the Cd adsorption than the TLM—inner sphere model (asymmetric inert electrolyte binding was allowed). The above electrostatic models treat the charges of specifically adsorbed ions as point charges.

In the recent CD [94] (charge distribution) model the charge of the specifically adsorbed ions is distributed between the surface plane and another electrostatic plane (whose distance from the surface corresponds roughly to the β-plane distance), thus, the center of charge of specifically adsorbed ions is located between these two planes. The CD concept was originally introduced to model adsorption of oxyanions. The negative charge of a specifically adsorbed oxoanion it attributed to its oxygen atoms. For geometrical reasons all four oxygen atoms of the phosphate or sulfate anion cannot be simultaneously located on the surface. Consequently only a part of the negative charge of a specifically adsorbed oxoanion is attributed to the surface plane. The rest of this negative charge is located in the "1"-plane within the Stern layer, i.e. closer to the surface than the distance of closest approach of the inert electrolyte ions. Thus, instead on one Stern capacitance, two capacitances are defined in the original CD model, and the overall Stern layer capacitance is calculated from the formula for two capacitors in series. In view of rather short distance between the "1"-plane and the Stern layer plane, the outer capacitance is rather high (about 5 F m^{-2}) and the overall Stern layer capacitance is nearly equal to that of the inner part. In addition to the above capacitances, another adjustable parameter was defined for each surface species, namely, f the fraction of the charge of a specifically adsorbed ion attributed to the surface plane. For instance, with $f = 0.5$, two oxygen atoms of the sulfate or phosphate anion are attributed to the surface, and with $f = 0.75$ three oxygen atoms of the sulfate or phosphate anion are attributed to the surface. The CD-MUSIC model (combination of the CD concept with the MUSIC model, cf. Section III.E) was tested using experimental data on phosphate adsorption on goethite from different sources, and obtained by different methods. The model involved mono- and bidentate surface complexes (in a bidentate surface complex two surface oxygen atoms bind one specifically adsorbed ion, cf. Section 3). Only indirect effect of the surface ($\equiv Fe_3$)OH$^{-1\ 2}$ groups on phosphate uptake was considered, namely, via their contribution to the surface charging. CD-MUSIC model was also used to model uptake of sulfate on goethite [95]. The CD-MUSIC concept was successful in explaining the spectroscopic and uptake data on selenate IV sorption on goethite [96]. The difficulty in explaining this data in terms of the 2-pK model was that the EXAFS data suggest a bidentate inner sphere complex, while the uptake data suggest co-adsorption of one proton with one adsorbed selenate IV anion, and this is rather consistent with monodenatate binding (in terms of 2-pK model). With MUSIC this discrepancy disappears, namely in the vicinity of PZC (which roughly coincides with selenate IV adsorption edge in the analyzed experimental data) the average proton stoichiometry is defined by the following reaction

$$\equiv FeOH^{-1/2} + \equiv FeOH_2^{+1/2} + H^+ + SeO_3^{2-} = (\equiv FeO)_2SeO^- + 2H_2O \quad (5.60)$$

whose proton stoichiometry leads to pH dependence close to that found experimentally. In the same publication [96] the MUSIC CD model was successfully applied to interpret the anion uptake in two-component systems. It should be emphasized that models derived from behavior of one-component systems often fail to properly explain the adsorption competition (cf. Chapter 4. VI).

A study of Cd adsorption on goethite was the first application of the CD model for cations [97]. Distribution of the positive charge of adsorbed cation between the surface and the "1" plane is explained by attributing this positive charge to the first shell of oxygen atoms around the cation. For a typical CN of 6, no more than three of six oxygen atoms can be simultaneously located on the surface for geometrical reasons.

The CD model [97] was used in combination with the MUSIC model. Contributions of the (110) face and (021) face to the total surface were estimated as 90% and 10%, respectively from the shape of the goethite crystals. Analysis of the surface composition of these faces led to the following site densities (nm^{-2}): 3 singly and 3 triply coordinated sites at the (110) face, and 7.5 singly and 3.75 doubly coordinated sites at the (021) face, respectively. Two Stern layer capacitances were 1.02 Fm^{-2} (inner) and 5 Fm^{-2} (outer) for the both faces, and protonation constants of singly and triply coordinated sites were both 9.3 (equal to the PZC). Counterion binding constant (cf. Section III. F) log K = -1 was chosen. The above parameters define the primary surface charging behavior of goethite. In order to obtain results consistent with EXAFS data two types of Cd surface complexes were considered, namely, $(\equiv FeOH)_2Cd^+$, log K 6.9, f 0.3 at the (110) face, and $[(\equiv FeOH)_2Fe_2OH]Cd$, (involving two singly and one doubly coordinated site) log K 9, f 0.58 at the (021) face. The fitted f values are consistent with the theoretical prediction, namely, for bidentate Cd surface complex two of six (1/3) and with tridentate three of six (1 2) ligands of Cd belong to the surface. Thus, only the log K values are fully adjustable parameters in this model. Very good agreement between calculated and experimental adsorption isotherms covering four decades of Cd concentrations, as well as between the calculated and experimental proton stiochiometry was obtained. The hypothesis that high affinity surface complexes are formed on the (021) face was also tested by performing similar experiments with another sample of goethite with shorter needles (the 021 face estimated as 20% of the surface). However, no significant difference in Cd uptake was observed between these two samples. In a subsequent paper [98] of the same authors an additional monodentate Cd complex log K 5.5, f 0.15 at the (110) face was introduced to explain enhanced uptake of Cd in the presence of phosphate. Binding constants and f values previously found for Cd and phosphate (in case of phosphate the binding constants and f values found for the entire surface were used for (110) and (021) faces) were used and ternary surface complexes (involving Cd and phosphate simultaneously) were not invoked.

The CD concept can be also combined with the Stern or TLM model without introduction of the additional electrostatic plane. This offers a possibility of formulation of problem in terms of the CD model within standard features offered by commercial speciation programs. Recent study of Gd and Ni adsorption on alumina in the framework of TLM model produced significantly better fit for the charge of the specifically adsorbed cations distributed 50–50 between the surface and the β-plane than for inner- or outer-sphere complexation [99].

C. Formulation of the Stability Constants of Surface Complexes

The values of stability constants of surface complexes and other parameters used in modeling of specific adsorption were reported in preceding section of this chapter and in Tables 4.1 and 4.2, but these stability constants were not yet defined. Explicit definition of the stability constant for each surface species separately would be rather space consuming, considering the number of surface species used in different publications to model specific adsorption, as reported in Chapter 4. Fortunately, there are generally accepted standards by formulation of the stability constants of surface complexes. Most stability constants whose values are reported in Tables 4.1 and 4.2 were defined according to these standards. Thus, the formula of the surface complex is sufficient to write the corresponding surface reaction, whose equilibrium constant is reported in Chapter 4, and for a few exceptions (stability constants defined in unusual way) the equations of surface reactions are written *in extenso*, or the stability constants were redefined to meet the standards, and their numerical values reported in Chapter 4 were recalculated.

The reaction responsible for the specific adsorption can be written in the following general form:

$$\text{surface site(s)} + \text{ions adsorbed from solution} = \text{surface species} +$$

$$\text{ions released into solution} \tag{5.61}$$

In speciation algorithms (Section II), reaction (5.61) is written in an equivalent from:

$$\text{surface site(s)} + \text{ions adsorbed from solution} - \text{ions released into solution}$$

$$= \text{surface species} \tag{5.62}$$

The equilibrium constant of reaction (5.62) is the stability constant of the surface species of interest, and values of this equilibrium constant are reported in Chapter 4. For monodentate surface complexes (one surface site on l.h.s. of reaction (5.62)):

$$K(\text{reaction }(5.62)) = [\text{surface species}] \times \Pi \{\text{ions released into solution}\} \times \Pi$$

$$\{\text{ions adsorbed from solution}\}^{-1} \times [\text{surface site}]^{-1} \times \text{Boltzmann factor,} \tag{5.63}$$

where the products of activities {} on the r.h.s. take into account the stoichiometric coefficients. The Boltzmann factor in Eq. (5.63) is defined as a product of the factors

$$\exp(\pm z_i e \psi_i / kT) \tag{5.64}$$

over all ionic reactants i whose solution activities appear in Eq. (5.63), taking into account the stoichiometric coefficients, where ψ_i is the electrostatic potential experienced by certain ion in the interfacial region (electrostatic positions and thus ψ_i can be different for particular ions) and the sign in (5.64) is plus for the ions adsorbed from solutions and minus for ions released into solution, and z_i is positive for cations and negative for anions.

For example, formation of the $\equiv\text{TiOCd}^+$ surface species (2-pK or TLM model) can be defined by the following reactions (among other):

$$\equiv \text{TiOH} + \text{Cd}^{2+} - \text{H}^+ = \equiv \text{TiOCd}^+ \tag{5.65}$$

$$\equiv \text{TiO}^- + \text{Cd}^{2+} = \equiv \text{TiOCd}^+ \tag{5.66}$$

$$\equiv \text{TiOH} + \text{CdOH}^+ - \text{H}_2\text{O} = \equiv \text{TiOCd}^+ \tag{5.67}$$

$$\equiv \text{TiO}^- + \text{CdOH}^+ - \text{OH}^- = \equiv \text{TiOCd}^+ \tag{5.68}$$

Each of the reactions (5.65)–(5.68) could be potentially used to define the stability constant of the surface complex $\equiv\text{TiOCd}^+$, but they have differently defined equilibrium constants (Eq. (5.63)), thus the numerical values of their equilibrium constants are also different. This would complicate comparison of results from different sources, unless some common standard is established. In most publications cited in Table 4.1 the following rules are observed.

- Neutral surface groups ($\equiv$TiOH in the present example) are the components (in the 2-pK or TLM model)
- Unhydrolyzed metal cations (Cd^{2+} in the present example) are the components (even when their actual concentration is low compared with hydroxycomplexes over the pH range of interest)
- Proton is the component (i.e. the OH^- ions do not appear in reactions defining the stability constants of surface complexes)

This leads to unique definition of the stability constants of monodentate surface complexes involved in the specific adsorption of cations. Thus, reaction (5.65) rather than reactions (5.66)–(5.68) should be chosen as a standard definition, according to the above standards. The exact definition of the stability constant of the $\equiv$TiOCd$^+$ species in the present example depends on the electrostatic position of Cd in the surface complex. It should be also emphasized that even with fixed electrostatic position of Cd, the numerical value of the equilibrium constant of reaction (5.65) depends on the choice of the model of primary surface charging. The details on the models of primary surface charging are not given in the tables in Chapter 4, and the reader is referred to the original publications.

Applications of the 1-pK or MUSIC models to surface complexation are rather rare, but a few examples can be found in Tables 4.1 and 4.2. These models require somewhat different standard definition of the stability constants of the surface species than outlined above for the 2-pK and TLM model, because neutral $\equiv$SOH groups do not occur 1-pK model. Deprotonated surface species, e.g. $\equiv$AlOH$^{1/2-}$ (cf. reaction(5.23)) are usually set as the components in the definition of the surface complexes responsible for the specific adsorption in the framework of the 1-pK and MUSIC models. There is some discord between such definition and the above definition in the framework of the 2-pK model. For example the $\equiv$SiOH species is used as the component in the 2-pK and TLM models, and the $\equiv$SiO$^-$ species is suggested by the definition proposed for the 1-pK and MUSIC models.

It would be beneficial to establish some common standard set of components also for specific adsorption of anions, as outlined above for cations, but unfortunately, such standard definition of the stability constant has not been set yet. The choice of neutral $\equiv$SOH groups (in the 2-pK and TLM models) and protons as the components (i.e. the same standards as for cations) is generally accepted, but everything from undissociated acid molecules to anions having the highest possible degree of deprotonation was selected as components in different publications (cf. Table 4.2). The values of stability constants of surface complexes depend on the selection of components, but the results can be easily re-calculated into stability constants defined by some standard approach when consecutive dissociation constants of the weak acid of inerest are known. This would facilitate comparison of results reported in different sources. However, in view of rather contradictory dissociation constants reported in literature for some important adsorbates (e.g. molybdic acid), this ambitious task was given up, and the numerical values of the stability constants of surface complexes reported in original publications are repeated in Table 4.2 without recalculation. For each publication the choice of the component (neutral acid molecule, monovalent anion, divalent anion, etc.) is specified. May the models of specific adsorption of anions be defined according to some common standard in future publications. Use of neutral acid molecules as the components seems to be a natural choice.

The above discussion was focussed on monodentate surface species. With bi- and tridentate surface species another controversy should be addressed, with respect to definition of stability constants of a surface complex (involving anion or cation). Let us consider the following surface reaction:

surface site (1) + surface site (2) + ions adsorbed from solution −

ions released into solution = bidentate surface complex   (5.69)

The stability constant of the bidentate surface complex in reaction (5.69) is often defined as:

K(reaction (5.69)) = [bidentate surface complex] × Π {ions released into

solution} × Π{ions adsorbed from solution}$^{-1}$ × {[surface site (1)] ×

[surface site (2)]}$^{-1}$ × Boltzmann factor   (5.70)

The product [surface species (1)] × [surface species (2)] in Eq. (5.70) expresses probability of finding the molecules of a two species of interest together. This is true for independent molecules or ions in solution but not for surface sites. Therefore such an approach to definition of stability constants of bi- and tridentate surface complexes has been widely criticized [100]. Nevertheless, definitions similar to Eq. (5.70), i.e. using the product (in case of different sites) or n-th power (in case of identical sites) of concentrations of surface sites are often encountered in the literature.

The values of stability constants defined in this manner are reported in Tables in Chapter 4 without any comment or correction, although Eq. (5.70) is not recommended as the definition of the stability constant for reaction (5.69).

The importance of using activity rather than concentration in definition of the equilibrium constants of chemical reactions was already emphasized in Section I. Nevertheless, in many original publications, the stability constants of surface species are defined in terms of concentrations. These stability constants are reported in Tables in Chapter 4 without any comment of correction, although the approach neglecting the activity coefficients in solution is not recommended.

D. Simple Models

Models of specific adsorption with single surface species (involving the specifically adsorbed ion) and with one site model of primary surface charging will be presented in this section. Many stability constants reported in Chapter 4 refer to such models. The present model calculations illustrate some aspects regarding the limitations of significance of the stability constants of surface complexes reported in Tables 4.1 and 4.2. The problem is similar as discussed in Section III for primary surface charging: many different models represent the experimental data nearly equally well, but publication of one set of best-fit model parameters may create an illusion that the unique model has been found.

In the present section the following questions will be discussed in detail:

- How the formula of the surface complex (including the electrostatic position of specifically adsorbed ion) affects the shapes of the calculated uptake (pH) curves at different ionic strengths.
- How the chosen model of primary surface charging affects the numerical value of the stability constant of the surface complex (models discussed in section III will be used as examples).

It is well known that parameters of the model of primary surface charging affect the best-fit value of the stability constant of the surface complex responsible for specific adsorption. For example Katz and Hayes [65] report the best fit log K for the $\equiv AlOCo^+$ surface complex (defined in standard way, cf. Section 3) ranging from -1.6 to -0.6 for different sets of TLM parameters (these TLM parameters produced practically identical charging curves). In contrast with the stability constant defined in standard way, the equilibrium constant of the surface reaction

$$\equiv AlO^- \cdots Y^+ + Me^{2+} = \equiv SOMe^+ + Y^+ \tag{5.71}$$

(Y is the cation of inert electrolyte) is rather insensitive to the choice of TLM parameters [101]. Certainly reaction (5.71) can be only defined for models with inert electrolyte binding, thus this approach does not apply to diffuse layer or Stern model.

The uptake(pH) curves were generated for two hypothetical materials, namely, alumina and silica whose surface area is 100 m^2/g, and whose pristine surface charging behavior corresponds to the model curves presented in Figs. 5.72 and 5.85, respectively. The model curves representing specific adsorption of Pb on these materials at low initial concentration ([surface sites] > > [total Pb]) were calculated to produce $pH_{50} = 5$ for 10 g solid/dm^3 and in the presence of 10^{-3} mol dm^{-3} inert electrolyte solution. For sufficiently low total Pb concentration the adsorption isotherms at constant pH are linear and the course of the calculated uptake(pH) curves is independent of the Pb initial concentration. The above solid to liquid ratio is typical for studies of specific adsorption, and $pH_{50} = 5$ is a realistic value for Pb adsorption on silica and alumina at this solid to liquid ratio, in view of the results of actual adsorption experiments compiled in Table 4.1. Lead has higher affinity to solid surfaces than most other divalent metal cations. The choice of the model curves from Fig. 5.72 (alumina) and 5.85 (silica), rather than model curves derived from any other set of experimental data analyzed in Section III, or calculated using any other model than TLM, or any other set of TLM parameters was arbitrary. This choice does not imply that TLM is favored over other models or that the experimental data used to derive these model curves are more reliable than other results used as examples in Section III.

The model representing Pb adsorption (within selected model of primary surface charging) involves the following variables:

- The electrostatic position of Pb
- The number of protons released per one adsorbed Pb

The value of the stability constant of the surface complex is defined by the condition: $pH_{50} = 5$ for 10 g solid/dm^3, in the presence of 10^{-3} mol dm^{-3} inert electrolyte. The analysis of the effect of the electrostatic position of Pb on the model uptake curves in limited to three possibilities: inner sphere (the adsorbed Pb experiences the surface potential), outer sphere (the adsorbed Pb experiences the β-plane potential), and CD model with $f=0.5$ (f is the fraction of the charge of specifically adsorbed ion attributed to the surface plane, in this case the charge of adsorbed Pb is distributed 50–50 between the surface and the β-plane. Two latter possibilities were only tested for TLM and Stern primary surface charging models. The model curves of specific adsorption change continuously between $f=0$ (outer sphere) to $f=1$ (inner sphere), thus the behavior of CD model with $f\neq0.5$ can be easily obtained by interpolation from the curves presented in this section. The number of protons released per one adsorbed Pb was set to 0 or 1 or 2. The activity coefficient of Pb species in solution was estimated from Davies formula.

Alumina carries high positive and silica carries negative surface charge at pH 5 (the pH_{50} assumed in the model calculations). Therefore the presented calculations represent specific adsorption at unfavorable electrostatic conditions on the one hand and favorable electrostatic conditions on the other. Detailed analysis is limited to specific adsorption of one divalent metal cation. One example for specific adsorption of trivalent metal cation, and one example for specific adsorption of an anion will be shown in the end of this section.

Many experimental studies of specific adsorption have been carried out at or near the PZC. Under such conditions the effect of choice of electrostatic model on the calculated specific adsorption is rather insignificant. Thus, experimental data obtained near the PZC are not particularly useful to test electrostatic models of specific adsorption.

The stability constant of the surface complex (calculated from the condition $pH_{50}=5$) depends on the assumed electrostatic position of Pb on the one hand and the number of protons released on the other.

Let us first consider specific adsorption at unfavorable electrostatic conditions (Pb adsorption on alumina). The numerical values of log K obtained for the TLM model (Fig. 5.72, Table 5.18) are summarized in Table 5.21.

Table 5.21 illustrates dramatic difference (almost eight decades per one proton released) in the stability constant calculated for different number of protons released per one adsorbed Pb assumed in the model calculations. The effect of the assumed electrostatic position of Pb is less significant, namely, only one order of magnitude in the stability constant between the inner and outer sphere complex. It should be emphasized that all these results were calculated using the same model for primary surface charging (one set of TLM parameters). Table 5.21 illustrates how limited is

TABLE 5.21 Logarithms of Stability Constant of the Alumina-Pb Surface Complex as the Function of the Electrostatic Position of Pb and the Number of Protons Released per One Adsorbed Pb

Protons released	0	1	2
Inner sphere	10.05	2.11	-5.81
CD f 0.5	9.58	1.65	-6.28
Outer sphere	9.11	1.18	-6.71

the significance of the stability constants of surface complexes reported without specified electrostatic position and the number of protons released.

The uptake (pH) curves calculated for the models summarized in Table 5.21 are presented in Figs. 5.105–5.116. All these models lead to s-shaped uptake curves

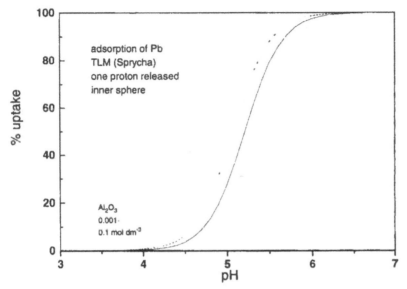

FIG. 5.105 Calculated uptake of Pb on alumina as the function of the pH and ionic strength: 0.1 mol dm^{-3} solid; 0.01 mol dm^{-3} dash; 0.001 mol dm^{-3} dot. Inner sphere complex, one proton released per one adsorbed Pb, TLM. For model parameters, cf. Tables 5.18 and 5.21.

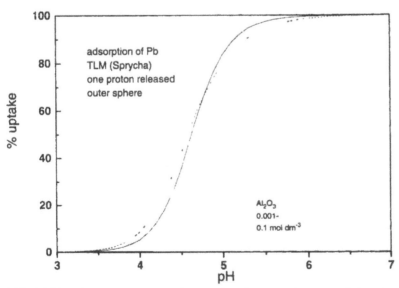

FIG. 5.106 Calculated uptake of Pb on alumina as the function of the pH and ionic strength: 0.1 mol dm^{-3} solid; 0.01 mol dm^{-3} dash; 0.001 mol dm^{-3} dot. Outer sphere complex, one proton released per one adsorbed Pb, TLM. For model parameters, cf. Tables 5.18 and 5.21.

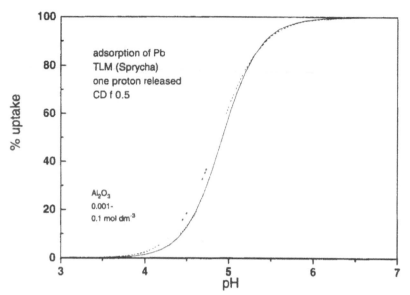

FIG. 5.107   Calculated uptake of Pb on alumina as the function of the pH and ionic strength: 0.1 mol dm^{-3} solid; 0.01 mol dm^{-3} dash; 0.001 mol dm^{-3} dot. CD, $f = 0.5$, one proton released per one adsorbed Pb, TLM. For model parameters, cf. Tables 5.18 and 5.21.

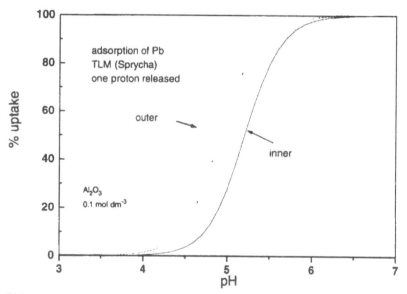

FIG. 5.108   Calculated uptake of Pb on alumina as the function of the pH and ionic strength 0.1 mol dm^{-3}: inner sphere complex solid; CD, $f = 0.5$ dash; outer sphere complex dot, one proton released per one adsorbed Pb, TLM. For model parameters, cf. Tables 5.18 and 5.21.

similar to that presented in Fig. 4.5 (A), but the ionic strength effects and the slopes of the curves differ from one set of model parameters to another. The uptake curves calculated for inner sphere binding of heavy metal cations are steeper than the curves calculated for outer sphere binding (Fig. 5.109), but the effect is not dramatic, and

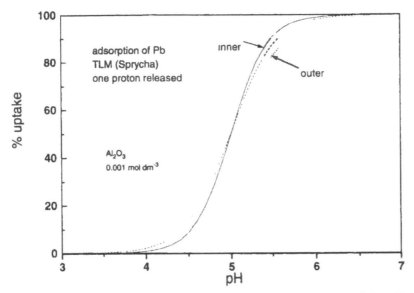

FIG. 5.109 Calculated uptake of Pb on alumina as the function of the pH at ionic strength 0.001 mol dm^{-3}: inner sphere complex solid; CD, $f = 0.5$ dash; outer sphere complex dot, one proton released per one adsorbed Pb, TLM. For model parameters, cf. Tables 5.18 and 5.21.

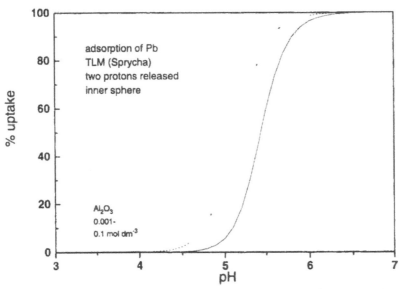

FIG. 5.110 Calculated uptake of Pb on alumina as the function of the pH and ionic strength: 0.1 mol dm^{-3} solid; 0.01 mol dm^{-3} dash; 0.001 mol dm^{-3} dot. Inner sphere complex, two protons released per one adsorbed Pb, TLM. For model parameters, cf. Tables 5.18 and 5.21.

with some scatter of experimental results, the f value can be only roughly estimated from experimental data. Figure 5.108 shows that the difference in slopes of the model curves calculated for 0.1 mol dm^{-3} of inert electrolyte for different f values is even less significant than for 0.001 mol dm^{-3} (Fig. 5.109).

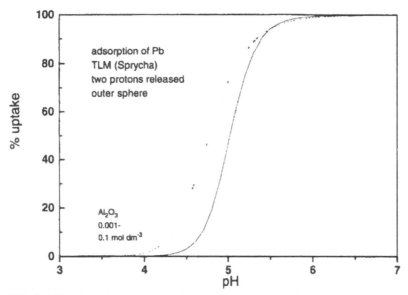

FIG. 5.111   Calculated uptake of Pb on alumina as the function of the pH and ionic strength: 0.1 mol dm^{-3} solid; 0.01 mol dm^{-3} dash; 0.001 mol dm^{-3} dot. Outer sphere complex, two protons released per one adsorbed Pb, TLM. For model parameters, cf. Tables 5.18 and 5.21.

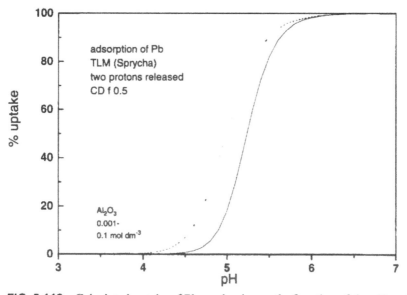

FIG. 5.112   Calculated uptake of Pb on alumina as the function of the pH and ionic strength: 0.1 mol dm^{-3} solid; 0.01 mol dm^{-3} dash; 0.001 mol dm^{-3} dot. CD, $f = 0.5$, two protons released per one adsorbed Pb, TLM. For model parameters, cf. Tables 5.18 and 5.21.

The ionic strength effect on the uptake curves calculated for various f values is significantly different. The uptake slightly decreases on increase of the ionic strength from 0.001 to 0.1 mol dm^{-3} for inner sphere complexation (Fig. 5.105), and increases for outer sphere complexation (Fig. 5.106), and with CD model (Fig. 5.107) the ionic

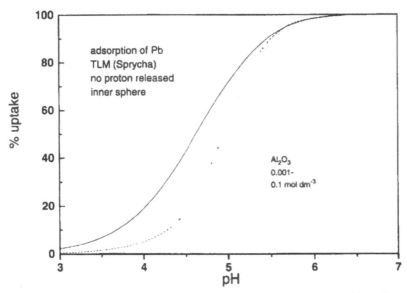

FIG. 5.113 Calculated uptake of Pb on alumina as the function of the pH and ionic strength: 0.1 mol dm⁻³ solid; 0.01 mol dm⁻³ dash; 0.001 mol dm⁻³ dot. Inner sphere complex, no proton released, TLM. For model parameters, cf. Tables 5.18 and 5.21.

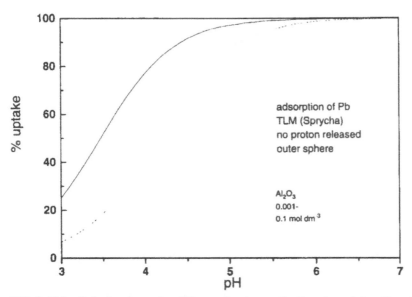

FIG. 5.114 Calculated uptake of Pb on alumina as the function of the pH and ionic strength: 0.1 mol dm⁻³ solid; 0.01 mol dm⁻³ dash; 0.001 mol dm⁻³ dot. Outer sphere complex, no proton released, TLM. For model parameters, cf. Tables 5.18 and 5.21.

strength effect is rather insignificant when release of one proton per one adsorbed Pb is assumed. On the other hand in models with two protons released per one adsorbed Pb the increase in the ionic strength from 0.001 to 0.1 mol dm⁻³ depresses the uptake (except for high pH with outer sphere complexation), and the effect is most

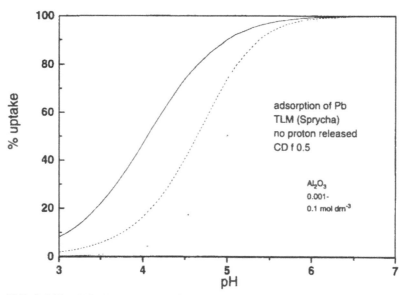

FIG. 5.115 Calculated uptake of Pb on alumina as the function of the pH and ionic strength: 0.1 mol dm^{-3} solid; 0.01 mol dm^{-3} dash; 0.001 mol dm^{-3} dot. CD, $f = 0.5$, no proton released, TLM. For model parameters, cf. Tables 5.18 and 5.21.

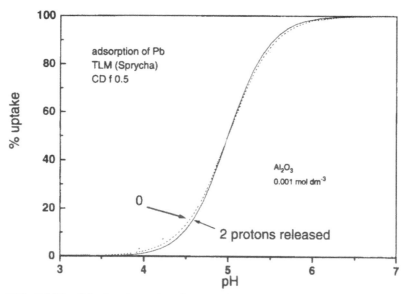

FIG. 5.116 Calculated uptake of Pb on alumina as the function of the pH at ionic strength 0.001 mol dm^{-3}: two protons released per one adsorbed Pb solid; 1 proton dash; no proton dot, CD, $f = 0.5$, TLM. For model parameters, cf. Tables 5.18 and 5.21.

significant with inner sphere complexation (Fig. 5.110) and least significant with outer sphere complexation (Fig. 5.111). Finally in a model assuming specific adsorption of Pb without proton release (Figs. 5.113–5.115) the increase in the ionic strength enhances the uptake, especially with outer sphere complexation. The calculated uptake curves are more steep for 2 protons released than with no protons released per one adsorbed Pb, but for low ionic strength the effect is not dramatic (Fig. 5.116). Thus, the ionic strength effect on uptake curves is more sensitive (than their slope) to the choice between models with different numbers of protons released per one adsorbed Pb. Then it may be surprising how many experimental studies of specific adsorption aimed at adsorption modeling were carried out at single ionic strength or without control of the ionic strength.

The above observations regarding the ionic strength effects (Figs 5.105–5.116) do not represent any general trends, namely, they are only valid for certain set of TLM parameters. Figures 5.112 and 5.117–5.119 show the ionic strength effect on the uptake curves calculated for the same electrostatic position of Pb (CD model, f 0.5), and the same number (two in the present example) of protons released per one adsorbed Pb, but using different sets of TLM parameters from Table 5.18. The stability constants of the alumina-Pb surface complex are presented in Table 5.22.

TABLE 5.22 The Stability Constants of Alumina-Pb Surface Complex (CD, f 0.5, 2 protons released) Used in Figs. 5.112 and 5.117–5.119. For TLM Parameters cf. Table 5.18

Fig.	TLM	Log K
5.117	Cox	-5.81
5.118	Csoban	-8.28
5.119	Mustafa	-6.3
5.112	Sprycha	-6.28

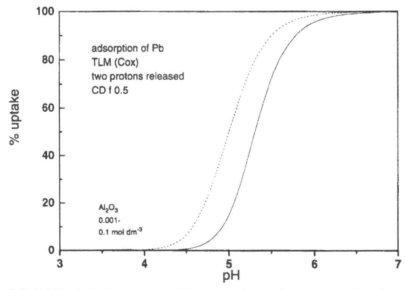

FIG. 5.117 Calculated uptake of Pb on alumina as the function of the pH and ionic strength: 0.1 mol dm^{-3} solid; 0.01 mol dm^{-3} dash; 0.001 mol dm^{-3} dot. CD, f= 0.5, two protons released per one adsorbed Pb, TLM. For model parameters, cf. Tables 5.18 and 5.22.

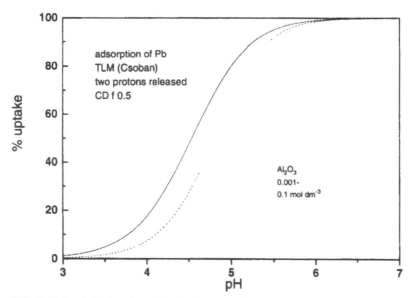

FIG. 5.118 Calculated uptake of Pb on alumina as the function of the pH and ionic strength: 0.1 mol dm^{-3} solid; 0.01 mol dm^{-3} dash; 0.001 mol dm^{-3} dot. CD, f=0.5, two proton released per one adsorbed Pb, TLM. For model parameters, cf. Tables 5.18 and 5.22.

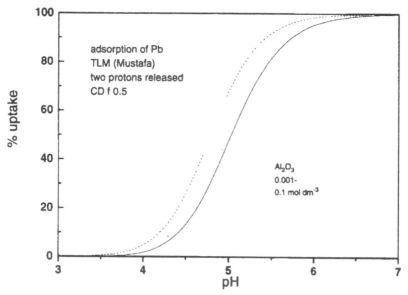

FIG. 5.119 Calculated uptake of Pb on alumina as the function of the pH and ionic strength: 0.1 mol dm^{-3} solid; 0.01 mol dm^{-3} dash; 0.001 mol dm^{-3} dot. CD, f=0.5, two proton released per one adsorbed Pb, TLM. For model parameters, cf. Tables 5.18 and 5.22.

For the same electrostatic position of Pb and the same number of protons released per one adsorbed Pb, two sets of TLM models (Figs. 5.112 and 5.117) give significant decrease in uptake on increase in the ionic strength from 0.001 to 0.1 mol dm^{-3}, with one set (Fig. 5.119) the effect was insignificant, and the fourth set (Fig. 5.118) predicts enhanced uptake at higher ionic strength. Similar discrepancies are obtained for different models of Pb adsorption. Table 5.22 illustrates the effect of the choice of the set of TLM parameters on the numerical value of the stability constant of the alumina-Pb surface complex defined in the same way (electrostatic position of Pb and the number of protons released). Three sets produced rather consistent K (difference by a factor of 3), but with the fourth set, K was lower by two orders of magnitude.

Thus, calculated K is only significant for certain set of TLM parameters.

The discrepancies between K values (characterizing specific adsorption) calculated for different sets of parameters of the model of primary surface charging are less significant with models having fewer adjustable parameters than TLM. The 1 pK-diffuse layer model was combined with the model of Pb adsorption assuming 1 proton released per one adsorbed Pb (inner sphere). In this model the ionic strength effect on the uptake curves is rather insignificant (Fig. 5.120).

Different parameters of diffuse layer model (cf. Table 5.12) produce rather consistent K (difference by a factor of 4, cf. Table 5.23).

TABLE 5.23 Calculated Stability Constants of Alumina-Pb Surface Complex (inner sphere, 1 proton released). For 1-pK-Diffuse Layer Model Parameters cf. Table 5.12

Diffuse layer model	log K
Cox	1.17
Csoban	1.4
Mustafa	0.83
Sprycha	0.98

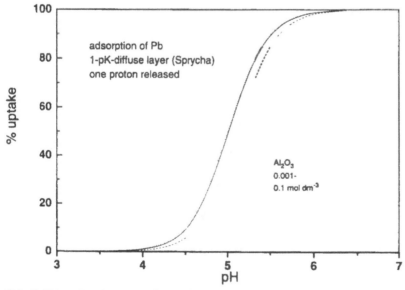

FIG. 5.120 Calculated uptake of Pb on alumina as the function of the pH and ionic strength: 0.1 mol dm^{-3} solid; 0.01 mol dm^{-3} dash; 0.001 mol dm^{-3} dot. One proton released per one adsorbed Pb, 1-pK-diffuse layer model. For model parameters, cf. Tables 5.12 and 5.23.

In view of different PZC in these four diffuse layer models, the K values in Table 5.23 are not fully correlated with N_s used in the diffuse layer model. Model uptake curves calculated for different diffuse layer models for 10^{-3} mol dm^{-3} inert electrolyte (only one such curve is shown in Fig. 5.120) are practically identical.

The 1-pK-Stern model was combined with the model of Pb adsorption assuming 1 proton released per one adsorbed Pb (inner sphere). In this model the ionic strength effect on the uptake curves is rather insignificant (Fig. 5.121). Different parameters of Stern model (cf. Tables 5.14 and 5.15) produce rather consistent K (difference by a factor of 4, cf. Table 5.24).

The K values in Tables 5.23 and 5.24 are not correlated, but the former are systematically higher. Model uptake curves calculated for different Stern models for 10^{-3} mol dm^{-3} inert electrolyte (only one such curve is shown in Fig. 5.121) are

TABLE 5.24 Calculated Stability Constants of Alumina-Pb Surface Complex (inner sphere, 1 proton released). For 1-pK-Stern Model Parameters cf. Tables 5.14 and 5.15

Stern model	log K
Cox	0.9
Csoban	0.8
Mustafa	0.29
Sprycha	0.91

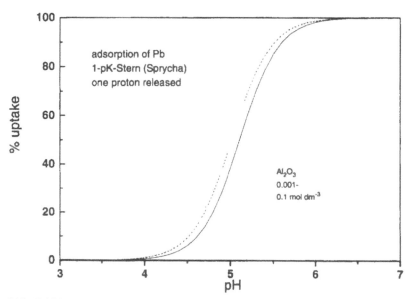

FIG. 5.121 Calculated uptake of Pb on alumina as the function of the pH and ionic strength: 0.1 mol dm^{-3} solid; 0.01 mol dm^{-3} dash; 0.001 mol dm^{-3} dot. One proton released per one adsorbed Pb, 1-pK-Stern model, inner sphere complex. For model parameters, cf. Tables 5.15 and 5.24.

practically identical. The stability constant of analogous alumina-Pb surface complex (inner sphere, 1 proton released) calculated with TLM derived from the same set of experimental primary charging curves (log K 2.11, Table 5.21) is significantly different from that calculated with Stern model (log K 0.91, Table 5.24) or diffuse layer model (log K 0.98, Table 5.23).

The 1-pK Stern model was also combined with the model of Pb adsorption assuming 1 proton released per one adsorbed Pb (CD, f 0.5). In this model the ionic strength effect on the uptake curves is rather insignificant (Fig. 5.122). Different parameters of Stern model (cf. Tables 5.14 and 5.15) produce rather consistent K (difference by a factor of 5, cf. Table 5.25).

Model uptake curves calculated for three Stern models for 10^{-3} mol dm^{-3} inert electrolyte can be hardly distinguished (cf. Fig. 5.123), and the fourth curve is slightly

TABLE 5.25 Calculated Stability Constants of Alumina-Pb Surface Complex (CD, f 0.5, 1 proton released). For Stern Model Parameters cf. Tables 5.14 and 5.15

Stern model	log K
Cox	0.66
Csoban	0.04
Mustafa	0.1
Sprycha	0.74

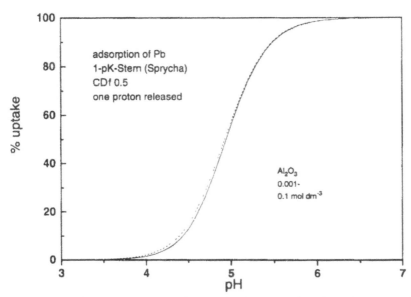

FIG. 5.122 Calculated uptake of Pb on alumina as the function of the pH and ionic strength: 0.1 mol dm^{-3} solid; 0.01 mol dm^{-3} dash; 0.001 mol dm^{-3} dot. One proton released per one adsorbed Pb, 1-pK-Stern model, CD, f=0.5. For model parameters, cf. Tables 5.15 and 5.25.

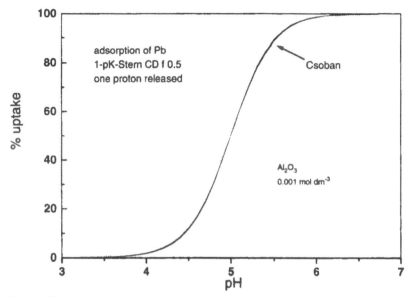

FIG. 5.123 Calculated uptake of Pb on alumina as the function of the pH at ionic strength 0.001 mol dm^{-3}: Cox, Mustafa and Sprycha solid; Csoban dot. One proton released per one adsorbed Pb, 1-pK-Stern model, CD, $f=0.5$. For model parameters, cf. Tables 5.14, 5.15, and 5.25.

TABLE 5.26 Logarithms of Stability Constant of the Silica-Pb Surface Complex as the Function of the Electrostatic Position of Pb and the Number of Protons Released per One Adsorbed Pb

Protons released	0	1	2
Inner sphere	1.36	-3.46	-8.25
Outer sphere	1.37	-3.44	-8.23

less steep. The stability constant of analogous alumina-Pb surface complex (CD, f 0.5, 1 proton released) calculated with TLM derived from the same set of primary charging curves (log K 1.65, Table 5.21) is significantly different from that calculated with Stern model (log K 0.74, Table 5.25).

The effect of the electrostatic model on the calculated uptake curves for favorable electrostatic conditions will be presented using the primary charging curves of silica from Section III.

The numerical values of log K obtained for the TLM model (Fig. 5.85, Table 5.19) combined with different models of Pb adsorption (electrostatic position of Pb, number of proton released per one adsorbed Pb), are summarized in Table 5.26.

Table 5.26 illustrates dramatic difference (almost five decades per one proton released) in the stability constant calculated for different number of protons released per one adsorbed Pb assumed in the model calculations. This effect is less significant than for alumina (Table 5.21), but still the significance of the stability constants of surface complexes reported without specified number of protons released is limited.

On the other hand, the effect of the assumed electrostatic position of Pb is rather insignificant. Therefore the results for CD, $f = 0.5$ are not shown. The absence of significant effect of the assumed electrostatic position of Pb on log K for silica is common for all analyzed TLM models for silica, cf. Table 5.27. On the other hand, all analyzed TLM models for alumina produced significant effect of the assumed electrostatic position of Pb on the calculated log K (this result is explicitly shown only for one TLM model in Table 5.21).

All model curves for model parameters from Table 5.26, presented in Figs. 5.124–5.129, predict decrease in uptake when the ionic strength increases. This effect is most significant for no proton released and least significant for two protons released per one adsorbed Pb, and more significant for outer sphere complex than for inner sphere complex. In this respect, silica is very different from alumina (cf. Figs. 5.105–5.119), for which the increase in the ionic strength can lead to increase or

TABLE 5.27 The Stability Constants of Silica-Pb Surface Complex (1 proton released) Used in Fig. 5.131. For TLM Parameters cf. Table 5.19

TLM	log K (inner sphere)	log K (outer sphere)
Kosmulski	-3.46	-3.44
Löbbus	-3.88	-3.82
Milonjic	-3.07	-3.04
Sidorova	-4.25	-4.18
Szekeres	-3.73	-3.67

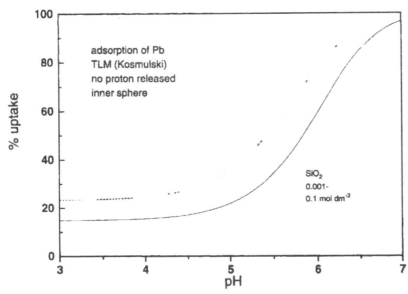

FIG. 5.124 Calculated uptake of Pb on silica as the function of the pH and ionic strength: 0.1 mol dm^{-3} solid; 0.01 mol dm^{-3} dash; 0.001 mol dm^{-3} dot. Inner sphere complex, no proton released, TLM. For model parameters, cf. Tables 5.19 and 5.26.

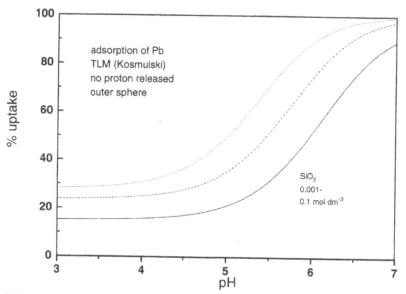

FIG. 5.125 Calculated uptake of Pb on silica as the function of the pH and ionic strength: 0.1 mol dm^{-3} solid; 0.01 mol dm^{-3} dash; 0.001 mol dm^{-3} dot. Outer sphere complex, no proton released, TLM. For model parameters, cf. Tables 5.19 and 5.26.

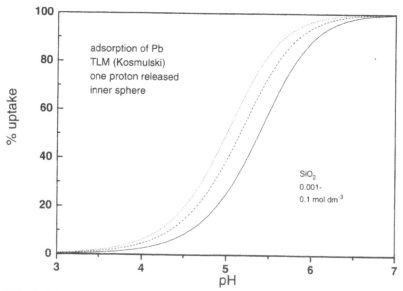

FIG. 5.126 Calculated uptake of Pb on silica as the function of the pH and ionic strength: 0.1 mol dm^{-3} solid; 0.01 mol dm^{-3} dash; 0.001 mol dm^{-3} dot. Inner sphere complex, one proton released per one adsorbed Pb, TLM. For model parameters, cf. Tables 5.19 and 5.26.

decrease in uptake (this depends on the assumed electrostatic position of Pb and on the number of protons released per one adsorbed Pb, and on the pH).

The model curves presented in Figs. 5.124 and 5.125 belong to the type shown in Fig. 4.6 B (0% uptake is not reached at unfavorable electrostatic conditions), but

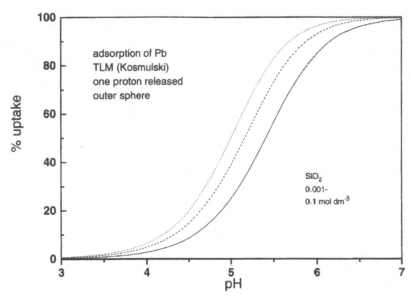

FIG. 5.127 Calculated uptake of Pb on silica as the function of the pH and ionic strength: 0.1 mol dm^{-3} solid; 0.01 mol dm^{-3} dash; 0.001 mol dm^{-3} dot. Outer sphere complex, one proton released per one adsorbed Pb, TLM. For model parameters, cf. Tables 5.19 and 5.26.

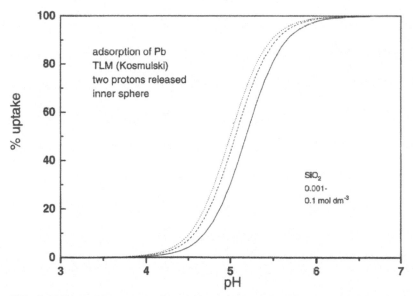

FIG. 5.128 Calculated uptake of Pb on silica as the function of the pH and ionic strength: 0.1 mol dm^{-3} solid; 0.01 mol dm^{-3} dash; 0.001 mol dm^{-3} dot. Inner sphere complex, two protons released per one adsorbed Pb, TLM. For model parameters, cf. Tables 5.19 and 5.26.

the other uptake curves for silica (and all curves for alumina) belong to the type shown in Fig. 4.5 (A) (typical adsorption edge). The model uptake curves for 2 protons released are significantly steeper than for no proton released (Fig. 5.130). Similar, but less significant effect was reported for alumina (Fig. 5.116).

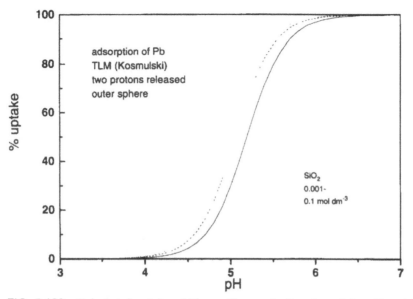

FIG. 5.129 Calculated uptake of Pb on silica as the function of the pH and ionic strength: 0.1 mol dm^{-3} solid; 0.01 mol dm^{-3} dash; 0.001 mol dm^{-3} dot. Outer sphere complex, two protons released per one adsorbed Pb, TLM. For model parameters, cf. Tables 5.19 and 5.26.

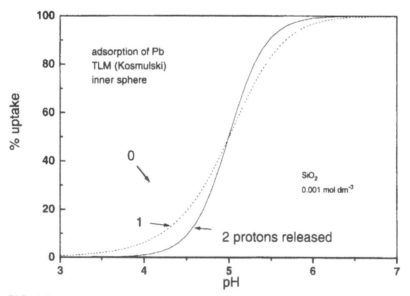

FIG. 5.130 Calculated uptake of Pb on silica as the function of the pH at ionic strength 0.001 mol dm^{-3}: two protons released per one adsorbed Pb solid; 1 proton dash; no proton dot. Inner sphere complex, TLM. For model parameters, cf. Tables 5.19 and 5.26.

The effect of the choice of the TLM model on the numerical value of the stability constant of silica-Pb surface complex (1 proton released) is illustrated in Table 5.27.

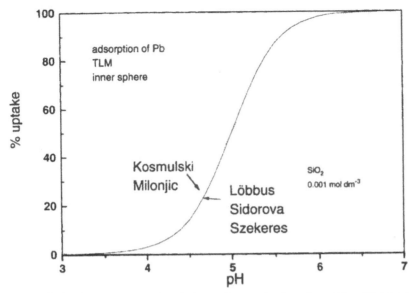

FIG. 5.131 Calculated uptake of Pb on silica as the function of the pH at ionic strength 0.001 mol dm^{-3}: Löbbus, Sidorova, and Szekeres solid; Kosmulski and Milonjic dot. Inner sphere complex, one proton released per one adsorbed Pb, TLM. For model parameters, cf. Tables 5.19 and 5.27.

The difference between the highest and the lowest stability constant of silica-Pb surface complex, calculated for different TLM models by a factor of 6 is less significant than the discrepancies reported for alumina (Table 5.22).

The model curves calculated for different TLM models (inner sphere complex, one proton released) are compared in Fig. 5.131. To avoid overcrowding, two groups of nearly identical curves were plotted as one curve each. The difference between the most and the least steep curve does not exceed typical scatter of experimental data points in adsorption measurements.

Figure 5.132 presents the ionic strength effect on the model uptake curves calculated for one proton released per one adsorbed Pb, using the diffuse layer model (Kosmulski, for model parameters cf. Table 5.13). The model curves are significantly steeper, and the ionic strength effect is less significant than in the analogous Pb adsorption model (inner sphere, one proton released) combined with TLM (Fig. 5.126). The calculated stability constant of the surface complex is higher by three orders of magnitude for the diffuse layer model (Table 5.28) than for TLM (Table 5.27).

Also for other diffuse layer models (cf. Table 5.13) the calculated stability constant of silica-Pb surface complex is considerably higher than for their TLM counterparts obtained from the same experimental data. The difference between the highest and the lowest K in Table 5.28 by almost two orders of magnitude is more significant than the discrepancies between the stability constants calculated for different diffuse layer models obtained for alumina (cf. Table 5.22). In spite of different K, the course of the calculated uptake curves obtained for different diffuse layer models and one proton released per one adsorbed Pb with

TABLE 5.28 The Stability Constants of Silica-Pb Surface Complex (1 proton released) for Diffuse Layer Model Parameters cf. Table 5.13

Diffuse layer model	log K
Kosmulski	0.17
Löbbus	-1.3
Milonjic	-0.07
Sidorova	-1.33
Szekeres	-1.58

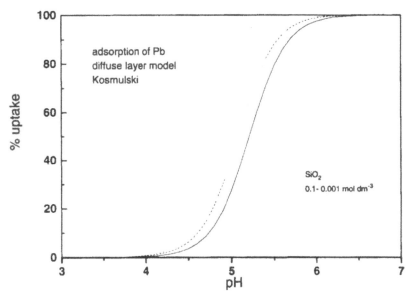

FIG. 5.132 Calculated uptake of Pb on silica as the function of the pH and ionic strength: 0.1 mol dm^{-3} solid; 0.01 mol dm^{-3} dash; 0.001 mol dm^{-3} dot. One proton released per one adsorbed Pb, diffuse layer model. For model parameters, cf. Tables 5.13 and 5.28.

10^{-3} mol dm^{-3} inert electrolyte is practically identical (one of these curves is shown in Fig. 5.132).

The stability constants of the silica-Pb surface complex (inner sphere, 1 proton released) calculated for different 1-pK-Stern models are summarized in Table 5.29.

Two best-fit N_s values in Table 5.29 are significantly higher than the highest N_s reported in Tables 5.1 and 5.3, and their physical sense is questionable. The log K ($\equiv$SiOPb$^+$) calculated for the best-fit Stern model (Table 5.29) are lower by about 0.2 than the corresponding log K for the diffuse layer model (Table 5.28), but higher by 2–3 than the log $K(\equiv$SiOPb$^+$) calculated for the best-fit TLM (Table 5.27). The calculated uptake curves presented in Fig. 5.132 (diffuse layer model) are practically identical as corresponding curves calculated for the Stern model (Kosmulski, model

TABLE 5.29 The Best-fit N_s in C/m^2, log K(reaction 5.25) and C in F/m^2 in the 1-pK-Stern Models, and Corresponding Stability Constants of Silica-Pb Surface Complex (inner sphere, 1 proton released)

Primary σ_o	N_s	Stern model parameters log K (5.25)	C	log K ($\equiv$SiOPb$^+$)
Kosmulski	1	8	4	-0.01
Löbbus	1	6.68	4	-1.46
Milonjic	1.6	7.98	2.1	-0.29
Sidorova	58	8.53	3.6	-1.54
Szekeres	92	8.56	4.6	-1.76

parameters from Table 5.29). The course of the calculated uptake curves obtained for different Stern models (inner sphere surface complex, one proton released per one adsorbed Pb) with 10^{-3} mol dm^{-3} inert electrolyte is almost identical (one of these curves is shown in Fig. 5.132). The uptake curves obtained for three Stern models (Löbbus, Sidorova, Szekeres) are slightly steeper than with two other Stern models, but the difference is less significant than for corresponding uptake curves obtained for TLM (cf. Fig. 5.131).

The discussed above examples are limited to adsorption of divalent metal cation. They indicate that the numerical value of the stability constant of the surface complex depends on the assumed model of primary surface charging. In this respect the significance of comparison of the stability constants of analogous surface complexes from different sources is questionable, when these stability constants were calculated using different models of primary surface charging. On the other hand the choice of the model of primary surface charging has rather limited effect on the shape of the calculated uptake curves. The shape of calculated uptake curves (slope, ionic strength effect) and the numerical value of the stability constant of the surface complex are both affected by the model of specific adsorption (electrostatic position of the specifically adsorbed cation and the number of protons released per one adsorbed cation).

The above conclusions are also valid for specific adsorption of cations whose valence is different from two, and for specific adsorption of anions. Figure 5.133 shows the calculated uptake curves of trivalent Gd on alumina. Identical model as in Fig. 5.120 was used: 1-pK-diffuse layer model (Sprycha) for primary surface charging, and one proton released per one adsorbed Gd, except the logarithm of the stability constant of Gd-alumina complex equals 3.76 (from the condition pH$_{50}$ = 5 at 10 g alumina/dm^3 and 10^{-3} mol dm^{-3} inert electrolyte). The uptake curves in Fig. 5.133 are steeper than corresponding curves in Fig. 5.120. This result is not surprising in view of higher valence of Gd (cf. Eq. (5.1)), and indeed, the experimental uptake curves are usually steeper for trivalent than for divalent cations.

The uptake curves for Gd (Fig. 5.133) and Pb (Figs. 5.105–5.132) were calculated over the pH range, where the hydrolysis of metal cations is negligible, i.e. Gd^{3+} and Pb^{2+} are the dominating solution species (cf. Fig. 4.2). This limited pH range was chosen on purpose. Adsorption experiments at higher pH (where hydrolysis plays a significant role) often result in precipitation of metal hydroxides,

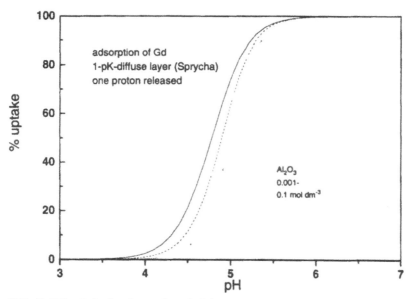

FIG. 5.133 Calculated uptake of Gd on alumina as the function of the pH and ionic strength: 0.1 mol dm^{-3} solid; 0.01 mol dm^{-3} dash; 0.001 mol dm^{-3} dot. One proton released per one adsorbed Gd, 1-pK-diffuse layer model. For model parameters cf. Table 5.12.

and application of surface complexation model to sorption in such systems would be questionable. For many common multivalent cations the hydrolysis is negligible at pH < 6, and significant fraction of these cations can be adsorbed on common materials at pH < 6 at sufficiently high solid to liquid ratio (cf. Figs. 4.28–4.59). Thus, the examples shown in Figs. 5.105–5.133 represent actual situations occurring by adsorption modeling.

In this respect the solution chemistry of common anions is very different. For example with phosphate (Fig. 4.4), HPO_4^{2-} and $H_2PO_4^-$ are the dominating solution species over the typically studied pH range, and fully dissociated anions occur only at extremely high pH values, which are of limited interest in adsorption studies, e.g. many common adsorbents are unstable at such a high pH (dissolution). On the other hand, the final products of hydrolysis on anions, i.e. fully protonated acid molecules are usually water soluble, thus, the applicability of surface complexation model is not limited by surface precipitation as it was discussed above for metal cations.

One example of calculated uptake curves representing anion adsorption is presented in Fig. 5.134. It should be emphasized that in contrast with Figs. 5.105–5.133, which reflect typical trends for di- and trivalent cations, the course of the curves shown in Fig. 5.134 is strongly influenced by the numerical values of the consecutive dissociation constants of the weak acid of interest (here: phosphoric acid), thus the conclusions from the analysis of such curves do not have general character. Figure 5.134 shows the calculated uptake curves of phosphoric acid on alumina. Identical model as in Figs. 5.120 and 5.133 was used: 1-pK-diffuse layer model (Sprycha) for primary surface charging, and one proton released per one adsorbed H_3PO_4 molecule, and the logarithm of the stability constant of H_3PO_4–alumina complex equals -1.86 (from the condition pH$_{50}$ = 5 at 10 g alumina/dm^3 and

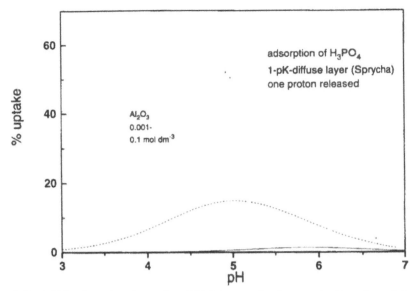

FIG. 5.134 Calculated uptake of H_3PO_4 on alumina as the function of the pH and ionic strength: 0.1 mol dm^{-3} solid; 0.01 mol dm^{-3} dash; 0.001 mol dm^{-3} dot. One proton released per one adsorbed H_3PO_4 molecule, 1-pK-diffuse layer model. For model parameters, cf. Table 5.12.

10^{-3} mol dm^{-3} inert electrolyte). The uptake curves in Fig. 5.134 represent the type shown in Fig. 4.6(C) (uptake curves with a maximum). Thus, this type of uptake curves, frequently observed for anions (Table 4.2), can be rationalized in terms of the surface complexation model. The width and height of the maximum depends on the electrostatic position and the number of proton released per one adsorbed molecule of phosphoric acid (results not shown here). Significant effect of the ionic strength on the calculated uptake curves in Fig. 5.134 (including the position of the uptake maximum) suggests that the experiments carried out at different ionic strength can be very valuable in testing the adsorption models for anions.

In principle the surface complexation model is suitable for small ions, but other applications can be also found in literature. For example, humic acid and related substances were modeled as a mixture of four to five monoprotic acids forming different types of surface complexes (electrostatic position, number of protons released) [102]. However, such approach can be only considered as data fitting model.

Surface species involving more than one surface site were not invoked in the discussed above examples. Let us consider two surface species, with the same electrostatic position of the specifically adsorbed ion, and the same number of proton released per one specifically adsorbed cation (or per one molecule of weak acid), but with different number of surface sites, e.g.

$$\equiv AlOH + H_2O + Pb^{2+} = \equiv AlOPbOH + 2H^+ \tag{5.72}$$

and

$$2 \equiv AlOH + Pb^{2+} = (\equiv AlO)_2Pb + 2H^+ \tag{5.73}$$

(TLM, inner sphere). It can be easily shown that models with these two surface species produce the same shape of uptake(pH) curves and the same ionic strength dependence. Thus, considering uptake data alone one cannot distinguish between surface species on r.h.s of reactions (5.72) and (5.73). In principle, spectroscopic data are necessary to choose between the reactions (5.72) or (5.73). However, in terms of the recent CD model (cf. Section B) the number of surface sites involved in the surface complex equals to the number of oxygen atoms (e.g. in the oxoanion) whose charge is attributed to the surface. For example with sulfate (four oxygen atoms), $f=0.25$ (one oxygen atom on the surface) corresponds to monodentate surface complex, $f=0.5$ (two oxygen atoms on the surface) corresponds to bidentate surface complex, etc. Thus, the number of surface sites involved in the surface complex can be inferred from the best-fit f value. In this approach, the number of surface sites involved in the surface complex cannot be treated as independent variable, but it is a consequence of the electrostatic position of the specifically adsorbed ion. It should be emphasized, however, that the effect of the assumed electrostatic position of the specifically adsorbed ion on the calculated uptake curves is often rather insignificant (cf. Figs. 5.124–5.129), and in such systems the f value cannot be even roughly estimated from the uptake data. Therefore, the spectroscopic data (EXAFS) are still practically the only confident method to distinguish between mono- and bidentate surface complexes.

E. Ternary Surface Complexes and Multiple Surface Species

The examples shown is Section D indicate that the shape of calculated uptake curves (slope, ionic strength effect) can be to some degree adjusted by the choice of the model of specific adsorption (electrostatic position of the specifically adsorbed species and the number of protons released per one adsorbed cation or coadsorbed with one adsorbed anion) on the one hand, and by the choice of the model of primary surface charging on the other. Indeed, in some systems, models with one surface species involving only the surface site(s) and the specifically adsorbed ion successfully explain the experimental results. For example, Rietra et al. [103] interpreted uptake, proton stoichiometry and electrokinetic data for sulfate sorption on goethite in terms of one surface species. Monodentate character of this species is supported by the spectroscopic data and by the best-fit charge distribution ($f\approx0.18$, vide infra).

However, the possibilities to adjust the course of model uptake curves by the selection of the simple model of specific adsorption (electrostatic position and the number of protons released per one specifically adsorbed cation or per one molecule of weak acid) are limited. It should be emphasized that the CD model was introduced quite recently, and in most publications summarized in Tables 4.1 and 4.2 only the choice between two electrostatic positions (inner- and outer-sphere) was considered. Thus, the ability to find a simple model properly simulating the actual uptake curves was even more limited than nowadays. But even when the charge distribution concept is taken into account, it often happens that the simple models fail to properly reflect the experimentally observed effects of the pH and ionic strength on the specific adsorption. This problem has been solved by

- Considering ternary surface complexes, i.e. surface species involving counterions (in addition to surface site(s) and the specifically adsorbed ions of interest).

• Considering simultaneous formation of multiple surface species involving the specifically adsorbed ion of interest.

The entries in Tables 4.1 and 4.2 with multiple formulae of surface species and logarithms of their stability constants usually indicate that the adsorption model assumed simultaneous formation of these surface species. Interpretation of such results requires carefulness. First, each surface species requires an additional adjustable parameter (its stability constant), and it is well known that simultaneous fitting of many adjustable parameters usually does not produce unequivocal results. Thus the reported set of values of stability constants should be rather considered as one of many sets (producing nearly identical model curves). This aspect of models with multiple surface species was discussed in more detail elsewhere [104].

Probably some sets of surface reactions reported in literature (outer sphere complex + inner sphere complex) can be replaced by single reaction formulated in terms of the CD model. A few publications report two sets of surface species along with their stability constants that produced equally good simulation of experimental data. These sets are separated by "or" in Tables in Chapter 4.

Some models with multiple surface reactions were formulated in the framework of MUSIC approach. For example uptake of fluoride on goethite was interpreted [105] in terms of formation of two surface species $\equiv FeF^{-1/2}$ and $\equiv Fe_2F$ involving two types of surface hydroxyl groups. The later species involves a $\equiv Fe_2OH$ group that does not participate in primary surface charging. This type of models should be clearly distinguished from models assuming one type surface sites, simultaneously responsible for primary surface charging and for formation of all types of surface species involved in specific adsorption.

In most models involving multiple surface species these species differ in the number of protons released per one specifically adsorbed cation (or molecule of weak acid). As discussed in Section II, such models have firm physical grounds, namely they can be interpreted in terms of stepwise deprotonation of adsorbed molecules of multivalent acids (or hydrolysis of adsorbed heavy metal cations) as the pH increases.

Adsorption of metal cations that form stable water soluble complexes with anions present in solution has been often interpreted in terms of formation of multiple surface species that differ in the number of anions coadsorbed with one adsorbed cation. When this number is greater than zero these species are termed ternary surface complexes. This can be interpreted as complexation of the adsorbed cation. This type of multiple surface species can be combined with the discussed above species that differ in the number of protons released per one specifically adsorbed cation. Less common (but also physically reasonable for ions that tend to form polymeric species in solution) is the idea of multiple polymeric surface species. For example, adsorption of cobalt on alumina was modeled in terms of formation of $\equiv AlOCo_2(OH)_2^+$ and $\equiv AlOCo_4(OH)_5^{2+}$ surface species [106].

Multiple species were also considered in interpretation of sorption of multivalent cations by materials that carry substantial amount of permanent charge (e.g. [107]). The uptake of ions by these materials is due to ion exchange on the one hand and to surface complexation on the other. A few stability constants of species responsible for sorption of metal cations listed in Table 4.1 were calculated in terms of such models.

V. SURFACE HETEROGENEITY

The discussions in Sections III and IV were focussed on the models assuming that the surface sites responsible for primary surface charging and for specific adsorption are all identical (homogeneous surface). These models often fail to properly reflect the experimental results. Many attempts to improve the goodness of the fit by allowing different types of surface sites (e.g. having different affinities to specifically adsorbing ions) have been published, and they are briefly discussed in this section. In such models one proton binding constant (cf. Eq. (5.24)) or one stability constant of surface complex is replaced by the distribution function. Koopal [108] emphasized the difference between patchwise heterogeneity and random heterogeneity (surface sites of different kinds evenly distributed over the surface). The modeling approach in which the distribution function is derived from adsorption isotherm (this section) should be clearly distinguished from the approach in which the distribution function is derived from crystallographic data (MUSIC).

A. Primary Surface Charging

The 1-pK-diffuse layer model has limited ability to properly reflect the surface charging of alumina (Figs. 5.39–5.42) and silica (Figs. 5.47–5.51) over the entire range of pH and ionic strength. Therefore it may be tempting to improve the goodness of fit by assuming a bimodal distribution of surface sites, i.e. to attribute the surface charge to two types of surface sites and try to adjust the best-fit site densities and proton binding constants. This increases the number of adjustable parameters (with respect to homogeneous model) by two for silica and by three for alumina. However, the best-fit distribution of surface sites for data presented in Figs. 5.39–5.42 and 5.47–5.51 gives rather insignificant improvement in the goodness of fit over the homogeneous model, thus introduction of new adjustable parameters is not justified. This result is in agreement with a theoretical study by van Riemsdijk et al. [52], who used Langmuir–Freundlich equation

$$\theta = (K_{ads}c)^m / [1 + (K_{ads}c)^m] \tag{5.74}$$

and two related equations to model charging curves of TiO_2 at three different ionic strengths. The heterogeneity parameter m ranges from 0 to 1, and with $m = 1$ Eq. (5.74) reduces to Langmuir adsorption isotherm (Eq. (5.4)). Equation (5.74) corresponds to nearly Gaussian (symmetrical) distribution function, and the heterogeneity parameter defines how broad is the distribution. The other two equations used by van Riemsdijk et al. represented asymmetrical distributions and they also reduce to Langmuir adsorption isotherm with $m = 1$. With the 1-pK model ($\theta = 0$ represents the surface with $\equiv TiOH^{1\,2-}$ groups only, and $\theta = 1/2$ at the PZC, cf. Eq. (5.23)), the calculated charging curves (Stern model) were rather insensitive to the parameter m (down to 0.25), i.e. the charging curves were rather useless in assessment of the value of parameter m in Eq. (5.74) and related equations. Also with the 2-pK model the effect of surface heterogeneity (considered in similar framework as discussed above for 1-pK model) on the calculated charging curves was rather insignificant according to van Riemsdijk et al. Also Rudzinski et al. [109] found that potentiometric titrations could be equally well modeled (using TLM) by homo- and heterogeneous models. On the other hand, significant improvement in the goodness of fit over homogeneous model (in the framework of diffuse layer model) was

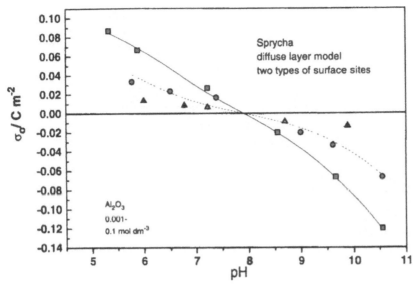

FIG. 5.135 Experimental (symbols) and calculated (lines) surface charge density of alumina as the function of the pH and ionic strength: 0.1 mol dm^{-3} squares, solid; 0.01 mol dm^{-3} circles, dash; 0.001 mol dm^{-3} triangles, dot. For model parameters see text.

observed when the 1-pK or 2-pK model (for alumina) was replaced by a heterogeneous model in which the positive charge (reaction (5.32)) and negative negative charge (reaction (5.33)) were due to two different types of surface sites, and densities of these sites and their protonation constants were the adjustable parameters. This approach ressembles MUSIC except the site densities and protonation constants are fitted (rather than derived from crystallographic data and bond valence, respectively). The results obtained for Sprycha data (0.19 C/m^2 sites responsible for negative charge, log protonation constant 8.65; 0.1 C/m^2 sites responsible for positive charge, log protonation constant 7.55) are presented in Fig. 5.135.

The present author is not aware of other publications reporting significant improvement in simulation of a set of primary charging curves (for at least three ionic strengths) when a homogeneous model is replaced by a heterogeneous model (within the same electrostatic model). On the other hand, numerous papers report modeling exercises with potentiometric data limited to one ionic strength.

Contescu et al. [110–112] derived continuous distributions of proton binding constant for γ-Al$_2$O$_3$ from their own potentiometric titration data and they obtained different distributions for different ionic strengths. These distributions had three or four peaks suggesting three or four types of surface sites. No attempt was made to find one distribution that could be used to calculate charging curves for different ionic strengths. In another paper of this series [113] distributions of proton binding constant are reported for different TiO$_2$ samples (rutile and anatase), but again no attempt was made to predict the charging behavior at different ionic strengths using one model.

Surface charging of rutile at one ionic strength was modeled using 3 kinds of surface sites (discrete distribution) having different proton binding constants [114].

The calculated distribution is again "conditional", i.e. it applies to one ionic strength.

The above results refer to metal oxides, but similar model exercises (discrete distributions of proton binding constants derived from one potentiometric curve) were reported for sparingly soluble phosphates [115] and for clay minerals [116].

Charmas [117] discussed "homogeneous" and "heterogeneous" version of TLM with different C_+ and C_-. The calculated charging curves did not significantly differ in these two versions, but a substantial difference in the calculated ζ potentials and adsorption of ions of inert electrolyte was reported.

B. Specific Adsorption

According to the models of specific adsorption discussed in Section IV, the uptake (pH) curves are independent of the total concentration of specifically adsorbing ions as long as this concentration is lower than about 1% of the concentration of surface hydroxyl groups. This corresponds to linear (Henry) adsorption isotherm at constant pH. Unfortunately this result is not confirmed by experimental data, namely, the actual adsorption isotherms show significant deviations from linearity and the uptake(pH) curves are dependent on the total concentration of specifically adsorbing ions, when their concentration is a small fraction of 1% of the concentration of surface hydroxyl groups (cf. Tables 4.1 and 4.2). In this respect, the stability constants of surface complexes calculated from uptake(pH) curves measured at certain total concentration of specifically adsorbing ions have conditional character, i.e. they are not necessarily applicable for other experimental conditions. This experimental observation resulted in a hypothesis (which was made very popular by Dzombak and Morel [118]) of "strong" and "weak" surface sites. In this model the concentration of "strong" sites is a few per cent of the concentration of "weak" surface sites, but their affinity to certain adsorbates is greater by many orders of magnitude (in terms of stability constant of surface complex). A few stability constants reported in Table 4.1 refer to models with strong sites (indicated by a subscript "s") and weak sites (subscript "w").

The bimodal distribution with strong and weak sites was widely criticized. Barrow et al. [119] emphasized lack of evidence for existence of "strong sites", large number of adjustable parameters compared with others models which fit the experimental results equally well, and numerical instability of the solutions. Barrow et al. favor continuous, e.g. normal distributions of binding energy. According to these authors the surface heterogeneity effects are more important for cation than for anion adsorption. Advantages and disadvantages of bimodal distribution of the surface sites on the one hand and of continuous Gaussian like distribution on the other, have been also debated by other authors [120,121].

Rudzinski et al. [122] represented the differential distribution of the adsorption (binding) energy E_i by the following function

$$\chi_i(E_i) = \begin{cases} 0 \text{ for } E_i < E_i^l \\ F_{iN}^{-1} c_i^{-1} \exp[(E_i - E_i^0)/c_i] \left\{ 1 + \exp[(E_i - E_i^0)/c_i] \right\}^{-2} \text{ for } E_i^l \leq E_i \leq E_i^m \\ 0 \text{ for } E_i > E_i^m \end{cases}$$

$$(5.75)$$

where the subscript i refers to the i-th adsorbed species, E_i^1, E_i^0, E_i^m, and c_i are the model parameters, and they are on the order of a few kJ and F_{iN} is the factor normalizing $\int \chi_i(E_i)dE_i$ to unity. TLM was used (four species responsible for surface charging) and the uptake of divalent metal cations was represented by the $\equiv$SOMe$^+$ surface complex. Two extreme cases were considered, namely full correlation of adsorption energies for different species, i.e.

$$E_j = E_i + \Delta_{ji} \qquad (5.76)$$

for given site on the one hand (the subscripts i and j refer to two different adsorbates), and complete lack of correlation on the other. The model without correlation was more successful in modeling the selected adsorption isotherms at constant pH. For the above distribution function the slope of log–log adsorption isotherm equals one for low surface coverage and kT/c_i for higher surface coverage. An extensive collection of log–log adsorption isotherms of cations at constant pH on various materials supporting this model was collected [123].

In addition to bimodal and continuous (Gaussian) distribution, many other types of distributions were discussed. Robertson and Leckie [124] analyzed effects of different (assumed) distributions of surface sites on the model uptake curves of mono-, di- and trivalent metal cations at various total concentrations. They combined different types of surface complexes (mono and bidentate, inner and outer sphere, different proton stoichiometries) with different electric interfacial layer models (diffuse layer, TLM, different site densities, etc.), and with three types of surface heterogeneity (in addition to one site model): 1% strong + 99% weak sites, 20 classes each containing 5% of total sites with log of cation binding constants differing by a factor 2 between the n-th and n + 1-th class, and 20 classes with the same binding constants as in the previous model but the concentrations differing by a factor of $\sqrt{2}$ between the n-th and n + 1-th class (highest affinity corresponds to lowest concentration). The results are presented as adsorption edges on the one hand, and log-log adsorption isotherms at constant pH on the other.

Uptake curves in Cd-rutile system (at four different total Cd concentrations and one ionic strength) were modeled using seven types of surface sites reacting with for Cd [114]. This model involves fourteen adjustable parameters (concentrations of surface sites and stability constants of corresponding surface complexes) and it is valid only for one ionic strength.

Cernik et al. [125] simultaneously fitted uptake (pH) curves of Sr (from Kolarik) for seven initial concentrations and charging curve of ferrihydrite (single ionic strength, taken from another source) to find a distribution of proton and Sr binding constants. The choice of the set of experimental data for this modeling exercise was rather unfortunate. Kolarik [126] used the term "sorption" in the title of his publication, abstract, and figure captions. However in the experimental part he wrote (free translation from German by Mrs. Izabela Kosmulska):

> "In all experiments iron III hydroxide was precipitated by means of sodium hydroxide in the presence of strontium ions."

It is clear, that Kolarik studied coprecipitation rather than adsorption, although this is mentioned only in this one sentence in the entire paper. This example illustrates a general tendency to mix up sorption and coprecipitation in older publications as discussed in previous chapters. One of the reasons why Kolarik did not consider the difference between sorption of Sr on previously formed hydroxides and coprecipitation worthy of special emphasizing is that in his earlier studies [127] similar uptakes by sorption on the one hand and coprecipitation on the other, were observed, by otherwise the same conditions. Typically uptake by sorption was lower only by a few per cent. Therefore, although the model itself is interesting, the physical sense of specific results obtained by Cernik et al. [125] is disputable.

Their strategy was to build a fixed grid (e.g. 1 log K unit of proton and Sr binding constants) and optimize concentrations of all surface sites representing the knots of the grid in some assumed computation window. Thus, in terms of Eq. (5.76), models without correlation of adsorption energies were discussed. This procedure resulted in a relatively high fraction of sites on the border of the computation window. The distribution can be regularized for a small number of sites (no fixed grid), on expense of the goodness of fit. The other possibility is regularization for smoothness (the concentrations of sites having similar proton and Sr binding constants should not differ too much), and this requires a finer grid. The later fitting still uses discrete sites but the distribution is quasi-continuous. Each of the three strategies gave different distribution, but almost identical model curves (proton and Sr adsorption). This result is a good illustration that attempts to find the distribution of surface sites from experimental adsorption data (or generally: simultaneous fitting of many adjustable parameter) inevitably result in multiple solutions producing almost identical model curves.

REFERENCES

1. M. Jaroniec and R. Madey. Physical Adsorption on Heterogeneous Surfaces, Elsevier, Amsterdam, 1988.
2. W. Rudzinski and D. H. Everett. Adsorption of Gases on Heterogeneous Surfaces, Academic Press, San Diego, 1991.
3. H. van Olphen, J. J. Fripat (Eds.) Data Handbook for Clay Materials, and Other Non-Metallic Minerals, Pergamon, Oxford, 1979.
4. M. Szekeres, I. Dekany and A. de Keizer. Colloids Surf. A 141: 327–336 (1998).
5. S. Kondo, M. Igarashi and K. Nakai. Colloids Surf. 63: 33–37 (1992).
6. B. I. Brookes, C. Kemball and H. F. Leach. J. Chem. Res. (S) 112–113 (1988).
7. C. M. Koretsky, D. A. Sverjensky and N. Sahai. Am. J. Sci. 298: 349–438 (1998).
8. S. Pivovarov. J. Colloid Int. Sci. 196: 321–323 (1997).
9. R. O. James and G. A. Parks. Surf. Colloid Sci. 12: 119–216 (1982).
10. G. W. Simmons and B. C. Beard. J. Phys. Chem. 91: 1143–1148 (1987).
11. E. McCafferty, J. P. Wightman and T. F. Cromer. J. Electrochem. Soc. 146: 2849–2852 (1999).
12. H. Tamura, A. Tanaka, K. Mita and R. Furuichi. J. Colloid Interf. Sci. 209: 225–231 (1999).
13. A. Herbelin and J. Westall. FITEQL ver. 3.1 Oregon State University, Corvallis, OR 1994.
14. J. O. Leckie. In: The Importance of Chemical Speciation in Environmental Processes M. Bernhard, F. E. Brinckman and P. J. Sadler, (Eds.), Springer Berlin, 1986, pp. 237–254.
15. W. H. van Riemsdijk, J. C. M. de Wit, S. L. J. Mous, L. K. Koopal and D. G. Kinniburgh, J. Colloid Interf. Sci. 183: 35–50 (1996).

16. I. Larson and P. Attard. J. Colloid Interf. Sci. 227: 152–163 (2000).
17. M. Kosmulski. Pol. J. Chem. 66: 1867–1878 (1992); Colloid Polym. Sci. 271: 1076–1082 (1993).
18. L. Bousse, N. F. de Rooij and P. Bergveld. Surf. Sci. 135: 479–496 (1983).
19. L. K. Koopal. Electrochim. Acta. 41: 2293–2306 (1996).
20. J. R. Rustad, E. Wasserman, A. R. Felmy and C. Wilke. J. Colloid Interf. Sci. 198: 119–129 (1998).
21. A. Seidel, M. Lobbus, W. Vogelsberger and J. Sonnefeld. Solid State Ionics. 101–103: 713–719 (1997).
22. M. Kosmulski, J. Colloid Interf. Sci. 208: 543–545 (1998).
23. G. H. Bolt and W. H. van Riemsdijk. In: Soil Chemistry B. Physico-Chemical Models, 2nd Edn. G. H. Bolt. (Ed.), Elsevier Amsterdam, 1982, p. 459–505.
24. W. Stumm, C. P. Huang and S. R. Jenkins. Croat. Chem. Acta 42: 223 (1970).
25. J. Westall and H. Hohl. Adv. Coll. Int. Sci. 12: 265–294 (1980).
26. R. Sprycha. J. Colloid Interf. Sci. 127: 1–11 (1989).
27. C. D. Cox and M. M. Ghosh. Water Res. 1181–1188 (1994).
28. S. Mustafa, B. Dilara, Z. Neelofer, A. Naeem and S. Tasleem. J. Colloid Interf. Sci. 204: 284–293 (1998).
29. K. Csoban and P. Joo. Colloids Surf. A 151: 97–112 (1999).
30. S. K. Milonjic. Colloids. Surf. 23: 301–312 (1987).
31. M. Kosmulski. J. Colloid Interf. Sci. 156: 305–310 (1993).
32. M. Löbbus, W. Vogelsberger, J. Sonnefeld and A. Seidel. Langmuir 14: 4386–4396 (1998).
33. M. P. Sidorova, H. Zastrow, L. E. Ermakova, N. F. Bogdanova and V. M. Smirnov. Koll. Zh. 61: 113–117 (1999).
34. J. Sonnefeld. J. Colloid Int. Sci. 155: 191–199 (1993).
35. J. Lutzenkirchen. J. Colloid Interf. Sci. 217: 8–18 (1999).
36. T. Hiemstra and W. H. van Riemsdijk. Colloids Surf. 59: 7–25 (1991).
37. J. Sonnefeld. Coll. Polym. Sci. 273: 932 938 (1995).
38. M. Kosmulski, R. Sprycha and J. Szczypa. In: Interfacial Dynamics, N. Kallay, (Ed.), Dekker, New York, 1999, pp. 163–223.
39. J. Lützenkirchen. J. Colloid Interf. Sci. 210: 384–390 (1999).
40. M. Borkovec. Langmuir. 13: 2608–2613 (1997).
41. Z. Tao and W. Dong. J. Colloid Interf. Sci. 208: 248 251 (1998).
42. T. Hiemstra, W. H. van Riemsdijk and G. H. Bolt. J. Colloid Interf. Sci. 133: 91–104 (1989).
43. T. Hiemstra, J. C. M. de Wit and W. H. van Riemsdijk. J. Colloid Interf. Sci. 133: 105–117 (1989).
44. T. Hiemstra, H. Yong and W. H. van Riemsdijk. Langmuir 15: 5942–5955 (1999).
45. W. F. Bleam. J. Colloid Interf. Sci. 159: 312–318 (1993).
46. T. Hiemstra, P. Venema and W. H. van Riemsdijk. J. Colloid Interf. Sci. 184: 680–692 (1996).
47. P. Venema, T. Hiemstra, P. G. Weidler and W. H. van Riemsdijk. J. Colloid Interf. Sci. 198: 282–295 (1998).
48. R. E. G. van Hall, J. C. T. Eijkel and P. Bergveld. Adv. Colloid Interf. Sci. 69: 31–62 (1996).
49. J. R. Rustad, E. Wasserman and A. R. Felmy. Surface Sci. 424: 28–35 (1999).
50. K. Nagashima and F. Blum. J. Colloid Interf. Sci. 217: 28 36 (1999).
51. P. A. Connor, K. D. Dobson and A. J. McQuillan. Langmuir 15: 2402–2408 (1999).
52. W. H. van Riemsdijk, G. H. Bolt, L. K. Koopal and J. Blaakmeer. J. Colloid Interf. Sci. 109: 219–228 (1986).

123. W. Rudzinski, R. Charmas and T. Borowiecki. In: Adsorption on New and Modified Inorganic Sorbents. A. Dabrowski and V. A. Tetrykh (Eds), Elsevier, 1996, pp. 357–409.
124. A. P. Robertson and J. O. Leckie, J. Colloid Interf. Sci. 188: 444–472 (1997).
125. M. Cernik, M. Borkovec and J. C. Westall. Langmuir 12: 6127–6137 (1996).
126. Z. Kolarik. Coll. Czechosl. Chem. Comm. 27: 938–950 (1962).
127. Z. Kolarik and V. Kourim. Coll. Czechosl. Chem. Comm. 25: 1000–100 (1960).

6

Sorption Properties of Selected Organic Materials

The surface properties of inorganic materials discussed in Chapters 2–5 can be tailored for specific applications by deposition of a thin layer of modifier (adsorption or grafting).

This leads to composite materials whose adsorption properties are chiefly defined by the modifier and the role of the chemistry of the core material is rather insignificant.

Therefore the relationship (emphasized in Chapters 2–5) between the bulk properties of the material and its adsorption properties is not applicable to modified adsorbents.

From this point of view all organic materials discussed in this chapter are in some sense modified materials, namely, their adsorption properties are defined by the surface treatment they underwent. Therefore, the adsorption properties of such materials reported in the literature, including the pH dependent surface charging, should be attributed to specific sample (whose surface treatment was more or less carefully controlled) rather than treated as generic properties of the studied core material.

Organic materials often occur as macroscopic specimens (reaction vessels, centrifuge tubes) whose surface area is too small to bring about substantial changes in solution over the usually studied range of adsorbate concentrations, but they can still affect adsorption experiments carried out at concentrations in the picomolar range. Several publications report the results of tests of potential uptake of the adsorbate of interest by the plasticware used in the experiments. For example Teflon is known as a very inert material.

On the other hand, activated carbon often has specific surface area in excess of 1000 m^2/g, i.e. higher then most inorganic materials. Thus, even with relatively low affinity to the adsorbate, activated carbon can bring about substantial uptake. Polymeric materials in form of dispersion of nearly spherical (and often nearly monodispersed) particles (size on the order of 100 nm) are termed latexes. Their specific surface area is typically on the order of a few m^2/g. The pH dependent surface charging of some activated carbons and latexes, and their reactions with other inorganic adsorbates will be presented in this chapter.

I. ACTIVATED CARBONS

Most experimental results reported in literature were obtained using commercially available activated carbons. A few other studies were carried out with self synthesized activated carbons derived from different natural organic materials. The preparation of activated carbons consists of two main stages termed carbonization and activation, and the choice of conditions under which these processes are carried out, gives practically infinite number of combinations. The temperature, time and atmosphere (the nature of the gas or gas mixture, and its pressure or flow rate) are the main variables. Moreover, different additives can be added before or between these processes to modify the final product. Wood and coal are the most common precursors, but sorption properties of activated carbons derived from other materials, e.g. coconut shells or plum kernels can be also found in literature. Although the precursor material certainly has some effect on the adsorption properties of activated carbon, these properties are chiefly defined by the conditions of thermal treatment and purification (*vide infra*). Therefore statements like, "*activated carbon from coconut shells has higher affinity to certain adsorbate than activated carbon from wood*" are misleading and they should be avoided.

Commercial products contain chiefly carbon, but substantial amount of mineral substances (e.g. metals and silicon) is always occluded in the pores. These impurities are difficult to remove (especially from micropores), and even multiple cycles of washing with hot hydrochloric and hydrofluoric acid or mixture thereof leave substantial amounts of impurities in the pores. Inorganic impurities affect the sorption properties of activated carbons. Therefore the elemental analysis of the activated carbons is often reported as the part of their characterization. In many studies the commercial materials were purified (usually using HCl and HF), but other authors did not pay attention to possible role of mineral impurities. Certainly the results obtained with relatively large and unknown amount of impurities are of limited significance.

Sorption properties of activated carbons depend on the degree of oxidation of the surface. The pH dependent surface charging and binding of metal cations by activated carbons are governed by oxygen containing surface groups (carboxylic, phenolic). In view of the effect of the neighboring atoms on the protonation of these groups (and the binding of metal cations to these groups), the surface should be rather considered in terms of some quasi continuous spectrum of pK than in terms of a few discrete values attributed to specific types of groups. The number of surface groups in certain ranges of protonation constant can be estimated from end-points of acid-base titration (using different titrants). Moreover, the surface groups containing heteroelements (phosphorus, sulfur) originating from the precursor material or introduced as modifiers exhibit acid-base and ion binding properties very different from the surface groups containing only carbon, oxygen and hydrogen.

In addition to acid-base and surface complexation reactions, different types of surface groups can undergo redox reactions during the adsorption process. Reductive adsorption of oxidants, e.g. Ag(I)→Ag(0) or Cr(VI)→Cr(III) is common-place with activated carbons. The redox potential is an important variable in such systems. For instance, presence of oxidants (nitrate V, chlorate VII) substantially depresses the uptake of Cr(VI) by activated carbon.

The oxygen containing surface groups are removed by evacuation at about 1000 K. On the other hand oxygen containing groups on the surface can be created when the original material is treated by H_2O_2 or other oxidants under controlled conditions. Many adsorption studies were carried out with such modified (outgased or oxidized) activated carbons. The oxidation is not necessarily limited to creation of surface groups. Application of strong oxidants may result in formation of compounds similar to humic acid. These compounds show high affinity to the surface thus they are difficult to remove from the adsorbent, and they substantially affect the sorption properties, e.g. binding of heavy metal ions. One should clearly distinguish between interactions with surface groups on the one hand, and interactions with macromolecular compounds (attached to the actual surface of activated carbon by adsorption forces) on the other. In extreme cases activated carbon acts rather as support for such macromolecular species (formed as the result of oxidation) than as actual adsorbent.

A. Primary Surface Charging

Activated carbons undergo pH dependent surface charging similar to that discussed in Chapter 3 for inorganic materials. Carboxyl and phenolic groups (among other) are responsible for the negative surface charge at high pH and amino groups are responsible for the positive surface charge at low pH. In this respect application of the 2-pK model which attributes positive and negative surface charge to one type of amphoteric surface groups (cf. Chapter 5) to surface charging of activated carbons is rather unfortunate.

Adsorption of cations and anions on activated carbons is affected by the electrostatic attraction or repulsion, thus it is pH dependent. However, many studies reported in literature, event quite recent (e.g. [1]) were carried out without pH control, probably at natural pH.

Babic et al. [2] collected some IEP and PZC of activated carbons reported in literature (five references, 12 samples). The PZC range from pH 2.2 to 10.4 and IEP range from pH 1.4 to 7.1. With a few materials for which both PZC and IEP were reported these two values did not match. These discrepancies are not surprising. Most studied materials were not purified from mineral constituents, thus the measurements were not carried out at pristine conditions. It should be emphasized that potentiometric titrations of activated carbons result in continuous drift of the pH (even after equilibration times of many hours), which is due to slow diffusion in micropores, and the apparent adsorption of acid or base depends on arbitrary selected equilibration time. Moreover, uncorrected PZC (cf. Chapter 3) obtained from titration at one ionic strength is usually reported.

Babic et al. [2] report a CIP of charging curves (25°C, 0.001–0.1 mol dm^{-3} KNO$_3$) at pH 7, for self synthesized active carbon obtained from carbonized viscose rayon cloth. Seco et al. [3] titrated commercial activated carbon at four different ionic strengths and attempted to determine the equilibrium constants of reactions (5.32) and (5.33) from these titrations. Only results obtained at extreme pH values (< 3 or > 11) were used, thus the apparent surface charge densities were obtained as differences of two large and almost equal numbers. On the other hand, at pH 4–10 the titration curve of carbon suspension and the blank curve were practically identical.

Recently Arafat et al. [4] reported uncorrected PZC (obtained from titration at one ionic strength, namely, 0.1 mol dm^{-3} NaCl) for a commercial activated carbon (from petroleum pitch). They obtained the following pH values: 7.15 for the original material washed with boiling water, 2.8 for oxidized sample (air treated at 350°C), and 9.2 for a sample treated with nitrogen at 850°C.

B. Adsorption of Metal Cations

Many publications report adsorption of heavy metal cations on activated carbon. Apparently, binding of metal cations is due to formation of surface complexes that are similar to chelate complexes in solution and some of them are very stable. The stability of these complexes is certainly dependent on the surface treatment. However, in some publications the sorption on activated carbons is treated as ion exchange.

Dobrowolski [5] emphasized potential analytical application, namely, activated carbon can be used to pre-concentrate trace amounts of toxic metals in chemical analysis of water. The heavy metal ions can be then desorbed, e.g. using complexing agents, or the dispersion of activated carbon can be directly analysed by AAS. This application requires activated carbon free of mineral impurities, especially the toxic metals of interest.

The uptake of six metal cations by commercial activated carbon (at constant pH) increases when the temperature increases [6]. This behavior is similar to typical results reported for inorganic materials (cf. Table 4.1), and very likely the increase in adsorption of metal cations when temperature increases is a generic property independent of the nature of the adsorbent. Therefore only the results obtained at controlled temperature are significant.

The uptake(pH) curves for the systems metal cation—activated carbon reported in the literature are very divergent, and they indicate significant role of the oxidation of the surface on the one hand and of the impurities on the other. These factors were often not controlled, thus it is difficult to quantify their effect on the adsorption.

Nearly linear increase in uptake of Zn by a commercial sample from 0% to 50% over the range 2–10 was reported [7] (no supporting electrolyte added). Inorganic adsorbents used in the same study (at the same solid to liquid ratio of 1 g/dm^{3}) turned out to be more efficient in removing Zn in spite of lower specific surface area. Moreover, the uptake of Zn by activated carbon at pH 8–10 was even lower than in control sample (without adsorbent). This suggests that some compounds leached out from the adsorbent stabilize Zn in solution.

Another commercial activated carbon was used to study the uptake of Cd and Zn as the function of pH [3]. The experiments were carried out at variable metal cation concentration and solid to liquid ratio, and at constant ionic strength (0.01 mol dm^{-3} NaCl). The uptake of the both metal cations increased with the pH but neither 0% uptake at low pH nor 100% uptake at high pH was reached (cf. Fig. 4.6(A) and (B)). Comparison of the uptake curves obtained at similar experimental conditions suggests that both metals had similar affinity to the surface, or even Cd had higher affinity. In contrast Zn has substantially higher affinity to mineral surfaces than Cd (cf. Section 4.II.C). The log–log adsorption isotherms of the both cations at constant pH were nearly linear with a slope of about 0.5 (pH dependent).

Biniak et al. [8] studied uptake of Cu by three de-ashed and modified active carbons from 0.05 mol dm^{-3} CuSO$_4$ at pH 0–6. The uptake of Cu was negligible at pH 0, but at pH 1 the uptake by all three studied materials was significant and it further increased when the pH increased. It should be emphasized that common inorganic materials do not adsorb Cu at pH 1 (Table 4.1).

C. Adsorption of Anions

Adsorption of borate on two types of commercial activated carbons (from deionized water and its mixtures with natural waters) resulted in uptake maximum at pH about 8 [9]. Nearly linear decrease in uptake of chromate by a commercial sample from 50% to 10% over the pH range 2–10 (no supporting electrolyte added) was reported [7]. In contrast Dobrowolski [5] reported uptake curves of chromate with a maximum at pH about 3 for three de-ashed and modified activated carbons. These two types of adsorption behavior are common in specific adsorption of weak acids on inorganic materials (cf. Table 4.2). Activated carbon was found to be an efficient adsorbent of iodides from synthetic clay water at pH as high as 8.5 [10].

D. Other Adsorbates

Commercial activated carbon was found to be more efficient as adsorbent of quinoline than common inorganic materials [11]. Adsorption of quinoline on activated carbon increases with pH up to pH 6 and at pH 6–11 the adsorption is nearly constant. In contrast the adsorption on inorganic materials has a sharp maximum at pH about 6.

Arafat et al. [4] studied the effect of the concentration of inorganic electrolyte on adsorption of benzene, toluene and phenol from aqueous solution at pH 11.6 on one commercial and two modified activated carbons and obtained very different results for these three adsorbates. The uptake of benzene was rather insensitive to the ionic strength. The uptake of toluene systematically decreased when the ionic strength increased. Finally the uptake of phenol was enhanced on addition of 0.5 mol dm^{-3} KCl, but further addition of salt depressed the uptake and with 0.8 mol dm^{-3} KCl the uptake dropped below that observed at low ionic strength. Adsorption of phenol on activated carbons was recently studied by other research groups [12,13], but without emphasis on the possible effects of pH dependent surface charging.

II. LATEXES

Polystyrene latex can be easily prepared on bench top, but many studies reported in the literature were carried out with commercially available samples. Polystyrene is the most common core material. Other polymers and copolymers (core materials) have different specific density (specific density of polystyrene is close to that of water), solubility in organic solvents and other physical properties, but the surface charging and adsorption of inorganics on latexes are due to their polar surface groups. Surface groups originate chiefly from the initiator of polymerization (e.g. persulfate is a source of $-OSO_3H$ groups), and from functional groups of the monomer/comonomer (e.g. methacrylic acid is a source of carboxyl groups) thus the presence of surface functional groups in latexes cannot be avoided. On the other hand there is no simple relationship between the nature and concentration of the surface groups in the resulting latex and the conditions of polymerization. Even with a monomer without functional groups (e.g. styrene) and simple initiator, the polymerization usually results in latex with multiple types of surface groups. In principle the polar end groups of the polymer chain tend to be exposed to the latex surface, but some polar groups are buried in the core material.

Once the polymerization is complete the modification of the surface groups is still possible, e.g. oxidation of hydroxyl groups results in carboxyl groups. Such modification can be carried out deliberately, in controlled manner or it can occur spontaneously during the storage of the dispersion. Post polymerization modification of surface groups by grafting was also considered. Although monomers with functional groups give latexes with certain type of surface groups, in principle the surface properties of latexes must not be attributed to the nature of the core material.

Latex particles having only strongly acidic surface groups can be treated as macroscopic multivalent ions with constant (pH independent) surface charge density and constant ζ potential at given ionic strength. For example ζ potential of latex of about -35 mV over a pH range 3–11 (in 10^{-3} mol dm^{-3} KNO$_3$) was reported [14]. The application of latexes discussed in Chapter 3 (cf. Fig. 3.9) is based upon this property. On the other hand, latex stabilized by weakly acidic carboxylic surface groups has also negative but strongly pH dependent ζ potential, namely the negative charge is more pronounced at high pH [15].

Positively charged latexes can be produced by polymerization of tertiary amines. Amphoteric latexes have surface amino groups on the one hand and carboxyl or other acidic groups on the other. For example IEP at pH about 3.5 was found for two commercial materials [16]. The same materials were modified by O_2-plasma etching, but the effect of this treatment on the IEP was rather insignificant.

Continuous decrease in the electrophoretic mobility when the ionic strength increases is observed for inorganic materials (Fig. 3.2) but not for latex. This anomalous behavior is attributed to the "hairy" structure of latex surfaces.

Temperature has dramatic effect on the electrokinetic properties of thermosensitive polymers. A pH independent electrophoretic mobility of about -0.5×10^{-8} m^2 V^{-1}s^{-1} (pH 3–11) in the presence of 0.1 mol dm^{-3} NaCl was reported for synthetic poly (N-isopropylacrylamide) latex (persulfate initiator) at 20°C [17]. The same material has a mobility of about -5×10^{-8} m^2 V^{-1}s^{-1} (ten times greater!) at 50°C. This dramatic change occurs when the lowest critical solubility temperature is exceeded, and it is accompanied by substantial decrease in particle size (factor about two,

TABLE 6.1 Isoelectric Points of Organic Materials

Material	Description	Salt	T	IEP	Ref.
High tensile carbon fibers	905 C treated, then exposed to oxygen at -42°C	0.001 mol dm^{-3} KCl		10	18
Polyamide 12		10^{-4}–10^{-3} mol dm^{-3} KCl		3.6–4.2	18
Polyamide 6–6	Nyltech Industries	0.001 mol dm^{-3} NaCl	23–27	5	19
Polycarbonate	MW 39 300	0.0003–0.001 mol dm^{-3} KCl		4	18
Polyetheretherketone		0.001 mol dm^{-3} KCl		4.4	18
Polyetherimide	MW 89 100	0.001 mol dm^{-3} KCl		4.1	18
Polystyrene	MW 198 000, free radical polymerization	0.001 mol dm^{-3} KCl		4	18
Polytetrafluoroethylene		0.01 mol dm^{-3} NaCl		3	20

measured by quasielastic light scattering) when the temperature increases from 20°C to 50°C.

Many publications are devoted to adsorption of surfactants, proteins and polymers on latexes, but adsorption of inorganic ions on latexes does not seem to be very active field. Latexes are also used as model colloids in studies of heterocoagulation, i.e. "adsorption" of colloids on colloids.

III. OTHER MATERIALS

Isoelectric points of a few organic materials reported in the literature are summarized in Table 6.1.

The IEP reported for different synthetic polymers are surprisingly consistent.

Interestingly all polymer surfaces had amphoteric character. The chemical nature of the core material is rather irrelevant here, and the surface charging is due to surface groups whose nature is unknown.

Recently, natural coals were found to strongly adsorb heavy metal cations [21] with the following affinity sequence $Pb > Cr(III) > Cd \approx Zn \approx Ni > Ca$. Typical adsorption edges (increase of uptake from practically 0% to practically 100% within 3 pH units, cf. Fig. 4.5 (A)) was observed.

REFERENCES

1. Y. F. Jia and K. M. Thomas. Langmuir 16: 1114–1122 (2000).
2. B. M. Babic, S. K. Milonjic, M. J. Polovina and B. V. Kaludierovic. Carbon 37: 477–481 (1999).

3.   A. Seco, P. Marzal, C. Gabaldon and J. Ferrer. Separ. Sci. Technol. 34: 1577–1593 (1999).
4.   H. A. Arafat, M. Franz and N. G. Pinto. Langmuir 15: 5997–6003 (1999).
5.   R. Dobrowolski. In: Adsorption and its Applications in Industry and Environmental Protection. A. Dabrowski, (Ed.), Elsevier, Amsterdam, 1998, pp. 777–805.
6.   R. Qadeer, J. Hanif, M. Saleem and M. Afzal. Coll. Polym. Sci. 271: 83–90 (1993).
7.   M. Lehmann, A. I. Zouboulis and K. A. Matis. Chemosphere 39: 881–892 (1999).
8.   S. Biniak, M. Pakula, G. S. Szymanski and A. Swiatkowski. Langmuir. 15: 6117–6122 (1999).
9.   W. W. Choi and K. Y. Chen. Environ. Sci. Technol. 13: 189–196 (1979).
10. N. Maes, A. Dierckx, X. Sillen, L. Wang, P. De Canniere, M. Put and J. Marivoet. Paper PB2-21, 7th International Conference on the Chemistry and Migration Behavior of Actinides and Fission Products in the Geosphere. Lake Tahoe, September, 1999.
11. A. K. Helmy, S. G. deBussetti and E. A. Ferreiro. Clays Clay Min. 31: 29–36 (1983).
12. B. Okolo, C. Park and M. A. Keane. J. Colloid Interf. Sci. 226: 308–317 (2000).
13. R. S. Juang, F. C. Wu and R. L. Tseng. J. Colloid Interf. Sci. 227: 437–444 (2000).
14. K. Furosawa, K. Nagashima and C. Anzai. Kobunshi Ronbuhshu 50: 343–347 (1993).
15. D. A. Antelmi and O. Spalla. Langmuir 15: 7478–7489 (1999).
16. C. J. Zimmermann, N. Ryde, N. Kallay, R. Partch and E. Matijevic. J. Mater. Res. 6: 855–860 (1991).
17. D. Duracher, A. Elaissari and C. Pichot. Colloid Polym. Sci. 277: 905–913 (1999).
18. A. Bismarck, M. E. Kumru and J. Springer. J. Colloid Interf. Sci. 217: 377–387 (1999).
19. P. Bouriat, P. Saulnier, P. Brochette, A. Graciaa and J. Lachaise. J. Colloid Interf Sci. 209: 445–448 (1999).
20. W. Smit, C. L. M. Holten, H. N. Stein, J. J. M. de Goeij and H. M. J. Theelen. J. Coll. Interf. Sci. 63: 120–128 (1978).
21. C. A. Burns, P. J. Cass, I. H. Harding and R. J. Crawford. Colloids Surf. A 155: 63–68 (1999).

Appendix A1

Abbreviations

This list of abbreviations does not cover

- Symbols of chemical elements, compounds and isotopes of hydrogen (cf. Fig. 1.1).
- SI units and their derivatives.
- Trade marks and names (cf. Appendix 2).
- Symbols used only in one equation and not referred to in further text.

Most symbols listed without subscript or superscript can be used with different subscripts and superscripts. Many symbols were used for different physical quantities. A few physical quantities were denoted by different symbols in different parts of the book, e.g. to avoid use of the same symbol for two different quantities in one equation or phrase.

0	(subscript) surface
a	activity
a	axis length (in crystallography)
a	stoichiometric coefficient of reagent A in reaction (2.1)
a	empirical coefficient in Eqs. (2.29), (3.14) and (4.16)
a	sphere radius
A	empirical coefficient in Eq. (3.19) and related equations
A	surface area, cross section

A	molecule of adsorbate
$A_{11(3)}$	Hamaker constant
AAS	atomic absorption spectroscopy
act	(superscript) defined in terms of activities
ads	(subscript) adsorption, adsorbed
AFM	atomic force microscopy
AMP	aminomethylphosphonic acid
AS	surface compound (adsorbate + surface site)
ATR	attenuated total reflection
b	axis length (in crystallography)
b	empirical coefficient in Eqs (2.29) and (4.16)
b	stoichiometric coefficient of reagent B in reaction (2.1)
b	(subscript) bulk
B	empirical coefficient in eq. (3.19) and related equations
BET	Brunauer-Emmett-Teller (adsorption isotherm)
bipy	2,2' dipirydyl
c	concentration
c	stoichiometric coefficient of reagent C in reaction 2.1
c	axis length (in crystallography)
c	empirical coefficient in eq. (2.29)
C	crystal field stabilization energy
C	capacitance
C_1	inner capacitance
C_2	outer capacitance
C_+	capacitance (for positive branch of charging curve)
C_-	capacitance (for negative branch of charging curve)
C*	vector of logarithms of concentrations
CCC	critical coagulation concentration
CD	charge distribution (model)
CDTA	trans-1,2,-diaminocyclohexanetetraacetic acid
CFSE	crystal field stabilization energy
c_i	molar concentration of species i
c_i	parameter in Eq (5.75)
CIP	common intersection point
CIR	cylindrical internal reflection
CMC	critical micellization concentration
CN	coordination number of metal cation
CN_O	coordination number of oxygen
C_p	heat capacity (at constant pressure)
d	stoichiometric coefficient of reagent D in reaction (2.1)
d	(subscript) dispersion
d	(subscript) diffuse
d	thickness (of capacitor)
d	day
d	empirical coefficient in Eq. (2.29)
d	specific density
D	diffusion coefficient
DCyTA	diaminocyclohexane N,N,N',N' tetraacetic acid

DLVO	Deraguin–Landau–Vervey–Overbeek (theory of colloid stability)
DM	cation binding energy from XPS spectrum
DMF	N,N dimethyl formamide
DMSO	dimethyl sulfoxide
DO	oxygen 1s binding energy from XPS spectrum
DOC	dissolved organic carbon
DR	diffuse reflectance (FTIR)
DRIFT	diffuse reflectance infrared Fourier transform
DSA	dimensionally stable electrode
DTPA	diethylenetriaminepentaacetic acid
DTPMP	diethylenetrinitrilopentakis-(methylenephosphonic acid)
e, e_0	charge of proton
E	field strength
E^0	normal potential
E_{act}	activation energy
EAP	equiadsorption point (of anion and cation of inert electrolyte)
EDTA	ethylenediaminetetraacetic acid
EDTMP	ethylenedinitrilotetrakis-(methylenephosphonic acid)
E_F	Fermi level
eff	(subscript) effective
E_g	band gap
EGME	specific surface area determined by means of 2-ethoxyethanol
EGTA	ethylenebis (oxyethylenenitrilo) tetraacetic acid
E_i, E_i^0, E_i^M	parameters in Eq. (5.75)
EMF	electromotive force
ENDOR	electron nuclear double resonance
EPR	electron paramagnetic resonance
ESR	electron spin resonance
EXAFS	extended X-ray absorption fine structure
f	fraction of charge of specifically adsorbed ion assigned to the surface in CD model
f	atomic fraction
f	volume fraction of solid
F	Faraday constant
F	exchange fraction (Eq. (4.26))
F	face centered
FA	fulvic acid
fi	(subscript) fictional
fix	(subscript) fixed (pH independent)
FTIR	Fourier transformed infrared
G	Gibbs energy
G^0	standard Gibbs energy
g_1, g_2	coefficients in Eq. (4.20)
Ger.	German
h	hour
H	enthalpy
H^0	standard enthalpy
HA	humic acid

HEDP	1-hydroxyethane-(1,1-diphosphonic acid) or 1-hydroxyethane-(1,2-diphosphonic acid)
HEDTA	1-(2-hydroxyethyl)ethylenediaminetriacetic acid
HEPES	4-(2-hydroxyethyl)-1 piperazine ethanesulfonic acid (buffer)
i	(subscript) component i
i	(radio)activity
I	ionic strength
ICP	inductively coupled plasma
IDMP	iminodi-(methylenephosphonic acid)
IEP	isoelectric point
IR	infrared (spectroscopy)
ISFET	ion sensitive field effect transistor
IUPAC	International Union of Pure and Applied Chemistry
k	rate constant
k	Boltzmann constant
K	equilibrium constant
K^*	vector of logarithms of equilibrium constants
K_{a1}, K_{a2}	surface acidity constants
K_D	distribution coefficient (cf. text under Eq. (4.9))
K_{sp}	solubility product (of sparingly soluble salt)
K_X, K_Y	stability constants of $\equiv SOH_2^+ \cdot\cdot X^-$ and $\equiv SO^- \cdot\cdot Y^+$ species in TLM model
L	metal–hydrogen distance (in surface OH group)
m	mass
m	parameter in Eq. (5.74)
M	(subscript) metal (usually divalent)
MA	methyl arsenate
Me	metal (usually divalent, in chemical reactions)
MES	4-morpholine ethanesulfonic acid (buffer)
min	minute
MK	Marek Kosmulski
MUSIC	multisite complexation (model)
MW	molecular weight (of polymers)
n	unknown or variable number of water molecules
n	number concentration
n	number of moles
n	refractive index
N	number
N	matrix of stoichiometric coefficients
N_A	Avogadro number
NMR	nuclear magnetic resonance
NOM	natural organic matter
N_s	total concentration of surface OH groups
NTA	nitrilotriacetic acid
NTMP	nitrilotris-(methylenephosphonic acid)
OC	organic carbon
Ox	oxalate
p	-log
p	pressure

p	number of elementary charges attributed to the surface in CD model
P	primitive
PAA	phosphonoacetic acid
pH_0	pristine point of zero charge
pH_{50}	the pH value for which the adsorbate is distributed 50–50 between the liquid and the solid
PPA	α-phosphonopropionic acid
PPZC	pristine point of zero charge
PVA	poly (vinyl alcohol)
PZC	point of zero charge
PZNPC	point of zero net proton charge
PZSE	point of zero salt effect
q	number of elementary charges attributed to the "1" plane in CD model
r	the number of protons released per one adsorbed metal cation (eq.4.14)
r	radius (of pore)
R	bond length
R	cation radius
R	rhombohedral
R	gas constant
R_f	chemical retardation (Eq. (4.10))
s	bond valence
s	(subscript) sublimation
s	(subscript) solid
s	(subscript) surface
s	(subscript) strong (site)
S	entropy
S^0	standard entropy
S, $\equiv$S	surface atom
SCE	saturated calomel electrode
SCM	surface complexation model
SDS	sodium dodecylsulfate
SEM	scanning electron microscopy
SFM	scanning force microscope
sp	(subscript) solubility product
SPM	suspended particulate matter
SXRF	synchrotron based micro X-ray fluorescence
t	time
t	transference number
T	temperature
T	total concentration of component
T	vector of total concentrations of components
TCNQ	7,7,8,8-tetracyanoquinodimethane
TDPAC	time differential perturbed angular correlation (spectroscopy)
TEA	tetraethyl ammonium
TEM	transmission electron microscopy
THF	tetrahydrofuran
TLM	triple layer model
TMA	tetramethyl ammonium

Appendix A2

Trade Names and Trademarks

The following names (trade names and trademarks) were used to characterize the materials in Tables in Chapters 3, 4 and 6. These products range from chemicals of high purity (e.g. reagent grade) to technical products of low purity. The purpose of this compilation is to facilitate a distinction of these trade names and trade marks from other names used to characterize the materials, i.e.

- Names of crystallographic forms (cf. Table 2.1).
- Geographic names (e.g. places where the mineralogical samples were collected).
- Last names (e.g. Stöber used as abbreviation for method of preparation of silica particles from tetraethoxysilane), etc.

The trade names and trade marks are spelled as in original publications, and this spelling is not necessarily identical with official names of companies and products.

3 M
Aerosil
Ajax
Akzo
Alcoa = Aluminum Company of America
Aldrich
Alfa
Alfa Aesar
Allied Signal

Alltech
Aluminum Co. of Canada
Amend
American Chemicals
American Cyanamid
American Petroleum Institute
Ashai
Baikalox
Baikowski
Baker
Barcroft
Bayer
Bayferrox
BDH = British Drug Houses
Buehler
Cabosil
Cabot
Carlo Erba
Catalyst and Chemical Ind.
CCI
China Clay Supreme
Chlorovinyl
Chromosorb
Condea
Crosfield
Custer
Cyanamid
Darvan
Davison
DBK
Degussa
Dia Showa
Du Pont
Duramax
Electrofact
English Clay Lovering Pochin
ES-gel
Eurospher
Fisher
Fluka
Fuji
Geltec
Geltech
Georgia Kaolin
Gregory, Bottley and Lloyd
Grillo
Harshaw
Herasil

High Purity Fine Chemical Inc.
Highways
Houdry
Huber
Hydral
Hypersil
IC 1; 12; 22
ICN
Ingold
INVAP
Japan Aerosil
Johnson–Matthey
JRC
Kaiser
Kanto
Kemika
Koch Light
Kristallhandel Kelpin
Kromasil
Kunimine
Laporte
La Roche
Lichrospher
Linde
Ludox
Machinery Nagel
Mager
Magnesium Elektron
Mallinckrodt
Mandoval
Marblehead
Martifin
Martinswerk
May and Baker
Merck
Min-U-Sil
Mitsui Toatsu
Monospher
Nanophase Technologies
Nikki
Nippon Silica Kogyo
Nishio
Nissan
NIST (National Institute of Standards and Technology)
Norton
Nucleosil
Nyltech
Pechiney

Pfalz and Bauer
Pfizer
Phenomenex
Plutonio Argentina
POCh
Polysciences
Porasil
Prolabo
Procatalyse
Puratronic
Purospher
Pyrex
Quarzwerke GmbH
Queensland Alumina
Reachim
Reanal
Research Organic Inorganic Chemical Co.
Reynols's
Rhone Poulenc
Riedel de Haen
Rohm and Haas
Sakei
Sarabhai
Schuchardt
Sifraco
Sigma
Sikron
Spherisorb
Spherosil
Starck
Sumitomo
Syowa Denko
Talcs de Luzenac
Tamei
TAMI
Thiokol
TICON
Tioxide
Titan
Tosoh Ceramics
Toyo Soda
Triton
Ube
UCAR
Union Carbide
Veb Lab
Ventron
Volclay

Wacker
Wako
Ward's
Wedron
Woelm Pharma
W.R. Grace
YMC
Zeo
Zeosyl
Zeothix
Z-Tech

Appendix A3

Points of Zero Charge

Tables A.1 and A.2 (added in proof) present the pristine points of zero charge that

- Were published after Chapter 3 was complete
- Overlooked in previous literature searches

and pristine points of zero charge of materials that do not fit any of the categories discussed in Chapters 3 and 6.

TABLE A.1 Zero Points of Simple Hydr(Oxides) and Other Materials

Oxide	Description	Salt	T	Method	pH_0	Source
Al_2O_3	Reynold's, 0.05% MgO	10^{-2} mol dm^{-3} KCl		iep	8.7	D.G. Goski and W.F. Caley. Ceram. Eng. Sci. Proc. 17: 187–194 (1996).
Al_2O_3	α, 99.9%, Sumitomo	10^{-4}–10^{-2} mol dm^{-3} NaCl	25	cip	8.5	A. El Ghzaoui. J. Appl. Phys. 86: 5894–5897 (1999).
Al_2O_3	fumed	none		iep	8.7	V.M. Gunko, V.I. Zarko, V.V. Turov, R. Leboda, E. Chibowski, E.M. Pakhlov, E.V. Goncharuk, M. Marciniak, E.F. Voronin and A.A. Chuiko, J. Colloid Interf. Sci. 220: 302–323 (1999).
				iep	9.8	
Al_2O_3	α, CL2500SG, Alcoa	10^{-3} mol dm^{-3} KCl	25	acousto	8.1	A. Pettersson, G. Marino, A. Pursiheimo and J.B. Rosenholm. J. Colloid Interf. Sci. 228: 73–81 (2000).
Al_2O_3	γ, from boehmite			pH	8.2	G.M.S. El Shafei, N.A. Moussa and C.A. Philip. J. Colloid Interf. Sci. 228: 105–113 (2000).
Al_2O_3	γ, Prolabo			pH	9.4	G.M.S. El Shafei, N.A. Moussa and C.A. Philip. J Colloid Interf. Sci. 228: 105–113 (2000).
Al_2O_3	α, Tamei	5×10^{-4} mol dm^{-3} LiCl in ethanol	20	iep	4.4	J. Widegren and L. Bergström. J. Eur. Cer. Soc. 20: 659–665 (2000).
Al_2O_3	α, reagent grade	1 mol dm^{-3} KCl		pH	7.1	E. Yu. Nevskaya, I.G. Gorichev, O.V. Kuchkovskaya, B.E. Zaitsev, A.I. Gorichev and A.D. Izotov. Zh. Fiz. Khim. 73: 1585-1591 (1999).
Al_2O_3	α, Pechiney, 3 samples	0.01 mol dm^{-3} $NaNO_3$		iep	8.7	A. Jacquet, L. Dupont and A. Foissy. Sil. ind. 64: 153–162 (1999).
					8.5	
					7	
AlOOH	boehmite, from gibbsite			pH	7.7	G.M.S. El Shafei, N.A. Moussa and C.A. Philip. J. Colloid Interf. Sci. 228: 105–113 (2000).
AlOOH	boehmite, Reanal	0.01–1 mol dm^{-3} KNO_3		cip	8.7	E. Tombacz, A. Dobos, M. Szekeres, H.D. Narres, E. Klumpp and I. Dekany. Colloid Polym.Sci. 278: 337 345 (2000).
AlOOH	α, reagent grade	0.01 mol dm^{-3} KCl		pH	8.8	E. Yu. Nevskaya, I.G. Gorichev, O.V. Kuchkovskaya, B.E. Zaitsev, A.I. Gorichev and A.D. Izotov. Zh. Fiz. Khim. 73: 1585–1591 (1999).

Material	Description	Electrolyte	T (°C)	Method	pzc	Reference
AlOOH	γ, reagent grade	0.0001–0.01 mol dm⁻³ KCl		iep	8.1	E. Yu. Nevskaya, I.G. Gorichev, O.V. Kuchkovskaya, B.E. Zaitsev, A.I. Gorichev and A.D. Izotov. Zh. Fiz. Khim. 73: 1585–1591 (1999).
Al(OH)₃	gibbsite, Prolabo			pH	5.4	G.M.S. El Shafei, N.A. Moussa and C.A. Philip. J. Colloid Interf. Sci. 228: 105–113 (2000).
Al(OH)₃	gibbsite, Alcoa	0.001–0.1 mol dm⁻³ NaNO₃		pH	8.7	R. Weerasooriya, B.B. Dharmasena and D. Aluthpatabendi. Colloids Surf.A. 170: 65–77 (2000).
Bi₂O₃·nH₂O	from nitrate	0.1 mol dm⁻³ NaNO₃		mass titration	6.8	B.S. Mathur and B. Venkataramani. Indian J. Chem. A 38: 1092–1099 (1999).
Fe₃O₄	natural magnetite, 2.4% SiO₂	10⁻², 10⁻¹ mol dm⁻³ NaCl	20	pH	9.9	S.K. Milonjic, A.L. Ruvarac and M.V. Susic. Bull. Soc. Chim. Beograd. 43: 207–210 (1978).
Fe₃O₄	Puratronic, reduced to remove Fe₂O₃	0.03 mol dm⁻³ Na triflate	50	intersection / inflection	6.3 / 6.5	D.J. Wesolowski, M.L. Machesky, D.A. Palmer and L.M. Anovitz. Chem. Geol. 167: 193–229 (2000). Temperature effect on the course of potentiometric titration was studied.
Fe₂O₃	hematite, spherical, from chloride	0.01 mol dm⁻³ NaNO₃	23 or 28	iep	7.6	Y. Zhang, N. Kallay and E. Matijevic. Langmuir 1: 201–206 (1985).
Fe₂O₃	hematite, spherical, from chloride	0.01 mol dm⁻³ NaClO₄	25	iep	7.3	R. Torres, N. Kallay and E. Matijevic. Langmuir 4: 706–710 (1988).
Fe₂O₃	hematite, Alfa	0.005 mol dm⁻³ KNO₃	25	cip / mass titration	6.3	T. Preocanin, Z. Brzovic, D. Dodlek and N. Kallay. Progr. Colloid Polym.Sci. 112: 76–79 (1999).
Fe₂O₃	hematite, Reachim	10⁻¹ mol dm⁻³ NaCl	25	pH	8.5	O.N. Karasyova, L.I. Ivanova, L.Z. Lakshtanov and L. Lövgren, J. Colloid Interf. Sci. 220: 419–428 (1999).
Fe₂O₃	hematite, spherical	10⁻³ mol dm⁻³		iep, salt titration	9.25	J.J. Lenhart and B.D. Honeyman. Geochim. Cosmochim. Acta. 63: 2891–2901 (1999).
Fe₂O₃	hematite, from chloride	0.005–0.5 mol dm⁻³ NaNO₃	25	cip	9.5	I. Christl and R. Kretzschmar. Geochim. Cosmochim. Acta. 63: 2929–2938 (1999).
Fe₂O₃	hematite, spherical	0.001–0.5 mol dm⁻³ NaNO₃		cip	8.1	K.K. Au, S. Yang and C.R.O' Melia. Environ. Sci. Technol. 32: 2900–2908 (1998); K.K. Au, A.C. Penisson, S. Yang and C.R.O' Melia. Geochim, Cosmochim. Acta 63: 2903–2917 (1999).
FeOOH	goethite from nitrate			cip / iep	8.5	D. Kovacevic, A. Pohlmeier, G. Ozbas, H.D. Narres and N. Kallay. Progr. Colloid Polym. Sci. 112: 183–187 (1999).

TABLE A.1 Continued

Oxide	Description	Salt	T	Method	pH$_0$	Source
FeOOH	goethite from nitrate, two samples	0.1 mol dm^{-3} NaNO$_3$	25	pH	9.5	J.F. Boily, P. Persson and S. Sjoberg. J. Colloid Interf. Sci. 227: 132–140 (2000).
FeOOH	goethite				8.9	J.R. Bargar, P. Persson and G.E. Brown. Geochim. Cosmochim. Acta. 63: 2957–2969 (1999).
FeOOH	goethite, Bayer	HNO$_3$ + KOH		iep	8.2	J. Subrt, L.A. Perez-Maqueda, J.M. Criado, C. Real, J. Bohacek and E. Vecernikova. J. Am. Cer. Soc. 83: 294–298 (2000).
FeOOH	from FeSO$_4$	HNO$_3$ + KOH		iep	8.2	J. Subrt, L.A. Perez-Maqueda, J.M. Criado, C. Real, J. Bohacek and E. Vecernikova. J. Am. Cer. Soc. 83: 294–298 (2000).
Mn$_3$O$_4$	from acetate	10^{-2} mol dm^{-3} NaCl		iep	5.8	M. Ocana. Colloid Polym. Sci. 278: 443–449 (2000).
MnOOH	α, from acetate	10^{-2} mol dm^{-3} NaCl		iep	9.6	M. Ocana. Colloid Polym. Sci. 278: 443–449 (2000).
SiO$_2$	precipitated, S 700, Z 175	KCl		salt addition	3.5 / 3.4	R. Zerouk, Thesis, quoted after A. Foissy and J. Persello in The Surface Properties of Silicas. A.P. Legrand. (Ed.), Wiley, New York, 1998, pp. 365–414.
SiO$_2$	precipitated, SiNa150	NaNO$_3$ 0.002–0.04 mol dm^{-3} NaCl		? / pH	2.5 / none	A. Foissy and J. Persello. In: The Surface Properties of Silicas. A.P. Legrand. (Ed.), Wiley, New York, 1998, pp. 365–414.
SiO$_2$	quartz, Sifraco C-600, washed	0.04–0.8 g/dm^3 NaCl / 8 g/dm^3 NaCl		iep	2 2.6 / 2.5–3.3	D.M. Nevskaia, A. Guerrero-Ruiz and J.D. Lopez-Gonzalez. J. Colloid Int. Sci. 205: 97–105 (1998).
SiO$_2$	Aerosil 200, Degussa	0.005 mol dm^{-3} KNO$_3$	25	cip / mass titration	4.4	T. Preocanin, Z. Brzovic, D. Dodlek and N. Kallay. Progr. Colloid Polym. Sci. 112: 76-79 (1999).
SiO$_2$	fumed	none		iep	2.2	V.M. Gunko, V.I. Zarko, V.V. Turov, R. Leboda, E. Chibowski, E.M. Pakhlov, E.V. Goncharuk, M. Marciniak, E.F. Voronin and A.A. Chuiko. J. Colloid Interf. Sci. 220: 302 323 (1999).
SiO$_2$	from ethoxide, with and without Ag core	0.001 mol dm^{-3} KNO$_3$		iep	< 3 if any	V.V. Hardikar and E. Matijevic. J. Colloid Interf. Sci. 221: 133–136 (2000).

Material	Description	Electrolyte	T	Method	PZC	Reference
$ThO_2 \cdot n\,H_2O$	from nitrate	$0.1\ \text{mol dm}^{-3}\ NaNO_3$		mass titration	7.8	B.S. Mathur and B. Venkataramani. Indian J. Chem. A 38: 1092–1099 (1999).
TiO_2	fumed	none		iep	6	V.M. Gunko, V.I. Zarko, V.V. Turov, R. Leboda, E. Chibowski, E.M. Pakhlov, E.V. Goncharuk, M. Marciniak, E.F. Voronin and A.A. Chuiko. J. Colloid Interf. Sci. 220: 302–323 (1999).
TiO_2	Anatase, Sigma			iep	5.3	J.P. Hsu and Y.T. Chang. Colloids Surf.A. 161: 423–437 (2000).
TiO_2	Degussa. P25	NaCl, conductivity $10^{-3.5}$ $10^{-1.5}\ \mu S/cm$	25	iep	6.4	L. Yezek, R.L. Rowell, L. Holysz and E. Chibowski. J. Colloid Interf. Sci. 225: 227–232 (2000).
TiO_2	Degussa. P25	none		iep	6.8	P. Fernandez-Ibanez, F.J. de las Nieves and S. Malato. J. Colloid Interf. Sci. 227: 510–516 (2000).
TiO_2	Nanophase Technologies, rutile + anatase	$5 \times 10^{-4}\ \text{mol dm}^{-3}\ LiCl$ in ethanol	20	iep	4.2	J. Widegren and L. Bergström. J. Eur. Cer. Soc. 20: 659–665 (2000).
TiO_2	fired and crushed xerogel	$0.01\ \text{mol dm}^{-3}\ NaCl$		iep	5	B.P. Nelson, R. Candal, R.M. Corn and M.A. Anderson. Langmuir 16: 6094–6101 (2000).
$Y(OH)_3$	thermal decomposition of basic carbonate	$NaNO_3$		coagulation	7.8–9.2	B.V. Eremenko, M.L. Malysheva, T.N. Bezuglaya, A.N. Savitskaya, I.S. Kozlov and L.G. Bogodist. Kolloid. Zh. 62: 58–64 (2000).
ZnO	Grillo	NaOH HCl		iep	9.5	A. Degen and M. Kosec. J. Eur. Ceram. Soc. 20: 667–673 (2000).
ZnO	Johnson Matthey	NaOH HCl		iep	$\zeta < 0$ at pH > 8.5; $\zeta = 0$ at pH 6–8.5	A. Degen and M. Kosec. J. Eur. Ceram. Soc. 20: 667–673 (2000).
ZrO_2	high purity, TSK, TZ-O, monoclinic	$10^{-4}\ 10^{-1}\ \text{mol dm}^{-3}\ NaCl$	25	cip; iep	5.5; 5.8	J. Zajac, M. Lindheimer and S. Partyka. Colloids Surf. A 98: 197 208 (1995).
ZrO_2	Ventron	$10^{-3}\ 10^{-1}\ \text{mol dm}^{-3}\ NaCl$	25	cip	7.6	W. Janusz. Fizykochemiczne Problemy Mineralurgii. 31: 125 135 (1997).
ZrO_2	high purity, TSK, TZ-O, monoclinic	$10^{-4}\ 10^{-2}\ \text{mol dm}^{-3}\ NaCl$	25	cip; iep	5.5; 5.8	A. El Ghzaoui. J. Appl. Phys. 86: 5894–5897 (1999).
ZrO_2	TZ-O, monoclinic, Tosoh	$10^{-3}\ \text{mol dm}^{-3}\ NaCl$	25	acousto	6.4	A. Pettersson, G. Marino, A. Pursiheimo and J.B. Rosenholm. J. Colloid Interf. Sci. 228: 73–81 (2000).
ZrO_2	Tosoh	$10^{-2}\ \text{mol dm}^{-3}\ NaCl$	26–27	iep	7–7.4	W.C. Wei, S.C. Wang and F.Y. Ho. J. Am. Cer. Soc. 82: 3385 3392 (1999). This paper contains also compilation of IEP of ZrO_2 and Y-doped ZrO_2 with 6 references.

TABLE A.1 Continued

Oxide	Description	Salt	T	Method	pH$_0$	Source
Zr(OH)$_4$	amorphous, precipitated from nitrate at different pH, values in brackets the same samples after γ-irradiation	10^{-3} mol dm^{-3} NaCl		iep	6.1 (5.8) 7 (7.7) 7 (6.2)	S. Music, G. Stefanic, N. Vdovic and A. Sekulic. J. Therm. Anal. 59: 837–846 (2000).
Mixed oxides containing Al$_2$O$_3$						
Al$_2$O$_3$·SiO$_2$	kyanite, from natural ore, impurities, % w/w FeO 0.3, CaO 0.29, K$_2$O 0.25, TiO$_2$ 0.17, MnO 0.01.	10^{-2} mol dm^{-3} KCl		iep	5.8	D.G. Goski and W.F. Caley. Ceram. Eng. Sci. Proc. 17: 187 194 (1996).
Clay minerals						
illite	Marblehead	10^{-2} 1 mol dm^{-3} KCl	25	pH	2.7	V.A. Sinitsyn, S.U. Aja, D.A. Kulik and S.A. Wood. Geochim. Cosmochim. Acta 64: 185 194 (2000).
kaolin	Sigma	0.04–8 g/dm^3 NaCl		iep	3 3.5	D.M. Nevskaia, A. Guerrero-Ruiz and J.D. Lopez-Gonzalez. J. Colloid Int. Sci. 205: 97–105 (1998).
Na-montmorillonite	From Wy2	10^{-2} mol dm^{-3} NaClO$_4$		iep	<2 if any	F. Thomas, L.J. Michot, D. Vantelon, E. Montarges, B. Prelot, M. Cruchaudet and J.F. Delon. Colloids Surf. A 159: 351 358 (1999).
Na$_x$Si$_{8-x}$Al$_x$Mg$_6$O$_{20}$(OH)$_4$	Synthetic saponites, 0.7 < x < 2, hydrothermal synthesis.	10^{-2} mol dm^{-3} NaClO$_4$		iep	<2 if any	F. Thomas, L.J. Michot, D. Vantelon, E. Montarges, B. Prelot, M. Cruchaudet and J.F. Delon. Colloids Surf. A 159: 351 358 (1999).
Silicate						
talc	Talcs de Luzenac	10^{-2} mol dm^{-3} NaClO$_4$		iep	3	F. Thomas, L.J. Michot, D. Vantelon, E. Montarges, B. Prelot, M. Cruchaudet and J.F. Delon. Colloids Surf. A 159: 351–358 (1999).

Mixed oxides containing ThO₂

Material	Description	Electrolyte	Temp	Method	Result	Reference
Bi_2O_3; $ThO_2 \cdot n\ H_2O$	different proportions, from nitrates	$0.1\ mol\ dm^{-3}\ NaNO_3$		mass titration	5–25% Th 5, 30% Th 6.6, 35% Th 7.3	B.S. Mathur and B. Venkataramani. Indian J. Chem. A 38: 1092–1099 (1999).

Mixed oxides containing ZrO₂

Material	Description	Electrolyte	Temp	Method	Result	Reference
$ZrO_2 + 3$ mol% Y_2O_3	HSY-3, 74% tetragonal + 26% monoclinic, Mandoval	$10^{-3}\ mol\ dm^{-3}\ NaCl$	25	acousto	6.6	A. Pettersson, G. Marino, A. Pursiheimo and J.B. Rosenholm. J. Colloid Interf. Sci. 228: 73–81 (2000).
$ZrO_2 + 3$–6 mol% Y_2O_3	Tosoh Y-TZP	$10^{-2}\ mol\ dm^{-3}\ NaCl$	26–27	iep	6.6–7.2	W.C. Wei, S.C. Wang and F.Y. Ho. J. Am. Cer. Soc. 82: 3385–3392 (1999). This paper contains also compilation of IEP of ZrO_2 and Y-doped ZrO_2 with 6 references.
Y-stabilized tetragonal ZrO_2	$ZrOCl_2 + Y(NO_3)_3 + NH_3$	$10^{-3}\ mol\ dm^{-3}\ KCl$		iep	5.8	J. Wang and L. Gao. Ceram. Int. 26: 187–191 (2000).

Sparingly soluble salts

Material	Description	Electrolyte	Temp	Method	Result	Reference
SiC	β, synthetic (SiO_2 + saccharose)	$10^{-1}\ mol\ dm^{-3}\ KNO_3$	25	pH	pH 8.5	L. Cerovic, S.K. Milonjic and L. Kostic-Gvozdenovic. J. Am. Cer. Soc. 78: 3093–3096 (1995). This paper contains also compilation of the IEP of SiC with 7 references.
SiC	β, Mitsui Toatsu	$5 \times 10^{-4}\ mol\ dm^{-3}\ LiCl$ in ethanol	20	iep	pH 7.5	J. Widegren and L. Bergström. J. Eur. Cer. Soc. 20: 659–665 (2000).
SiC	review with 14 references.			iep	pH 2–9	S.K. Milonjic, L.S. Cerovic, D.P. Uskokovic. In: Materials Science of Carbides, Nitrides, and Borides. Y.G. Gogotsi and R.A. Andrievski, (Eds.), Kluwer, Dordrecht 1999, pp. 343–358.
Si_3N_4	review with 23 references.			iep	pH 3.2–9.2	S.K. Milonjic, L.S. Cerovic, D.P. Uskokovic. In: Materials Science of Carbides, Nitrides, and Borides. Y.G. Gogotsi and R.A. Andrievski, (Eds.) Kluwer, Dordrecht 1999, pp. 343–358.
$Ca_{10}(PO_4)_6(OH)_2$	synthetic ($Ca(OH)_2$ + H_3PO_4)	$10^{-1}\ mol\ dm^{-3}\ KNO_3$	25	pH	pH 4.1–6.1	I.D. Smiciklas, S.K. Milonjic, P. Pfendt and S. Raicevic. Separ. Purif. Technol. 18: 185–194 (2000).

TABLE A.1 Continued

Oxide	Description	Salt	T	Method	pH_0	Source
$Ca_{10}(PO_4)_6(OH)_2$	review with 8 references				pH 4.3–8.6	I.D. Smiciklas, S.K. Milonjic, P. Pfendt and S. Raicevic. Separ. Purif.Technol. 18: 185–194 (2000).
$Ni(OH)_{1.4}(SO_4)_{0.3}$	Needle like particles	10^{-2} mol dm^{-3} NaCl		iep	pH 12.5	M. Ocana. J. Colloid Interf. Sci. 228: 259–262 (2000).
$BaTiO_3$	two commercial samples, TiO_2:BaO ratio 1.05	10^{-3} mol dm^{-3} KCl		iep	pH 6.5	M.C. Blanco Lopez, B. Rand and F.L. Riley. J. Eur. Cer. Soc. 20: 107–118 (2000).

Glass

glass		HCl/NaOH		iep	<2.5 if any	Y. Gu and D. Li. J. Colloid Interf. Sci. 226: 328–339 (2000).

Other materials

red mud	hematite 37%, Al_2O_3 18%, SiO_2 17%, TiO_2 6%, Na_2O 8%, CaO 4% by weight	10^{-3} 1 mol dm^{-3} NaCl, CsCl	25	cip	8	G. Atun and G. Hisarli. J. Colloid Interf. Sci. 228: 40–45 (2000).
tourmaline dravite		0.001 mol dm^{-3} KNO_3		iep	pH5	L.J. Warren. Colloids Surf. 5: 301–319 (1982).
ice	H_2O, D_2O	0–10^{-3} mol dm^{-3} NaCl	1–4	pH iep	pH 7 pH 3.5	J. Drzymala, Z. Sadowski, L. Holysz and E. Chibowski, J. Colloid Interf. Sci. 220: 229–234 (1999).

Table A.2 Zero Points of Simple Hydr(Oxides) and Other Materials

Material	Description	Salt	T	Method	pH$_0$	Source
Al$_2$O$_3$	Fluka 507 C	10^{-3} 10^{-1} mol dm^{-3} KNO$_3$		cip	7	A.K.Helmy, and E.A.Ferreiro. Z. phys. Chem. (Leipzig) 257: 881 892 (1976).
Al$_2$O$_3$	A16SG, Alcoa	10^{-2} mol dm^{-3} NaCl	20	acousto	8.5	C. Pagnoux, M.Serantoni, R.Laucournet, T.Chartier, and J.F.Baumard. J. Eur. Cer. Soc. 19: 1935 1948 (1999).
Al$_2$O$_3$	AKP30, Sumitomo	10^{-2} mol dm^{-3} NaCl	20	acousto	9	C. Pagnoux, M.Serantoni, R.Laucournet, T.Chartier, and J.F.Baumard. J. Eur. Cer. Soc. 19: 1935-1948 (1999). Temperature effect on the IEP was also studied.
Al$_2$O$_3$	AKP30, Sumitomo	0.006 mol dm^{-3} NaCl	22	acousto	8.7	J.K.Beattie, and A.Djerdjev. J. Am. Ceram. Soc. 83: 2360-2364 (2000).
Al$_2$O$_3$	Sumitomo AA 05	0-0.05 mol dm^{-3} KCl		acousto	9 10 (hysteresis)	R.Greenwood, and K.Kendall. Powder Techn. 113: 148 157 (2000).
Al$_2$O$_3$	SDK 161 (Whitfield) SDK 160 (Whitfield)	0-0.01 mol dm^{-3} KCl	25	acousto	7.5-8.2 8.1-9 (hysteresis)	R.Greenwood, and K.Kendall. J.Eur.Cer.Soc. 20: 77-84 (2000).
Al$_2$O$_3$	Aldrich	10^{-3} 10^{-1} mol dm^{-3} NaCl NaClO$_4$	20	cip	8.2 7.8	J.Jablonski, W.Janusz, M.Reszka, R.Sprycha, and J.Szczypa. Pol.J.Chem. 74: 1399-1409 (2000).
Al$_2$O$_3$		10^{-4} mol dm^{-3} NaNO$_3$		iep	8.5	M.Pattanaik, and S.K.Bhaumik. Mater. Letters 44: 352 360 (2000).
Al$_2$O$_3$	Wusong	10^{-3} mol dm^{-3} KCl		iep	8.3	Y.Liu, L.Gao, and J.Guo. Colloids Surf. A 174: 349-356 (2000).
Al$_2$O$_3$	Riedel de Haen	10^{-2} mol dm^{-3} NaCl	25	iep	9	J.D.G.Duran, M.M.Ramos-Tejada, F.J.Arroyo, and F.Gonzalez-Caballero. J.Colloid Interf. Sci. 229: 107 117 (2000).
Al$_2$O$_3$	amorphous, from chloride	10^{-2} mol dm^{-3} NaCl		iep	9.4	S.Goldberg, and C.T.Johnston. J.Colloid Interf. Sci. 234: 204-216 (2001).
Al$_2$O$_3$	A16SG, Alcoa	10^{-3}-10^{-1} mol dm^{-3} KCl	20	pH	9.9-10.1	G.Tari, S.M.Olhero, and J.M.F.Ferreira. J. Colloid Interf. Sci. 231: 221-227 (2000).
Al$_2$O$_3$	Degussa C, δ, heated at 1000 °C	0.005-0.5 mol dm^{-3} NaNO$_3$ 0.005-0.5 mol dm^{-3} NaCl	room	cip	8 9.2 (original) 8.6 (purified)	E. Tombacz, M.Szekeres, I.Kertesz and L.Turi. Prog. Coll. Polym. Sci. 160:98 (1995). E. Tombacz and M.Szekeres. Langmuir 17: 1411-1419 (2001).
Al$_2$O$_3$	Degussa, γ	10^{-1} mol dm^{-3} NaNO$_3$		iep	9.3	Y.Arai, E.J.Elzinga and D.L.Sparks. J. Colloid Interf. Sci. 235: 80-88 (2001).

Table A.2 Continued

Material	Description	Salt	T	Method	pH$_0$	Source
Cu$_2$O·n H$_2$O	from nitrate or sulfate, in presence or absence of phosphate, 5 different samples	10^{-2} mol dm^{-3} NaNO$_3$		iep	4–5	P.McFadyen, and E.Matijevic. J. Inorg. Nucl. Chem. 35: 1883–1893 (1973).
CuO	from nitrate	NaOH + HNO$_3$		iep	7.5	S.H.Lee, Y.S.Her, and E.Matijevic. J. Colloid Interf. Sci. 186: 193- 202 (1997).
CuO	precipitated	10^{-3} mol dm^{-3} NaNO$_3$	25	iep	9.4	K.Subramaniam, S.Yiacoumi, and C.Tsouris. Colloids Surf. A 177: 133–146 (2001).
Cu(OH)$_2$	needle-like, from sulfate or acetate, different samples	10^{-2} mol dm^{-3} NaClO$_4$		iep	9.5- 10.2	L.Durand-Keklikian, and E.Matijevic. Colloid Polym. Sci. 268: 1151–1158 (1990).
Fe$_2$O$_3$	Johnson Matthey	NaNO$_3$		from acidity constants	8.1	N.Marmier, J.Dumonceau, J.Chupeau, and F.Fromage. C.R. Acad. Sci. Paris Ser. II 317: 311 317 (1993).
Fe$_2$O$_3$	natural hematite	NaOH + HCl		iep	2.8	K.B.Quast. Miner. Eng. 13: 1361 1376 (2000).
Fe$_2$O$_3$		10^{-1} mol dm^{-3} KNO$_3$		pH	6.6	
Fe$_2$O$_3$		10^{-4} mol dm^{-3} NaNO$_3$		iep	7.8	M.Pattanaik, and S.K.Bhaumik. Mater. Letters 44: 352–360 (2000).
Fe$_2$O$_3$	hematite, from chloride	0.05- 1 mol dm^{-3} NaNO$_3$	25	cip acousto	8.5 8.3	M.Gunnarsson, A.M.Jakobsson, S.Ekberg, Y.Albinsson, and E. Ahlberg. J. Colloid Interf. Sci. 231: 326- 336 (2000).
Fe$_2$O$_3$	Polysciences	10^{-3} mol dm^{-3} NaNO$_3$	25	iep pH	< 3 3	K.Subramaniam, S.Yiacoumi, and C.Tsouris. Colloids Surf. A 177: 133 146 (2001).
Fe$_2$O$_3$	amorphous	10^{-2} mol dm^{-3} NaCl		iep	8.5	S.Goldberg, and C.T.Johnston. J. Colloid Interf. Sci. 234: 204- 216 (2001).
FeOOH	goethite, from nitrate	0.001–0.1 mol dm^{-3} NaNO$_3$ NaCl	25	cip	9 9.2	M.Villalobos, and J.O.Leckie. Geochim. Cosmochim. Acta 64: 3787 3802 (2000). M.Villalobos, and J.O.Leckie. J. Colloid Interf. Sci. 235:15 32 (2001).
FeOOH	goethite, from nitrate, two samples	0.01 2 mol dm^{-3} NaNO$_3$	25	merge acousto	9.4	J.F.Boily, J.Lutzenkirchen, O.Balmes, J.Beattie, and S.Sjoberg. Colloids Surf. A 179: 11 27 (2001).
Ga$_2$O$_3$	reagent grade, washed	NaNO$_3$	25	iep	9	M.Kosmulski. J. Colloid Interf. Sci. 238: 225 227 (2001).
In$_2$O$_3$	reagent grade, washed	NaNO$_3$	25	cip iep	8.7	M.Kosmulski. J. Colloid Interf. Sci. 238: 225 227 (2001).

Material	Description	Electrolyte	T	Method	pzc	Reference
NiO	Joung Dong			iep	8.2	J.W.Moon, H.L.Lee, J.D.Kim, G.D.Kim, D.A.Lee, and H.W.Lee. Materials Lett. 38: 214-220 (1999).
SiO_2	3 samples: gel, opal, quartz	0.1, 0.7 mol dm⁻³ NaCl, KCl		pH	<6 if any	F.A.Rodrigues, P.J.M.Monteiro, and G.Sposito. Cement Concrete Res. 29: 527-530 (1999).
SiO_2	Riedel de Haen	10^{-2} mol dm⁻³ NaCl	25	iep	<3 if any	J.D.G.Duran, M.M.Ramos-Tejada, F.J.Arroyo, and F.Gonzalez-Caballero. J. Colloid Interf. Sci. 229: 107-117 (2000).
SiO_2	flame hydrolysis deposition, amorphous	0.001-0.1 mol dm⁻³ NaCl		authors'	3.5	R.Sabia, and L.Ukrainczyk J. Non-Crystal. Solids 277: 1-9 (2000).
SiO_2	Sigma	10^{-3} mol dm⁻³ $NaNO_3$	25	iep	<3	K.Subramaniam, S.Yiacoumi, and C.Tsouris. Colloids Surf. A 177: 133-146 (2001).
SiO_2	Quartz, Ward's	0.002 mol dm⁻³ KNO_3		pH	4.7	P.Huang, and D.W.Fuerstenau. Colloids Surf. A 177: 147-156 (2001).
SiO_2				iep	<4	
SiO_2	Machinery Nagel	0.1 mol dm⁻³ KNO_3	25	iep	3.7	M.Berka, and I.Banyai. J. Colloid Interf. Sci. 233: 131-135 (2001).
SiO_2				pH	4.8	
SiO_2	Degussa A-300	0.1 mol dm⁻³ NaCl		pH	<4 if any	N.N.Vlasova. J. Colloid Interf. Sci. 233: 227-233 (2001).
SnO_2	from chloride	10^{2} mol dm⁻³ $NaClO_4$	20	iep	4.2	M.Ocana, and E.Matijevic. J. Mater. Res. 5:1083-1091 (1990).
TiO_2	Degussa P 25			iep	7.5	A.Fernandez-Nieves, C.Richter, and F.J. de las Nieves. Progr. Colloid Polym. Sci. 110: 21-24 (1998).
TiO_2	from isopropoxide (two samples)	$NaNO_3$		iep	5.9; 7.2	P.A.Venz, R.L.Frost, J.T.Kloprogge. J. Non-Cryst. Solids 276: 95-112 (2000).
TiO_2	rutile	10^{-4} mol dm⁻³ $NaNO_3$		iep	5.9	M.Pattanaik, and S.K.Bhaumik. Mater. Letters 44: 352-360 (2000).
TiO_2	Degussa P 25	0.01 mol dm⁻³ KCl	25	iep	7.0	L.A.G.Rodenas, A.D.Weisz, G.A.Magaz, and M.A.Blesa. J. Colloid Interf. Sci. 230: 181-185 (2000).
TiO_2	Du Pont	0.001 mol dm⁻³ KCl		iep	6	W.P.Hsu, M.Kosmulski, and E.Matijevic. submitted.
Y_2O_3	monodispersed, spherical, calcination of basic carbonate	10^{-3} mol dm⁻³ $NaNO_3$		iep	7.6	B.Aiken, W.P.Hsu, and E.Matijevic. J. Am. Ceram. Soc. 71: 845-853 (1988).
ZrO_2	monodispersed, spherical, calcination of basic carbonate	10^{-2} mol dm⁻³ $NaNO_3$		iep	4.2	B.Aiken, W.P.Hsu, and E.Matijevic. J. Mater. Sci. 25:1886-1894 (1990).
ZrO_2	monodispersed, hollow particles, tetragonal + monoclinic	10^{-2} mol dm⁻³ $NaNO_3$		iep	4.5	N.Kawahashi, C.Persson, and E.Matijevic. J. Mater. Chem. 1:577-582 (1991).

Table A.2 Continued

Material	Description	Salt	T	Method	pH_0	Source
Mixed oxides containing Al						
$Al_2O_3 + TiO_2$	membrane	0.001, 0.005 mol dm^{-3} KCl	20	iep	6.5	A.Szymczyk, P.Fievet, J.C.Reggiani, and J.Pagetti. Desalination 115:129 134 (1998).
$Al_2O_3 + TiO_2 +$ 9% SiO_2	membrane	0.001 mol dm^{-3} KCl	20	iep	4.5	A.Szymczyk, P.Fievet, J.C.Reggiani, and J.Pagetti. Desalination 115:129 134 (1998).
Clay minerals						
kaolinite		10^{-4} mol dm^{-3} NaNO$_3$		iep	2.5	M.Pattanaik, and S.K.Bhaumik. Mater. Letters 44: 352 360 (2000).
Na-montmori-lonite	from natural bentonite	10^{-2} mol dm^{-3} NaCl	25	iep	<3 if any	J.D.G.Duran, M.M.Ramos-Tejada, F.J.Arroyo, and F. Gonzalez-Caballero. J. Colloid Interf. Sci. 229: 107 117 (2000).
Basic carbonates						
$EuOHCO_3$	monodispersed, from chloride and urea	10^{-3} mol dm^{-3} NaNO$_3$		iep	7.6	E.Matijevic, and W.P.Hsu. J. Colloid Interf. Sci. 118: 506 523 (1987).
$GdOHCO_3$	monodispersed, from chloride and urea	10^{-3} mol dm^{-3} NaNO$_3$		iep	7.6	E.Matijevic, and W.P.Hsu. J. Colloid Interf. Sci. 118: 506 523 (1987).
$NiCO_3 \cdot Ni(OH)_2 \cdot H_2O$	NiSO$_4$ + urea	10^{-3} mol dm^{-3}		iep	10	I.ul Haq, E.Matijevic, and K.Akhtar. Chem. Materials 9: 2659 2665 (1997).
$TbOHCO_3$	monodispersed, from chloride and urea	10^{-3} mol dm^{-3} NaNO$_3$		iep	7.6	E.Matijevic, and W.P.Hsu. J. Colloid Interf. Sci. 118: 506 523 (1987).
$YOHCO_3 \cdot H_2O$	monodispersed, spherical	10^{-3} mol dm^{-3} NaNO$_3$		iep	7.6	B.Aiken, W.P.Hsu, and E.Matijevic. J. Am. Ceram. Soc. 71: 845 853 (1988).
$Y_xCe_{1-x}OH$ $CO_3 \cdot H_2O$	monodispersed, spherical, x 0.25 or 0.75	10^{-3} mol dm^{-3} NaNO$_3$		iep	7.1	B.Aiken, W.P.Hsu, and E.Matijevic. J. Am. Ceram. Soc. 71: 845 853 (1988).
$Zr_2O_3(OH)_2$ CO_3	monodispersed, spherical	10^{-2} mol dm^{-3} NaNO$_3$		iep	3.4	B.Aiken, W.P.Hsu, and E.Matijevic. J. Mater. Sci. 25:1886 1894 (1990).
$Zr_yY_{0.8}(OH)_{3.8}$ $(CO_3)_{1.3}$	monodispersed, spherical	10^{-2} mol dm^{-3} NaNO$_3$		iep	3	B.Aiken, W.P.Hsu, and E.Matijevic. J. Mater. Sci. 25:1886 1894 (1990).

Mixed oxides containing Ce

Material	Description	Electrolyte	T	Method	pzc	Reference
CeO + Y$_2$O$_3$	monodispersed, spherical, calcination of mixed basic carbonate	10^{-3} mol dm^{-3} NaNO$_3$		iep	3.9 (+ maximum in ζ at pH≈7)	B.Aiken, W.P.Hsu, and E.Matijevic. J. Am. Ceram. Soc. 71: 845 853 (1988).

Ferrates

Material	Description	Electrolyte	T	Method	pzc	Reference
CoFe$_{3.3}$O$_{11.9}$	monodispersed, from sulfate	10^{-4} 10^{-2} mol dm^{-3} NaNO$_3$		iep	6.5	J.de Vicente, A.V.Delgado, R.C.Plaza, J.D.G.Duran, and F.Gonzalez-Caballero. Langmuir 16: 7954 7961 (2000).

Mixed oxides containing Pb

Material	Description	Electrolyte	T	Method	pzc	Reference
Pb$_2$Ru$_{1.75}$ Pb$_{0.25}$O$_{7-y}$				authors'	3.9	J.M.Zen, A.S.Kumar, and J.C.Chen. J. Molec. Cat. A 165: 177 188 (2001).

Silicates and mixed oxides containing Si

Material	Description	Electrolyte	T	Method	pzc	Reference
talc	Ward's	0.002 mol dm^{-3} KNO$_3$		iep	2.5	P.Huang, D.W.Fuerstenau. Colloids Surf. A 177: 147 156 (2001).
7% TiO$_2$ + 93 %SiO$_2$	flame hydrolysis deposition, amorphous	0.001–0.1 mol dm^{-3} NaCl		authors'	2.5	R.Sabia, and L.Ukrainczyk J. Non-Crystal. Solids 277: 1 9 (2000).

Zirconates

Material	Description	Electrolyte	T	Method	pzc	Reference
ZrY$_{0.8}$O$_{3.2}$	monodispersed, spherical, calcination of basic carbonate	10^{-2} mol dm^{-3} NaNO$_3$		iep	positive ζ at pH < 5; negative ζ at pH > 7	B.Aiken, W.P.Hsu, and E.Matijevic. J. Mater. Sci. 25: 1886-1894 (1990).
ZrO$_2$ + 8 mol % Y$_2$O$_3$	MEL	0 0.01 mol dm^{-3} KCl	25	acousto	6.7 7.5	R.Greenwood, and K.Kendall. J. Eur. Cer. Soc. 20: 77 84 (2000).
	Tosoh TZ8Y				9.2 9.6 (hysteresis)	
	Daiichi HSY8				6.5	

Sparingly soluble salts

Material	Description	Electrolyte	T	Method	pzc	Reference
MnCO$_3$	MnSO$_4$ + urea	10^{-3} mol dm^{-3}		iep	5.8	I.ul Haq, E.Matijevic, and K.Akhtar. Chem. Materials 9: 2659 2665 (1997).
Si$_3$N$_4$	Kema Nord, original and etched by KOH, HF, or HCl	0.01 mol dm^{-3} KCl	25	iep acousto pH	6 9	B.V.Zhmud, J.Sonnefeld, and L.Bergström. Colloids Surf. A 158: 327–341 (1999).

Table A.2 Continued

Material	Description	Salt	T	Method	pH$_0$	Source
Clays						
kaolin	Engelhard			acousto	<4 if any	A.S.Dukhin, P.J.Goetz, and S.Truesdail. Langmuir 17: 964–968 (2001).
Coatings						
ZrO$_2$ on NiO	from ZrO(NO$_3$)$_2$			iep	6.2	J.W.Moon, H.L.Lee, J.D.Kim, G.D.Kim, D.A.Lee, and H.W.Lee. Materials Lett. 38: 214–220 (1999)

Index